Mathematical
Surveys
and
Monographs

Volume 202

Asymptotic Geometric Analysis, Part I

Shiri Artstein-Avidan
Apostolos Giannopoulos
Vitali D. Milman

American Mathematical Society
Providence, Rhode Island

2010 *Mathematics Subject Classification.* Primary 52Axx, 46Bxx, 60Dxx, 28Axx, 46B20, 46B09, 52A20, 52A21, 52A23, 68-02.

For additional information and updates on this book, visit
www.ams.org/bookpages/surv-202

Library of Congress Cataloging-in-Publication Data

Artstein-Avidan, Shiri, 1978–
Asymptotic geometric analysis / Shiri Artstein-Avidan, Apostolos Giannopoulos, Vitali D. Milman.
pages cm. – (Mathematical surveys and monographs ; volume 202)
Includes bibliographical references and index.
ISBN 978-1-4704-2193-9 (pt. 1: alk. paper)
1. Geometric analysis. 2. Functional analysis. I. Giannopoulos, Apostolos, 1963– II. Milman, Vitali D., 1939– III. Title.

QA360.A78 2015
515′.1–dc23 2014049152

♾ The paper used in this book is acid-free and falls within the guidelines
established to ensure permanence and durability.
Visit the AMS home page at `http://www.ams.org/`

10 9 8 7 6 5 4 3 2 1   20 19 18 17 16 15

Contents

Preface

In this book we present the theory of *asymptotic geometric analysis*, a theory which stands at the midpoint between geometry and functional analysis. The theory originated from functional analysis, where one studied Banach spaces, usually of infinite dimensions. In the first few decades of its development it was called "local theory of normed spaces", which stood for investigating infinite dimensional Banach spaces via their finite dimensional features, for example subspaces or quotients. Soon, geometry started to become central. However, as we shall explain below in more detail, the study of "isometric" problems, a point of view typical for geometry, had to be substituted by an "isomorphic" point of view. This became possible with the introduction of an asymptotic approach to the study of high-dimensional spaces (asymptotic with respect to dimensions increasing to infinity). Finally, these finite but very high-dimensional questions and results became interesting in their own right, influential on other mathematical fields of mathematics, and independent of their original connection with infinite dimensional theory. Thus the name asymptotic geometric analysis nowadays describes an essentially new field.

Our primary object of study will be a finite dimensional normed space X; we may assume that X is $\mathbb{R}^n$ equipped with a norm $\|\cdot\|$. Such a space is determined by its unit ball $K_X = \{x \in \mathbb{R}^n : \|x\| \leqslant 1\}$, which is a compact convex set with non-empty interior (we call this type of set "a convex body"). Conversely, if K is a centrally symmetric convex body in $\mathbb{R}^n$, then it is the unit ball of a normed space $X_K = (\mathbb{R}^n, \|\cdot\|_K)$. Thus, the study of finite dimensional normed spaces is in fact a study of centrally symmetric convex bodies, but again, the low-dimensional type questions and the corresponding intuition are very different from what is needed when the emphasis is on high-dimensional asymptotic behaviour. An example that clarifies this difference is given by the following question: does there exist a universal constant $c > 0$ such that every convex body of volume one has a hyperplane section of volume more than c? In any fixed dimension n, simple compactness arguments show that the answer is affirmative (although the question to determine the optimal value of the corresponding constant c_n may remain interesting and challenging). However, this is certainly not enough to conclude that a constant $c > 0$ exists which applies to any body of volume one in any dimension. This is already an *asymptotic type question.* In fact, it is unresolved to this day and will be discussed in Chapter 10.

Classical geometry (in a fixed dimension) is usually an isometric theory. In the field of asymptotic geometric analysis, one naturally studies *isomorphic geometric objects* and derives *isomorphic geometric results.* By an "isomorphic" geometric object we mean a family of objects in different spaces of increasing dimension and by an "isomorphic" geometric property of such an "isomorphic" object we mean a property shared by the high-dimensional elements of this family. One is interested in the asymptotic behaviour with respect to some parameter (most often it is the dimension n) and in the control of how the geometric quantities involved depend on this parameter. The appearance of such an isomorphic geometric object is a new feature of asymptotic high-dimensional theory. Geometry and analysis meet here in a non-trivial way. We will encounter throughout the book many geometric inequalities in isomorphic form. Basic examples of such inequalities are the "isomorphic isoperimetric inequalities" that led to the discovery of the "concentration phenomenon", one of the most powerful tools of the theory, responsible for many counterintuitive results. Let us briefly describe it here, through the primary example of the sphere. A detailed account is given in Chapter 3. Consider the Euclidean unit sphere in $\mathbb{R}^n$, denoted S^{n-1}, equipped with the Lebesgue measure, normalized to have total measure 1. Let A be a subset of the sphere of measure $1/2$. Take an ε-extension of this set, with respect to Euclidean or geodesic distance, for some fixed but small ε; this is the set of all points which are at a distance of at most ε from the original set (usually denoted by A_ε). It turns out that the remaining set (that is, the set $S^{n-1} \setminus A_\varepsilon$ of all points in the sphere which are at a distance more than ε from A) has, in high dimensions, a very small measure, decreasing to zero exponentially fast as the dimension n grows. This type of statement has meaning only in asymptotic language, since in fact we are considering a sequence of spheres of increasing dimensions, and a sequence of subsets of these spheres, each of measure one half of its corresponding sphere, and the sequence of the measures of the ε-extensions (where ε is fixed for all n) is a sequence tending to 1 exponentially fast with dimension. We shall see how the above statement, which is proved very easily using the isoperimetric inequality on the sphere, plays a key role in some of the very basic theorems in this field.

We return to the question of changing intuition. The above paragraph shows that, for example, an ε-neighbourhood of the equator $x_1 = 0$ on S^{n-1} already contains an exponentially close to 1 part of the total measure of the sphere (since the sets $x_1 \leqslant 0$ and $x_1 \geqslant 0$ are both of measure $1/2$, and this set is the intersection of their ε-neighbourhoods). While this is again easy to prove (say, by integration) once it is observed, it does not correspond to our three-dimensional intuition. In particular, the far reaching consequences of these observations are hard to anticipate in advance. So, we see that in high dimension some of the intuition which we built for ourselves from what we know about three-dimensional space fails, and this "break" in intuition is the source of what one may call "surprising phenomena" in high dimensions. Of course, the surprise is there until intuition corrects itself, and the next surprise occurs only with the next break of intuition.

Here is a very simple example: The volume of the Euclidean ball B_2^n of radius one seems to be increasing with dimension. Indeed, denote this by κ_n and compute:

$$\kappa_1 = 2 < \kappa_2 = \pi < \kappa_3 = \frac{4\pi}{3} < \kappa_4 < \kappa_5 < \kappa_6.$$

However, a simple computation which is usually performed in Calculus III classes shows that

$$\mathrm{Vol}_n(B_2^n) = \kappa_n = \frac{\pi^{n/2}}{\Gamma(\frac{n}{2}+1)} = \left(c_n/\sqrt{n}\right)^n$$

where $c_n \to \sqrt{2\pi e}$. We thus see that in fact the volume of the Euclidean unit ball decreases like $n^{-n/2}$ with dimension (and one has the recursion formula $\kappa_n = \frac{2\pi}{n}\kappa_{n-2}$). So, for example, if one throws a point into the cube circumscribing the ball, at random, the chance that it will fall inside the ball, even in dimension 20, say, is practically zero. One cannot find this ball inside the cube.

Let us try to develop an intuition of high-dimensional spaces. We illustrate, with another example, how changing the intuition can help us understand, and anticipate, results. To begin, we should understand how to draw "high-dimensional" pictures, or, in other words, to try and imagine what do high-dimensional convex bodies "look like". The first non-intuitive fact is that the volume of parallel hypersections of a convex body decays exponentially after passing the median level (this is a consequence of the Brunn-Minkowski inequality, see Section 3.5). If we want to capture this property, it means that our two- or three-dimensional pictures of a high-dimensional convex body should have a "hyperbolic" form! Thus, K is a convex set but, as the rate of volume decay has a crucial influence on the geometry, we should find a way to visualize it in our pictures. For example, one may draw the convex set K as follows:

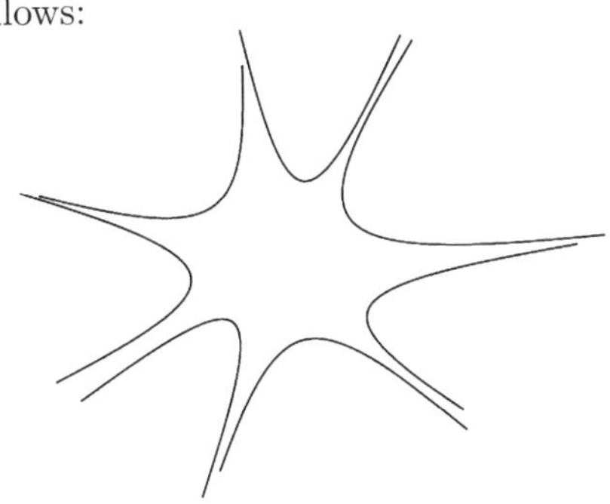

The convexity is no longer seen in the picture, but the volumetric properties are apparent. Next, with such a picture in mind, we may intuitively understand the following fact (it is a special case of Theorem 5.5.4 in Section 5.5): Consider the convex body $K = \sqrt{n}B_1^n := \mathrm{conv}(\pm\sqrt{n}e_i)$ (also called the unit ball of L_1^n). Take a random rotation UK of K and intersect it with the original body.

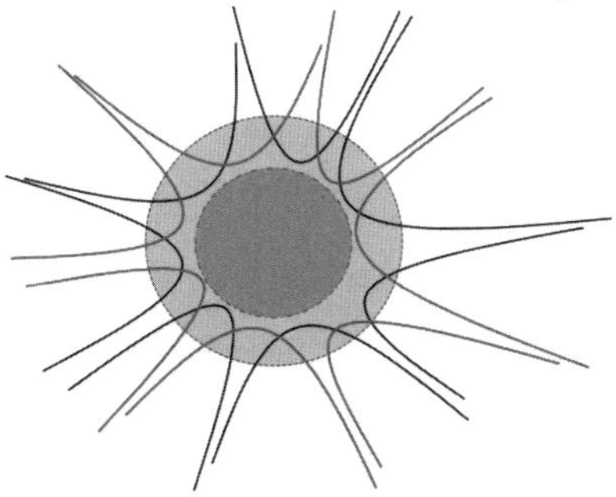

The resulting body, $K \cap UK$ is, with high probability over the choice of the random rotation U, contained in a Euclidean ball of radius C, where C is a universal constant independent of the dimension. Note that the original body, which contains a Euclidean ball of radius 1 (as does the intersection), has points in distance $\sqrt{n}$ from the origin. That is, the smallest Euclidean ball containing K is $\sqrt{n}B_2^n$. However, the simple (random) procedure of rotation and intersection, with high probability cuts out all these "remote regions" and regularizes the body completely so that it becomes an isomorphic Euclidean ball.

This was an example of a very concrete body, but it turns out the same property holds for a large class of bodies (called "finite volume ratio" bodies, see Section 5.5). Actually, if one allows more rotations, $\frac{n}{\log n}$ of them in dimension n, one may always regularize any body by the same process to become an isomorphic Euclidean ball. This last claim needs a small correction to be completely true: we have not explained how one chooses a random rotation. To this end one considers the Haar probability measure on the space of orthogonal rotations. To consider orthogonal rotations one must first fix a Euclidean structure, and the above statement is true after fixing the "right" structure corresponding to the body in question. The story of choosing a Euclidean structure, which is the same as choosing a "position" for the body, is an important topic, and for different goals different structures should be chosen. This topic is covered in Chapter 2.

Let us emphasize that while the geometric picture is what helps us understand which phenomena may occur, the picture is of course not a proof, and in each case a proof should be developed and is usually non-trivial.

This last example brings us to another important point which will be a central theme in this book, and this is the way in which, in this theory, *randomness* and *patterns* appear together. A perceived random nature of high dimensions is at the root of the reasons for the patterns produced and the unusual phenomena observed in high dimensions. In the dictionary, "randomness" is the exact opposite of "pattern". Randomness means "no pattern". But, in fact, objects created by independent identically distributed random processes, while being different from one another, are many times indistinguishable and similar in the statistical sense. Consider for example the unit cube, $[0,1]^n$. Choosing a random point inside it with respect to the uniform distribution means simply picking the n coordinates independently and uniformly at random in $[0,1]$. We know that such a point has some very special statistical properties (the simplest of which is the law of large numbers and the central limit theorem regarding the behaviour of the sum of these coordinates). It turns out that similar phenomena occur when the unit cube is replaced by a general convex body (again, a position should be specified). It is a challenge to uncover these similarities, a pattern, in very different looking objects. When we discover very similar patterns in arbitrary, and apparently very diverse convex bodies or normed spaces, we interpret them as a manifestation of the randomness principle mentioned above.

On the one hand, high dimension means many variables and many "possibilities", so one may expect an increase in the diversity and complexity as dimension increases. However, the concentration of measure and similar effects caused by the convexity assumption imply in fact a reduction of the diversity with increasing dimension, and the collapse of many different possibilities into one, or, in some cases, a few possibilities only. We quote yet another simple example which is a version of the "global Dvoretzky-type theorem". For details see Section 5.6. (The Minkowski sum of two sets is defined by $A+B=\{a+b : a\in A, b\in B\}$.)

> Let $n\in\mathbb{N}$ and let $K\subset\mathbb{R}^n$ be a convex body such that the Euclidean ball B_2^n is the ellipsoid of maximal volume inside K. Then, for $N=Cn/\log n$ random orthogonal transformations $U_i\in O(n)$, with probability at least $1-e^{-cn}$ we have that

$$B_2^n \subset \frac{1}{N}\big(U_1K+U_2K+\cdots+U_NK\big)\subset C'B_2^n,$$

where $0 < c, C, C' < \infty$ are universal constants (independent of K and of n).

One way in which diversity is compensated and order is created in the mixture caused by high dimensionality, is the concentration of measure phenomenon. As the dimension n increases, the covering numbers of a generic body of the same volume as the unit Euclidean ball, say, by the Euclidean ball itself (this means the number of translates of the ball needed to cover the body, see Sections 4.1 and 4.2) become large, usually exponentially so, meaning e^{cn} for some constant $c > 0$, and so seem impossible to handle. The concentration of measure is, however, of exponential order too (this time $e^{-c'n}$ for some constant $c' > 0$), so that in the end proofs become a matter of comparison of different constants in the various exponents (this is, of course, a very simplistic description of what is going on).

Let us quote from the preface of P. Lévy to the second edition of his book of 1951 [**380**]:

> "It is quite paradoxical, that an increase in the number of variables might cause simplifications. In reality, any law of large numbers presupposes the existence of some rule governing the influence of sequential variables; starting with such a rule, we often obtain simple asymptotic results. Without such a rule, complete chaos ensues, and since we are unable to describe, for instance, an infinite sequence of numbers, without resorting to an exact rule, we are unable to find order in the chaos, where, as we know, one can find mysterious non-measurable sets, which we can never truly comprehend, but which nevertheless will not cease to exist."

As we shall see below, the above facts reflect the probabilistic nature of high dimensions. We mean by this more than just the fact that we are using probabilistic techniques in many steps of the proofs. Let us mention one more very concrete example to illustrate this "probabilistic nature": Assume you are given a body $K \subset \mathbb{R}^n$, and you know that there exist 3 orthogonal transformations $U_1, U_2, U_3 \in O(n)$ such that the intersection of U_1K, U_2K and U_3K is, up to constant 2, say, a Euclidean ball. Then, for a *random* choice of 10 rotations, $\{V_i\}_{i=1}^{10} \subset O(n)$, with high probability on their choice, one has that $\bigcap_{i=1}^{10} V_iK$ is up to constant C (which depends on the numbers 2, 3 and 10, not on the dimension n, and may be computed) a Euclidean ball. This is a manifestation of a principle which is sometimes called "random is the best", namely that in various situations the results obtained by a random method cannot be substantially improved if the random choice is replaced by the best choice for the specific goal.

There are a number of reasons for this observed ordered behaviour. One may mention "repetition", which creates order, as statistics demonstrates. What we explain here and shall see throughout the book is that very high dimensions, or more generally, high parametric families, are another source of order.

We mention at this point that historically we observe the study of finite, but very high-dimensional spaces and their asymptotic properties as dimension increases already in Minkowski's work, who for the purposes of analytic number theory considered n-dimensional space from a geometric point of view. Before him, as well as long after him, geometry had to be two- or three-dimensional, see, e.g., the works of Blaschke. A paper of von Neumann from 1942 also portrays the same asymptotic point of view. We quote below from Sections 4 and 5 of the introduction of [**601**]. Here E_n denotes n-dimensional Euclidean space and M_n denotes the space of all $n \times n$ matrices. Whatever is in brackets is the present authors' addition.

> "Our interest will be concentrated in this note on the conditions in E_n and M_n - mainly M_n - when n is *finite*, but *very great*. This is an approach to the study of the infinite dimensional, which differs essentially from the usual one. The usual approach consists in studying an actually infinite dimensional unitary space, i.e. the Hilbert space E_∞. We wish to investigate instead the *asymptotic* behaviour of E_n and M_n for finite n, when $n \to \infty$.
>
> We think that the latter approach has been unjustifiably neglected, as compared with the former one. It is certainly not contained in it, since it permits the use of the notions $\|A\|$ and $t(A)$ (normalized Hilbert Schmidt norm, and trace) which, owing to the factors $1/n$ appearing in (their definitions) possess no analogues in E_∞.
>
> Since Hilbert space E_∞ was conceived as a limiting case of the E_n for $n \to \infty$, we feel that such a study is necessary in order to clarify to what extent E_∞ is or is not the only possible limiting case. Indeed we think that it is not, and that investigations on operator rings by F. J. Murray and the author show that other limiting cases exist, which under many aspects are more natural ones.
>
> Our present investigations originated in fact mainly from the desire to solve certain questions... We hope, however, that the reader will find that they also have an interest of their own, mainly in the sense indicated above: as a study of the asymptotic behaviour of E_n and M_n for finite n, when $n \to \infty$.
>
> From the point of view described (above) it seems natural to ask this question: How much does the character of E_n and M_n change when n increases - especially if n has already assumed very great values?"

Let us turn to a short description of the various chapters of the book; this will give us the opportunity to comment on additional fundamental ideas of the theory.

In Chapter 1 we recall basic notions from classical convexity. In fact, a relatively large portion of this book is dedicated to convexity theory, since a large part of the development of asymptotic geometric analysis is connected strongly with the classical theory. We present several proofs of the Brunn-Minkowski inequality and some of its fundamental applications. We have chosen to discuss in detail those proofs as they allow us to introduce fruitful ideas which we shall revisit throughtout the book. In the appendices we provide a more detailed exposition of basic facts from elementary convexity, convex analysis and the theory of mixed volumes. In particular, we describe the proof of Minkowski's theorem on the polynomiality of the volume of the sum of compact convex sets, and of the Alexandrov-Fenchel inequality, one of the most beautiful, non-trivial and profound theorems in convexity, which is linked with algebraic geometry and number theory. We emphasize the functional analytic point of view into classical convexity. This point of view opened a new field which is sometimes called "*functionalization of geometry*" or "*geometrization of probability*": It turns out that almost any notion or inequality connected with convex bodies has an analogous notion or inequality in the world of convex *functions*. This analogy between bodies and functions is fruitful in many different ways. On the one hand, it allows to predict functional inequalities which then are interesting in their own right. On the other hand, the generalization into the larger world of convex functions enables one to see the bigger picture and better understand what is going on. Finally, the results for functions may sometimes have implications back in the convex bodies world. This general idea is considered in parallel with the classical theory throughout the book.

In Chapter 2 we introduce the most basic and classical positions of convex bodies: Given a convex body K in $\mathbb{R}^n$, the family of its *positions* is the family of its affine images $\{x_0 + T(K)\}$ where $x_0 \in \mathbb{R}^n$ and $T \in GL_n$. In the context of functional analysis, one is given a norm (whose unit ball is K) and the choice of a position reflects a choice of a Euclidean structure for the linear space $\mathbb{R}^n$. Note that the choice of a Euclidean structure specifies a unit ball of the Euclidean norm, which is an ellipsoid. Thus, we may equivalently see a "position" as a choice of a special ellipsoid. The different ellipsoids connected with a convex body (or the different positions, corresponding to different choices of a Euclidean structure) that we consider in this chapter reflect different traces of symmetries which the convex body has. We introduce John position (also called maximal volume ellipsoid position), minimal surface area position and minimal mean width position. It turns out that when a position is extremal then some differential must vanish, and its vanishing is connected with isotropicity of some connected measure.

We also discuss some applications, mainly of John position, and introduce a main tool, which is useful in many other results in the theory, called the Brascamp-Lieb inequality. We state and prove one of its most useful forms, which is the so-called "normalized form" put forward by K. Ball, together with its reverse form, using F. Barthe's transportation of measure argument. In the second volume of this book we shall discuss the general form of the Brascamp-Lieb inequality, its various versions, proofs, and reverse form, as well as further applications to convex geometric analysis.

In Chapter 3 we discuss the concentration of measure phenomenon, first put forward in V. Milman's version of Dvoretzky theorem. Concentration is the central phenomenon that is responsible for the main results in this book. We present a number of approaches, all leading to the same type of behaviour: in high parametric families, under very weak assumptions of various types, a function tends to concentrate around its mean or median. Classical isoperimetric inequalities for metric probability spaces, such as the sphere, Gauss space and the discrete cube, are at the origin of measure concentration, and we start our exposition with these examples. Once the extremal sets (the solutions of the isoperimetric problem) are known, concentration inequalities come as a consequence of a simple computation. However, in very few examples are the exteremal sets known. We therefore do not focus on extremal sets but mainly on different ways to get concentration inequalities. We explore various such ways, and determine the different sources for concentration. In the second volume of this book we shall come back to this subject and study its functional aspects: Sobolev and logarithmic Sobolev inequalities, tensorization techniques, semi-group approaches, Laplace transform and infimum convolutions, and investigate in more detail the subject of transportation of measure.

In Chapter 4 we introduce the covering numbers $N(A,B)$ and the entropy numbers $e_k(A,B)$ as a way of measuring the "size" of a set A in terms of another set B. As we will see in the next chapters, they are a very useful tool and play an important role in the theory. Here, we explain some of their properties, derive relations and duality between these numbers, and estimate them in terms of other parameters of the sets involved — estimates which shall be useful in the sequel.

Chapter 5 is the starting point for our exposition of the asymptotic theory of convex bodies. It is devoted to the Dvoretzky-Milman theorem and to the main developments around it. In geometric language the theorem states that every high-dimensional centrally symmetric convex body has central sections of high dimension which are almost ellipsoidal. The dependence of the dimension k of these sections on the dimension n of the body is as follows: for every n-dimensional normed space $X = (\mathbb{R}^n, \|\cdot\|)$ and every $\varepsilon \in (0,1)$ there exist an integer $k \geqslant c\varepsilon^2 \log n$ and a k-dimensional subspace F of X which satisfies $d_{BM}(F, \ell_2^k) \leqslant 1+\varepsilon$, where d_{BM} denotes Banach-Mazur distance, a natural geometric distance between two normed spaces, and c is some absolute constant. The proof of the Dvoretzky-Milman theorem exploits the concentration of measure phenomenon for the Euclidean sphere S^{n-1}, in the form of a deviation inequality for Lipschitz functions $f : S^{n-1} \to \mathbb{R}$, which implies that the values of $\|\cdot\|$ on S^{n-1} concentrate near their average

$$M = \int_{S^{n-1}} \|x\|\, d\sigma(x).$$

A remarkable fact is that in Milman's proof, a formula for such a k is given in terms of n, M and the Lipschitz constant (usually called b) of the norm, and that this formula turns out to be sharp (up to a universal constant) in full generality. This gives us the opportunity to introduce one more new idea of the theory, which is *universality*. In different fields, and also in the origins of asymptotic geometric analysis, for a long time one knew how to write very precise estimates reflecting different asymptotic behaviour of certain specific high-dimensional (or high parametric) objects (say, for the spaces ℓ_p^n). Usually, one could show that these estimates are sharp, in an isomorphic sense at least. However, an accumulation of results indicates that, in fact, available estimates are exact for *every* sequence of spaces in increasing dimension (and thus one is tempted to say "for every space"). These kinds of estimates are called "asymptotic formulae". Let us demonstrate another such formula, concerning the diameter of a random projection of a convex body. All constants appearing in the statement below (C, c_1, C_2, c') are universal and do not depend on the body or on the dimension. Let $K \subset \mathbb{R}^n$ be a centrally symmetric convex body. One denotes by $h_K(u)$ the support function of K in direction u, defined as half the width of the minimal slab orthogonal to u which includes K, that is,

$$h_K(u) = \max\{\langle x, u\rangle : x \in K\}.$$

Denote by $d = d(K)$ the smallest constant such that $K \subset dB_2^n$, that is, half of the diameter of K, and actually $d = \max_{u \in S^{n-1}} h_K(u)$. Denote by $M^* = M^*(K)$ the average of h_K over S^{n-1}, that is,

$$M^*(K) = \int_{S^{n-1}} h_K(u) d\sigma(u)$$

where σ is the Haar probability measure on S^{n-1}. It turns out that for dimensions larger than $k^* = C(M^*/d)^2 n$, the diameter of the projection of K onto a random k-dimensional subspace is, with high probability, approximately $d\sqrt{k/n}$. That is, between $c_1 d\sqrt{k/n}$ and $C_2 d\sqrt{k/n}$. Around the critical dimension $k^* = k^*(K)$, the projection becomes already (with high probability on the choice of a subspace) a Euclidean ball of radius approximately $M^*(K)$, and this will be, again up to constants, the diameter (and the inner-radius) of a random projection onto dimension $c'k^*$ and less. In this result the isomorphic nature of the result is very apparent. Indeed, the diameter need not be ε-isometrically close to $d\sqrt{k/n}$ for k in the range between k^* and n, but only isomorphically. Isometric results are known in the regime $k \leqslant c'k^*$ when the projection is already with high probability a Euclidean ball (this is actually the Dvoretzky-Milman theorem). We describe this result in detail in Section 5.7.1. Another property of this last example is a threshold behaviour of the function $f(t)$ giving the average diameter of a projection into dimension tn. The function, which is monotone, attains its maximum, d, at $t = 1$, behaves like $d\sqrt{t}$ in the range $[C(M^*/d)^2, 1]$, and like a constant, close to M^*, in the range $[0, c'(M^*/d)^2]$. Threshold phenomena have been known for a long time in many areas of mathematics, for example, in mathematical physics. Here we see that these occur in complete generality (for *any* convex body, the same type of threshold). More examples of threshold behaviour in asymptotic geometric analysis shall be demonstrated in the book.

Before moving to the description of Chapters 6–10 we mention another point of view one should keep in mind when reading the book: the comparison between *local* and *global* type results. The careful readers may have already noted the similarity of two of the statements given so far in this preface: a part of the statement about decrease of diameter in fact said that after some critical dimension, a random projection of a convex body is with high probability close to a Euclidean ball (this also follows from the Dvoretzky-Milman theorem by duality of projections and sections). This is called a "local" statement. Two other theorems quoted above regarded what happens when one intersects random rotations of a convex body (for example, B_1^n), or when one takes the Minkowski sum (average) of random rotations of a convex body (for example, the cube). Again the results were that after a suitable (and not very large) number of such rotations, the resulting body is an isomorphic Euclidean ball. These types of results, pertaining to the body as a whole and not its sections or projections, are called "global" results. At the heart of the global results presented in this book, which have convex geometric flavor, stand methods which come from functional analysis (considering norms, their averages, etc). Again, by global properties we refer to properties of the original body or norm in question, while the local properties pertain to the structure of lower dimensional sections and projections of the body or normed space. From the beginning of the 1970's the need for geometric functional analysis led to a deep investigation of the linear structure of finite dimensional normed spaces (starting with Dvoretzky theorem). However, it had to develop a long way before this structure was understood well enough to be used for the study of the global properties of a space. The culmination of this study was an understanding of the fact that subspaces (and quotient spaces) of proportional dimension behave very predictably. An example is the theorem quoted above regarding the decay of diameter. This understanding formed a bridge between the problems of functional analysis and the global asymptotic properties of convex sets, and is the reason the two fields of convexity and of functional analysis work together nowadays.

In Chapter 6 we discuss upper bounds for the parameter $M(K)M^*(K)$, or equivalently, the product of the mean width of K and the mean width of its polar, the main goal being to minimize this parameter over all positions of the convex body. (The polar of a convex body K is the closed convex set generating the norm given by h_K, and is denoted K°.) We will see that the quantity MM^* can be bounded from above by a parameter of the space $(X, \|\cdot\|_K)$ which is called its K-convexity constant, and which in turn can be bounded from above, for X of dimension n, by $c[\log(d_{BM}(X, \ell_2^n)) + 1] \leqslant c' \log n$ for universal c, c'. This estimate for the K-convexity constant is due to G. Pisier and as we will see it is one of the fundamental facts in the asymptotic theory. The estimate for $M(K)M^*(K)$ brings us to one more main point, which concerns *duality*, or *polarity*. In many situations two dual operations performed one after the other already imply complete regularization. That is, one operation cancels a certain type of "bad behaviour", and the dual operation cancels the "opposite" bad behaviour. Other examples include the quotient of a subspace theorem (see Chapter 7) or its corresponding global theorem: if one takes the sum of a body and a random (in the right coordinate system) rotation of it, then considers the polar of this set, to which again one applies a random rotation and takes the sum, the resulting body will be with high probability on the choice of rotations, an isomorphic Euclidean ball. If one uses just one of these two operations, there may be a need for $n/\log n$ such operations.

Chapter 7 is devoted to results about proportional subspaces and quotients of an n-dimensional normed space, i.e., of dimension λn, where the "proportion" $\lambda \in (0,1)$ can sometimes be very close to 1. The first step in this direction is Milman's M^*-estimate. In a geometric language, it says that there exists a function $f : (0,1) \to \mathbb{R}^+$ such that, for every centrally symmetric convex body K in $\mathbb{R}^n$ and every $\lambda \in (0,1)$, a random $\lfloor \lambda n \rfloor$-dimensional section $K \cap F$ of K satisfies the inclusion

$$K \cap F \subseteq \frac{M^*(K)}{f(\lambda)} B_2^n \cap F.$$

In other words, the diameter of a random "proportional section" of a high-dimensional centrally symmetric convex body K is controlled by the mean width $M^*(K)$ of the body. We present several proofs of the M^*-estimate; based on these, we will be able to say more about the best possible function f for which the theorem holds true and about the corresponding estimate for the probability of subspaces in which this occurs. As an application of the M^* estimate we obtain Milman's quotient of a subspace theorem. We also complement the M^* estimate by a lower bound for the outer-radius of sections of K, which holds for *all* subspaces, we compare "best" sections with "random" ones of slightly lower dimension, and we provide a linear relation between the outer-radius of a section of K and the outer-radius of a section of K°.

In Chapter 8 we present one of the deepest results in asymptotic geometric analysis: the existence of an M-position for every convex body K. This position can be described "isometrically" (if, say, K has volume 1) as minimizing the volume of $T(K) + B_2^n$ over all $T \in SL_n$. However, such a characterization hides its main properties and advantages that are in fact of an "isomorphic" nature. The isomorphic formulation of the result states that there exists an ellipsoid of the same volume as the body K, which can replace K, in many computations, up to universal constants. This result, which was discovered by V. Milman, leads to the reverse Santaló inequality and the reverse Brunn-Minkowski inequality. The reverse Santaló inequality concerns the volume product, sometimes called the Mahler product, of K which is defined by

$$s(K) := \mathrm{Vol}_n(K)\mathrm{Vol}_n(K^\circ).$$

The classical Blaschke-Santaló inequality states that, given a centrally symmetric convex body K in $\mathbb{R}^n$, the volume product $s(K)$ is less than or equal to the volume product $s(B_2^n) = \kappa_n^2$, and that equality holds if and only if K is an ellipsoid. In the opposite direction, a well-known conjecture of Mahler states that $s(K) \geqslant 4^n/n!$ for every centrally symmetric convex body K (i.e., the cube is a minimizer for $s(K)$ among centrally symmetric convex bodies) and that $s(K) \geqslant (n+1)^{n+1}/(n!)^2$ in the not necessarily symmetric case, meaning that in this case the simplex is a minimizer. The reverse Santaló inequality of Bourgain and Milman verifies this conjecture in the asymptotic sense: there exists an absolute constant $c > 0$ such that

$$\left(\frac{s(K)}{s(B_2^n)}\right)^{1/n} \geqslant c$$

for every centrally symmetric convex body K in $\mathbb{R}^n$. Milman's reverse Brunn-Minkowski inequality states that for any pair of convex bodies K and T that are in M-position, one has

$$\mathrm{Vol}_n(K+T)^{1/n} \leqslant C\left[\mathrm{Vol}_n(K)^{1/n} + \mathrm{Vol}_n(T)^{1/n}\right].$$

(The reverse inequality, with constant 1, is simply the Brunn-Minkowski inequality of Chapter 1.)

Another way to define the M-position of a convex body is through covering numbers, as was presented in Milman's proof. Pisier has proposed a different approach to these results, which allows one to find a whole family of special M-ellipsoids satisfying stronger entropy estimates. We describe his approach in the last part of Chapter 8.

In Chapter 9 we introduce a "Gaussian approach" to some of the main results which were presented in previous chapters, including sharp versions of the Dvoretzky-Milman theorem and of the M^*-estimate. The proof of these results is based on comparison principles for Gaussian processes, due to Gordon, which extend a theorem of Slepian. The geometric study of random processes, and especially of Gaussian processes, has strong connections with asymptotic geometric analysis. The tools presented in this chapter will appear again in the second volume of the book.

In the last chapter of this volume, Chapter 10, we discuss more recent discoveries on the distribution of volume in high-dimensional convex bodies, together with the unresolved "slicing problem" which was mentioned briefly at the beginning of this preface, with some of its equivalent formulations. A natural framework for this study is the *isotropic position* of a convex body: a convex body $K \subset \mathbb{R}^n$ is called isotropic if $\mathrm{Vol}_n(K) = 1$, its barycenter (center of mass) is at the origin and its inertia matrix is a multiple of the identity, that is, there exists a constant $L_K > 0$

such that

$$\int_K \langle x,\theta\rangle^2 dx = L_K^2$$

for every θ in the Euclidean unit sphere S^{n-1}. The number L_K is then called the isotropic constant of K. The isotropic position arose from classical mechanics back in the 19th century. It has a useful characterization as a solution of an extremal problem: the isotropic position $\tilde{K}=T(K)$ of K minimizes the quantity

$$\int_{\tilde{K}} |x|^2 dx$$

over all $T\in GL_n$ such that $\text{Vol}_n(\tilde{K})=1$ and $\int_{\tilde{K}} x dx = 0$.

The central theme in Chapter 10 is the *hyperplane conjecture* (or *slicing problem*): it asks whether there exists an absolute constant $c>0$ such that $\max_{\theta\in S^{n-1}} \text{Vol}_{n-1}(K\cap\theta^{\perp})\geqslant c$ for every n and every convex body K of volume 1 in $\mathbb{R}^n$ with barycenter at the origin. We will see that an affirmative answer to this question is equivalent to the fact that there exists an absolute constant $C>0$ such that

$$L_n := \max\{L_K : K \text{ is an isotropic convex body in } \mathbb{R}^n\} \leqslant C.$$

We shall work in the more general setting of a finite log-concave measure μ, where a corresponding notion of isotropicity is defined via the covariance matrix $\text{Cov}(\mu)$ of μ. We present the best known upper bounds for L_n. Around 1985-86, Bourgain obtained the upper bound $L_n\leqslant c\sqrt[4]{n}\log n$ and, in 2006, this estimate was improved by Klartag to $L_n\leqslant c\sqrt[4]{n}$. In fact, Klartag obtained a solution to an isomorphic version of the hyperplane conjecture, the "isomorphic slicing problem", by showing that, for every convex body K in $\mathbb{R}^n$ and any $\varepsilon\in(0,1)$, one can find a centered convex body $T\subset\mathbb{R}^n$ and a point $x\in\mathbb{R}^n$ such that $(1+\varepsilon)^{-1}T\subseteq K+x\subseteq(1+\varepsilon)T$ and $L_T\leqslant C/\sqrt{\varepsilon}$ for some absolute constant $C>0$. An additional essential ingredient in Klartag's proof of the bound $L_n\leqslant c\sqrt[4]{n}$, which is a beautiful and important result in its own right, is the following very useful deviation inequality of Paouris: if μ is an isotropic log-concave probability measure on $\mathbb{R}^n$, then

$$\mu(\{x\in\mathbb{R}^n : |x|\geqslant ct\sqrt{n}\})\leqslant \exp\left(-t\sqrt{n}\right)$$

for every $t\geqslant 1$, where $c>0$ is an absolute constant. The proof is presented in Section 10.4 along with the basic theory of the L_q-centroid bodies of an isotropic log-concave measure. Another important result regarding isotropic log-concave measures is the *central limit theorem* of Klartag, which states that the 1-dimensional marginals of high-dimensional isotropic log-concave measures μ are approximately Gaussian with high probability. We will come back to this result and related ones in the second volume of the book and we will see that precise quantitative relations exist between the hyperplane conjecture, the optimal answer to the central limit problem, and other conjectures regarding volume distribution in high dimensions.

Acknowledgements. This book is based on material gathered over a long period of time with the aid of many people. We would like to mention two ongoing working seminars in which many of the ideas and results were presented and discussed: these are the Asymptotic Geometric Analysis seminars at the University of Athens and at Tel Aviv University. The active participation of faculty members, students and visitors in these seminars, including many discussions and collaborations, have made a large contribution to the possibility of this book. We would like to mention the names of some people whose contribution was especially important, whether in offering us mathematical and technical advice, in reading specific chapters of the book, in allowing us to make use of their research notes and material, and for sending us to correct and less known references and sources. We thank S. Alesker, S. Bobkov, D. Faifman, B. Klartag, H. König, A. Litvak, G. Pisier, R. Schneider, B. Slomka, S. Sodin and B. Vritsiou. Finally, we would like to thank S. Gelfand and the AMS team for offering their publishing house as a home for this manuscript, and for encouraging us to complete this project.

The second named author would like to acknowledge partial support from the ARISTEIA II programme of the General Secretariat of Research and Technology of Greece during the final stage of this project. The first and third named authors would like to acknowledge partial support from the Israel Science Foundation.

November 2014

CHAPTER 1

Convex bodies: classical geometric inequalities

The first section of this chapter introduces basic notions and background material from classical convexity. In the appendices we provide a more detailed exposition of those facts from elementary convexity, convex analysis and the theory of mixed volumes that will be used in this book. The main part of the chapter is devoted to the Brunn-Minkowski inequality and some of its applications. We present several proofs of the Brunn-Minkowski inequality; besides the classical ones, we have chosen to discuss in detail several other proofs because they introduce very fruitful ideas that we will revisit. Among them, the Prékopa-Leindler inequality is a first example of the interplay between geometric and functional inequalities, while two proofs introduce important volume preserving transformations: the Knothe map and the Brenier map. All the applications that are outlined in Section 1.5 are crucial for the development of the asymptotic theory of convex bodies.

1.1. Basic convexity

1.1.1. Convex sets

A set $A \subseteq \mathbb{R}^n$ is called *convex* if $(1-\lambda)x + \lambda y \in A$ for any $x, y \in A$ and any $\lambda \in [0,1]$. The *Minkowski sum* of two sets $A, B \subseteq \mathbb{R}^n$ is defined by

$$A + B := \{a + b : a \in A, b \in B\}.$$

Clearly the sum of two convex sets is convex, and a set is convex if and only if $\lambda K + (1-\lambda)K = K$ for every $\lambda \in (0,1)$.

The notion of convexity has been known since ancient times. For example, it was noted by Archimedes that the area between a chord and a "convex" line is monotone with respect to changing the line, a property which clearly does not hold without the convexity assumption:

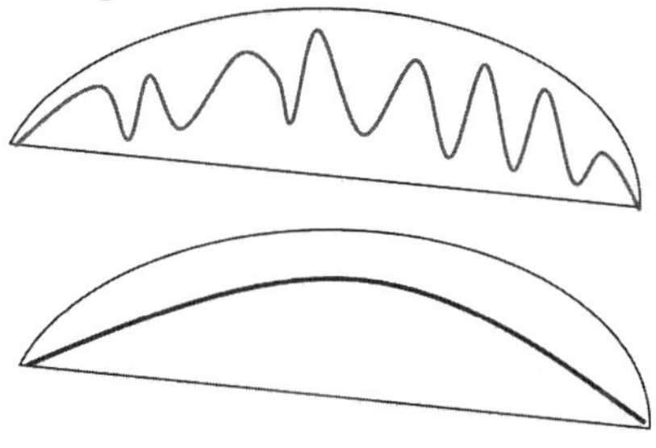

DEFINITION 1.1.1. A *convex body* is a convex set $K \subseteq \mathbb{R}^n$ which is compact and has non-empty interior. We say that K is centrally symmetric if $K = -K$. We

denote the class of convex bodies in $\mathbb{R}^n$ by $\mathcal{K}^n$. When 0 is assumed to be in the interior of the body, we write $K \in \mathcal{K}^n_{(0)}$.

Convexity theory is a rich field with many useful results, and there are several very good books dedicated to the topic (see the notes and remarks section at the end of the chapter). In this book we have chosen to include a brief account of those parts that are central to convexity and needed for the theory of asymptotic geometric analysis. So as not to distract the reader from the main line of the book, we have included several of these parts and most of the longer proofs in an appendix. In this first section we recall those notions and results which are needed for our discussion, and refer the reader to the appendix or to other sources when needed.

1.1.2. Convex functions

A function $\varphi : \mathbb{R}^n \to (-\infty, +\infty]$ is called *convex* if its *epigraph*, defined by

$$\mathrm{epi}(\varphi) = \{(x,t) \in \mathbb{R}^n \times \mathbb{R} : \varphi(x) \leqslant t\},$$

is a convex subset of $\mathbb{R}^{n+1}$. Of course, this is equivalent to the function satisfying the inequality

$$\varphi((1-\lambda)x + \lambda y) \leqslant (1-\lambda)\varphi(x) + \lambda\varphi(y)$$

for every $\lambda \in [0,1]$ and $x, y \in \mathbb{R}^n$. Let us also mention that a convex function φ has the property that its sub-level sets $\{x \in \mathbb{R}^n : \varphi(x) \leqslant t\}$ are convex.

A convex function $\varphi : \mathbb{R}^n \to \mathbb{R}$ is called *proper* if it is not identically $+\infty$. The *domain* of φ is the convex set $\mathrm{dom}\,(\varphi) = \{x : \varphi(x) < +\infty\}$. If A is the interior of $\mathrm{dom}\,(\varphi)$, then φ is continuous on A. The function is called *closed* if its epigraph is a closed (convex) set, and this is equivalent to lower semicontinuity of φ. Moreover, we can modify the values of a convex function φ on the boundary of $\mathrm{dom}\,(\varphi)$ so that φ will become closed. In this book the convex functions we shall consider will always be lower semicontinuous.

With regards to derivative, it is known that, given a convex function φ, $\nabla\varphi$ exists almost everywhere in A. Even when it does not exist, one may replace it with the following notion (having fixed a scalar product $\langle\cdot,\cdot\rangle$ in advance).

DEFINITION 1.1.2. Let $\varphi : \mathbb{R}^n \to (-\infty, +\infty]$ be convex and $x \in \mathrm{dom}\,(\varphi)$. The *subdifferential* of φ at x is defined by

$$\partial\varphi(x) = \{u \in \mathbb{R}^n : \varphi(y) \geqslant \varphi(x) + \langle u, y - x\rangle \text{ for all } y\}.$$

Any $u \in \partial\varphi(x)$ is called a *subgradient* of φ at x.

The subdifferential parametrizes the supporting hyperplanes of φ. It is easy to see that $\nabla\varphi(x)$ exists if and only if $\partial\varphi(x) = \{\nabla\varphi(x)\}$. Let $x \in \mathrm{int}\,(\mathrm{dom}\,(\varphi))$. Then the set $\partial\varphi(x)$ is non-empty, since there is at least one supporting hyperplane at $(x, \varphi(x))$, this hyperplane being the graph of an affine function which lies below φ and touches it at $(x, \varphi(x))$.

The set $\partial\varphi(x)$ is convex. For $x \in \mathrm{int}\,(\mathrm{dom}\,(\varphi))$, it is also compact (namely it is closed, and at the same time $\partial\varphi(x)$ cannot include a ray since x is in the interior of the domain, thus it is also bounded).

Again, the topic of convex functions is a rich one; in particular, regularity and smoothness issues have been studied very deeply, because it turns out that the convexity assumption rules out many pathological behaviours of functions. Although this is not the main topic of this book, given that convex functions do play an important role in the theory, we have included some more of the basic facts on them in the appendix.

Finally, we define a family of functions that are in one-to-one correspondence with convex functions, and which throughout the book serve as a functional generalization of convex bodies: these are the logarithmically-concave functions.

DEFINITION 1.1.3 (Log-concave function). A function $f : \mathbb{R}^n \to [0, \infty)$ is called *log-concave* if $f = \exp(-\varphi)$ for some convex function $\varphi : \mathbb{R}^n \to (-\infty, +\infty]$.

This log-concave re-normalization of a convex function φ has some advantages over φ, one of them being that the log-concave functions can often be integrated. In many cases, statements about convex bodies can be generalized to log-concave functions, with volume replaced by integral. Log-concave functions are intimately connected with log-concave measures, which will be defined in Chapter 3 and will be investigated later on.

1.1.3. The support function

The notions of the support function and of the polar of a convex set are central in convexity, and are useful in many other fields of mathematics, for example linear programming.

DEFINITION 1.1.4. Given $K \subseteq \mathbb{R}^n$ non-empty and convex, we define the *support function* corresponding to K by

$$h_K(u) = \sup_{x \in K} \langle x, u \rangle$$

where $u \in \mathbb{R}^n$.

One may check that h_K is positively homogeneous, lower semicontinuous and convex. Oppositely, every positively homogeneous lower semicontinuous convex $h : \mathbb{R}^n \to (-\infty, \infty]$ is the support function of some (unique) closed convex set K. The operation $K \mapsto h_K$ is order preserving: $K \subseteq T$ implies $h_K \leqslant h_T$ and $h_K \leqslant h_T$ implies $\overline{K} \subset \overline{T}$ (here $\overline{A}$ stands for the closure of A).

The quantity $h_K(u) + h_K(-u)$ is called the *width* of K in direction u.

A simple but important property of the support function regards the way it interacts with Minkowski addition. For $\lambda, \mu \geqslant 0$ one has for every $u \in \mathbb{R}^n$

$$h_{\lambda K + \mu T}(u) = \lambda h_K(u) + \mu h_T(u).$$

One also has $h_{-K}(u) = h_K(-u)$, and so K is centrally symmetric if and only if h_K is even, while $0 \in K$ if and only if $h_K \geqslant 0$. In the latter case, we may consider the support function as a gauge function on $\mathbb{R}^n$ or, when K is centrally symmetric, as a norm. The unit ball of this norm is called the polar of K and is denoted by K°. This set may be defined even without the symmetry assumption, as long as 0 resides inside the set K.

DEFINITION 1.1.5. Given $K \subseteq \mathbb{R}^n$ convex with $0 \in K$, we define the *polar* of K to be the set

$$K^\circ = \left\{ y \in \mathbb{R}^n : \sup_{x \in K} \langle x, y \rangle \leqslant 1 \right\}.$$

Note that even without any assumption on K, the set K° is a closed convex set. It is compact if $0 \in \operatorname{int}(K)$, while the mapping $K \mapsto K^\circ$ reverses order, and is an involution both when it is defined on the class of convex bodies that contain 0 in their interior, as well as on the bigger class of all closed convex bodies containing 0. As a mapping of elements in these classes, it exchanges intersection with closure of convex hull.

A *polytope* is the convex hull of a finite set of points. The support function of a polytope is very simple since, if $P = \operatorname{conv}\{x_i\}_{i=1}^m$, by the above remarks

$$h_P(u) = \max_{1 \leqslant i \leqslant m} \langle x_i, u \rangle.$$

So, we see that the support function of a polytope is piecewise linear. In fact h_K is piecewise linear if and only if K is a polytope.

1.1.4. Hausdorff distance

The space of convex bodies in $\mathbb{R}^n$ is endowed with a natural topology. It can be introduced as the topology induced by the *Hausdorff metric* δ^H:

DEFINITION 1.1.6. For $K, T \in \mathcal{K}^n$ we define

$$\delta^H(K,T) = \max\left\{ \max_{x \in K} \min_{y \in T} |x-y|, \max_{x \in T} \min_{y \in K} |x-y| \right\},$$

where $|x|$ denotes the Euclidean norm of $x \in \mathbb{R}^n$.

It is not hard to check that

$$\begin{aligned} \delta^H(K,T) &= \inf\{\delta \geqslant 0 : K \subseteq T + \delta B_2^n, T \subseteq K + \delta B_2^n\} \\ &= \max\{|h_K(u) - h_T(u)| : u \in S^{n-1}\}, \end{aligned}$$

where B_2^n is the Euclidean unit ball in $\mathbb{R}^n$ and S^{n-1} is the $(n-1)$-dimensional unit sphere. Thus, the embedding $K \mapsto h_K$ from $\mathcal{K}^n$ to the space $C(S^{n-1})$ of continuous functions on the sphere is an isometry between $(\mathcal{K}^n, \delta^H)$ and a subset of $C(S^{n-1})$ endowed with the supremum norm. Note that this mapping is positively linear (mapping Minkowski addition to sum of functions) and order-preserving (between inclusion and point-wise inequality). Note that there is no linearity for negative scalars, namely $h_{K-T} \neq h_K - h_T$ (where by $K - T$ we mean $K + (-T)$).

We mention a useful corollary from these simple facts.

COROLLARY 1.1.7. *Given $K, L, M \in \mathcal{K}^n$ we have*

$$\delta^H(K, L) = \delta^H(K + M, L + M).$$

In particular, we have the cancellation law

$$K + M = L + M \textit{ implies } K = L.$$

It is useful to know that any convex set can be approximated by polytopes:

PROPOSITION 1.1.8. *Let $K \in \mathcal{K}^n$ and $\varepsilon > 0$.*

(i) *There exists a polytope $P \subseteq K$ with $\delta^H(P, K) \leqslant \varepsilon$.*
(ii) *There exists a polytope $P \supseteq K$ with $\delta^H(P, K) \leqslant \varepsilon$.*
(iii) *If $0 \in \operatorname{int}(K)$ then there exists a polytope $P \subseteq \operatorname{int}(K)$ with $K \subseteq \operatorname{int}((1 + \varepsilon)P)$.*

We remark that actually, when approximating several sets simultaneously, one can additionally require that the approximating polytopes are strongly isomorphic, namely that their faces (of all dimensions) lie in the same directions. For a proof of the above proposition, and for a complete statement of this last remark, see the appendix.

Finally, it is useful to recall the following compactness theorem of Blaschke.

THEOREM 1.1.9 (Blaschke selection). *Any sequence of convex bodies $K_j \in \mathcal{K}^n$ all of whose elements are contained in some fixed ball, has a convergent subsequence.*

1.1.5. Mixed Volumes

The history of mixed volumes is an interesting one and will be elaborated upon in the Notes and Remarks section as well as in the appendix, where many of the proofs are given. For this section we quote the main theorem of Minkowski on the polynomiality of volume with respect to Minkowski addition. Note that the formula applies even when the sets are of dimension lower than n.

THEOREM 1.1.10 (Minkowski polarization). *Let $K_1, \dots, K_m$ be non-empty compact convex subsets of $\mathbb{R}^n$. For any n-tuple $1 \leqslant i_n, \dots, i_n \leqslant m$ there exists non-negative coefficients $V(K_{i_1}, \dots, K_{i_n})$, $1 \leqslant i_1, \dots, i_n \leqslant m$, that are symmetric with respect to the indices $i_1, \dots, i_n$, such that*

$$\mathrm{Vol}_n\,(t_1K_1 + \cdots + t_mK_m) = \sum_{i_1,\dots,i_n=1}^{m} V(K_{i_1}, \dots, K_{i_n})\, t_{i_1} \cdots t_{i_n}$$

for all $t_1, \dots, t_m \geqslant 0$. The coefficient $V(K_{i_1}, \dots, K_{i_n})$ is the mixed volume of $K_{i_1}, \dots, K_{i_n}$ and depends only on these n bodies. We use $V(K; m, T; n-m)$ to denote the mixed volume of m copies of K with $n-m$ copies of T.

A special case of the above is the formula for $\mathrm{Vol}_n(K+sT)$, that is a polynomial of degree n in $s > 0$. Steiner's formula, which was already known in 1840, is the even more special case of that where T is the Euclidean ball. The volume of $K + tB_2^n$, $t > 0$, can be expanded as a polynomial in t:

$$\mathrm{Vol}_n\,(K + tB_2^n) = \sum_{i=0}^{n} \binom{n}{i} W_i(K) t^i$$

where $W_i(K) = V(K; n-i, B_2^n; i)$ is called the *i-th quermassintegral* of K. It is easy to see (depending on the definition one gives for surface area, see Section 1.5.3 for details) that the surface area $\partial(K) := \mathrm{Vol}_{n-1}(\partial K)$ of K is given by

$$\partial(K) = nW_1(K).$$

We present the proof of Minkowski's theorem in the appendix, and there we also prove many useful and simple facts about mixed volumes, which we now list for convenience.

(i) We have a polarization formula for mixed volumes, given by

$$V(K_1, \dots, K_n) = \frac{1}{n!} \sum_{k=1}^{n} (-1)^{n+k} \sum_{1 \leqslant j_1 < \cdots < j_k \leqslant n} \mathrm{Vol}_n\Big(\sum_{m=1}^{k} K_{j_m} \Big).$$

(ii) It is easy to see by the definitions that V is multilinear in each argument and that $V(AK_1, AK_2, \dots, AK_n) = V(K_1, K_2, \dots, K_n)$ for $A \in SL_n$.

(iii) V is continuous with respect to each of its arguments. As stated in the theorem, $V \geqslant 0$ and furthermore it is monotone increasing in each argument.

(iv) In the case of polytopes one may write an inductive formula for the coefficients $V(K_1, \dots, K_n)$, which involves the support function of, say, K_1, and the $(n-1)$-dimensional mixed volumes of the $(n-1)$-dimensional facets of the other bodies. In the case of general bodies the respective formula involves mixed measures, that we do not need at this stage at all. However, in the case of $W_i(K)$ there is a beautiful *integral formula of Kubota* that allows one to compute the coefficients:

$$W_i(K) = \frac{\kappa_n}{\kappa_{n-i}} \int_{G_{n,n-i}} \mathrm{Vol}_{n-i}(P_F(K))\, d\nu_{n,n-i}(F), \tag{1.1.1}$$

where integration is with respect to the Haar probability measure $\nu_{n,n-i}$ on $G_{n,n-i}$, namely on the family of i-codimensional subspaces of $\mathbb{R}^n$, and where κ_m denotes the volume of the Euclidean unit ball in $\mathbb{R}^m$. We prove this formula in the appendix as well. A special case of this formula is the *Cauchy formula* for the surface area:

$$\partial(K) = \frac{n\kappa_n}{\kappa_{n-1}} \int_{S^{n-1}} \mathrm{Vol}_{n-1}(P_{u^\perp}(K))\, d\sigma(u),$$

where $P_{u^\perp}(K)$ is the orthogonal projection of K onto the subspace $u^\perp = \{x \in \mathbb{R}^n : \langle x, u\rangle = 0\}$.

There are many interesting and mysterious inequalities which hold for mixed volumes, the most important one being the Alexandrov-Fenchel inequality, from which one may deduce many others. Let us quote it here, leaving the proof, and many implications, for the appendix.

THEOREM 1.1.11 (Alexandrov-Fenchel). *Let $K, L, K_3, \dots, K_n$ be non-empty compact convex subsets of $\mathbb{R}^n$. Then,*

$$V(K, L, K_3, \dots, K_n)^2 \geqslant V(K, K, K_3, \dots, K_n) \cdot V(L, L, K_3, \dots, K_n).$$

As an application of this inequality one may show the following two facts: for all $0 \leqslant i < n$ one has

$$W_i(K+L)^{\frac{1}{n-i}} \geqslant W_i(K)^{\frac{1}{n-i}} + W_i(L)^{\frac{1}{n-i}},$$

and for all $n > i > j \geqslant 0$,

$$\left(\frac{W_i(K)}{\mathrm{Vol}_n(B_2^n)}\right)^{\frac{1}{n-i}} \geqslant \left(\frac{W_j(K)}{\mathrm{Vol}_n(B_2^n)}\right)^{\frac{1}{n-j}}.$$

These are called Alexandrov inequalities.

1.1.6. Mixed discriminants

The Alexandrov-Fenchel inequalities are the most advanced representatives of a series of very important inequalities. They should perhaps be called "hyperbolic" inequalities in contrast to the more often used in analysis "elliptic" inequalities: Cauchy-Schwarz, Hölder, and their consequences (various triangle inequalities). The reasons for calling these inequalities "hyperbolic" will become clearer in the appendix, where we discuss these inequalities and their relatives in detail. A consequence of "hyperbolic" inequalities is *concavity* of some important quantities. There are similar inequalities going back to Newton which hold for symmetric functions. We next describe a number of very interesting inequalities concerning mixed discriminants, which are quite similar to Newton's inequalities. They were first discovered by Alexandrov in one of his approaches to the Alexandrov-Fenchel inequalities quoted above.

Let $A_1, \dots, A_m$ be symmetric matrices. Clearly, $\det(t_1 A_1 + \cdots + t_m A_m)$ is a homogeneous polynomial of degree n in $t_1, \dots, t_m$, where $t_j \in \mathbb{R}$. Its coefficients (which are chosen so that they are symmetric with respect to permutations) are denoted by $D(A_{i_1}, \dots, A_{i_n})$ and are called *mixed discriminants*. We therefore have

$$\det(t_1 A_1 + \cdots + t_m A_m) = \sum_{i_1=1}^{m} \cdots \sum_{i_n=1}^{m} D(A_{i_1}, \dots, A_{i_n})\, t_{i_1} \dots t_{i_n}.$$

Furthermore, it is easy to show the polarization formula

$$D(A_1, \dots, A_n) = \frac{1}{n!} \sum_{\varepsilon \in \{0,1\}^n} (-1)^{n+\sum \varepsilon_i} \det\left(\sum \varepsilon_i A_i\right),$$

which should be compared with polarization for mixed volumes. Indeed, this type of formula is simply a fact about polynomials. If the matrices are assumed to be positive definite, then the coefficients $D(A_1, \dots, A_n)$ are non-negative. The fact that the polynomial $P(t) = \det(A + tI)$ has only real roots for any $A \geqslant 0$ plays a central role in the proof of this fact.

Let us mention some of the inequalities which apply to mixed discriminants of positive semi-definite matrices; a detailed account of inequalities for mixed discriminants will be given in the appendix. A main inequality of Alexandrov from which all others follow is

THEOREM 1.1.12 (Alexandrov). *Let $A, B, C_3, \dots, C_n$ be positive semi-definite (symmetric) $n \times n$ matrices, Then*

$$D(A, B, C_3, \dots, C_n)^2 \geqslant D(A, A, C_3, \dots, C_n) \cdot D(B, B, C_3, \dots, C_n).$$

From this it follows that

$$D(A_1, A_2, \dots, A_n) \geqslant \prod_{i=1}^{n} [\det A]^{\frac{1}{n}},$$

and also that the function $[\det A]^{1/n}$, defined on the cone of positive definite matrices, is concave, namely that the following inequality, called Minkowski's inequality, is true:

$$[\det(A_1 + A_2)]^{\frac{1}{n}} \geqslant [\det A_1]^{\frac{1}{n}} + [\det A_2]^{\frac{1}{n}}.$$

This last fact will be useful for us and has many simple self-contained proofs, one of which we shall present in Section 2.1.2. We explain more on mixed discriminants in the appendix.

1.1.7. Steiner symmetrization

To end this introductory section, let us describe the process of *successive Steiner symmetrizations* that one can apply to a convex body so as to turn it into a ball. This method is very useful in proving geometric inequalities, as we shall see below.

DEFINITION 1.1.13. Let $K \subseteq \mathbb{R}^n$ be a convex body and $u \in S^{n-1}$. The *Steiner symmetral* $S_u(K)$ of K in the direction of u is defined so that for any $x \in u^{\perp}$

$$\mathrm{Vol}\left((x + \mathbb{R}u) \cap K\right) = \mathrm{Vol}\left((x + \mathbb{R}u) \cap S_u(K)\right),$$

and so that the right hand term is an interval centered at x.

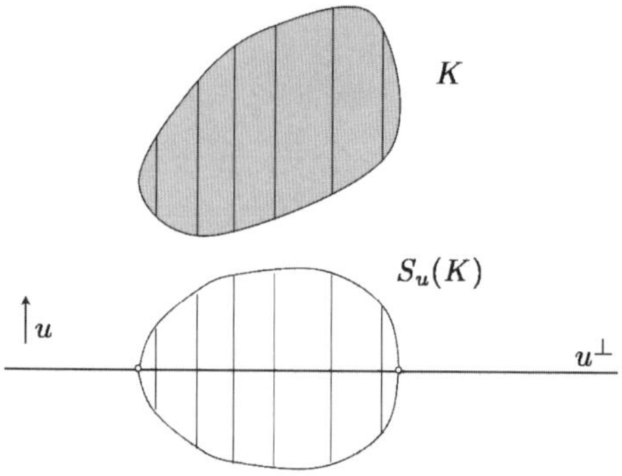

The most useful fact about Steiner symmetrization, which was observed by Brunn in [**124**], is that it preserves convexity, namely that, if $K \subseteq \mathbb{R}^n$ is a convex body and $u \in S^{n-1}$, the body $S_u(K)$ is convex. The proof follows easily by considering trapezoids. Indeed, convexity is a two-dimensional notion, so it is enough to consider x and y in $u^\perp$ and check that for any $0 < \lambda < 1$ the interval centered at $z = (1-\lambda)x + \lambda y$ with length $l_z = \mathrm{Vol}(K \cap (z + \mathbb{R}u))$ has length greater than or equal to $(1-\lambda)l_x + \lambda l_y$; the latter follows from the fact that, by convexity, $(1-\lambda)(K \cap (x + \mathbb{R}u)) + \lambda(K \cap (y + \mathbb{R}u)) \subseteq K \cap (z + \mathbb{R}u)$.

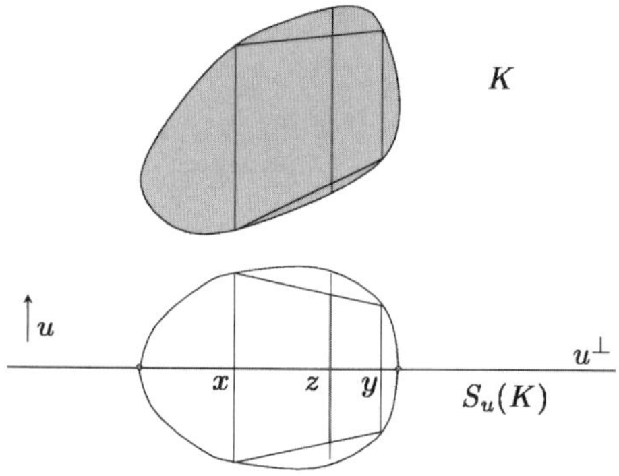

Steiner symmetrization has many useful properties, some of which we collect in the following list. For us the most important features are the fact that it preserves volume (which follows easily from Fubini's theorem) and the way it interacts with Minkowski addition, namely that the symmetral of $K + T$ in a direction u contains $S_u(K) + S_u(T)$. Let $K, T \in \mathcal{K}^n$ and $u \in S^{n-1}$. Then:

(i) $\lambda S_u(K) = S_u(\lambda K)$.
(ii) $S_u(K) + S_u(T) \subseteq S_u(K + T)$.
(iii) $K \subseteq T$ implies $S_u(K) \subseteq S_u(T)$.
(iv) S_u is continuous with respect to the Hausdorff metric.
(v) $\mathrm{Vol}_n(S_u(K)) = \mathrm{Vol}_n(K)$.
(vi) $\partial(S_u(K)) \leqslant \partial(K)$, where $\partial(C)$ denotes the surface area of C.
(vii) $\mathrm{diam}(S_u(K)) \leqslant \mathrm{diam}(K)$.
(viii) $\mathrm{inradius}(S_u(K)) \geqslant \mathrm{inradius}(K)$ and $\mathrm{circumrad}(S_u(K)) \leqslant \mathrm{circumrad}(K)$.

REMARK 1.1.14. In fact not only is surface area monotone with respect to Steiner symmetrizations, but so are all the quermassintergrals W_i.

Finally, we shall use the following monotonicity-of-dual-volume result. We quote it here for completeness, but return to it in Section 1.5.4 where we provide its proof.

PROPOSITION 1.1.15. Let K be a centrally symmetric convex body in $\mathbb{R}^n$ and let $u \in S^{n-1}$. Then

$$\mathrm{Vol}_n(K^\circ) \leqslant \mathrm{Vol}_n((S_u(K))^\circ).$$

Having seen that various quantities are constant, or monotone, with respect to Steiner symmetrization, our next step is to explain the fact that applying successive symmetrizations one approaches the Euclidean ball. We shall not prove this claim in detail but indicate roughly how it is done.

THEOREM 1.1.16. *Let $K \subset \mathbb{R}^n$ be a convex body such that $\mathrm{Vol}_n(K) = \mathrm{Vol}_n(B_2^n) =: \kappa_n$. Then there exists a sequence of vectors $u_j \in S^{n-1}$ such that, setting $K_j = S_{u_j}(K_{j-1})$, we have $K_j \to B_2^n$ (in the Hausdorff metric).*

Proof. Consider the class of all bodies obtained by successive symmetrizations of the original body K. They are all included in any ball around 0 that contains K. Take the infimal circumradius of bodies in this family, R_0. Take a sequence of symmetrals whose circumradius converges to R_0, and by Theorem 1.1.9, a convergent subsequence of it, $K_j \to L$. First we shall explain how one shows that the limiting body L is a ball (its circumradius must be the infimal circumradius of course). Assume the contrary, then L is included in the ball with the infimal radius, and misses some cap of this ball (say a cap or radius $\delta > 0$). By compactness one may cover the boundary of this ball with finitely many mirror images of this cap with respect to a sequence of hyperplanes.

Pick a body in the sequence that is very close to the limit, which implies in particular that so are all of its symmetrizations. One can check that symmetrizing the limit body with respect to all these hyperplanes will produce a body contained in a ball which is strictly inside the original one, contradicting minimality of its circumradius.

We are almost done, as we have shown that for any ε we can find a sequence of symmetrizations which bring us close to a ball up to ε. We may now build our sequence inductively. Note that this proves in fact that we can find a sequence of symmetrizations with respect to which several bodies, together, converge to a ball. □

REMARK 1.1.17. The question of how fast these symmetrizations approach B_2^n (given the best choice of u_j, or a random choice, or a semi-random one) is of high importance in this field. The reader is referred to Part II, where it is shown that one needs $O(n)$ symmetrizations in order to approach a Euclidean ball within small distance.

1.2. Brunn–Minkowski inequality

The Brunn-Minkowski inequality provides a fundamental relation between volume in $\mathbb{R}^n$ and Minkowski addition.

Brunn-Minkowski inequality. *Let K and T be two non-empty compact subsets of $\mathbb{R}^n$. Then,*

$$\mathrm{Vol}_n(K+T)^{1/n} \geqslant \mathrm{Vol}_n(K)^{1/n} + \mathrm{Vol}_n(T)^{1/n}. \tag{1.2.1}$$

If we make the additional hypothesis that K and T are convex bodies, then we can have equality in (1.2.1) only if K and T are homothetical.

In this section we provide many different proofs for this inequality, each of which is beautiful and reflects different ideas of symmetrizations.

The inequality expresses in a sense the fact that volume, to the correct power, is a *concave* function with respect to Minkowski addition. For this reason, it is often written in the following form: if K and T are non-empty compact subsets of $\mathbb{R}^n$ and $\lambda \in (0,1)$, then

$$\mathrm{Vol}_n(\lambda K + (1-\lambda)T)^{1/n} \geqslant \lambda \mathrm{Vol}_n(K)^{1/n} + (1-\lambda)\mathrm{Vol}_n(T)^{1/n}. \tag{1.2.2}$$

Using (1.2.2) and the arithmetic-geometric means inequality we can also write

$$\mathrm{Vol}_n(\lambda K + (1-\lambda)T) \geqslant \mathrm{Vol}_n(K)^{\lambda}\mathrm{Vol}_n(T)^{1-\lambda}. \tag{1.2.3}$$

This weaker form of the Brunn-Minkowski inequality has the advantage (or, for some problems, the disadvantage) of being dimension-free. It is actually equivalent to (1.2.1) in the following sense: if we know that (1.2.3) holds for all K, T and λ, we can recover (1.2.1) as follows.

Consider non-empty compact sets K and T (we may assume that both $\mathrm{Vol}_n(K) > 0$ and $\mathrm{Vol}_n(T) > 0$, otherwise there is nothing to prove), and define

$$K_1 = \mathrm{Vol}_n(K)^{-1/n}K,\ T_1 = \mathrm{Vol}_n(T)^{-1/n}T \text{ and } \lambda = \frac{\mathrm{Vol}_n(K)^{1/n}}{\mathrm{Vol}_n(K)^{1/n} + \mathrm{Vol}_n(T)^{1/n}}.$$

Then, K_1 and T_1 have volume 1, and hence (1.2.3) gives

$$\mathrm{Vol}_n(\lambda K_1 + (1-\lambda)T_1) \geqslant 1.$$

Since

$$\lambda K_1 + (1-\lambda)T_1 = \frac{K+T}{\mathrm{Vol}_n(K)^{1/n} + \mathrm{Vol}_n(T)^{1/n}},$$

we immediately get (1.2.1).

1.2.1. Brunn's concavity principle

Historically, the first proof of the Brunn-Minkowski inequality was based on Brunn's concavity principle:

THEOREM 1.2.1. *Let K be a convex body in $\mathbb{R}^n$ and let F be a k-dimensional subspace. Then, the function $f : F^{\perp} \to \mathbb{R}$ defined by*

$$f(x) = \mathrm{Vol}_k(K \cap (F+x))^{1/k}$$

is concave on its support.

Sketch of the proof. The proof goes by way of Steiner symmetrizations of K in directions $u \in F$. As a small variation of the argument in Theorem 1.1.16 explained above, one can find a sequence of successive Steiner symmetrizations in directions $u \in F$ so that the limiting convex body $\tilde{K}$ has the following property:

> For every $x \in F^{\perp}$, $\tilde{K} \cap (F+x)$ is a ball with centre at x and radius $r(x)$ such that $\mathrm{Vol}_k(\tilde{K} \cap (F+x)) = \mathrm{Vol}_k(K \cap (F+x))$.

Now, the proof of the theorem is immediate. Convexity of $\tilde{K}$ implies that the function r is concave on its support, and this shows that f is also concave. □

Proof of the Brunn-Minkowski inequality. Brunn's concavity principle implies the Brunn-Minkowski inequality for convex bodies as follows. If K and T are convex bodies in $\mathbb{R}^n$, we define

$$K_1 = K \times \{0\} \quad \text{and} \quad T_1 = T \times \{1\}$$

in $\mathbb{R}^{n+1}$ and consider their convex hull L. If

$$L(t) = \{x \in \mathbb{R}^n : (x,t) \in L\} \qquad (t \in [0,1])$$

we easily check that $L(0) = K$, $L(1) = T$ and

$$L(1/2) = \frac{K+T}{2}.$$

Then, Brunn's concavity principle for $F = \mathbb{R}^n$ shows that

$$\mathrm{Vol}_n\Big(\frac{K+T}{2}\Big)^{1/n} \geqslant \frac{1}{2}\mathrm{Vol}_n(K)^{1/n} + \frac{1}{2}\mathrm{Vol}_n(T)^{1/n},$$

and (1.2.1) follows. □

REMARK 1.2.2. A similar proof using another property of Steiner symmetrizations goes as follows: pick a sequence of Steiner symmetrizations $(S_{u_m})_{m\in\mathbb{N}}$ such that the symmetrals of K, of T, and of $K+T$ all converge to balls, say of radii r_K, r_T and r_{K+T} respectively. We have

$$S_{u_2}(S_{u_1}(K+T)) \supseteq S_{u_2}(S_{u_1}(K) + S_{u_1}(T)) \supseteq S_{u_2}(S_{u_1}(K)) + S_{u_2}(S_{u_1}(T)),$$

and by induction this inclusion remains valid for all symmetrals in the sequence. In the limit we get

$$r_{K+T}B_2^n \supseteq (r_K + r_T)B_2^n,$$

which is equivalent to the inequality $r_{K+T} \geqslant r_K + r_T$. By writing out the radii explicitly we get the Brunn-Minkowski inequality.

We proceed with yet another, alternative, proof of Theorem 1.2.1, which is due to Gromov and Milman. Their argument proves a more general statement.

DEFINITION 1.2.3. Let K be a convex set in $\mathbb{R}^n$ and let $f : K \to \mathbb{R}^+$. We say that f is *α-concave* for some $\alpha > 0$ if $f^{1/\alpha}$ is concave on K. Equivalently, if

$$f^{1/\alpha}(\lambda x_1 + \mu x_2) \geqslant \lambda f^{1/\alpha}(x_1) + \mu f^{1/\alpha}(x_2)$$

for all $x_1, x_2 \in K$ and $\lambda, \mu > 0$ with $\lambda + \mu = 1$.

Note that as $\alpha \to 0^+$ we get Definition 1.1.3 of log-concave functions.

LEMMA 1.2.4. *Let $f, g : K \to \mathbb{R}^+$. If f is α-concave and g is β-concave, then fg is $(\alpha+\beta)$-concave.*

Proof. Applying Hölder's inequality (for a two-point probability space) with $p = (\alpha+\beta)/\alpha$ and $q = (\alpha+\beta)/\beta$, we get

$$\begin{aligned}
&\lambda\big(f(x_1)g(x_1)\big)^{\frac{1}{\alpha+\beta}} + \mu\big(f(x_2)g(x_2)\big)^{\frac{1}{\alpha+\beta}} \\
&\qquad \leqslant \Big(\lambda f(x_1)^{\frac{1}{\alpha}} + \mu f(x_2)^{\frac{1}{\alpha}}\Big)^{\frac{\alpha}{\alpha+\beta}} \Big(\lambda g(x_1)^{\frac{1}{\beta}} + \mu g(x_2)^{\frac{1}{\beta}}\Big)^{\frac{\beta}{\alpha+\beta}} \\
&\qquad \leqslant (f(\lambda x_1 + \mu x_2))^{\frac{1}{\alpha+\beta}} (g(\lambda x_1 + \mu x_2))^{\frac{1}{\alpha+\beta}}
\end{aligned}$$

for all $x_1, x_2 \in K$ and $\lambda, \mu > 0$ with $\lambda + \mu = 1$, by the α-concavity of f and the β-concavity of g. $\square$

Now, let K be a convex body in $\mathbb{R}^n$ and let $\theta \in S^{n-1}$. For every $y \in P_{\theta^\perp}(K)$ we write I_y for the set $\{t \in \mathbb{R} : y + t\theta \in K\}$. From the convexity of K it follows that I_y is an interval for every $y \in P_{\theta^\perp}(K)$. Let $f : K \to \mathbb{R}^+$ be a continuous function. We consider the marginal $P_\theta f : P_{\theta^\perp}(K) \to \mathbb{R}^+$ of f with respect to θ, defined by

$$(P_\theta f)(y) := \int_{I_y} f(y + t\theta)dt, \qquad y \in P_{\theta^\perp}(K).$$

LEMMA 1.2.5. *If f is α-concave, then $P_\theta f$ is $(1 + \alpha)$-concave.*

Proof. Concavity of a function is a "two-dimensional" notion, so we may clearly assume that $K \subseteq \mathbb{R}^2$, in which case $P_{\theta^\perp}(K)$ is an interval. Let $y_1, y_2 \in P_{\theta^\perp}(K)$ and write $I_{y_i} = [a_i, b_i]$, $i = 1, 2$. For every $\lambda, \mu > 0$ with $\lambda + \mu = 1$, we set $y_\lambda = \lambda y_1 + \mu y_2$. Then,

$$I_{y_\lambda} \supseteq [\lambda a_1 + \mu a_2, \lambda b_1 + \mu b_2].$$

We define $c_i \in I_{y_i}$ by the equations

$$\int_{a_i}^{b_i} f(y_i + t\theta)dt = 2\int_{a_i}^{c_i} f(y_i + t\theta)dt = 2\int_{c_i}^{b_i} f(y_i + t\theta)dt.$$

If K' is the convex hull of the intervals $[y_i + c_i\theta, y_i + b_i\theta]$, $i = 1, 2$ and K'' is the convex hull of the intervals $[y_i + a_i\theta, y_i + c_i\theta]$, $i = 1, 2$, we define $f' = f|_{K'}$ and $f'' = f|_{K''}$. By the definition of c_i it is not hard to check that, if $P_\theta f'$ and $P_\theta f''$ are $(1+\alpha)$-concave, then $P_\theta f$ satisfies the $(1+\alpha)$-concavity condition at y_1, y_2 and y_λ for every λ. Thus, we only need to prove that $P_\theta f'$ and $P_\theta f''$ are $(1 + \alpha)$-concave.

For every $n \geqslant 2$ we define partitions $a_i = t_{0,i} < t_{1,i} < \cdots < t_{n-1,i} < t_{n,i} = b_i$ of $[a_i, b_i]$ such that

$$\int_{a_i}^{b_i} f(y_i + t\theta)dt = n\int_{t_{k-1,i}}^{t_{k,i}} f(y_i + t\theta)dt$$

for all $k \in \{1, \ldots, n\}$. The same observation as above shows that it suffices to check that $P_\theta f^k$ is $(1+\alpha)$-concave for every k, where $f^k = f|_{K^{(k)}}$ and $K^{(k)}$ is the convex hull of the intervals $[y_i + t_{k-1,i}\theta, y_i + t_{k,i}\theta]$.

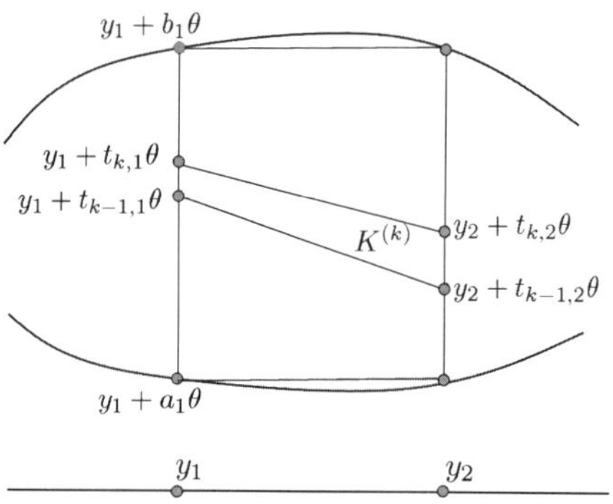

Passing to the limit we see that we have to check the following infinitesimal claim:

Let $t_i \in I_{y_i}$ and $d_i > 0$, $i = 1, 2$. Given $\lambda \in (0, 1)$, set $y_\lambda = \lambda y_1 + (1-\lambda) y_2$, $t(\lambda) = \lambda t_1 + (1-\lambda) t_2$ and $d(\lambda) = \lambda d_1 + (1-\lambda) d_2$. Then, the function $\lambda \mapsto f(y_\lambda + t(\lambda)\theta) \cdot d(\lambda)$ is $(1+\alpha)$-concave.

This claim follows from Lemma 1.2.4, since $\lambda \mapsto f(y_\lambda + t(\lambda)\theta)$ is α-concave and the linear function $\lambda \mapsto d(\lambda)$ is 1-concave. □

Finally, with Lemma 1.2.5 we can give another proof of Brunn's Theorem 1.2.1 as follows. The indicator function of K is constant on K, and hence it is α-concave for every $\alpha > 0$. We choose an orthonormal basis $\{\theta_1, \dots, \theta_k\}$ of F and perform successive projections in the directions of θ_i. Lemma 1.2.5 shows that the function $x \mapsto \mathrm{Vol}_k(K \cap (F + x))$ is $(\alpha + k)$-concave on $P_{F^\perp}(K)$, for every $\alpha > 0$. It follows that $f(x) = \mathrm{Vol}_k(K \cap (F + x))^{1/k}$ is concave. □

1.2.2. Elementary sets

Next, we give a proof of the Brunn-Minkowski inequality for non-empty compact sets in $\mathbb{R}^n$ that are not necessarily convex. It is due to Hadwiger and Ohmann. The argument is based on *elementary sets.* An elementary set is a finite union of non-overlapping boxes whose edges are parallel to the coordinate axes.

THEOREM 1.2.6. *Let A and B be elementary sets in $\mathbb{R}^n$. Then,*

$$\mathrm{Vol}_n(A + B)^{1/n} \geqslant \mathrm{Vol}_n(A)^{1/n} + \mathrm{Vol}_n(B)^{1/n}.$$

Proof. We first examine the case where both A and B are boxes. Assume that $a_1, \dots, a_n > 0$ are the lengths of the edges of A, and $b_1, \dots, b_n > 0$ are the lengths of the edges of B. Then, $A + B$ is also a box, whose edges have lengths $a_1 + b_1, \dots, a_n + b_n$. Thus, we have to show that

$$((a_1 + b_1) \cdots (a_n + b_n))^{1/n} \geqslant (a_1 \cdots a_n)^{1/n} + (b_1 \cdots b_n)^{1/n}.$$

This is equivalent to the inequality

$$\left(\frac{a_1}{a_1 + b_1} \cdots \frac{a_n}{a_n + b_n} \right)^{1/n} + \left(\frac{b_1}{a_1 + b_1} \cdots \frac{b_n}{a_n + b_n} \right)^{1/n} \leqslant 1.$$

But, the arithmetic-geometric means inequality shows that the left hand side is less than or equal to

$$\frac{1}{n} \left(\frac{a_1}{a_1 + b_1} + \cdots + \frac{a_n}{a_n + b_n} \right) + \frac{1}{n} \left(\frac{b_1}{a_1 + b_1} + \cdots + \frac{b_n}{a_n + b_n} \right) = 1.$$

For every pair of elementary sets A and B, we define the *complexity* of (A, B) as the total number of boxes in A and B. We will prove the theorem by induction on the complexity m of (A, B). The case $m = 2$ was our first step.

Assume then that $m \geqslant 3$ and that the statement holds true for all pairs of elementary sets with complexity less than or equal to $m - 1$. Since $m \geqslant 3$, we may assume that A consists of at least two boxes. Let I_1 and I_2 be two of them. Since they are non-overlapping we can separate them by a coordinate hyperplane which, without loss of generality, may be described by the equation $x_n = \rho$ for some $\rho \in \mathbb{R}$. We define

$$A^+ = A \cap \{x \in \mathbb{R}^n : x_n \geqslant \rho\} \text{ and } A^- = A \cap \{x \in \mathbb{R}^n : x_n \leqslant \rho\}.$$

Then, A^+ and A^- are non-overlapping elementary sets, and each one of them consists of a smaller number of boxes than A. We now pass to B, and find a hyperplane $x_n = s$ such that if $B^+ = B \cap \{x \in \mathbb{R}^n : x_n \geqslant s\}$ and $B^- = B \cap \{x \in \mathbb{R}^n : x_n \leqslant s\}$, then

$$\frac{\mathrm{Vol}_n(A^+)}{\mathrm{Vol}_n(A)} = \frac{\mathrm{Vol}_n(B^+)}{\mathrm{Vol}_n(B)} =: \lambda. \tag{1.2.4}$$

Then, B^+ and B^- are elementary sets, we clearly have $0 < \lambda < 1$, and the complexity of each pair $(A^\pm, B^\pm)$ is less than m. It is clear that

$$A + B = (A^+ + B^+) \cup (A^+ + B^-) \cup (A^- + B^+) \cup (A^- + B^-).$$

On the other hand, since $A^+ + B^+ \subseteq \{x : x_n \geqslant \rho + s\}$ and $A^- + B^- \subseteq \{x : x_n \leqslant \rho + s\}$, the sets $A^+ + B^+$ and $A^- + B^-$ have disjoint interiors. Therefore,

$$\mathrm{Vol}_n(A + B) \geqslant \mathrm{Vol}_n(A^+ + B^+) + \mathrm{Vol}_n(A^- + B^-).$$

Our inductive hypothesis can be applied to the right hand side, giving

$$\mathrm{Vol}_n(A^+ + B^+)^{\frac{1}{n}} \geqslant \mathrm{Vol}_n(A^+)^{\frac{1}{n}} + \mathrm{Vol}_n(B^+)^{\frac{1}{n}}$$

and

$$\mathrm{Vol}_n(A^- + B^-)^{\frac{1}{n}} \geqslant \mathrm{Vol}_n(A^-)^{\frac{1}{n}} + \mathrm{Vol}_n(B^-)^{\frac{1}{n}}.$$

Taking the definition of λ into account we get

$$\begin{aligned}\mathrm{Vol}_n(A + B) &\geqslant \lambda\big(\mathrm{Vol}_n(A)^{\frac{1}{n}} + \mathrm{Vol}_n(B)^{\frac{1}{n}}\big)^n + (1-\lambda)\big(\mathrm{Vol}_n(A)^{\frac{1}{n}} + \mathrm{Vol}_n(B)^{\frac{1}{n}}\big)^n \\ &= \big(\mathrm{Vol}_n(A)^{\frac{1}{n}} + \mathrm{Vol}_n(B)^{\frac{1}{n}}\big)^n,\end{aligned}$$

which proves the theorem. □

Since every compact set in $\mathbb{R}^n$ can be approximated by a decreasing sequence of elementary sets in the Hausdorff metric, Theorem 1.2.6 is easily extended to non-empty compact sets A and B in $\mathbb{R}^n$.

THEOREM 1.2.7. *Let C_1 and C_2 be non-empty compact sets in $\mathbb{R}^n$. Then,*

$$\mathrm{Vol}_n(C_1 + C_2)^{1/n} \geqslant \mathrm{Vol}_n(C_1)^{1/n} + \mathrm{Vol}_n(C_2)^{1/n}.$$

1.2.3. Induction on the dimension

Our next proof of the Brunn-Minkowski inequality is by induction on the dimension and is due to Kneser and Süss. The argument works only for convex bodies. One of the advantages of the proof is that it clarifies the cases of equality in the convex case. A second one is that a suitable modification of the argument leads to "stability results" (see notes and remarks at the end of the chapter).

THEOREM 1.2.8. *Let K_0 and K_1 be convex bodies in $\mathbb{R}^n$. For every $\lambda \in [0, 1]$,*

$$\mathrm{Vol}_n((1-\lambda)K_0 + \lambda K_1)^{1/n} \geqslant (1-\lambda)\mathrm{Vol}_n(K_0)^{1/n} + \lambda \mathrm{Vol}_n(K_1)^{1/n}$$

with equality only if K_0 and K_1 are homothetical.

Proof. We prove the theorem by induction on n. The statement is clearly true when $n = 1$, therefore we may assume that $n \geqslant 2$ and that the theorem has been proved in dimension $n - 1$. We may also assume, by the same argument as the one given at the beginning of Section 1.2, that $\mathrm{Vol}_n(K_0) = \mathrm{Vol}_n(K_1) = 1$ and by translation invariance of volume, that the origin is the barycenter of both K_0 and K_1. Given $\theta \in S^{n-1}$ we define

$$f_i(t) = \mathrm{Vol}_{n-1}(\{x \in K_i : \langle x, \theta\rangle = t\})$$
$$g_i(t) = \mathrm{Vol}_n(\{x \in K_i : \langle x, \theta\rangle \leqslant t\})$$

for $i = 0, 1$ and $t \in \big(-h_{K_i}(-\theta), h_{K_i}(\theta)\big)$. Then,

$$g_i(t) = \int_{-h_{K_i}(-\theta)}^{t} f_i(s)ds.$$

Since f_i is continuous in this domain and strictly positive, we have

$$g_i'(t) = f_i(t), \qquad -h_{K_i}(-\theta) < t < h_{K_i}(\theta)$$

and g_i is strictly increasing. Now, if $h_i : (0,1) \to \mathbb{R}$ is the inverse function of g_i, we have

$$h_i'(u) = \frac{1}{g_i'(h_i(u))} = \frac{1}{f_i(h_i(u))}, \qquad 0 < u < 1.$$

Let $K_\lambda = (1-\lambda)K_0 + \lambda K_1$ and write $K_\lambda(u) = K_\lambda \cap \{y + h_\lambda(u)\theta : y \in \theta^\perp\}$ for every $\lambda \in [0,1]$ and $0 < u < 1$, where $h_\lambda = (1-\lambda)h_0 + \lambda h_1$. We can easily check that for every $u \in [0,1]$

$$K_\lambda(u) \supseteq (1-\lambda)K_0(u) + \lambda K_1(u).$$

Therefore, if we make the change of variables $t = h_\lambda(u)$ we get

$$\begin{aligned}
\mathrm{Vol}_n(K_\lambda) &= \int_{-h_{K_\lambda}(-\theta)}^{h_{K_\lambda}(\theta)} \mathrm{Vol}_{n-1}(K_\lambda \cap (\theta^\perp + t\theta))dt \\
&\geqslant \int_0^1 \mathrm{Vol}_{n-1}(K_\lambda(u))h_\lambda'(u)du \\
&\geqslant \int_0^1 \mathrm{Vol}_{n-1}((1-\lambda)K_0(u) + \lambda K_1(u)) \left(\frac{1-\lambda}{f_0(h_0(u))} + \frac{\lambda}{f_1(h_1(u))}\right) du.
\end{aligned}$$

The inductive hypothesis shows that this is greater than or equal to

$$\int_0^1 \left((1-\lambda)f_0(h_0(u))^{\frac{1}{n-1}} + \lambda f_1(h_1(u))^{\frac{1}{n-1}}\right)^{n-1} \left(\frac{1-\lambda}{f_0(h_0(u))} + \frac{\lambda}{f_1(h_1(u))}\right) du.$$

By the arithmetic-geometric means inequality applied to each term, the integrand is greater than or equal to 1 on $(0,1)$ with equality only if $f_0(h_0(u)) = f_1(h_1(u))$. This shows that $\mathrm{Vol}_n(K_\lambda) \geqslant 1$ and completes the inductive step.

Suppose that for some $\lambda \in (0,1)$ we have $\mathrm{Vol}_n((1-\lambda)K_0 + \lambda K_1) = 1$. Then, $f_0(h_0(u)) = f_1(h_1(u))$ for every $u \in (0,1)$, which implies $h_0' = h_1'$. Thus, $h_1 - h_0$ is a constant function. Since 0 is the barycenter of K_i,

$$\begin{aligned}
0 &= \int_{K_i} \langle x, \theta\rangle dx = \int t f_i(t)\, dt = \int_0^1 h_i(u) f_i(h_i(u)) h_i'(u)\, du \\
&= \int_0^1 h_i(u)du,
\end{aligned}$$

which shows that $h_0 \equiv h_1$. In particular, their value as $u \to 1$ is the same, so that $h_{K_0}(\theta) = h_{K_1}(\theta)$. Since this should hold for every $\theta \in S^{n-1}$, we must have $K_0 = K_1$. $\square$

1.3. Volume preserving transformations

Let K and T be two open convex bodies in $\mathbb{R}^n$. By a *volume preserving transformation* we mean a map $\phi : K \to T$ which is one to one, onto and has a Jacobian with constant determinant equal to $\mathrm{Vol}_n(T)/\mathrm{Vol}_n(K)$. In this section we describe two such maps, the Knothe map and the Brenier map. Applying each one of them we may obtain alternative proofs of the Brunn-Minkowski inequality.

1.3.1. Knothe map

We fix an orthonormal basis $\{e_1, \dots, e_n\}$ in $\mathbb{R}^n$, and consider two open convex bodies K and T. The properties of the *Knothe map* from K to T with respect to the given coordinate system are described in the following theorem.

THEOREM 1.3.1. *There exists a map $\phi : K \to T$ with the following properties:*

(a) *ϕ is triangular: the i-th coordinate function of ϕ depends only on $x_1, \dots, x_i$. That is,*

$$\phi(x_1, \dots, x_n) = (\phi_1(x_1), \phi_2(x_1, x_2), \dots, \phi_n(x_1, \dots, x_n)).$$

(b) *The partial derivatives $\frac{\partial \phi_i}{\partial x_i}$ are positive on K, and the determinant of the Jacobian matrix $J(\phi)$ of ϕ is constant. More precisely, for every $x \in K$*

$$|\det J(\phi)(x)| = \prod_{i=1}^{n} \frac{\partial \phi_i}{\partial x_i}(x) = \frac{\mathrm{Vol}_n(T)}{\mathrm{Vol}_n(K)}.$$

Proof. For each $i = 1, \dots, n$ and $s = (s_1, \dots, s_i) \in \mathbb{R}^i$ we consider the section

$$K_s = \{y \in \mathbb{R}^{n-i} : (s, y) \in K\}$$

of K (similarly for T). We shall define a one to one and onto map $\phi : K \to T$ as follows.

Let $x = (x_1, \dots, x_n) \in K$. Then, $K_{x_1} \neq \emptyset$ and we can define $\phi_1(x) = \phi_1(x_1)$ by

$$\frac{1}{\mathrm{Vol}_n(K)} \int_{-\infty}^{x_1} \mathrm{Vol}_{n-1}(K_{s_1}) ds_1 = \frac{1}{\mathrm{Vol}_n(T)} \int_{-\infty}^{\phi_1(x_1)} \mathrm{Vol}_{n-1}(T_{t_1}) dt_1.$$

In other words, we move in the direction of e_1 until we "catch" a percentage of T which is equal to the percentage of K occupied by $K \cap \{s = (s_1, \dots, s_n) : s_1 \leqslant x_1\}$. Note that ϕ_1 is defined on K but $\phi_1(x)$ depends only on the first coordinate of $x \in K$. Also,

$$\frac{\partial \phi_1}{\partial x_1}(x) = \frac{\mathrm{Vol}_n(T)}{\mathrm{Vol}_n(K)} \frac{\mathrm{Vol}_{n-1}(K_{x_1})}{\mathrm{Vol}_{n-1}(T_{\phi_1(x_1)})}.$$

We continue by induction. Assume that we have defined $\phi_1(x) = \phi_1(x_1)$, $\phi_2(x) = \phi_2(x_1, x_2)$ and $\phi_{j-1}(x) = \phi_{j-1}(x_1, \dots, x_{j-1})$ for some $j \geqslant 2$. If $x = (x_1, \dots, x_n) \in K$ then $K_{(x_1, \dots, x_{j-1})} \neq \emptyset$, and we define $\phi_j(x) = \phi_j(x_1, \dots, x_j)$ by

$$\begin{aligned} \frac{\mathrm{Vol}_{n-j+1}(T_{(\phi_1(x_1), \dots, \phi_{j-1}(x_1, \dots, x_{j-1}))})}{\mathrm{Vol}_{n-j+1}(K_{(x_1, \dots, x_{j-1})})} & \int_{-\infty}^{x_j} \mathrm{Vol}_{n-j}(K_{(x_1, \dots, x_{j-1}, s_j)}) ds_j \\ = \int_{-\infty}^{\phi_j(x_1, \dots, x_j)} & \mathrm{Vol}_{n-j}(T_{(\phi_1(x_1), \dots, \phi_{j-1}(x_1, \dots, x_{j-1}), t_j)}) dt_j. \end{aligned}$$

It is clear that

$$\frac{\partial \phi_j}{\partial x_j}(x) = \frac{\mathrm{Vol}_{n-j+1}(T_{(\phi_1(x),\dots,\phi_{j-1}(x))})}{\mathrm{Vol}_{n-j+1}(K_{(x_1,\dots,x_{j-1})})} \frac{\mathrm{Vol}_{n-j}(K_{(x_1,\dots,x_j)})}{\mathrm{Vol}_{n-j}(T_{(\phi_1(x),\dots,\phi_j(x))})}.$$

Continuing in this way, we obtain a map $\phi = (\phi_1, \dots, \phi_n) : K \to T$. It is easy to check that ϕ is one to one and onto. Note that

$$\frac{\partial \phi_n}{\partial x_n}(x) = \frac{\mathrm{Vol}_1(T_{(\phi_1(x),\dots,\phi_{n-1}(x))})}{\mathrm{Vol}_1(K_{(x_1,\dots,x_{n-1})})}.$$

By construction, ϕ has properties (a) and (b). □

Remark. Observe that each choice of coordinate system in $\mathbb{R}^n$ produces a different Knothe map from K onto T.

Using the Knothe map we can give one more proof of the Brunn-Minkowski inequality for convex bodies. We may clearly assume that K and T are open. Consider the Knothe map $\phi : K \to T$. It is clear that

$$(\mathrm{Id} + \phi)(K) \subseteq K + \phi(K) = K + T,$$

and, since $J(\mathrm{Id} + \phi)$ is triangular, we get

$$\begin{aligned} \mathrm{Vol}_n(K+T) &\geqslant \int_{(\mathrm{Id}+\phi)(K)} dx \\ &= \int_K |\det J(\mathrm{Id}+\phi)(x)| dx \\ &= \int_K \prod_{j=1}^n \Big(1 + \frac{\partial \phi_j}{\partial x_j}(x)\Big) dx \\ &\geqslant \int_K \Big(1 + \Big(\prod_{j=1}^n \frac{\partial \phi_j}{\partial x_j}(x)\Big)^{1/n}\Big)^n dx \\ &= \mathrm{Vol}_n(K)\Big(1 + \Big(\frac{\mathrm{Vol}_n(T)}{\mathrm{Vol}_n(K)}\Big)^{1/n}\Big)^n \\ &= \big(\mathrm{Vol}_n(K)^{1/n} + \mathrm{Vol}_n(T)^{1/n}\big)^n. \end{aligned}$$

1.3.2. Brenier map

In this section we show yet another way to produce a mapping $\psi : K \to T$ which is volume preserving. This time we will make sure that the Jacobian of ψ is positive definite, by ensuring $\psi = \nabla f$ for a (twice differentiable) convex function f. Using this map (called the *Brenier map*) we shall have one more proof of the Brunn-Minkowski inequality for convex bodies as follows: Since $(\mathrm{Id} + \psi)(K) \subseteq K + T$,

$$\begin{aligned} \mathrm{Vol}_n(K+T) &\geqslant \int_K |\det J(\mathrm{Id}+\psi)(x)| dx = \int_K |\det\,(\mathrm{Id} + \mathrm{Hess} f)\,(x)| dx \\ &= \int_K \prod_{i=1}^n (1 + \lambda_i(x)) dx, \end{aligned}$$

where $\lambda_i(x)$ are the non-negative eigenvalues of $\mathrm{Hess} f$. Moreover, by the ratio-of-volumes-preserving property of ψ, we have $\prod_{i=1}^n \lambda_i(x) = \mathrm{Vol}_n(T)/\mathrm{Vol}_n(K)$ for every $x \in K$. Therefore, the arithmetic-geometric means inequality gives

$$\mathrm{Vol}_n(K+T) \geqslant \int_K \Big(1 + \Big[\prod_{i=1}^n \lambda_i(x)\Big]^{1/n}\Big)^n dx = \big(\mathrm{Vol}_n(K)^{1/n} + \mathrm{Vol}_n(T)^{1/n}\big)^n.$$

This proof has the advantage of providing a description for the equality cases: either K or T must be a point, or K must be homothetical to T.

So, one must find a mapping which is of the form $\psi = \nabla f$ for a convex function f defined on K, the image of which is T. We shall first generalize the notion of "volume preserving" to mappings that push forward one measure into another. The special case of uniform measures on convex sets will be of particular interest.

DEFINITION 1.3.2. Consider the space $\mathcal{P}(\mathbb{R}^n)$ of Borel probability measures on $\mathbb{R}^n$ as a subset of the unit ball of $C_\infty(\mathbb{R}^n)^*$ (the dual of the space of infinitely differentiable functions which vanish uniformly at infinity). Let $\mu,\ \nu \in \mathcal{P}(\mathbb{R}^n)$. We say that a probability measure $\gamma \in \mathcal{P}(\mathbb{R}^n \times \mathbb{R}^n)$ has *marginals* μ and ν if for every bounded Borel measurable functions $f, g : \mathbb{R}^n \to \mathbb{R}$ we have

$$\int_{\mathbb{R}^n} f(x)d\mu(x) = \int_{\mathbb{R}^n\times\mathbb{R}^n} f(x)d\gamma(x,y)$$

and

$$\int_{\mathbb{R}^n} g(y)d\nu(y) = \int_{\mathbb{R}^n\times\mathbb{R}^n} g(y)d\gamma(x,y).$$

DEFINITION 1.3.3 (Rockafellar). Let $G \subseteq \mathbb{R}^n \times \mathbb{R}^n$. We say that G is *cyclically monotone* if for every $m \geqslant 2$ and $(x_i, y_i) \in G$, $i \leqslant m$, we have

$$\langle y_1, x_2 - x_1\rangle + \langle y_2, x_3 - x_2\rangle + \cdots + \langle y_m, x_1 - x_m\rangle \leqslant 0.$$

PROPOSITION 1.3.4. *Let μ and ν be Borel probability measures on $\mathbb{R}^n$. There exists a joint probability measure γ on $\mathbb{R}^n \times \mathbb{R}^n$ which has cyclically monotone support and marginals μ, ν.*

The proof of Proposition 1.3.4 is a consequence of the next discrete lemma and a limiting argument.

LEMMA 1.3.5. *Let $x_i, y_i \in \mathbb{R}^n$, $i = 1, \dots, m$ and consider the measures*

$$\mu = \frac{1}{m}\sum_{i=1}^m \delta_{x_i} \text{ and } \nu = \frac{1}{m}\sum_{i=1}^m \delta_{y_i}.$$

There exists a probability measure γ on $\mathbb{R}^n \times \mathbb{R}^n$ which has cyclically monotone support and marginals μ, ν.

Proof. For every permutation σ of $\{1, \dots, m\}$ we consider the measure

$$\gamma_\sigma = \frac{1}{m}\sum_{i=1}^m \delta_{(x_{\sigma(i)}, y_i)}.$$

It is clear that γ_σ has marginals μ and ν for every σ. Let

$$F(\sigma) = \sum_{i=1}^m \langle y_i, x_{\sigma(i)}\rangle.$$

We will show that if $F(\sigma)$ is maximal, then the support $(x_{\sigma(i)}, y_i)$ of γ_σ is cyclically monotone.

Without loss of generality we may assume that $F(I)$ is maximal, where I denotes the identity permutation. We want to show that $G = \{(x_i, y_i) : i \leqslant m\}$ is cyclically monotone. Let $k \leqslant m$, $i_1, \ldots, i_k$ be distinct indices and consider the points $(x_{i_s}, y_{i_s}) \in G$. If σ is the permutation defined by $\sigma(i_s) = i_{s+1}$ if $s < k$, $\sigma(i_k) = i_1$ and $\sigma(i) = i$ otherwise, we have

$$\begin{aligned} 0 \geqslant F(\sigma) - F(I) &= \sum_{s=1}^{k} \left(\langle y_{i_s}, x_{\sigma(i_s)} \rangle - \langle y_{i_s}, x_{i_s} \rangle \right) \\ &= \langle y_{i_1}, x_{i_2} - x_{i_1} \rangle + \langle y_{i_2}, x_{i_3} - x_{i_2} \rangle + \cdots + \langle y_{i_k}, x_{i_1} - x_{i_k} \rangle. \end{aligned}$$

This proves the lemma. □

Proof of Proposition 1.3.4. Given μ and ν we construct discrete measures $\mu_n \to \mu$, $\nu_n \to \nu$ which converge in the weak-$*$ topology. By Lemma 1.3.5, for every n there exists γ_n with cyclically monotone support and marginals μ_n, ν_n. By a standard compactness argument (Arzéla-Ascoli) there exists a weak-$*$ subsequential limit γ of γ_n. It clearly has cyclically monotone support and μ, ν as its marginals. □

We shall next link the property of cyclical monotonicity with that of being contained in the sub-gradient of some convex function.

PROPOSITION 1.3.6 (Rockafellar). *Let $G \subseteq \mathbb{R}^n \times \mathbb{R}^n$. Then, G is contained in the subdifferential of a proper convex function $f : \mathbb{R}^n \to \mathbb{R}$ if and only if G is cyclically monotone.*

Proof. It is easy to check that the subdifferential of a proper convex function is cyclically monotone. Let $(x_i, y_i) \in \partial(f)$, $i = 1, \ldots, m$ (namely let $y_i \in \partial(f)(x_i)$). Then,

$$\begin{aligned} \langle y_1, x_2 - x_1 \rangle &\leqslant f(x_2) - f(x_1) \\ \langle y_2, x_3 - x_2 \rangle &\leqslant f(x_3) - f(x_2) \\ &\vdots \\ \langle y_m, x_1 - x_m \rangle &\leqslant f(x_1) - f(x_m) \end{aligned}$$

by the definition of the subdifferential. Adding the inequalities we get

$$\langle y_1, x_2 - x_1 \rangle + \langle y_2, x_3 - x_2 \rangle + \cdots + \langle y_m, x_1 - x_m \rangle \leqslant 0.$$

It follows that every $G \subseteq \partial(f)$ is cyclically monotone.

For the opposite direction, assume that G is non-empty and cyclically monotone, and fix $(x_0, y_0) \in G$. We define $f : \mathbb{R}^n \to \mathbb{R}$ by

$$f(x) = \sup\left\{ \langle y_m, x - x_m \rangle + \langle y_{m-1}, x_m - x_{m-1} \rangle + \cdots + \langle y_0, x_1 - x_0 \rangle \right\},$$

where the supremum is taken over all $m \geqslant 0$ and $(x_i, y_i) \in G$, $1 \leqslant i \leqslant m$. The function f is convex since it is the supremum of a family of affine functions. Using the cyclical monotonicity of G we easily check that $f(x_0) = 0$. This shows that f is proper. Finally, G is contained in the subdifferential of f: Let $(x, y) \in G$. We will show that

$$t + \langle z - x, y \rangle < f(z)$$

for every $t < f(x)$ and every $z \in \mathbb{R}^n$. This implies that $(x, y) \in \partial(f)$. Since $t < f(x)$, there exist $(x_1, y_1), \ldots, (x_m, y_m) \in G$ such that

$$t < \langle y_m, x - x_m \rangle + \cdots + \langle y_0, x_1 - x_0 \rangle.$$

By the definition of f again,

$$\begin{aligned} f(z) &\geqslant \langle y, z - x \rangle + \langle y_m, x - x_m \rangle + \cdots + \langle y_0, x_1 - x_0 \rangle \\ &> \langle y, z - x \rangle + t. \end{aligned}$$

This completes the proof. □

DEFINITION 1.3.7. Let μ, $\nu \in \mathcal{P}(\mathbb{R}^n)$. Let $T : \mathbb{R}^n \to \mathbb{R}^n$ be a measurable function which is defined μ-almost everywhere and satisfies

$$\nu(B) = \mu(T^{-1}(B))$$

for every Borel subset B of $\mathbb{R}^n$. We then say that T *pushes forward* μ to ν and write $T\mu = \nu$. It is easy to see that $T\mu = \nu$ if and only if for every bounded Borel measurable $g : \mathbb{R}^n \to \mathbb{R}$ we have

$$\int_{\mathbb{R}^n} g(y) d\nu(y) = \int_{\mathbb{R}^n} g(T(x)) d\mu(x).$$

Given μ and $\nu \in \mathcal{P}(\mathbb{R}^n)$, we have the next theorem establishing the existence of a map that is the gradient of a convex function and pushes forward μ to ν. Brenier proved its existence and uniqueness under some integrability assumptions on the moments of μ and ν; these were later removed by McCann. The precise formulation is as follows.

THEOREM 1.3.8 (Brenier-McCann). *Let μ, $\nu \in \mathcal{P}(\mathbb{R}^n)$ and assume that μ is absolutely continuous with respect to the Lebesgue measure. Then, there exists a convex function $f : \mathbb{R}^n \to \mathbb{R}$ such that $\nabla f : \mathbb{R}^n \to \mathbb{R}^n$ is defined μ-almost everywhere, and $(\nabla f)\mu = \nu$.*

Proof. Proposition 1.3.4 shows that there exists a probability measure γ on $\mathbb{R}^n \times \mathbb{R}^n$ which has cyclically monotone support and marginals μ, ν. By Proposition 1.3.6, the support of γ is contained in the subdifferential of a proper convex function $f : \mathbb{R}^n \to \mathbb{R}$.

Since f is convex and μ is absolutely continuous with respect to the Lebesgue measure, f is differentiable μ-almost everywhere. Since $\mathrm{supp}(\gamma) \subset \partial(f)$, by the definition of the subdifferential we have $y = \nabla f(x)$ for almost all pairs (x, y) with respect to γ. Then, for every bounded Borel measurable $g : \mathbb{R}^n \to \mathbb{R}$ we see that

$$\begin{aligned} \int g(y) d\nu(y) &= \int g(y) d\gamma(x, y) = \int g(\nabla f(x)) d\gamma(x, y) \\ &= \int g(\nabla f(x)) d\mu(x), \end{aligned}$$

which shows that $(\nabla f)\mu = \nu$. □

Moreover, regularity results from the theory of elliptic equations also provide smoothness of the Brenier map. In the particular case that μ is the normalized Lebesgue measure on some convex body K and ν is the normalized Lebesgue measure on some other convex body T, a regularity result of Caffarelli shows that f may be assumed twice continuously differentiable. More precisely, we have the following theorem, from which one can deduce the Brunn-Minkowski inequality as explained at the beginning of this subsection.

THEOREM 1.3.9. *Let K and T be open convex bodies in $\mathbb{R}^n$. There is a convex function $f \in C^2(K)$ such that $\psi = \nabla f : K \to T$ is one to one, onto and volume preserving.* □

An interesting variant of Theorem 1.3.9 was given in a paper of Alesker, Dar and V. Milman. They built a volume preserving map between convex bodies which has the following remarkable property.

THEOREM 1.3.10. *Let K and T be open convex bounded subsets of $\mathbb{R}^n$ of volume 1. Then, there exists a C^1-diffeomorphism $F : K \to T$ preserving the Lebesgue measure, such that, for every $\lambda > 0$, $\{x + \lambda F(x) : x \in K\} = K + \lambda T$.*

As an illustrative example for the meaning of Theorem 1.3.10 consider two ε-extensions of two orthogonal intervals in $\mathbb{R}^2$. Then the map $x \mapsto x + F(x)$ tends to a Peano type curve as $\varepsilon \to 0$. The function F in Theorem 1.3.10 cannot be in general the Brenier map, as the example of K being a ball and T an ellipsoid shows. Indeed, in such a case the Brenier map is linear, but then $\mathrm{Id} + F$ is also linear, whereas the sum of a ball and an ellipsoid is not, in general, an ellipsoid. However, F in this theorem is the composition of two Brenier maps, mapping the uniform measure on K to the Gaussian measure (say) and mapping the Gaussian measure to the uniform measure on T.

The key element in the proof of Theorem 1.3.10 is a lemma that can be found in Gromov's paper [**268**] and in the book of Rockafellar [**521**]:

LEMMA 1.3.11. *If $f_i : \mathbb{R}^n \to \mathbb{R}^n$ are C^2-smooth convex functions with strictly positive Hessian, then the convex sets K_i which are the images of the gradient maps $\mathrm{Im}(\nabla f_i) =: K_i$ are open convex sets satisfying*

$$\mathrm{Im}(\nabla f_1 + \nabla f_2) = K_1 + K_2.$$

This result is then paired with a corresponding result of Cafarelli on the regularity of the Brenier map from the Gaussian measure on $\mathbb{R}^n$ onto the uniform measure on a convex set.

It is worthwhile to notice that the fact that $\mathrm{Vol}_n(\sum t_i K_i)$ is a polynomial of degree n in the scalars t_i can also be directly deduced from this line of reasoning. Indeed, fix a measure on $\mathbb{R}^n$, say the Gaussian measure γ_n again, and consider the Brenier maps $\nabla f_i : \mathbb{R}^n \to K_i$. Then

$$\sum_{i=1}^{m} t_i K_i = \mathrm{Im}\Big(\sum_{i=1}^{m} t_i \nabla f_i\Big).$$

Therefore, denoting $\nabla^2 := \mathrm{Hess}$, we have

$$\begin{aligned}\mathrm{Vol}_n\Big(\sum_{i=1}^{m} t_i K_i\Big) &= \int_{\mathbb{R}^n} \det\Big(\sum_{i=1}^{m} t_i \nabla^2 f_i(x)\Big)\, dx \\ &= \sum_{i_1=1}^{m} \cdots \sum_{i_n=1}^{m} \int_{\mathbb{R}^n} \prod_{j=1}^{n} t_{i_j} D(\nabla^2 f_{i_1}(x), \ldots, \nabla^2 f_{i_n}(x))dx\end{aligned}$$

where we have used the polynomiality of determinant with mixed-discriminant coefficients as explained in Section 1.1.6, and elaborated upon in the appendix. This gives the formula for mixed volume

$$V(K_1,\dots,K_n)=\int_{\mathbb{R}^n} D(\nabla^2 f_1(x),\dots,\nabla^2 f_n(x))dx.$$

1.4. Functional forms

A *functional form* of the Brunn-Minkowski inequality is an integral inequality which reduces to the Brunn-Minkowski inequality by appropriate choice of the functions involved. Such functional inequalities can be applied in different contexts and to different ends: for example, in subsequent chapters we will see how the Prékopa-Leindler inequality may be applied to yield the logarithmic Sobolev inequality and several important concentration results in Gauss space.

1.4.1. Prékopa-Leindler inequality

The inequality of Prékopa and Leindler is the following statement.

THEOREM 1.4.1 (Prékopa-Leindler). *Let $f,g,h:\mathbb{R}^n\to\mathbb{R}^+$ be measurable functions, and let $\lambda\in(0,1)$. We assume that f and g are integrable, and that for every $x,y\in\mathbb{R}^n$*

$$h(\lambda x+(1-\lambda)y)\geqslant f(x)^\lambda g(y)^{1-\lambda}. \tag{1.4.1}$$

Then,

$$\int_{\mathbb{R}^n} h\geqslant\left(\int_{\mathbb{R}^n} f\right)^\lambda\left(\int_{\mathbb{R}^n} g\right)^{1-\lambda}.$$

Proof. The proof goes by induction on the dimension n.

(*a*) $n=1$: We may assume that f and g are continuous and strictly positive.
We define $x,y:(0,1)\to\mathbb{R}$ by the equations

$$\int_{-\infty}^{x(t)} f=t\int_{\mathbb{R}} f \quad\text{and}\quad \int_{-\infty}^{y(t)} g=t\int_{\mathbb{R}} g.$$

In view of our assumptions, x and y are differentiable, and for every $t\in(0,1)$ we have

$$x'(t)f(x(t))=\int_{\mathbb{R}} f \quad\text{and}\quad y'(t)g(y(t))=\int_{\mathbb{R}} g.$$

We now define $z:(0,1)\to\mathbb{R}$ by

$$z(t)=\lambda x(t)+(1-\lambda)y(t).$$

Since x and y are strictly increasing, z is also strictly increasing, and the arithmetic-geometric means inequality shows that

$$z'(t)=\lambda x'(t)+(1-\lambda)y'(t)\geqslant(x'(t))^\lambda(y'(t))^{1-\lambda}.$$

Hence, we can estimate the integral of h making the change of variables $s = z(t)$, as follows:

$$\begin{aligned}\int_{\mathbb{R}} h &= \int_0^1 h(z(t))z'(t)dt \\ &\geqslant \int_0^1 h(\lambda x(t) + (1-\lambda)y(t))(x'(t))^{\lambda}(y'(t))^{1-\lambda}dt \\ &\geqslant \int_0^1 f^{\lambda}(x(t))g^{1-\lambda}(y(t))\left(\frac{\int f}{f(x(t))}\right)^{\lambda}\left(\frac{\int g}{g(y(t))}\right)^{1-\lambda}dt \\ &= \left(\int_{\mathbb{R}} f\right)^{\lambda}\left(\int_{\mathbb{R}} g\right)^{1-\lambda}.\end{aligned}$$

(*b*) *Inductive step:* We assume that $n \geqslant 2$ and the assertion of the theorem has been proved in all dimensions $k \in \{1, \dots, n-1\}$. Let f, g and h be as in the theorem. For every $s \in \mathbb{R}$ we define $h_s : \mathbb{R}^{n-1} \to \mathbb{R}^+$ setting $h_s(w) = h(w, s)$, and $f_s, g_s : \mathbb{R}^{n-1} \to \mathbb{R}^+$ in an analogous way. From (1.4.1) it follows that if $x, y \in \mathbb{R}^{n-1}$ and $s_0, s_1 \in \mathbb{R}$ then

$$h_{\lambda s_1+(1-\lambda)s_0}(\lambda x + (1-\lambda)y) \geqslant f_{s_1}(x)^{\lambda}g_{s_0}(y)^{1-\lambda},$$

thus our inductive hypothesis gives

$$\begin{aligned}H(\lambda s_1 + (1-\lambda)s_0) := \int_{\mathbb{R}^{n-1}} h_{\lambda s_1+(1-\lambda)s_0} &\geqslant \left(\int_{\mathbb{R}^{n-1}} f_{s_1}\right)^{\lambda}\left(\int_{\mathbb{R}^{n-1}} g_{s_0}\right)^{1-\lambda} \\ &=: F^{\lambda}(s_1)G^{1-\lambda}(s_0).\end{aligned}$$

Applying the inductive hypothesis once again, this time with $n = 1$, to the functions F, G and H, we get

$$\int_{\mathbb{R}^n} h = \int_{\mathbb{R}} H \geqslant \left(\int_{\mathbb{R}} F\right)^{\lambda}\left(\int_{\mathbb{R}} G\right)^{1-\lambda} = \left(\int_{\mathbb{R}^n} f\right)^{\lambda}\left(\int_{\mathbb{R}^n} g\right)^{1-\lambda},$$

where we have used Fubini's theorem as well. □

Proof of the Brunn-Minkowski inequality. Let K and T be non-empty compact subsets of $\mathbb{R}^n$, and $\lambda \in (0,1)$. We define $f = \mathbf{1}_K$, $g = \mathbf{1}_T$, and $h = \mathbf{1}_{\lambda K+(1-\lambda)T}$, where $\mathbf{1}_A$ denotes the indicator function of a set A. It is easily checked that the assumptions of Theorem 1.4.1 are satisfied, therefore

$$\mathrm{Vol}_n(\lambda K + (1-\lambda)T) = \int_{\mathbb{R}^n} h \geqslant \left(\int_{\mathbb{R}^n} f\right)^{\lambda}\left(\int_{\mathbb{R}^n} g\right)^{1-\lambda} = \mathrm{Vol}_n(K)^{\lambda}\mathrm{Vol}_n(T)^{1-\lambda}.$$

We also state another useful corollary of Theorem 1.4.1, which says that log-concavity of a function (see Definition 1.1.3) is preserved under convolutions.

COROLLARY 1.4.2. *Let $F, G : \mathbb{R}^n \to \mathbb{R}^+$ be integrable log-concave functions. Then, their convolution*

$$(F * G)(x) = \int_{\mathbb{R}^n} F(x-y)G(y)dy$$

is also a log-concave function.

Proof. Let $\lambda \in (0,1)$ and $x_1, x_2 \in \mathbb{R}^n$. We should show that

$$(F * G)(\lambda x_1 + (1-\lambda)x_2) \geqslant ((F * G)(x_1))^{\lambda} ((F * G)(x_2))^{(1-\lambda)}.$$

Setting $h(y) = F(\lambda x_1 + (1-\lambda)x_2 - y)G(y)$, $f(y) = F(x_1 - y)G(y)$ and $g(y) = F(x_2 - y)G(y)$ we easily check that they satisfy the condition of Theorem 1.4.1, by log-concavity of F and G. Applying the conclusion of the theorem, we get an inequality for their integrals which is exactly the inequality needed. $\square$

In order to state the next result, we introduce some notation.

DEFINITION 1.4.3. Let $p \neq 0$ and $\lambda \in (0,1)$. For all $x, y > 0$ we set

$$M_p^{\lambda}(x,y) = (\lambda x^p + (1-\lambda) y^p)^{1/p}.$$

If $x, y \geqslant 0$ and $xy = 0$ we set $M_p^{\lambda}(x,y) = 0$. Observe that

$$\lim_{p \to 0^+} M_p^{\lambda}(x,y) = x^{\lambda} y^{1-\lambda}.$$

By Hölder's inequality, if $x, y, z, w \geqslant 0$, $a, b, \gamma > 0$ and $\frac{1}{a} + \frac{1}{b} = \frac{1}{\gamma}$, then

$$M_a^{\lambda}(x,y) \cdot M_b^{\lambda}(z,w) \geqslant M_{\gamma}^{\lambda}(xz, yw). \tag{1.4.2}$$

PROPOSITION 1.4.4. *Let $f, g, h : \mathbb{R}^n \to \mathbb{R}^+$ be measurable functions, and let $p > 0$ and $\lambda \in (0,1)$. We assume that f and g are integrable, and for every $x, y \in \mathbb{R}^n$*

$$h(\lambda x + (1-\lambda) y) \geqslant M_p^{\lambda}(f(x), g(y)). \tag{1.4.3}$$

Then,

$$\int_{\mathbb{R}^n} h \geqslant M^{\lambda}_{\frac{p}{pn+1}} \left(\int_{\mathbb{R}^n} f, \int_{\mathbb{R}^n} g \right).$$

Proof. We will consider only the case $n = 1$. As in the proof of the Prékopa-Leindler inequality, we define $x, y : (0,1) \to \mathbb{R}$ by the equations

$$\int_{-\infty}^{x(t)} f = t \int_{\mathbb{R}} f \quad \text{and} \quad \int_{-\infty}^{y(t)} g = t \int_{\mathbb{R}} g.$$

Then,

$$x'(t) f(x(t)) = \int_{\mathbb{R}} f \quad \text{and} \quad y'(t) g(y(t)) = \int_{\mathbb{R}} g.$$

We define $z : (0,1) \to \mathbb{R}$ by

$$z(t) = \lambda x(t) + (1-\lambda) y(t).$$

Then, z is strictly increasing, and

$$z'(t) = \lambda x'(t) + (1-\lambda) y'(t) = M_1^{\lambda}(x'(t), y'(t)).$$

Hence, using (1.4.2) and (1.4.3) we can estimate the integral of h making the change of variables $s = z(t)$, as follows:

$$\begin{aligned}\int_{\mathbb{R}} h &= \int_0^1 h(z(t))z'(t)dt \\ &\geqslant \int_0^1 h(\lambda x(t) + (1-\lambda)y(t))M_1^{\lambda}(x'(t), y'(t))dt \\ &\geqslant \int_0^1 M_p^{\lambda}(f(x(t)), g(y(t)))M_1^{\lambda}(x'(t), y'(t))dt \\ &\geqslant \int_0^1 M_{\frac{p}{p+1}}^{\lambda}\left(f(x(t))x'(t), g(y(t))y'(t)\right) dt \\ &= M_{\frac{p}{p+1}}^{\lambda}\left(\int_{\mathbb{R}} f, \int_{\mathbb{R}} g\right).\end{aligned}$$

The inductive step is exactly as in Theorem 1.4.1. □

REMARK 1.4.5. Using Proposition 1.4.4 we may give an alternative proof of Lemma 1.2.4. The claim was the following: Assume that K is a two-dimensional convex body and $f : K \to \mathbb{R}^+$ is an α-concave function. If

$$(Pf)(y) := \int_{\mathbb{R}} \mathbf{1}_K(y,t)f(y,t)dt,$$

then Pf is $(1+\alpha)$-concave.

Proof. Let $F_y(t) = \mathbf{1}_K(y,t)f(y,t)$, $y \in PK$. Then, for all $y, z \in PK$ we have

$$F_{\lambda y+(1-\lambda)z}(\lambda t + (1-\lambda)s) \geqslant M_{1/\alpha}^{\lambda}(F_y(t), F_z(s))$$

by the α-concavity of f and the convexity of K. The claim follows from Proposition 1.4.4. □

We close this section with one more functional inequality that will be used in Chapter 10.

THEOREM 1.4.6. *Let $\gamma > 0$ and $\lambda, \mu > 0$ with $\lambda + \mu = 1$. Let $w, g, h : \mathbb{R}^+ \to \mathbb{R}^+$ be integrable functions such that*

$$h(M_{-\gamma}^{\lambda}(r,s)) \geqslant w(r)^{\frac{\lambda s^{\gamma}}{\lambda s^{\gamma}+\mu r^{\gamma}}} g(s)^{\frac{\mu r^{\gamma}}{\lambda s^{\gamma}+\mu r^{\gamma}}} \tag{1.4.4}$$

for every pair $(r,s) \in \mathbb{R}^+ \times \mathbb{R}^+$. Then,

$$\int_0^{\infty} h \geqslant M_{-\gamma}^{\lambda}\left(\int_0^{\infty} w, \int_0^{\infty} g\right). \tag{1.4.5}$$

Proof. We may assume that w and g are continuous and strictly positive. We define $r, s : [0,1] \to \mathbb{R}^+$ by the equations

$$\int_0^{r(t)} w = t\int_0^{\infty} w \quad \text{and} \quad \int_0^{s(t)} g = t\int_0^{\infty} g.$$

Then, r and s are differentiable, and for every $t \in (0,1)$ we have

$$r'(t)w(r(t)) = \int_0^{\infty} w \quad \text{and} \quad s'(t)g(s(t)) = \int_0^{\infty} g.$$

Next, we define $z : [0,1] \to \mathbb{R}^+$ by

$$z(t) = M_{-\gamma}^{\lambda}(r(t), s(t)).$$

Note that

$$\begin{aligned} z' &= z^{\gamma+1}\left(\frac{\lambda r'}{r^{\gamma+1}}+\frac{\mu s'}{s^{\gamma+1}}\right) \\ &= \lambda\frac{\int w}{w(r)}\left(\frac{s^\gamma}{\lambda s^\gamma+\mu r^\gamma}\right)^{\frac{\gamma+1}{\gamma}}+\mu\frac{\int g}{g(s)}\left(\frac{r^\gamma}{\lambda s^\gamma+\mu r^\gamma}\right)^{\frac{\gamma+1}{\gamma}} \\ &= \frac{\lambda s^\gamma}{\lambda s^\gamma+\mu r^\gamma}\left(\frac{\int w}{w(r)}\left(\frac{s^\gamma}{\lambda s^\gamma+\mu r^\gamma}\right)^{\frac{1}{\gamma}}\right)+\frac{\mu r^\gamma}{\lambda s^\gamma+\mu r^\gamma}\left(\frac{\int g}{g(s)}\left(\frac{r^\gamma}{\lambda s^\gamma+\mu r^\gamma}\right)^{\frac{1}{\gamma}}\right) \\ &\geqslant \left(\frac{\int w}{w(r)}\left(\frac{s^\gamma}{\lambda s^\gamma+\mu r^\gamma}\right)^{\frac{1}{\gamma}}\right)^{\frac{\lambda s^\gamma}{\lambda s^\gamma+\mu r^\gamma}}\left(\frac{\int g}{g(s)}\left(\frac{r^\gamma}{\lambda s^\gamma+\mu r^\gamma}\right)^{\frac{1}{\gamma}}\right)^{\frac{\mu r^\gamma}{\lambda s^\gamma+\mu r^\gamma}}, \end{aligned}$$

by the arithmetic-geometric means inequality. Now, making a change of variables we write

$$\begin{aligned} \int_0^\infty h &= \int_0^1 h(z)z'dz \\ &\geqslant \int_0^1 M_0^{\frac{\lambda s^\gamma}{\lambda s^\gamma+\mu r^\gamma}}\left(\left(\frac{s^\gamma}{\lambda s^\gamma+\mu r^\gamma}\right)^{\frac{1}{\gamma}}\int_0^\infty w,\left(\frac{r^\gamma}{\lambda s^\gamma+\mu r^\gamma}\right)^{\frac{1}{\gamma}}\int_0^\infty g\right). \end{aligned}$$

Since $M_0^\delta(a,b)\geqslant M_{-\gamma}^\delta(a,b)$ we finally get

$$\begin{aligned} \int_0^\infty h &\geqslant \int_0^1 M_{-\gamma}^{\frac{\lambda s^\gamma}{\lambda s^\gamma+\mu r^\gamma}}\left(\left(\frac{s^\gamma}{\lambda s^\gamma+\mu r^\gamma}\right)^{\frac{1}{\gamma}}\int_0^\infty w,\left(\frac{r^\gamma}{\lambda s^\gamma+\mu r^\gamma}\right)^{\frac{1}{\gamma}}\int_0^\infty g\right) \\ &= \int_0^1 M_{-\gamma}^\lambda\Big(\int_0^\infty w,\int_0^\infty g\Big) = M_{-\gamma}^\lambda\Big(\int_0^\infty w,\int_0^\infty g\Big). \end{aligned}$$

This completes the proof. □

1.5. Applications of the Brunn-Minkowski inequality

In this section we give a first list of important geometric applications of the Brunn-Minkowski inequality and its relatives. This list is far from being complete but shows the fundamental character and strength of the Brunn-Minkowski inequality. We will be returning to more advanced applications throughout the book.

1.5.1. An inequality of Rogers and Shephard

DEFINITION 1.5.1. The *difference body* of a convex body K is the centrally symmetric convex body

$$K-K=\{x-y : x,y\in K\}.$$

From the Brunn-Minkowski inequality it is clear that $\mathrm{Vol}_n(K-K)\geqslant 2^n\mathrm{Vol}_n(K)$, with equality if and only if K has a centre of symmetry. Rogers and Shephard gave a sharp upper bound for the volume of the difference body.

THEOREM 1.5.2 (Rogers-Shephard). *Let K be a convex body in $\mathbb{R}^n$. Then,*

$$\mathrm{Vol}_n(K-K)\leqslant\binom{2n}{n}\mathrm{Vol}_n(K).$$

Proof. The Brunn-Minkowski inequality enters the proof through the observation that $f(x) = \mathrm{Vol}_n(K \cap (x+K))^{1/n}$ is a concave function supported on $K - K$. This can be proved directly, using the inclusion

$$K \cap ((1-\lambda)x + \lambda y + T) \supset (1-\lambda)(K \cap (x+T)) + \lambda(K \cap (y+T))$$

and the Brunn-Minkowski inequality, or by using Brunn's concavity principle for the body $K \times T \subset \mathbb{R}^{2n}$ and the n-dimensional subspace $\{(x,x) : x \in \mathbb{R}^n\}$ (in both cases one employs the special case $T = K$).

We define a second function $g : K - K \to \mathbb{R}^+$ as follows: each $x \in K - K$ can be written in the form $x = r\theta$, where $\theta \in S^{n-1}$ and $0 \leqslant r \leqslant \rho_{K-K}(\theta)$. Recall that ρ_W denotes the *radial function* of W:

$$\rho_W(\theta) = \max\{t > 0 : t\theta \in W\}, \qquad \theta \in S^{n-1}.$$

We then set $g(x) = f(0)(1 - r/\rho_{K-K}(\theta))$. By definition, g is linear on the interval $[0, \rho_{K-K}(\theta)\theta]$, it vanishes on the boundary of $K - K$, and $g(0) = f(0)$. Since f is concave, we see that $f \geqslant g$ on $K - K$. Therefore, we can write

$$\begin{aligned}
\int_{K-K} \mathrm{Vol}_n(K \cap (x+K))dx &= \int_{K-K} f^n(x)dx \geqslant \int_{K-K} g^n(x)dx \\
&= [f(0)]^n n\kappa_n \int_{S^{n-1}} \int_0^{\rho_{K-K}(\theta)} r^{n-1}(1 - r/\rho_{K-K}(\theta))^n dr\, d\sigma(\theta) \\
&= n\kappa_n \mathrm{Vol}_n(K) \int_{S^{n-1}} \rho_{K-K}^n(\theta)\, d\sigma(\theta) \int_0^1 t^{n-1}(1-t)^n dt \\
&= \mathrm{Vol}_n(K)\mathrm{Vol}_n(K-K) \frac{n\Gamma(n)\Gamma(n+1)}{\Gamma(2n+1)} \\
&= \binom{2n}{n}^{-1} \mathrm{Vol}_n(K)\mathrm{Vol}_n(K-K),
\end{aligned}$$

using integration in polar coordinates (we denote by σ the rotationally invariant probability measure on the Euclidean unit sphere S^{n-1}, and use the fact that its surface area is $\mathrm{Vol}_{n-1}(S^{n-1}) = n\kappa_n$). On the other hand, Fubini's theorem gives

$$\begin{aligned}
&\int_{K-K} \mathrm{Vol}_n(K \cap (x+K))dx = \int_{\mathbb{R}^n} \mathrm{Vol}_n(K \cap (x+K))dx \\
&= \int_{\mathbb{R}^n}\int_{\mathbb{R}^n} \mathbf{1}_K(y)\mathbf{1}_{x+K}(y)dy\,dx = \int_{\mathbb{R}^n} \mathbf{1}_K(y)\left(\int_{\mathbb{R}^n} \mathbf{1}_{y-K}(x)dx\right)dy \\
&= \int_K \mathrm{Vol}_n(y-K)dy = \mathrm{Vol}_n(K)^2.
\end{aligned}$$

Combining the above, we conclude the proof. □

REMARK 1.5.3. If we take a closer look at the argument and take into account the equality case in the Brunn-Minkowski inequality, we see that we can have equality in Theorem 1.5.2 if and only if K has the following property: if $(rK+x) \cap (sK+y) \neq \emptyset$ for some $r, s > 0$ and $x, y \in \mathbb{R}^n$, then

$$(rK+x) \cap (sK+y) = tK + w$$

for some $t \geqslant 0$ and $w \in \mathbb{R}^n$. That is, the non-empty intersection of homothetical copies of K is again homothetical to K or a point. Rogers and Shephard proved that this property characterizes the simplex.

REMARK 1.5.4. It is worth noting that the main ingredient we have used is that $f(x) = \mathrm{Vol}_n(K \cap (x+K))^{1/n}$ is a concave function supported on $K-K$. Given two convex bodies, K and T, the same argument shows that $f(x) = \mathrm{Vol}_n(K \cap (x-T))^{1/n}$ is a concave function supported on $K+T$. So one can apply the same arguments and get the more general inequality

$$\mathrm{Vol}_n(K+T)\,\mathrm{Vol}_n(K \cap (-T)) \leqslant \binom{2n}{n} \mathrm{Vol}_n(K)\,\mathrm{Vol}_n(T).$$

A similar argument shows that if K and T are convex bodies in $\mathbb{R}^n$ then

$$\mathrm{Vol}_n(K)\,\mathrm{Vol}_n(T) \leqslant \mathrm{Vol}_n(K+T)\,\sup_x \mathrm{Vol}_n(K \cap (x-T)).$$

As in the proof of the Rogers-Shephard inequality, we write

$$\begin{aligned}\mathrm{Vol}_n(K)\mathrm{Vol}_n(T) &= \int_{\mathbb{R}^n} \mathrm{Vol}_n(K \cap (x-T))dx \\ &= \int_{K+T} \mathrm{Vol}_n(K \cap (x-T))dx \\ &\leqslant \mathrm{Vol}_n(K+T) \cdot \max_x \mathrm{Vol}_n(K \cap (x-T)).\end{aligned}$$

In particular, this implies that

$$\max_x \mathrm{Vol}_n(K \cap (x-K)) \geqslant 2^{-n}\mathrm{Vol}_n(K).$$

Thus, every convex body K contains a symmetric convex body L (with centre of symmetry some $x' \in K$) such that $\mathrm{Vol}_n(K) \leqslant 2^n \mathrm{Vol}_n(L)$.

The usefulness of Theorem 1.5.2 rests upon the fact that the volume of $K-K$ is not much larger than the volume of K. We have

$$\mathrm{Vol}_n(K-K)^{1/n} \leqslant 4\mathrm{Vol}_n(K)^{1/n},$$

which means that every convex body (that contains the origin) is contained in a centrally symmetric convex body with more or less the same volume radius; the *volume radius* of a convex body is the radius of the Euclidean ball centered at 0 that has the same volume, or, in other words, it is the quantity $(\mathrm{Vol}_n(K)/\kappa_n)^{1/n}$.

We mention another curious consequence, which is a simple consequence of Theorem 1.5.2 but also reminiscent of inequalities we shall encounter in Chapter 8.

COROLLARY 1.5.5. *Let $K, L \subset \mathbb{R}^n$ be convex bodies, then*

$$\mathrm{Vol}_n(K-L) \leqslant \binom{2n}{n} \mathrm{Vol}_n(K+L).$$

Proof. Indeed, we simply use inclusion and Theorem 1.5.2 to get that

$$\mathrm{Vol}_n(K-L) \leqslant \mathrm{Vol}_n(K+L-(K+L)) \leqslant \binom{2n}{n} \mathrm{Vol}_n(K+L).$$

□

It is good to note that Rogers and Shephard also proved that, when the barycenter of K is 0, we additionally have that

$$\mathrm{Vol}_n(K \cap (-K)) \geqslant 2^{-n}\mathrm{Vol}_n(K).$$

So, every convex body with barycenter at the origin is not only contained in a centrally symmetric convex body of similar volume radius, but also contains a centrally symmetric convex body with the same volume radius up to a constant. Again this result has a generalization as follows:

$$\mathrm{Vol}_n(K)\,\mathrm{Vol}_n(L) \leqslant \mathrm{Vol}_n(K+L)\,\mathrm{Vol}_n(K\cap(-L)).$$

When the bodies are symmetric this easily follows from integration. In the general case one needs to assume that their barycenters coincide (say, both are at the origin) and the result is due to Milman and Pajor. A proof of this result will be given in Chapter 4. Also, in Chapter 8 we will use one more, connected, inequality of Rogers and Shephard, which we next state and prove.

LEMMA 1.5.6. *Let K be a centrally symmetric convex body in $\mathbb{R}^n$. Let $F\in G_{n,k}$ and write $F^{\perp}$ for the orthogonal complement of F. Then,*

$$(1.5.1)\qquad \binom{n}{k}^{-1}\mathrm{Vol}_k(K\cap F)\,\mathrm{Vol}_{n-k}(P_{F^{\perp}}(K)) \leqslant \mathrm{Vol}_n(K) \leqslant \mathrm{Vol}_k(K\cap F)\,\mathrm{Vol}_{n-k}(P_{F^{\perp}}(K)).$$

Proof. Using Fubini's theorem we write

$$(1.5.2)\qquad \mathrm{Vol}_n(K) = \int_{P_{F^{\perp}}(K)} \mathrm{Vol}_k(K\cap(x+F))\,dx.$$

From the symmetry of K we have

$$\mathrm{Vol}_k(K\cap(x+F)) = \mathrm{Vol}_k(K\cap(-x+F))$$

for every $x\in P_{F^{\perp}}(K)$. Since

$$\frac{1}{2}K\cap(x+F) + \frac{1}{2}K\cap(-x+F) \subseteq K\cap F,$$

by the Brunn-Minkowski inequality it follows that

$$\mathrm{Vol}_k(K\cap(x+F)) \leqslant \mathrm{Vol}_k(K\cap F)$$

for all $x\in P_{F^{\perp}}(K)$, which proves the right hand side inequality in (1.5.1).

For the left hand side inequality we first observe, using the convexity of K, that if $x\in tP_{F^{\perp}}(K)$ for some $0\leqslant t\leqslant 1$ then there exists $z\in F$ such that $x+tz\in tK$, and hence

$$K\cap(x+F) \supseteq (1-t)(K\cap F) + x + tz.$$

Taking volumes, we get

$$\mathrm{Vol}_k(K\cap(x+F)) \geqslant (1-\|x\|_{F,K})^k\mathrm{Vol}_k(K\cap F)$$

for every $x\in P_{F^{\perp}}(K)$, where $\|\cdot\|_{F,K}$ is the norm induced by $P_{F^{\perp}}(K)$ on $F^{\perp}$. Thus, the right hand side of (1.5.2) is greater than or equal to

$$\mathrm{Vol}_k(K\cap F)\,\mathrm{Vol}_{n-k}(P_{F^{\perp}}(K)) \int_0^1 k\,t^{n-k}(1-t)^{k-1}dt.$$

Computing the last integral we complete the proof of the lemma. $\square$

1.5.2. Borell's lemma

Borell's lemma implies *concentration of volume* in convex bodies in $\mathbb{R}^n$: if $A \cap K$ captures more than half of the volume of K, then the percentage of K that stays outside tA, when $t > 1$, decreases exponentially with respect to t as $t \to \infty$, at a rate that does not depend at all on the body K or the dimension n.

THEOREM 1.5.7 (Borell's Lemma). *Let K be a convex body in $\mathbb{R}^n$ with volume $\mathrm{Vol}_n(K) = 1$, and let A be a closed, convex and centrally symmetric set such that $\mathrm{Vol}_n(K \cap A) = \delta > \frac{1}{2}$. Then, for every $t > 1$ we have*

$$\mathrm{Vol}_n(K \cap (tA)^c) \leqslant \delta \left(\frac{1-\delta}{\delta} \right)^{\frac{t+1}{2}}.$$

Proof. We first show that

$$A^c \supseteq \frac{2}{t+1}(tA)^c + \frac{t-1}{t+1}A.$$

If this were not so, we could write some $a \in A$ in the form

$$a = \frac{2}{t+1}y + \frac{t-1}{t+1}a_1,$$

for some $a_1 \in A$ and $y \notin tA$. But then we would have

$$\frac{1}{t}y = \frac{t+1}{2t}a + \frac{t-1}{2t}(-a_1) \in A,$$

because of the convexity and symmetry of A. This means that $y \in tA$, which is a contradiction.

Since K is convex, we have

$$A^c \cap K \supseteq \frac{2}{t+1}\big((tA)^c \cap K\big) + \frac{t-1}{t+1}\big(A \cap K\big).$$

An application of the Brunn-Minkowski inequality gives

$$\begin{aligned} 1 - \delta = \mathrm{Vol}_n(A^c \cap K) &\geqslant \mathrm{Vol}_n((tA)^c \cap K)^{\frac{2}{t+1}} \mathrm{Vol}_n(A \cap K)^{\frac{t-1}{t+1}} \\ &= \mathrm{Vol}_n((tA)^c \cap K)^{\frac{2}{t+1}} \delta^{\frac{t-1}{t+1}}. \end{aligned}$$

This proves the theorem. □

REMARK 1.5.8. One might wonder whether the number 1/2 appearing in Borell's lemma is really crucial. In other words, assume e.g. that $\mathrm{Vol}_n(K \cap A) = \delta = \frac{1}{3}$. Does this also imply that $\mathrm{Vol}_n(K \cap (tA)^c)$ is exponentially small when t is large? The answer is of course yes, and can be given in two steps as follows. Choose $t_0 > 0$ so that the set $B := t_0 A$ satisfies $\mathrm{Vol}(t_0 A \cap K) = \frac{7}{12}$, and set $K_1 := K \cap B$. Then $\mathrm{Vol}_n(K_1 \cap A) = \frac{1}{3} = \frac{4}{7}\frac{7}{12}$, and so we see that A and K_1 satisfy the conditions in Borell's lemma with $\delta = \frac{4}{7}$. As a result, for every $t > 1$,

$$\mathrm{Vol}_n(K_1 \cap (tA)^c) \leqslant \frac{7}{12} \cdot \frac{4}{7} \left(\frac{1-(4/7)}{(4/7)} \right)^{\frac{t+1}{2}} = \frac{1}{3} \cdot \left(\frac{3}{4} \right)^{\frac{t+1}{2}}.$$

In particular, for some finite t_1, that can be explicitly computed, we have that $\mathrm{Vol}_n(K_1 \cap t_1 A) > \frac{13}{24}$. We then use $A_1 = t_1 A$ as the set in Borell's lemma, and see that for $t > t_1$ we have

$$\mathrm{Vol}_n(K \cap (tA)^c) = \mathrm{Vol}_n\Big(K \cap \Big(\frac{t}{t_1}A_1\Big)^c\Big) \leqslant \frac{13}{24}\left(\frac{11}{13}\right)^{\frac{(t/t_1)+1}{2}}.$$

In fact, a more precise estimate is known, see the notes and remarks section.

1.5.3. The isoperimetric inequality for convex bodies

There are several ways to define the surface area of a convex body K, and they are all equivalent. One may think of ∂K as an $(n-1)$-dimensional manifold in $\mathbb{R}^n$ and look at its $(n-1)$-dimensional Hausdorff measure. It is not hard to show (see the appendix) that this coincides with "naively" defining the surface area of polytopes as the sum of the $(n-1)$-dimensional volumes of their facets, and then taking the limit for a sequence of convex polytopes converges to the given body. Indeed, to this end it is useful to note that under a convexity assumption, surface area is a monotone quantity (as are all the mixed volumes; recall that in Section 1.1.5 we saw that the surface area of a body K appears as (a multiple of) one of the quermassintegrals of K). This can be seen for polytopes as follows: the generalized triangle inequality for polytopes states that the $(n-1)$-dimensional volume of any facet is smaller than the sum of the $(n-1)$-dimensional volumes of all other facets. This fact can be directly seen by taking the projection onto this facet. Next, given polytopes $P_1 \supseteq P_2$ one may construct a sequence of polytopes, each of them being the intersection of the previous one with a halfspace, with the initial element in the sequence being P_1, the final one being P_2. By the generalized triangle inequality, the surface area decreases along this sequence. With this monotonicity in hand, the definition of surface area of a convex body as the infimum of the surface areas of the polytopes containing it, or as the supremum of the surface areas of the polytopes contained in it, both make sense and agree with the previous definition.

Yet another definition which coincides with these (see the appendix for details) can be given as follows:

$$\partial(K) := \lim_{t\to 0^+} \frac{\mathrm{Vol}_n(K + tB_2^n) - \mathrm{Vol}_n(K)}{t}.$$

It is a well-known fact that among all convex bodies of a given volume the ball has minimal surface area. This fact has many proofs (the history is also quite interesting here, with many erroneous proofs). However, it also follows as an immediate consequence of the Brunn-Minkowski inequality: if K is a convex body in $\mathbb{R}^n$ with $\mathrm{Vol}_n(K) = \mathrm{Vol}_n(rB_2^n)$, then for every $t > 0$

$$\mathrm{Vol}_n(K + tB_2^n)^{1/n} \geqslant \mathrm{Vol}_n(K)^{1/n} + t\mathrm{Vol}_n(B_2^n)^{1/n} = (r+t)\mathrm{Vol}_n(B_2^n)^{1/n}.$$

We thus see that the surface area $\partial(K)$ of K satisfies

$$\begin{aligned}\partial(K) &= \lim_{t\to 0^+} \frac{\mathrm{Vol}_n(K + tB_2^n) - \mathrm{Vol}_n(K)}{t} \\ &\geqslant \lim_{t\to 0^+} \frac{(r+t)^n - r^n}{t}\mathrm{Vol}_n(B_2^n) = n\mathrm{Vol}_n(B_2^n)^{\frac{1}{n}}\mathrm{Vol}_n(K)^{\frac{n-1}{n}},\end{aligned}$$

with equality if $K = rB_2^n$. The question of uniqueness in the equality case is more delicate.

Note that a stronger isoperimetric statement holds true: if, say, $\mathrm{Vol}_n(K) = \mathrm{Vol}_n(B_2^n)$ then

$$\mathrm{Vol}_n(K + tB_2^n) \geqslant \mathrm{Vol}_n(B_2^n + tB_2^n)$$

for every $t > 0$. In other words, for fixed volume and for every $t > 0$, the t-extension

$$K_t = \{y : \mathrm{dist}(y, K) \leqslant t\}$$

has minimal volume if K is a ball.

One immediately sees that in the proof the Euclidean ball had no special role. That is, if one defines (with some abuse of notation)

$$V(\partial K, T) = \lim_{t\to 0^+} \frac{\mathrm{Vol}_n(K + tT) - \mathrm{Vol}_n(K)}{t},$$

then the same reasoning gives that

$$\begin{aligned}\mathrm{Vol}_n(K + tT)^{1/n} &\geqslant \mathrm{Vol}_n(K)^{1/n} + t\mathrm{Vol}_n(T)^{1/n}\\ &= \Big(\Big(\frac{\mathrm{Vol}_n(K)}{\mathrm{Vol}_n(T)}\Big)^{1/n} + t\Big)\mathrm{Vol}_n(T)^{1/n},\end{aligned}$$

so that

$$V(\partial K, T) \geqslant n\mathrm{Vol}_n(T)^{\frac{1}{n}}\mathrm{Vol}_n(K)^{\frac{n-1}{n}}.$$

This inequality is called Minkowski's first inequality.

1.5.4. Blaschke-Santaló inequality

DEFINITION 1.5.9. Let K be a centrally symmetric convex body in $\mathbb{R}^n$. The *volume product* $s(K)$ of K (called also "Mahler volume" or "Mahler product") is the product $\mathrm{Vol}_n(K)\mathrm{Vol}_n(K^\circ)$ of the volumes of K and its polar.

It is easily seen that the volume product is an invariant of the linear class of K: if $T \in GL_n$, then

$$s(K) = \mathrm{Vol}_n(K)\,\mathrm{Vol}_n(K^\circ) = \mathrm{Vol}_n(TK)\,\mathrm{Vol}_n((TK)^\circ) = s(TK).$$

The Blaschke-Santaló inequality asserts that $s(K)$ is maximized when K is an ellipsoid.

THEOREM 1.5.10 (Blaschke-Santaló). *Let K be a centrally symmetric convex body in $\mathbb{R}^n$. Then, $\mathrm{Vol}_n(K)\mathrm{Vol}_n(K^\circ) \leqslant \kappa_n^2$ with equality if and only if K is an ellipsoid.*

In fact, central symmetry is not as crucial a point as it seems in the beginning; given a general convex body one may define the following quantity:

$$s(K) = \inf_x\{\mathrm{Vol}(K)\mathrm{Vol}((K - x)^\circ) : x \in \mathrm{int}(K)\}.$$

In the case that K is centrally symmetric, this definition coincides with the previous one. The non-symmetric Santaló inequality states that, in general, $s(K) \leqslant \kappa_n^2$.

We shall explain here how the inequality in Theorem 1.5.10 can be proven, without analysing the equality case. We shall need Proposition 1.1.15 that we quoted in Section 1.1.7.

Proposition 1.1.15. *Let K be a centrally symmetric convex body in $\mathbb{R}^n$ and let $u \in S^{n-1}$. Then*

$$\mathrm{Vol}(K^\circ) \leqslant \mathrm{Vol}((S_u(K))^\circ).$$

Indeed, this fact together with Theorem 1.1.16 proves Theorem 1.5.10. So, it suffices to prove here the proposition, and we can do so by means of the Brunn-Minkowski inequality.

Proof of Proposition 1.1.15. Without loss of generality we shall symmetrize with respect to the hyperplane $x_n = 0$, that is, $u = e_n$. We may write

$$S_u(K) = \left\{ \left(x, \frac{s-t}{2} \right) : (x,s), (x,t) \in K \right\}$$

and then,

$$(S_u K)^\circ = \{(y,r) : \langle x, y\rangle + r(s-t)/2 \leqslant 1 \text{ for all } (x,s), (x,t) \in K\}.$$

For a set $A \subseteq \mathbb{R}^n$ let $A(r)$ denote the $(n-1)$-dimensional section of A at height r, that is, $\{x \in \mathbb{R}^{n-1} : (x,r) \in A\}$. Then,

$$\begin{aligned}
&\frac{1}{2}\left(K^\circ(r) + K^\circ(-r)\right) \\
&\quad = \left\{ \frac{y+z}{2} : \langle x,y\rangle + sr \leqslant 1, \langle w,z\rangle - tr \leqslant 1 \text{ for all } (x,s),(w,t) \in K \right\} \\
&\quad \subseteq \left\{ \frac{y+z}{2} : \langle x,y\rangle + sr \leqslant 1, \langle x,z\rangle - tr \leqslant 1 \text{ for all } (x,s),(x,t) \in K \right\} \\
&\quad \subseteq \left\{ \frac{y+z}{2} : \frac{1}{2}\langle x, y+z\rangle + \frac{(s-t)}{2} r \leqslant 1 \text{ for all } (x,s),(x,t) \in K \right\} \\
&\quad = \left\{ v : \langle x, v\rangle + \frac{(s-t)}{2} r \leqslant 1 \text{ for all } (x,s),(x,t) \in K \right\} \\
&\quad = (S_u(K))^\circ(r).
\end{aligned}$$

Note that, for $A = K^\circ$, we have $A = -A$ and thus

$$A(-r) = \{x : (x,-r) \in A\} = \{x : (-x,r) \in A\} = \{-y : (y,r) \in A\} = -A(r),$$

so that both have the same volume. Thus, by the Brunn-Minkowski inequality we get that

$$\begin{aligned}
\mathrm{Vol}_{n-1}\left(\frac{K^\circ(r) + K^\circ(-r)}{2} \right) &\geqslant \mathrm{Vol}_{n-1}(K^\circ(r))^{1/2} \mathrm{Vol}_{n-1}(K^\circ(-r))^{1/2} \\
&= \mathrm{Vol}_{n-1}(K^\circ(r)).
\end{aligned}$$

Putting these together we see that

$$\begin{aligned}
\mathrm{Vol}_n(S_u(K))^\circ &= \int_{-\infty}^{\infty} \mathrm{Vol}_{n-1}((S_u(K))^\circ(r)) dr \geqslant \int_{-\infty}^{\infty} \mathrm{Vol}_{n-1}(K^\circ(r)) dr \\
&= \mathrm{Vol}_n(K^\circ),
\end{aligned}$$

as claimed. □

1.5.5. Urysohn's inequality

Let us recall the definition of the support function h_K and of the width function of a convex body K: given $u \in S^{n-1}$, we have $h_K(u) = \max_{x \in K}\langle x, u\rangle$ and $w_K(u) = h_K(u) + h_K(-u)$. The *mean width* $w(K)$ of K is the average

$$w(K) = \int_{S^{n-1}} \frac{w_K(u)}{2}\, d\sigma(u) = \int_{S^{n-1}} h_K(u)\, d\sigma(u).$$

Note that by Kubota's formula (1.1.1)

$$W_{n-1}(K) = \frac{\kappa_n}{2}\int_{S^{n-1}} \mathrm{Vol}_1(P_uK)d\sigma(u) = \kappa_n\, w(K),$$

which shows that we are dealing in fact with one of the quermassintegrals. All these quantities are minimized, for fixed volume, at the Euclidean ball. In the case of mean width, this fact is a classical inequality of Urysohn.

THEOREM 1.5.11 (Urysohn). *Let K be a convex body in $\mathbb{R}^n$. Then,*

$$w(K) \geqslant \left(\frac{\mathrm{Vol}_n(K)}{\mathrm{Vol}_n(B_2^n)}\right)^{1/n}.$$

We will give a simple proof of this fact using Steiner symmetrization. By continuity of the mean width with respect to the Hausdorff metric, it suffices to prove the following

PROPOSITION 1.5.12. *Let K be a convex body in $\mathbb{R}^n$ and let $\theta \in S^{n-1}$. Then,*

$$w(S_\theta(K)) \leqslant w(K).$$

REMARK 1.5.13. Again this is a particular case of the more general statement that all W_i are decreasing with respect to Steiner symmetrization.

Proof. We may assume that $\theta = e_n$ and write

$$S_\theta(K) = K_1 = \left\{\left(x, \frac{t_1 - t_2}{2}\right) : x \in P_{\theta^\perp}K, (x,t_1) \in K, (x,t_2) \in K\right\}.$$

Let $u = (u_1, \dots, u_n) \in S^{n-1}$. We set $u' = (u_1, \dots, u_{n-1}, -u_n)$. Then,

$$\begin{aligned} h_{S_\theta K}(u) &= \max\left\{\left\langle \left(x, \frac{t_1 - t_2}{2}\right), u\right\rangle : (x, t_i) \in K\right\} \\ &\leqslant \frac{1}{2}\max\{\langle (x,t_1), u\rangle : (x,t_1) \in K\} + \frac{1}{2}\max\{\langle (x,-t_2), u\rangle : (x,t_2) \in K\} \\ &= \frac{1}{2}\max\{\langle (x,t_1), u\rangle : (x,t_1) \in K\} + \frac{1}{2}\max\{\langle (x,t_2), u'\rangle : (x,t_2) \in K\} \\ &= \frac{1}{2}h_K(u) + \frac{1}{2}h_K(u'). \end{aligned}$$

Since u and u' have the same distribution in S^{n-1}, we see that

$$\begin{aligned} w(S_\theta(K)) &= \int_{S^{n-1}} h_{S_\theta(K)}(u)d\sigma(u) \\ &\leqslant \frac{1}{2}\int_{S^{n-1}} h_K(u)d\sigma(u) + \frac{1}{2}\int_{S^{n-1}} h_K(u')d\sigma(u) \\ &= w(K). \end{aligned}$$

REMARK 1.5.14. A related inequality, that will be also used later on, states that if $0 \in \mathrm{int}(K)$ then

$$w(K) \geqslant \left(\frac{\kappa_n}{\mathrm{Vol}_n(K^\circ)}\right)^{1/n}. \tag{1.5.3}$$

To see this, use Hölder's inequality and the fact that $h_K(\theta) = \rho_{K^\circ}^{-1}(\theta)$ to write

$$\begin{aligned} w(K) &= \int_{S^{n-1}} h_K(\theta) d\sigma(\theta) \geqslant \left(\int_{S^{n-1}} (h_K(\theta))^{-n} d\sigma(\theta) \right)^{-1/n} \\ &= \left(\int_{S^{n-1}} (\rho_{K^\circ}(\theta))^n d\sigma(\theta) \right)^{-1/n} = \left(\frac{\kappa_n}{\mathrm{Vol}_n(K^\circ)} \right)^{1/n}. \end{aligned}$$

REMARK 1.5.15. It is interesting to mention here two more proofs of Urysohn's inequality. The first is by averaging over orthogonal images of K. One easily checks that $K_N = \frac{1}{N}\sum_{i=1}^N U_i(K)$ has the same mean width as K. Clearly, by taking U_i to be a net in $O(n)$, one can make sure that K_N converges to a multiple of the Euclidean ball, and hence that it must converge to $w(K)B_2^n$. But, by the Brunn-Minkowski inequality, the volume of K_N is greater than the volume of K, and we get the inequality in the limit.

Yet another proof can be obtained by another method of symmetrization, which is interesting in its own right, called *Minkowski symmetrization*. Given a direction $\theta \in S^{n-1}$ we set

$$\tau_\theta(K) := \frac{K + \pi_\theta(K)}{2}$$

where $\pi_\theta x := x - 2\langle \theta, x\rangle\theta$ is the reflection across $\theta^\perp$. It is not hard to check that $S_\theta(K) \subseteq \tau_\theta(K)$ (we will see this in Part II), that $\tau_\theta(K)$ has larger volume than K and the same mean width as K. Finally, it is also not hard to show that there exists a sequence of directions θ_j such that, letting $K_j = \tau_{\theta_j}(K_{j-1})$, these bodies converge to a Euclidean ball. Putting all these facts together completes the proof.

1.5.6. Grünbaum's lemma

We close this section with the following beautiful proposition which claims that, in any direction, a hyperplane passing through the barycenter of a convex body K in $\mathbb{R}^n$ divides the body in two parts of more or less the same volume.

PROPOSITION 1.5.16 (Grünbaum). *Let K be a convex body of volume 1 in $\mathbb{R}^n$ with barycenter at the origin. For every $\theta \in S^{n-1}$ we have*

$$\mathrm{Vol}_n(\{x \in K : \langle x, \theta\rangle \geqslant 0\}) \geqslant 1/e.$$

In fact, a more precise estimate depending on the dimension is known; one has

$$\mathrm{Vol}_n(\{x \in K : \langle x, \theta\rangle \geqslant 0\}) \geqslant \left(\frac{n}{n+1}\right)^n.$$

Proof. The argument that we present below is from [**398**]. Since K is compact there exists $M > 0$ such that

$$\mathrm{Vol}_n(\{x \in K : |\langle x, \theta\rangle| > M\}) = 0.$$

Let $G(t) = \mathrm{Vol}_n(\{x \in K : \langle x, \theta\rangle \geqslant -t\})$. Then, G is a log-concave increasing function and we have $G(t) = 0$ for $t \leqslant -M$ and $G(t) = \mathrm{Vol}_n(K) = 1$ for $t \geqslant M$. Since the barycenter of K is at the origin, we have

$$\int_{-M}^{M} tG'(t)dt = 0,$$

and applying integration by parts we see that

$$\int_{-M}^{M} G(t)dt = M.$$

Our task is to prove that $G(0) \geqslant 1/e$. Observe that, since $\log G$ is a concave function, we can write

$$G(t) \leqslant G(0)e^{\alpha t}$$

with $\alpha = G'(0)/G(0)$. We may also choose M large enough so that $1/\alpha < M$. Then, using the fact that $G(t) \leqslant G(0)e^{\alpha t}$ if $t \leqslant 1/\alpha$ and that, trivially, $G(t) \leqslant \mathrm{Vol}_n(K) = 1$ if $t > 1/\alpha$, we can write

$$M = \int_{-M}^{M} G(t)dt \leqslant \int_{-\infty}^{1/\alpha} G(0)e^{\alpha t}dt + \int_{1/\alpha}^{M} \mathbf{1}\,dt = \frac{eG(0)}{\alpha} + M - \frac{1}{\alpha}.$$

We conclude that $G(0) \geqslant 1/e$ as claimed. □

1.6. Minkowski's problem

The *area measure* σ_K of a convex body K in $\mathbb{R}^n$ is defined on S^{n-1} and corresponds to the usual surface measure on K via the Gauss map: First we define the reverse normal image with respect to K in $\mathbb{R}^n$. For a Borel subset B of the sphere S^{n-1} we define

$$n_K^{-1}(B) = \{x \in \partial K : \exists \text{ exterior normal } n_K(x) \in B\}.$$

We measure $(n-1)$-dimensional Lebesgue volume on ∂K as usual (say, with its $(n-1)$ dimensional Hausdorff measure). One may check that for Borel B, $n_K^{-1}(B)$ is measurable (to see this, observe that if B is compact then so is its inverse image). Then, we define the area measure σ_K on S^{n-1} by

$$\sigma_K(B) = \mathrm{Vol}_{n-1}(n_K^{-1}(B)).$$

Clearly $\sigma_K(S^{n-1}) = \mathrm{Vol}_{n-1}(\partial K)$.

For a polytope P with normal vectors u_j and facets F_j, the area measure σ_P is a discrete measure which assigns mass $\mathrm{Vol}_{n-1}(F_j)$ at each normal vector u_j.

The following theorem of Minkowski gives necessary and sufficient conditions for a measure on the sphere to be the surface area measure of a polytope.

THEOREM 1.6.1 (Minkowski existence). *Given a set of m vectors $u_j \in S^{n-1}$ and weights $\lambda_j > 0$, the following are equivalent:*

(i) $\sum_{j=1}^{m} \lambda_j u_j = 0$ *and not all u_j lie in a halfspace whose boundary hyperplane contains the origin.*
(ii) *The vectors u_j and the weights λ_j are exterior normals and areas of a (unique, up to translation) polytope.*

Proof. We first show that (ii) implies (i). Assuming (ii) the polytope is

$$P = \{x : \langle x, u_j\rangle \leqslant h_P(u_j),\ j = 1, \ldots, m\}.$$

Assume that there exists $v \neq 0$ such that $\langle u_j, v\rangle \leqslant 0$ for all j. Then, we easily see that for every $x \in P$ we also have $x + tv \in P$ for all $t \in \mathbb{R}$; this contradicts the assumption that P is bounded. Next, for any $v \in S^{n-1}$ we write

$$\Big\langle v, \sum_{j=1}^{m} \lambda_j u_j \Big\rangle = \sum_{\{j:\langle u_j,v\rangle>0\}} \lambda_j \langle v, u_j\rangle + \sum_{\{j:\langle u_j,v\rangle<0\}} \lambda_j \langle v, u_j\rangle.$$

The absolute value of both sums on the right hand side is equal to the volume of the projection of P to $v^{\perp}$. Therefore, the expression is 0 for all v, thus $\sum_{j=1}^{m} \lambda_j u_j = 0$.

Next, we show that (i) implies (ii). For the existence we shall use Lagrange multipliers. For every m-tuple of non-negative s_i's, define

$$P(s) = \{x : \langle x, u_j \rangle \leqslant s_j,\ j = 1, \dots, m\}.$$

This is clearly a convex polyhedral set. It is even a polytope - for this one has to check boundedness, which follows from the conditions (i); otherwise it would contain a ray, in the direction of v, say, which means the defining u_i's would all satisfy $\langle tv, u_j \rangle \leqslant s_j$ for all $t > 0$, and hence $\langle v, u_j \rangle \leqslant 0$, but this would mean that they are contained in a halfspace whose boundary hyperplane contains the origin, which cannot be by assumption.

Let $P(s)$ be a polytope as above, with facets $F_j(s) = \{x \in P(s) : \langle x, u_j \rangle = s_j\}$, $j \leqslant m$ (it may happen that $F_j(s) = \emptyset$ for some j's). The first observation is that the volume of $P(s)$ as a function of s is differentiable for $s_1, \dots, s_m > 0$, and its derivative with respect to s_j is

$$\frac{\partial \mathrm{Vol}_n(P(s))}{\partial s_j} = \mathrm{Vol}_{n-1}(F_j(s)).$$

The function $s \mapsto \mathrm{Vol}_n(P(s))$ attains a maximum on the simplex $S = \{s \geqslant 0 : \sum_{j=1}^{m} \lambda_j s_j = 1\}$. If it is attained at a boundary point $\hat{s}$ of the simplex, one of the s_j's is 0, and then the origin is on the boundary of $P(\hat{s})$. We choose a vector $t = (t_1, \dots, t_m)$ so that $0 \in \mathrm{int}(t + P(\hat{s}))$. Note that

$$t + P(\hat{s}) = \{x : \langle x, u_j \rangle \leqslant \hat{s}_j + \langle u_j, t \rangle, j = 1, \dots m\} = P(s'),$$

where $s' = \hat{s} + (\langle u_1, t \rangle, \dots, \langle u_m, t \rangle)$, and that $0 \in \mathrm{int}(P(s')$, which implies that $s'_j > 0$. We also have that

$$\sum_{j=1}^{m} \lambda_j s'_j = \sum_{j=1}^{m} \lambda_j \hat{s}_j + \sum_{j=1}^{m} \lambda_j \langle u_j, t \rangle = 1,$$

because we have assumed that $\sum_{j=1}^{m} \lambda_j u_j = 0$. Therefore, $s' \in \mathrm{relint}(S)$. This argument allows us to assume that the volume of $P(s)$ is maximized at a (relative) interior point $\hat{s}$ of S.

Then, by Lagrange multipliers theory, there exists $\alpha \in \mathbb{R}$ such that

$$\frac{\partial \big(\mathrm{Vol}_n(P(s)) - \alpha \sum_{j=1}^{m} \lambda_j s_j\big)}{\partial s_j}(\hat{s}) = \mathrm{Vol}_{n-1}(F_j(\hat{s})) - \alpha \lambda_j = 0,\ j = 1, \dots, m.$$

Since $\hat{s}_j > 0$ for all j, we see that $P(\hat{s})$ has non-empty interior; in particular, $\mathrm{Vol}_{n-1}(F_j(\hat{s})) \neq 0$ for at least one j, which shows that $\alpha \neq 0$ and $\hat{s}_j = \alpha^{-1} \mathrm{Vol}_{n-1}(F_j(s))$ for all $j = 1, \dots, m$. Then, the polytope $\alpha^{-\frac{1}{n-1}} P(\hat{s})$ satisfies (ii).

Uniqueness follows from equality conditions in Minkowski's first inequality: if P and Q are two polytopes with exterior normals u_j and the areas of their corresponding facets are equal to λ_j then

$$V(P, Q, \dots, Q) = \frac{1}{n} \sum_{j=1}^{m} \lambda_j h_P(u_j) = V(P, P, \dots, P) = \mathrm{Vol}_n(P).$$

Then,

$$\mathrm{Vol}_n(P) = V(P, Q, \ldots, Q) \geqslant (\mathrm{Vol}_n(P))^{\frac{1}{n}} (\mathrm{Vol}_n(Q))^{\frac{n-1}{n}},$$

which shows that $\mathrm{Vol}_n(P) \geqslant \mathrm{Vol}_n(Q)$. By symmetry, we also get $\mathrm{Vol}_n(Q) \geqslant \mathrm{Vol}_n(P)$. Then, we have equality in Minkowski's first inequality, and since the volumes of P and Q are equal we conclude that Q is a translate of P. □

The continuous case is also of interest, since it is intimately related with the following classical "Minkowski's problem": Given a positive function κ on the sphere, can one find a convex body whose Gauss curvature at the point with normal u is $\kappa(u)$?

There is a simple connection between the Gauss curvature and the area measure:

$$\kappa_K(u) = \lim_{B \downarrow u} \frac{\mathrm{Vol}_{n-1}(B)}{\sigma_K(B)}.$$

That is $d\mathrm{Vol}_{n-1}(u) = \kappa(u) d\sigma_K(u)$. Here we obviously assume that Gauss curvature is well-defined, namely that the convex body is C^2. One sometimes writes (in the case that κ is nowhere 0) that

$$\sigma_K(B) = \int_B \frac{1}{\kappa(u)} d\mathrm{Vol}_{n-1}(u).$$

In order to be able to use discrete approximation of a general body by a polytope we shall need the following lemma.

LEMMA 1.6.2. *If a sequence $\{K_i\}$ of convex bodies converges to K in the Hausdorff metric, then the corresponding area measures weakly converge: $\sigma_{K_i} \to \sigma_K$.*

This enables one to prove the next theorem.

THEOREM 1.6.3. *Let σ be a Borel measure on S^{n-1}. Assume it satisfies $\int_{S^{n-1}} u d\sigma(u) = 0$ and it is not concentrated on a great circle. Then there is a convex body K, unique op to translation, such that $\sigma = \sigma_K$. In fact, the converse is also true.*

Note that in particular this solves Minkowski's problem stated above.

Sketch of the proof. The idea is to approximate the measure by a discrete one satisfying the same requirements, for it to use Minkowski's existence theorem for polytopes, and to show convergence of the polytopes to a body with the required property. Uniqueness will follow from equality conditions in Minkowski's inequality.

The fact that the converse is also true can be shown using approximation by polytopes. □

1.7. Notes and remarks

Basic Convexity

We have included a section in the appendix in which we elaborate on the basic definitions and structures from convexity. To read about convex geometry in more detail we refer the reader to the monographs by Schneider [**556**], Gruber [**277**], Schneider and Weil [**557**]. We have also included a section on convex functions in the appendix. A recommended source in which one may read in depth about convex functions is Rockafellar's book [**521**].

Brunn–Minkowski inequality

The history of the Brunn-Minkowski inequality starts with the work of Brunn [**124**], [**125**] who discovered it in dimensions 2 and 3. Minkowski considered the n-dimensional case (the proof of Brunn is obviously valid for any dimension n) and characterized the case of equality in [**474**] (see also [**473**] for dimensions $n = 2$ and 3, and [**126**]). One should emphasize here that the study of geometry during that time (end of 19th century and beginning of 20th century) concerned only dimensions 2 or 3; for example, almost all the work of Blaschke was restricted to 2 or 3 dimensional objects. It was Minkowski who realized the importance of the geometric study of higher dimensional spaces as a way of solving some outstanding problems of analytic number theory. This was the starting point of n-dimensional convexity theory for any dimension n.

Alternative proofs of the Brunn-Minkowski inequality were given by Blaschke, Hilbert, Bonnesen. Schwarz symmetrization was introduced in [**560**] and the proof of the Brunn-Minkowski inequality via this method is due to Blaschke [**75**]. Theorem 1.2.8, that provides a proof of the inequality using induction on the dimension, is due to Kneser and Süss, [**350**]. These classical proofs can be found in the book by Bonnesen and Fenchel [**93**].

Lusternik [**404**] extended the inequality to the class of compact sets. An alternative proof of Lusternik's result was obtained by Henstock and Macbeath in [**302**]. The proof, using elementary sets, that is presented in Theorem 1.2.6 is due to Hadwiger and Ohmann [**297**], who also clarified Lusternik's conditions for equality.

Lemma 1.2.5 appears in an appendix of the article [**273**] of Gromov and V. Milman. In this paper they introduced the idea of "localization" and proved very general isoperimetric type inequalities on the sphere S^n equipped with some absolutely continuous measure and metric, continuous with respect to Euclidean metric (see also [**6**] and [**269**]). As an application they proved a concentration result for the unit sphere of a uniformly convex space. This idea of bisection was later developed by Lovász and Simonovits in [**397**], who presented it in a very useful form and applied it in a joint work with Kannan [**325**] to obtain isoperimetric inequalities for log-concave measures. We will come back to the Lovász-Simonovits localization theorem in Volume II.

The survey articles by Gardner [**218**], Barthe [**53**] and Maurey [**420**] provide beautiful expositions of the Brunn-Minkowski inequality and its importance from different perspectives.

Alexandrov-Fenchel inequality

The Alexandrov-Fenchel inequalities are among the most deep inequalities in Mathematics. The history behind them is non-trivial. It seems that as a question it was not known to Hilbert (and the general n-dimensional Minkowski's inequality covers them only in the 2 and 3-dimensional case). On the other hand, it also seems that in the 1930's it was a well known folklore type problem. Fenchel's papers don't give clear historical information about who first conjectured the general Alexandrov-Fenchel inequalities. In his note [**195**] Fenchel first mentions that Minkowski's inequalities between mixed volumes are consequences of Brunn's theorem, and then he writes the following:

> However, one has good reasons to presume that not all inequalities between mixed volumes can be obtained in this way; in particular, one has suggested the truth of the inequality $V_{123\ldots n}^2 \geqslant V_{113\ldots n}V_{223\ldots n}$, which one has not been able to deduce, neither from Brunn's theorem nor from its improvement due to Bonnesen.

In his article [**196**], Fenchel first mentions Bonnesen's book of 1929, and then he continues:

> Transferred to n-dimensional space, Bonnesen's methods lead in certain cases to inequalities of the kind considered by Minkowski, which are no longer accessible by Brunn-Minkowski methods. After this,

> it was rather clear that for $n \geqslant 4$ one was by no means in possession of all inequalities of the considered kind, and one was led to conjecture the existence of a much more general inequality, which would contain all known and many new inequalities of Minkowski type as special cases. This conjecture could indeed be verified a few years ago, in different ways, on one hand by direct treatment of a related minimum problem (footnote 4), on the other hand by taking the pattern from a proof for the Minkowski inequalities (footnote 5) given by Hilbert.

Footnote 4 refers to Fenchel's two Comptes Rendus notes, and footnote 5 refers to Alexandrov's papers. So, it appears that the conjecture of the general inequality was "folklore" at that time, as we say today. We cite from Fenchel's artice [**197**]:

> Minkowski considered in this context only dimensions 2 and 3. The main part of the theory, including the two uniqueness and existence theorems, could easily be generalized to arbitrary dimension. It was, however, clear that there must be valid inequalities between the mixed volumes which are not consequences of the Brunn-Minkowski theorem. Another challenge was to subordinate the theorems on the determination of convex polyhedra and smooth convex bodies to one theorem valid for arbitrary convex bodies. Satisfactory results in both directions were achieved in the second half of the thirties, mainly through the efforts of A. D. Alexandrov.

Fenchel presented a geometric idea in [**195**] with a sketch of proof, which does not seem to work, and Alexandrov in four long papers (mainly in [**1**] and [**2**]) gave two different and complete proofs. Confirming this situation, Leichtweiss in his book [**376**] calls these results "quadratic Alexandrov's inequalities".

In the 1970's and 80's efforts of many people have led to the discovery that the Alexandrov-Fenchel inequalities are parallel to the Hodge index theorem (the so called "Hodge inequalities") and may be deduced from these together with the theorem on the number of roots of a system of polynomial equations. This was done independently, in 1978, by Khovanskii [**331**] and Teissier [**590**], see Section 27 of [**127**].

The starting point for this development was the formula for the number of complex roots of a system of polynomial equations, say, $P_1(x) = \cdots = P_n(x) = 0$. To briefly explain this connection let us introduce the notion of Newton polyhedron. Given a monomial of the form $cz_1^{m_1} \cdots z_n^{m_n}$ in n complex variables with a complex coefficient $c \neq 0$ consider the point with integer coordinates $m = (m_1, ..., m_n)$ in $\mathbb{R}^n$. We consider Laurent polynomials which are finite sums of monomials with integers m_i which may be positive or negative. The Newton polyhedron of such a polynomial is the convex hull in $\mathbb{R}^n$ of all integer points corresponding to the monomials participating in the given polynomial. Given a system of polynomial equations $P_1(x) = \cdots = P_n(x) = 0$ we have built n convex bodies $K_1, \ldots, K_n$. For simplicity of the formulation of the result, let us consider only polynomials with non-zero constant terms. The relation between algebraic geometry and the theory of mixed volumes now follows from the theorem on the number of (complex) roots of such a system, which was initiated by the result of Kushnirenko [**364**] and completed in the work by D. Bernstein [**67**]. They show that the number of complex roots of the general ("typical", i.e., in general position) system of polynomial equations $P_1(x) = \cdots = P_n(x) = 0$ with Newton polyhedra $K_1, \ldots, K_n$ is equal to the mixed volume of these bodies multiplied by $n!$.

There is a very interesting connection of this result with the following coincidence, going back to the middle of 19th century. Actually, the roots of the results of Kushnirenko and Bernstein may be found in a 1840 theorem of Minding. It is a very curious historical coincidence that around the same time that Steiner proved his result on polynomiality of volume (actually, in his case, area) of the expression $K + tB$ for the Euclidean disc B, Minding [**472**] considered the problem of finding a formula for the number N of solutions of two polynomial equations $P_1 = P_2 = 0$ of two variables (today we would say polynomials in "general position"). As we know today (through the Kushnirenko-Bernstein theory) the mixed volume $V(K_1, K_2)$ of the Newton polyhedra K_1 and K_2 (in $\mathbb{R}^2$) of these polynomials is responsible for the number N. And this mixed volume $V(K_1, K_2)$ is the only non-trivial coefficient in the decomposition

$$Area(K_1 + tK_2) = Area(K_1) + 2tV(K_1, K_2) + t^2 Area(K_2).$$

However, Steiner proved this formula only in the case $K_2 = B$ (the Euclidean disc), which was a most interesting case in geometry but which was, on the contrary, not an interesting case for Minding because B is never the Newton polyhedron of any polynomial. So, the connection was not observed until the 1970's. The Euclidean ball was the only interesting case for Steiner because the isoperimetric inequality was his main target. It is in fact another very curious coincidence regarding Alexandrov-Fenchel inequalities from 1936 and Hodge theory from the same year: it took forty years to understand that they describe very similar mathematical structures.

Minkowski polarization

Following Theorem 1.1.10 we discussed the properties of mixed volumes. Slightly weaker properties than (ii) and (iii) in the list are that the mixed volume $V(K_1, \dots, K_n)$ is multi-linear and monotone by every argument with respect to set inclusion. It was shown by Milman and Schneider in [**467**] that any such functional defined on n-tuples of centrally symmetric convex bodies which is annihilated when any two of these bodies are parallel intervals, must be proportional to the mixed volume functional. As a corollary they derived the minimality property of mixed volumes: if some multi-linear and monotone functional $F(K_1, \dots, K_n)$ is bounded from above by the corresponding mixed volume of these (centrally symmetric convex) bodies, then $F = aV$ for some constant $a \geqslant 0$. Similarly, the mixed discriminant was characterized by Florentin, Milman and Schneider in [**213**] as the unique (up to a multiplicative constant) multi-linear monotone functional on n-tuples of positive semi-definite $n \times n$ matrices which is annihilated when any two of these matrices are proportional rank one matrices.

Minkowski addition: generalizations and additional information

Using the language of the support function one may generalize the usual Minkowski addition, corresponding to $h_{K+T} = h_K + h_T$ as follows. For $p \geq 1$ define $L = K +_p T$ by

$$h_L^p = h_K^p + h_T^p.$$

As p tends to ∞ the corresponding body tends to the convex hull of K and T. This type of addition, called p-Minkowski-addition or p-sum, was introduced by Firey [**207**] and intensively studied by Lutwak [**405**]. In a recent paper [**463**], Milman and Rotem show that p-sums represent all existing summations which satisfy a small and natural set of requirements. Under even weaker conditions they show that the "polynomiality" of volume with respect to summation implies that the summation is the usual Minkowski summation. These questions, under different assumptions, were studied by Gardner, Hug and Weil [**220**].

Let us note here one extremely curious recent observation by I. Molchanov from [**479**]. Let $K \subset \mathbb{R}^n$ be any convex body such that $rB_2^n \subset K$ for some $r > 1$. Then there exists a convex body Z such that $Z^\circ = Z + K$. It follows by similar methods that the same is true for p-summation, namely there exists some (other) Z such that $Z^\circ = Z +_p K$ (the case $p = \infty$ is trivial, taking $Z = K^\circ$ which is contained in K).

Refinements of the Brunn-Minkowski inequality

A linear refinement of the Brunn-Minkowski inequality holds true for pairs of convex bodies K and T that have projections of the same volume onto some hyperplane: if $\mathrm{Vol}_{n-1}(P_\theta(K)) = \mathrm{Vol}_{n-1}(P_\theta(T))$, then

$$\mathrm{Vol}_n((1-\lambda)K + \lambda T) \geqslant (1-\lambda)\mathrm{Vol}_n(K) + \lambda \mathrm{Vol}_n(T)$$

for every $\lambda \in (0,1)$. To see this one first proves the same assertion under the stronger assumption that the two projections coincide and then uses Steiner symmetrization of the bodies K and T in the direction of θ and an additional sequence of Steiner symmetrizations in directions perpendicular to θ that leads to bodies of revolution with the same projection onto $\theta^\perp$ (see [**93**]).

A second result of this type concerns convex bodies K and T that have equal maximal hyperplane sections in some direction: if $\sup_r \mathrm{Vol}_{n-1}(K \cap (\theta^\perp + r\theta)) = \sup_s \mathrm{Vol}_{n-1}(T \cap (\theta^\perp + s\theta))$ for some $\theta \in S^{n-1}$, then for any $\lambda \in (0,1)$ we have

$$\mathrm{Vol}_n\Big((1-\lambda)K + \lambda T\Big) \geqslant (1-\lambda)\mathrm{Vol}_n(K) + \lambda \mathrm{Vol}_n(T).$$

This result goes back to Bonnesen [**92**]. The case $\lambda = 1/2$, for example, is a consequence of the next more general result: if $\gamma = \sup_r \mathrm{Vol}_{n-1}(K \cap (\theta^\perp + r\theta))^{\frac{1}{n-1}}$ and $\delta = \sup_s \mathrm{Vol}_{n-1}(T \cap (\theta^\perp + s\theta))^{\frac{1}{n-1}}$, then

$$\mathrm{Vol}_n(K+T) \geqslant (\gamma+\delta)^{n-1}\left(\frac{\mathrm{Vol}_n(K)}{\gamma^{n-1}} + \frac{\mathrm{Vol}_n(T)}{\delta^{n-1}}\right).$$

Stability results

The argument of Kneser and Süss for the proof of the Brunn-Minkowski inequality allows one to obtain stability results. Assume that $\mathrm{Vol}_n(K_0) = \mathrm{Vol}_n(K_1) = 1$ and that both bodies have their barycenter at the origin. Let $D = \max\{\mathrm{diam}(K_0), \mathrm{diam}(K_1)\}$. Groemer [**265**] proved (see also the works of Diskant [**173**], [**174**] and the article [**104**] of Bourgain and Lindenstrauss) that if

$$\mathrm{Vol}_n((1-\lambda)K_0 + \lambda K_1)^{1/n} \leqslant 1 + \varepsilon$$

for some $\varepsilon > 0$ and some $\lambda \in (0,1)$, then

$$\delta^H(K_0, K_1) \leqslant Cn\left(\frac{1}{\sqrt{\lambda(1-\lambda)}} + 2\right) D\varepsilon^{1/(n+1)}$$

where $C \sim 6$. An equivalent formulation is that

$$\mathrm{Vol}_n((1-\lambda)K_0 + \lambda K_1)^{1/n} \geqslant 1 + \left(\frac{\sqrt{\lambda(1-\lambda)}}{CnD(1+2\sqrt{\lambda(1-\lambda)})}\right)^{n+1} \delta^H(K_0,K_1)^{n+1}$$

for all $0 \leqslant \lambda \leqslant 1$.

Using a new proof of the Brunn-Minkowski inequality, based on mass transportation arguments, Figalli, Maggi and Pratelli showed in [**201**] and [**202**] that the relative asymmetry parameter

$$A(K,T) = \inf_{x \in \mathbb{R}^n} \frac{\mathrm{Vol}_n(K \triangle (\lambda T + x))}{\mathrm{Vol}_n(K)}$$

of K and T (where $\lambda^n = \mathrm{Vol}_n(K)/\mathrm{Vol}_n(T)$) satisfies

$$A(K,T) \leqslant Cn^{7/2}\sqrt{\beta(K,T)}$$

where

$$\beta(K,T) = \frac{\mathrm{Vol}_n(K+T)^{1/n}}{\mathrm{Vol}_n(K)^{1/n} + \mathrm{Vol}_n(T)^{1/n}} - 1.$$

The exponent 7/2 is due to Segal [**561**], who actually showed that it could be improved to 1 modulo a conjecture of Dar [**164**]: according to this conjecture the inequality

$$\mathrm{Vol}_n(K+T)^{1/n} \geqslant M(K,T)^{1/n} + \left(\frac{\mathrm{Vol}_n(K)\mathrm{Vol}_n(T)}{M(K,T)}\right)^{1/n}$$

holds for every pair of convex bodies K and T in $\mathbb{R}^n$, where

$$M(K,T) = \max_{x\in\mathbb{R}^n} \mathrm{Vol}_n(K\cap(x+T)).$$

This statement is stronger than the Brunn-Minkowski inequality and has been verified only in some special cases.

Volume preserving transformations

The origin of Brenier's map is in the theory of optimal transport, that starts with the question of Monge [**478**] on "how to transport a certain amount of soil, extracted from the ground, to places where it should be incorporated in a construction". A modern L_2-version of the problem is the following: given two finite measures μ and ν on $\mathbb{R}^n$, minimize $\int_{\mathbb{R}^n} |x - T(x)|^2 d\mu(x)$ over all transformations T that push forward μ to ν. Kantorovich expressed the optimal transport problem as a question of convex programming. For more information we refer the reader to the books of Villani [**598**] and [**599**]. Brenier proved in [**121**], [**122**] that the optimal transport map T from a measure μ that is absolutely continuous with respect to the Lebesgue measure to a measure ν is given by the gradient of a convex function and that it is essentially unique. Brenier's theorem was proved under some assumptions that were later removed by McCann [**424**], [**425**]. The proof of the uniqueness employs a refinement of Alexandrov's argument (see [**3**]) for the uniqueness of convex surfaces with prescribed integral curvature. The notion of cyclical monotonicity was put forward by Rockafellar as a direct generalization of usual monotonicity of a function of one variable, we refer to his monograph [**521**]. A nice pedagogical article of Ball [**45**] shows the existence of Brenier map via Brouwer's fixed point theorem.

Regularity results of Caffarelli (see [**130**], [**131**] and [**132**]) from the theory of elliptic equations provide smoothness of the Brenier map that allow one to use it in our setting. Instances of such use can be found e.g. in Alesker, Dar and V. Milman [**9**], where Theorem 1.3.10 is taken from. Lemma 1.3.11 appears in Gromov [**268**].

The proof of the existence of Knothe's map is from [**351**]. This construction is dependent on the choice of a basis (note that also Brenier's map depends on the choice of a Euclidean structure, needed in order to define the gradient); on the other hand Knothe's map avoids questions of regularity. It is worthwhile to note that Knothe's map may be perceived as a degeneration of Brenier's map in the sense that one can consider rescaled optimal transport problems whose solutions converge to the Knothe map; see Garlier, Galichon and Santambrogio [**142**].

Functional forms

Theorem 1.4.1 is usually attributed to Prékopa and Leindler; it was proved in [**377**] and [**515**] (see also [**514**]). Das Gupta offers in [**165**] detailed information on the historical background and on related results by several authors. Corollary 1.4.2 was first proved by Davidovic, Hacet and Korenbljum in [**166**].

Proposition 1.4.4 is from the paper of Henstock and Macbeath [**302**]. Various extensions (including Theorem 1.4.6) appear e.g. in Uhrin [**595**] and Barthe [**49**].

Steiner Symmetrization

Steiner introduced symmetrization in [**571**], [**572**] in connection with the isoperimetric problem. More precisely, he showed that perimeter or surface area decreases under symmetrization in dimensions $n = 2$ and 3 respectively. Blaschke studied symmetrization systematically in [**75**] and found several applications of this method. An interesting result of Falconer [**191**] asserts that if C is a compact subset of $\mathbb{R}^n$ with the property that all its Steiner symmetrals are convex, then C itself is convex. He also showed that if C has concave marginals onto a.e. hyperplane then C is convex. Figalli and Jerison [**200**] showed that this is still true under weaker assumptions on the marginals: if the marginals have convex support and are uniformly Lipschitz strictly inside their support, then the set is convex. In particular, this implies that if a set has log-concave marginals onto a.e. hyperplane then it is convex.

The fact that for any convex body K in $\mathbb{R}^n$ there exists a sequence of directions $u_j \in S^{n-1}$ such that $(S_{u_n} \circ \cdots \circ S_{u_1})(K)$ converges to a ball in the Hausdorff metric was observed by Gross [**274**] who also studied symmetrizations of bodies that are not necessarily convex. Mani has proved in [**413**] that if we choose an infinite random sequence of directions $u_j \in S^{n-1}$ and apply successive Steiner symmetrizations S_{u_j} to a body K in these directions, then we almost surely get a sequence of convex bodies converging to a ball. See also the work of Volčič [**600**] who proves almost sure convergence, in the L_1 distance, of sequences of random Steiner symmetrizations of measurable sets having finite measure to the ball having the same measure and that sequences of random symmetrizations of a compact set converge almost surely in the Hausdorff distance to the ball having the same measure.

The upper bounds for the number of successive Steiner symmetrizations that are required in order to bring a convex body at a uniformly bounded distance from a Euclidean ball of equal volume were exponential in the dimension (see e.g. [**294**]) until Bourgain, Lindenstrauss and Milman proved in [**107**] that there exist absolute constants $c, c_1, c_2 > 0$ with the following property: if K is a convex body in $\mathbb{R}^n$, then there are $k \leqslant cn \log n$ unit vectors u_j such that successive Steiner symmetrizations in the directions of u_j transform K into a convex body K_1 with $c_1 \rho B_2^n \subseteq K_1 \subseteq c_2 \rho B_2^n$, where B_2^n is the Euclidean unit ball and $\mathrm{Vol}_n(K) = \mathrm{Vol}_n(\rho B_2^n)$. An essentially best possible result was obtained by Klartag and V. Milman in [**346**]: for every $\varepsilon > 0$ there exist constants $c_1(\varepsilon), c_2(\varepsilon) > 0$ such that, given a convex body K in $\mathbb{R}^n$ with $\mathrm{Vol}_n(K) = \mathrm{Vol}_n(B_2^n)$, there are $k \leqslant (2+\varepsilon)n$ unit vectors u_j with the property that successive Steiner symmetrizations in the directions of u_j transform K into a convex body K' with $c_1(\varepsilon) B_2^n \subseteq K' \subseteq c_2(\varepsilon) B_2^n$. Later, Klartag [**334**] gave also an almost isometric version of this result. We will present these results in Part II. In fact, that whole chapter is devoted to analogous questions regarding other classical symmetrization methods.

There is a continuous way to look at Steiner symmetrization, letting the various intervals move in constant velocity (that depends of course on the interval). This is called a *system of moving shadows*. This point of view was suggested by Rogers and Shephard in a more general framework [**523**]. For a base set $B \subset \mathbb{R}^n$ (which is usually compact) one is given a velocity $v(x)$ for any element in B, and then, given a vector $u \in S^{n-1}$ too, one constructs the system

$$K_t = \mathrm{conv}\{x + tv(x)u : x \in B\}.$$

One can show that $\mathrm{Vol}(K_t)$ is a convex function of t, as is $\mathrm{Vol}_E(P_E K_t)$ for any subspace E. From this, together with Kubota's formula (1.1.1), it follows that all quermassintegrals $W_i(K_t)$ are convex functions of t.

Since the Steiner symmetral $S_u(K)$ can be viewed as the body $S_u(K) = K_0$ where $K_{-1} = K$ and $K_1 = R_u(K)$ (here R_u is the reflection with respect to $u^{\perp}$ of K), this immediately shows that Steiner symmetrization decreases surface area, and all the quermassintegrals. Finally, Campi and Gronchi [**134**] showed that $1/\mathrm{Vol}(K_t^\circ)$ is also a convex function of t.

Applications of the Brunn-Minkowski inequality

This book presents a limited number of the applications of the Brunn-Minkowski inequality in geometry and analysis. In this chapter we described the ones that will be of fundamental importance for the development of the theory we are concerned with.

The inequality of Rogers and Shephard on the volume of the difference body was proved in [**522**] (some special cases are discussed in the book of Bonnesen and Fenchel). The slightly different proof of Theorem 1.5.2 that we describe is due to Chakerian [**143**]. Other variants of the proof as well as extensions of this result can be found in [**523**], [**524**]. The equality case, as explained in Remark 1.5.3, follows from a refinement of a characterization of the simplex by Choquet [**149**]. A shorter proof was given by Eggleston, Grünbaum, and Klee [**181**]. Recently some functional versions of the Rogers-Shephard inequality were considered, see Colesanti [**152**] and Artstein-Avidan, Florentin, Gutman and Ostrover [**32**]. In the latter, the authors use a functional version of the Rogers-Shephard inequality to prove the following curious new volume inequality

$$\mathrm{Vol}_n(\mathrm{conv}(K \cup -L))\mathrm{Vol}_n(K^\circ + L^\circ)^\circ) \leq \mathrm{Vol}_n(K)\mathrm{Vol}_n(L).$$

Borell's Lemma is from [**94**]. In the more general version described in Remark 1.5.8 the constants can be improved, and it is known that for any log-concave probability measure μ and any centrally symmetric convex set A one has for all $t > 1$ that

$$1 - \mu(tA) \leqslant (1 - \mu(A))^{\frac{t+1}{2}}.$$

Furthermore, this inequality is best possible as follows from the example of the measure $d\mu = c\exp(-|x - x_0|)$ and $A = [-1, 1]$ on $\mathbb{R}$. The case $A = B_2^n$ is proved by Lovász and Simonovits [**397**] and the general case is a result of Guédon [**287**], where he proves a Borell-type lemma for s-concave measures.

The Blaschke-Santaló inequality was proved by Blaschke [**76**] in dimension $n = 3$ and by Santaló [**533**] in all dimensions. Saint-Raymond proved the equality case in [**531**]. Proposition 1.1.15 appears in the form above in Ball [**37**] and in Meyer and Pajor [**434**]; the not necessarily symmetric case is studied in Meyer and Pajor [**435**]. In the opposite direction, a well-known conjecture of Mahler states that $s(K) \geqslant 4^n/n!$ for every centrally symmetric convex body K (with one of the minimizers being the n-dimensional cube), while in the not necessarily symmetric case it declares that $s(K) \geqslant (n+1)^{n+1}/(n!)^2$ (with the minimum being attained at an n-dimensional simplex). We will discuss this question in Chapter 8, where we will describe the proof of a relaxed version of Mahler's conjecture that has been verified, namely the Bourgain-Milman inequality: there exists an absolute constant $c > 0$ such that

$$\left(\frac{s(K)}{s(B_2^n)}\right)^{1/n} = \left(\frac{s(K)}{\kappa_n^2}\right)^{1/n} \geqslant c$$

for every convex body K in $\mathbb{R}^n$. The inequality was first proved in [**110**] and answers the question of Mahler in the asymptotic sense: for every convex body K in $\mathbb{R}^n$, the affine invariant $s(K)^{1/n}$ is of the order of $1/n$.

Urysohn's inequality appears in [**596**]. The isodiametric inequality of Bieberbach (see [**73**]), which says that

$$\mathrm{Vol}_n(K) \leqslant \left(\frac{\mathrm{diam}(K)}{2}\right)^n \mathrm{Vol}_n(B_2^n),$$

can also be obtained as a consequence of the Brunn-Minkowski inequality (see e.g. [**277**, p. 153]).

Minkowski problem

Theorem 1.6.1 was proved by Minkowski for three-dimensional polytopes in [**473**] and [**475**]. The proof that we present is due to Alexandrov [**4**] (and can be also found in [**277**]). The general version of Minkowski's theorem was obtained by Fenchel and Jessen [**198**]. The literature on variants of the Minkowski problem, as well as stability and regularity versions of Minkowski's existence theorem, is quite extensive. Further information and references may be found in the articles of Cheng and Yau[**148**], Gluck [**246**] and the book of Pogorelov [**512**].

Closely related to the ideas behind Minkowski's existence theorem are also the works of Jerison [**309**], where electrostatic capacity plays the role of volume and the Minkowski problem is studied for the measure derived by its first variation, and [**310**] where similar problems are treated with methods of the calculus of variations. Colesanti continued this line of research (see [**151**], and [**153**] where the Minkowski problem for the torsional rigidity is studied).

The Christoffel problem asks for necessary and sufficient conditions for a measure μ on S^{n-1} so that it will be the first surface area measure of a convex body (see the appendix for the definition of the j-th area measure; Minkowski problem corresponds to the case $j = n-1$). The problem started with Christoffel's [**150**] work and it was answered by Firey (see [**209**] and [**210**]) and Berg [**65**]; however, their characterizations of the solution are not easy to apply in concrete situations. An alternative answer was given by Schneider [**554**] for the case of polytopes. The article [**253**] of Goodey and Weil establishes a connection between the Christoffel problem and spherical Radon transforms in the case of centrally symmetric convex bodies. Goodey, Yaskin and Yaskina [**254**] used Fourier transform techniques for a new approach to Berg's solution of the Christoffel problem.

A lot of work has been done for the general Christoffel-Minkowski problem regarding the j-th surface area measure. Equivalently, the question is to characterize, for any given $1 \leqslant j \leqslant n-1$, those functions that correspond to the j-th elementary symmetric function of the principal radii of curvature of a convex body. The interested reader will find information in the works of B. Guan, P. Guan, Lin, Ma and Zhou [**281**], [**282**], [**283**], [**284**], [**285**] and in Schneider's book [**556**].

CHAPTER 2

Classical positions of convex bodies

We fix a Euclidean structure $\langle\cdot,\cdot\rangle$ in $\mathbb{R}^n$. Given a convex body K in $\mathbb{R}^n$, we consider the family of its *positions*, meaning its affine images $\{x_0 + T(K)\}$ where $x_0 \in \mathbb{R}^n$ and $T \in GL_n$. In the context of functional analysis, one is given a norm (whose unit ball is K) and the choice of a position reflects a choice of a Euclidean structure for the linear space $\mathbb{R}^n$. In this case, when the body is centrally symmetric, one usually disregards translations and works only with linear images of K. In this language, the choice of a Euclidean structure specifies a unit ball of the Euclidean norm, which is an ellipsoid. Thus, one may regard a "position" as a choice of a special ellipsoid (though one usually thinks of the image of K by the linear map mapping this ellpsoid into a euclidean ball as the "position" of choice). An ellipsoid in $\mathbb{R}^n$ is a convex body of the form

$$\mathcal{E} = \left\{ x \in \mathbb{R}^n : \sum_{i=1}^{n} \frac{\langle x, v_i\rangle^2}{\alpha_i^2} \leqslant 1 \right\},$$

where $\{v_i\}_{i\leqslant n}$ is an orthonormal basis of $\mathbb{R}^n$ with respect to $\langle\cdot,\cdot\rangle$, and $\alpha_1,\ldots,\alpha_n$ are positive reals (the directions and lengths of the semiaxes of $\mathcal{E}$, respectively). It is easy to check that $\mathcal{E} = T(B_2^n)$, where T is the positive definite linear transformation of $\mathbb{R}^n$ defined by $T(v_i) = \alpha_i v_i$, $i = 1,\ldots,n$. Therefore, the volume of $\mathcal{E}$ is equal to

$$\mathrm{Vol}_n(\mathcal{E}) = \kappa_n \prod_{i=1}^{n} \alpha_i$$

(where, as before, $\kappa_n = \mathrm{Vol}_n(B_2^n)$). Note that by standard linear algebra, for any $A \in GL_n$, the body $A(B_2^n)$ is also an ellipsoid, and its volume is $\kappa_n|\det(A)|$.

The different ellipsoids connected with a convex body (or the different positions, corresponding to different choices of a Euclidean structure) that we shall consider in this chapter reflect different traces of symmetries which the convex body has. In some special cases, when the body has enough symmetries, meaning that there is a unique ellipsoid that satisfies all these symmetries, most (or all) of these positions coincide. This is the case, for example, when the normed space has a *symmetric basis* (i.e., a basis with respect to which the norm is invariant under permutations of the coordinates and changes of signs in the coordinates). However, in the general case a convex body has several different, useful, positions.

In this chapter we shall discuss the most basic and classical positions of convex bodies, which arise as solutions of extremal problems. These include John position (also called maximal volume ellipsoid position), minimal surface area position and minimal mean width position. The attentive reader will notice that the ideas behind the proofs are quite similar. Indeed, when a position is extremal then clearly some differential must vanish, and its vanishing is connected with isotropicity of some connected measure.

We shall also consider some direct applications, mainly of John position, and introduce a main tool, which is useful in many other results in the theory, the Brascamp-Lieb inequality. We state and prove one of its most useful forms, which is the so-called "normalized form" put forward by K. Ball. We prove it together with its reverse form, using Barthe's argument which is an ingenious transportation of measure argument similar to the one used in the proof of the Prékopa-Leindler inequality. In Part II of this book we shall discuss the general form of the Brascamp-Lieb inequality, its various versions, proofs, and reverse form, as well as further applications to convex geometric analysis.

We would like to mention that there are two other very important positions which are even more useful in terms of the subject of this book, and each will have its own chapter dedicated to its investigation.

The first of the two, which arose from classical mechanics back in the 19th century, is the *isotropic* position. It is defined to be the (unique, up to orthogonal transformations) position in which the body has its barycenter at the origin, the inertia matrix of the convex body is scalar and the volume of the body is one. Namely, given a convex body K in $\mathbb{R}^n$, the isotropic position of K is defined as $\tilde{K} = T(K) + x_0$ with $T \in GL_n$ and $x_0 \in \mathbb{R}^n$ so that $\mathrm{Vol}_n(\tilde{K}) = 1$, $\int_{\tilde{K}} x dx = 0$ and

$$\int_{\tilde{K}} \langle x, \theta\rangle^2 dx = L_K^2$$

for any $\theta \in S^{n-1}$, where $L_K > 0$ is a constant called the *isotropic constant* of K. In Chapter 10 (and later on, in Part II) we describe the main results and research directions connected with this very central position. Let us mention that the isotropic position has a useful characterization as a solution of an extremal problem: indeed, the isotropic position $\tilde{K} = T(K) + x_0$ of K minimizes the quantity

$$\int_{\tilde{K}} |x|^2 dx$$

over all $T \in GL_n$ and $x_0 \in \mathbb{R}^n$ such that $\mathrm{Vol}_n(\tilde{K}) = 1$ and $\int_{\tilde{K}} x dx = 0$.

The second position which is very useful in many results and which dates to the mid 1980's is the so called Milman position, or M-position of a convex body. It can be described "isometrically" as the position $\tilde{K} = T(K) + x_0$ with the properties that $\mathrm{Vol}_n(\tilde{K}) = \mathrm{Vol}_n(B_2^n)$ and $\int_{\tilde{K}} x dx = 0$, which in addition minimizes the volume $\mathrm{Vol}_n(\tilde{K} + B_2^n)$ among all such $T \in GL_n$. However, such a characterization hides its main properties and advantages that are in fact of an "isomorphic" nature. It turns out that the ellipsoid connected with the M-position (which is called the Milman ellipsoid or in short M-ellipsoid of K) can replace the original body in volume estimates. Given a body that is in M-position, the covering numbers (which we will describe in Chapter 4) of the body *by* a Euclidean ball and *of* a Euclidean ball by the body are not large. Furthermore, not only does this hold for the body, but also for its polar. Each of the two above facts imply that, for any pair of convex bodies K and T that are in M-position, a *reverse* Brunn-Minkowski type inequality is true, namely there exists a universal constant C such that for any n, and any $K, T \in \mathcal{K}^n$ that are in M-position, one has

$$\mathrm{Vol}_n(K+T)^{1/n} \leqslant C\left[\mathrm{Vol}_n(K)^{1/n} + \mathrm{Vol}_n(T)^{1/n}\right].$$

The details will appear in Chapter 8.

2.1. John's theorem

We shall show that for every convex body K in $\mathbb{R}^n$ there is a unique ellipsoid $\mathcal{E}$ of maximal volume that is contained in K. We will say that $\mathcal{E}$ is the *maximal volume ellipsoid* of K.

John's theorem gives a characterization of the maximal volume ellipsoid, or, more precisely, a characterization for a body being in "John position".

DEFINITION 2.1.1. Given a body $K \in \mathcal{K}^n$ that has the origin as an interior point (for such a body we write $K \in \mathcal{K}^n_{(0)}$) we shall say that it is in "John position" if B_2^n is a maximal volume ellipsoid of K.

It is easy to see by standard compactness arguments (which we illustrate below for completeness) that every $K \in \mathcal{K}^n_{(0)}$ has some linear image $A(K)$ which is in John position. We shall prove below that the maximal volume ellipsoid is unique, and thus the John position of a convex body is unique up to orthogonal transformations. One interesting application of the John position is the fact that, for any centrally symmetric convex body K, we can find an ellipsoid $\mathcal{E}$ such that

$$\mathcal{E} \subseteq K \subseteq \sqrt{n}\mathcal{E}. \tag{2.1.1}$$

In particular, if K is in John position, then $\mathcal{E} = B_2^n$. In the language of geometric functional analysis this means that the Banach-Mazur distance between any n-dimensional normed space and ℓ_2^n is at most $\sqrt{n}$.

More precisely, we define

DEFINITION 2.1.2 (Banach-Mazur distance). Let $X = (\mathbb{R}^n, \|\cdot\|_K)$ and $Y = (\mathbb{R}^n, \|\cdot\|_T)$. The Banach-Mazur distance between X and Y, denoted $d_{BM}(X,Y)$ is the infimal constant $C > 0$ such that there exists a linear operator $A : X \to Y$ satisfying

$$\|x\|_K \leqslant \|Ax\|_T \leqslant C\|x\|_K$$

for all $x \in X$. Equivalently we have $C^{-1}A(K) \subset T \subset A(K)$. For simplicity we will usually write $d(X,Y)$ instead of $d_{BM}(X,Y)$ unless there is some danger of confusion.

Given two centrally symmetric convex bodies K and T in $\mathbb{R}^n$ we agree to write

$$\begin{aligned} d(K,T) &= d_{BM}(K,T) := d_{BM}(X_K, X_T) \\ &= \min\{t > 0 : A(K) \subseteq T \subseteq tA(K) \text{ for some } A \in GL_n.\} \end{aligned}$$

We also extend the definition of the Banach-Mazur distance to the setting of not necessarily symmetric convex bodies as follows: if K and T are convex bodies in $\mathbb{R}^n$ then the distance of K and T is defined by

$$d(K,T) = d_{BM}(K,T) = \min\{t > 0 : A(K) + y \subseteq T + x \subseteq t(A(K) + y)\}$$

where the minimum is over all $x, y \in \mathbb{R}^n$ and $A \in GL_n$. Finally, it will also be useful to have the notion of the *geometric* distance between two spaces (or bodies), which is denoted $d_G(K,T)$, and defined as follows

$$d_G(K,T) = \inf\{ab : a, b > 0\, \exists x_0, y_0 \in \mathbb{R}^n, \frac{1}{b}(T + y_0) \subset K + x_0 \subset a(T + y_0)\}.$$

Of course we then have

$$d_{BM}(K,T) = \inf\{d_G(K, AT) : A \in GL_n\}.$$

We may now reformulate (2.1.1) as $d(X, \ell_2^n) \leqslant \sqrt{n}$ for any $X = (\mathbb{R}^n, \|\cdot\|_K)$. Since this fact has an elementary proof we shall devote the next subsection to it. We shall then discuss minimal and maximal volume ellipsoids of a convex body. Finally, we shall state and prove John's theorem regarding the contact points of K and B_2^n when K is in John position and the corresponding decomposition of the identity.

2.1.1. The "simple" John's theorem

THEOREM 2.1.3 (John). *Let K be a centrally symmetric convex body in John position. Then $K \subseteq \sqrt{n}B_2^n$.*

Proof. Assume towards a contradiction that there exists some $x_0 \in K$ with $|x_0| = R > \sqrt{n}$. Without loss of generality assume that $x_0 = R\, e_1$. From symmetry, $\mathrm{conv}(\pm R\, e_1, B_2^n) \subseteq K$. Consider an ellipsoid

$$\mathcal{E} = \left\{ x : \frac{x_1^2}{a^2} + \sum_{i=2}^{n} \frac{x_i^2}{b^2} \leqslant 1 \right\},$$

where we set $b^2 = (1 - \varepsilon)$ for a small ε, and $a > 1$ is to be determined later in such a way that we will have $\mathcal{E} \subseteq K$. The volume of $\mathcal{E}$ is $ab^{n-1}\kappa_n$. To have it included in the aforementioned convex hull, note that we need just a two-dimensional argument, since we are working with revolution bodies. We shall need the following two drawings: the first is the picture of the convex hull, the second is the same picture, squeezed in one axis by b/a.

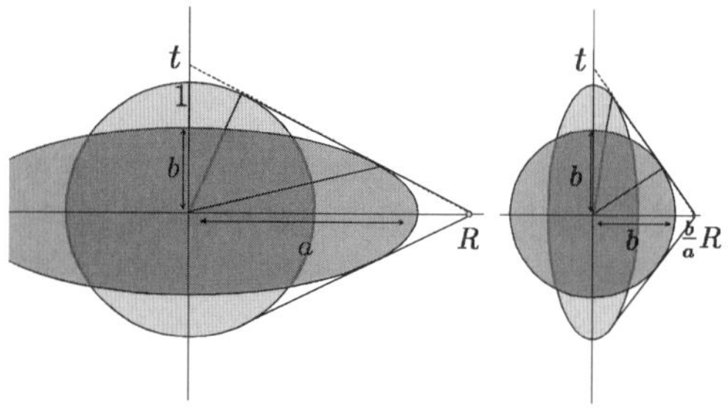

We see that $1/R = t/\sqrt{R^2 + t^2}$ from the first picture. Thus $t^2R^2 = R^2 + t^2$ and $t^2 = R^2/(R^2 - 1)$.

From the second picture we see that $a/R = b/(Rb/a) = t/\sqrt{R^2b^2/a^2 + t^2}$. That is, $R^2b^2 + t^2a^2 = R^2t^2$, and by replacing t^2 with $R^2/(R^2 - 1)$ we get

$$a^2 = b^2 + R^2(1 - b^2).$$

Writing also $1 - \varepsilon$ instead of b^2, we see that $a = 1 + (R^2 - 1)\varepsilon$ and that the square of the volume of the whole ellipsoid is equal to

$$(1-\varepsilon)^{n-1}(1+\varepsilon(R^2-1)) = (1-(n-1)\varepsilon+o(\varepsilon))(1+(R^2-1)\varepsilon) = 1+(R^2-n)\varepsilon+o(\varepsilon).$$

Therefore, if it held that $R > \sqrt{n}$, then, for small enough ε, this volume would indeed be greater than 1, a contradiction. □

REMARK 2.1.4. We remark that a similar argument can be given in the non-centrally symmetric case, resulting in the factor n instead of $\sqrt{n}$. This is clearly the best possible constant in the non-symmetric case, as is seen by considering the example of the simplex.

2.1.2. Maximal and minimal volume ellipsoids

The fact that there cannot be two different maximal volume ellipsoids inside a convex body follows easily from the following result on determinants of matrices (this result goes back to Minkowski).

LEMMA 2.1.5. *Let $A_1, A_2 \in GL_n$ be symmetric and positive definite. Then*

$$\det(A_1 + A_2)^{1/n} \geqslant \det(A_1)^{1/n} + \det(A_2)^{1/n}.$$

Equality holds if and only if $A_1 = \mu A_2$ for some $\mu > 0$. Equivalently,

$$\det((1-\lambda)A_1 + \lambda A_2) \geqslant \det(A_1)^{1-\lambda}\det(A_2)^{\lambda}$$

for all $\lambda \in [0,1]$. Here the equality condition is that $A_1 = A_2$.

This inequality can be proved as a consequence of the arithmetic-geometric means inequality in the same way as in the proof of Theorem 1.2.6 from Chapter 1. One only has to notice the fact from linear algebra that two positive definite linear transformations may be brought to diagonal form simultaneously by an SL_n transform. We give here a different proof using Hölder's inequality.

Proof. We shall use the well known formula

$$\int_{\mathbb{R}^n} e^{-\langle Ax,x\rangle}dx = \frac{(2\pi)^{n/2}}{\det(A)^{1/2}}.$$

Then we write

$$\begin{aligned}\frac{(2\pi)^{n/2}}{\det((1-\lambda)A+\lambda B)^{1/2}} &= \int_{\mathbb{R}^n} e^{-\langle((1-\lambda)A+\lambda B)x,x\rangle}dx \\ &= \int_{\mathbb{R}^n}\left(e^{-\langle Ax,x\rangle}\right)^{(1-\lambda)}\left(e^{-\langle Bx,x\rangle}\right)^{\lambda}dx \\ &\underset{\text{(Hölder)}}{\leqslant} \left(\int_{\mathbb{R}^n} e^{-\langle Ax,x\rangle}dx\right)^{1-\lambda}\left(\int_{\mathbb{R}^n} e^{-\langle Bx,x\rangle}dx\right)^{\lambda} \\ &= \frac{(2\pi)^{n/2}}{(\det A)^{\frac{1-\lambda}{2}}(\det B)^{\frac{\lambda}{2}}}\end{aligned}$$

and the lemma follows. □

PROPOSITION 2.1.6. *Given a convex body $K \subset \mathbb{R}^n$, there exists a unique ellipsoid of maximal volume inscribed in K.*

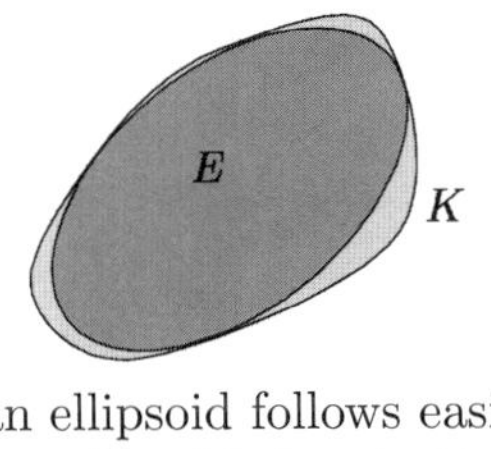

Proof. The existence of such an ellipsoid follows easily from standard compactness arguments in $\mathbb{R}^{n(n+1)/2}$, or from Blaschke's selection theorem. Next assume that there are two different ellipsoids of maximal volume inside K. Without loss of generality one of them is B_2^n, and the other can be written as $x_0 + A(B_2^n)$, with A

a positive definite matrix which then must satisfy $\det(A) = 1$. By convexity of K we also have

$$\frac{x_0}{2} + \frac{A + \mathrm{Id}}{2} B_2^n \subseteq K.$$

This is an ellipsoid, whose volume is equal to $\det\left(\frac{A+\mathrm{Id}}{2}\right)\kappa_n$, with the last determinant being at least 1 by Lemma 2.1.5. The equality condition implies that $A = \mathrm{Id}$ which means that K contains B_2^n and a translate of it, $x_0 + B_2^n$ for some $x_0 \neq 0$ (as we assumed the ellipsoids were different). But then it contains their convex hull, which is easily seen to contain a slightly more elongated ellipsoid, and we get a contradiction to B_2^n having the maximum volume among all ellipsoids in K. □

We next show that there is a unique ellipsoid $\mathcal{E}$ which contains a convex body K in $\mathcal{K}_{(0)}^n$ and has minimal volume (the *minimal volume ellipsoid* of K). When this ellipsoid is a multiple of the Euclidean ball we say that L is in Löwner position.

PROPOSITION 2.1.7. *Let K be a convex body in $\mathcal{K}_{(0)}^n$. There exists a unique ellipsoid $\mathcal{E} \supseteq K$ with minimal volume.*

In fact, Proposition 2.1.7 and Proposition 2.1.6 are equivalent because of the following duality property:

LEMMA 2.1.8. *Let K be a convex body in $\mathbb{R}^n$. A centrally symmetric ellipsoid $\mathcal{E}$ is the ellipsoid of maximal volume inside K if and only if $\mathcal{E}^\circ$ is the ellipsoid of minimal volume outside K°.*

Proof. Polarity reverses order so that for every ellipsoid $\mathcal{E} \subseteq K$ there corresponds an ellipsoid $\mathcal{E}^\circ \supseteq K^\circ$ (and vice versa given that polarity is an involution). Furthermore, by the computation $(A(K))^\circ = (A^*)^{-1}(K^\circ)$ or the linear invariance of the volume product we have that $\mathrm{Vol}(\mathcal{E})\mathrm{Vol}(\mathcal{E}^\circ) = \kappa_n^2$, and thus, if $\mathrm{Vol}(\mathcal{E}_1) \leqslant \mathrm{Vol}(\mathcal{E}_2)$, then $\mathrm{Vol}(\mathcal{E}_1^\circ) \geqslant \mathrm{Vol}(\mathcal{E}_2^\circ)$. Therefore the volume of the maximal ellipsoid in K is κ_n^2/V where V is the volume of the minimal ellipsoid outside K°. □

REMARK 2.1.9. *We leave it to the reader to generalize the above to the case where the center of the ellipsoids is not assumed to be necessarily at the origin.*

2.1.3. Contact points and John's theorem

Assume that a convex body K is in John position. We will say that $x \in \mathbb{R}^n$ is a *contact point* of K and B_2^n if $|x| = \|x\|_K = \|x\|_{K^\circ} = 1$. Note that under the assumption that $B_2^n \subseteq K$ the first two conditions ($|x| = \|x\|_K = 1$) imply the third one ($\|x\|_{K^\circ} = 1$) since both B_2^n and K have $x + x^\perp$ as a supporting hyperplane at x (indeed, the former conditions imply that any supporting hyperplane of K at x must also be a supporting hyperplane of B_2^n at x, and thus this can only be $x + x^\perp$). John's theorem describes the distribution of contact points on the unit sphere S^{n-1}.

THEOREM 2.1.10 (John). *If B_2^n is the maximal volume ellipsoid of a centrally symmetric convex body in $\mathbb{R}^n$, there exist contact points $x_1, \ldots, x_m$ of K and B_2^n, and positive real numbers $c_1, \ldots, c_m$ with $\sum_{j=1}^m c_j = n$ such that*

$$x = \sum_{j=1}^m c_j \langle x, x_j \rangle x_j$$

for every $x \in \mathbb{R}^n$. Moreover, one may choose $m \leqslant \binom{n+1}{2} + 1$.

REMARK 2.1.11. Theorem 2.1.10 says that the identity operator Id of $\mathbb{R}^n$ can be represented in the form

$$\mathrm{Id} = \sum_{j=1}^{m} c_j x_j \otimes x_j,$$

where $x_j \otimes x_j$ is the projection in the direction of x_j: $(x_j \otimes x_j)(x) = \langle x, x_j \rangle x_j$. We call this form a decomposition of identity. Note that, if such a representation exists, then for every $x \in \mathbb{R}^n$

$$|x|^2 = \langle x, x \rangle = \sum_{j=1}^{m} c_j \langle x, x_j \rangle^2.$$

Also, if we choose $x = e_i$, $i = 1, \ldots, n$, where $\{e_i\}$ is the standard orthonormal basis of $\mathbb{R}^n$, we have

$$\begin{aligned} n = \sum_{i=1}^{n} |e_i|^2 &= \sum_{i=1}^{n} \sum_{j=1}^{m} c_j \langle e_i, x_j \rangle^2 = \sum_{j=1}^{m} c_j \sum_{i=1}^{n} \langle e_i, x_j \rangle^2 \\ &= \sum_{j=1}^{m} c_j |x_j|^2 = \sum_{j=1}^{m} c_j, \end{aligned}$$

so this assertion of Theorem 2.1.10 follows from the others.

Proof of Theorem 2.1.10. The preceding remark shows that, if such a representation exists, we must have $\sum_{j=1}^{m} \frac{c_j}{n} = 1$. Our purpose is then to show that Id/n is a convex combination of matrices of the form $x \otimes x$, where x is a contact point of K and B_2^n. To this end, we define

$$\mathcal{C} = \{x \otimes x : |x| = \|x\|_K = 1\}$$

and show that $\mathrm{Id}/n \in \mathrm{conv}(\mathcal{C})$. Note that $\mathrm{conv}(\mathcal{C})$ is a non-empty compact convex subset of $\mathbb{R}^{n^2}$ (actually, of the cone of positive semi-definite matrices, which is a subset of an $\binom{n+1}{2}$-dimensional subspace of $\mathbb{R}^{n^2}$).

Let us assume towards contradiction that Id/n may be separated from $\mathrm{conv}(\mathcal{C})$. Equivalently, we assume that there is a linear functional ϕ such that

$$\langle \phi, \mathrm{Id}/n \rangle < r \leqslant \langle \phi, u \otimes u \rangle$$

for all $u \in S^{n-1} \cap \partial K$ and some $r \in \mathbb{R}$. We may also assume without loss of generality that $\phi = \phi^*$ since otherwise we can set $\psi = \frac{\phi + \phi^*}{2}$ and then, because both $u \otimes u$ and Id/n are symmetric, ψ will satisfy the same inequality. (Equivalently, one may right away restrict to the space of linear symmetric matrices which is isometric to its dual space, and which Id/n as well as the elements of $\mathcal{C}$ belong to, and separate there.)

Note that the traces of both $u \otimes u$ and of Id/n are 1, so that the "scalar product" with the identity matrix is equal for both. In other words, we may subtract from ϕ any linear functional that assigns to each matrix its trace times any constant c as a value, and still get separation, with r changed to $r - c$. If this constant is chosen to be $\mathrm{tr}(\phi)/n$ we get that the new functional takes the value 0 at Id/n.

So far we have shown that there exists a linear functional B, given by a symmetric matrix $B = B^*$, such that $\langle B, \mathrm{Id} \rangle = \mathrm{tr}(B) = 0$ and $\langle B, u \otimes u \rangle > s > 0$ for all $u \in S^{n-1} \cap \partial K$.

The idea is to use B in order to define an ellipsoid of greater volume than B_2^n which still resides in K; this ellipsoid shall be defined as

$$\mathcal{E}_\delta = \{x : \langle (\mathrm{Id} + \delta B)x, x\rangle \leqslant 1\}$$

for a suitably chosen $\delta > 0$. Note that for small enough δ the above set is indeed an ellipsoid as $\mathrm{Id} + \delta B$ is still positive definite (simply take $\delta < 1/\|B\|$ (where $\|B\| = \max\{|B(x)| : x \in S^{n-1}\}$) and then $\mathrm{Id} + \delta B$ will have a square root, say S_δ, so that $\mathcal{E}_\delta = S_\delta^{-1}(B_2^n)$).

Next, we shall show two things: first, that for small enough δ this ellipsoid is indeed inside K, and second, that the volume of this ellipsoid is greater than the volume of B_2^n (because the trace of B is 0). To show the former, we first consider the set of contact points of K and B_2^n; denote this set by U. We separate the points of S^{n-1} to points that are close to U (namely points $v \in S^{n-1}$ with $\mathrm{dist}(v, U) \leqslant s/(2\|B\|)$) and points that are far from it.

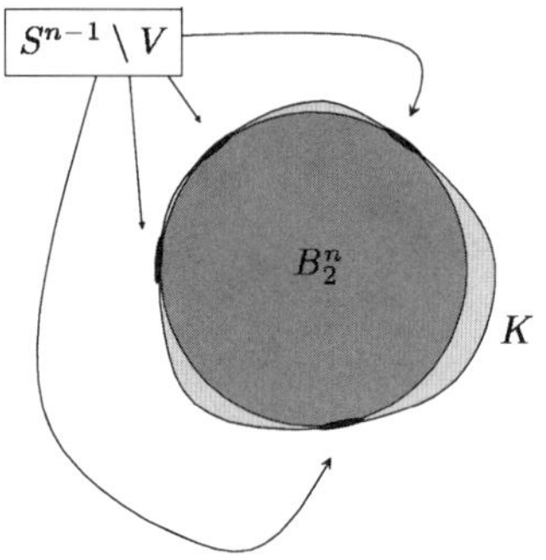

Let us look first at points of S^{n-1} which are far from contact points: $V = \{v \in S^{n-1} : \mathrm{dist}(v, U) \geqslant s/(2\|B\|)\}$. This is a compact set and thus it has positive distance from the boundary of K; in particular, $\alpha = \max\{\|x\| : x \in V\} < 1$. We claim that, for small enough δ, $v/\|v\| \notin \mathcal{E}_\delta$ for all $v \in V$ (geometrically this means that $\mathcal{E}_\delta \cap \mathbb{R}^+ v \subseteq K$ for all $v \in V$).

Indeed, let $\lambda = \min_{v\in V}\langle Bv, v\rangle$. Note that this is a negative number: since the trace of B is zero (and given that the quantity $\langle Bx, x\rangle$ is not identically zero on $\mathbb{R}^n$), there must be *some* w such that $\langle Bw, w\rangle < 0$. However, for all $u \in U$ we have $\langle Bu, u\rangle > s > 0$ and hence we must also have

$$\langle Bv, v\rangle = \langle Bu, u\rangle + \langle Bu, v-u\rangle + \langle B(v-u), v\rangle > s - 2|v-u|\|B\| > 0$$

for every v such that $|v-u| < s/(2\|B\|)$. Therefore, $w \in V$ and we conclude that $\lambda \leqslant \langle Bw, w\rangle < 0$.

Now is the time to choose $\delta < (1-\alpha^2)/|\lambda|$. Then for $v \in V$

$$\Big\langle (\mathrm{Id} + \delta B)\Big(\frac{v}{\|v\|}\Big), \frac{v}{\|v\|}\Big\rangle = \frac{1 + \delta\langle Bv, v\rangle}{\|v\|^2} \geqslant \frac{1+\delta\lambda}{\alpha^2} > 1.$$

This completes the assertion in the case of V.

For $S^{n-1}\setminus V$ the same assertion follows easily by our choice of neighborhood of U: for every $u \in U$ we know that $\langle Bu, u\rangle = \langle B, u\otimes u\rangle > s$, and thus $\langle (\mathrm{Id} + \delta B)u, u\rangle > 1 + \delta s$. Thus $u \notin \mathcal{E}_\delta$. Furthermore, if $v \in S^{n-1}$ then

$$\begin{aligned}|\langle (\mathrm{Id}+\delta B)v, v\rangle - \langle (\mathrm{Id}+\delta B)u, u\rangle| &= \delta|(\langle Bv, v\rangle - \langle Bu, u\rangle)| \leqslant \\ &\delta|\langle Bv, v\rangle - \langle Bv, u\rangle| + \delta|\langle Bv, u\rangle - \langle Bu, u\rangle| \leqslant 2\delta\|B\||u-v|.\end{aligned}$$

This shows that for an $s/(2\|B\|)$-neighbourhood of U on S^{n-1}, we have $\langle(\mathrm{Id}+\delta B)v,v\rangle>1$ and in particular $v\notin\mathcal{E}_\delta$. However $v\in B_2^n\subseteq K$, and hence $\mathcal{E}_\delta\cap\mathbb{R}^+v\subseteq K$ for all $v\in S^{n-1}\setminus V$ too.

We have completed the first part, namely to show that $\mathcal{E}_\delta\subseteq K$ for small enough δ. As for the volume of this ellipsoid, we have

$$\mathrm{Vol}(\mathcal{E}_\delta)=\kappa_n/\det(\mathrm{Id}+\delta B)^{1/2}.$$

Note now that, since $\det(\mathrm{Id}+\delta B)^{1/n}\leqslant\mathrm{tr}(\mathrm{Id}+\delta B)/n=1$ by the arithmetic-geometric means inequality and since B_2^n has the maximum volume among the ellipsoids contained in K, we ought to have that the ellipsoid $\mathcal{E}_\delta$ has the same volume as B_2^n, and hence that it coincides with the maximal volume ellipsoid in K which, by our assumption, is B_2^n. We thus arrive at a contradiction since $B\neq 0$. We conclude that $\mathrm{Id}/n\in\mathrm{conv}(\mathcal{C})$. In addition, by Carathéodory's theorem from classical convexity (Theorem A.1.3 in Appendix A), applied to the set $\mathcal{C}$ which is a subset of an $\binom{n+1}{2}$-dimensional subspace of $\mathbb{R}^{n^2}$, we may write Id/n as a combination of at most $m=\binom{n+1}{2}+1$ points in $\mathrm{conv}(\mathcal{C})$. $\square$

DEFINITION 2.1.12. A Borel measure μ on S^{n-1} is called *isotropic* if

$$\int_{S^{n-1}}\langle x,\theta\rangle^2d\mu(x)=\frac{\mu(S^{n-1})}{n}\tag{2.1.2}$$

for every $\theta\in S^{n-1}$. We will make frequent use of the next standard lemma.

LEMMA 2.1.13. *Let μ be a Borel measure on S^{n-1}. The following are equivalent:*

(i) *μ is isotropic.*

(ii) *For every $i,j=1,\dots,n$,*

$$\int_{S^{n-1}}x_ix_jd\mu(x)=\frac{\mu(S^{n-1})}{n}\delta_{i,j}.\tag{2.1.3}$$

(iii) *For every linear transformation $T:\mathbb{R}^n\to\mathbb{R}^n$,*

$$\int_{S^{n-1}}\langle x,T(x)\rangle d\mu(x)=\frac{\mathrm{tr}(T)}{n}\mu(S^{n-1}).\tag{2.1.4}$$

Proof. Setting $\theta=e_i$ and $\theta=\frac{e_i+e_j}{\sqrt{2}}$ in (2.1.2) we get (2.1.3). Moreover, from the observation that if $T=(t_{ij})_{i,j=1}^n$ then $\langle x,T(x)\rangle=\sum_{i,j=1}^n t_{ij}x_ix_j$, we readily see that (2.1.3) implies (2.1.4). Finally, note that applying (2.1.4) with $T(x)=\langle x,\theta\rangle\theta$ we get (2.1.2). $\square$

Note. Theorem 2.1.10 implies that

$$\sum_{j=1}^m c_j\langle x_j,\theta\rangle^2=1$$

for every $\theta\in S^{n-1}$. In our terminology, the measure μ on S^{n-1} that gives mass c_j to the point x_j, $j=1,\dots,m$, is *isotropic*.

Conversely, we have the following proposition which shows that if a body K that contains B_2^n has enough points in $\partial K\cap S^{n-1}$, enough meaning that an isotropic measure with support on this set can be constructed, then the body must be in John position. This is very useful in that we can sometimes immediately determine whether a body is in John position by observing the distribution of its contact points.

PROPOSITION 2.1.14 (Ball). *Let K be a centrally symmetric convex body in $\mathbb{R}^n$ that contains the Euclidean unit ball B_2^n. Assume that the set of contact points of K and B_2^n is non-empty and assume that there exists an isotropic Borel measure μ on S^{n-1} whose support lies in this set. Then, B_2^n is the maximal volume ellipsoid of K.*

Proof. Instead of K consider the set

$$L = \{y : \langle x, y\rangle \leqslant 1 \text{ for all } x \in \operatorname{supp}(\mu).\}$$

(Note: this is the polar of $\operatorname{conv}(\operatorname{supp}(\mu))$.)

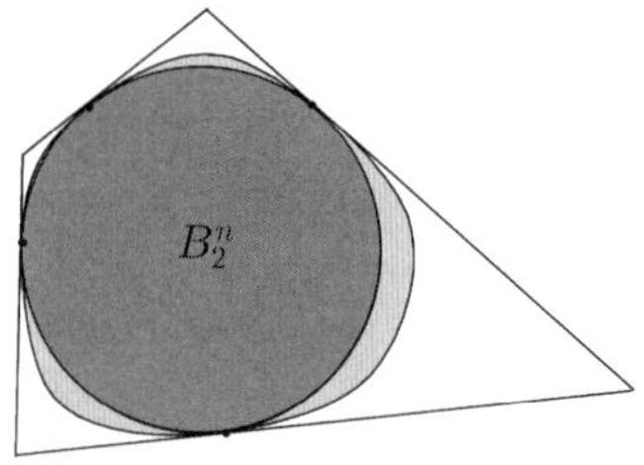

Clearly $K \subseteq L$ (since $\|x\|_{K^\circ} = 1$ for every contact point x of K) and so it is enough to show that B_2^n is maximal for L. Pick some ellipsoid $\mathcal{E} \subseteq L$ given by

$$\mathcal{E} = \left\{ x \in \mathbb{R}^n : \sum_{i=1}^n \frac{\langle x, v_i\rangle^2}{\alpha_i^2} \leqslant 1 \right\},$$

where $\{v_i : i = 1, \dots, n\}$ is some orthonormal basis and α_i are the respective lengths of the semi-axes.

If $u \in \operatorname{supp}(\mu)$ then the vector $y \equiv y(u)$ defined as

$$y(u) = \sum_{j=1}^n \alpha_j \langle u, v_j\rangle v_j$$

belongs to $\mathcal{E} \subseteq L$. Indeed, $y(u)$ has "j-th coordinate" (that is, with respect to the basis $\{v_i\}$) equal to $\alpha_j\langle u, v_j\rangle$, so that the sum of the squares weighted by α_j^{-2} is $|u|^2 = 1$. Since $y(u) \in \mathcal{E} \subset L$, we see that $\langle u, y(u)\rangle \leqslant 1$, which gives $\sum_{j=1}^n \alpha_j\langle u, v_j\rangle^2 \leqslant 1$. This holds for all $u \in \operatorname{supp}(\mu)$.

Isotropicity means that for every x, $\int_{S^{n-1}} \langle x, u\rangle^2 d\mu(u) = |x|^2\mu(S^{n-1})/n$. Using this for the v_j's, we may now write

$$\begin{aligned}\mu(S^{n-1}) = \int_{S^{n-1}} d\mu(u) &\geqslant \int_{S^{n-1}} \langle y(u), u\rangle d\mu(u) \\ &= \int_{S^{n-1}} \sum_{j=1}^n \alpha_j \langle v_j, u\rangle^2 d\mu(u) = \frac{\mu(S^{n-1})}{n} \sum_{j=1}^n \alpha_j.\end{aligned}$$

By the arithmetic-geometric means inequality, $\sum_{j=1}^n \alpha_i \leqslant n$ implies $\prod_{j=1}^n \alpha_i \leqslant 1$, or, in other words, that $\mathrm{Vol}_n(\mathcal{E}) \leqslant \kappa_n$ as claimed. □

Theorem 2.1.10 and Proposition 2.1.14 provide the following characterization of the John position:

THEOREM 2.1.15. *Let K be a centrally symmetric convex body in $\mathbb{R}^n$ that contains the Euclidean unit ball B_2^n. Then, B_2^n is the maximal volume ellipsoid of K if and only if there exists an isotropic measure μ supported by the contact points of K and B_2^n.*

REMARK 2.1.16. John's theorem about the distribution of contact points still holds for a general convex body $K \subseteq \mathbb{R}^n$; namely we have:

Theorem. Let K be a convex body in $\mathbb{R}^n$. The Euclidean unit ball B_2^n is the ellipsoid of maximal volume in K if and only if K contains B_2^n and there exist contact points $(x_j)_{j=1}^m$ of B_2^n and ∂K and positive numbers $(c_j)_{j=1}^m$ so that

$$\sum_{j=1}^m c_j x_j = 0$$

and

$$\mathrm{Id} = \sum_{j=1}^m c_j x_j \otimes x_j.$$

The reader may verify that the proof carries over to this case as well.

Finally let us show how Theorem 2.1.10 implies Theorem 2.1.3 above which states that, if K is a centrally symmetric convex body in $\mathbb{R}^n$ and $\mathcal{E}$ is the maximal volume ellipsoid of K, then $K \subseteq \sqrt{n}\mathcal{E}$. This is equivalent to the statement that, if B_2^n is the maximal volume ellipsoid of K, then $K \subseteq \sqrt{n} B_2^n$.

Proof of Theorem 2.1.3. Consider the representation of the identity

$$x = \sum_{j=1}^m c_j \langle x, x_j\rangle x_j$$

of Theorem 2.1.10. We will use the fact that $\|x_j\|_K = \|x_j\|_{K^\circ} = |x_j| = 1$, $j = 1, \ldots, m$. For every $x \in K$ we have

$$|x|^2 = \sum_{j=1}^m c_j \langle x, x_j\rangle^2 \leqslant \sum_{j=1}^m c_j \big(\|x\|_K \|x_j\|_{K^\circ}\big)^2 \leqslant \sum_{j=1}^m c_j = n.$$

This shows that $|x| \leqslant \sqrt{n}$. Therefore, $B_2^n \subseteq K \subseteq \sqrt{n} B_2^n$. $\square$

REMARK 2.1.17. In the non-symmetric case a similar argument shows that $K \subseteq nB_2^n$. Indeed, let K be in John position, that is, assume that $B_2^n \subseteq K$ and that there are contact points $x_j \in \partial K \cap S^{n-1}$ such that $\mathrm{Id} = \sum_{j=1}^m c_j x_j \otimes x_j$ and $\sum_{j=1}^m c_j x_j = 0$ for some positive real numbers c_j. To show that $K \subseteq nB_2^n$ we shall show that $h_K(u) \leqslant n$ for all $u \in S^{n-1}$. Note that $\mathrm{conv}\{x_1, \ldots, x_m\} \subseteq K^\circ$ and hence, for every $u \in \mathbb{R}^n$,

$$h_K(u) = \|u\|_{K^\circ} \leqslant \inf \left\{ \sum_{j=1}^m t_j : u = \sum_{j=1}^m t_j x_j,\ t_j \geqslant 0,\ j = 1, \ldots, m \right\}.$$

(We use $\|\cdot\|$ to denote a gauge function, not necessarily symmetric.) Since

$$u = \sum_{j=1}^m c_j \langle u, x_j\rangle x_j = \sum_{j=1}^m c_j \big(\langle u, x_j\rangle - \min_{1\leqslant j\leqslant m} \langle u, x_j\rangle\big) x_j$$

and $\sum_{j=1}^m c_j = n$, we conclude that

$$h_K(u) \leqslant \sum_{j=1}^m c_j\langle u, x_j\rangle - n \min_{1\leqslant j\leqslant m} \langle u, x_j\rangle = -n \min_{1\leqslant j\leqslant m} \langle u, x_j\rangle \leqslant n|u|\,|x_j| = n.$$

Another nice application of John's theorem is the following theorem of Kadets and Snobar.

THEOREM 2.1.18 (Kadets-Snobar). *Let X be a finite dimensional Banach space and let Y be a subspace of X with $\dim(Y) = n$. Then there exists a projection $P : X \to Y$ with $\|P\| \leqslant \sqrt{n}$. Here, projection means a linear operator with $P = P^*$ and $P^2 = \mathrm{Id}$.*

Proof. Choose a Euclidean structure on Y such that $K_Y = \{y \in Y : \|y\| \leqslant 1\}$ is in John position. From John's theorem there exist contact points $u_j \in \partial K_Y$ and $c_j > 0$ such that on Y we have $\mathrm{Id}_Y = \sum_{j=1}^m c_j u_j \otimes u_j$. In particular, $\|u_j\|_{K_Y} = |u_j| = \|u_j\|_{K_{Y^*}} = 1$.

Extend each functional $\langle u_j, \cdot\rangle$ to $\tilde{u}_j : X \to \mathbb{R}$ using the Hahn-Banach theorem so that $\|\tilde{u}_j\|_{X^*} = 1$, and define $P : X \to Y$ by

$$P(x) = \sum_{j=1}^m c_j\tilde{u}_j(x)u_j.$$

Then,

$$\begin{aligned}\|P\| &= \sup_{\|x\|=1} \Big\|\sum_{j=1}^m c_j\tilde{u}_j(x)u_j\Big\|_Y = \sup_{\|x\|=1}\sup_{y^*\in K_{Y^*}} \sum_{j=1}^m c_j\tilde{u}_j(x)\langle u_j, y^*\rangle \\ &\leqslant \sup_{y^*\in K_{Y^*}} \sum_{j=1}^m c_j|\langle u_j, y^*\rangle|,\end{aligned}$$

where we have used the facts that $|\tilde{u}_j(x)| \leqslant 1$ and that, as Y is finite dimensional, we can represent each $y^* \in Y^*$ as a vector in Y acting by scalar product. Note that, since $B_2^n \subseteq K_Y$, we have $|y^*| \leqslant 1$ for all $y^* \in K_{Y^*}$. Combining all these, we can continue as follows:

$$\begin{aligned}\|P\| &\leqslant \sup_{y^*\in K_{Y^*}} \sum_{j=1}^m c_j|\langle u_j, y^*\rangle| = \sup_{y^*\in K_{Y^*}} \sum_{j=1}^m \sqrt{c_j}\sqrt{c_j}|\langle u_j, y^*\rangle| \\ &\leqslant \sup_{y^*\in K_{Y^*}} \Big(\sum_{j=1}^m c_j\Big)^{1/2}\Big(\sum_{j=1}^m c_j|\langle u_j, y^*\rangle|^2\Big)^{1/2} \\ &= \sqrt{n}\sup_{y^*\in K_{Y^*}} \sqrt{\langle y^*, y^*\rangle} \leqslant \sqrt{n},\end{aligned}$$

thus completing the proof. □

2.1.4. Contact points and the Dvoretzky-Rogers lemmas

Let $X = (\mathbb{R}^n, \|\cdot\|)$ be an n-dimensional normed space. Assume that B_2^n is the ellipsoid of maximal volume contained in the unit ball K of X. Recall that from John's theorem there exist $x_1, \ldots, x_m \in \mathbb{R}^n$ with $\|x_j\| = |x_j| = \|x_j\|_* = 1$ and

$c_1, \ldots, c_m > 0$ such that $\mathrm{Id} = \sum_{j=1}^m c_j x_j \otimes x_j$. In other words, for every $x \in \mathbb{R}^n$,

$$x = \sum_{j=1}^{m} c_j \langle x, x_j \rangle x_j. \tag{2.1.5}$$

We also have $\sum_{j=1}^m c_j = n$, and we may assume that $m \leqslant \binom{n+1}{2} + 1$.

In this subsection we collect several useful results on the distribution of the contact points x_j on S^{n-1}, known as "Dvoretzky-Rogers type lemmas".

LEMMA 2.1.19. *If B_2^n is the maximal volume ellipsoid of a centrally symmetric convex body K in $\mathbb{R}^n$, then for every $T \in L(\mathbb{R}^n)$ we can find a contact point y of K and B_2^n with the property*

$$\langle y, T(y) \rangle \geqslant \frac{\mathrm{tr}(T)}{n}.$$

Proof. We have

$$\mathrm{tr}\,(T) = \langle T, \mathrm{Id} \rangle = \sum_{j=1}^{m} c_j \langle T, x_j \otimes x_j \rangle.$$

Since $\sum_{j=1}^m c_j = n$, we may find y among the x_j that satisfies

$$\langle y, T(y) \rangle = \langle T, y \otimes y \rangle \geqslant \frac{\mathrm{tr}(T)}{n}$$

as desired. □

THEOREM 2.1.20 (Dvoretzky-Rogers). *If B_2^n is the maximal volume ellipsoid of a centrally symmetric convex body K in $\mathbb{R}^n$, there exists an orthonormal sequence $z_1, \ldots, z_n$ in $\mathbb{R}^n$ such that*

$$\left(\frac{n-i+1}{n} \right)^{1/2} \leqslant \|z_i\| \leqslant |z_i| = 1$$

for all $i = 1, \ldots, n$.

Proof. We define the z_i inductively. Choose z_1 to be any contact point of K and B_2^n, and assume that $z_1, \ldots, z_k$ have been chosen for some $k < n$.

We set $F_k = \mathrm{span}\{z_1, \ldots, z_k\}$. Then, $\mathrm{tr}(P_{F_k^\perp}) = n - k$, and applying Lemma 2.1.19 we may find a contact point y_{k+1} of K and B_2^n with

$$|P_{F_k^\perp}(y_{k+1})|^2 = \langle y_{k+1}, P_{F_k^\perp}(y_{k+1}) \rangle \geqslant \frac{n-k}{n}.$$

It follows that $\|P_{F_k}(y_{k+1})\| \leqslant |P_{F_k}(y_{k+1})| \leqslant \sqrt{k/n}$.

We define $z_{k+1} = P_{F_k^\perp}(y_{k+1}) / |P_{F_k^\perp}(y_{k+1})|$. Then, since $h_K(y_{k+1}) = 1$, we have

$$1 = |z_{k+1}| \geqslant \|z_{k+1}\| \geqslant \langle y_{k+1}, z_{k+1} \rangle = \left| P_{F_k^\perp}(y_{k+1}) \right| \geqslant \left(\frac{n-k}{n} \right)^{1/2},$$

and the inductive step is complete. □

COROLLARY 2.1.21. *Assume that B_2^n is the maximal volume ellipsoid of a centrally symmetric convex body K in $\mathbb{R}^n$. If $k = \lfloor n/2 \rfloor + 1$, we can find orthonormal vectors $z_1, \ldots, z_k$ such that*

$$\frac{1}{\sqrt{2}} \leqslant \|z_j\| \leqslant 1$$

for all $j = 1, \ldots, k$. □

REMARK 2.1.22. It is useful to note that one also has a version of Corollary 2.1.21 that provides a bound for the norms of *all* of the orthonormal vectors z_i:

PROPOSITION 2.1.23. *Assume that B_2^n is the maximal volume ellipsoid of a centrally symmetric convex body K in $\mathbb{R}^n$. We can find an orthonormal basis $w_1, \dots, w_n$ such that*

$$\frac{1}{4} \leqslant \|w_j\| \leqslant \sqrt{2}$$

for all $j = 1, \dots, n$.

Proof. We use Theorem 2.1.20 to produce an orthonormal sequence $z_1, \dots, z_n$ satisfying that theorem, and we rearrange the vectors in the sequence in decreasing order in terms of $\|z_i\|$ so that the inequality in the statement of the theorem will continue to hold. Denote by i_0 the largest index such that $\|z_i\| \geqslant 1/4$. By Corollary 2.1.21 we have $i_0 \geqslant n/2$. This in particular implies that $\|z_i\| \geqslant \|z_{n-i_0}\| \geqslant \|z_{\lfloor n/2 \rfloor+1}\| \geqslant 1/\sqrt{2}$ for all $1 \leqslant i \leqslant n - i_0$, while for the vectors $z_{i_0+1}, \dots, z_n$ we clearly have $\|z_i\| < 1/4$.

We shall construct the basis w_i as follows. Its first $2(n - i_0)$ vectors will be

$$\frac{z_1 + z_{i_0+1}}{\sqrt{2}}, \frac{z_1 - z_{i_0+1}}{\sqrt{2}}, \dots, \frac{z_j + z_{i_0+j}}{\sqrt{2}}, \frac{z_j + z_{i_0+j}}{\sqrt{2}}, \dots, \frac{z_{n-i_0} + z_n}{\sqrt{2}}, \frac{z_{n-i_0} + z_n}{\sqrt{2}}.$$

They are all orthonormal, and by the triangle inequality satisfy

$$\sqrt{2} \geqslant \frac{\|z_j \pm z_{i_0+j}\|}{\sqrt{2}} \geqslant \frac{1}{\sqrt{2}} \left[\frac{1}{\sqrt{2}} - \frac{1}{4} \right] > \frac{1}{4}.$$

Thus, we have defined the first $2(n - i_0)$ vectors of an orthonormal sequence. The other $2i_0 - n$ vectors will simply be $z_{n-i_0+1}, \dots, z_{i_0}$, which (by the choice of i_0) have the property that $1 \geqslant \|z_j\| \geqslant 1/4$ and are orthogonal to the already defined w_i. □

The original Dvoretzky-Rogers lemma is the following refined version of Theorem 2.1.20.

THEOREM 2.1.24. *Assume that B_2^n is the maximal volume ellipsoid of a centrally symmetric convex body K in $\mathbb{R}^n$. There exist $y_1, \dots, y_n \in \mathbb{R}^n$ and an orthonormal basis $\{z_1, \dots, z_n\}$ which satisfy the following:*

(i) $\|y_i\| = |y_i| = \|y_i\|_* = 1$ *for all* $i = 1, \dots, n$.

(ii) $y_i = \sum_{j=1}^{i} a_{ij} z_j$ *for all* $i = 1, \dots, n$.

(iii) $a_{ii}^2 = 1 - \sum_{j<i} a_{ij}^2 \geqslant \frac{n-i+1}{n}$ *and* $a_{ii} > 0$ *for all* $i = 1, \dots, n$.

(iv) $\|y_i - z_i\|^2 \leqslant |y_i - z_i|^2 \leqslant 2\left(1 - \sqrt{(n-i+1)/n}\right)$ *for all* $i = 1, \dots, n$.

Proof. We start with John's decomposition: $\mathrm{Id} = \sum_{j=1}^{m} c_j x_j \otimes x_j$ where $x_1, \dots, x_m$ are contact points of K and B_2^n and $c_j > 0$. We set $z_1 = y_1 = x_1$.

Assume that, for some $1 \leqslant k < n$, we have defined $z_1, \dots, z_k$ and $y_1, \dots, y_k$ so that (i), (ii) and (iii) hold true. We set $F_k = \mathrm{span}\{z_1, \dots, z_k\}$ and consider the orthogonal projection $P := P_{F_k^\perp}$ onto $F_k^\perp$. We have $P(z_j) = P(y_j) = 0$ for all $j = 1, \dots, k$. There exists $s \leqslant m$ such that

$$|Px_s| = \max\{|Px_j| : 1 \leqslant j \leqslant m\}.$$

We define $y_{k+1} = x_s$. Note that $|Px_s| > 0$. We define

$$z_{k+1} = \frac{Px_s}{|Px_s|}.$$

Then, $|z_{k+1}| = 1$ and $\langle z_{k+1}, z_j\rangle = 0$ for all $j = 1, \ldots, k$. Since y_{k+1} is a contact point, it satisfies (i). Observe that

$$y_{k+1} - |Px_s| z_{k+1} = y_{k+1} - Px_s = x_s - Px_s \in F_k.$$

Therefore,

$$y_{k+1} = |Px_s| z_{k+1} + \sum_{j=1}^{k} a_{k+1,j} z_j,$$

which proves (ii). To check (iii) observe that $P = \sum_{j=1}^{m} c_j x_j \otimes Px_j$, and hence

$$n - k = \mathrm{tr}(P) = \sum_{j=1}^{m} c_j \langle x_j, Px_j\rangle = \sum_{j=1}^{m} c_j |Px_j|^2 \leqslant |Px_s|^2 \sum_{j=1}^{m} c_j = n|Px_s|^2.$$

It follows that

$$a_{k+1,k+1}^2 = |Px_s|^2 \geqslant \frac{n-k}{n}.$$

Finally, for every $i = 1, \ldots, n$ we can write

$$|y_i - z_i|^2 = |1 - a_{ii}|^2 + \sum_{j<i} a_{ij}^2 = 2 - 2a_{ii} \leqslant 2\left(1 - \sqrt{\frac{n-i+1}{n}}\right),$$

thus concluding (iv) and completing the proof. □

The next result provides a bound for the dual norm of the orthonormal vectors z_j in Theorem 2.1.24 as long as $j \leqslant \sqrt{n}/4$.

THEOREM 2.1.25. *Assume that B_2^n is the maximal volume ellipsoid of the centrally symmetric convex body K in $\mathbb{R}^n$. Let $s = \lfloor \sqrt{n}/4 \rfloor$. There exist orthonormal vectors $z_1, \ldots, z_s$ such that*

$$\sqrt{\frac{n-i+1}{n}} \leqslant \|z_i\| \leqslant |z_i| = 1 \leqslant \|z_i\|_* \leqslant 2$$

for all $i = 1, \ldots, s$.

Proof. Consider the orthonormal basis of Theorem 2.1.24. For every $i = 1, \ldots, s$,

$$\|z_i\| = \|z_i\| \cdot \|y_i\|_* \geqslant \langle z_i, y_i\rangle = a_{ii} \geqslant \sqrt{\frac{n-i+1}{n}}.$$

We have $\|z_1\|_* = 1 \leqslant 2$. Assume that $1 \leqslant k \leqslant \sqrt{n}/4 - 1$ and that $\|z_i\|_* \leqslant 2$ for all $i = 1, \ldots, k$. We write

$$\begin{aligned}\left\|\sum_{j=1}^{k} a_{k+1,j} z_j\right\|_* &\leqslant \sum_{j=1}^{k} |a_{k+1,j}|\, \|z_j\|_* \leqslant \left(\sum_{j=1}^{k} a_{k+1,j}^2\right)^{1/2} \left(\sum_{j=1}^{k} \|z_j\|_*^2\right)^{1/2} \\ &\leqslant \sqrt{k/n} \cdot \sqrt{k} \max_{1\leqslant j\leqslant k} \|z_j\|_* \leqslant 2k/\sqrt{n}.\end{aligned}$$

Then,

$$\begin{aligned}\|z_{k+1}\|_* &\leqslant \frac{1}{a_{k+1,k+1}} \left(\|y_{k+1}\|_* + \|y_{k+1} - a_{k+1,k+1} z_{k+1}\|_*\right) \\ &= \frac{1}{a_{k+1,k+1}} \left(1 + \left\|\sum_{j=1}^{k} a_{k+1,j} z_j\right\|_*\right) \leqslant \sqrt{\frac{n}{n-k}} \left(1 + \frac{2k}{\sqrt{n}}\right).\end{aligned}$$

Since $k \leqslant \sqrt{n}/4 - 1$, this last quantity is smaller than 2. □

A consequence of Theorem 2.1.25 is the next statement which is known as Dvoretzky-Rogers factorization.

THEOREM 2.1.26. *Assume that B_2^n is the maximal volume ellipsoid of the centrally symmetric convex body K in $\mathbb{R}^n$. There exist $s \simeq \sqrt{n}$ and orthonormal vectors $z_1, \ldots, z_s$ in $\mathbb{R}^n$ such that for all $a_1, \ldots, a_s \in \mathbb{R}$,*

$$\frac{1}{2}\max_{i\leqslant s}|a_i| \leqslant \left\|\sum_{i=1}^s a_i z_i\right\| \leqslant \left(\sum_{i=1}^s a_i^2\right)^{1/2}.$$

Proof. From Theorem 2.1.25, if $s = \lfloor \sqrt{n}/4 \rfloor$ then

$$\sqrt{\frac{n-i+1}{n}} \leqslant \|z_i\| \leqslant 1 \leqslant \|z_i\|_* \leqslant 2$$

for all $i = 1, \ldots, s$. It follows that for all $a_1, \ldots, a_s \in \mathbb{R}$,

$$\left\|\sum_{i=1}^s a_i z_i\right\| \leqslant \left|\sum_{i=1}^s a_i z_i\right| = \left(\sum_{i=1}^s a_i^2\right)^{1/2},$$

and for every $k = 1, \ldots, s$,

$$|a_k| = \left|\left\langle \sum_{i=1}^s a_i z_i, z_k\right\rangle\right| \leqslant \left\|\sum_{i=1}^s a_i z_i\right\| \|z_k\|_* \leqslant 2\left\|\sum_{i=1}^s a_i z_i\right\|.$$

This completes the proof. □

2.2. Minimal mean width position

The next position we discuss is the position in which the mean width of the body is minimized. We recall that, given a convex body $K \subseteq \mathbb{R}^n$, the mean width $w(K)$ of K is the quantity

$$w(K) = \int_{S^{n-1}} h_K(u)\, d\sigma(u).$$

(without loss of generality we may assume that $0 \in \mathrm{int}(K)$). We have already encountered this parameter in Section 1.5 where we proved Urysohn's inequality, stating that among all convex bodies of a given volume, the Euclidean ball has minimal mean width.

We say that a convex body K of volume 1 is in *minimal mean width position* if $w(K) \leqslant w(TK)$ for every $T \in SL_n$.

REMARK 2.2.1. Analogous positions can be defined if we wish to minimize other integrals of the form

$$\int_{S^{n-1}} \|u\|^p d\sigma(u)$$

for other powers $p \geqslant 1$ and with $\|\cdot\|$ being either h_K or $\|\cdot\|_K$. We emphasize that, up to a universal constant depending only on p (and not on the dimension of the space nor on the norm involved), the numbers thus attained for different values of p are all equivalent. This is a consequence of Kahane's inequality which appears in Section 3.5 as Theorem 3.5.2. When $p = 2$ and $\|\cdot\| = \|\cdot\|_K$, the position minimizing the corresponding expression is called the ℓ-position and will be discussed in Chapter 6.

It seems that the choice of h_K and of the power 1 that we consider in this section leads to a more "geometric" position (as we shall see below) whereas the power 2 (combined with the norms) corresponds to the functional analytic point of view, as we shall explain in Chapter 6. Because of the equivalence of the relevant parameters for different values of p up to constants depending only on p, these different choices of powers for a given norm are interchangeable in a large part of the theory.

Our purpose in this section is to find necessary and sufficient conditions for a body K to have minimal mean width. We assume for simplicity that h_K is twice continuously differentiable (we then say that K is *smooth enough*).

THEOREM 2.2.2. *A smooth enough convex body K in $\mathbb{R}^n$ has minimal mean width if and only if*

$$\int_{S^{n-1}} \langle \nabla h_K(u), Tu\rangle d\sigma(u) = \frac{\mathrm{tr}(T)}{n} w(K) \tag{2.2.1}$$

for every $T \in L(\mathbb{R}^n)$. Moreover, this minimal mean width position is unique up to orthogonal transformations.

Proof. Assume first that K has minimal mean width. Let $T \in L(\mathbb{R}^n)$ and $\varepsilon > 0$ be small enough. Then $(\mathrm{Id} + \varepsilon T)^* / [\det(\mathrm{Id} + \varepsilon T)]^{1/n}$ is volume preserving, hence

$$\int_{S^{n-1}} h_K(u + \varepsilon Tu) d\sigma(u) \geqslant [\det(\mathrm{Id} + \varepsilon T)]^{1/n} \int_{S^{n-1}} h_K(u) d\sigma(u).$$

Since $h_K(u + \varepsilon Tu) = h_K(u) + \varepsilon \langle \nabla h_K(u), Tu\rangle + O(\varepsilon^2)$ and $[\det(\mathrm{Id} + \varepsilon T)]^{1/n} = 1 + \varepsilon \frac{\mathrm{tr}(T)}{n} + O(\varepsilon^2)$, letting $\varepsilon \to 0^+$ we obtain

$$\int_{S^{n-1}} \langle \nabla h_K(u), Tu\rangle d\sigma(u) \geqslant \frac{\mathrm{tr}(T)}{n} w(K). \tag{2.2.2}$$

Replacing T by $-T$ in (2.2.2) we see that there must be equality in (2.2.2) for every $T \in L(\mathbb{R}^n)$.

Conversely, assume that (2.2.1) is satisfied and let $T \in SL_n$. Up to an orthogonal transformation we may assume that T is symmetric positive-definite (which allows us to write $\mathrm{tr}(T) \geqslant n \cdot \det(T)^{1/n} = n$ by Hadamard's and the arithmetic-geometric means inequalities). Then,

$$w(TK) = \int_{S^{n-1}} h_{TK}(u) d\sigma(u) = \int_{S^{n-1}} h_K(Tu) d\sigma(u). \tag{2.2.3}$$

It is a known fact (see [**556**, p. 54]) that $\nabla h_K(u)$ is the unique point on the boundary of K at which u is the outer normal to K. In particular, $\nabla h_K(u) \in K$, which implies that

$$\langle \nabla h_K(u), z\rangle \leqslant h_K(z) \tag{2.2.4}$$

for every $z \in \mathbb{R}^n$. Therefore, by (2.2.1), (2.2.3) and (2.2.4) we get

$$w(TK) \geqslant \int_{S^{n-1}} \langle \nabla h_K(u), Tu\rangle d\sigma(u) = \frac{\mathrm{tr}(T)}{n} w(K) \geqslant w(K). \tag{2.2.5}$$

This shows that K has minimal mean width. Moreover, we can have equality in (2.2.5) only if T is the identity. This proves uniqueness of the minimal mean width position up to $U \in O(n)$. □

Consider the measure ν_K on S^{n-1} that has density h_K with respect to σ. Giannopoulos and Milman proved that a smooth enough convex body K has minimal mean width if and only if ν_K is isotropic. We first fix some notation: If f is a real or vector-valued function on $\mathbb{R}^n\backslash\{0\}$ then we write $\widehat{f}$ for the restriction of f to S^{n-1}. If F is defined on S^{n-1}, the radial extension f of F to $\mathbb{R}^n\backslash\{0\}$ is given by $f(x)=F(x/|x|)$. If F is a twice differentiable real function on S^{n-1}, we define

$$\Delta_{\circ}F=\widehat{(\Delta f)}\quad\text{and}\quad\nabla_{\circ}F=\widehat{(\nabla f)},$$

where f is the radial extension of F. The operator $\Delta_{\circ}$ is usually called the *Laplace-Beltrami operator*, while $\nabla_{\circ}$ is referred to as the *gradient*. As a consequence of Green's formula we have

$$\int_{S^{n-1}}F\cdot\Delta_{\circ}G\,d\sigma=\int_{S^{n-1}}G\cdot\Delta_{\circ}F\,d\sigma=-\int_{S^{n-1}}\langle\nabla_{\circ}F,\nabla_{\circ}G\rangle\,d\sigma.$$

LEMMA 2.2.3. *Let K be a smooth enough convex body in $\mathbb{R}^n$. We define*

$$I_K(\theta)=\int_{S^{n-1}}\langle\nabla h_K(u),\theta\rangle\langle u,\theta\rangle d\sigma(u),\qquad\theta\in S^{n-1}.$$

Then,

$$w(K)+I_K(\theta)=(n+1)\int_{S^{n-1}}h_K(u)\langle u,\theta\rangle^2d\sigma(u)\tag{2.2.6}$$

for every $\theta\in S^{n-1}$.

Proof. Let $\theta\in S^{n-1}$ and consider the function $f(x)=\langle x,\theta\rangle^2/2$. A direct computation shows that

$$(\nabla_{\circ}\widehat{f})(u)=\langle u,\theta\rangle\theta-\langle u,\theta\rangle^2u\tag{2.2.7}$$

and

$$(\Delta_{\circ}\widehat{f})(u)=1-n\langle u,\theta\rangle^2.\tag{2.2.8}$$

Since h_K is positively homogeneous of degree 1, we have $(\nabla_{\circ}\widehat{h_K})(u)=\nabla h_K(u)-h_K(u)u$ and $h_K(u)=\langle\nabla h_K(u),u\rangle$, $u\in S^{n-1}$. Taking into account (2.2.7) we obtain

$$\langle(\nabla_{\circ}\widehat{f})(u),(\nabla_{\circ}\widehat{h_K})(u)\rangle=\langle\nabla h_K(u),\theta\rangle\langle u,\theta\rangle-h_K(u)\langle u,\theta\rangle^2.$$

Integrating on the sphere and using Green's formula we get

$$I_K(\theta)-\int_{S^{n-1}}h_K(u)\langle u,\theta\rangle^2d\sigma(u)=-\int_{S^{n-1}}h_K(u)(\Delta_{\circ}\widehat{f})(u)d\sigma(u),$$

which is equal to

$$-w(K)+n\int_{S^{n-1}}h_K(u)\langle u,\theta\rangle^2d\sigma(u)$$

by (2.2.8). This proves (2.2.6). □

THEOREM 2.2.4 (Giannopoulos-Milman). *A convex body K has minimal mean width if and only if*

$$\int_{S^{n-1}}h_K(u)\langle u,\theta\rangle^2d\sigma(u)=\frac{w(K)}{n}$$

for every $\theta\in S^{n-1}$ (equivalently, if ν_K is isotropic).

Proof. We may assume by an approximation argument that the body is smooth enough. It is not hard to check (in the same way that Lemma 2.1.13 is proven, for example) that (2.2.1) is true for every $T \in L(\mathbb{R}^n)$ if and only if

$$I_K(\theta) = \frac{w(K)}{n}$$

for every $\theta \in S^{n-1}$. The result now follows from Theorem 2.2.2 and Lemma 2.2.3. □

2.3. Minimal surface area position

Next we discuss the position of a convex body in which surface area is minimized. The isoperimetric inequality in $\mathbb{R}^n$ states that among bodies of a given volume, the Euclidean balls have least surface area. Measurable sets of finite volume may have infinite surface area and, if $n \geqslant 2$, even convex bodies of a given volume may have arbitrarily large surface area if they are very flat. Nevertheless, if we consider classes of affinely equivalent convex bodies rather than individual bodies one may show an inequality between volume and surface area for one representative of each class: the "least flat" member of that class.

Recall that the area measure σ_K of K is defined on S^{n-1} and corresponds to the usual surface measure on K via the Gauss map: for every Borel $A \subseteq S^{n-1}$, we have

$$\sigma_K(A) = \mathrm{Vol}_{n-1}(\{x \in \mathrm{bd}(K) : \text{ the outer normal to } K \text{ at } x \text{ is in } A\}).$$

We obviously have $\partial(K) = \sigma_K(S^{n-1})$.

We say that K is in *minimal surface area position* if $\partial(K) \leqslant \partial(TK)$ for every $T \in SL_n$. The existence of this minimal position is easily checked by compactness arguments. A characterization of the minimal surface area position through the area measure was given by Petty.

THEOREM 2.3.1. *A convex body K in $\mathbb{R}^n$ is in minimal surface area position if and only if σ_K satisfies the isotropic condition*

$$\int_{S^{n-1}} \langle u, \theta\rangle^2 d\sigma_K(u) = \frac{\partial(K)}{n}, \qquad \theta \in S^{n-1}. \tag{2.3.1}$$

Moreover, this minimal surface area position is unique up to orthogonal transformations.

Proof. First we show that if K has minimal surface area then σ_K is isotropic. We use the fact that

$$\partial((A^{-1})^* K) = \int_{S^{n-1}} |A(u)|\, d\sigma_K(u) \tag{2.3.2}$$

for every $A \in SL_n$. Consider any $T \in L(\mathbb{R}^n)$ and let $\varepsilon > 0$ be small enough. Then, $(\mathrm{Id} + \varepsilon T)/[\det(\mathrm{Id} + \varepsilon T)]^{1/n}$ is volume preserving, and hence the assumption on K and (2.3.2) give

$$\int_{S^{n-1}} |(\mathrm{Id} + \varepsilon T)(u)|\, d\sigma_K(u) \geqslant [\det(\mathrm{Id} + \varepsilon T)]^{\frac{1}{n}} \partial(K).$$

Observe that $|u+\varepsilon Tu| = 1+\varepsilon\langle u, Tu\rangle + O(\varepsilon^2)$ and $[\det(\mathrm{Id}+\varepsilon T)]^{1/n} = 1+\varepsilon\frac{\mathrm{tr}(T)}{n} + O(\varepsilon^2)$. Letting $\varepsilon \to 0^+$ we get

$$\int_{S^{n-1}} \langle u, Tu\rangle \, d\sigma_K(u) \geqslant \frac{\mathrm{tr}(T)}{n}\partial(K),$$

and then by replacing T with $-T$ we conclude that

$$\int_{S^{n-1}} \langle u, Tu\rangle \, d\sigma_K(u) = \frac{\mathrm{tr}(T)}{n}\partial(K). \tag{2.3.3}$$

This implies (2.3.1).

Next we show that if K satisfies (2.3.1), or equivalently (2.3.3), then it is in minimal surface area position. Given $A \in SL_n$, we can write $A = UT$, where T^{-1} is symmetric positive definite and $U \in O(n)$, and then

$$\begin{aligned}\partial(AK) &= \int_{S^{n-1}} |((UT)^{-1})^* u| \, d\sigma_K(u) = \int_{S^{n-1}} |(T^{-1})^* u| \, d\sigma_K(u) \\ &\geqslant \int_{S^{n-1}} \langle u, T^{-1}u\rangle \, d\sigma_K(u) = \frac{\mathrm{tr}(T^{-1})}{n}\partial(K) \geqslant \partial(K),\end{aligned}$$

because $\mathrm{tr}(T^{-1})/n \geqslant [\det(T^{-1})]^{1/n} = 1$.

The latter argument shows also that the minimal surface area position is unique up to orthogonal transformations. Assume that K has minimal surface area, and $\partial(AK) = \partial(K)$ for some $A = UT \in SL_n$ (where T^{-1} is symmetric positive definite and $U \in O(n)$). Then arguing as above we obtain

$$\partial(K) = \partial(UT(K)) = \partial(TK) \geqslant \frac{\mathrm{tr}(T^{-1})}{n}\partial(K),$$

which gives that $\mathrm{tr}(T^{-1}) \leqslant n$, and hence, since T^{-1} is symmetric and positive definite, that $T = \mathrm{Id}$. This proves the theorem. □

Definition 2.3.2. Let K be a convex body in $\mathbb{R}^n$. The *minimal surface invariant* of K is defined as

$$\partial_K = \min\left\{ \frac{\partial(TK)}{\mathrm{Vol}_n(TK)^{\frac{n-1}{n}}} : T \in GL_n \right\}.$$

Clearly, a convex body K of volume 1 has minimal surface area if $\partial(K) = \partial_K$.

Note that an equivalent formulation of the isoperimetric inequality for convex bodies is:

$$\partial_K \geqslant \partial_{B_2^n} = n\kappa_n^{1/n}.$$

This last quantity is of the order of $\sqrt{n}$ as $n \to \infty$. With the notion of minimal surface area position in hand, a natural question to ask next is which body K has maximal "minimal surface invariant". This is the topic we study in the following section.

2.4. Reverse isoperimetric inequality

Modulo affine transformations, among all convex bodies of a given volume in $\mathbb{R}^n$, the n-dimensional simplex has "largest" surface area, while among centrally symmetric convex bodies, the cube is the extremal body.

THEOREM 2.4.1 (Ball). *Let K be a convex body in $\mathbb{R}^n$ and T a regular n-dimensional solid simplex. Then there is an affine image $\tilde{K}$ of K satisfying*

$$\mathrm{Vol}(\tilde{K}) = \mathrm{Vol}(T) \qquad \text{and} \qquad \partial(\tilde{K}) \leqslant \partial(T).$$

If K is a centrally symmetric convex body in $\mathbb{R}^n$ and Q an n-dimensional cube then there is a linear image $\tilde{K}$ of K satisfying

$$\mathrm{Vol}(\tilde{K}) = \mathrm{Vol}(Q) \qquad \text{and} \qquad \partial(\tilde{K}) \leqslant \partial(Q).$$

A main tool in the proof of the above theorem is the Brascamp-Lieb inequality. In the next subsection we state and prove its "normalized form" put forward by K. Ball together with its reverse form, which is due to Barthe.

2.4.1. Ball's normalized form for the Brascamp-Lieb inequality

THEOREM 2.4.2 (Ball). *Let $u_1, \ldots, u_m \in S^{n-1}$ and $c_1, \ldots, c_m > 0$ satisfy*

$$\mathrm{Id} = \sum_{j=1}^m c_j u_j \otimes u_j.$$

If $f_1, \ldots, f_m : \mathbb{R} \to \mathbb{R}^+$ are measurable functions, then

$$\int_{\mathbb{R}^n} \prod_{j=1}^m f_j^{c_j}(\langle x, u_j\rangle)dx \leqslant \prod_{j=1}^m \left(\int_{\mathbb{R}} f_j(t)dt\right)^{c_j}.$$

The advantage of the normalization condition $\mathrm{Id} = \sum_{j=1}^m c_j u_j \otimes u_j$ is that the constant in the inequality, which is (always) attained by Gaussian functions, is 1.

PROPOSITION 2.4.3. *Let $u_1, \ldots, u_m \in S^{n-1}$ and $c_1, \ldots, c_m > 0$ satisfy* $\mathrm{Id} = \sum_{j=1}^m c_j u_j \otimes u_j$. *Then,*

$$\sup\left\{ \frac{\int_{\mathbb{R}^n} \prod_{j=1}^m g_j^{c_j}(\langle x, u_j\rangle)dx}{\prod_{j=1}^m \left(\int_{\mathbb{R}} g_j\right)^{c_j}} : g_j(t) = e^{-\lambda_j t^2}, \lambda_j > 0\right\}$$
$$= \inf\left\{ \frac{\det\left(\sum_{j=1}^m c_j\lambda_j u_j \otimes u_j\right)}{\prod_{j=1}^m \lambda_j^{c_j}} : \lambda_j > 0\right\} = 1.$$

Proof. Recall that $(u \otimes u)(x) = \langle x, u\rangle u$. Let $g_j(t) = \exp(-\lambda_j t^2)$, $j = 1, \ldots, m$, where λ_j are positive reals. Then,

$$\begin{aligned}
\int_{\mathbb{R}^n} \prod_{j=1}^m g_j^{c_j}(\langle x, u_j\rangle)dx &= \int_{\mathbb{R}^n} \exp\left(-\sum_{j=1}^m c_j\lambda_j \langle x, u_j\rangle^2\right)dx \\
&= \int_{\mathbb{R}^n} \exp\left(-\Big\langle\Big(\sum_{j=1}^m c_j\lambda_j u_j \otimes u_j\Big)(x), x\Big\rangle\right)dx \\
&= \frac{\pi^{n/2}}{\sqrt{\det\left(\sum_{j=1}^m c_j\lambda_j u_j \otimes u_j\right)}}.
\end{aligned}$$

On the other hand,

$$\prod_{j=1}^m \left(\int_{\mathbb{R}} g_j\right)^{c_j} = \prod_{j=1}^m \left(\int_{\mathbb{R}} \exp(-\lambda_j t^2)dt\right)^{c_j} = \prod_{j=1}^m \left(\frac{\sqrt{\pi}}{\sqrt{\lambda_j}}\right)^{c_j}$$
$$= \frac{\pi^{n/2}}{\sqrt{\prod_{j=1}^m \lambda_j^{c_j}}}$$

since $c_1 + \cdots + c_m = n$. It follows that

$$\inf\left\{\left(\frac{\prod_{j=1}^m \left(\int_{\mathbb{R}} g_j\right)^{c_j}}{\int_{\mathbb{R}^n}\prod_{j=1}^m g_j^{c_j}(\langle x,u_j\rangle)dx}\right)^2 : g_j(t) = e^{-\lambda_j t^2},\ \lambda_j > 0\right\}$$
$$= \inf\left\{\frac{\det\left(\sum_{j=1}^m c_j\lambda_j u_j\otimes u_j\right)}{\prod_{j=1}^m \lambda_j^{c_j}} : \lambda_j > 0\right\}.$$

We turn to showing that this constant is 1.

Let $\lambda_j > 0$, $j = 1, \ldots, m$. For every $I \subseteq \{1, \ldots, m\}$ with cardinality $|I| = n$ we define

$$\lambda_I = \prod_{i\in I}\lambda_j \quad \text{and} \quad U_I = \left(\det\left(\sum_{j\in I} c_j u_j\otimes u_j\right)\right)^2.$$

By the Cauchy-Binet formula we have

$$\det\left(\sum_{j=1}^m c_j\lambda_j u_j\otimes u_j\right) = \det\left(\sum_{j=1}^m \lambda_j(\sqrt{c_j}u_j)\otimes(\sqrt{c_j}u_j)\right) = \sum_{|I|=n}\lambda_I U_I. \tag{2.4.1}$$

Setting $\lambda_j = 1$ in (2.4.1) we see that

$$\sum_{|I|=n} U_I = 1.$$

By the arithmetic-geometric means inequality,

$$\sum_{|I|=n}\lambda_I U_I \geqslant \prod_{|I|=n}\lambda_I^{U_I} = \prod_{j=1}^m \lambda_j^{\sum_{\{I:j\in I\}} U_I}. \tag{2.4.2}$$

Applying the Cauchy-Binet formula again, we have

$$\sum_{\{I:j\in I\}} U_I = \sum_{|I|=n} U_I - \sum_{\{I:j\notin I\}} U_I = 1 - \det\left(\mathrm{Id} - (\sqrt{c_j}u_j)\otimes(\sqrt{c_j}u_j)\right)$$
$$= 1 - (1 - c_j|u_j|^2) = c_j.$$

Going back to (2.4.1) and (2.4.2) we see that

$$\det\left(\sum_{j=1}^m c_j\lambda_j u_j\otimes u_j\right) \geqslant \prod_{j=1}^m \lambda_j^{c_j} \tag{2.4.3}$$

and thus

$$\inf\left\{\frac{\det\left(\sum_{j=1}^m c_j\lambda_j u_j\otimes u_j\right)}{\prod_{j=1}^m \lambda_j^{c_j}} : \lambda_j > 0\right\} \geqslant 1.$$

The choice $\lambda_j = 1$ gives equality in (2.4.3), which completes the proof. □

We set

$$I(f_1,\dots,f_m)=\int_{\mathbb{R}^n}\prod_{j=1}^m f_j^{c_j}(\langle x,u_j\rangle)dx.$$

By considering the Gaussians, we have shown that,

$$\sup\left\{I(f_1,\dots,f_m):\int_{\mathbb{R}} f_j=1\ ,\ j=1,\dots,m\right\}\geqslant 1. \tag{2.4.4}$$

The following is a reverse form of Theorem 2.4.2.

THEOREM 2.4.4 (Barthe). *Let $u_1,\dots,u_m\in S^{n-1}$ and $c_1,\dots,c_m>0$ satisfy $\mathrm{Id}=\sum_{j=1}^m c_ju_j\otimes u_j$. If $h_1,\dots,h_m:\mathbb{R}\to\mathbb{R}^+$ are measurable functions, we set*

$$K(h_1,\dots,h_m)=\int_{\mathbb{R}^n}^{*}\sup\left\{\prod_{j=1}^m h_j^{c_j}(\theta_j):\theta_j\in\mathbb{R}\ ,\ x=\sum_{j=1}^m\theta_jc_ju_j\right\}dx.$$

Then,

$$\inf\left\{K(h_1,\dots,h_m):\int_{\mathbb{R}} h_j=1\ ,\ j=1,\dots,m\right\}=1.$$

Testing centered Gaussian functions we get the easy part of this *reverse Brascamp-Lieb inequality.*

PROPOSITION 2.4.5. *With the notation of Theorem 2.4.4 we have*

$$\inf\left\{K(h_1,\dots,h_m):\int_{\mathbb{R}} h_j=1\ ,\ j=1,\dots,m\right\}\leqslant 1.$$

Proof. Let $\lambda_j>0$, $j=1,\dots,m$ and consider the functions $h_j(t)=\exp(-t^2/\lambda_j)$. Then, the function

$$m(x):=\sup\left\{\prod_{j=1}^m h_j^{c_j}(\theta_j):x=\sum_{j=1}^m\theta_jc_ju_j\right\}$$

is given by

$$m(x)=\exp\left(-\inf\left\{\sum_{j=1}^m\frac{c_j}{\lambda_j}\theta_j^2:x=\sum_{j=1}^m\theta_jc_ju_j\right\}\right).$$

Define

$$\|x\|^2=\sum_{j=1}^m c_j\lambda_j\langle x,u_j\rangle^2=\langle Ax,x\rangle$$

where A is the symmetric positive-definite operator $A:=\sum_{j=1}^m c_j\lambda_ju_j\otimes u_j$. It is not hard to check that the dual norm is exactly

$$\|y\|_*^2=\inf\left\{\sum_{j=1}^m\frac{c_j}{\lambda_j}\theta_j^2:y=\sum_{j=1}^m\theta_jc_ju_j\right\}.$$

Therefore,

$$\|y\|_*^2=\langle By,y\rangle,$$

where $B=A^{-1}$. It follows that

$$\int_{\mathbb{R}^n}m(x)dx=\frac{\pi^{n/2}}{\sqrt{\det B}}=\pi^{n/2}\sqrt{\det A}.$$

On the other hand,

$$\prod_{j=1}^{m}\left(\int_{\mathbb{R}}\exp(-t^2/\lambda_j)dt\right)^{c_j}=\pi^{n/2}\prod_{j=1}^{m}\lambda_j^{c_j/2}.$$

This shows that

$$\inf\left\{K^2(h_1,\dots,h_m):\int_{\mathbb{R}}h_j=1\right\}$$
$$\leqslant \inf\left\{\frac{\det\left(\sum_{j=1}^m c_j\lambda_j u_j\otimes u_j\right)}{\prod_{j=1}^m\lambda_j^{c_j}}:\lambda_j>0\right\}=1$$

and the proof is complete. □

The main step in Barthe's argument is the following proposition.

PROPOSITION 2.4.6. *Let* $f_1,\dots,f_m:\mathbb{R}\to\mathbb{R}^+$ *and* $h_1,\dots,h_m:\mathbb{R}\to\mathbb{R}^+$ *be integrable functions with*

$$\int_{\mathbb{R}}f_j(t)dt=\int_{\mathbb{R}}h_j(t)dt=1,\qquad j=1,\dots,m.$$

Then,

$$I(f_1,\dots,f_m)\leqslant K(h_1,\dots,h_m).$$

Proof. We may assume that f_j,h_j are continuous and strictly positive. We use the transportation of measure idea that was used for the proof of the Prékopa-Leindler inequality: For every $j=1,\dots,m$ we define $T_j:\mathbb{R}\to\mathbb{R}$ by the equation

$$\int_{-\infty}^{T_j(t)}h_j(s)ds=\int_{-\infty}^{t}f_j(s)ds.$$

Then, each T_j is strictly increasing, 1-1 and onto, and

$$T_j'(t)h_j(T_j(t))=f_j(t),\quad t\in\mathbb{R}. \tag{2.4.5}$$

We now define $W:\mathbb{R}^n\to\mathbb{R}^n$ by

$$W(y)=\sum_{j=1}^{m}c_jT_j(\langle y,u_j\rangle)u_j. \tag{2.4.6}$$

A simple computation shows that

$$J(W)(y)=\sum_{j=1}^{m}c_jT_j'(\langle y,u_j\rangle)u_j\otimes u_j.$$

This implies

$$\langle[J(W)(y)](v),v\rangle>0\quad\text{if}\quad v\neq 0$$

and hence W is injective. Consider the function

$$m(x)=\sup\left\{\prod_{j=1}^{m}h_j^{c_j}(\theta_j):x=\sum_{j=1}^{m}\theta_jc_ju_j\right\}.$$

Then, (2.4.6) shows that

$$m(W(y))\geqslant\prod_{j=1}^{m}h_j^{c_j}(T_j(\langle y,u_j\rangle))$$

for every $y \in \mathbb{R}^n$. It follows that

$$\begin{aligned}\int_{\mathbb{R}^n} m(x)dx &\geqslant \int_{W(\mathbb{R}^n)} m(x)dx \\ &= \int_{\mathbb{R}^n} m(W(y)) \cdot |\det J(W)(y)|\, dy \\ &\geqslant \int_{\mathbb{R}^n} \prod_{j=1}^m h_j^{c_j}(T_j(\langle y, u_j\rangle)) \det\left(\sum_{j=1}^m c_j T_j'(\langle y, u_j\rangle) u_j \otimes u_j\right) dy \\ &\geqslant \int_{\mathbb{R}^n} \prod_{j=1}^m h_j^{c_j}(T_j(\langle y, u_j\rangle)) \prod_{j=1}^m \left(T_j'(\langle y, u_j\rangle)\right)^{c_j} dy,\end{aligned}$$

where in the last inequality we have used Proposition 2.4.3. Therefore, taking (2.4.5) into account we have

$$\int_{\mathbb{R}^n} m(x)dx \geqslant \int_{\mathbb{R}^n} \prod_{j=1}^m f_j^{c_j}(\langle y, u_j\rangle)dy = I(f_1, \dots, f_m).$$

In other words, $I(f_1, \dots, f_m) \leqslant K(h_1, \dots, h_m)$. □

Proof of Theorem 2.4.4 and Theorem 2.4.2. Let $f_1, \dots, f_m$ and $h_1, \dots, h_m : \mathbb{R} \to \mathbb{R}^+$ be integrable functions with

$$\int_{\mathbb{R}} f_j(t)dt = \int_{\mathbb{R}} h_j(t)dt = 1, \quad j = 1, \dots, m.$$

Then,

$$I(f_1, \dots, f_m) \leqslant K(h_1, \dots, h_m).$$

Taking the supremum over all such functions f_j and the infimum over all such functions h_j we get that

$$1 \leqslant \sup\left\{I(f_1, \dots, f_m)\right\} \leqslant \inf\left\{K(h_1, \dots, h_m)\right\} \leqslant 1,$$

so that there must be equality throughout. □

2.4.2. Maximal volume ratio

DEFINITION 2.4.7. The *volume ratio* of a convex body K is defined to be

$$\mathrm{vr}(K) = \inf_{\mathcal{E} \subseteq K} \left(\frac{\mathrm{Vol}_n(K)}{\mathrm{Vol}_n(\mathcal{E})}\right)^{1/n}$$

where the infimum is taken over all ellipsoids $\mathcal{E}$ inside K.

THEOREM 2.4.8 (Ball). (i) *Among centrally symmetric convex bodies in $\mathbb{R}^n$, the cube has the largest volume ratio.*
(ii) *Among convex bodies in $\mathbb{R}^n$, the simplex has the largest volume ratio.*

Proof. (i) One has to show that if a centrally symmetric convex body K is in John position then $\mathrm{Vol}_n(K) \leqslant 2^n$. But when a body is in John position, by Theorem 2.1.10 there exist contact points u_j of it and B_2^n and $c_j > 0$ such that

$$\mathrm{Id} = \sum_{j=1}^m c_j u_j \otimes u_j.$$

Since K is symmetric, for every contact point u_j of K and B_2^n we have that $-u_j$ is also a contact point. Thus the body is contained in the intersection of the strips $C := \{x : |\langle x, u_j\rangle| \leqslant 1\}$ (because for all contact points u_j we have that $h_K(u_j) = 1$), and its volume is at most

$$\mathrm{Vol}_n(C) = \int_{\mathbb{R}^n} \prod_{j=1}^m \mathbf{1}_{[-1,1]}(\langle x, u_j\rangle)^{c_j}\,dx \leqslant \prod_{j=1}^m \left(\int_{\mathbb{R}} \mathbf{1}_{[-1,1]}\right)^{c_j} = 2^n$$

by the Brascamp-Lieb inequality and the fact that $\sum_{j=1}^m c_j = n$.

(ii) Moving to the non-symmetric case, we note that the n-dimensional simplex that circumscribes B_2^n has volume

$$\frac{n^{n/2}(n+1)^{(n+1)/2}}{n!}$$

(it is easier to check this working in $\mathbb{R}^{n+1}$ on the hyperplane $\sum x_i = 1$, say). We thus need to establish this bound for the volume of a body in John position. Use John's theorem in the non-symmetric case to find contact points u_j and $c_j > 0$ such that $\sum_j c_j u_j = 0$ and $\mathrm{Id} = \sum_j c_j u_j \otimes u_j$. As above, the body K lies inside the possibly larger body

$$L := \{x : \langle x, u_j\rangle \leqslant 1, i = j, \dots, m\}$$

(this is a bounded set because $\sum_j c_j u_j = 0$), so it would suffice to bound this body's volume by the desired bound.

We construct a new sequence of vectors $(v_i)_{i=1}^m$ in $\mathbb{R}^{n+1}$ which would be orthogonal in the extreme case where K is a simplex. The estimate follows from an application of the Brascamp-Lieb inequality to a family of functions whose product is supported on a cone in $\mathbb{R}^{n+1}$ that has cross-sections similar to K: regard $\mathbb{R}^{n+1}$ as $\mathbb{R}^n \times \mathbb{R}$ and for each j, let

$$v_j = \sqrt{\frac{n}{n+1}}\left(-u_j, \frac{1}{\sqrt{n}}\right) \in \mathbb{R}^{n+1} \quad \text{and} \quad d_j = \frac{n+1}{n} c_j \in \mathbb{R}^+.$$

Direct computation shows that

$$\mathrm{Id}_{n+1} = \sum_{j=1}^m d_j v_j \otimes v_j.$$

Take $f_j = e^{-t}, t \geqslant 0$. Apply the Brascamp-Lieb inequality to get

$$\int_{\mathbb{R}^{n+1}} \prod f_j^{d_j}(\langle x, v_j\rangle)dx \leqslant \prod_{j=1}^m \left(\int_{\mathbb{R}} f_j\right)^{d_j} = 1, \tag{2.4.7}$$

because $\int f_j = 1$. Next, we compute the same integral over each hyperplane $x_{n+1} = r$. One can check that the function $\prod f_j^{d_j}(\langle x, v_j\rangle)$ is non-zero precisely when $r \geqslant 0$ and the point x is in $\frac{r}{\sqrt{n}} L \times \{r\}$ (given that each f_j is defined to be non-zero precisely at the non-negative $t \in \mathbb{R}$), and in that case it equals exactly $e^{-r\sqrt{n+1}}$

(because we know that $\sum_j c_j u_j = 0$). In other words,

$$\int_{\{x_{n+1}=r\}} \prod f_j^{d_j}(\langle x, v_j\rangle)dx = e^{-r\sqrt{n+1}}\mathrm{Vol}_n\left(\frac{r}{\sqrt{n}}L\right) = e^{-r\sqrt{n+1}}\left(\frac{r}{\sqrt{n}}\right)^n \mathrm{Vol}_n(L).$$

Then, (2.4.7) implies

$$1 \geqslant \mathrm{Vol}_n(L)\int_0^\infty e^{-r\sqrt{n+1}}\left(\frac{r}{\sqrt{n}}\right)^n dr = \frac{\mathrm{Vol}_n(L)\cdot n!}{n^{n/2}(n+1)^{(n+1)/2}}.$$

We conclude the proof noticing that $\mathrm{Vol}_n(K) \leqslant \mathrm{Vol}_n(L)$ and that the upper bound for $\mathrm{Vol}_n(K)$ is no other than the number we had equality for in the case of the simplex. □

2.4.3. Proof of the reverse isoperimetric inequality

It is perhaps surprising that in proving a statement about the maximal "minimal surface invariant", we shall not use the minimal surface area position but rather the John position. Of course, in the case of the extremal bodies for the required inequality, these two positions coincide. We shall prove only the second part of Theorem 2.4.1, as the proof of the first part is identical.

Proof of Theorem 2.4.1. We have to find a linear image of a given centrally symmetric convex body K such that

$$\partial(\tilde{K}) \leqslant c_n \mathrm{Vol}(\tilde{K})^{\frac{n-1}{n}}$$

where c_n is determined so that for the (say, side-length-1) cube there is equality, that is, $c_n = 2n$. This position will be the John position of K: indeed, for this position we have

$$\begin{aligned}\partial(K) &= \lim_{t\to 0}\frac{\mathrm{Vol}_n(K+tB_2^n) - \mathrm{Vol}_n(K)}{t} \\ &\leqslant \lim_{t\to 0}\frac{\mathrm{Vol}_n((1+t)K) - \mathrm{Vol}_n(K)}{t} \\ &= n\mathrm{Vol}_n(K) \leqslant 2n\mathrm{Vol}(K)^{\frac{n-1}{n}},\end{aligned}$$

where we make use of Theorem 2.4.8 for the last inequality. □

2.5. Notes and remarks

Positions of convex bodies

The other two classical positions that are mentioned at the beginning of Chapter 2 will be presented in detail in subsequent chapters. The isotropic position is discussed in Chapter 10. We mention here that its first appearance in asymptotic geometric analysis is in the paper of Milman and Pajor [**459**]. Similarly, the M-position will be discussed and remarked upon in another chapter, Chapter 8; its existence was first established in [**448**].

John's theorem

The idea of a maximal volume ellipsoid, and of the extremal properties associated with it, appears in Fritz John's paper [**313**] where it was given as an example of how to work with Lagrange multipliers in a non-smooth setting. His approach was later used a lot in partial differential equations. Nevertheless, the application of Theorem 2.1.3 was close to his heart because in the 1930's he presented several papers in convexity where he considered the problem of estimating the distance from the Euclidean space in low dimensions. An interesting historical detail is that John's article [**313**], which is one of the most cited papers of the local theory of normed spaces, was rejected from the journal where it had first been submitted, and then the author offered it to the Courant anniversary volume. The main achievement is not so much the upper bound $\sqrt{n}$ for the distance of an n-dimensional space from the Euclidean space but rather the isotropic decomposition of the identity into the discrete sum of rank one projections given by contact points. Our presentation of the proof of Theorem 2.1.10 follows the one by K. Ball in his survey article [**43**]. Proposition 2.1.14 is due to Ball, see [**42**].

Given a pair of convex bodies K and T in $\mathbb{R}^n$ we say that K is in a *position of maximal volume* in T if $K \subseteq T$ and $\mathrm{Vol}_n(AK) \leqslant \mathrm{Vol}_n(K)$ for every affine image AK of K which is also contained in T. It was observed by V. Milman (see [**593**, Theorem 14.5]) that if both K and T are centrally symmetric then one can have an analogue of John's decomposition of the identity. It was presented in a very useful form by Lewis [**381**] in the language of operator norms and we discuss it in Chapter 6. This fact was later proved in the not-necessarily symmetric context. The precise statement is that if K is in a position of maximal volume in T and if $0 \in \mathrm{int}(T)$ then there exist $m \leqslant n(n+1)$ contact points x_j of K and T, contact points y_j of K° and T° and $c_j > 0$ such that: $\langle x_j, y_j\rangle = 1$,

$$\sum_{j=1}^m c_j y_j = 0 \quad \text{and} \quad \mathrm{Id} = \sum_{j=1}^m c_j x_j \otimes y_j.$$

Moreover, one can find $z \in \mathrm{int}(K)$ and $m \leqslant n(n+1)$ reals $c_j > 0$ and contact pairs (x_j, y_j) of $K - z$ and $T - z$ as above so that

$$\sum_{j=1}^m c_j x_j = \sum_{j=1}^m c_j y_j = 0 \quad \text{and} \quad \mathrm{Id} = \sum_{j=1}^m c_j x_j \otimes y_j.$$

These results were first obtained by Giannopoulos, Perissinaki and Tsolomitis in [**244**]; see also Dilworth [**171**] and later Bastero and Romance [**56**] for an extension to the non-convex case and, finally, the article of Gordon, Litvak, Meyer and Pajor [**262**] for a proof without any smoothness or other additional assumptions on K and/or T, in the spirit of John's original argument. A related result of Giannopoulos and Hartzoulaki [**224**] concerns the volume ratio of two convex bodies: if K is in a position of maximal volume in T then $(\mathrm{Vol}_n(T)/\mathrm{Vol}_n(K))^{1/n} \leqslant c\sqrt{n}\log n$; the proof will be discussed in Part II.

The inequality in Remark 2.1.17 implies that the Banach-Mazur distance between any convex body K in $\mathbb{R}^n$ and B_2^n is bounded by n. Leichtweiss in 1959 [**375**], and later Palmon [**489**], showed that one can have equality in $d(K, B_2^n) \leqslant n$ only if K is a simplex. The equality in the centrally symmetric case is much more delicate. Milman and Wolfson [**471**] showed that when a centrally symmetric convex body K has extremal distance $d(K, B_2^n) = \sqrt{n}$ from the Euclidean ball, then it has a section of dimension $k \geqslant c\log n$ which is isometric to the unit ball of ℓ_1^k (this estimate is exact, up to the constant c, as one can see by the example of a cube). In the same paper, an isomorphic result is obtained under the assumption that $d(K, B_2^n) \geqslant c\sqrt{n}$; this will be discussed in Part II.

A modified version of the Banach-Mazur distance was introduced by Grünbaum in [**280**] as follows:

$$\tilde{d}_{BM}(K,T) = \inf\{|\lambda| : \lambda \in \mathbb{R},\ K - w \subseteq A(T - z) \subseteq \lambda(K - w)\}$$

where the infimum is over all $z, w \in \mathbb{R}^n$ and all $A \in GL_n$. Grünbaum conjectured that $\tilde{d}_{BM}(K,T) \leqslant n$ for every pair of convex bodies K and T. An affirmative answer was given by Gordon, Litvak, Meyer and Pajor in [**262**], where they also proved that if S is a simplex then $\tilde{d}_{BM}(K,S) = n$ for every centrally symmetric convex body K in $\mathbb{R}^n$. The analogue of the Leichtweiss-Palmon result for Grünbaum's distance was studied by Jiménez and Naszódi in [**311**]; they showed that if K is strictly convex or smooth, then $\tilde{d}_{BM}(K,T) = n$ implies that T is a simplex.

Gruber proved in [**276**] that the closure of the set of those convex bodies in $\mathbb{R}^n$ that have less than $N_n = \frac{1}{2}n(n+3)$ contact points with their maximal volume ellipsoid is nowhere dense in the class of all convex bodies in $\mathbb{R}^n$ (in the topology induced by the Banach-Mazur distance). However, Rudelson showed in [**526**] and [**527**] that for every convex body K and any $\varepsilon > 0$ there exists a convex body T such that $d_{BM}(K,T) \leqslant 1+\varepsilon$ and the number of contact points of T with its John ellipsoid is less than $Cn\log n/\varepsilon^2$. The $\log n$-term is not necessary; this was proved by Srivastava [**568**], who showed the following: given a centrally symmetric convex body K, one can find a convex body T with Banach-Mazur distance from K as above, $d_{BM}(K,T) \leqslant 1+\varepsilon$, which has $m \leqslant 32n/\varepsilon^2$ contact points with its John ellipsoid, while in the not necessarily symmetric case, given K one can find T so that $d_{BM}(K,T) \leqslant 2.24$ and T has $m \leqslant Cn$ contact points with its John ellipsoid. Srivastava's argument is based on a powerful linear algebraic method (see [**58**] and [**567**]) that will be discussed in Part II.

Theorem 2.1.18 is due to Kadets and Snobar, [**321**]. The estimate is not optimal but it is "almost optimal" (see the survey article [**353**] of Koldobsky and König for more information). The theorem is sharp in the following sense: there exists a construction by Szarek [**581**] of a sequence of spaces E_n of dimension n such that, for *all* projections $P : E_n \to E_n$ with rank $n/2$, one has $\|P\| \geqslant c\sqrt{n}$ for some fixed $c > 0$. The same result, with $\sqrt{n/\log n}$ instead of $\sqrt{n}$, was obtained, before Szarek, by Gluskin [**248**].

Dvoretzky-Rogers Lemmas

We use the terminology "Dvoretzky-Rogers lemma" for a series of statements on the distribution (on the unit sphere) of the contact points of a convex body in John position with the Euclidean unit ball. Theorem 2.1.24 was proved by Dvoretzky and Rogers in [**179**] and plays a key role in the proof of their main result that every infinite dimensional Banach space X contains an unconditionally convergent series that is not absolutely convergent. More precisely, they showed that if (a_n) is a sequence of positive reals with $\sum_{n=1}^{\infty} a_n < \infty$ then one may find a sequence (x_n) in X so that $\|x_n\| = \sqrt{a_n}$ for all n and the series $\sum_{n=1}^{\infty} x_n$ is unconditionally convergent.

The results of this section play an important role in the sequel. Theorem 2.1.26 was the starting point for Grothendieck's question that led to Dvoretzky theorem; moreover, Theorem 2.1.25 is used in all the proofs of Dvoretzky theorem that we present in this book.

Proposition 2.1.23 is a remark by W. Johnson.

Minimal mean width position

The isotropic description of the minimal mean width position, Theorem 2.2.4, is due to Giannopoulos and Milman [**232**]. For details on the Laplace Beltrami operator we refer the reader to the book of Groemer [**266**]. In the same article, the question to characterize the minimal positions with respect to other quermassintegrals was also studied. Naturally, one says that a convex body K *minimizes* W_i if $W_i(K) \leqslant W_i(TK)$ for every $T \in SL_n$. Minimal mean width corresponds to $i = n-1$ and minimal surface area to $i = 1$. The situation is less clear if $1 < i < n-1$. A necessary condition is obtained in [**232**], namely, if K minimizes W_i then the area measure $S_{n-i}(K,\cdot)$ is isotropic. This line of thought was continued by several authors who studied positions of convex bodies arising as solutions of extremal problems (see e.g. the paper [**57**] of Bastero and Romance).

Recall that a way to introduce M-position is given by the next result of V. Milman: there exists an absolute constant $\beta > 0$ such that every convex body K in $\mathbb{R}^n$ with barycenter at the origin has a linear image $\tilde{K}$ with $\mathrm{Vol}_n(\tilde{K}) = 1$ that satisfies $\mathrm{Vol}_n(K + D_n)^{1/n} \leqslant \beta$, where D_n is the Euclidean ball of volume 1. Then, we say that K is in *M-position with constant β*. The question whether the minimal surface area position is an M-position in the above sense was posed in [**232**] and was answered in the negative by Saroglou in [**534**]. An alternative, more natural, proof was given by Markessinis, Paouris and Saroglou in [**415**] where it was proved that for every $n \in \mathbb{N}$ there exists an unconditional convex body K of volume 1 in $\mathbb{R}^n$ which is in minimal surface area position and satisfies $\mathrm{Vol}_n(K + D_n)^{1/n} \geqslant c\sqrt[8]{n}$. The exponent 1/8 in this statement is optimal (up to the value of the isotropic constant L_K of a body) in the symmetric case. The same authors proved that if K is a centrally symmetric convex body of volume 1 in $\mathbb{R}^n$ that has minimal surface area, then $\mathrm{Vol}_n(K + D_n)^{1/n} \leqslant C\sqrt[8]{n}L_K$ for some absolute constant C. The method in [**415**] shows that it is not true that the minimal mean width position is always an M-position.

Minimal surface area position

Theorem 2.3.1 is due to Petty [**498**]. See also the paper of Giannopoulos and Papadimitrakis [**243**] for a proof and applications of the fact that when K has minimal surface area then its area measure σ_K is isotropic. As an example, we describe bounds for the volume of the *projection body* ΠK of K, which is defined by

$$h_{\Pi K}(x) := \mathrm{Vol}_{n-1}(P_{x^\perp}(K)) = \frac{1}{2}\int_{S^{n-1}} |\langle x, z\rangle| d\sigma_K(z).$$

For simplicity, consider a polytope K with facets F_j and normals u_j, $j = 1, \dots, m$. If K is in minimal surface area position, then Petty's theorem states that σ_K is isotropic; this statement is equivalent to the representation of the identity

$$\mathrm{Id} = \sum_{j=1}^{m} c_j u_j \otimes u_j$$

and $\Pi K = \frac{\partial(K)}{2n}\sum_{j=1}^m c_j[-u_j, u_j]$ where $c_j = \frac{n\mathrm{Vol}_{n-1}(F_j)}{\partial(K)}$. Using Barthe's reverse Brascamp-Lieb inequality one can give a lower bound for the volume of ΠK. Namely,

$$\mathrm{Vol}_n(\Pi K) \geqslant 2^n\left(\frac{\partial(K)}{2n}\right)^n.$$

The example of the cube shows that this inequality is sharp for bodies with minimal surface area.

Combined with Theorem 2.2.4 this volume estimate leads to a sharp reverse Urysohn inequality for zonoids (this class consists of the limits of Minkowski sums of line segments in the Hausdorff metric, and coincides with the class of the projection bodies of convex bodies; see Bourgain and Lindenstrauss [**104**] and Bolker [**91**]). The next inequality appears in the paper of Giannopoulos, Milman and Rudelson[**236**]: if Z is a zonoid in $\mathbb{R}^n$ with volume 1 and minimal mean width, then $w(Z) \leqslant w(Q_n) = \frac{2\kappa_{n-1}}{\kappa_n}$ where Q_n is the cube of volume 1. To see this, write Z as the projection body of a convex body K and note that from the characterizations of minimal surface area (applied to K) and minimal mean width (applied to Z), K has minimal surface area. Then, one can check that

$$w(Z) = \frac{\kappa_{n-1}}{n\kappa_n}\partial(K),$$

and using the reverse isoperimetric inequality one concludes that $w(Z) \leqslant 2\kappa_{n-1}/\kappa_n$. Equality holds when K is a cube, and this corresponds to the case $Z = Q_n$.

Reverse isoperimetric inequality

Reverse isoperimetric inequalities for affine classes of convex bodies were considered by Behrend [**60**] and Green [**264**] in the planar case. The sharp reverse isoperimetric inequality of Theorem 2.4.1 was proved by Ball [**41**] in 1991. Closely related is the isomorphic reverse Brunn-Minkowski inequality proved by V. Milman [**448**] in 1986 (see Chapter 8 of this book).

Theorem 2.4.2, the normalized form of the Brascamp-Lieb inequality, is due to Ball. The proof of the inequality and its reverse form that we present in the text is due to Barthe [**50**] (see also [**51**] and [**49**]). We will devote a significant part of a chapter in Part II to rearrangement inequalities and the history of the Brascamp-Lieb inequality; we will also present there the statement and the proof of the reverse form of the multidimensional Brascamp-Lieb inequality, in which the Brenier map plays a key role.

Here we only mention a few applications of the Brascamp-Lieb inequality and its reverse form to sharp geometric inequalities. Ball first used it in [**39**] (see also [**44**]) to obtain estimates on the volume of sections and projections of the unit cube. Barthe [**52**] proved the following extremal property of the simplex: if K is a convex body whose maximal volume ellipsoid is B_2^n then $w(K) \leqslant w(S_n)$, where S_n is the regular simplex circumscribing B_2^n. In the symmetric case one has $w(K) \leqslant w(B_1^n)$ (this is much simpler and was observed by Schechtman and Schmuckenschläger in [**543**]). The proof of both inequalities makes use of the reverse Brascamp-Lieb inequality. In Löwner position (that is, position when the ellipsoid of minimal volume inclusing K is a multiple of the Euclidean ball)) the simplex and the cube are the extremal bodies for $w(K)$.

CHAPTER 3

Isomorphic isoperimetric inequalities and concentration of measure

Concentration of measure is the central phenomenon at the base of the field of asymptotic geometric analysis. Since its discovery as a general phenomenon in the end of the 1960's, numerous approaches have been found, all leading to the same type of behaviour: in high parametric families, under very weak assumptions of various types, a function tends to concentrate around its mean or median. This simple fact, which is perhaps first encountered (under very specialized conditions) by undergraduate or high school students in the study of the law of large numbers, is applicable in a variety of cases and has far reaching consequences. The concept of concentration has first been put forward in Milman's version of Dvoretzky theorem, that will appear in Chapter 5 and will be our starting point for developing the theory. Concentration proved to be a powerful tool with numerous applications in geometry, analysis, probability theory and discrete mathematics.

In this chapter we shall explore the most basic approaches towards concentration. Classical isoperimetric inequalities for metric probability spaces are at the origin of measure concentration, so we will start with the most basic examples such as the sphere, Gauss space and the cube. In these (and not many other) examples, the extremal sets for isoperimetric questions are known and can be easily described. However, in all the applications which we will consider, the extremal sets are not important but only the concentration estimates which one derives from them. We thus do not concentrate our attention on extremal sets but rather on the many different ways to get concentration inequalities. In this chapter we shall explore various such ways, and we try each time to determine the different source for concentration. Of course, probability and independence are a well known source for concentration. We shall see that also curvature or convexity is a source. (Recall Borell's Lemma in Section 1.5.2 which is also a form of concentration). But also, a spectral gap is a source of concentration as we will see in Part II, martingales can be such a source as we show in Section 3.3, and concentration can be an aspect of conditional probability as we show in Section 3.5. We shall investigate approaches which do not need a metric, and other approaches which do not need a measure. In Part II we discuss functional aspects of the subject such as Sobolev and logarithmic Sobolev inequalities, tensorization techniques, semi-group approaches, Laplace transform and infimum convolutions, transportation of measure, and more.

3.1. An approach through extremal sets, and the basic terminology

The classical setting for the concentration phenomenon is on a metric probability space.

DEFINITION 3.1.1. Let (X, d) be a metric space. We denote the Borel σ-algebra of X by $\mathcal{B}(X)$. For each non-empty $A \in \mathcal{B}(X)$ and any $t > 0$, the *t-extension* of A is the set

$$A_t = \{x \in X : d(x, A) < t\}.$$

Informally, given a Borel probability measure μ on X, we shall say that there is "concentration of measure" if, given a set of some fixed measure (say, 1/2), its t-extension for small values of $t > 0$ must have large measure. The formal definitions for the different types of concentration will be given in Section 3.1.7. We shall then also see the simple but incredibly useful fact that when there is concentration, Lipschitz functions are close to constant on a large measure.

We shall begin with three examples.

3.1.1. Three basic examples

EXAMPLES 3.1.2. (i) *The sphere* S^{n-1}. We write $|\cdot|$ for the Euclidean norm on $\mathbb{R}^n$. We consider the unit sphere $S^{n-1} = \{x \in \mathbb{R}^n : |x| = 1\}$ equipped with the geodesic metric ρ; if $x, y \in S^{n-1}$ then $\rho(x, y)$ is the convex angle $\widehat{xoy}$ in the plane determined by the origin o and x, y. The sphere S^{n-1} becomes a probability space with the unique rotationally invariant measure σ, which can be realized as follows: For any Borel set $A \subseteq S^{n-1}$ we set

$$\sigma(A) := \frac{\mathrm{Vol}_n(C(A))}{\mathrm{Vol}_n(B_2^n)},$$

where

$$C(A) := \{sx : x \in A \text{ and } 0 \leqslant s \leqslant 1\}.$$

One can check that σ coincides with the normalized Lebesgue measure on S^{n-1}. One may also check that ρ is indeed a metric, and $|x - y| = 2\sin\frac{\rho(x,y)}{2}$. Therefore,

$$\frac{2}{\pi}\rho(x, y) \leqslant |x - y| \leqslant \rho(x, y).$$

We shall see in Subsection 3.1c that under these definitions the sphere exhibits very strong concentration, namely for A with $\sigma(A) \geqslant 1/2$ one has

$$\sigma(A_t) \geqslant 1 - \sqrt{\pi/8}\,\exp(-t^2 n/2).$$

(ii) *Gauss space.* We consider the measure γ_n on $\mathbb{R}^n$ with density

$$g_n(x) = (2\pi)^{-n/2} e^{-|x|^2/2}.$$

In other words, if A is a Borel subset of $\mathbb{R}^n$ then

$$\gamma_n(A) = \frac{1}{(2\pi)^{n/2}} \int_A e^{-|x|^2/2} dx.$$

The metric probability space $(\mathbb{R}^n, |\cdot|, \gamma_n)$ is the n-dimensional Gauss space.

The standard Gaussian measure γ_n has three important properties: it is a product measure, more precisely $\gamma_n = \gamma_1 \otimes \cdots \otimes \gamma_1$, and it is invariant under orthogonal transformations: if $U \in O(n)$ and A is a Borel subset of $\mathbb{R}^n$ then $\gamma_n(U(A)) = \gamma_n(A)$. Moreover, orthogonally projecting the measure γ_n onto a subspace of dimension k (that is, taking a marginal) one gets exactly γ_k.

We shall see in Subsection 3.1.4 that for A with $\gamma_n(A) \geqslant 1/2$ one has

$$\gamma_n(A_t) \geqslant 1 - \frac{1}{2}\exp(-t^2/2).$$

(iii) *The discrete cube.* We consider the set $E_2^n = \{-1,1\}^n$, which we identify with the set of vertices of the unit cube $Q_n = [-1,1]^n$ in $\mathbb{R}^n$. We equip E_2^n with the uniform probability measure μ_n which gives mass 2^{-n} to each point, and with the Hamming metric

$$d_n(x,y) = \frac{1}{n}\mathrm{card}\{i \leqslant n : x_i \neq y_i\} = \frac{1}{2n}\sum_{i=1}^n |x_i - y_i|.$$

We shall see in Subsection 3.1.5 that for A with $\mu_n(A) \geqslant 1/2$ one has

$$\mu_n(A_t) \geqslant 1 - \frac{1}{2}\exp(-2t^2 n).$$

3.1.2. The isoperimetric problem

DEFINITION 3.1.3 (Minkowski content). Let (X,d) be a metric space and let μ be a (not necessarily finite) measure on the Borel σ-algebra $\mathcal{B}(X)$. The surface area (or Minkowski content) of a non-empty $A \in \mathcal{B}(X)$ is defined by

$$\mu^+(A) = \liminf_{t\to 0^+} \frac{\mu(A_t \setminus A)}{t},$$

where A_t is the t-extension of A. If $\mu(A) < \infty$ then

$$\mu^+(A) = \liminf_{t\to 0^+} \frac{\mu(A_t) - \mu(A)}{t}.$$

On a metric probability space one can formulate the isoperimetric problem:

(i) Given $0 < \alpha < 1$ and $t > 0$, find

$$\inf\{\mu(A_t) : A \in \mathcal{B}(X), \mu(A) \geqslant \alpha\}$$

and identify (if they exist) those sets A for which this infimum is attained.

(ii) Given $0 < \alpha < 1$ find

$$\inf\{\mu^+(A) : A \in \mathcal{B}(X), \mu(A) \geqslant \alpha\}$$

and identify (if they exist) those sets A for which this infimum is attained.

The answer to the first question may vary with t. Nevertheless, in most classical examples, the minimizers are independent from t and quite symmetric subsets of X. Therefore, we can easily compute the measure of their t-extension as well as their surface area. We emphasize, again, that for applications it is important mainly to know an estimate on concentration with the right asymptotic behaviour (in terms of t, and often also in terms of the parameter n which has a meaning of dimension). The extremal sets themselves are, for our applications, usually of lesser importance. We shall thus, in the three cases discussed above, omit long detailed proofs for extremal sets when simple direct proofs for isoperimetric inequalities are available.

3.1.3. The spherical isoperimetric inequality

The isoperimetric problem for S^{n-1} is formulated as follows: Let $\alpha \in (0,1)$ and $t > 0$. Among all Borel subsets A of the sphere which satisfy $\sigma(A) = \alpha$, determine the ones for which the surface area $\sigma(A_t)$ of their t-extension is minimal.

The answer to this question is given by the next theorem.

THEOREM 3.1.4 (Lévy, Schmidt). *Let $\alpha \in (0,1)$ and let $B(x,r)$ be a geodesic ball of radius $r > 0$ in S^{n-1} such that $\sigma(B(x,r)) = \alpha$. Then, for every $A \subseteq S^{n-1}$ with $\sigma(A) = \alpha$ and every $t > 0$, we have*

$$\sigma(A_t) \geqslant \sigma(B(x,r)_t) = \sigma(B(x,r+t)).$$

In other words, for any given value of α and any $t > 0$, the spherical caps of measure α provide the solution to the isoperimetric problem.

A proof of the spherical isoperimetric inequality can be given with spherical symmetrization and induction on the dimension. Let us consider the special case $\alpha = 1/2$. If $\sigma(A) = 1/2$ and $t > 0$, then we can estimate the size of A_t using Theorem 3.1.4: we have

$$\sigma(A_t) \geqslant \sigma\left(B\left(x, \tfrac{\pi}{2} + t\right)\right) \tag{3.1.1}$$

for every $t > 0$ and $x \in S^{n-1}$. Then, (3.1.1) leads to the following inequality.

THEOREM 3.1.5. *Let $A \subseteq S^{n+1}$ with $\sigma(A) = 1/2$ and let $0 < t < \pi/2$. Then,*

$$\sigma(A_t) \geqslant 1 - \sqrt{\pi/8}\, \exp(-t^2 n/2). \tag{3.1.2}$$

Proof. By (3.1.1), it is enough to give a lower bound for $\sigma\left(B\left(x, \frac{\pi}{2} + t\right)\right)$. We write

$$\sigma\left(B\left(x, \tfrac{\pi}{2} + t\right)\right) = \frac{\int_0^{\frac{\pi}{2}+t} \sin^n \theta \, d\theta}{\int_0^{\pi} \sin^n \theta \, d\theta},$$

and setting $h(t,n) = 1 - \sigma\left(B\left(x, \frac{\pi}{2} + t\right)\right)$, we need an upper bound for

$$h(t,n) = \frac{\int_{\frac{\pi}{2}+t}^{\pi} \sin^n \theta \, d\theta}{\int_0^{\pi} \sin^n \theta \, d\theta} = \frac{\int_t^{\frac{\pi}{2}} \cos^n \phi \, d\phi}{2I_n},$$

where $I_n = \int_0^{\pi/2} \cos^n \phi \, d\phi$. The change of variables $s = \phi\sqrt{n}$ gives

$$h(t,n) = \frac{1}{2\sqrt{n} I_n} \int_{t\sqrt{n}}^{\frac{\pi}{2}\sqrt{n}} \cos^n(s/\sqrt{n}) \, ds. \tag{3.1.3}$$

Differentiation of the function $s \mapsto e^{s^2/2} \cos s$ shows that it is decreasing on $[0, \pi/2]$, and hence

$$\cos s \leqslant \exp(-s^2/2)$$

for all $0 \leqslant s \leqslant \pi/2$. Then, (3.1.3) gives

$$\begin{aligned} h(t,n) &\leqslant \frac{1}{2\sqrt{n}I_n} \int_{t\sqrt{n}}^{\frac{\pi}{2}\sqrt{n}} \exp(-s^2/2)\, ds \\ &= \frac{1}{2\sqrt{n}I_n} \int_0^{(\frac{\pi}{2}-t)\sqrt{n}} \exp(-(s+t\sqrt{n})^2/2)\, ds \\ &\leqslant \frac{\exp(-t^2n/2)}{2\sqrt{n}I_n} \int_0^{\infty} \exp(-s^2/2)\, ds \\ &= \frac{\sqrt{\pi/8}}{\sqrt{n}I_n} \exp(-t^2n/2). \end{aligned}$$

The proof of the theorem will be complete if we check that $\sqrt{n}I_n \geqslant 1$ for every $n \geqslant 1$. To this end, we observe that from the recursive formula $(n+2)I_{n+2} = (n+1)I_n$ we have

$$\sqrt{n+2}I_{n+2} = \sqrt{n+2}\frac{n+1}{n+2}I_n = \frac{n+1}{\sqrt{n+2}}I_n \geqslant \sqrt{n}I_n,$$

and this means that we only have to check the cases $n=1$ and $n=2$: we have

$$I_1 = \int_0^{\pi/2} \cos\phi\, d\phi = 1 \geqslant 1$$

and

$$\sqrt{2}I_2 = \sqrt{2}\int_0^{\pi/2} \cos^2\phi d\phi = \frac{\sqrt{2}\pi}{4} \geqslant 1.$$

This completes the proof. □

REMARK 3.1.6. Looking closely at the proof one sees that the constant $\sqrt{\pi/8}$ can be improved when the dimension n is high, since asymptotically

$$\sqrt{n}I_n = \sqrt{n}\frac{\sqrt{\pi}}{2}\frac{\Gamma(\frac{n+1}{2})}{\Gamma(\frac{n}{2}+1)} \to_{n\to\infty} \sqrt{\pi/2}.$$

Thus for large n we can get as close as needed to a constant $1/2$ in front of the exponential term.

The proof of Theorem 3.1.5 is heavily based on the spherical isoperimetric inequality. However, in the applications, we don't really need the precise solution to the isoperimetric problem; we only need an inequality providing an estimate of the form (3.1.2). One can give a very simple proof of an analogous exponential estimate using the Brunn-Minkowski inequality. The key point is the following lemma of Arias de Reyna, Ball and Villa.

LEMMA 3.1.7. *Consider the probability measure* $\mu(A) = \mathrm{Vol}_n(A)/\mathrm{Vol}_n(B_2^n)$ *on the Euclidean unit ball* B_2^n. *If* A, B *are Borel subsets of* B_2^n *with* $\mu(A) \geqslant \alpha$ *and* $\mu(B) \geqslant \alpha$, *and if* $\rho(A,B) = \inf\{|a-b| : a \in A, b \in B\} = \rho > 0$, *then*

$$\alpha \leqslant \exp(-\rho^2 n/8).$$

In other words, if two disjoint Borel subsets of B_2^n have positive distance ρ, then at least one of them must have small volume (depending on ρ) when the dimension n is high.

Proof. We may assume that A and B are closed. By the Brunn-Minkowski inequality, $\mu(\frac{A+B}{2}) \geqslant \alpha$. On the other hand, the parallelogram law shows that if $a \in A$ and $b \in B$ then

$$|a+b|^2 = 2|a|^2 + 2|b|^2 - |a-b|^2 \leqslant 4 - \rho^2.$$

It follows that $\frac{A+B}{2} \subseteq \sqrt{1-\frac{\rho^2}{4}}B_2^n$, hence

$$\mu\left(\frac{A+B}{2}\right) \leqslant \left(1-\frac{\rho^2}{4}\right)^{n/2} \leqslant \exp(-\rho^2 n/8)$$

as claimed. □

Proof of Theorem 3.1.5 (with weaker constants). Assume that $A \subseteq S^{n-1}$ with $\sigma(A) = 1/2$. Let $t > 0$ and define $B = S^{n-1} \setminus A_t$. We fix $\lambda \in (0,1)$ and consider the subsets

$$\tilde{A} = \cup\{sA : \lambda \leqslant s \leqslant 1\} \quad \text{and} \quad \tilde{B} = \cup\{sB : \lambda \leqslant s \leqslant 1\}$$

of B_2^n. These are disjoint with distance $2\lambda\sin(t/2) \geqslant \frac{2}{\pi}\lambda t$. Lemma 3.1.7 shows that $\mu(\tilde{B}) \leqslant \exp(-c\lambda^2t^2n)$, and since $\mu(\tilde{B}) = (1-\lambda^n)\sigma(B)$ we obtain

$$\sigma(A_t) \geqslant 1 - \frac{1}{1-\lambda^n}\exp(-c\lambda^2t^2n).$$

We conclude the proof by choosing $\lambda = 1/2$. □

Remark 3.1.8. For future use we provide lower and upper bounds for the measure of spherical caps. There are two ways in which a cap may be defined. First, given $u \in S^{n-1}$ and $\varepsilon \in (0,1)$ we may set $C(u,\varepsilon) = \{\theta \in S^{n-1} : \langle u,\theta\rangle \geqslant \varepsilon\}$. Then, one can show that

$$\sigma(C(u,\varepsilon)) \leqslant \exp(-\varepsilon^2 n/2).$$

To see this, at least when ε is small enough, note that the measure $\sigma(C(u,\varepsilon))$ is equal to the percentage of B_2^n which is occupied by the *spherical cone* which corresponds to $C(u,\varepsilon)$ and that if $0 < \varepsilon < \frac{1}{\sqrt{2}}$ then this cone is contained in a Euclidean ball of radius $(1-\varepsilon^2)^{1/2}$; it follows that

$$\sigma(C(u,\varepsilon)) \leqslant (1-\varepsilon^2)^{n/2} \leqslant \exp(-\varepsilon^2 n/2).$$

Note also that $C(u,\varepsilon)$ can be written in the form $B(u,r) = \{\theta \in S^{n-1} : |\theta - u| \leqslant r\}$, where $r^2 = 2(1-\varepsilon)$. By the previous estimate we get

$$\sigma(B(u,r)) \leqslant \exp(-c(1-r^2/2)n)$$

for all $r \geqslant \sqrt{2-\sqrt{2}}$. For a lower bound, one can easily check that $\sigma(B(x,r)) \geqslant (r/3)^n$. For this claim, we start with an r-net $\mathcal{N}$ of S^{n-1} of cardinality $|\mathcal{N}| \leqslant \left(1+\frac{2}{r}\right)^n$. Then, $S^{n-1} \subseteq \bigcup_{x\in\mathcal{N}} B(x,r)$ and we must have $\sigma(B(x,r))|\mathcal{N}| \geqslant 1$. This implies that

$$\sigma(B(x,r)) \geqslant \left(\frac{r}{r+2}\right)^n \geqslant \left(\frac{r}{3}\right)^n.$$

3.1.4. Isoperimetric inequality in Gauss space

The isoperimetric inequality in "Gauss space" is the following statement.

THEOREM 3.1.9 (Borell, Sudakov-Tsirelson). *Let $\alpha \in (0,1)$ and $\theta \in S^{n-1}$, and let $H = \{x \in \mathbb{R}^n : \langle x, \theta\rangle \leqslant \lambda\}$ be a half-space in $\mathbb{R}^n$ with $\gamma_n(H) = \alpha$. Then, for every $t > 0$ and every Borel $A \subseteq \mathbb{R}^n$ with $\gamma_n(A) = \alpha$ we have*

$$\gamma_n(A_t) \geqslant \gamma_n(H_t).$$

In particular, among sets with fixed gaussian measure, γ^+ is minimized for half-spaces.

COROLLARY 3.1.10. *If $\gamma_n(A) \geqslant 1/2$ then, for every $t > 0$,*

$$1 - \gamma_n(A_t) \leqslant \frac{1}{2}\exp(-t^2/2). \tag{3.1.4}$$

Proof. From Theorem 3.1.9 we know that

$$1 - \gamma_n(A_t) \leqslant 1 - \gamma_n(H_t)$$

where H is a half-space of measure $1/2$. Since γ_n is invariant under orthogonal transformations, we may assume that $H = \{x \in \mathbb{R}^n : x_1 \leqslant 0\}$, and then, it follows that

$$1 - \gamma_n(H_t) = \frac{1}{\sqrt{2\pi}}\int_t^\infty e^{-s^2/2}ds.$$

One may check by direct computations and differentiation that the function

$$F(x) = e^{x^2/2}\int_x^\infty e^{-s^2/2}ds$$

is decreasing on $[0, +\infty)$. The fact that $F(t) \leqslant F(0)$ completes the proof. $\square$

As in the case of the sphere, the proof given above of the isomorphic isoperimetric inequality (3.1.4) requires knowing the exact solution of the Gaussian isoperimetric problem. We shall not give a proof of Theorem 3.1.9, although it is not difficult and can be done using symmetrization, because there are many simple direct proofs for gaussian concentration. For example, Maurey observed that the Prékopa-Leindler inequality may be used to give a simple proof of the isomorphic isoperimetric inequality in Gauss space.

THEOREM 3.1.11. *Let A be a non-empty Borel subset of $\mathbb{R}^n$. Then,*

$$\int_{\mathbb{R}^n} e^{d(x,A)^2/4}d\gamma_n(x) \leqslant \frac{1}{\gamma_n(A)},$$

where $d(x,A) = \inf\{|x-y| : y \in A\}$. Therefore, if $\gamma_n(A) = \frac{1}{2}$ we have

$$1 - \gamma_n(A_t) \leqslant 2\exp(-t^2/4)$$

for every $t > 0$, where $A_t = \{y : d(y,A) < t\}$.

Proof. Consider the functions

$$f(x) = e^{d(x,A)^2/4}g_n(x)\,, \quad g(x) = \mathbf{1}_A(x)g_n(x)\,, \quad m(x) = g_n(x),$$

where g_n is the density of γ_n. For every $x \in \mathbb{R}^n$ and $y \in A$ we see that

$$\begin{aligned}(2\pi)^n f(x)g(y) &= e^{d(x,A)^2/4}e^{-|x|^2/2}e^{-|y|^2/2} \leqslant \exp\left(\frac{|x-y|^2}{4} - \frac{|x|^2}{2} - \frac{|y|^2}{2}\right) \\ &= \exp\left(-\frac{|x+y|^2}{4}\right) = \left(\exp\left(-\frac{1}{2}\left|\frac{x+y}{2}\right|^2\right)\right)^2 \\ &= (2\pi)^n \left(m\left(\frac{x+y}{2}\right)\right)^2,\end{aligned}$$

by the parallelogram law and the fact that $d(x,A) \leqslant |x-y|$. Since $g(y) = 0$ whenever $y \notin A$, this implies that f, g and m satisfy the assumptions of the Prékopa-Leindler inequality with $\lambda = 1/2$. Therefore,

$$\left(\int_{\mathbb{R}^n} e^{d(x,A)^2/4}\, d\gamma_n(x)\right)\gamma_n(A) = \left(\int_{\mathbb{R}^n} f\right)\left(\int_{\mathbb{R}^n} g\right) \leqslant \left(\int_{\mathbb{R}^n} m\right)^2 = 1.$$

This proves the first assertion of the theorem. For the second one, observe that if $\gamma_n(A) = \frac{1}{2}$ then

$$e^{t^2/4}\gamma_n(\{x : d(x,A) \geqslant t\}) \leqslant \int_{\mathbb{R}^n} e^{d(x,A)^2/4}\, d\gamma_n(x) \leqslant \frac{1}{\gamma_n(A)} = 2.$$

This shows that $\gamma_n(A_t^c) \leqslant 2\exp(-t^2/4)$. □

3.1.5. Isoperimetric inequality in the discrete cube

The discrete cube $E_2^n = \{-1,1\}^n$, the set of vertices of the cube $Q_n = [-1,1]^n$ in $\mathbb{R}^n$, becomes a metric probability space with the uniform measure μ_n and the Hamming metric

$$d_n(x,y) = \frac{1}{n}\mathrm{card}\{i \leqslant n : x_i \neq y_i\} = \frac{1}{2n}\sum_{i=1}^n |x_i - y_i|.$$

The metric d_n takes finitely many values: $0, 1/n, 2/n, \ldots, 1$. These are the values of $t > 0$ for which the t-extension $A_t = \{y \in E_2^n : d_n(y,A) \leqslant t\}$ of A is of some interest, because A_t does not change as t varies in an interval of the form $[k/n, (k+1)/n)$. Note the slight difference in the definition of the t-extension that we adopt here; this is done for notational convenience.

The isoperimetric problem for the discrete cube can be formulated as follows. We are given a natural number $N = 1, 2, \ldots, 2^n$ and some $t = k/n$, $k = 1, \ldots, n$ and we want to find those $A \subseteq E_2^n$ with cardinality N for which the k/n-extension of A has the smallest cardinality. Harper proved that the solution to the problem is given by the d_n-balls (the so-called Hamming balls of E_2^n) in the case where N is the cardinality of some d_n-ball. The minimizer for the isoperimetric problem is known in the general case too. The isoperimetric inequality for E_2^n is the following statement.

Theorem 3.1.12 (Harper). *Let $A \subseteq E_2^n$ with $N = \sum_{k=0}^m \binom{n}{k}$ elements. Then, for every $s = 1, \ldots, n-m$ we have*

$$\mu_n(A_{s/n}) \geqslant \frac{1}{2^n}\sum_{k=0}^{m+s}\binom{n}{k} = \mu_n(B(x, m/n)_{s/n}) = \mu_n(B(x,(m+s)/n)) \tag{3.1.5}$$

where x is an arbitrary element of E_2^n. □

REMARK 3.1.13. For N which is not of this form, the extremal sets are, up to symmetries, simply the first N vectors in lexicographical order.

Theorem 3.1.12 leads to an isomorphic isoperimetric inequality for E_2^n.

COROLLARY 3.1.14. *If* $\mu_n(A) \geqslant 1/2$ *and* $t > 0$, *then*

$$\mu_n(A_t^c) \leqslant \frac{1}{2}\exp(-2t^2 n). \tag{3.1.6}$$

To prove the corollary observe that in order to estimate $\mu_n(A_t)$ it is enough to choose $m = n/2$ and $s = tn$ in (3.1.5), assuming without loss of generality that both m and s are integers. Then, we see that

$$\mu_n(A_t^c) \leqslant \frac{1}{2^n} \sum_{j=(\frac{1}{2}+t)n}^{n} \binom{n}{j},$$

which decreases exponentially fast to 0 as $n \to \infty$. Indeed, when n is large, this sum is approximately

$$\frac{1}{\sqrt{2\pi}} \int_{2t\sqrt{n}}^{\infty} \exp(-x^2/2)\, dx \leqslant \frac{1}{2}\exp(-2t^2 n).$$

We will not prove Theorem 3.1.12. The proof is combinatorial and can be done by induction on n. We will present a direct proof of an isomorphic isoperimetric inequality of the same form as (3.1.6). The proof is based on the following theorem of Talagrand which we study in the next subsection. The theorem actually applies to the much more general setting of product measures, and has many applications in combinatorial approximation.

THEOREM 3.1.15 (Talagrand). *Let* A *be a non-empty subset of* $E_2^n \subset \mathbb{R}^n$. *We consider its convex hull* $\mathrm{conv}(A)$ *in* $\mathbb{R}^n$ *and for every* $x \in E_2^n$ *we define*

$$\phi_A(x) = \min\{|x-y| : y \in \mathrm{conv}(A)\}.$$

Then,

$$\int_{E_2^n} \exp(\phi_A^2(x)/8)\, d\mu_n(x) \leqslant \frac{1}{\mu_n(A)}.$$

Now, let A be a non-empty subset of E_2^n. The function ϕ_A of Theorem 3.1.15 and the function

$$d_n(x, A) = \min\left\{\frac{1}{2n}\sum_{i=1}^{n} |x_i - y_i| : y \in A\right\}$$

which measures the distance from x to A are related as follows.

LEMMA 3.1.16. *For every non-empty* $A \subseteq E_2^n$ *and every* $x \in E_2^n$,

$$2\sqrt{n}\, d_n(x, A) \leqslant \phi_A(x).$$

Proof. Let $x \in E_2^n$. For every $y \in A$ we have

$$\langle x - y, x\rangle = \sum_{i=1}^{n} x_i(x_i - y_i) = 2n d_n(x, y) \geqslant 2n d_n(x, A). \tag{3.1.7}$$

Clearly this inequality applies also to all $y \in \mathrm{conv}(A)$. We thus see from (3.1.7) that, for every $y \in \mathrm{conv}(A)$,

$$\sqrt{n}|x - y| \geqslant \langle x - y, x\rangle \geqslant 2n d_n(x, A).$$

This proves the lemma. □

Combining the above with Theorem 3.1.15 we get the isomorphic isoperimetric inequality for E_2^n.

THEOREM 3.1.17. *Let $A \subseteq E_2^n$ with $\mu_n(A) \geqslant 1/2$. Then, for every $t > 0$ we have*
$$\mu_n(A_t) \geqslant 1 - 2\exp(-t^2 n/2).$$

Proof. If $x \notin A_t$, then $d_n(x, A) \geqslant t$ and Lemma 3.1.16 shows that $\phi_A(x) \geqslant 2t\sqrt{n}$. But, from Theorem 3.1.15 we have
$$e^{t^2 n/2} \mu_n(\{x : \phi_A(x) \geqslant 2t\sqrt{n}\}) \leqslant \int_{E_2^n} \exp(\phi_A^2(x)/8)\, d\mu_n(x) \leqslant \frac{1}{\mu_n(A)} = 2,$$
and this gives
$$\mu_n(A_t^c) \leqslant \mu_n(\{x : \phi_A(x) \geqslant 2t\sqrt{n}\}) \leqslant 2\exp(-t^2 n/2).$$
$\square$

3.1.6. Talagrand's inequality for the discrete cube.

Recall that $E_2^n = \{-1, 1\}^n$, and that $\phi_A(x) = \min\{|x - y| : y \in \mathrm{conv}(A)\}$. We have denoted by μ_n the uniform probability measure on E_2^n, and shall use $\mathbb{E}(f)$ to denote $\int_{E_2^n} f\, d\mu_n$. Note that this is simply the expectation with respect to a finite sequence $\{\epsilon_i\}_{i=1}^n$ of Bernoulli random variables.

In this subsection we prove Theorem 3.1.15: for every $A \subseteq E_2^n$,
$$\mathbb{E}\big(\exp(\phi_A^2/8)\big) \leqslant \frac{1}{\mu_n(A)}.$$

The proof goes by induction on the cardinality of A. If $A = \{y\}$ is a singleton then $\phi_A(x) = |x - y|$ for all $x \in E_2^n$. Thus,
$$\mathbb{E}\big(\exp(\phi_A^2/8)\big) = \frac{1}{2^n} \sum_{x \in E_2^n} e^{|x-y|^2/8}.$$

The set of those $x \in E_2^n$ which differ from y in exactly i coordinates has cardinality $\binom{n}{i}$, and $|x - y|^2 = 4i$ in this case. It follows that
$$\mathbb{E}\big(e^{|x-y|^2/8}\big) = \frac{1}{2^n} \sum_{i=0}^n \binom{n}{i} e^{i/2} = \frac{1}{2^n}\big(1 + e^{1/2}\big)^n$$
$$= \left(\frac{1 + e^{1/2}}{2}\right)^n \leqslant 2^n = \frac{1}{\mu_n(A)},$$
because $e^{1/2} \leqslant e \leqslant 3$.

Assume that $\mathrm{card}(A) \geqslant 2$. The case $n = 1$ is trivial: we have $A = E_2^1$, and hence $\phi_A(x) = 0$ for every $x \in E_2^1$. Then, $\mathbb{E}\big(\exp(\phi_A^2/8)\big) = 1 = 1/\mu_1(A)$.

For the inductive step we consider $A \subseteq E_2^{n+1}$ with $\mathrm{card}(A) \geqslant 2$. Without loss of generality we may assume that
$$A = (A_1 \times \{1\}) \cup (A_{-1} \times \{-1\})$$
where $A_1, A_{-1} \neq \emptyset$ and $\mathrm{card}(A_{-1}) \leqslant \mathrm{card}(A_1)$.

LEMMA 3.1.18. *For every $x \in E_2^n$,*
$$\phi_A((x, 1)) \leqslant \phi_{A_1}(x).$$

Proof. This is immediate once we observe that $\mathrm{conv}\,(A_1) \times \{1\} \subset \mathrm{conv}\,(A)$. □

LEMMA 3.1.19. *For every $x \in E_2^n$ and every $0 \leqslant a \leqslant 1$,*

$$\phi_A^2((x,-1)) \leqslant 4a^2 + a\phi_{A_1}^2(x) + (1-a)\phi_{A_{-1}}^2(x).$$

Proof. Let $z_i \in \mathrm{conv}\,(A_i)$ $(i = 1, -1)$ such that $\phi_{A_i}(x) = |x - z_i|$. Since $(z_i, i) \in \mathrm{conv}(A)$ and $\mathrm{conv}(A)$ is convex, we get

$$z := a(z_1, 1) + (1-a)(z_{-1}, -1) = (az_1 + (1-a)z_{-1}, 2a-1) \in \mathrm{conv}(A).$$

We have

$$\begin{aligned}|(x,-1) - z|^2 &= |(x - az_1 - (1-a)z_{-1}, -2a)|^2 \\ &\leqslant (a|x - z_1| + (1-a)|x - z_{-1}|)^2 + 4a^2 \\ &\leqslant a|x - z_1|^2 + (1-a)|x - z_{-1}|^2 + 4a^2 \\ &= a\phi_{A_1}^2(x) + (1-a)\phi_{A_{-1}}^2(x) + 4a^2.\end{aligned}$$

Thus we get the lemma. □

Using the lemmas, we write

$$\begin{aligned}\mathbb{E}\big(e^{\phi_A^2/8}\big) &= \frac{1}{2^{n+1}} \sum_{x \in E_2^{n+1}} e^{\phi_A^2(x)/8} \\ &= \frac{1}{2^{n+1}} \sum_{x \in E_2^n} e^{\phi_A^2((x,1))/8} + \frac{1}{2^{n+1}} \sum_{x \in E_2^n} e^{\phi_A^2((x,-1))/8} \\ &\leqslant \frac{1}{2^{n+1}} \sum_{x \in E_2^n} e^{\phi_{A_1}^2(x)/8} + \frac{1}{2^{n+1}} e^{a^2/2} \sum_{x \in E_2^n} e^{a\phi_{A_1}^2(x)/8 + (1-a)\phi_{A_{-1}}^2(x)/8} \\ &\leqslant \frac{1}{2^{n+1}} \sum_{x \in E_2^n} e^{\phi_{A_1}^2(x)/8} \\ &\quad + \frac{1}{2^{n+1}} e^{a^2/2} \Big(\sum_{x \in E_2^n} e^{\phi_{A_1}^2(x)/8} \Big)^a \Big(\sum_{x \in E_2^n} e^{\phi_{A_{-1}}^2(x)/8} \Big)^{1-a} \\ &= \frac{1}{2}\mathbb{E}(e^{\phi_{A_1}^2/8}) + \frac{1}{2} e^{a^2/2} \Big(\mathbb{E}\big(e^{\phi_{A_1}^2/8}\big)\Big)^a \Big(\mathbb{E}\big(e^{\phi_{A_{-1}}^2/8}\big)\Big)^{1-a}.\end{aligned}$$

We set

$$u_1 = \mathbb{E}\big(e^{\phi_{A_1}^2/8}\big), v_1 = \frac{1}{\mu_n(A_1)} \quad \text{and} \quad u_{-1} = \mathbb{E}\big(e^{\phi_{A_{-1}}^2/8}\big), v_{-1} = \frac{1}{\mu_n(A_{-1})}.$$

By the inductive hypothesis we have $u_1 \leqslant v_1$ and $u_{-1} \leqslant v_{-1}$ (also, the assumption $\mathrm{card}(A_{-1}) \leqslant \mathrm{card}(A_1)$ is equivalent to $v_1 \leqslant v_{-1}$). Then, the previous inequality takes the form

$$\begin{aligned}\mathbb{E}\left(e^{\phi_A^2/8}\right) &\leqslant \frac{1}{2}u_1 + \frac{1}{2}e^{a^2/2}(u_1)^a(u_{-1})^{1-a} \leqslant \frac{1}{2}v_1 + \frac{1}{2}e^{a^2/2}(v_1)^a(v_{-1})^{1-a} \\ &\leqslant \frac{v_1}{2}[1 + e^{a^2/2}(v_1/v_{-1})^{a-1}].\end{aligned}$$

The minimum of this last quantity is attained when $a = -\ln(v_1/v_{-1})$. The value $-\ln(v_1/v_{-1})$ is roughly equal to $1 - v_1/v_{-1}$. We choose $a_0 = 1 - v_1/v_{-1}$. Since $v_1 \leqslant v_{-1}$, we have $0 \leqslant a_0 \leqslant 1$ and then we can write

$$\mathbb{E}(e^{\phi_A^2/8}) \leqslant \frac{v_1}{2}[1 + e^{a_0^2/2}(1 - a_0)^{a_0 - 1}].$$

LEMMA 3.1.20. *For every $0 \leqslant a \leqslant 1$ we have*

$$1 + e^{a^2/2}(1-a)^{a-1} \leqslant \frac{4}{2-a}.$$

Proof. Simple calculations show that the inequality is equivalent to

$$g(a) = \ln(2+a) - \ln(2-a) - a^2/2 - (a-1)\ln(1-a) \geqslant 0.$$

We differentiate and check that $g'' \geqslant 0$ and $g'(0) = 0$. So, g is increasing on $[0,1]$. Since $g(0) = 0$, the claim follows. □

Using Lemma 3.1.20 we get

$$\mathbb{E}(e^{\phi_A^2/8}) \leqslant \frac{v_1}{2}\frac{4}{2-a_0} = \frac{2v_1}{1+v_1/v_{-1}} = \frac{2}{1/v_1 + 1/v_{-1}} = \frac{2}{\mu_n(A_1) + \mu_n(A_{-1})}$$
$$= \frac{1}{\mu_{n+1}(A)}.$$

This completes the inductive step and the proof of Theorem 3.1.15. □

Theorem 3.1.15 and a direct application of Markov's inequality give:

COROLLARY 3.1.21. *For every $t > 0$ we have*

$$\mu_n(\phi_A \geqslant t) \leqslant \frac{1}{\mu_n(A)} e^{-t^2/8}.$$

REMARK 3.1.22. In Lemma 3.1.16 we checked that, for every non-empty $A \subseteq E_2^n$ and for every $x \in E_2^n$, one has $2\sqrt{n}d_n(x,A) \leqslant \phi_A(x)$. So,

$$\{d_n(x,A) \geqslant t\} \subseteq \{\phi_A \geqslant 2t\sqrt{n}\}.$$

The isoperimetric inequality for E_2^n gives the estimate

$$\mu_n(\{d_n(x,A) \geqslant t\} \leqslant \frac{1}{2} e^{-2t^2 n}$$

if $\mu_n(A) = 1/2$. From Corollary 3.1.21 we obtain the estimate

$$\mu_n(\phi_A \geqslant 2t\sqrt{n}) \leqslant 2e^{-t^2 n/2}$$

for the measure of $\{\phi_A \geqslant 2t\sqrt{n}\}$, which is a larger set.

REMARK 3.1.23. Talagrand's inductive argument applies in a much more general setting than that of the discrete cube. Let $\Omega = \prod_{i=1}^n \Omega_i$, where each Ω_i is a probability space, and endow Ω with the product measure μ_n. Given a subset $A \subset \Omega$ and an $x \in \Omega$ we let $U(x,A) \subset \{-1,1\}^n$ be defined so that if $s \in U(x,A)$ then there is some $y \in A$ which differs from x only in the coordinates for which $s_i = 1$. We then let $\phi_A(x) = \min\{|s| : s \in U(x,A)\}$. As in the statement of Theorem 3.1.15, in such a situation we have that

$$\int_\Omega \exp(\phi_A^2/8) \leqslant \frac{1}{\mu_n(A)}.$$

This inequality has many applications in combinatorics.

3.1.7. Concentration function

A convenient way to keep track of the different concentration behaviour in different spaces is to introduce the concentration function and the concept of a Lévy family of spaces.

DEFINITION 3.1.24 (concentration function). *Let (X, d, μ) be a metric probability space. The concentration function of (X, d, μ) is defined on $(0, \infty)$ by*

$$\alpha_\mu(t) := \sup\{1 - \mu(A_t) : \mu(A) \geqslant 1/2\}.$$

The function α_μ is obviously decreasing and, as the next proposition shows, $\lim_{t\to\infty} \alpha_\mu(t) = 0$.

PROPOSITION 3.1.25. *On every metric probability space (X, d, μ) one has*

$$\lim_{t\to\infty} \alpha_\mu(t) = 0.$$

Proof. We fix $x \in X$ and $0 < \varepsilon \leqslant 1/2$. From the fact that $X = \bigcup_{n=1}^{\infty} B(x, n)$ it follows that there exists $r \in \mathbb{N}$ such that

$$\mu(B(x, r)) > 1 - \varepsilon.$$

Then, for every $A \in \mathcal{B}(X)$ with $\mu(A) \geqslant 1/2$ we have $A \cap B(x, r) \neq \emptyset$, which implies that $B(x, r) \subseteq A_{2r}$. Then, for every $t \geqslant 2r$ we get

$$1 - \mu(A_t) \leqslant 1 - \mu(A_{2r}) \leqslant 1 - \mu(B(x, r)) < \varepsilon,$$

and this shows that $\alpha_\mu(t) \leqslant \varepsilon$. □

DEFINITION 3.1.26 (concentration of measure). We say that we have measure concentration on the metric probability space (X, d, μ) if $\alpha_\mu(t)$ decreases fast as $t \to \infty$. More precisely:

(i) We say that μ has *normal concentration* on (X, d) if there exist constants $C, c > 0$ such that, for every $t > 0$,

$$\alpha_\mu(t) \leqslant Ce^{-ct^2}.$$

(ii) We say that μ has *exponential concentration* on (X, d) if there exist constants $C, c > 0$ such that, for every $t > 0$,

$$\alpha_\mu(t) \leqslant Ce^{-ct}.$$

We shall say that a family (X_n, d_n, μ_n) is a *Lévy family* if for any $t > 0$

$$\alpha_{\mu_n}(t \cdot \mathrm{diam}(X_n)) \to_{n\to\infty} 0$$

and we shall say that it is a *normal Lévy family* with constants c, C if for any $t > 0$

$$\alpha_{\mu_n}(t) \leqslant Ce^{-ct^2 n}.$$

The results of the previous section (more precisely, the estimates that follow from the solutions of the corresponding isoperimetric problems) show that this is the case for the sphere, the discrete cube and the Gauss space. According to the definition 3.1.24 of the concentration function, in these spaces we have:

(i) For the sphere (S^{n-1}, ρ, σ) one has

$$\alpha_\sigma(t) \leqslant \sqrt{\pi/8}\, \exp(-t^2 n/2).$$

(ii) For the Gauss space $(\mathbb{R}^n, |\cdot|, \gamma_n)$ one has

$$\alpha_{\gamma_n}(t) \leqslant \frac{1}{2}\exp(-t^2/2).$$

(iii) For the discrete cube (E_2^n, d_n, μ_n) one has

$$\alpha_{\mu_n}(t) \leqslant \frac{1}{2}\exp(-2t^2 n).$$

It turns out that there are many more normal Lévy families than perhaps one would originally expect. We introduce and discuss a number of important such examples in Section 3.3.

REMARK 3.1.27. Let φ be a 1-Lipschitz function between two metric spaces (X_1, d_1) and (X_2, d_2). If μ_1 is a probability measure on (X_1, d_1) then we define $\mu_2 = \varphi(\mu_1)$, that is, $\mu_2(A) = \mu_1(\varphi^{-1}(A))$. Note that $A_t \supseteq \varphi([\varphi^{-1}(A)]_t)$ for all $A \in \mathcal{B}(X_2)$. It follows that the concentration functions satisfy

$$\alpha_{\mu_2} \leqslant \alpha_{\mu_1}.$$

In particular, this applies to projections.

Many of the applications of the concentration of measure are based on the next theorem.

THEOREM 3.1.28. *Let (X, d, μ) be a metric probability space. If $f : X \to \mathbb{R}$ is a 1-Lipschitz function, i.e. $|f(x) - f(y)| \leqslant d(x, y)$ for all $x, y \in X$, then*

$$\mu\left(\{x \in X : |f(x) - \mathrm{med}(f)| > t\}\right) \leqslant 2\alpha_\mu(t),$$

where $\mathrm{med}(f)$ *is a median, also called a Lévy mean, of f.*

Note. A median, also called Lévy mean, $\mathrm{med}(f)$ of f is a number such that

$$\mu(\{f \geqslant \mathrm{med}(f)\}) \geqslant 1/2 \text{ and } \mu(\{f \leqslant \mathrm{med}(f)\}) \geqslant 1/2.$$

Proof of Theorem 3.1.28. We set $A = \{x : f(x) \geqslant \mathrm{med}(f)\}$ and $B = \{x : f(x) \leqslant \mathrm{med}(f)\}$. For every $y \in A_t$ there exists $x \in A$ with $d(x, y) \leqslant t$, and then

$$f(y) = f(y) - f(x) + f(x) \geqslant -d(y, x) + \mathrm{med}(f) \geqslant \mathrm{med}(f) - t$$

because f is 1-Lipschitz. Similarly, if $y \in B_t$ then there exists $x \in B$ with $d(x, y) \leqslant t$, and then

$$f(y) = f(y) - f(x) + f(x) \leqslant d(y, x) + \mathrm{med}(f) \leqslant \mathrm{med}(f) + t.$$

It follows that if $y \in A_t \cap B_t$ then $|f(x) - \mathrm{med}(f)| \leqslant t$. In other words,

$$\{x \in X : |f(x) - \mathrm{med}(f)| > t\} \subseteq (A_t \cap B_t)^c = A_t^c \cup B_t^c.$$

From the definition of the concentration function we have $\mu(A_t) \geqslant 1 - \alpha_\mu(t)$ and $\mu(B_t) \geqslant 1 - \alpha_\mu(t)$. It follows that

$$\mu\left(\{|f - \mathrm{med}(f)| > t\}\right) \leqslant (1 - \mu(A_t)) + (1 - \mu(B_t)) \leqslant 2\alpha_\mu(t)$$

as claimed. □

COROLLARY 3.1.29. *Let (X, d, μ) be a metric probability space. If $f : X \to \mathbb{R}$ is a Lipschitz function with constant $\|f\|_{\mathrm{Lip}} = \sigma$, i.e. if $|f(x) - f(y)| \leqslant \sigma d(x, y)$ for all $x, y \in X$, then*

$$\mu\left(\{x \in X : |f(x) - \mathrm{med}(f)| > t\}\right) \leqslant 2\alpha_\mu(t/\sigma),$$

where $\mathrm{med}(f)$ *is a Lévy mean of f.*

When the concentration function of the space decreases fast, Theorem 3.1.28 shows that 1-Lipschitz functions are "almost constant" and "equal" to their Lévy mean on "almost all of the space". This is the source for the name "concentration phenomenon". A converse statement is also valid.

THEOREM 3.1.30. *Let (X, d, μ) be a metric probability space. Assume that for some $\eta > 0$ and some $t > 0$ one has*

$$\mu\left(\{x \in X : |f(x) - \mathrm{med}(f)| > t\}\right) \leqslant \eta$$

for every 1-Lipschitz function $f : X \to \mathbb{R}$. Then, $\alpha_\mu(t) \leqslant \eta$.

Proof. Let A be a Borel subset of X with $\mu(A) \geqslant 1/2$. We consider the function $f(x) = d(x, A)$. Then, f is 1-Lipschitz and $\mathrm{med}(f) = 0$ because f is non-negative and $\mu(\{x : f(x) = 0\}) \geqslant 1/2$. From the assumption we get

$$\mu(\{x \in X : d(x, A) > t\}) \leqslant \eta,$$

that is $1 - \mu(A_t) \leqslant \eta$. It follows that $\alpha_\mu(t) \leqslant \eta$. □

A useful variant of Theorem 3.1.30 holds true if we replace the Lévy mean by the expectation. This formulation has some advantages since the expectation of a function is often easier to determine.

PROPOSITION 3.1.31. *Let (X, d, μ) be a metric probability space and let $\alpha : \mathbb{R}^+ \to \mathbb{R}^+$ be a function such that, for every bounded 1-Lipschitz function $f : (X, d) \to \mathbb{R}$ and for every $t > 0$,*

$$\mu\Big(\Big\{f \geqslant \int f d\mu + t\Big\}\Big) \leqslant \alpha(t).$$

Then, for every Borel set $A \subseteq X$ with $\mu(A) > 0$ and for every $t > 0$,

$$1 - \mu(A_t) \leqslant \alpha(\mu(A)t).$$

In particular,

$$\alpha_\mu(t) \leqslant \alpha(t/2) \quad \textit{for all}\, t > 0.$$

Proof. We fix $t > 0$ and $A \in \mathcal{B}(X)$ with $\mu(A) > 0$. We consider the function $f_t(x) = \min\{d(x, A), t\}$. Note that $\|f_t\|_{\mathrm{Lip}} \leqslant 1$ and

$$\int f_t\, d\mu \leqslant (1 - \mu(A))t. \tag{3.1.8}$$

From the assumption we have

$$\begin{aligned} 1 - \mu(A_t) = \mu(\{f_t = t\}) &\leqslant \mu\Big(\Big\{f_t \geqslant \int f_t\, d\mu + \mu(A)t\Big\}\Big) \\ &\leqslant \alpha(\mu(A)t). \end{aligned}$$

For the last claim of the proposition we observe that if $\mu(A) \geqslant 1/2$ then from (3.1.8) we have $\int f_t\, d\mu \leqslant t/2$ and, repeating the previous argument, we see that $\alpha_\mu(t) \leqslant \alpha(t/2)$. □

3.2. Deviation inequalities for Lipschitz functions on classical metric probability spaces

A direct consequence of Corollary 3.1.29 and of our estimate for the concentration function of the sphere, Theorem 3.1.5, is the next deviation inequality for Lipschitz functions $f : S^{n-1} \to \mathbb{R}$.

THEOREM 3.2.1. *Let $f : S^{n-1} \to \mathbb{R}$ be a Lipschitz continuous function with constant b. Then, for every $t > 0$,*

$$\sigma\left(\{x \in S^{n-1} : |f(x) - L_f| \geqslant bt\}\right) \leqslant 2\exp(-c_1 t^2 n).$$

where L_f is the Lévy mean of f and $c_1 > 0$ is an absolute constant.

A very useful observation is that one can replace the Lévy mean of f by the expectation of f in Theorem 3.2.1.

THEOREM 3.2.2. *Let $f : S^{n-1} \to \mathbb{R}$ be a Lipschitz continuous function with constant b. Then, for every $t > 0$,*

$$\sigma\left(\{x \in S^{n-1} : |f(x) - \mathbb{E}(f)| \geqslant bt\}\right) \leqslant 4\exp(-c_1 t^2 n),$$

where $\mathbb{E}(f)$ is the expectation of f and $c_1 > 0$ is an absolute constant.

One can give a direct proof of Theorem 3.2.2 starting from the corresponding result in Gauss space that will be described next. However, there is also a general argument which relates deviation from the expectation to deviation from the Lévy mean, and allows us to deduce Theorem 3.2.2 from Theorem 3.2.1.

Proof of Theorem 3.2.2. We may clearly assume that $b = 1$. Let $\tilde{f}$ be an independent copy of f on S^{n-1}. Note that

$$\begin{aligned}(\sigma\otimes\sigma)(\{(x,y) : |f(x) - \tilde{f}(y)| \geqslant t\}) &\leqslant \sigma(\{x : |f(x) - L_f| \geqslant t/2\})\\ &\quad + \sigma(\{y : |\tilde{f}(y) - L_f| \geqslant t/2\})\\ &\leqslant 4\exp(-c_1 t^2 n/4)\end{aligned}$$

for every $t > 0$. Then, we may write

$$\begin{aligned}\mathbb{E}_{\sigma\otimes\sigma}\left(\exp(\lambda^2|f - \tilde{f}|^2)\right) &= 2\int_0^\infty \lambda^2 t e^{\lambda^2 t^2}(\sigma\otimes\sigma)(\{(x,y) : |f(x) - \tilde{f}(y)| \geqslant t\})\,dt\\ &\leqslant 8\lambda^2\int_0^\infty t e^{\lambda^2 t^2 - c_2 t^2 n}dt\end{aligned}$$

for any $\lambda > 0$, where $c_2 = c_1/4$. Choosing $\lambda = \sqrt{c_2 n/2}$ we get

$$\mathbb{E}_{\sigma\otimes\sigma}\left(\exp(c_3 n|f - \tilde{f}|^2)\right) \leqslant 4,$$

where $c_3 = c_2/2$. Since $t \mapsto \exp(c_3 n t^2)$ is a convex function, Jensen's inequality implies that

$$\mathbb{E}_\sigma\left(\exp(c_3 n|f - \mathbb{E}(f)|^2)\right) \leqslant 4.$$

Then, Markov's inequality gives

$$\begin{aligned}\sigma(\{x : |f(x) - \mathbb{E}(f)| \geqslant t\}) &\leqslant e^{-c_3 t^2 n}\mathbb{E}_\sigma\left(\exp(c_3 n|f - \mathbb{E}(f)|^2)\right)\\ &\leqslant 4e^{-c_3 t^2 n}\end{aligned}$$

for every $t > 0$. □

A simple proof of the corresponding result in Gauss space may be given by introducing infimum convolution and property (τ) that will be discussed in more detail in Part II.

DEFINITION 3.2.3 (infimum convolution). Let f and g be (Borel) measurable functions on $\mathbb{R}^n$. We denote by $f\square g$ the *infimum convolution* of f and g, defined by

$$(f\square g)(x) = \inf\{f(x-y)+g(y) : \ y\in\mathbb{R}^n\}.$$

If μ is a probability measure on $\mathbb{R}^n$ and φ is a non-negative measurable function on $\mathbb{R}^n$, we say that the pair (μ,φ) has *property* (τ) if for every bounded measurable function f on $\mathbb{R}^n$ we have

$$\left(\int_{\mathbb{R}^n} e^{f\square\varphi}\,d\mu\right)\left(\int_{\mathbb{R}^n} e^{-f}\,d\mu\right)\leqslant 1.$$

THEOREM 3.2.4 (Maurey). *The pair $(\gamma_n, |x|^2/4)$ has property (τ).*

Proof. A simple proof can be given using the Prékopa-Leindler inequality. Let f be a bounded measurable function on $\mathbb{R}^n$. We define $\varphi(y)=|y|^2/4$ and $\psi=f\square\varphi$. If

$$m(x)=f(x)+\frac{|x|^2}{2}, g(y)=-\psi(y)+\frac{|y|^2}{2} \text{ and } h(z)=\frac{|z|^2}{2},$$

then we easily check that

$$h\left(\frac{x+y}{2}\right)\leqslant\frac{m(x)+g(y)}{2}.$$

So,

$$\left(\int_{\mathbb{R}^n} e^{-m(x)}dx\right)\left(\int_{\mathbb{R}^n} e^{-g(x)}dx\right)\leqslant\left(\int_{\mathbb{R}^n} e^{-h(x)}dx\right)^2.$$

In other words,

$$\left(\int_{\mathbb{R}^n} e^{-f}\,d\gamma_n\right)\left(\int_{\mathbb{R}^n} e^{f\square\varphi}\,d\gamma_n\right)\leqslant 1$$

which proves the theorem. $\square$

As an application, we give a proof of a result known to us from Maurey and Pisier, see the notes and remarks section.

THEOREM 3.2.5. *Let $f:\mathbb{R}^n\to\mathbb{R}$ be a Lipschitz function with constant 1 with respect to the Euclidean norm. Then, for every $t>0$,*

$$\int_{\mathbb{R}^n}\int_{\mathbb{R}^n}\exp\left(t(f(x)-f(y))\right)d\gamma_n(x)\,d\gamma_n(y)\leqslant e^{t^2}.$$

Proof. We fix $t>0$, consider the function $\varphi(y)=|y|^2/4$ and define

$$\psi_t=(tf)\square\varphi.$$

Let $x\in\mathbb{R}^n$ and $y=y(x,t)\in\mathbb{R}^n$ such that

$$\psi_t(x)=tf(y)+\frac{|x-y|^2}{4}.$$

Since $\|f\|_{\text{Lip}}\leqslant 1$, we get

$$\psi_t(x)\geqslant tf(x)-t|x-y|+\frac{|x-y|^2}{4}\geqslant tf(x)-t^2. \tag{3.2.1}$$

From Theorem 3.2.4,

$$\left(\int_{\mathbb{R}^n} e^{\psi_t}\,d\gamma_n\right)\left(\int_{\mathbb{R}^n} e^{-tf}\,d\gamma_n\right) \leqslant 1.$$

Using (3.2.1) we get

$$\left(\int_{\mathbb{R}^n} e^{tf}\,d\gamma_n\right)\left(\int_{\mathbb{R}^n} e^{-tf}\,d\gamma_n\right) \leqslant e^{t^2},$$

or equivalently,

$$\int_{\mathbb{R}^n}\int_{\mathbb{R}^n} \exp\left(t(f(x)-f(y))\right) d\gamma_n(x)\,d\gamma_n(y) \leqslant e^{t^2}$$

as claimed. □

COROLLARY 3.2.6. *Let $f : \mathbb{R}^n \to \mathbb{R}$ be a Lipschitz function with $\|f\|_{\mathrm{Lip}} \leqslant 1$. Then,*

$$\gamma_n\left(\left\{x : \left|f(x) - \int_{\mathbb{R}^n} f\,d\gamma_n\right| \geqslant s\right\}\right) \leqslant 2e^{-s^2/4}$$

for every $s > 0$.

Proof. Let $s > 0$. From Theorem 3.2.5 and Jensen's inequality we obtain

$$\mathbb{E}\left(\exp\left(t(f - \mathbb{E}(f))\right)\right) \leqslant e^{t^2}.$$

for every $t > 0$. This shows that

$$\gamma_n\left(\{x : f(x) - \mathbb{E}(f) \geqslant s\}\right) \leqslant \exp\left(t^2 - ts\right)$$

for every $t > 0$. Minimizing with respect to t and applying the same argument to $-f$, we conclude the proof. □

We end this section with a useful corollary of Talagrand's theorem which describes the concentration of the values of convex Lipschitz functions, when restricted to the discrete cube, around their Lévy mean.

THEOREM 3.2.7. *Let $f : \mathbb{R}^n \to \mathbb{R}$ be a convex Lipschitz function with Lipschitz constant σ. Let L be a Lévy mean of f on E_2^n, with respect to the uniform measure μ_n. Then, for every $t > 0$ we have*

$$\mu_n(\{|f - L| \geqslant t\}) \leqslant 4e^{-t^2/8\sigma^2}.$$

Proof. By the definition of L we have $\mu_n(\{f \geqslant L\}) \geqslant 1/2$ and $\mu_n(\{f \leqslant L\}) \geqslant 1/2$. We set $A = \{f \leqslant L\}$. Since f is convex, for every $y \in \mathrm{conv}(A)$ we have $f(y) \leqslant L$. It follows that if $f(x) \geqslant L + t$ for some $x \in E_2^n$, then

$$f(x) \geqslant L + t \geqslant f(y) + t$$

for every $y \in \mathrm{conv}(A)$. This implies that $\sigma|x - y| \geqslant |f(x) - f(y)| \geqslant t$, and hence

$$\phi_A(x) \geqslant t/\sigma.$$

From Corollary 3.1.21 and from $\mu_n(A) \geqslant 1/2$ we have

$$\mu_n(\{f \geqslant L + t\}) \leqslant \mu_n(\{\phi_A \geqslant t/\sigma\}) \leqslant \frac{1}{\mu_n(A)} e^{-t^2/8\sigma^2}$$
$$\leqslant 2e^{-t^2/8\sigma^2}.$$

Let $t > 0$ and $B = \{f \leqslant L - t\}$. As before, assuming that B is non-empty, we check that $f(x) \geqslant L$ implies that $\phi_B(x) \geqslant t/\sigma$. Using Corollary 3.1.21 we get
$$\mu_n(\{f(x) \geqslant L\}) \leqslant \mu_n(\{\phi_B \geqslant t/\sigma\}) \leqslant \frac{1}{\mu_n(B)} e^{-t^2/8\sigma^2}.$$
Since $1/2 \leqslant \mu_n(\{f(x) \geqslant L\})$, this shows that
$$\mu_n(B) \leqslant 2e^{-t^2/8\sigma^2}.$$
Combining the above, we have
$$\mu_n(\{|f - L| \geqslant t\}) = \mu_n(\{f \geqslant L + t\}) + \mu_n(\{f \leqslant L - t\})$$
$$\leqslant 2e^{-t^2/8\sigma^2} + 2e^{-t^2/8\sigma^2}$$
$$= 4e^{-t^2/8\sigma^2}.$$

$\square$

3.3. Concentration on homogeneous spaces

Many examples of normal Lévy families, other than the ones discussed in the previous section, have found applications in the asymptotic theory of finite dimensional normed spaces. For some important metric probability spaces (X, d, μ), the exact solution to the isoperimetric problem was (and still is) unknown: new and very interesting techniques were invented in order to estimate the concentration function $\alpha_\mu(t)$. In this section we will consider families of some compact groups and their homogeneous spaces. We start by introducing some basic notions and facts.

Let (M, ρ) be a compact metric space and let G be a group whose members act as isometries on M, i.e. for $g \in G$, $t, s \in M$, $\rho(gt, gs) = \rho(t, s)$. We begin by introducing the Haar measure.

THEOREM 3.3.1. *There exists a regular measure μ on the Borel subsets of M which is invariant under the action of members of G, i.e. $\mu(A) = \mu(gA)$ for all $A \subseteq M$, $g \in G$. Alternatively,*
$$\int_M f(t) d\mu(t) = \int_M f(gt) d\mu(t)$$
for all $g \in G$ and $f \in C(M)$ (where $C(M)$ is the linear space of all real, continuous functions on M).

If the action of G on M is transitive, i.e. for all $t, s \in M$ there exists $g \in G$ such that $gt = s$, then M is called a *homogeneous space* of G.

Fix $t \in M$ and let
$$G_\circ = \{g \in G : gt = t\}.$$
Then $G_\circ$ is a subgroup of G (called an *isotropic subgroup*) and $M = G/G_\circ$ where any $s \in M$ is identified with the equivalence class $gG_\circ$ with g such that $gt = s$.

To illustrate the definition we give a simple example: fix an inner product $\langle \cdot, \cdot \rangle$ on $\mathbb{R}^n$, let $|x| = \langle x, x \rangle^{1/2}$ for $x \in \mathbb{R}^n$ and let $G = O(n)$ be the orthogonal group on $(\mathbb{R}^n, |\cdot|)$. We identify $O(n)$ with the set of n-tuples $(e_1, \ldots, e_n)$ of orthonormal vectors (note that, if we fix one orthonormal basis $(e_1^\circ, \ldots, e_n^\circ)$, then any orthogonal operator A uniquely determines another such n-tuple: $(Ae_1^\circ, \ldots, Ae_n^\circ)$). Let $M = S^{n-1} = \{x \in \mathbb{R}^n : |x| = 1\}$ and $\phi : O(n) \to S^{n-1}$ be defined by $\phi(e_1, \ldots, e_n) = e_1$.

Then clearly, for any $t \in S^{n-1}$, $\phi^{-1}(t)$ can be identified with $O(n-1)$. Thus, $S^{n-1} = O(n)/O(n-1)$.

THEOREM 3.3.2. *If (M, ρ) is a compact metric homogeneous space of the group G, then the measure of Theorem 3.3.1 is unique up to a constant.*

Proof. Define a semi-metric on G by $d(g, h) = \sup_{t \in M} \rho(gt, ht)$. Identifying elements whose distance from each other is zero, we get a group H which still acts as isometries on M and also on itself (there are two ways in which H acts on itself - we choose right multiplication, namely $hh' = h' \cdot h$ where "$\cdot$" is the multiplication of the group G). One checks that H is compact (actually, in all our applications $G = H$ will be given as a compact group). Let μ on M and ν on H be measures that are invariant under the action of G. Then, for all $f \in C(M)$,

$$\nu(1)\mu(f) = \int_G \int_M f(gt)\, d\mu(t)\, d\nu(g) = \int_M \int_G f(gt)\, d\nu(g)\, d\mu(t). \tag{3.3.1}$$

By the transitivity of G on M and the invariance of ν, the inner integral on the right depends on f but not on t, so we may call it $\bar{\nu}(f)$. Then

$$\nu(1)\mu(f) = \bar{\nu}(f)\mu(1).$$

It follows that if μ' is another invariant measure on M, then

$$\mu(f)\mu'(1) = \mu'(f)\mu(1),$$

whence we are done. □

REMARKS 3.3.3. (i) The proof shows also that any right invariant probability measure on a compact metric group G is equal to any normalized (i.e., probability) left invariant measure.
(ii) It is easily checked that the unique normalized invariant measure on G is also invertible invariant, which means that $\int_G f(t)\, d\mu(t) = \int_G f(t^{-1})\, d\mu(t)$ for every $f \in C(G)$.

In what follows, μ will denote the normalized Haar measure on the space in question, so that it may appear twice in the same formula denoting measures on different spaces. We pass now to several examples of homogeneous spaces of the group $O(n)$ of all $n \times n$ real orthogonal matrices.

EXAMPLES 3.3.4. (i) $S^{n-1} = \{(x_1, \ldots, x_n) \in \mathbb{R}^n : \sum_{i=1}^n x_i^2 = 1\}$ with either the Euclidean or geodesic metric is easily seen to be equivalent to $O(n)/O(n-1)$, as was discussed above.
(ii) The *Stiefel manifolds*. For $1 \leqslant k \leqslant n$,

$$W_{n,k} := \{e = (e_1, \ldots, e_k) : e_i \in \mathbb{R}^n, \langle e_i, e_j \rangle = \delta_{ij}, 1 \leqslant i, j \leqslant k\}$$

with the metric $\rho(e, f) := \left(\sum_{i=1}^k d(e_i, f_i)^2\right)^{1/2}$, d being either the Euclidean or the geodesic metric. Note that $W_{n,n} = O(n)$, $W_{n,1} = S^{n-1}$ and $W_{n,n-1} = SO(n) = \{T \in O(n) : \det T = 1\}$. In general, $W_{n,k}$ may be identified with $O(n)/O(n-k)$ via the map $\phi : O(n) \to W_{n,k}$ defined by $\phi(e_1, \ldots, e_n) = (e_1, \ldots, e_k)$.
(iii) The *Grassmann manifold* $G_{n,k}$, $1 \leqslant k \leqslant n$, consists of all k-dimensional subspaces of $\mathbb{R}^n$ with the metric being the Hausdorff distance between the unit balls of the two subspaces:

$$\rho(F, H) = \sup_{x \in S^{n-1} \cap H} \rho(x, S^{n-1} \cap F).$$

The equivalence $G_{n,k} = O(n)/(O(k) \times O(n-k))$ is again easily verified.

(iv) If G is any group with invariant metric ρ and $G_\circ$ is a subgroup, we may define a metric d on $M = G/G_\circ$ by

$$d(t,s) = \inf\{\rho(g,h) : \phi(g) = t, \phi(h) = s\},$$

where ϕ is the quotient map. In this way M becomes a homogeneous space of G. Note that in all the previous examples the metric given on the homogeneous space of $O(n)$ is equivalent, up to a universal constant (not depending on n), to the metric described here.

The uniqueness of the normalized Haar measure allows us to deduce several interesting consequences.

(1) The first remark is that, for any $A \subseteq S^{n-1}$ and $x_0 \in S^{n-1}$,

$$\mu(\{T \in O(n) : T(x_0) \in A\}) = \mu(A).$$

(2) Next, we give two identities. Fix $1 \leqslant k \leqslant n$ and, for $F \in G_{n,k}$, denote the $(k-1)$-dimensional sphere $S^{n-1} \cap F$ of F by S_F. Then

$$\int_{S^{n-1}} f\,d\mu = \int_{G_{n,k}} \int_{S_F} f(t)\,d\mu_F(t)\,d\mu(F)$$

for all $f \in C(S^{n-1})$, where μ_F is the normalized Haar measure on S_F (by our convention, μ on the left is the normalized Haar measure on S^{n-1} and on the right the one on $G_{n,k}$).

We identify $\mathbb{R}^{2n}$ with $\mathbb{C}^n$ (by introducing a complex structure in one of the possible ways). For each k we denote the collection of complex k-dimensional subspaces of $\mathbb{C}^n$ by $\mathbb{C}G_{n,k}$ and the unit sphere of any $F \in \mathbb{C}G_{n,k}$ by $\mathbb{C}S_F$ (which can be identified with S^{2k-1}). Again, $\mathbb{C}G_{n,k}$ is a homogeneous space and we get an identity similar to the previous one:

$$\int_{S^{2n-1}} f\,d\mu = \int_{\mathbb{C}G_{n,k}} \int_{\mathbb{C}S_F} f(t)\,d\mu_F(t)\,d\mu(F).$$

Note that here one integrates on a much smaller space, $\mathbb{C}G_{n,k}$, than the one in the first identity which, adjusted to the dimensions here, would be $G_{2n,2k}$.

The following few examples of Lévy families are consequences of a general isoperimetric inequality for connected Riemannian manifolds due to Gromov. Let μ_X be the normalized Riemannian volume element on a connected Riemannian manifold X which has no boundary, and let $R(X)$ be the Ricci curvature of X.

THEOREM 3.3.5 (Lévy-Gromov). *Let $A \subseteq X$ be measurable and let $\varepsilon > 0$. Then*

$$\mu_X(A_\varepsilon) \geqslant \mu(B_\varepsilon)$$

for (any) ball B on the sphere $r \cdot S^n$ with the property that

$$\mu_X(A) = \mu(B),$$

where $n = \dim X$, r is such that

$$R(X) = R(r \cdot S^n) (= (n-1)/r^2),$$

and μ is the normalized Haar measure on $r \cdot S^n$.

The value of $R(X)$, known in some examples, together with the computation for the measure of a cap in the proof of Theorem 3.1.5 leads to the following examples.

THEOREM 3.3.6 (Gromov-Milman). *The family* $SO(n) = \{T \in O(n) : \det T = 1\}$, $n = 1, 2, \ldots$, *with the metric described in Example* 3.3.4(ii) *and the normalised Haar measure is a normal Lévy family with constants* $c_1 = \sqrt{\pi/8}$, $c_2 = 1/8$.

Similarly, for each m the family

$$X_n = \prod_{i=1}^{m} S^n, \qquad n = 1, 2, \ldots,$$

with the product measure and the metric

$$d(x, y) = \Big(\sum_{i=1}^{m} \rho(x_i, y_i)^2\Big)^{1/2}, \quad x = (x_1, \ldots, x_m), y = (y_1, \ldots, y_m) \in X_n$$

(where ρ is the geodesic metric on S^n) is a normal Lévy family with constants $c_1 = \sqrt{\pi/8}$, $c_2 = 1/2$.

Next, we show that homogeneous spaces on which $SO(n)$ acts inherit the property of being Lévy families. Let G be a subgroup of $SO(n)$ and let $V = SO(n)/G$. Let μ be the Haar measure on V and let d_n be a metric defined as

$$d_n(t, s) = \inf\{\rho(g, h) : \varphi g = t, \varphi h = s\},$$

where φ is the quotient map.

Clearly, $\mu(A \subseteq V) = \mu(\varphi^{-1}(A) \subseteq SO(n))$. By the definition of d_n,

$$\varphi^{-1}(A_\varepsilon) \supseteq (\varphi^{-1}(A))_\varepsilon.$$

Therefore, if $\mu(A \subseteq V) \geqslant 1/2$, then $\mu(\varphi^{-1}(A) \subseteq SO(n)) \geqslant 1/2$ and $\mu(A_\varepsilon) \geqslant \mu((\varphi^{-1}(A))_\varepsilon)$. We conclude

THEOREM 3.3.7 (Gromov-Milman). *Given* $n \in \mathbb{N}$, *let* G_n *be a subgroup of* $SO(n)$ *with the metric described above and with the normalized Haar measure* μ_n. *Then*

$$(SO(n)/G_n, d_n, \mu_n), \quad n \geqslant 1,$$

is a normal Lévy family with constants $c_1 = \sqrt{\pi/8}$, $c_2 = 1/8$.

The above theorem, together with Examples 3.3.4(ii) and 3.3.4(iii), implies immediately that the following families are normal Lévy families with constants $c_1 = \sqrt{\pi/8}$, $c_2 = 1/8$:

(i) Any family of Stiefel manifolds $\{W_{n,k_n}\}_{n=1}^{\infty}$ where $1 \leqslant k_n \leqslant n$, $n \geqslant 1$.
(ii) Any family of Grassmann manifolds $\{G_{n,k_n}\}_{n=1}^{\infty}$ where $1 \leqslant k_n \leqslant n$, $n \geqslant 1$.

3.4. An approach through conditional expectation and martingales

3.4.1. Martingales

DEFINITION 3.4.1. Let $(\Omega, \mathcal{F}, P)$ be a probability space. If $\mathcal{G}$ is a sub-σ-algebra of $\mathcal{F}$ and if $f \in L_1(\Omega, \mathcal{F}, P)$, then the function

$$\mu(A) = \int_A f dP, \quad A \in \mathcal{G}$$

defines a measure on $\mathcal{G}$, which is absolutely continuous with respect to $P|_{\mathcal{G}}$. By the Radon–Nikodym theorem, there exists a unique $h \in L_1(\Omega, \mathcal{G}, P)$ with the property

$$\int_A h dP = \int_A f dP$$

for every $A \in \mathcal{G}$. The function h is the *conditional expectation* of f with respect to $\mathcal{G}$ and it is denoted by $h = \mathbb{E}(f|\mathcal{G})$.

The next lemma gives the basic properties of the conditional expectation.

LEMMA 3.4.2. (i) *The operator* $f \mapsto \mathbb{E}(f|\mathcal{G})$ *is positive, linear and has norm* 1 *on every* L_p, $1 \leqslant p \leqslant \infty$.

(ii) *If* $\mathcal{G}_1$ *is a sub-*σ*-algebra of* $\mathcal{G}$, *then* $\mathbb{E}(\mathbb{E}(f|\mathcal{G})|\mathcal{G}_1) = \mathbb{E}(f|\mathcal{G}_1)$.

(iii) *If* $g \in L_\infty(\Omega, \mathcal{G}, P)$ *then* $\mathbb{E}(f \cdot g|\mathcal{G}) = g \cdot \mathbb{E}(f|\mathcal{G})$.

(iv) *If* $\mathcal{G} = \{\emptyset, \Omega\}$ *is the trivial* σ*-algebra, then* $\mathbb{E}(f|\mathcal{G})$ *is constant and equal to the expectation of* f:

$$\mathbb{E}(f|\mathcal{G}) = \mathbb{E}(f) = \int f dP.$$

Proof. (i) Linearity is clear from the definition of conditional expectation. We prove positivity: if $f \in L_1(\Omega, \mathcal{A}, P)$ and $f \geqslant 0$, then there exists $h \in L_1(\Omega, \mathcal{G}, P)$ such that

$$\int_A h\, dP = \int_A f\, dP \geqslant 0$$

for every $A \in \mathcal{G}$. If we set $E_n = \{\omega : h(\omega) \leqslant -\frac{1}{n}\}$ we have that $E_n \in \mathcal{G}$ and

$$0 \leqslant \int_{E_n} f\, dP = \int_{E_n} h\, dP \leqslant -\frac{1}{n} P(E_n),$$

which gives $P(E_n) = 0$. Therefore,

$$P(\{\omega : h(\omega) < 0\}) = P\Big(\bigcup_{n=1}^{\infty} E_n\Big) = 0.$$

The fact that the operator $T(f) = \mathbb{E}(f|\mathcal{G})$ is positive and linear implies that it is monotone: if $f, g \in L_1(\Omega, \mathcal{A}, P)$ and $f \leqslant g$, then $\mathbb{E}(f|\mathcal{G}) \leqslant \mathbb{E}(g|\mathcal{G})$. In particular,

$$|\mathbb{E}(f|\mathcal{G})| \leqslant \mathbb{E}(|f|\,|\mathcal{G})$$

for every $f \in L_1(\Omega, \mathcal{A}, P)$. Then, the conditional expectation $T : L_1 \to L_1$ is a bounded operator of norm 1. To see this, note that

$$\|\mathbb{E}(f|\mathcal{G})\|_1 = \int |\mathbb{E}(f|\mathcal{G})|\, dP \leqslant \int \mathbb{E}(|f|\,|\mathcal{G})\, dP = \int |f|\, dP = \|f\|_1.$$

We also have $\mathbb{E}(\mathbf{1}|\mathcal{G}) = \mathbf{1}$. From Hölder's inequality it follows that $L_p \subseteq L_1$ for every $1 \leqslant p \leqslant \infty$. Therefore, if $f \in L_\infty$, then

$$|\mathbb{E}(f|\mathcal{G})| \leqslant \mathbb{E}(|f||\mathcal{G}) \leqslant \mathbb{E}(\|f\|_\infty|\mathcal{G}) = \|f\|_\infty.$$

It follows that for every $f \in L_\infty$ we also have $\mathbb{E}(f|\mathcal{G}) \in L_\infty$; in fact, $\|\mathbb{E}(f|\mathcal{G})\|_\infty \leqslant \|f\|_\infty$. In other words, the conditional expectation $T : L_\infty \to L_\infty$ is a well-defined bounded operator of norm 1. It remains to prove that $T : L_p \to L_p$ is also well-defined. This follows from the next claim, which can be thought of as a generalization of Jensen inequality and admits a similar proof.

Claim. Let $f \in L_1$ and let $\varphi : \mathbb{R} \to \mathbb{R}$ be a convex function such that $\mathbb{E}|\varphi(f)| < \infty$. Then,

$$\varphi(\mathbb{E}(f|\mathcal{G})) \leqslant \mathbb{E}(\varphi(f)|\mathcal{G}).$$

Proof of the Claim. It is known that there exist sequences of real numbers (a_n), (b_n) such that $\varphi(x) = \sup_n (a_n x + b_n)$ for every $x \in \mathbb{R}$. Then, for every $n \in \mathbb{N}$ we have

$$a_n f(x) + b_n \leqslant \varphi(f(x)).$$

It follows that for every $n \in \mathbb{N}$ there exists $E_n \in \mathcal{G}$ with $P(E_n) = 0$ and

$$a_n \mathbb{E}(f|\mathcal{G}) + b_n \leqslant \mathbb{E}(\varphi(f)|\mathcal{G})$$

for all $x \in \Omega \setminus E_n$ (since we are in L_1, all knowledge about the behaviour of f can be only in the sense of "almost everywhere"). If we set $E = \bigcup_{n=1}^{\infty} E_n$, then $P(E) = 0$ and for all $x \in \Omega \setminus E$ we have

$$a_n \mathbb{E}(f|\mathcal{G}) + b_n \leqslant \mathbb{E}(\varphi(f)|\mathcal{G})$$

for every $n \in \mathbb{N}$. Taking the supremum with respect to n we see that

$$\varphi(\mathbb{E}(f|\mathcal{G})) \leqslant \mathbb{E}(\varphi(f)|\mathcal{G})$$

for almost all $x \in \Omega$. □

Applying this inequality to $\varphi(t) = |t|^p$ we get

$$|\mathbb{E}(f|\mathcal{G})|^p \leqslant \mathbb{E}(|f|^p|\mathcal{G})$$

and integration shows that $\|\mathbb{E}(f|\mathcal{G})\|_p \leqslant \|f\|_p$ for every $f \in L_p$.

(ii) Let $g = \mathbb{E}(f|\mathcal{G})$. Then, for every $A \in \mathcal{G}$ we have $\int_A f\,dP = \int_A g\,dP$. If $B \in \mathcal{G}_1 \subseteq \mathcal{G}$ then $\int_B g\,dP = \int_B f\,dP$. From the uniqueness in the definition of conditional expectation it follows that $\mathbb{E}(g|\mathcal{G}_1) = \mathbb{E}(f|\mathcal{G}_1)$.

(iii) It suffices to check the case of $\mathcal{G}$-measurable indicator functions. If $g = \mathbf{1}_A$ and $A, B \in \mathcal{G}$ then

$$\int_B \mathbb{E}(fg|\mathcal{G})\,dP = \int_B fg\,dP = \int_{A\cap B} f\,dP = \int_{A\cap B} \mathbb{E}(f|\mathcal{G})\,dP = \int_B g\mathbb{E}(f|\mathcal{G})\,dP.$$

Therefore, because $\mathbf{1}_A\mathbb{E}(f|\mathcal{G}) \in L_1(\Omega, \mathcal{G}, P)$, we have $\mathbb{E}(fg|\mathcal{G}) = g\mathbb{E}(f|\mathcal{G})$.

(iv) Clear from the definition. □

DEFINITION 3.4.3. Let $\mathcal{F}_0 \subseteq \mathcal{F}_1 \subseteq \cdots \subseteq \mathcal{F}$ be a sequence of σ-algebras. A sequence $f_0, f_1, \ldots$ of functions $f_i \in L_1(\Omega, \mathcal{F}_i, P)$ is called a *martingale* with respect to $\{\mathcal{F}_i\}$ if $\mathbb{E}(f_i|\mathcal{F}_{i-1}) = f_{i-1}$ for every $i \geqslant 1$.

A simple but good example to keep in mind is that of a finite atomic probability space, and a finite sequence of σ-algebras generated by a partition, a refinement of this partition, etc.

3.4.2. Azuma's inequality

THEOREM 3.4.4 (Azuma). *Let $f \in L_\infty(\Omega, \mathcal{F}, P)$ and let $\{\emptyset, \Omega\} = \mathcal{F}_0 \subseteq \mathcal{F}_1 \subseteq \cdots \subseteq \mathcal{F}_n = \mathcal{F}$ be a sequence of σ-algebras. We set $d_j = \mathbb{E}(f|\mathcal{F}_j) - \mathbb{E}(f|\mathcal{F}_{j-1})$, $j = 1, \ldots, n$. Then, for every $t > 0$ we have*

$$P(|f - \mathbb{E}(f)| \geqslant t) \leqslant 2\exp\left(-\frac{t^2}{4\sum_{j=1}^n \|d_j\|_\infty^2}\right).$$

Proof. We first observe that the sequence $\{\mathbb{E}(f|\mathcal{F}_j)\}_{j=0}^n$ is a martingale with respect to $\{\mathcal{F}_j\}_{j=0}^n$. We have

$$\mathbb{E}(\mathbb{E}(f|\mathcal{F}_j)|\mathcal{F}_{j-1}) = \mathbb{E}(f|\mathcal{F}_{j-1})$$

and $\mathbb{E}(f|\mathcal{F}_j) \in L_1(\Omega, \mathcal{F}_j, P)$ from the definition of conditional expectation. Moreover, we have $\mathbb{E}(d_j|\mathcal{F}_{j-1}) = 0$ for every $j \geqslant 1$. Comparing the Taylor series of e^x and $e^{x^2/2}$ we see that $e^x \leqslant x + e^{x^2}$ for every $x \in \mathbb{R}$. Since the operator $f \mapsto \mathbb{E}(f|\mathcal{F})$ is positive, we conclude that for every $\lambda \in \mathbb{R}$ and any $k = 1, \dots, n$

$$\mathbb{E}(e^{\lambda d_k}|\mathcal{F}_{k-1}) \leqslant \mathbb{E}(\lambda d_k|\mathcal{F}_{k-1}) + \mathbb{E}(e^{\lambda^2 d_k^2} \mid \mathcal{F}_{k-1}) = \mathbb{E}(e^{\lambda^2 d_k^2} \mid \mathcal{F}_{k-1}).$$

Since we have assumed that $f \in L_\infty(\Omega, \mathcal{F}, P)$, every $d_k \in L_\infty(\Omega, \mathcal{F}_k, P)$. It follows that

$$\mathbb{E}(e^{\lambda d_k}|\mathcal{F}_{k-1}) \leqslant \mathbb{E}(e^{\lambda^2 d_k^2} \mid \mathcal{F}_{k-1}) \leqslant \mathbb{E}(e^{\lambda^2 \|d_k\|_\infty^2}|\mathcal{F}_{k-1}) = e^{\lambda^2 \|d_k\|_\infty^2}.$$

We use this inductively to show

$$\mathbb{E}\left(e^{\sum_{j=1}^n \lambda d_j}\right) \leqslant e^{\lambda^2 \sum_{j=1}^n \|d_j\|_\infty^2}.$$

Indeed, since $e^{\sum_{j=1}^k \lambda d_j} \in L_\infty(\Omega, \mathcal{F}_k, P)$, from Lemma 3.4.2 (iii) we get

$$\mathbb{E}(e^{\sum_{j=1}^k \lambda d_j} e^{\lambda d_{k+1}}|\mathcal{F}_k) = e^{\sum_{j=1}^k \lambda d_j}\mathbb{E}(e^{\lambda d_{k+1}}|\mathcal{F}_k).$$

so that we may write

$$\begin{aligned}\mathbb{E}\left(e^{\lambda \sum_{j=1}^k d_j}\right) &= \mathbb{E}\left[\mathbb{E}\left(e^{\sum_{j=1}^{k-1} \lambda d_j} e^{\lambda d_k} \mid \mathcal{F}_{k-1}\right)\right] = \mathbb{E}\left[e^{\sum_{j=1}^{k-1} \lambda d_j}\mathbb{E}\left(e^{\lambda d_k} \mid \mathcal{F}_{k-1}\right)\right]\\ &\leqslant \mathbb{E}\left[e^{\sum_{j=1}^{k-1} \lambda d_j} e^{\lambda^2 \|d_k\|_\infty^2}\right] = e^{\lambda^2 \|d_k\|_\infty^2}\mathbb{E}\left(e^{\sum_{j=1}^{k-1} \lambda d_j}\right).\end{aligned}$$

Continuing by induction we get that

$$\mathbb{E}\left(e^{\lambda \sum_{j=1}^k d_j}\right) \leqslant e^{\lambda^2 \sum_{j=1}^k \|d_j\|_\infty^2}.$$

For every $\lambda > 0$ we have

$$\begin{aligned}P(f - \mathbb{E}f \geqslant t) &= P(\mathbb{E}(f|\mathcal{F}_n) - \mathbb{E}(f|\mathcal{F}_0) \geqslant t) = P\Big(\sum_{j=1}^n d_j \geqslant t\Big)\\ &\leqslant \mathbb{E}\left(e^{\lambda \sum_{j=1}^n d_j - \lambda t}\right) \leqslant e^{\lambda^2 \sum_{j=1}^n \|d_j\|_\infty^2 - \lambda t}.\end{aligned}$$

Minimizing with respect to λ we see that

$$P(f - \mathbb{E}(f) \geqslant t) \leqslant \exp\left(-\frac{t^2}{4\sum_{j=1}^n \|d_j\|_\infty^2}\right).$$

Applying the same argument to $-f$, we get the desired inequality. □

3.4.3. Concentration of measure on Π_n

We consider the family Π_n of permutations of the set $[n] := \{1, 2, \dots, n\}$ as a probability metric space, with the distance $d(\sigma, \tau) = \frac{1}{n}|\{i : \sigma(i) \neq \tau(i)\}|$ and with the uniform probability measure P which gives mass $\frac{1}{n!}$ to every $\sigma \in S_n$. The purpose of this section is to provide a proof of the following theorem of Maurey.

THEOREM 3.4.5 (Maurey). *Let $f : \Pi_n \to \mathbb{R}$ be a 1-Lipschitz function. Then,*

$$P(|f - \mathbb{E}f| \geqslant t) \leqslant 2e^{-t^2 n/16}$$

for all $t > 0$.

In view of Proposition 3.1.31 this in turn implies an estimate for the concentration function of Π_n. The proof of Theorem 3.4.5 is based on Azuma's inequality.

Proof of Theorem 3.4.5. Let $\mathcal{F}_j$ be the algebra which is generated by the sets

$$A_{i_1,\ldots,i_j} = \{\sigma : \sigma(1) = i_1, \ldots, \sigma(j) = i_j\}$$

where $i_1, \ldots, i_j$ are distinct elements of $\{1, \ldots, n\}$. This is a partition of Π_n, which corresponds to conditioning on the event of knowing what the first j entries of σ are. We consider the sequence

$$\{\emptyset, \Pi_n\} = \mathcal{F}_0 \subseteq \mathcal{F}_1 \subseteq \cdots \subseteq \mathcal{F}_n = 2^{\Pi_n}.$$

Then, $\mathcal{F}_j \subseteq \mathcal{F}_{j+1}$ for all $j = 0, 1, \ldots, n-1$. To see this, note that any element of $\mathcal{F}_j$ can be written in the form

$$A_{i_1,i_2,\ldots,i_j} = \bigcup_{k \in [n] \setminus \{i_1 \ldots, i_j\}} A_{i_1,i_2,\ldots,i_j,k},$$

and hence, it belongs to $\mathcal{F}_{j+1}$.

Let $f : \Pi_n \to \mathbb{R}$ be a 1-Lipschitz function and let $(f_j)_{j=0}^n$ be the martingale $f_j = \mathbb{E}(f|\mathcal{F}_j)$ induced by f.

LEMMA 3.4.6. *For every atom $A = A_{i_1,i_2,\ldots,i_j}$ of $\mathcal{F}_j$ and any pair of atoms $B = A_{i_1,i_2,\ldots,i_j,r}$ and $C = A_{i_1,\ldots,i_j,s}$ of $\mathcal{F}_{j+1}$ which are contained in A, we can find a $1-1$ and onto map $\phi : B \to C$ such that $d(b, \phi(b)) \leqslant \frac{2}{n}$ for every $b \in B$.*

Proof. Let π be the permutation which commutes r and s and fixes all other elements of $\{1, \ldots, n\}$. We define $\phi : B \to C$ by $\phi(\sigma) = \pi \circ \sigma$.

Then, $\phi(\sigma)(i) = \sigma(i)$ for $i \neq j+1$ and $i \neq \sigma^{-1}(s)$. If $i = j+1$ then $\phi(\sigma)(j+1) = \pi \circ \sigma(j+1) = \pi(r) = s$ and if $i = \sigma^{-1}(s)$ then $\phi(\sigma)(i) = \pi(s) = r$. It follows that

$$d(b, \phi(b)) \leqslant \frac{2}{n}.$$

It is clear that ϕ is 1-1 and the fact that $|B| = |C|$ implies that ϕ is onto. □

We fix A, B, C as in Lemma 3.4.6. Since B, C are atoms of $\mathcal{F}_{j+1}$, we have that f_{j+1} is constant on B, C. We have

$$\int_B \mathbb{E}(f|\mathcal{F}_{j+1}) dP = \int_B f dP = \frac{1}{n!} \sum_{\sigma \in B} f(\sigma).$$

Since $f_{j+1} = \mathbb{E}(f|\mathcal{F}_{j+1})$ is constant on B, we get

$$f_{j+1}|_B \equiv \frac{1}{P(B)} \frac{1}{n!} \sum_{\sigma \in B} f(\sigma) = \frac{1}{|B|} \sum_{\sigma \in B} f(\sigma).$$

Similarly $f_{j+1}|_C = \frac{1}{|C|} \sum_{\sigma \in C} f(\sigma)$. We write

$$f_{j+1}|C = \frac{1}{|C|} \sum_{\sigma \in C} f(\sigma) = \frac{1}{|\phi(B)|} \sum_{\sigma \in B} f(\phi(\sigma)) = \frac{1}{|B|} \sum_{\sigma \in B} f(\phi(\sigma)),$$

where ϕ is the function from Lemma 3.4.6. Since f is 1-Lipschitz,

$$|f_{j+1}|B - f_{j+1}|C| \leqslant \frac{1}{|B|}\sum_{\sigma\in B}|f(\sigma) - f(\phi(\sigma))| \leqslant \frac{1}{|B|}\sum_{\sigma\in B} d(\sigma,\phi(\sigma))$$
$$\leqslant \frac{1}{|B|}\sum_{\sigma\in B}\frac{2}{n} = \frac{2}{n}.$$

It follows that $|f_{j+1}|_B - f_j|_A| \leqslant \frac{2}{n}$. Indeed, we have

$$|A| = \Big|\bigcup_{s\notin\{i_1,\dots,i_j\}} A_{i_1,\dots,i_j,s}\Big| = \sum_{s\notin\{i_1,\dots,i_j\}} |A_{i_1,i_2,\dots,i_j,s}| = (n-j)|C|.$$

Then,

$$f_j|_A = \frac{1}{|A|}\sum_{\sigma\in A} f(\sigma) = \frac{1}{|A|}\sum_{s\notin\{i_1,\dots,i_j\}}\ \sum_{\sigma\in A_{i_1,\dots,i_j,s}} f(\sigma)$$
$$= \frac{1}{(n-j)}\sum_{C\subseteq A}\frac{1}{|C|}\sum_{\sigma\in C} f(\sigma) = \frac{1}{n-j}\sum_{C\subseteq A} f_{j+1}|_C$$

where the sum is over all $C\subset A$ of the form $C = A_{i_1,i_2,\dots,i_j,s}$. Therefore,

$$|f_{j+1}|_B - f_j|_A| = \Big|f_{j+1}|_B - \frac{1}{n-j}\sum_{C\subseteq A} f_{j+1}|_C\Big|$$
$$= \Big|\sum_{C\subseteq A}\frac{1}{n-j}(f_{j+1}|_B - f_{j+1}|_C)\Big|$$
$$\leqslant \sum_{C\subseteq A}\frac{1}{n-j}|f_{j+1}|_B - f_{j+1}|_C|$$
$$\leqslant \sum_{C\subseteq A}\frac{1}{n-j}\frac{2}{n} = \frac{2}{n}.$$

We will show that $|d_{j+1}|_{B_i}| \leqslant \frac{2}{n}$ for all atoms B_i of $\mathcal{F}_{j+1}$, where $d_j = f_j - f_{j-1}$ are defined as in Azuma's inequality: we have

$$|d_{j+1}|B_i| = |f_{j+1}|_{B_i} - f_j|_{B_i}| = |f_{j+1}|_{B_i} - f_j|_{A_i}| \leqslant \frac{2}{n}$$

where A_i is the atom of F_j which contains B_i. Therefore,

$$\|d_{j+1}\|_\infty \leqslant \frac{2}{n}.$$

It is clear that f is in $L_\infty(\Pi_n,\mathcal{F}_n,P)$, and hence, the previous inequality and Azuma's inequality give

$$P(|f-\mathbb{E}f|\geqslant t)\leqslant 2e^{-t^2n/16}$$

for every $t>0$. □

3.4.4. Some more examples

The proof of Theorem 3.4.5 suggests the next general definition: suppose that (X,d) is a finite metric space, viewed also as a probability space with the normalized counting measure P. We say that (X,d) has *length* at most ℓ if there exist positive reals $a_1,\dots,a_n$ with $\ell^2 = \sum_{i=1}^n a_i^2$ and a sequence $\{P_k\}_{k=0}^n$ of partitions of X such that $P_0 = \{X\}$, $P_n = \{\{x\} : x \in X\}$, P_k is a refinement of P_{k-1} for all $1 \leqslant k \leqslant n$, and the following holds true:

> If $A \in P_{k-1}$, $B, C \in P_k$ and $B, C \subseteq A$ then we may find a bijection $\varphi : B \to C$ such that $d(x, \varphi(x)) \leqslant a_k$ for all $x \in B$.

Following the proof of Maurey's theorem one can check that if $f : (X, d) \to \mathbb{R}$ is a 1-Lipschitz function then, for all $t > 0$,

$$P(\{|f - \mathbb{E}(f)| \geqslant t\}) \leqslant 2\exp(-t^2/4\ell^2). \tag{3.4.1}$$

This applies in particular to the discrete cube E_2^n thus providing an alternative proof of the estimate of its concentration function. Note that the natural choice of partitions P_k (that we obtain if we fix k coordinates of the points $\epsilon \in E_2^n$) shows that E_2^n has length at most $1/\sqrt{n}$.

In the same spirit, consider a compact metric group (G, d) where d is translation invariant, that is, $d(hg_1, hg_2) = d(g_1, g_2) = d(g_1h, g_2h)$ for all $g_1, g_2, h \in G$. We denote by μ the unique translation invariant (Haar) probability measure on G.

Given a closed subgroup, we may endow G/H with a natural metric, denoted by $\bar{d}$, letting

$$\bar{d}(g_1H, g_2H) = d(g_2^{-1}g_1, H).$$

It is easy to check that $\bar{d}$ is indeed a metric. The following theorem is in fact a generalization of Theorem 3.4.5.

THEOREM 3.4.7. *Let (G, d) be a compact group with a translation invariant metric. Let*

$$G = G_0 \supseteq G_1 \supseteq \cdots \supseteq G_n = \{1\}$$

be a decreasing sequence of closed subgroups of G. Let a_k denote the diameter of G_{k-1}/G_k, $k = 1, \ldots, n$. Assume $f : G \to \mathbb{R}$ is a 1-Lipschitz function. Then for any $t > 0$ we have

$$\mu(\{x : |f(x) - \mathbb{E}(f)| > t\}) \leqslant 2\exp\Big(-t^2/\Big(4\sum_{k=1}^{n} a_k^2\Big)\Big).$$

In particular, $\alpha_\mu(t) \leqslant 2\exp(-t^2/(16\sum_{k=1}^n a_k^2))$.

Proof. We denote by $\mathcal{F}_k$ the σ-algebra generated by the sets $\{gG_k\}_{g\in G}$. Denote $f_k = \mathbb{E}(f|\mathcal{F}_k)$. To use Theorem 3.4.4 we should estimate $\|d_k\|_\infty$ for $d_k = \mathbb{E}(f|\mathcal{F}_k) - \mathbb{E}(f|\mathcal{F}_{k-1})$. As in the proof of Theorem 3.4.5, we shall bound the oscillation of f_k on each atom of $\mathcal{F}_{k-1}$. To this end assume that both $h_1G_k, h_2G_k \subset gG_{k-1}$. Then both $g^{-1}h_1, g^{-1}h_2 \in G_{k-1}$, so that also $h_2^{-1}h_1 \in G_{k-1}$. We shall define a mapping $\varphi : h_1G_k \to h_2G_k$ satisfying $d(x, \varphi(x)) \leqslant a_k$, and continue in the same way as we did in the proof of Theorem 3.4.5. To define the mapping, we pick a special $s \in G_k$ which satisfies that

$$d(h_2^{-1}h_1, G_k) = d(h_2^{-1}h_1, s)$$

and define for $g \in G_k$

$$\varphi(h_1g) = h_2sg.$$

Clearly the mapping is $\varphi : h_1G_k \to h_2G_k$. Further we see that

$$d(h_1g, \varphi(h_1g)) = d(h_1g, h_2sg) = d(h_2^{-1}h_1, s) = d(h_2^{-1}h_1, G_k) \leqslant a_k.$$

We thus get that the oscillation of f_k on $\mathcal{F}_{k-1}$ is bounded by a_k, so that $\|d_k\|_\infty \leqslant a_k$, and by Theorem 3.4.4, we are done. $\square$

3.5. Khintchine type inequalities

We saw in Section 3.1 that if (X, d, μ) is a metric probability space with a concentration function that decays fast then every 1-Lipschitz function on X concentrates around its Lévy mean. More precisely

$$\mu(\{x \in X : |f(x) - \mathrm{med}(f)| \geqslant t) \leqslant 2\alpha_\mu(t).$$

This type of concentration implies equivalence of the L_p-norms for Lipschitz functions on X, that is, inverse Hölder inequalities of the form

$$\|f\|_{L_p(\mu)} \le c(p,\mu)\|f\|_{L_1(\mu)},$$

where the order of the constant $c(p,\mu)$ as $p \to \infty$ reflects the degree of concentration.

Such inverse Hölder inequalities appear often in the context of probability spaces, and reflect a different kind of concentration which involves measures but not distances, see for example [**233**] for a discussion of this point of view. A classical example is Khintchine's inequality and its generalization by Kahane, which will be discussed in Section 3.5.1. Another example is the fact that linear functionals on a convex body K of volume 1 satisfy the inequality

$$\|f\|_{L_p(K)} \leqslant cp\|f\|_{L_1(K)}$$

where $c > 0$ is an absolute constant. More generally, Bourgain has shown that if $f : K \to \mathbb{R}$ is a polynomial of degree m, then $\|f\|_{L_p(K)} \leqslant c(p,m)\|f\|_{L_2(K)}$ for every $p > 2$, where $c(p,m)$ depends only on p and on the degree m of f. In Section 3.5.2 we introduce the ψ_α-norm of a measurable function; this notion provides the right framework for a systematic discussion of the *level of concentration of a class of functions* on a probability space. In Section 3.5.3 we introduce logarithmically concave measures as a more general setting than that of convex bodies, and we establish Khintchine type inequalities for the class of seminorms. These estimates play a key role in the asymptotic theory of convex bodies, especially in that part of the theory which concerns volume distribution in high dimensions; see Chapter 10.

We close this section with a discussion of S. Bernstein type inequalities about sums of independent random variables. The fact that they exhibit good ψ_α-behaviour (more precisely, the corresponding tail estimates) is a very useful tool in the applications of the probabilistic method to asymptotic geometric analysis.

3.5.1. Khintchine-Kahane inequality

The Rademacher functions $r_i : [0,1] \to \mathbb{R}$, $i \geqslant 1$, are defined by

$$r_i(t) = \mathrm{sign}\,\sin(2^{i-1}\pi t).$$

They are ± 1-valued (if we ignore a set of measure zero) independent random variables on $[0,1]$ and they form an orthonormal sequence in $L_2[0,1]$. An equivalent way to introduce a sequence with these properties is to consider $E_2 = \prod_{i=1}^\infty \{-1,1\}$ endowed with the standard product measure and to define, for every $\epsilon = (\epsilon_i)_{i=1}^\infty$,

$$r_i(\epsilon) = \epsilon_i.$$

The classical Khintchine inequality states that for every $p > 0$ there exist constants $A_p, B_p > 0$ with the following property: for every $n \geqslant 1$ and any n-tuple of real numbers $a_1, \ldots, a_n$,

$$A_p\Big(\sum_{i=1}^n a_i^2\Big)^{1/2} \leqslant \Big(\int_{E_2^n}\Big|\sum_{i=1}^n a_i\epsilon_i\Big|^p d\mu_n(\epsilon)\Big)^{1/p} \leqslant B_p\Big(\sum_{i=1}^n a_i^2\Big)^{1/2}. \tag{3.5.1}$$

Since

$$\Big(\sum_{i=1}^n a_i^2\Big)^{1/2} = \Big(\int_{E_2^n}\Big|\sum_{i=1}^n a_i\epsilon_i\Big|^2 d\mu_n(\epsilon)\Big)^{1/2}$$

for all $a_1, \ldots, a_n$, an equivalent way to state Khintchine inequality is the following.

THEOREM 3.5.1 (Khintchine). *For every $p > 0$ there exist $A_p, B_p > 0$ such that for every $n \geqslant 1$ and any $a = (a_1, \ldots, a_n) \in \ell_2^n$,*

$$A_p\Big\|\sum_{i=1}^n a_i\epsilon_i\Big\|_{L_2(E_2^n)} \leqslant \Big\|\sum_{i=1}^n a_i\epsilon_i\Big\|_{L_p(E_2^n)} \leqslant B_p\Big\|\sum_{i=1}^n a_i\epsilon_i\Big\|_{L_2(E_2^n)}. \tag{3.5.2}$$

Let A_p^*, B_p^* denote the best constants for which the statement of Theorem 3.5.1 is valid. From Hölder's inequality it is clear that $A_p^* = 1$ if $p \geqslant 2$ and $B_p^* = 1$ if $0 < p \leqslant 2$. The exact values of A_p^* and B_p^* have been determined by Szarek ($A_1^* = 1/\sqrt{2}$) and Haagerup (for all p).

One can give a proof of Khintchine's inequality using the martingales approach. We are primarily interested in the behaviour of B_p^* as $p \to \infty$. The argument below gives $B_p^* = O(\sqrt{p})$; comparing with Haagerup's exact result we see that this is the correct order of B_p^*.

Proof of Theorem 3.5.1. We describe the proof in the case $p \geqslant 1$. It will be convenient to work with the Rademacher functions $r_1, \ldots, r_n$ defined on $([0,1], \lambda)$, where λ denotes Lebesgue measure. By homogeneity, we may assume that $\{a_i\}_{i=1}^n$ satisfy $\sum_{i=1}^n a_i^2 = 1$. For $1 \leqslant k \leqslant n$ we consider the algebra F_k of finite unions of intervals of the form $[s/2^k, (s+1)/2^k)$, $s = 0, 1, \ldots, 2^k - 1$. Note that $r_1, \ldots, r_k$ are F_k-measurable. It is clear that $F_1 \subseteq F_2 \subseteq \cdots \subseteq F_n$ and we can easily check that $\{\sum_{i=1}^k a_ir_i\}_{k=1}^n$ is a martingale with respect to $\{F_k\}_{k=1}^n$.

We set $f = \sum_{i=1}^n a_ir_i$. Then, $\mathbb{E}(f|F_k) = \sum_{i=1}^k a_ir_i$ and if we define $d_k = \mathbb{E}(f|F_k) - \mathbb{E}(f|F_{k-1})$ we see that

$$\|d_k\|_\infty = \|a_kr_k\|_\infty = |a_k|. \tag{3.5.3}$$

Since $f \in L_\infty(E_2^n)$ and $\mathbb{E}(f) = 0$, applying Azuma's inequality we get

$$\mu_n\Big(\Big|\sum_{i=1}^n a_ir_i\Big| > t\Big) \leqslant 2\exp\Big(-\frac{t^2}{4\sum_{i=1}^n \|d_i\|_\infty^2}\Big) = 2e^{-\frac{t^2}{4}}$$

for every $t > 0$. Then, we write

$$\begin{aligned}\int_0^1 |\sum a_ir_i|^p d\lambda &= \int_0^\infty pe^{p-1}\lambda\Big(\Big|\sum_{i=1}^n a_ir_i\Big| > t\Big)dt \\ &\leqslant 2\int_0^\infty pt^{p-1}e^{-\frac{t^2}{4}}\,dt \\ &= 2^pp\int_0^\infty e^{-x}x^{p/2-1}dx \leqslant (C\sqrt{p})^p,\end{aligned}$$

using the fact that $(\Gamma(p/2))^{1/p} \leqslant c\sqrt{p}$ by Stirling's formula. This proves the right hand side of (3.5.2) for $p \geqslant 2$, with $B_p = O(\sqrt{p})$ as $p \to \infty$.

Next, we consider the case $1 \leqslant p \leqslant 2$ (a similar argument works for $0 < p < 1$). From Hölder's inequality we have

$$1 = \int_0^1 \Big|\sum_{i=0}^n a_i r_i\Big|^2 d\lambda = \int_0^1 \Big|\sum_{i=1}^n a_i r_i\Big|^{2/3} \Big|\sum_{i=1}^n a_i r_i\Big|^{4/3} d\lambda$$
$$\leqslant \Big(\int_0^1 \Big|\sum_{i=1}^n a_i r_i\Big| d\lambda\Big)^{2/3} \Big(\int_0^1 \Big|\sum_{i=1}^n a_i r_i\Big|^4 d\lambda\Big)^{1/3}.$$

It follows that

$$B_4^{-2} \leqslant \int_0^1 \Big|\sum_{i=1}^n a_i r_i\Big| d\lambda.$$

Combining the above we conclude that for all $1 \leqslant p < \infty$ there exist positive constants B_p, A_p such that

$$A_p^{-1} \leqslant \Big\|\sum_{i=1}^n a_i r_i\Big\|_p \leqslant B_p$$

for all $a_1, \dots, a_n$ with $\sum_{i=1}^n a_i^2 = 1$. □

Kahane's inequality generalizes Khintchine's inequality.

THEOREM 3.5.2 (Kahane). *There exists $K > 0$ such that for every normed space X, for any $n \geqslant 1$, for any $x_1, \dots, x_n \in X$ and any $p \geqslant 1$,*

$$\Big(\mathbb{E}\Big\|\sum_{i=1}^n \epsilon_i x_i\Big\|^p\Big)^{1/p} \leqslant 2\mathbb{E}\Big\|\sum_{i=1}^n \epsilon_i x_i\Big\| + K\sigma\sqrt{p}, \tag{3.5.4}$$

where

$$\sigma^2 = \sup\Big\{\sum_{i=1}^n |x^*(x_i)|^2 : x^* \in X^*, \|x^*\| \leqslant 1\Big\}. \tag{3.5.5}$$

In particular, there exists $K > 0$ such that for every normed space X,

$$\Big(\mathbb{E}\Big\|\sum_{i=1}^n \epsilon_i x_i\Big\|^p\Big)^{1/p} \leqslant (2 + K\sqrt{2}\sqrt{p})\mathbb{E}\Big\|\sum_{i=1}^n \epsilon_i x_i\Big\|. \tag{3.5.6}$$

The "in particular" follows from Khintchine's inequality, since if $\|x^*\| \leqslant 1$ then

$$\Big(\sum_{i=1}^n |x^*(x_i)|^2\Big)^{1/2} \leqslant \sqrt{2}\,\mathbb{E}\Big|\sum_{i=1}^n \epsilon_i x^*(x_i)\Big| \leqslant \sqrt{2}\,\mathbb{E}\Big\|\sum_{i=1}^n \epsilon_i x_i\Big\|,$$

which shows that

$$\sigma \leqslant \sqrt{2}\mathbb{E}\Big\|\sum_{i=1}^n \epsilon_i x_i\Big\|.$$

One may derive Theorem 3.5.2 from Borell's Lemma (Theorem 1.5.7) applied to the full cube, and this gives an estimate with a constant of the order of p instead of $\sqrt{p}$. We will give a proof of Theorem 3.5.2 using concentration of the values of convex Lipschitz functions around their Lévy mean, a fact which follows from Talagrand's Theorem 3.1.21 and which we proved as Theorem 3.2.7 above. We use it to get

PROPOSITION 3.5.3. *Let X be a normed space and let $(x_i)_{i=1}^n$ be a sequence of vectors in X. We set*

$$\sigma^2 = \sup\Big\{\sum_{i=1}^n |x^*(x_i)|^2 : x^* \in X^*, \|x^*\| \leqslant 1\Big\}.$$

Let L be a Lévy mean of $\|\sum_{i=1}^n \epsilon_i x_i\|$ on E_2^n. Then, for all $t \geqslant 0$,

$$\mu_n\Big(\{|\, \|\sum_{i=1}^n \epsilon_i x_i\| - L\,| \geqslant t\}\Big) \leqslant 4e^{-t^2/8\sigma^2}.$$

Proof. Define $f(u) = \|\sum_{i=1}^n u_i x_i\|$. Note that f is a convex function. We shall use Theorem 3.2.7. We claim that σ in the statement of the proposition is an upper bound for the Lipschitz constant of f. Indeed let $x^* \in X^*$ with $\|x^*\| \leqslant 1$ and $u, v \in \mathbb{R}^n$. From the Cauchy-Schwarz inequality we have

$$\begin{aligned}\Big|x^*\Big(\sum_{i=1}^n u_i x_i - \sum_{i=1}^n v_i x_i\Big)\Big| &= \Big|\sum_{i=1}^n (u_i - v_i) x^*(x_i)\Big| \\ &\leqslant \Big(\sum_{i=1}^n |x^*(x_i)|^2\Big)^{1/2} \Big(\sum_{i=1}^n (u_i - v_i)^2\Big)^{1/2} \\ &\leqslant \sigma |u - v|.\end{aligned}$$

By the Hahn-Banach theorem we conclude that

$$|f(u) - f(v)| \leqslant \Big\|\sum_{i=1}^n u_i x_i - \sum_{i=1}^n v_i x_i\Big\| \leqslant \sigma |u - v|,$$

and hence f is Lipschitz with constant σ. Applying Theorem 3.2.7 for f we get the result. $\square$

Now, we can give a proof of the Khintchine-Kahane inequality with the best possible dependence on p.

Proof of Theorem 3.5.2. We define $f : E_2^n \to \mathbb{R}^+$ by

$$f(\epsilon_1, \ldots, \epsilon_n) = \Big|\, \Big\|\sum_{i=1}^n \epsilon_i x_i\Big\| - L\, \Big|,$$

where L is a Lévy mean of $\|\sum_{i\leqslant n} \epsilon_i x_i\|$ on E_2^n. From Proposition 3.5.3, making the change of variables $s = t^2/8\sigma^2$ we have

$$\begin{aligned}\int_{E_2^n} \Big|\, \Big\|\sum_{i=1}^n \epsilon_i x_i\Big\| - L\, \Big|^p d\mu_n &= p \int_0^\infty t^{p-1} \mu_n\Big(\Big\{\epsilon : |\|\sum_{i=1}^n \epsilon_i x_i\| - L| \geqslant t\Big\}\Big) dt \\ &\leqslant 4p \int_0^\infty t^{p-1} e^{-t^2/8\sigma^2} dt \\ &= 2^{p+1} p (\sqrt{2}\sigma)^p \int_0^\infty e^{-s} s^{p/2-1} ds \leqslant (K\sigma\sqrt{p})^p\end{aligned}$$

for some absolute constant $K > 0$. Thus,

$$\Big(\int |\, \|\sum_{i=1}^n \epsilon_i x_i\| - L|^p d\mu_n\Big)^{1/p} \leqslant K\sigma\sqrt{p}.$$

By the triangle inequality in $L_p(E_2^n,\mu_n)$,

$$\Big(\int_{E_2^n}\|\sum_{i=1}^n \epsilon_i x_i\|^p d\mu_n\Big)^{1/p} \leqslant L + K\sigma\sqrt{p} \tag{3.5.7}$$

for every $p\geqslant 1$. Finally, we observe that $L\leqslant 2\mathbb{E}\,\|\sum_{i\leqslant n}\epsilon_i x_i\|$ by Markov's inequality. □

3.5.2. ψ_α-estimates

DEFINITION 3.5.4 (Orlicz-norm). Let $(\Omega,\mathcal{A},\mu)$ be a probability space and let $\psi(t)$ be a strictly increasing convex function on $[0,\infty)$ with $\psi(0)=0$ and $\lim_{t\to\infty}\psi(t)=\infty$. We denote by $L_\psi(\mu)$ the space of all real valued measurable functions on Ω which satisfy $\int_\Omega \psi(|f|/\lambda)d\mu<\infty$ for some $\lambda>0$ and we define

$$\|f\|_{L_\psi(\mu)} = \inf\Big\{\lambda>0 : \int_\Omega \psi(|f|/\lambda)d\mu\leqslant 1\Big\}.$$

One can easily check that $\|\cdot\|_{L_\psi(\mu)}$ is a norm on $L_\psi(\mu)$.

For any $\alpha\geqslant 1$ we consider the function $\psi_\alpha(t)=e^{t^\alpha}-1$. Then, the ψ_α-norm of $f\in L_{\psi_\alpha(\mu)}$ is defined as follows:

$$\|f\|_{\psi_\alpha} := \inf\Big\{\lambda>0 : \int_\Omega \exp\Big(\Big(\frac{|f(\omega)|}{\lambda}\Big)^\alpha\Big)\,d\mu(\omega)\leqslant 2\Big\}.$$

We shall be concerned mainly with the two functions

$$\psi_1(t)=e^t-1 \quad\text{and}\quad \psi_2(t)=e^{t^2}-1$$

(besides of course the functions t^p, $1\leqslant p<\infty$, which give rise to the usual L_p spaces).

The next lemma gives an equivalent expression for the ψ_α-norm of a function in terms of its L_q-norms.

LEMMA 3.5.5. *Let $(\Omega,\mathcal{A},\mu)$ be a probability space. Let $\alpha\geqslant 1$ and let $f:\Omega\to\mathbb{R}$ be a function in $L_{\psi_\alpha}(\mu)$. Then,*

$$\|f\|_{\psi_\alpha}\simeq \sup\Big\{\frac{\|f\|_{L_p(\mu)}}{p^{1/\alpha}} : p\geqslant\alpha\Big\}.$$

Proof. First we show that there exists an absolute constant $C>0$ such that for any $p\geqslant\alpha$ we have

$$\|f\|_p\leqslant Cp^{1/\alpha}\|f\|_{\psi_\alpha}.$$

Indeed, we set $A:=\|f\|_{\psi_\alpha}$ and using the elementary inequality $1+\frac{t^k}{k!}\leqslant e^t$, which is true for any $t>0$, we obtain

$$1+\int_\Omega \frac{|f(\omega)|^{k\alpha}}{k!A^{k\alpha}}\,d\mu(\omega)\leqslant \int_\Omega \exp\left((|f|/A)^\alpha\right)d\mu = 2,$$

which implies

$$\int_\Omega |f|^{k\alpha}\,d\mu\leqslant k!A^{k\alpha}.$$

Let $p \geqslant \alpha$. There exists a unique $k \in \mathbb{N}$ such that $k\alpha \leqslant p < (k+1)\alpha$. Then, using Hölder's inequality we get

$$\begin{aligned}\|f\|_p &\leqslant \|f\|_{(k+1)\alpha} \leqslant [(k+1)!]^{\frac{1}{(k+1)\alpha}} A \leqslant (2k)^{1/\alpha} A \\ &\leqslant \left(\frac{2p}{\alpha}\right)^{1/\alpha} A \leqslant 2p^{1/\alpha} A.\end{aligned}$$

Conversely, if $\gamma := \sup_{p\geqslant\alpha} \frac{\|f\|_p}{p^{1/\alpha}}$, then $\int_\Omega |f|^p \, d\mu \leqslant \gamma^p p^{p/\alpha}$ for all $p \geqslant \alpha$. Then, for any $c > 0$, we write

$$\begin{aligned}\int_\Omega \exp\left((|f|/c\gamma)^\alpha\right) d\mu &= 1 + \sum_{k=1}^\infty \frac{1}{(c\gamma)^{k\alpha} k!} \int_\Omega |f|^{k\alpha} \, d\mu \leqslant 1 + \sum_{k=1}^\infty \frac{(k\alpha)^k}{k! c^{k\alpha}} \\ &\leqslant 1 + \sum_{k=1}^\infty \left(\frac{e\alpha}{c^\alpha}\right)^k,\end{aligned}$$

where we have used the elementary inequality $k! \geqslant (k/e)^k$. If we choose $c_\alpha := (2e\alpha)^{1/\alpha} \leqslant 2e \cdot e^{1/e} := c$, then we have $\|f\|_{\psi_\alpha} \leqslant c_\alpha \gamma \leqslant c\gamma$. □

DEFINITION 3.5.6. Let μ be a probability measure on $\mathbb{R}^n$, and let $\alpha \geqslant 1$ and $\theta \in S^{n-1}$. We say that μ satisfies a *ψ_α-estimate with constant $b_\alpha = b_\alpha(\theta)$ in the* direction of θ if we have

$$\|\langle \cdot, \theta\rangle\|_{\psi_\alpha} \leqslant b_\alpha \|\langle \cdot, \theta\rangle\|_2.$$

We say that μ is a ψ_α-measure with constant $B_\alpha > 0$ if

$$\sup_{\theta\in S^{n-1}} \frac{\|\langle \cdot, \theta\rangle\|_{\psi_\alpha}}{\|\langle \cdot, \theta\rangle\|_2} \leqslant B_\alpha.$$

Using Lemma 3.5.5 we see that μ satisfies a ψ_α-estimate with constant b_α in the direction of $\theta \in S^{n-1}$ if

$$\|\langle \cdot, \theta\rangle\|_p \leqslant c b_\alpha p^{1/\alpha} \|\langle \cdot, \theta\rangle\|_2$$

for all $p \geqslant \alpha$. The next lemma gives one more description of the ψ_α-norm for linear functionals.

LEMMA 3.5.7. *Let μ be a probability measure on $\mathbb{R}^n$ and let $\alpha \geqslant 1$ and $\theta \in S^{n-1}$.*

(i) *If μ satisfies a ψ_α-estimate with constant b in the direction of θ then for all $t > 0$ we have $\mu(\{x : |\langle x, \theta\rangle| \geqslant t\|\langle \cdot, \theta\rangle\|_2\}) \leqslant 2e^{-t^\alpha/b^\alpha}$.*
(ii) *If we have $\mu(\{x : |\langle x, \theta\rangle| \geqslant t\|\langle \cdot, \theta\rangle\|_2\}) \leqslant 2e^{-t^\alpha/b^\alpha}$ for some $b > 0$ and for all $t > 0$ then μ satisfies a ψ_α-estimate with constant $\leqslant cb$ in the direction of θ, where $c > 0$ is an absolute constant.*

Proof. The first assertion is a direct application of Markov's inequality. For the second, it suffices to prove that

$$\left(\int_{\mathbb{R}^n} |\langle x, \theta\rangle|^p \, d\mu(x)\right)^{1/p} \leqslant cbp^{1/\alpha} \|\langle \cdot, \theta\rangle\|_2,$$

for any $p \geqslant \alpha$, where $c > 0$ is an absolute constant. We write

$$\begin{aligned}\int_{\mathbb{R}^n} |\langle x, \theta\rangle|^p \, d\mu(x) &= \int_0^\infty pt^{p-1}\mu(\{x : |\langle x, \theta\rangle| \geqslant t\})\, dt \\ &= \|\langle \cdot, \theta\rangle\|_2^p \int_0^\infty pt^{p-1}\mu(\{x : |\langle x, \theta\rangle| \geqslant t\|\langle \cdot, \theta\rangle\|_2\})\, dt \\ &\leqslant 2\|\langle \cdot, \theta\rangle\|_2^p \int_0^\infty pt^{p-1}e^{-t^\alpha/b^\alpha}\, dt,\end{aligned}$$

using the tail estimate. Making the change of variables $s = (t/b)^\alpha$, we arrive at

$$\begin{aligned}\int_{\mathbb{R}^n} |\langle x, \theta\rangle|^p \, d\mu(x) &\leqslant 2(b\|\langle, \theta\rangle\|_2)^p \int_0^\infty \frac{p}{\alpha} s^{p/\alpha - 1} e^{-s}\, ds \\ &= 2(b\|\langle \cdot, \theta\rangle\|_2)^p \Gamma\Big(\frac{p}{\alpha} + 1\Big).\end{aligned}$$

Using Stirling's formula, we get the result. □

We close this section with a useful estimate on the expectation of the maximum of a finite set of ψ_α-random variables.

PROPOSITION 3.5.8. *Let $\alpha \geqslant 1$ and assume that the random variables $\{X_i\}_{i=1}^N$, $N \geqslant 2$, satisfy the ψ_α-estimate*

$$\|X_i\|_{\psi_\alpha} \leqslant b$$

for all $i = 1, \ldots, N$. Then

$$\mathbb{E} \max_{1 \leqslant i \leqslant N} |X_i| \leqslant Cb(\log N)^{1/\alpha},$$

where $C > 0$ is an absolute constant.

Proof. By the definition of the ψ_α-norm and Markov's inequality, for every $t > 0$ we have

$$\begin{aligned}\text{Prob}\left(\max_{1 \leqslant i \leqslant N} |X_i| \geqslant t\right) &\leqslant \sum_{i=1}^N \text{Prob}\,(x \in K : |\langle x, \theta_i\rangle| \geqslant t) \\ &\leqslant 2N \exp\left(-(t/b)^\alpha\right).\end{aligned}$$

Then, given $A > 0$ we may write

$$\begin{aligned}\mathbb{E} \max_{1 \leqslant i \leqslant N} |X_i| &= \int_0^\infty \text{Prob}\left(\max_{1 \leqslant i \leqslant N} |X_i| \geqslant t\right) dt \\ &\leqslant A + \int_A^\infty \text{Prob}\left(\max_{1 \leqslant i \leqslant N} |X_i| \geqslant t\right) dt \\ &\leqslant A + 2N \int_A^\infty \exp\left(-(t/b)^\alpha\right) dt.\end{aligned}$$

Choosing $A = 4b(\log N)^{1/\alpha}$ we get

$$\begin{aligned}\int_A^\infty \exp\left(-(t/b)^\alpha\right) dt &= 4b(\log N)^{1/\alpha}\int_1^\infty \exp(-4^\alpha s^\alpha \log N) ds\\ &\leqslant 4b(\log N)^{1/\alpha}\int_1^\infty \exp(-4s\log N) ds\\ &\leqslant 4b(\log N)^{1/\alpha}\exp(-2\log N)\int_1^\infty e^{-s} ds\\ &\leqslant 4b(\log N)^{1/\alpha}N^{-2},\end{aligned}$$

where we have used the fact that

$$\exp(-4s\log N) \leqslant \exp(-2\log N)\cdot e^{-s}$$

is valid for all $s \geqslant 1$. It follows that

$$\mathbb{E}\max_{1\leqslant i\leqslant N}|X_i| \leqslant Cb(\log N)^{1/\alpha}$$

with $C = 8$. □

3.5.3. Log-concave probability measures

DEFINITION 3.5.9. A Borel probability measure μ on $\mathbb{R}^n$ is called *log-concave* if for all non-empty compact subsets A, B of $\mathbb{R}^n$ and all $0 < \lambda < 1$ we have

$$\mu((1-\lambda)A + \lambda B) \geqslant \mu(A)^{1-\lambda}\mu(B)^\lambda.$$

A function $f : \mathbb{R}^n \to [0, \infty)$ is called *log-concave* if

$$f((1-\lambda)x + \lambda y) \geqslant f(x)^{1-\lambda}f(y)^\lambda$$

for all $x, y \in \mathbb{R}^n$ and any $0 < \lambda < 1$.

Let $\mathcal{P}_n$ denote the class of Borel probability measures on $\mathbb{R}^n$ which are absolutely continuous with respect to the Lebesgue measure. By a theorem of C. Borell, if μ is a log-concave probability measure on $\mathbb{R}^n$ such that $\mu(H) < 1$ for any hyperplane H, then $\mu \in \mathcal{P}_n$ and has a log-concave density f, that is $d\mu(x) = f(x)\,dx$. The opposite direction follows from the Prékopa-Leindler inequality, namely that a measure with log-concave density is log-concave.

The barycenter of a density f is defined as

$$\text{bar}(f) = \int_{\mathbb{R}^n} xf(x)\,dx.$$

In particular, f has barycenter (or center of mass) at the origin if

$$\int_{\mathbb{R}^n}\langle x, \theta\rangle f(x)\,dx = 0$$

for all $\theta \in S^{n-1}$. A non-degenerate log-concave probability measure μ will be called centered if its density f has barycenter at the origin (if so, we will say that f is centered as well).

As we have seen in Chapter 1, for a convex body of volume 1, $K \subset \mathbb{R}^n$ setting

$$\mu_K(A) = \text{Vol}_n(K \cap A) = \int_A \mathbf{1}_K(x)dx$$

for every Borel $A \subseteq \mathbb{R}^n$, the Brunn-Minkowski inequality implies that μ_K is a log-concave probability measure. Similarly for every $c > 0$, the function $f_c(x) = \exp(-c|x|^2)$ is even and log-concave on $\mathbb{R}^n$ so the measure

$$\gamma_{n,c}(A) = \frac{1}{I(c)} \int_A \exp(-c|x|^2)dx$$

where $I(c) = \int_{\mathbb{R}^n} \exp(-c|x|^2)dx$, is a symmetric log-concave probability measure. In particular, this holds true for the standard Gaussian measure γ_n.

Borell's lemma (Theorem 1.5.7) remains valid in the more general context of log-concave probability measures.

LEMMA 3.5.10 (Borell). *Let μ be a log-concave measure in $\mathcal{P}_n$. Then, for any symmetric convex set A in $\mathbb{R}^n$ with $\mu(A) = \alpha \in (1/2, 1)$ and any $t > 1$ we have*

$$1 - \mu(tA) \leqslant \alpha\Big(\frac{1-\alpha}{\alpha}\Big)^{\frac{t+1}{2}}. \tag{3.5.8}$$

Proof. Using the symmetry and convexity of A we check that

$$\frac{2}{t+1}\mathbb{R}^n \setminus (tA) + \frac{t-1}{t+1}A \subseteq \mathbb{R}^n \setminus A.$$

for every $t > 1$. Then, we apply the log-concavity of μ to get the result. □

Using Borell's lemma we see that there exists an absolute constant $C > 0$ such that every log-concave measure $\mu \in \mathcal{P}_n$ is a ψ_1-measure with constant C.

THEOREM 3.5.11. *Let μ be a non-degenerate log-concave probability measure on $\mathbb{R}^n$. If $f : \mathbb{R}^n \to \mathbb{R}$ is a seminorm then, for any $q > p \geqslant 1$, we have*

$$\Big(\int_{\mathbb{R}^n} |f|^p \, d\mu\Big)^{1/p} \leqslant \Big(\int_{\mathbb{R}^n} |f|^q \, d\mu\Big)^{1/q} \leqslant c\frac{q}{p}\Big(\int_{\mathbb{R}^n} |f|^p \, d\mu\Big)^{1/p},$$

where $c > 0$ is an absolute constant.

Proof. We write $\|f\|_p^p := \int |f|^p \, d\mu$. Then, the set

$$A = \{x \in \mathbb{R}^n : |f(x)| \leqslant 3\|f\|_p\}$$

is symmetric and convex with $\mu(A) \geqslant 1 - 3^{-p}$, which implies $\frac{1-\alpha}{\alpha} \leqslant \frac{3^{-p}}{1-3^{-p}} \leqslant e^{-p/2}$. Note that for any $t > 0$ we have

$$tA = \{x \in \mathbb{R}^n : |f(x)| \leqslant 3t\|f\|_p\}.$$

From Borell's lemma we see that

$$\mu(\{x : |f(x)| \geqslant 3t\|f\|_p\}) \leqslant e^{-c_1 p(t+1)} \leqslant e^{-c_1 pt}$$

for any $t > 1$, with $c_1 = \frac{1}{4}$. Now, we write

$$\begin{aligned}\int |f|^q \, d\mu &= \int_0^\infty qs^{q-1}\mu(\{x : |f(x)| \geqslant s\})\, ds \\ &\leqslant \int_0^{3\|f\|_p} qs^{q-1} dx + \int_{3\|f\|_p}^\infty qs^{q-1}\mu(\{x : |f(x)| \geqslant s\})\, ds \\ &\leqslant (3\|f\|_p)^q + (3\|f\|_p)^q \int_1^\infty qt^{q-1}e^{-c_1pt}\, dt \\ &\leqslant (3\|f\|_p)^q + (3\|f\|_p)^q \int_0^\infty qt^{q-1}e^{-c_1pt}\, dt \\ &\leqslant (3\|f\|_p)^q + \Big(\frac{3\|f\|_p}{c_1p}\Big)^q \Gamma(q+1).\end{aligned}$$

Stirling's formula and the fact that $(a+b)^{1/q} \leqslant a^{1/q} + b^{1/q}$ for all $a, b > 0$ and $q \geqslant 1$, imply that $\|f\|_{L_q(\mu)} \leqslant c\frac{q}{p}\|f\|_{L_p(\mu)}$. □

REMARKS 3.5.12. (i) Any map $x \mapsto |\langle x, \theta\rangle|$ (where $\theta \in S^{n-1}$) satisfies the hypothesis of Theorem 3.5.11. Therefore,

$$\|\langle\cdot, \theta\rangle\|_q \leqslant c_1 q \|\langle\cdot, \theta\rangle\|_1$$

for all $\theta \in S^{n-1}$ and $q \geqslant 1$, where $c_1 > 0$ is an absolute constant. It follows that

$$\|\langle\cdot, \theta\rangle\|_{\psi_1} \leqslant c_2 \|\langle\cdot, \theta\rangle\|_1$$

for all $\theta \in S^{n-1}$. By Lemma 3.5.7 we have that, for all $t > 0$,

$$\mu(\{x : |\langle x, \theta\rangle| \geqslant t\|\langle\cdot, \theta\rangle\|_1\}) \leqslant 2e^{-t/c_2}.$$

(ii) Since the n-dimensional Gaussian measure is a log-concave probability measure, any seminorm f satisfies the conclusion of Theorem 3.5.11. On the other hand, integrating in polar coordinates we get

$$\Big(\int |f(x)|^q \, d\gamma_n(x)\Big)^{1/q} \simeq \sqrt{n+q}\Big(\int_{S^{n-1}} |f(\theta)|^q \, d\sigma(\theta)\Big)^{1/q},$$

for any $q \geqslant 1$. Combining these estimates we obtain

$$\Big(\int_{S^{n-1}} |f(\theta)|^q \, d\sigma(\theta)\Big)^{1/q} \leqslant c\frac{q}{p}\sqrt{\frac{n+p}{n+q}}\Big(\int_{S^{n-1}} |f(\theta)|^p \, d\sigma(\theta)\Big)^{1/p}$$

for any $1 \leqslant p \leqslant q$, where $c > 0$ is an absolute constant.

3.5.4. Bernstein type inequalities

In this subsection we discuss a number of Bernstein type inequalities about sums of independent random variables $X_1, \dots, X_N$ on some probability space (Ω, μ). For the original versions of these results see the notes and remarks at the end of this chapter.

Estimates for the probability

$$\mathbb{P}\Big(\Big\{|\sum_{j=1}^N X_j| > tN\Big\}\Big), \qquad t > 0$$

naturally depend on our information on the X_j's; the results that we present below have proved to be an extremely useful tool in applications of the probabilistic method to asymptotic geometric analysis. A sample of such applications will appear in subsequent chapters. We shall present three main cases of the inequality: bounded random variables, ψ_1 random variables and ψ_2 random variables.

THEOREM 3.5.13 (Bernstein). *Let $\{X_j\}_{j=1}^N$ be a sequence of independent random variables with mean 0 on some probability space (Ω,μ). Assume that $\|X_j\|_\infty \leqslant M$ for all j and some constant M. Let $\sigma^2 = \frac{1}{N}\sum_{j=1}^N \mathbb{E}(X_j^2)$. Then, for all $t>0$,*

$$\mathbb{P}\Big(\Big\{\sum_{j=1}^N X_j \geqslant tN\Big\}\Big) \leqslant \exp\Big(-\frac{\sigma^2 N}{M^2}F\Big(\frac{Mt}{\sigma^2}\Big)\Big),$$

where $F(u) = (1+u)\log(1+u) - u$, $u>0$.

Proof. Let $t>0$. From Markov's inequality and by independence we have

$$\begin{aligned}\mathbb{P}\Big(\sum_{j=1}^N X_j \geqslant tN\Big) &= \mathbb{P}\Big(\Big\{\exp\Big(\frac{\lambda}{N}\sum_{j=1}^N X_j\Big) \geqslant e^{\lambda t}\Big\}\Big)\\ &\leqslant e^{-\lambda t}\mathbb{E}\exp\Big(\frac{\lambda}{N}\sum_{j=1}^N X_j\Big)\\ &= e^{-\lambda t}\prod_{j=1}^N \mathbb{E}\exp(\lambda X_j/N)\end{aligned}$$

for every $\lambda>0$. Next, observe that since $\mathbb{E}(X_j)=0$,

$$\begin{aligned}\mathbb{E}\exp(\lambda X_j/N) &= 1 + \sum_{k=2}^\infty \frac{\lambda^k \mathbb{E}(X_j^k)}{N^k k!} \leqslant 1 + \mathbb{E}(X_j^2)\sum_{k=2}^\infty \frac{\lambda^k M^{k-2}}{N^k k!}\\ &= 1 + \frac{\mathbb{E}(X_j^2)}{M^2}\Big(e^{\frac{\lambda M}{N}} - \frac{\lambda M}{N} - 1\Big).\end{aligned}$$

Since $e^u \geqslant 1+u$ for all $u\in\mathbb{R}$, we get

$$\prod_{j=1}^N \mathbb{E}\exp(\lambda X_j/N) \leqslant \exp\Big(\frac{\sum_{j=1}^N \mathbb{E}(X_j)^2}{M^2}\Big(e^{\frac{\lambda M}{N}} - \frac{\lambda M}{N} - 1\Big)\Big).$$

By the definition of σ^2 this shows that

$$\mathbb{P}\Big(\sum_{j=1}^N X_j \geqslant tN\Big) \leqslant \exp\Big(\frac{\sigma^2 N}{M^2}\Big(e^{\frac{\lambda M}{N}} - \frac{\lambda M}{N} - 1\Big) - \lambda t\Big)$$

for all $\lambda>0$. We choose λ so that $\exp(\lambda M/N) = 1 + tM/\sigma^2$ to conclude the proof. $\square$

One can check that $F(u) \geqslant u^2/(2+2u/3)$ for all $u>0$, and hence, $F(u)\geqslant 3u/8$ if $u\geqslant 1$ and $F(u)\geqslant 3u^2/8$ if $0<u\leqslant 1$. Thus, we can state Theorem 3.5.13 in the following form.

THEOREM 3.5.14 (Bernstein). *Let $\{X_j\}_{j=1}^N$ be a sequence of independent random variables with mean 0 on some probability space (Ω,μ). Assume that $\|X_j\|_\infty \leqslant$*

M for all j and some constant M. Let $\sigma^2 = \frac{1}{N}\sum_{j=1}^N \mathbb{E}(X_j^2)$. Then, for all $t > 0$,

$$\mathbb{P}\Big(\Big\{\sum_{j=1}^N X_j \geqslant tN\Big\}\Big) \leqslant \exp\Big(-\frac{3N}{8}\min\Big\{\frac{t^2}{\sigma^2}, \frac{t}{M}\Big\}\Big).$$

In the theorem below we have some uniform bound on the L_1 and the L_∞-norm of the X_j's.

THEOREM 3.5.15 (Bernstein). *Let $\{X_j\}_{j=1}^\infty$ be a sequence of independent random variables with mean 0 on some probability space (Ω, μ). Assume that X_j are all bounded and that $\mathbb{E}|X_j| \leqslant 2$ and $\|X_j\|_\infty \leqslant M$ for all j and some constant M. Then, for $0 < t < 1$,*

$$\mathbb{P}\Big(\Big\{\Big|\sum_{j=1}^N X_j\Big| \geqslant tN\Big\}\Big) \leqslant 2\exp(-t^2N/(8M)).$$

Proof. Using that $e^x \leqslant 1 + x + x^2$ for $0 \leqslant x \leqslant 1$ and that $\mathbb{E}(X_j) = 0$ and $|X_j| \leqslant M$ we get for $0 < \lambda \leqslant 1/M$ that

$$\mathbb{E}\exp(\lambda X_j) \leqslant 1 + \lambda^2\mathbb{E}(X_j^2) \leqslant 1 + \lambda^2\|X_j\|_1\|X_j\|_\infty \leqslant \exp(2\lambda^2 M).$$

By independence,

$$\mathbb{E}\exp(\sum_{j=1}^N \lambda X_j) = \prod_{j=1}^N \mathbb{E}\exp(\lambda X_j) \leqslant \exp(2\lambda^2 MN).$$

Finally use, for $\lambda = t/(4M)$,

$$\begin{aligned}\mathbb{P}\Big(\Big\{\sum_{j=1}^N X_j > tN\Big\}\Big) &\leqslant \exp(-\lambda tN)\mathbb{E}\exp\Big(\lambda\sum_{j=1}^N X_j\Big)\\ &\leqslant \exp(-\lambda tN)\exp(2\lambda^2 MN)\\ &= \exp(2\lambda^2 MN - \lambda tN) = \exp(-t^2N/(8M)).\end{aligned}$$

By considering the symmetric case we get the required estimate. □

Next we consider the case of independent ψ_1-random variables.

THEOREM 3.5.16 (Bernstein). *Let $\{X_j\}_{j=1}^N$ be a sequence of independent random variables with mean 0 on some probability space (Ω, μ). Assume that $\|X_j\|_{L_{\psi_1}} \leqslant M$ for all j and some constant M. Let $\sigma^2 = \frac{1}{N}\sum_{j=1}^N \|X_j\|_{L_{\psi_1}}^2$. Then, for all $t > 0$,*

$$\mathbb{P}\Big(\Big\{\Big|\sum_{j=1}^N X_j\Big| \geqslant tN\Big\}\Big) \leqslant \exp\Big(-cN\min\Big\{\frac{t^2}{\sigma^2}, \frac{t}{M}\Big\}\Big),$$

where $c > 0$ is an absolute constant.

Proof. We start with the observation that

$$\mathbb{E}|X_j|^k \leqslant k!\|X_j\|_{\psi_1}^k$$

for all j and all $k \geqslant 1$. Since we also have $\mathbb{E}(X_j) = 0$, we get

$$\begin{aligned}\mathbb{E}\exp(\lambda X_j/N) &\leqslant 1 + \sum_{k=2}^{\infty} \frac{\lambda^k \mathbb{E}|X_j|^k}{N^k k!} \leqslant 1 + \frac{\lambda^2 \|X_j\|_{\psi_1}^2}{N^2\left(1 - \frac{\lambda}{N}\|X_j\|_{\psi_1}\right)} \\ &\leqslant 1 + \frac{\lambda^2 \|X_j\|_{\psi_1}^2}{N^2\left(1 - \frac{\lambda M}{N}\right)}\end{aligned}$$

for all $0 < \lambda < N/M$. It follows that

$$\mathbb{E}\exp(\lambda Y) \leqslant \exp\Big(\frac{\lambda^2}{N^2\left(1 - \frac{\lambda M}{N}\right)} \sum_{j=1}^{N} \|X_j\|_{\psi_1}^2\Big) = \exp\Big(\frac{\lambda^2 \sigma^2}{N - \lambda M}\Big),$$

where $Y = \frac{1}{N}\sum_{j=1}^{N} X_j$.

Let $t > 0$. From Markov's inequality we see that

$$\mathbb{P}(Y \geqslant t) \leqslant \exp\Big(-\lambda t + \frac{\lambda^2 \sigma^2}{N - \lambda M}\Big)$$

for all $0 < \lambda < N/M$. We consider two cases:

(i) If $t \leqslant \sigma^2/M$ we choose $\lambda = tN/(4\sigma^2) \leqslant N/(4M)$ to get

$$\mathbb{P}(Y \geqslant t) \leqslant \exp\Big(-\frac{t^2 N}{6\sigma^2}\Big).$$

(ii) If $t > \sigma^2/M$ we choose $\lambda = N/(4M)$ to get

$$\mathbb{P}(Y \geqslant t) \leqslant \exp\Big(-\frac{tN}{6M}\Big).$$

Repeating the same argument for $-Y$ we get the theorem. □

Finally, we consider the case where $\|X_j\|_{L_{\psi_2}} \leqslant A$.

THEOREM 3.5.17 (Bernstein). *Let $\{X_j\}_{j=1}^{\infty}$ be a sequence of independent random variables with mean 0 on some probability space (Ω, μ). Assume that X_j belong to $L_{\psi_2}(\mu)$ and that $\|X_j\|_{L_{\psi_2}} \leqslant A$ for all j and some constant $A > 0$. Then, for all $t > 0$,*

$$\mathbb{P}\Big(\Big\{\Big|\sum_{j=1}^{N} X_j\Big| \geqslant tN\Big\}\Big) \leqslant 2\exp(-t^2 N/(8A)).$$

Note that, in contrast to the previous cases, in Theorem 3.5.16 we do not have any restriction on the size of t.

Proof. Without loss of generality we may assume that $\|X_j\|_{L_{\psi_2}} \leqslant 1$ and in particular for all $k \in \mathbb{N}$ we have

$$\mathbb{E}(|X_j|^{2k}) \leqslant k!.$$

Since $|X_j|^{2k-1} \leqslant |X_j|^{2k}$ if $|X_j| \geqslant 1$ and $|X_j|^{2k-1} \leqslant 1$ otherwise, we also have $\mathbb{E}(|X_j|^{2k-1}) \leqslant k! + 1$ for all $k \geqslant 1$. We deduce

$$\begin{aligned}\mathbb{E}(\exp(\lambda X_j)) &\leqslant 1 + \sum_{k=2}^{\infty} \frac{1}{k!}\mathbb{E}(|\lambda X_j|^k) \\ &\leqslant 1 + \lambda^2 + \sum_{k=2}^{\infty} \lambda^{2k}\Big((k!+1)\frac{1}{(2k-1)!} + k!\frac{1}{(2k)!}\Big) \\ &\leqslant e^{2\lambda^2}\end{aligned}$$

for all $\lambda > 0$, where the last inequality may be checked using Taylor expansion for $\exp(2\lambda^2)$ and a term-by-term comparison. Consequently,

$$\begin{aligned}\mathbb{P}\Big(\Big\{\sum_{j=1}^{N} X_j > tN\Big\}\Big) &= \mathbb{P}\Big(\Big\{\exp\Big(\lambda\sum_{j=1}^{N} X_j\Big) > \exp(\lambda tN)\Big\}\Big)\\ &\leqslant \exp(-\lambda tN)\mathbb{E}\exp\Big(\lambda\sum_{j=1}^{N} X_j\Big)\\ &= \exp(-\lambda tN)\prod_{j=1}^{N}\mathbb{E}(\exp(\lambda X_j))\\ &\leqslant \exp(-\lambda tN)\exp(2\lambda^2 N) = \exp(2\lambda^2 N - \lambda tN).\end{aligned}$$

By taking $\lambda = t/4$ we deduce

$$\mathbb{P}\Big(\Big\{\sum_{j=1}^{N} X_j > tN\Big\}\Big) \leqslant \exp(-t^2N/8)$$

and by considering the symmetric case we get the estimate. □

3.6. Raz's Lemma

Recall that $\nu_{n,k}$ is the normalized Haar measure on $G_{n,k}$ and, as usual, σ is the normalized Lebesgue measure on S^{n-1}. Given a subspace $V \in G_{n,k}$ we write σ_V for the normalized Lebesgue measure on $S^{n-1} \cap V$ and for $A \subset S^{n-1}$ we consider the conditional probability measure $\sigma|_V(A) = \sigma_V(A \cap V)$. Of course we have that

$$\sigma(A) = \int_{G_{n,k}} \sigma|_V(A)d\nu_{n,k}(V).$$

The function $V \mapsto \sigma|_V(A)$ need not be continuous. However, the following result of R. Raz states that it is very well concentrated around its mean $\sigma(A)$.

THEOREM 3.6.1 (Raz). *Let $A \subset S^{n-1}$ and denote $\sigma(A) = \alpha$. Then for any $k < n$ and $\varepsilon > 0$ one has*

$$\nu_{n,k}(\{V : |\sigma|_V(A) - \alpha| \geqslant \varepsilon\}) \leqslant 2e^{-\varepsilon^2 k/2}.$$

The proof is a careful analysis of conditional probability. The main difficulty lies in the following observation: picking randomly a two dimensional subspace, and in it, randomly and independently, two points, their joint distribution is not the same as that of simply choosing independently two random points on the sphere. We shall use the following definition and notation for the conditional probability of an "event" (which is a subset $S \subset S^{n-1} \times \cdots \times S^{n-1}$, k times) given $V \in G_{n,k}$:

$$P(S|V) = \int_{x_1,\dots,x_n \in S^{n-1}} \int_{T \in O(\mathrm{span}\,\{x_1,\dots,x_k\}:V)} \mathbf{1}_S(T(x_1),\dots,T(x_k)),$$

where the integral is calculated with respect to Haar measures, and $O(U : V)$ stands for the collection of unitary transformations from U to V. Clearly, $P(\cdot|V)$ it is concentrated on $(S^{n-1} \cap V) \times \cdots \times (S^{n-1} \cap V)$, k times.

Note that under this definition, by a standard change of variables,

$$\int_{V\in G_{n,k}} P(S|V)d\nu_{n,k} = P(S),$$

P being the product of k Haar measures on S^{n-1}. Indeed,

$$\begin{aligned}
&\int_{V\in G_{n,k}}\int_{x_1,\dots,x_n\in S^{n-1}}\int_{T\in O(\mathrm{span}\,\{x_1,\dots,x_k\}:V)} \mathbf{1}_S(T(x_1),\dots,T(x_k))\\
&= \int_{x_1,\dots,x_n\in S^{n-1}}\int_{V\in G_{n,k}}\int_{T\in O(\mathrm{span}\,\{x_1,\dots,x_k\}:V)} \mathbf{1}_S(T(x_1),\dots,T(x_k))\\
&= \int_{x_1,\dots,x_n\in S^{n-1}}\int_{T\in O(n)} \mathbf{1}_S(T(x_1),\dots,T(x_k))\\
&= \int_{T\in O(n)}\int_{x_1,\dots,x_n\in S^{n-1}} \mathbf{1}_S(T(x_1),\dots,T(x_k))\\
&= \int_{T\in O(n)}\int_{x_1,\dots,x_n\in S^{n-1}} \mathbf{1}_S(x_1,\dots,x_k)\\
&= \int_{x_1,\dots,x_n\in S^{n-1}} \mathbf{1}_S(x_1,\dots,x_k) = P(S).
\end{aligned}$$

We define, for $B\subset G_{n,k}$,

$$P(S|B) = \frac{1}{\nu_{n,k}(B)}\int_B P(S|V)\,d\nu_{n,k}(V),$$

and clearly for any S we have that $P(S) = P(S|B)\nu_{n,k}(B) + P(S|B^c)(1-\nu_{n,k}(B))$ so that

$$\nu_{n,k}(B) \leqslant \frac{P(S)}{P(S|B)}.$$

Proof of Theorem 3.6.1. Let $y_1,\dots,y_k$ be independent random points, uniformly distributed on S^{n-1}. We study the events

$$B_1 = \left\{(y_i)_{i=1}^k \mid \sigma|_{\mathrm{span}\,\{y_i\}_{i=1}^k}(A) \geqslant \alpha + 2\varepsilon\right\}$$

and

$$B_2 = \left\{(y_i)_{i=1}^k \mid \sigma|_{\mathrm{span}\,\{y_i\}_{i=1}^k}(A) \leqslant \alpha - 2\varepsilon\right\}.$$

We will estimate their probability from above, using the conditional probability of the events

$$C_1 = \left\{(y_i)_{i=1}^k \Big| \frac{\sum_{i=1}^k \mathbf{1}_A(y_i)}{k} - \alpha \geqslant \varepsilon\right\},\ C_2 = \left\{(y_i)_{i=1}^k \Big| \frac{\sum_{i=1}^k \mathbf{1}_A(y_i)}{k} - \alpha \leqslant -\varepsilon\right\}.$$

We know (see Section 3.2) that $P(C_i) \leqslant e^{-2\varepsilon^2 k}$ (computer scientists call this sampling estimate a Chernoff-type bound, but it goes back to Kolmogorov). In order to estimate $P(C_1|B_1)$, we fix a subspace V where B_1 holds. By Chebychev's inequality

$$P\left(C_1\Big|\ \mathrm{span}\,\{y_i\}_{i=1}^k = V\right) \geqslant \frac{\mathbb{E}\left(\frac{\sum \mathbf{1}_A(y_i)}{k} - (\alpha+\varepsilon)\Big|\ \mathrm{span}\,\{y_i\}_{i=1}^k = V\right)}{\max\left(\frac{\sum \mathbf{1}_A(y_i)}{k} - (\alpha+\varepsilon)\Big|\mathrm{span}\,\{y_i\}_{i=1}^k = V\right)}.$$

Similarly, fixing this time V where B_2 holds,

$$P\left(C_2\Big|\ \mathrm{span}\,\{y_i\}_{i=1}^k = V\right) \geqslant \frac{\mathbb{E}\left((\alpha-\varepsilon) - \frac{\sum \mathbf{1}_A(y_i)}{k}\Big|\ \mathrm{span}\,\{y_i\}_{i=1}^k = V\right)}{\max\left(\alpha-\varepsilon-\frac{\sum \mathbf{1}_A(y_i)}{k}\Big|\mathrm{span}\,\{y_i\}_{i=1}^k = V\right)}.$$

Since each y_i is uniform in span $\{y_i\}_{i=1}^k$, and regardless of their conditional dependence, we get for every $V \in B_1$ that

$$P\left(C_1\Big|\mathrm{span}\,\{y_i\}_{i=1}^k = V\right) \geqslant \frac{\sigma|_V(A) - (\alpha+\varepsilon))}{1-(\alpha+\varepsilon)} \geqslant \varepsilon$$

and for every $V \in B_2$,

$$P\left(C_2\Big|\mathrm{span}\,\{y_i\}_{i=1}^k = V\right) \geqslant \frac{(\alpha-\varepsilon-\sigma|_V(A)}{\alpha-\varepsilon} \geqslant \varepsilon.$$

We deduce that in both cases

$$P(C_i|B_i) \geqslant \varepsilon \tag{3.6.1}$$

and conclude that

$$P(B_i) \leqslant \frac{P(C_i)}{P(C_i|B_i)} \leqslant \frac{e^{-2\varepsilon^2 k}}{\varepsilon}.$$

If we rescale ε we find

$$\nu_{n,k}\{V : |\sigma|_V(A) - \alpha| \geqslant \varepsilon\} \leqslant \frac{4e^{-\varepsilon^2 k/2}}{\varepsilon}.$$

This was the form of the original Raz's lemma. We next follow a simple tensorization argument, due to Milman and Wagner, to dismiss the $4/\varepsilon$ factor. We do this only for one side of the estimate, the other is completely symmetric as was seen in the above proof.

Fix $k < n$ and arbitrary m. Consider $y_1, \dots y_{mk}$ independent Haar-uniform variables in S^{n-1}. Now the sequence $\{\mathrm{span}\,\{y_i\}_{i=(j-1)k+1}^{jk}\}_{j=1}^m$ is Haar-uniform and independent in $G_{n,k}$, and for every $1 \leqslant \ell \leqslant k$ the variable $y_{(j-1)k+\ell}$ is Haar-uniform in span $\{y_i\}_{i=(j-1)k+1}^{jk}$ (we make no claim as to their conditional independence here).

We consider the events

$$C = \left\{(y_i)_{i=1}^{mk}\ \Big|\ \frac{\sum_{i=1}^{mk}\mathbf{1}_A(y_i)}{mk} - \alpha > \varepsilon\right\}$$

and

$$B = B_1 \cap B_2 \cap \dots \cap B_m\ ,$$

where

$$B_j = \left\{(y_i)_{i=1}^{mk}\ |\ \sigma|_{\mathrm{span}\,\{y_i\}_{i=(j-1)k+1}^{jk}}(A) \geqslant \alpha + 2\varepsilon\right\}.$$

Just as before, $P(C) \leqslant e^{-2\varepsilon^2 mk}$. Furthermore, by Chebychev's inequality,

$$\begin{aligned} &P(C \mid \forall j : \mathrm{span}\,\{y_i\}_{i=(j-1)k+1}^{jk} = V_j) \\ &\geqslant \frac{\mathbb{E}\left(\frac{\sum \mathbf{1}_A(y_i)}{mk}\ \Big|\ \forall j : \mathrm{span}\,\{y_i\}_{i=(j-1)k+1}^{jk} = V_j\right) - (\alpha+\varepsilon)}{\max\left(\frac{\sum \mathbf{1}_A(y_i)}{mk}\ \Big|\ \forall j : \mathrm{span}\,\{y_i\}_{i=(j-1)k+1}^{jk} = V_j\right) - (\alpha+\varepsilon)} \\ &\geqslant \frac{\frac{1}{m}\sum_j \sigma|_{V_j}(A)\ - (\alpha+\varepsilon)}{1-(c+\varepsilon)}. \end{aligned}$$

As before, we conclude that $P(C|B) \geqslant \varepsilon$.

Since the original y_i's are independent, so are the events B_i, and we get $P(B) = P(B_1)\cdots P(B_m) = P(B_1)^m$. It follows that

$$P(B_1)^m \leqslant \frac{P(C)}{P(C|B)} \leqslant \frac{4e^{-2\varepsilon^2 km}}{\varepsilon}.$$

Letting m go to infinity, we find:

$$P(B_1) \leqslant e^{-2\varepsilon^2 k}\ .$$

Again, rescaling and repeating for the lower tail, the proof of the theorem is complete. □

REMARK 3.6.2. The case in which $\varepsilon = t\sigma(A)$ for some fixed $t \in (0,1)$, and $\sigma(A)$ is small, is of particular interest for computer scientists (it is related to the so called vector-in-subspace problem). In such a case, Theorem 3.6.1 allows the value $\sigma(A) \approx 1/\sqrt{k}$ and not less. Klartag and Regev improved this bound, and showed (for $t = 1/10$, for example) that

$$P\Big\{\Big|\frac{\sigma(A\cap E)}{\sigma(A)} - 1\Big| \geqslant \frac{1}{10}\Big\} \leqslant ce^{-ck/\log^2(\sigma(A))}.$$

Thus, one may consider sets A of exponentially small measure and still get a similar answer to the one in Raz's lemma.

3.7. Notes and remarks

The development of the concept of concentration should be traced in Emile Borel's geometric interpretation of the central limit theorem for n independent random variables uniformly distributed on the unit interval $[-1/2, 1/2]$. Obviously, it means that the volume of the n-dimensional cube is concentrated in a small slice around the hyperplane section of the cube orthogonal to the diagonal which joins the vertices $(-1/2, \dots, -1/2)$ and $(1/2, \dots, 1/2)$. "Small" here means relatively to the length of this diagonal. Borel asked what is the meaning of this observation. P. Lévy reacted on this question by demonstrating in 1919 a similar behaviour on the Euclidean n-dimensional ball and published it in 1922 in a book devoted, in part, to this answer. Let us cite P. Lévy from the preface to the second addition of his book [**380**], published in 1951:

> "When calculating the average value of a function $\Phi(M)$ over a region V, it is often the case that some value μ is dominant over all others, since the surface $\Phi(M) = \mu$ fills almost all the volume: for an arbitrarily small number $\epsilon > 0$, the part of the region where the inequality $|\Phi(M) - \mu| > \epsilon$ holds, is negligibly small. For that reason, the average is not the true average; having calculated that unique value, one can claim that the function under observation is approximately equal to that value nearly everywhere.
>
> Quite a long time ago, in his work "A geometric introduction to some physical theories", Emile Borel noted this phenomenon in some particular cases. In the lectures given by the author in 1919, as well as in the book published in 1922, those results of Borel were extended to the case of the average of a uniformly continuous function over a ball or a sphere, and then to the case where, instead of a ball, one takes a region bounded by a convex closed surface with radii of curvature bounded from above."

Let us emphasize that although P. Lévy computed the deviation (concentration) inequality for the Euclidean sphere S^n, he actually never used its strength. Any level of decay, i.e. the weakest concentration inequality, would be sufficient for the applications he considered in his book. Again, our citation from P. Lévy comes from the same preface:

> "In a ball inside a Hilbert space, a uniformly continuous function is approximately equal to its average value nearly everywhere. For arbitrarily small positive constants ϵ, ϵ', a uniformly continuous function, for n large enough, will differ, inside an n-dimensional section of the ball, from its average value m_n by less than ϵ, excluding a part of that ball of measure not greater than ϵ'."

This property was used by P. Lévy to study the Laplace operator on the infinite dimensional Hilbert sphere. All the applications of this ingenious observation by P. Lévy that appeared in the next fifty years (from 1919/1922 till 1969) focused on the same kind of problems outlined by P. Lévy in these two editions of his book.

The picture changed with the proof of the Dvoretzky-Milman theorem (see Section 5.2) using the idea of concentration. The role of the concentration inequality became clear. The full use of its strength led to the best estimates in many results of the asymptotic theory of finite dimensional normed spaces (see Chapter 5). This led V. Milman to the idea of "oscilation stability" ("spectrum", as he called it in the 1970's) and to the realization that this phenomenon should be a broadly spread property of high-dimensional structures. Already in 1971 he proved and used it for Stieffel and Grassmann manifolds ([**442**] and [**443**]), for complex spaces and some other situations. The next stage was the concept of the "concentration phenomenon" for metric probability spaces as we understand it today. It is sometimes called "Lévy-Milman" concentration (see e.g. [**269**]) to be distinguished from two other concentration type concepts. One of these concepts is applied to metric spaces in situations where no natural measure may be introduced (e.g. for infinite dimensional spaces), and another one for probability spaces in situations where the metric we use is not natural for the problem and not connected with concentration. For the first concept (that was considered by Gromov and V. Milman) see the subsection on concentration for infinite dimensional spaces below. Typical results of the second type of concentration (put forward by Giannopoulos and V. Milman in [**233**] as explained in the introduction of Section 3.5) are Borell's lemma and Khintchine-Kahane inequalities, as well as any kind of "reverse Hölder" inequalities.

Approach through extremal sets

The solution for the isoperimetric problem on the sphere is given by Paul Lévy in [**380**] and by Schmidt in [**547**]. For a proof using spherical symmetrization see Figiel, Lindenstrauss and V. Milman [**205**], where the inequality is presented and applications to the local theory of normed spaces are given. Lemma 3.1.7, which leads to a very simple proof of the isomorphic isoperimetric inequality for the sphere, is due to Arias de Reyna, Ball and Villa [**21**]. A similar type of reasoning was also used in Gromov and V. Milman [**271**] for proving concentration on the class of uniformly convex normed spaces.

The isoperimetric inequality for Gauss space was discovered by Sudakov-Tsirelson [**576**] and Borell [**95**] who used the isoperimetric theorem on the sphere and the observation that projections of uniform measures on N-dimensional spheres of radius $\sqrt{N}$ when projected to $\mathbb{R}^n$ approximate Gaussian measure as $N \to \infty$. This is sometimes called "Poincaré lemma" (see [**573**] and [**372**]) but was known already to Maxwell, see, e.g. [**169**]. Borell [**95**] among other things obtained a Brunn-Minkowski inequality in Gauss space. The Gaussian isoperimetric inequality was also proved by Erhard [**182**] who developed a rearrangement of sets argument in Gauss space. Let us note that Bobkov [**80**] proved an isoperimetric inequality on the discrete cube from which he also derived the

Gaussian isoperimetric inequality (see also [**78**] and [**79**]). His ingenious argument provides the first purely analytic (probabilistic) proof, without any geometry involved, of the Gaussian isoperimetric inequality and automatically implies the isoperimetric inequality for the Euclidean sphere. Theorem 3.1.11, which establishes the isomorphic isoperimetric inequality in Gauss space as a direct application of the Prékopa-Leindler inequality, is due to Maurey [**418**].

Harper's paper with the extremal sets is [**300**]. We also mention the result of Wang and Wang [**604**] where the setting of the cube is extended to that of the family of n vectors in which the ith coordinate is assumed to belong to the set $\{0, 1, \dots, t_i\}$, and the answer is in the same spirit.

Talagrand's Theorem 3.1.15 appears in [**587**]. For its many applications and its relatives see Steele [**569**, Chapter 6], Alon-Spencer [**13**] and the chapter on concentration by McDiarmid in [**426**].

The notion of a Lévy-family and of the concentration function were introduced in the works of Gromov, Milman [**271**] and Amir and Milman [**16**]; the formal definitions that we present in the text were put forward in [**271**] and in the article of Alon and Milman [**12**] which introduces the notion of a "concentrated Levy family", meaning only exponential decay of the concentration function, a situation that is very often met in combinatorics and theoretical computer science (see also [**11**]). The connection between concentration and deviation inequalities for Lipschitz functions was emphasized by Lévy and V. Milman [**452**]. Our exposition here follows the books of Milman-Schechtman [**464**] and Ledoux [**373**]. For other introductions to the concentration of measure phenomenon we refer to the survey articles of Milman [**452**], Schechtman [**538**], Talagrand [**589**] and Gromov [**269**, Chapter $3\frac{1}{2}$].

Deviation inequalities

The proof of Theorem 3.2.2 follows [**464**, Appendix V]. Property (τ) was introduced by Maurey in [**418**], who emphasized its relation with concentration properties of product measures and gave a short proof for Talagrand's concentration inequality (see [**588**]) for the product exponential measure. We will come back to it in Part II. The applications of property (τ) to concentration in Gauss space that we saw in Section 3.2 come from the same paper. Theorem 3.2.7 is due to Talagrand and appears in [**587**]. The result stated as Theorem 3.2.5 was popularized by Maurey and Pisier but somewhat equivalent results were later found in earlier references due to Borell, Tsirelson, Ibragimov and Sudakov, see [**576**] [**308**].

Concentration on homogeneous spaces

Theorem 3.3.1 is well known, and its proof may be found in various books, for example [**464**].

Theorem 3.3.5 is due to Gromov, [**267**] and then published in the book of Milman and Schechtman [**464**] as an appendix written by Gromov; see also Gromov [**269**] or Burago and Zalgaller [**127**]. In the same appendix of [**464**], parts of the proofs of the results of this section are explained in detail. The interested reader may consult the books of Cheeger and Ebin [**146**], Gallot, Hulin and Lafontaine [**217**], Chavel [**144**] for a through introduction to curvature in Riemannian geometry and for bounds on the Ricci curvature in the classical examples that are mentioned in Section 3.2. Theorem 3.3.6 is due to Gromov and Milman from [**271**] as is Theorem 3.3.7. Concentration properties of $W_{n,k}$ and $G_{n,k}$ had been also applied by Milman in [**442**] and [**443**].

An approach through conditional expectation and martingales

Azuma's inequality appears in [**35**]. Yurinskii used it in [**610**] and [**611**] to give bounds for the deviation of the norm of sums of independent random vectors from its mean. The martingale method as presented here, and in particular the concentration inequality for the group of permutations, goes back to Maurey [**417**] who used it to study symmetric basic sequences in Banach spaces. Schechtman developed Maurey's ideas in [**535**] and [**536**]. Theorem 3.4.7 appears in V. Milman and Schechtman [**464**].

Khintchine type inequalities

Khintchine inequality first appears in [**330**], but it was first stated in this form by Littlewood [**386**]. Its usefulness for the study of L_p-spaces was understood much later, in the 1970's. The exact values of A_p^* and B_p^* have been determined by Szarek [**578**] who showed that $A_1^* = 1/\sqrt{2}$, and by Haagerup [**293**] who determined the best constants for all p. They are:

$$A_p^* = 2^{1/2-1/p} \text{ if } 0 < p \leqslant p_0 \quad \text{and} \quad A_p^* = \sqrt{2}(\Gamma((p+1)/2)/\sqrt{\pi})^{1/p} \text{ if } p_0 < p < 2$$

and

$$B_p^* = \sqrt{2}(\Gamma((p+1)/2)/\sqrt{\pi})^{1/p} \text{ for all } p > 2,$$

where $p_0 \simeq 1.84742$ is the solution of the equation $\Gamma((p+1)/2) = \sqrt{\pi/2}$.

Kahane's inequality was proved by Kahane in [**322**] with constant proportional to p as $p \to \infty$; the dependence $\sqrt{p}$ is due to Kwapien, see [**365**]. The proof that we present here comes from [**587**].

Bourgain, answering a question of V. Milman, showed in [**100**] that for any $n \geqslant 1$, any convex body K of volume 1 in $\mathbb{R}^n$ and any polynomial $f : \mathbb{R}^n \to \mathbb{R}$ of degree m, one has $\|f\|_\psi \leqslant C\|f\|_1$, where $\psi(t) = e^{t^{c/m}} - 1$ and $c, C > 0$ are absolute constants. Later, Bobkov [**82**] using a localization technique of Kannan, Lovász and Simonovits [**325**] (based on the "localization lemma" of Lovász and Simonovits [**397**]) proved that one can have $c = 1$ in Bourgain's theorem. Nazarov, M. Sodin and Volberg proved in [**483**] a "geometric Kannan-Lovász-Simonovits lemma" stating that $\mu(F_t^c) \leqslant \mu(F^c)^{\frac{t+1}{2}}$, $t > 0$, for every log-concave probability measure μ and any Borel set F in $\mathbb{R}^n$, where F_t is the t-dilation of F defined by

$$F_t = \left\{ x \in \mathbb{R}^n : \text{ there exists an interval } I \text{ s. t. } x \in I \text{ and } |I| < \frac{t+1}{2}|F \cap I| \right\}.$$

In [**483**] they used this inequality to obtain sharp large and small deviations, as well as Khintchine-type inequalities for functions that satisfy a Remez type inequality. Using the original localization lemma of Lovász and Simonovits, Carbery and Wright (see [**136**]) determined the sharp constants between the L_p and L_q norms of polynomials over convex bodies in $\mathbb{R}^n$.

The theorem of C. Borell, regarding the characterization of log-concave probability measures on $\mathbb{R}^n$ is from [**94**]. In fact, it is part of a more general theorem about α-concave measures corresponding to α-concave functions, as in Proposition 1.4.4 (with $f = g = h$).

Bernstein type inequalities

We would like to thank Sasha Sodin for the following text which he wrote by our request and which provides information about Bernstein's original work. The first paper of Bernstein [**68**] on the subject appeared in 1924. It contained a ψ_1 theorem stated below as Theorem A, and certain inequalities for the special case of (biased) Bernoulli random variables, discussed below. It was probably the first time Chebyshev's inequality was applied to an exponential test function.

Theorem A. *Let X_j be independent random variables with zero mean, and suppose*

$$\left|\mathbb{E}(X_j^k)\right| \leqslant \frac{k!}{4!}\left(\frac{L}{5}\right)^{k-4}\mathbb{E}(X_j^4)\ , \quad k \geqslant 4\ .$$

Let

$$B_k = \sum_{j=1}^{N} \mathbb{E}(X_j^k)\ .$$

Then

$$\mathbb{P}\left\{\left|\sum_{j=1}^{N} X_j - \frac{t^2}{6}\frac{B_3}{B_2}\right| \geqslant t\sqrt{B_2}\left[1 + \frac{B_4 t^2}{12B_2^2}\right]\right\} \leqslant 2e^{-t^2/2}\ , \quad 0 \leqslant t \leqslant \frac{5}{2L}\sqrt{B_2}\ .$$

Compared to, say, Theorem 3.5.16, Theorem A has several interesting features. First, the constant in front of the quadratic term in the exponent is sharp (i.e. equal to the one given by the central limit theorem). Apparently, the theorem gives good numerical bounds already for small values of N. Second, the deviations are not about zero.

In the textbook [**69**], published in 1927, Bernstein gave two less precise but also less technical results, a ψ_1 result (Theorem B) and a ψ_2 result (Theorem C).

Theorem B. *Suppose X_j are independent random variables with zero mean, and suppose*

$$\left|\mathbb{E}(X_j^k)\right| \leqslant \frac{\beta_j}{2}L^{k-2}k!\ , \quad k \geqslant 2\ .$$

Then, for $B = \sum_{j=1}^N \beta_j$,

$$\mathbb{P}\left\{\sum X_j \geqslant t\sqrt{B}\right\} \leqslant e^{-t^2/4}\ , \quad 0 \leqslant t \leqslant \frac{\sqrt{B}}{2L}\ .$$

Unlike Theorem A, the constant in the exponent is off by a factor of 2.

Theorem C. *Suppose X_j are independent random variables with even distribution such that*

$$\mathbb{E}(X_j^{2k}) \leqslant \left(\frac{\beta_j}{2}\right)^k \frac{(2k)!}{k!}\ .$$

Then, for $B = \sum_{j=1}^N \beta_j$,

$$\mathbb{P}\left\{\sum_{j=1}^{N} X_j \geqslant t\sqrt{B}\right\} \leqslant e^{-t^2/2}\ , \quad 0 \leqslant t\ .$$

Here the constant is sharp (and the proof is very simple due to the symmetry restriction). Bernstein remarks that a similar result without the symmetry condition can be proved, but the statement and proof are more cumbersome. The general case can be reduced to the symmetric case if we agree to lose a constant in the exponent.

In 1937, Bernstein published a short note [**70**] where he observed that the arguments can be extended to dependent random variables under certain conditions. In modern terminology, these results are Bernstein type inequalities for martingales (which were first studied by Bernstein and by P. Lévy, more or less simultaneously, although the term was coined later by Jean Ville). The results of Bernstein are stronger in several respects than the much later work by Azuma (in particular, the conditions are weaker, and the results incorporate a maximal inequality in the spirit of earlier results by Kolmogorov and Bernstein).

A significant part of the first paper [**68**] is devoted to the special case of independent Bernoulli random variables with

$$\mathbb{P}\{X_j = 1\} = p = 1 - q\ .$$

Bernstein obtains very precise asymptotics for the number $w = w(t)$ such that

$$\mathbb{P}\left\{\sum_{j=1}^{N} X_j \geqslant w(t)\right\} = \int_t^{\infty} e^{-s^2/2} \frac{ds}{\sqrt{2\pi}} .$$

This may be thought of as monotone transportation from the Gaussian distribution to the binomial one. The asymptotic expression $w^*(t)$ is so precise that $|w(t) - w^*(t)| < 1$ for $Npq \geqslant 365$. It is not clear how this compares, e.g., to Chernoff's bounds, though it seems that bounds in the spirit of the latter appear inside the argument of Bernstein.

Theorem 3.5.13 was probably first proved by Bennett (see [**61**] and [**62**]; also Hoeffding [**304**]). Bennett admits that Bernstein's original papers are inaccessible for him, and that he learned about the work of Bernstein from the papers of Craig [**155**] and Godwin [**251**]. Both Craig and Godwin write that they learned about the work of Bernstein from secondary sources. Thus, neither of these three papers quotes the results of Bernstein in their original form.

Concentration on infinite dimensional spaces

For a more precise description of *concentration without measure* assume that $X = (X, \rho, \mu)$ is a metric probability space which is also a G-space: a group G of metric and measure preserving maps act on X. Note that if $A \subseteq X$ has not too small measure, say $\mu(A) \geqslant 1/10$, but $\alpha_\mu(t)$ is very small then it is easy to see that $\mu(A_t)$ will be close to 1. Assume for example that $\alpha_\mu(t)$ is so small that $\mu(A_t) > 1 - \frac{1}{100}$. Then, for any $\{g_i\}_{i=1}^{100} \subset G$ the intersection $\bigcap_{i=1}^{100} g_i(A_t) \neq \emptyset$. Therefore, for *any* partition of $X = \bigcup_{i=1}^{100} A_i$ one may find A_{i_0} such that for any set of 100 "rotations" g_i from G the intersection $\bigcap_{i=1}^{100} g_i((A_{i_0})_t)$ is non-empty. This is one of the important schemes of how the concentration property is used. But it essentially deals only with a metric property of X and of the group G that is acting on it.

Now let (X, ρ) be a metric G-space (without any measure structure on it). The next example provides a way to introduce a notion of concentration on X with respect to G. Let $X = S^\infty = S(H)$ be the unit sphere of a Hilbert space H. Let G be a subgroup of orthogonal operators U. We call a set $A \subset S^\infty$ *essential* with respect to G if for all $\varepsilon > 0$ and for all n and $g_1, \dots, g_n$ in G we have

$$\bigcap_{i=1}^{n} g_i(A_t) \neq \emptyset.$$

We say that (S^∞, G) has a "*concentration property*" if for every finite partition $S^\infty = \bigcup_{i=1}^{N} A_i$ of S^∞ there exists i_0 such that A_{i_0} is essential. Let us mention some concrete examples:

(i) Fix an orthonormal basis $\{e_i\}_{i=1}^{\infty}$ and let $U(n)$ be the subgroup generated by the unitary group on $\text{span}\{e_1, \dots, e_n\}$ and the identity action on $\text{span}\{e_i : i > n\}$. Let $G = \bigcup_{i=1}^{n} U(n)$. Then, (S^∞, G) has the concentration property.

(ii) Let U be any unitary operator on H and $G = \{U^n\}_{-\infty}^{\infty}$. Then (S^∞, G) has the concentration property.

(iii) Let $\mathcal{A}$ be a family of pairwise commuting unitary operators on H. Then $(S^\infty, \mathcal{A})$ has the concentration property.

These notions were considered by Gromov and V. Milman in the end of the 1970's and beginning of the 1980's; see [**452**]. Later Pestov (see [**496**] and [**497**]) connected this property with the *amenability* of the G-action on X (i.e. with the existence of an invariant mean, which may be viewed as a substitute for measure). For example, Pestov proved that a locally compact group G is amenable if and only if the uniform action $(S_{L_2(G)}, G)$ has the concentration property.

Raz's Lemma

Raz's Lemma in its original form appears in [**516**]. It was improved to the form given here by V. Milman and Wagner in [**470**]. The most advanced results so far are due to Klartag and Regev [**347**] where the relation to quantum communication is explained in detail.

CHAPTER 4

Metric entropy and covering numbers estimates

Entropy numbers and covering numbers are a way of measuring the "size" of a set in terms of another set. They are useful tools in various areas of mathematics, and play an important role in asymptotic geometric analysis. Their developement is usually attributed to Kolmogorov [**356**] and Kolmogorov-Tikhomirov [**357**], and they were used with several other approximation numbers in approximation theory. However, in the metric setting they have already appeared in the paper of Pontrjagin and Schnirelman [**513**], and in the paper [**601**] quoted in the introduction, von Neumann uses coverings of certain high-dimensional sets of matrices by balls, bounding these by volumetric arguments similar to the ones we shall encounter below.

In this chapter we introduce covering numbers and entropy numbers, explain some of their properties, derive relations and duality between these numbers, volume ratios and other parameters of the sets involved.

4.1. Covering numbers

4.1.1. The basics

Definition 4.1.1. Let A and B be two convex bodies in $\mathbb{R}^n$. The *covering number* $N(A,B)$ of A by B is the least number of translates of B that are needed in order to cover A:

$$N(A,B) = \min\left\{ N \in \mathbb{N} \mid \exists\, x_1, \dots, x_N \in \mathbb{R}^n : A \subseteq \bigcup_{i=1}^{N} (x_i + B) \right\}. \tag{4.1.1}$$

A variant of $N(A,B)$ is defined if we require that the centers x_i belong to A; then we set

$$\overline{N}(A,B) = \min\left\{ N \in \mathbb{N} \mid \exists\, x_1, \dots, x_N \in A : A \subseteq \bigcup_{i=1}^{N} (x_i + B) \right\}. \tag{4.1.2}$$

Remark 4.1.2. We assumed above that the convex bodies are in $\mathbb{R}^n$, that is, in a finite dimensional linear space. The definition makes sense in the more general setting of sets in a linear space X, possibly infinite dimensional. However, in such a case the numbers involved may turn out to be infinite (the minimum of an empty set). We shall work in the infinite dimensional setting in Section 4.3, where it is more convenient. Indeed, classically these numbers are used to study measures of compactness of certain operators between Banach spaces. The reader is advised to check that many of the relations below (the ones which do not involve the dimension, or volume) hold also for this more general case.

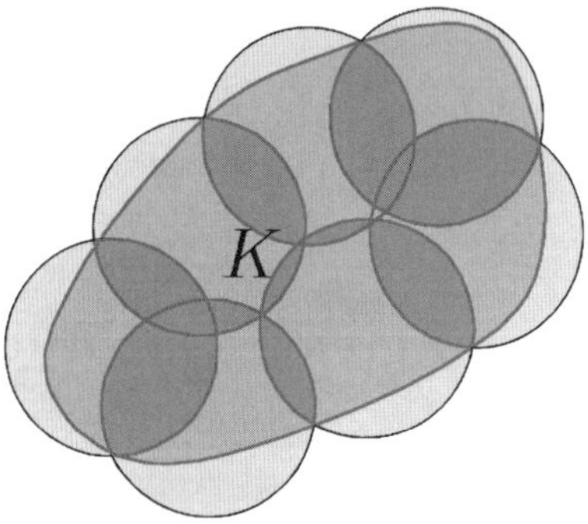

FACT 4.1.3. *For any convex bodies A, B, C in $\mathbb{R}^n$ with $A \subseteq B$, we have*

$$N(A, C) \leqslant N(B, C), \ N(C, B) \leqslant N(C, A) \text{ and } \overline{N}(C, B) \leqslant \overline{N}(C, A).$$

Proof. Immediate. □

FACT 4.1.4. *For any convex bodies A, B in $\mathbb{R}^n$ one has*

$$\overline{N}(A, B - B) \leqslant N(A, B) \leqslant \overline{N}(A, B).$$

In the case of covering by a multiple of the Euclidean ball B_2^n we have

$$N(A, RB_2^n) = \overline{N}(A, RB_2^n).$$

Proof. Note first that for any $x \in \mathbb{R}^n$ with $(x + B) \cap A \neq \emptyset$ we may find $y \in A$ such that $(x + B) \cap A \subseteq y + B - B$. Indeed, take any $y \in (x + B) \cap A$, then $y - x \in B$ so that $x + B \subseteq y + (B - B)$. In the case where $B = RB_2^n$ the factor 2 may be omitted. Indeed, we then take as y the metric projection of x onto A, that is, the closest point to x on A. Then for any $z \in (x + B) \cap A$ we must have

$$R^2 \geqslant |z - x|^2 = |z - y|^2 + |y - x|^2 + 2\langle z - y, y - x\rangle \geqslant |z - y|^2$$

so that $(x + B) \cap A \subseteq y + B$. Finally given a covering of A by B change each x_i to a $y_i \in A$ satisfying the above property. □

In the remainder of this section we gather some more of the basic well known properties of covering numbers which will be often used in the sequel. Their proofs, which are straightforward, are included for completeness.

FACT 4.1.5. *For any convex bodies A, B in $\mathbb{R}^n$ and every invertible linear operator $T : \mathbb{R}^n \to \mathbb{R}^n$ one has $N(A, B) = N(T(A), T(B))$.*

Proof. Follows from the fact that $A \subseteq \bigcup_{i=1}^N (x_i + B)$ if and only if $T(A) \subseteq \bigcup_{i=1}^N (Tx_i + T(B))$. □

FACT 4.1.6. *For any two ellipsoids $\mathcal{E}_1, \mathcal{E}_2$ one has $N(\mathcal{E}_1, \mathcal{E}_2) = N(\mathcal{E}_2^\circ, \mathcal{E}_1^\circ)$.*

Proof. This follows from the fact that for any two ellipsoids there exists a linear operator such that $\mathcal{E}_2^\circ = T(\mathcal{E}_1)$ and $\mathcal{E}_1^\circ = T(\mathcal{E}_2)$, together with the previous fact. Indeed, since any two ellipsoids are positive definite linear maps of B_2^n, we pick some T with $\mathcal{E}_2^\circ = T(\mathcal{E}_1)$ and $T = T^*$. Then

$$T(\mathcal{E}_2) = T((T(\mathcal{E}_1))^\circ) = TT^{-*}(\mathcal{E}_1^\circ) = \mathcal{E}_1^\circ.$$

□

FACT 4.1.7. *For any convex bodies A, B and C one has*

$$N(A,B) \leqslant N(A,C)N(C,B).$$

Proof. Follows from the fact that if $A \subseteq \bigcup_{i=1}^N (x_i + C)$ and $C \subseteq \bigcup_{j=1}^M (y_j + B)$ then

$$A \subseteq \bigcup_{i=1}^N \bigcup_{j=1}^M ((x_i + y_j) + B).$$

□

FACT 4.1.8. *For any convex bodies A, B and C one has $N(A+C, B+C) \leqslant N(A,B)$.*

Proof. Follows from the fact that if $A \subseteq \bigcup_{i=1}^N (x_i + B)$ then

$$A + C \subseteq \bigcup_{i=1}^N (x_i + B + C).$$

□

FACT 4.1.9. *For any convex bodies A and B one has*

$$\overline{N}(A, (A-A) \cap B) = \overline{N}(A,B). \tag{4.1.3}$$

In particular we have that: for centrally symmetric A,

$$\overline{N}(A, 2A \cap B) = \overline{N}(A,B) \ \text{ and } \ N(A, 2(A \cap B)) \leqslant N(A,B).$$

For covering by a Euclidean ball, $N(A, RB_2^n \cap (A-A)) = N(A, RB_2^n)$, and in the general not necessarily symmetric case we have $N(A, (A-A) \cap (B-B)) \leqslant N(A,B)$.

Proof. Given a minimal covering of A by B with centres $x_i \in A$, meaning $A \subseteq \bigcup_{i=1}^N (x_i + B)$, we intersect it with A to get

$$A \subseteq \bigcup_{i=1}^N ((x_i + B) \cap A).$$

For every $y \in (x_i + B) \cap A$ we have $y - x_i \in B \cap (A - A)$ so that $A \subseteq \bigcup_{i=1}^N (x_i + [B \cap (A-A)])$. This gives $\overline{N}(A,B) \geqslant \overline{N}(A, (A-A) \cap B)$. By Fact 4.1.3 we get the opposite inequality, and thus equality. For $N(A,B)$ we use the same reasoning as in Fact 4.1.4. We start with a minimal covering of A by B $A \subseteq \bigcup_{i=1}^N (x_i + B)$ and we intersect it with A. Since N was minimal, for every i there exists some $y_i \in (x_i + B) \cap A$ and further $(x_i + B) \cap A \subseteq y_i + ((B - B) \cap (A - A))$. Indeed, let $z \in (x_i + B) \cap A$, then $z - y_i \in A - A$ and also, as $y_i = x_i + b$ for some $b \in B$ and $z - x_i \in B$ we have that $z - y_i = (z - x_i) + (x_i - y_i) \in B - B$. We get that $z - y_i \in (A - A) \cap (B - B)$ as claimed. Thus

$$N(A, (A-A) \cap (B-B)) \leqslant N(A,B).$$

In the centrally symmetric case we change $B - B = 2B$ and $A - A = 2A$. In the case where $B = RB_2^n$ is a multiple of the Euclidean ball we know that $N(A, RB_2^n) = \overline{N}(A, RB_2^n)$ so that (4.1.3) implies

$$N(A, (A-A) \cap RB_2^n) \leqslant \overline{N}(A, (A-A) \cap RB_2^n) = N(A, RB_2^n)$$

and from monotonicity we get equality again. □

DEFINITION 4.1.10. Given a convex set A and a centrally symmetric convex set B the *separation number* $M(A,B)$ is the maximal cardinality of a B-separated set in A, that is,

$$M(A,B) = \max\{N \in \mathbb{N} : \exists\, x_1,\dots,x_N \in A \text{ s.t. } \forall j \neq i, x_j \notin (x_i + B)\}.$$

Equivalently one may ask

$$\max\left\{N \in \mathbb{N} : \exists\, x_1,\dots,x_N \in A \text{ s.t. } \forall j \neq i, \left(x_j + \frac{1}{2}B\right) \cap \left(x_i + \frac{1}{2}B\right) = \emptyset\right\}.$$

Yet another equivalent formulation, say in the case of closed B, is that $\|x_i - x_j\|_B > 1$.

The separation number is closely related to the covering number of A by B:

FACT 4.1.11. $M(A,2B) \leqslant N(A,B) \leqslant \overline{N}(A,B) \leqslant M(A,B)$.

Proof. For the right hand side inequality assume we are given a maximal B-separated set $\{x_i\}_{i=1}^M$ in A. From the fact that it is maximal we know that no point $x \in A$ exists such that $x \notin \bigcup_{i=1}^M (x_i + B)$ which means $A \subseteq \bigcup_{i=1}^M (x_i + B)$, thus $\overline{N}(A,B) \leqslant M$. For the left hand side inequality, let $A \subseteq \bigcup_{i=1}^N (y_i + B)$ and assume we are given some $2B$-separated set $\{x_j\}_{j=1}^{M'}$ in A. Because $x_j \in A$ there exists some $i(j)$ such that $x_j \in y_{i(j)} + B$. On the other hand, if $j \neq j'$ then $i(j) \neq i(j')$ otherwise $y_{i(j)} \in (x_j + B) \cap (x_{j'} + B)$ which contradicts the separation property of $\{x_j\}_{j=1}^{M'}$. Thus we have found an injective mapping from $\{1,\dots,M'\}$ to $\{1,\dots,N\}$ meaning $M' \leqslant N$ as needed. □

REMARK 4.1.12. We note that we have not provided here a clear definition for separation number in the case of covering by a non-symmetric set B, since in that case the different forms given in Definition 4.1.10 are no longer equivalent. We also remark that the term "packing number" is frequently used instead of "separation number", however packing can also refer to the maximal number of disjoint copies of a certain body which can be put inside another, that is, not only their centers but the whole set, and to avoid confusion we do not use this terminology in this book.

4.1.2. Volume bounds

In this section we collect some technical statements which will be used throughout the book. The reader may glance through them and return to them when she encounters their use.

The simplest bounds on covering numbers are given by volumetric arguments.

THEOREM 4.1.13. *Let K, T be convex bodies in $\mathbb{R}^n$. Then,*

$$\frac{\mathrm{Vol}_n(K)}{\mathrm{Vol}_n(T)} \leqslant N(K,T),$$

and if T is centrally symmetric then

$$N(K,T) \leqslant 2^n \frac{\mathrm{Vol}_n(K + \frac{T}{2})}{\mathrm{Vol}_n(T)}.$$

Proof. Clearly if $K \subseteq \bigcup_{i=1}^N (x_i + T)$ then $\mathrm{Vol}_n(K) \leqslant N\mathrm{Vol}_n(T)$ which proves the first inequality. For the second we first use Fact 4.1.11 to bound $N(K,T) \leqslant M(K,T)$. We use the equivalent definition of $M(K,T)$ in which all the copies $x_i + \frac{T}{2}$ are disjoint, which means that $\mathrm{Vol}_n\left(\bigcup_{i=1}^M (x_i + \frac{T}{2})\right) = M\frac{1}{2^n}\mathrm{Vol}_n(T)$. Finally, as $x_i \in K$ we have that $\bigcup_{i=1}^M (x_i + \frac{T}{2}) \subseteq K + \frac{T}{2}$ and thus we get

$$M\frac{1}{2^n}\mathrm{Vol}_n(T) \leqslant \mathrm{Vol}_n\left(K + \frac{T}{2}\right)$$

which completes the proof. □

An important instance in which this bound is very useful is that of bodies for which a significant part of the volume is included in the largest inscribed homothetic copy of the body with which they are being covered. The main case in which the following is interesting is when T is a Euclidean ball, and in this case one says that the body K satisfying the condition of the statement has bounded volume ratio, a notion to which we shall return in Chapter 5, Section 5.5.

COROLLARY 4.1.14. *Let $T \subseteq K \subset \mathbb{R}^n$ convex, $T = -T$, and assume that for some $A > 0$*

$$\left(\frac{\mathrm{Vol}_n(K)}{\mathrm{Vol}_n(T)}\right)^{1/n} \leqslant A.$$

Then, for any $\varepsilon > 0$,

$$N(K, \varepsilon T) \leqslant A^n\left(1 + \frac{2}{\varepsilon}\right)^n.$$

In particular we have

COROLLARY 4.1.15. *Let $K \subset \mathbb{R}^n$ be a centrally symmetric convex body. Then, for $0 < \varepsilon < 1$,*

$$\left(\frac{1}{\varepsilon}\right)^n \leqslant N(K, \varepsilon K) \leqslant \left(1 + \frac{2}{\varepsilon}\right)^n.$$

REMARK 4.1.16. The question of the best covering of the Euclidean ball B_2^n by εB_2^n has attracted much attention, due to its close connection with coding theory. The best known constants up to date are given by Kabatianski and Levenshtein in [**320**]. They actually provide an estimate for the maximum density δ_n of packings of translates of B_2^n, and not covering directly. Their bound is $\delta_n \leqslant 2^{-0.599n+o(n)}$.

Another related question is that of Hadwiger, whether for any K there exists $\lambda < 1$ such that $N(K, \lambda K) \leqslant 2^n$. This is clearly best possible since for parallelepipeds there is equality.

The following lemma is useful, and in particular we shall make use of it in Chapter 8.

LEMMA 4.1.17. *Let $K, P \subset \mathbb{R}^n$ be convex and centrally symmetric. Then*

$$N(K, sB_2^n) \geqslant \frac{\mathrm{Vol}_n(K + P)}{\mathrm{Vol}_n((sB_2^n \cap K) + P)} \tag{4.1.4}$$

Proof. Our first observation is that if D, K, P are centrally symmetric convex bodies in $\mathbb{R}^n$ then

$$\max_{x \in \mathbb{R}^n} \mathrm{Vol}_n(\{(x + D) \cap K\} + P) = \mathrm{Vol}_n((D \cap K) + P). \tag{4.1.5}$$

This follows from the Brunn-Minkowski inequality in a similar way to that explained in the proof of Theorem 1.5.2. Indeed, set $T_x = \{(x+D) \cap K\} + P$ and use the symmetry and convexity of D, K, P to check that
$$\frac{T_{-x} + T_x}{2} \subseteq T_0.$$
Then, apply the Brunn-Minkowski inequality (and the fact that $\text{Vol}_n(T_x) = \text{Vol}_(T_{-x})$) to get
$$\text{Vol}_n(T_x) \leqslant \text{Vol}_n\left(\frac{T_x + T_{-x}}{2}\right) \leqslant \text{Vol}_n(T_0)$$
for every $x \in \mathbb{R}^n$.

Next note that from the definition of $N = N(K, sB_2^n)$ there exist $x_1, \ldots, x_n \in \mathbb{R}^n$ such that $K \subseteq \cup_{i=1}^N [(x_i + sB_2^n) \cap K]$, and hence
$$K + P \subseteq \bigcup_{i=1}^N [\{(x_i + sB_2^n) \cap K\} + P].$$
From (4.1.5) we get
$$\text{Vol}_n(K+P) \leqslant N(K, sB_2^n) \max_{1 \leqslant i \leqslant N} \text{Vol}_n(\{(x_i + sB_2^n) \cap K\} + P)$$
$$\leqslant N(K, sB_2^n) \text{Vol}_n((sB_2^n \cap K) + P).$$
$\square$

For bodies which are not centrally symmetric we have similar estimates.

THEOREM 4.1.18. *Let $K, L \subset \mathbb{R}^n$ be convex bodies. Then*
$$4^{-n} \frac{\text{Vol}_n(K-L)}{\text{Vol}_n(L)} \leqslant N(K, L) \leqslant 4^n \frac{\text{Vol}_n(K-L)}{\text{Vol}_n(L)}.$$

As a consequence we get the following

COROLLARY 4.1.19. *Let K be a convex body in $\mathbb{R}^n$ with $0 \in \text{int}(K)$. Then,*
$$N(K-K, K) \leqslant 4^n \frac{|K-2K|}{|K|} \leqslant 8^n \frac{|K-K|}{|K|} \leqslant 32^n,$$
and in particular
$$N(K, \varepsilon K) \leqslant N(K-K, \varepsilon K) \leqslant N(K-K, \varepsilon(K-K)) \cdot N(K-K, K)$$
$$\leqslant \left(32 + \frac{64}{\varepsilon}\right)^n.$$

The proof of Theorem 4.1.18 is based on the following volume estimate of Milman and Pajor.

THEOREM 4.1.20. *Let K and L be two convex bodies with the same barycenter in $\mathbb{R}^n$. Then,*
$$\text{Vol}_n(K) \text{Vol}_n(L) \leqslant \text{Vol}_n(K-L) \text{Vol}_n(K \cap L).$$
In particular, if K has its barycenter at the origin then
$$\text{Vol}_n(K \cap (-K)) \geqslant 2^{-n} \text{Vol}_n(K).$$

Proof of Theorem 4.1.18. Assume that the barycenter of L is at the origin. Then on the one hand we have that if $K \subseteq \bigcup_{i=1}^N (x_i + L)$ then $K - L \subseteq \bigcup_{i=1}^N (x_i + (L - L))$ and thus by Rogers-Shephard inequality (Theorem 1.5.2)

$$\mathrm{Vol}_n(K - L) \leqslant N(K, L)\mathrm{Vol}_n(L - L) \leqslant 4^n N(K, L)\mathrm{Vol}_n(L).$$

On the other hand,

$$\begin{aligned} N(K, L) \leqslant N(K, L \cap (-L)) &\leqslant \frac{\mathrm{Vol}_n(2K + (L \cap (-L)))}{\mathrm{Vol}_n(L \cap (-L))} \\ &\leqslant 2^n \frac{\mathrm{Vol}_n(K - L)}{\mathrm{Vol}_n(L \cap (-L))} \leqslant 4^n \frac{\mathrm{Vol}_n(K - L)}{\mathrm{Vol}_n(L)}, \end{aligned}$$

where we have used the estimate $\mathrm{Vol}_n(L \cap (-L)) \geqslant 2^{-n}\mathrm{Vol}_n(L)$ from Theorem 4.1.20. □

The proof of Theorem 4.1.20 makes use of the next lemma.

LEMMA 4.1.21. *Let μ be a probability measure on $\mathbb{R}^n$ and let $\psi : \mathbb{R}^n \to \mathbb{R}$ be a non-negative log-concave function with finite, positive integral. Then,*

$$\int_{\mathbb{R}^n} \psi(x)\, d\mu(x) \leqslant \psi\left(\int_{\mathbb{R}^n} x \frac{\psi(x)}{\int \psi\, d\mu}\, d\mu(x) \right).$$

Proof. Applying Jensen's inequality to the convex function $t \mapsto t \log t$, $t > 0$, we may write

$$\int_{\mathbb{R}^n} \psi(x) \log \psi(x)\, d\mu(x) \geqslant \Big(\int_{\mathbb{R}^n} \psi(x)\, d\mu(x) \Big) \log \Big(\int_{\mathbb{R}^n} \psi(x)\, d\mu(x) \Big).$$

Equivalently,

$$\int_{\mathbb{R}^n} \log \psi(x) \frac{\psi(x)}{\int \psi\, d\mu}\, d\mu(x) \geqslant \log \Big(\int_{\mathbb{R}^n} \psi(x)\, d\mu(x) \Big).$$

Using the assumption that $\log \psi$ is concave on $\{\psi > 0\}$ we get (using Jensen again, this time for a different probability measure)

$$\begin{aligned} \log \left[\psi \left(\int_{\mathbb{R}^n} x \frac{\psi(x)}{\int \psi\, d\mu}\, d\mu(x) \right) \right] &\geqslant \int_{\mathbb{R}^n} \log \psi(x) \frac{\psi(x)}{\int \psi\, d\mu}\, d\mu(x) \\ &\geqslant \log \Big(\int_{\mathbb{R}^n} \psi(x)\, d\mu(x) \Big), \end{aligned}$$

which is the assertion of the lemma. □

Proof of Theorem 4.1.20. For any integrable (possibly vector valued) function f on $\mathbb{R}^n$ we set

$$I(f) := \int_{K \times L} f\Big(\frac{x - y}{\sqrt{2}} \Big)\, dx\, dy.$$

Also, given $v \in \mathbb{R}^n$ we set

$$C_v := (\sqrt{2}K - v) \cap (\sqrt{2}L + v).$$

Making the change of variables $u = \frac{x+y}{\sqrt{2}}$ and $v = \frac{x-y}{\sqrt{2}}$ we see that $dxdy = dudv$ and

$$I(f) = \int_{\mathbb{R}^n} f(v)\, \mathrm{Vol}_n(C_v)\, dv. \tag{4.1.6}$$

Note that the function $\psi(v) := \mathrm{Vol}_n(C_v)$ is log-concave and supported on the set $M := \frac{K-L}{\sqrt{2}}$. We denote the uniform measure on M by μ. Then

$$\mathrm{Vol}_n(K)\mathrm{Vol}_n(L) = \int_M \mathrm{Vol}_n(C_v)\, dv = \mathrm{Vol}(M) \int_M \psi(y) d\mu$$

(this amounts to choosing $f = \mathbf{1}_M$ in the above equation). The choice $f(v) = v\mathbf{1}_M(v)$ in (4.1.6) gives (after division by $\mathrm{Vol}(K)\mathrm{Vol}(L)$) that

$$\begin{aligned} 0 &= \frac{1}{\sqrt{2}\mathrm{Vol}_n(K)} \int_K x\, dx - \frac{1}{\sqrt{2}\mathrm{Vol}_n(L)} \int_L y\, dy \\ &= \frac{1}{\mathrm{Vol}_n(K)\mathrm{Vol}_n(L)} \int_M v\mathrm{Vol}_n(C_v)\, dv. \end{aligned}$$

Thus,

$$0 = \frac{\mathrm{Vol}_n(M)}{\mathrm{Vol}_n(K)\mathrm{Vol}_n(L)} \int v\psi(v)\, d\mu(v) = \int_{\mathbb{R}^n} v \frac{\psi(v)}{\int \psi\, d\mu}\, d\mu(v),$$

and applying Lemma 4.1.21 for the log-concave function ψ we get

$$\frac{\mathrm{Vol}_n(K)\mathrm{Vol}_n(L)}{\mathrm{Vol}_n(M)} \leqslant \psi(0) = 2^{n/2}\mathrm{Vol}_n(K \cap L). \tag{4.1.7}$$

Assuming that K has its barycenter at the origin and setting $L = -K$ we have $K - L = 2K$ and we obtain the second assertion of the theorem. □

The following three lemmas will be useful later; they connect covering numbers and volumes of two bodies with those of their convex hull.

LEMMA 4.1.22. *Let K, L be centrally symmetric convex bodies in $\mathbb{R}^n$ and assume that $L \subseteq bK$ for some $b \geqslant 1$. Then,*

$$N(\mathrm{conv}(K \cup L), (1 + 1/n)K) \leqslant 2bnN(L, K).$$

Proof. From the definition of $N = N(L, K)$, there exist $x_1, \ldots, x_N \in \mathbb{R}^n$ such that $L \subseteq \bigcup_{i=1}^N (x_i + K)$. We may assume that $L \cap (x_i + K) \neq \emptyset$, and hence $x_i \in L + K \subseteq (b + 1)K$.

If $0 \leqslant \lambda \leqslant 1$ then

$$\lambda L + (1 - \lambda)K \subseteq \bigcup_{i=1}^N (\lambda x_i + \lambda K) + (1 - \lambda)K = \bigcup_{i=1}^N (\lambda x_i + K),$$

so

$$\mathrm{conv}\,(L \cup K) \subseteq \bigcup_{i=1}^N \bigcup_{0 \leqslant \lambda \leqslant 1} (\lambda x_i + K).$$

We define $\lambda_j = \frac{j}{2bn}, j = 1, \ldots, \lfloor 2bn \rfloor$ in $[0, 1]$. For every $z \in \mathrm{conv}\,(L \cup K)$ there exists i and $\lambda \in [0, 1]$ such that $z \in \lambda x_i + K$ and so, taking λ_j such that $|\lambda - \lambda_j| \leqslant \frac{1}{2bn}$ we get that $z \in \lambda_j x_i + \left(\frac{b+1}{2bn} + 1\right) K$ and hence

$$\begin{aligned} \mathrm{conv}\,(L \cup K) &\subseteq \bigcup_{j=1}^{\lfloor 2bn \rfloor} \bigcup_{i=1}^N \left(\lambda_j x_i + \left(\frac{b+1}{2bn} + 1\right) K\right) \\ &\subseteq \bigcup_{j=1}^{\lfloor 2bn \rfloor} \bigcup_{i=1}^N \left(\lambda_j x_i + \left(\frac{1}{n} + 1\right) K\right), \end{aligned}$$

as $\frac{b+1}{2b} \leqslant 1$. □

As a simple consequence we get the following two volume estimates.

LEMMA 4.1.23. *Let L be a convex body and let K be a centrally symmetric convex body in $\mathbb{R}^n$. Assume that $L \subseteq bK$ for some $b \geqslant 1$. Then*

$$\mathrm{Vol}_n\Big(\mathrm{conv}\big(K \cup L\big)\Big) \leqslant 2ebn\, N(L, K)\mathrm{Vol}_n(K).$$

Proof. By the previous lemma we have

$$N = N\left(\mathrm{conv}(K \cup L), (1 + 1/n)K\right) \leqslant 2bnN(L, K)$$

and therefore

$$\mathrm{Vol}_n(\mathrm{conv}(K \cup L)) \leqslant N(1 + 1/n)^n \mathrm{Vol}_n(K) \leqslant 2ebn\, N(L, K)\mathrm{Vol}_n(K)$$

as claimed. □

LEMMA 4.1.24. *Assume that $B_2^n \subseteq sbK$ for some $s > 0$ and some $b \geqslant 1$. For every centrally symmetric convex body P in $\mathbb{R}^n$ one has*

$$\mathrm{Vol}_n\left(\mathrm{conv}\left(K \cup \frac{1}{s}B_2^n\right) + P\right) \leqslant 2ebnN(B_2^n, sK)\,\mathrm{Vol}_n(K + P).$$

Proof. Setting $L = \frac{1}{s}B_2^n$ by Lemma 4.1.22 there exist $x_1, \ldots, x_M \in \mathbb{R}^n$ with $M = N(B_2^n, sK)\lfloor 2bn \rfloor$ such that

$$\begin{aligned}\mathrm{conv}\left(K \cup \frac{1}{s}B_2^n\right) + P &\subseteq \bigcup_{j=1}^{M} (x_j + (1 + 1/n)K) + P \\ &\subseteq \bigcup_{j=1}^{M} (x_j + (1 + 1/n)(K + P)).\end{aligned}$$

Therefore,

$$\begin{aligned}\mathrm{Vol}_n\Big(\mathrm{conv}\Big(K \cup \frac{1}{s}B_2^n\Big) + P\Big) &\leqslant M\mathrm{Vol}_n((1 + 1/n)(K + P)) \\ &\leqslant eM\, \mathrm{Vol}_n(K + P)\end{aligned}$$

and the result follows. □

4.2. Sudakov's inequality and its dual

Let K be a centrally symmetric convex body in $\mathbb{R}^n$. In this section we give an estimate for the covering numbers $N(K, tB_2^n)$ and $N(B_2^n, tK)$ using the mean width (and dual mean width) of the body, which was defined in 1.5.5 and is denoted by $w(K)$. The bound for $N(K, tB_2^n)$ is called Sudakov's inequality. The bound for $N(B_2^n, tK)$ was originally proved by Pajor and Tomczak-Jaegermann. At the present state of the theory one follows from the other by the duality Theorem 4.4.4 which we present in Section 4.4. In Section 9.1 of Chapter 9 we shall return to Sudakov's inequality from a different, probabilistic, viewpoint.

THEOREM 4.2.1 (Sudakov). *Let K be a centrally symmetric convex body in $\mathbb{R}^n$. For every $t > 0$,*

$$\log N(K, tB_2^n) \leqslant cn\left(\frac{w(K)}{t}\right)^2,$$

where $c > 0$ is an absolute constant.

We will deduce Sudakov's inequality from its dual, for which we shall present the proof of Talagrand.

THEOREM 4.2.2 (Pajor-Tomczak). *Let K be a centrally symmetric convex body in $\mathbb{R}^n$. For every $t > 0$,*

$$\log \overline{N}(B_2^n, tK) \leqslant cn \left(\frac{w(K^\circ)}{t}\right)^2,$$

where $c > 0$ is an absolute constant.

Talagrand's argument, similarly to bounds presented in the previous sections, is a volumetric argument, but it makes use of the Gaussian measure γ_n instead of Lebesgue measure. We need a simple lemma.

LEMMA 4.2.3. *Let K be a centrally symmetric convex body in $\mathbb{R}^n$. For every $z \in \mathbb{R}^n$ we have*

$$\gamma_n(K+z) \geqslant \exp(-|z|^2/2)\gamma_n(K).$$

Proof. Using the symmetry of K we see that

$$\begin{aligned}\gamma_n(K+z) &= \frac{1}{(2\pi)^{n/2}}\int_K \exp(-|z+x|^2/2)dx \\ &= \frac{1}{(2\pi)^{n/2}}\int_K \exp(-|z-x|^2/2)dx,\end{aligned}$$

and hence

$$\gamma_n(K+z) = \frac{1}{(2\pi)^{n/2}}\int_K \frac{\exp(-|z+x|^2/2)+\exp(-|z-x|^2/2)}{2}dx.$$

Since the exponential function is convex,

$$\begin{aligned}\gamma_n(K+z) &\geqslant \frac{1}{(2\pi)^{n/2}}\int_K \exp\big(-\big(|x+z|^2+|x-z|^2\big)/4\big)dx \\ &= \frac{1}{(2\pi)^{n/2}}\int_K \exp(-|x|^2/2-|z|^2/2)dx \\ &= \exp(-|z|^2/2)\frac{1}{(2\pi)^{n/2}}\int_K \exp(-|x|^2/2)dx \\ &= \exp(-|z|^2/2)\gamma_n(K).\end{aligned}$$

This proves the lemma. □

Proof of Theorem 4.2.2. Let $x_1, \ldots, x_N$ form a set of points in B_2^n which is maximal with respect to the condition

$$i \neq j \Longrightarrow \|x_i - x_j\|_K \geqslant t. \tag{4.2.1}$$

Then, the sets $x_i + \frac{t}{2}K$ have disjoint interiors. It follows that for every $\lambda > 0$ the sets $\lambda x_i + \frac{\lambda t}{2}K$ have disjoint interiors, and since γ_n is a probability measure, we have

$$\sum_{i=1}^N \gamma_n\left(\lambda x_i + \frac{\lambda t}{2}K\right) = \gamma_n\left(\bigcup_{i=1}^N \left(\lambda x_i + \frac{\lambda t}{2}K\right)\right) \leqslant 1.$$

Since $x_i \in B_2^n$, we obviously have $|\lambda x_i| \leqslant \lambda$, $i = 1, \dots, N$. From Lemma 4.2.3 we get

$$\gamma_n\left(\lambda x_i + \frac{\lambda t}{2}K\right) \geqslant \exp(-\lambda^2/2)\gamma_n\left(\frac{\lambda t}{2}K\right) \qquad i = 1, \dots, N.$$

This gives an upper bound for N: for every $\lambda > 0$,

$$N \leqslant \frac{\exp(\lambda^2/2)}{\gamma_n\left(\frac{\lambda t}{2}K\right)}.$$

In order to choose a suitable $\lambda > 0$ we write

$$\begin{aligned}\int_{\mathbb{R}^n} \|x\|_K \gamma_n(dx) &= \frac{1}{(2\pi)^{n/2}} \int_{\mathbb{R}^n} \|x\|_K e^{-|x|^2/2} dx \\ &= \frac{n\mathrm{Vol}_n(B_2^n)}{(2\pi)^{n/2}} \int_{S^{n-1}} \int_0^\infty \|\theta\|_K u^n e^{-u^2/2} du\, d\sigma(\theta) \\ &= \frac{n}{2^{n/2}\Gamma\left(\frac{n}{2}+1\right)} \int_0^\infty u^n e^{-u^2/2} du \cdot \int_{S^{n-1}} h_{K^\circ}(\theta)\, d\sigma(\theta) \\ &= \frac{\sqrt{2}\Gamma\left(\frac{n+1}{2}\right)}{\Gamma\left(\frac{n}{2}+1\right)} \cdot w(K^\circ) \leqslant c\sqrt{n} w(K^\circ).\end{aligned}$$

Here c is a universal constant (in fact $c = 2$ works).

Applying Markov's inequality we get

$$\gamma_n(\|x\|_K \geqslant \lambda t/2) \leqslant \frac{2}{\lambda t} \int_{\mathbb{R}^n} \|x\|_K \gamma_n(dx) \leqslant \frac{2c\sqrt{n}}{\lambda t} w(K^\circ),$$

that is,

$$1 - \gamma_n\left(\frac{\lambda t}{2}K\right) \leqslant \frac{2c\sqrt{n}}{\lambda t} w(K^\circ).$$

If we choose $\lambda = 4c\sqrt{n}w(K^\circ)/t$, we have

$$\gamma_n\left(2c\sqrt{n}w(K^\circ)K\right) \geqslant \frac{1}{2},$$

and hence

$$N \leqslant 2\exp\left(8c^2 n w^2(K^\circ)/t^2\right).$$

Using the connection between covering and separation, namely Fact 4.1.11, we see that

$$\log N(B_2^n, tK) \leqslant \log N \leqslant 8c^2 n \left(\frac{w(K^\circ)}{t}\right)^2$$

as claimed. □

REMARK 4.2.4. An equivalent way of stating Theorem 4.2.2 is the following:

$$\sup_{t>0} t\left(\log \overline{N}(B_2^n, tK)\right)^{1/2} \leqslant c\sqrt{n} w(K^\circ).$$

If we apply this inequality for K°, we obtain

$$A := \sup_{t>0} t\left(\log \overline{N}(B_2^n, tK^\circ)\right)^{1/2} \leqslant c\sqrt{n} w(K).$$

Tomczak-Jaegermann observed that $N(K, tB_2^n)$ and $N(B_2^n, tK^\circ)$ are equivalent in the following sense.

LEMMA 4.2.5. *Let K be a centrally symmetric convex body in $\mathbb{R}^n$. Then,*

$$N(K, tB_2^n) \leqslant N(K, 2tB_2^n) N\left(B_2^n, \frac{t}{8}K^\circ\right).$$

Proof. We first observe that

$$2K \cap \left(\frac{t^2}{2}K^\circ\right) \subseteq tB_2^n.$$

Because, if $x \in 2K$ and $x \in (t^2/2)K^\circ$, then

$$|x|^2 = \langle x, x\rangle \leqslant \|x\|_K \|x\|_{K^\circ} \leqslant 2\frac{t^2}{2} = t^2.$$

It follows that

$$N(K, tB_2^n) \leqslant N\left(K, (2K) \cap \left(\frac{t^2}{2}K^\circ\right)\right).$$

Next we use Fact 4.1.9, Fact 4.1.7 and Fact 4.1.3 to get that

$$\begin{aligned} N(K, tB_2^n) &\leqslant N\Big(K, \frac{t^2}{2}K^\circ \cap 2K\Big) \leqslant N\Big(K, \frac{t^2}{4}K^\circ\Big) \\ &\leqslant N(K, 2tB_2^n) N\Big(B_2^n, \frac{t}{8}K^\circ\Big). \end{aligned}$$

□

There are various equivalences which follow from this lemma. The first will show that Theorem 4.2.2 implies Theorem 4.2.1.

THEOREM 4.2.6. *Let K be a centrally symmetric convex body in $\mathbb{R}^n$. If*

$$B = \sup_{t>0}\, t\left(\log N(K, tB_2^n)\right)^{1/2},$$

then $B \leqslant 10A$.

Proof. The inequality of Lemma 4.2.5 and simple calculations give

$$\begin{aligned} t^2 \log N(K, tB_2^n) &\leqslant \frac{1}{4}(2t)^2 \log N(K, 2tB_2^n) + 64(t/8)^2 \log N\left(B_2^n, \frac{t}{8}K^\circ\right) \\ &\leqslant \frac{1}{4}(2t)^2 \log N(K, 2tB_2^n) + 64A^2, \end{aligned}$$

and taking sup over all $t > 0$ we arrive at $3B^2 \leqslant 256A^2$. □

Proof of Theorem 4.2.1. From Theorem 4.2.6 and Remark 4.2.4 for every $t > 0$ we have

$$t^2 \log N(K, tB_2^n) \leqslant 100A^2 \leqslant cnw^2(K),$$

where $c > 0$ is an absolute constant. □

Finally, let us mention that Sudakov's inequality holds true for not necessarily symmetric convex bodies as well.

PROPOSITION 4.2.7. *Let K be a convex body in $\mathbb{R}^n$. For every $t > 0$,*

$$t^2 \log N(K, tB_2^n) \leqslant cnw^2(K),$$

where $c > 0$ is an absolute constant.

Proof. Consider the difference body $K-K$ of K. Then,

$$w(K-K) = \int_{S^{n-1}} h_{K-K}(u)\sigma(du) = \int_{S^{n-1}} [h_K(u) + h_{-K}(u)]\sigma(du)$$
$$= \int_{S^{n-1}} [h_K(u) + h_K(-u)]\sigma(du) = 2w(K).$$

Since there is a translate of K which is contained in $K-K$, Sudakov's inequality gives

$$t^2 \log N(K, tB_2^n) \leqslant t^2 \log N(K-K, tB_2^n) \leqslant cnw^2(K-K) = 4cnw^2(K).$$

□

4.3. Entropy numbers and approximation numbers

In this section we shall work in general Banach spaces, not necessarily finite dimensional ones. There are several reasons for slightly changing the point of view. One reason is the origin of the theory of approximation numbers, which came from the need to understand operators between Banach spaces. For example, compactness of an operator $u : X \to Y$ where X and Y are two Banach spaces with unit balls B_X and B_Y, amounts to the fact that for every $\varepsilon > 0$ the covering number $N(uB_X, \varepsilon B_Y)$ is finite. To get a quantitative statement about "how compact" the operator is, one may consider the whole function $f(\varepsilon) = N(uB_X, \varepsilon B_Y)$, or, as is more customary, the "entropy numbers" sequence associated to u, defined below. The questions which come up via these considerations usually have a finite dimensional meaning (usually about some inequalities with constants not depending on the dimension), but the intuition, at least to begin with, comes from the infinite dimensional point of view.

Another, more "down to earth" reason for moving to the Banach space language here in our book is that in the proofs of some of the results below one needs to use some approximation procedure which amounts, in the finite dimensional settings, to an increase in dimension which is sometimes so big that the finiteness of the dimension becomes almost irrelevant. For example, every centrally symmetric polytope can be realized, up to a linear transformation, as a section of some higher dimensional cube. However, the dimension of the cube might be as high as half the number of facets of the polytope, and if this polytope itself came from approximating a general convex body by a polytope, the resulting dimension might be enormously high. By working in the infinite dimensional setting, we avoid dimension considerations altogether.

However, in the results we shall consider we still will try to convey the finite dimensional meanings of the relevant statements, and our main interest will of course be in tools useful for asymptotic geometric analysis.

Finally, we think it is good also for the reader to be able to adopt, at times, the Banach space theory point of view, to know this "second language", as it turns out that sometimes questions and directions are natural in one realm and very much not so in another, which is a source for finding interesting questions, connections and phenomena.

Let X and Y be two Banach spaces with unit balls B_X and B_Y, and let $u : X \to Y$ be a compact linear operator. In this section we introduce the sequences of entropy numbers, Gelfand numbers, Kolmogorov numbers and approximation numbers of u. We start with entropy numbers, which were first formally introduced by Pietsch in his work on operator theory; they are in a sense an inverse function for the covering numbers, defined as follows:

DEFINITION 4.3.1 (entropy numbers). For $k \geqslant 1$ we let

$$e_k(u) = \inf\{\varepsilon > 0 : N(uB_X, \varepsilon B_Y) \leqslant 2^{k-1}\}.$$

Note that $e_1(u) = \|u\|$. It is clear that $(e_k(u))$ is decreasing and, since u is compact, we have $e_k(u) \to 0$ as $k \to \infty$ (in fact, this property is equivalent to the compactness of u). Since the sequence $(e_k(u))_{k\geqslant 1}$ is an "inverse function" of the covering numbers $N(uB_X, \varepsilon B_Y)$, most of the facts quoted in Section 4.1 have their counterparts for entropy numbers. From the elementary properties of the covering numbers described in Section 4.1.1, one can check that for every $k, n \geqslant 1$ and any $u : X \to Y$, $v : Y \to Z$,

$$e_{k+n-1}(vu) \leqslant e_k(v)e_n(u).$$

Also, if $u_1, u_2 : X \to Y$, we have

$$e_{k+n-1}(u_1 + u_2) \leqslant e_k(u_1) + e_n(u_2).$$

It is well known that if an operator is compact then so is its dual operator. This implies a connection between $e_k(u)$ and $e_k(u^*)$ which will be made clear by Theorem 4.3.7 at the end of this section. In Section 4.4 we shall see that in many cases these two sequences are comparable in a much stronger sense.

DEFINITION 4.3.2. Let X and Y be two Banach spaces and let $u : X \to Y$ be a compact linear operator.

(i) The *approximation numbers* of u are defined for each $k \geqslant 1$ by

$$a_k(u) = \inf\{\|u - v\| \,:\, v : X \to Y,\ \operatorname{rank}(v) < k\}.$$

(ii) The *Gelfand numbers* of u are defined for each $k \geqslant 1$ by

$$c_k(u) = \inf\{\|u|_F\| \ :\ F \subset X\ ,\ \operatorname{codim}(F) < k\}.$$

(iii) For every closed subspace S of Y we consider the quotient map $Q_S : Y \to Y/S$. The *Kolmogorov numbers* of u are defined for each $k \geqslant 1$ by

$$d_k(u) = \inf\{\|Q_E u\| \,:\, E \subset Y,\ \dim(E) < k\}.$$

We denote by s_k any of the numbers a_k, c_k and d_k. Then, the sequence $(s_k(u))_{k\geqslant 1}$ is decreasing and $s_k(u) \to 0$ as $k \to \infty$. We also have $s_1(u) = \|u\|$.

REMARK 4.3.3. Let X, Y and Z be Banach spaces and let $u, u_1, u_2 : X \to Y$ and $v : Y \to Z$ be compact linear operators. We will need some basic properties of the various approximation numbers of u and v. First, since the polar of a section is a projection (the dual of a subspace is a quotient) and vice versa, we have
(i) For every $k \geqslant 1$

(4.3.1) $$c_k(u) = d_k(u^*) \quad \text{and} \quad d_k(u) = c_k(u^*).$$

Denoting $s_k = c_k$ or d_k, for every $k, n \geqslant 1$ we have the following two properties:
(ii) For the composition of two operators,

(4.3.2) $$s_{k+n-1}(vu) \leqslant s_k(v)s_n(u).$$

(iii) For the sum of two operators,

(4.3.3) $$s_{k+n-1}(u_1 + u_2) \leqslant s_k(u_1) + s_n(u_2).$$

The next theorem of Carl shows that the sequence $(e_k(u))_{k\geqslant 1}$ is dominated, in a sense, by any of the sequences $(a_k(u))_{k\geqslant 1}$, $(c_k(u))_{k\geqslant 1}$ and $(d_k(u))_{k\geqslant 1}$. We will make special use of this theorem in Chapter 8, where an estimate for covering numbers will follow from corresponding estimates for diameters of sections and projections of convex bodies.

THEOREM 4.3.4 (Carl). *We use s_k to denote one of the numbers a_k, c_k and d_k. For every $\alpha > 0$ there exists a constant ρ_α with the following property: for every $u : X \to Y$ and every $n \in \mathbb{N}$,*

$$\sup_{k\leqslant n} k^\alpha e_k(u) \leqslant \rho_\alpha \sup_{k\leqslant n} k^\alpha s_k(u).$$

For the proof we need a simple lemma, which in turn uses two well known facts about the spaces $\ell_\infty(I)$ and $\ell_1(I)$ where I is some index set.

FACT 4.3.5. *Let I be some index set. Then*
(i) $\ell_\infty(I)$ has the "extension property", namely any bounded linear operator defined on a closed linear subspace S of some Banach space X and taking values in $\ell_\infty(I)$, has an extension to the whole Banach space X, with the same norm.
(ii) $\ell_1(I)$ has the "lifting property", namely for every pair of Banach spaces X and Y and every operator S from Y onto X, the following condition holds: for every operator $T : \ell_1(I) \to X$ there is an operator $\tilde{T} : \ell_1(I) \to Y$ such that $T = S\tilde{T}$. Moreover, whenever $S : Y \to X$ is a quotient map then, for every $\varepsilon > 0, \tilde{T} : \ell_1(I) \to Y$ may be chosen so that $\|\tilde{T}\| \leq (1+\varepsilon)\|T\|$.

The proofs of these two (dual) facts are standard, and essentially follow from the Hahn-Banach theorem and duality. We shall use these together with the fact that every Banach space X is isometric to a quotient space of $\ell_1(I)$ and every Banach space Y can be isometrically embedded into $\ell_\infty(I)$ for some set I. This corresponds to the finite dimensional facts that any centrally symmetric polytope $P \subset \mathbb{R}^n$ is a linear image of B_1^m for some $m \geqslant n$ ($m/2$ being the number of vertices of P) and is also an n-dimensional section of some linear image of the cube B_∞^m for some $m \geqslant n$ (here $m/2$ is the number of facets of P).

LEMMA 4.3.6. *Let X, X_1, Y, Y_1 be Banach spaces. Assume that X is isometric to a quotient space of X_1 and that Y is isometric to a subspace of Y_1. Let $q : X_1 \to X$ be the quotient map and let $j : Y \to Y_1$ be an isometric embedding. Then, for every operator $u : X \to Y$ we have*

$$e_k(u) = e_k(uq)$$

and

$$e_k(u)/2 \leqslant e_k(ju) \leqslant e_k(u)$$

for all $k \geqslant 1$. Moreover, if $X_1 = \ell_1(I)$ then $d_k(u) = a_k(uq)$, while if $Y_1 = \ell_\infty(I)$ then $a_k(ju) = c_k(u)$.

Proof. The first claim follows immediately from the fact that $qB_{X_1} = B_X$. Note also that for every $k \geqslant 1$ one has

$$e_k(ju) \leqslant \|j\| e_k(u) = e_k(u).$$

Next, observe that if we cover $ju(B_X)$ by 2^{k-1} balls of radius ε in Y_1, then the set $u(B_X)$ is the union of 2^{k-1} subsets of diameter 2ε because j is an isometry, and then by the definition of the entropy numbers we get

$$e_k(u) \leqslant 2e_k(ju).$$

For the last two claims we use the lifting property of $\ell_1(I)$ and the extension property of $\ell_\infty(I)$ respectively. $\square$

We mention that the covering numbers analogue of the left hand side in the second inequality is that for centrally symmetric convex K and T and any subspace E we have

$$N(K \cap E, 2(T \cap E)) \leqslant N(K, T)$$

which follows since for any x there exists y such that $(x+T)\cap E \subseteq y+2(T\cap E)$; this in turn easily follows from the convexity and central symmetry of T. The geometric formulation of the other inequalities is trivial (for example, $N(K \cap E, T \cap E) \leqslant N(K \cap E, T)$).

Proof of Theorem 4.3.4. As every Banach space X is isometric to a quotient space of $\ell_1(I)$ and every Banach space Y can be isometrically embedded into $\ell_\infty(I)$ for some set I, by the previous lemma it is enough to prove the theorem for the case $s_k = a_k$.

Let $u : X \to Y$. We may assume that $n = 2^N$ for some $N \geqslant 0$ and that $\sup_{k\leqslant n} k^\alpha a_k(u) \leqslant 1$. Then, for every $m \leqslant N$ there exists an operator $v_m : X \to Y$ with $\mathrm{rank}(v_m) < 2^m$ such that

$$\|u - v_m\| < 2^{-m\alpha}.$$

We set $T_0 = v_0$, $T_1 = v_1 - v_0, \dots, T_N = v_N - v_{N-1}$. Then,

$$u = \sum_{m=0}^{N} T_m + (u - v_N),$$

and we have

$$\|T_m\| \leqslant 2^{\alpha+1}2^{-m\alpha} \quad \text{and} \quad \mathrm{rank}(T_m) < 2^{m+1}.$$

We set $K = u(B_X)$, $K_m = T_m(B_X)$ and consider $t_m > 0$ which will be determined later on. A standard covering argument and the fact that $K_m \subseteq \|T_m\|B_Y$ show that

$$N(K_m, t\|T_m\|B_Y) \leqslant \left(1 + \frac{2}{t}\right)^{\mathrm{rank}(T_m)} \leqslant \left(1 + \frac{2}{t}\right)^{2^{m+1}}$$

for every $t > 0$.

Let $r \in (0,1)$. Using the elementary inequality $(1+y)^d \leqslant (1+y^r)^{d/r}$ we write

$$\left(1 + \frac{2}{t}\right)^{2^{m+1}} < \exp\left(\left(\frac{2}{t}\right)^r \frac{2^{m+1}}{r}\right).$$

We set $N(t) := N(K, tB_Y)$. Then,

$$K \subseteq \sum_{m=0}^{N} K_m + (u - v_N)(B_X)$$

and taking into account the fact that $\|u - v_N\| < 2^{-N\alpha}$ we obtain

$$N\left(\sum_{m=0}^{N} t_m\|T_m\| + 2^{-N\alpha}\right) \leqslant \prod_{m=0}^{N} N(K_m, t_m\|T_m\|B_Y)$$

$$\leqslant \exp\left(\sum_{m=0}^{N}\left(\frac{2}{t_m}\right)^r \frac{2^{m+1}}{r}\right).$$

We choose $\beta > \alpha$ and $r \in (0,1)$ such that $r < 1/\beta$. For every $s > 0$ we define $t_m = s2^{m\beta}2^{-N\beta}$. With this choice,

$$\sum_{m=0}^{N} t_m\|T_m\| + 2^{-N\alpha} \leqslant 2^{-N\alpha}(c_1 s + 1)$$

and

$$\sum_{m=0}^{N}\left(\frac{2}{t_m}\right)^r \frac{2^{m+1}}{r} \leqslant c_2 s^{-r}2^N,$$

where c_1, c_2 are constants depending only on α, β and r. Finally, we choose s large enough (depending only on α, β and r) so that

$$2\exp(c_2 s^{-r}2^N) \leqslant 2^{2^N} = 2^n.$$

Then,

$$N(2^{-N\alpha}(c_1 s + 1)) \leqslant 2^{n-1},$$

that is,

$$e_n(u) \leqslant 2^{-N\alpha}(c_1 s + 1) = n^{-\alpha}(c_1 s + 1).$$

This proves that, for $n = 2^N$,

$$n^\alpha e_n(u) \leqslant (c_1 s + 1)\sup_{k\leqslant n} k^\alpha a_k(u),$$

which is the result in the case $n = 2^N$. Extending the result to any n is simple. □

To end this section we present a consequence of Lemma 4.2.5 in terms of entropy numbers; we have the following theorem of Tomczak-Jaegermann on the connection between the two sequences $e_k(u)$ and $e_k(u^*)$ for an operator u between a normed space and the Euclidean space of the same dimension.

THEOREM 4.3.7 (Tomczak). *Let $X = (\mathbb{R}^n, \|\cdot\|_K)$ be a normed space with unit ball K. Let $u : X \to \ell_2^n$ be a linear operator. Then,*

$$\sum_{j=1}^{\infty} e_j(u) \leqslant 64\sum_{j=1}^{\infty} e_j(u^*).$$

REMARK 4.3.8. A similar inequality, with a different universal constant, holds also in the opposite direction, which will follow from a similar lemma given as Lemma 4.4.1 below, but as we shall show a much stronger duality relation in Section 4.4, we do not state it here. We also mention that the condition that one of the spaces is Euclidean can be much relaxed, for example to one of the spaces being, say, uniformly convex, and then the constant instead of 64 will depend on the modulus of convexity. We also remark that the extension to infinite dimensional Banach spaces (and compact operators) is straightforward, and will be elaborated upon in the next section.

Proof. As in Lemma 4.2.5 we have that

$$N(K, tB_2^n) \leqslant N\left(K, \frac{t^2}{4}K^\circ\right) \leqslant N(K, 4tB_2^n)N\left(B_2^n, \frac{t}{16}K^\circ\right).$$

We shall without loss of generality assume that $u : X \to \ell_2^n$ is the formal identity operator in $\mathbb{R}^n$, and so $u^* : \ell_2^n \to X$ is the identity operator as well. Plugging in $t = \frac{1}{4}e_k(u) + 16e_k(u^*)$ we see that

$$e_{2k}(u) \leqslant e_{2k-1}(u) \leqslant \frac{1}{4}e_k(u) + 16e_k(u^*)$$

and thus

$$\sum_{j=1}^\infty e_k(u) = \sum_{j=1}^\infty e_{2k}(u) + \sum_{j=1}^\infty e_{2k-1}(u) \leqslant \frac{1}{2}\sum_{j=1}^\infty e_k(u) + 32\sum_{j=1}^\infty e_k(u^*),$$

from which it follows that

$$\sum_{j=1}^\infty e_k(u) \leqslant 64\sum_{j=1}^\infty e_k(u^*),$$

$\square$

4.4. Duality of entropy

4.4.1. Statement and background

If T is the unit ball of a normed space Y, and K is the image of the unit ball of another space X under some given compact linear operator u, the number $N(K, tT)$ as a function of the parameter t quantifies in some sense the compactness of $u : X \to Y$.

One of the main subjects of research in the field of geometric operator theory was the duality of entropy numbers, which in the language of covering numbers states that in an appropriate sense, as functions of t, the expressions $N(K, tT)$ and $N(T^\circ, tK^\circ)$ are equivalent, where K° and T° denote the polar bodies of K and T respectively. The most important case (which is the one appearing in applications to probability theory, ergodic theory, learning theory and other fields) is when one of the two bodies is assumed to be the Euclidean ball. In the language of operators, this means that either the domain or the range of the operator u is a Hilbert space.

A first result in this direction follows from the analysis given in the proof of Theorem 4.2.6. Indeed, let us restate part of the proof, and a dual argument, as a lemma which will be useful for us later on.

LEMMA 4.4.1. *Let K and T be centrally symmetric convex bodies in $\mathbb{R}^n$. Then,*

$$N(K, tB_2^n) \leqslant N(K, 4tB_2^n)N\left(B_2^n, \frac{t}{16}K^\circ\right).$$

and

$$N(B_2^n, tT) \leqslant N(B_2^n, 4tT)N\left(T^\circ, \frac{t}{16}B_2^n\right).$$

Proof. The first inequality appears already in the beginning of the proof of Theorem 4.3.7 (and follows by a simple modification of the last step in the proof of Lemma 4.2.5). For the second inequality we need the dual fact, which one gets by applying polarity to the equation $\frac{t}{2}T \cap \frac{2}{t}T^\circ \subseteq B_2^n$: we have

$$B_2^n \subseteq \mathrm{conv}\Big(\frac{t}{2}T \cup \frac{2}{t}T^\circ\Big) \subseteq \frac{t}{2}T + \frac{2}{t}T^\circ.$$

Therefore

$$\begin{aligned} N(B_2^n, tT) &\leqslant N\Big(\frac{t}{2}T + \frac{2}{t}T^\circ, tT\Big) \leqslant N\Big(\frac{2}{t}T^\circ, \frac{t}{2}T\Big) \\ &\leqslant N\Big(\frac{2}{t}T^\circ, \frac{1}{8}B_2^n\Big) N\Big(\frac{1}{8}B_2^n, \frac{t}{2}T\Big) \\ &= N(B_2^n, 4tT) N\left(T^\circ, \frac{t}{16}B_2^n\right), \end{aligned}$$

where we have used Fact 4.1.8 in the second inequality. □

As a corollary we get a very simple form of duality, namely

Corollary 4.4.2. *Let K be a centrally symmetric convex body in $\mathbb{R}^n$.*

(i) *If $K \subseteq 4B_2^n$ then*

$$N(K, B_2^n) \leqslant N\Big(B_2^n, \frac{1}{16}K^\circ\Big). \tag{4.4.1}$$

(ii) *If $B_2^n \subseteq 4K$ then*

$$N(B_2^n, K) \leqslant N\Big(K^\circ, \frac{1}{16}B_2^n\Big). \tag{4.4.2}$$

It turns out that also without the inclusion assumption of Corollary 4.4.2 a very strong connection between the covering number of K by B_2^n and the covering number of B_2^n by K° exists. In fact, under an additional assumption on T a more general duality holds connecting $N(K, tT)$ and $N(T^\circ, tK^\circ)$.

Remark 4.4.3. Note that in the case of two ellipsoids, Fact 4.1.6 is an exact duality relation.

In the next subsection we prove the following duality result of Artstein-Avidan, V. Milman and Szarek for covering numbers.

Theorem 4.4.4 (Artstein-Milman-Szarek). *There exist two absolute constants α and β such that for any dimension n and any centrally symmetric convex body K in $\mathbb{R}^n$ one has*

$$N(B_2^n, \alpha^{-1}K^\circ)^{\frac{1}{\beta}} \leqslant N(K, B_2^n) \leqslant N(B_2^n, \alpha K^\circ)^{\beta}. \tag{4.4.3}$$

For instance one may take $\beta = 10$ and $\alpha = 1/128$. This theorem establishes a strong connection between the geometry of a set and its polar or, equivalently, between a normed space and its dual. Notice that since the theorem is true for any K, we can actually infer that for any $t > 0$

$$\beta^{-1}\log N(B_2^n, \alpha^{-1}tK^\circ) \leqslant \log N(K, tB_2^n) \leqslant \beta \log N(B_2^n, \alpha t K^\circ). \tag{4.4.4}$$

For the proof of theorem 4.4.4 we shall use a reduction scheme which will allow us to consider only "bounded bodies", to which we may apply Corollary 4.4.2.

4.4.2. A reduction scheme

The following proposition enables us, while trying to prove a duality theorem for two general convex bodies K and T (not requiring, yet, that one of them be a Euclidean ball), to restrict our attention to the special case in which $K \subseteq 4T$. In the case where T is Euclidean, Corollary 4.4.2 above together with the following proposition complete the proof of Theorem 4.4.4.

PROPOSITION 4.4.5. *Let T be a centrally symmetric convex body in a Euclidean space such that, for some given constants $c, C > 0$, the following holds: for any K which is a centrally symmetric convex body with $K \subseteq 4T$, one has*

$$N(K,T) \leqslant N(T^\circ, cK^\circ)^C.$$

Then there exist some other constants $c', C' > 0$ (depending only on c, C) such that for any centrally symmetric convex body K,

$$N(K,T) \leqslant N(T^\circ, c'K^\circ)^{C'}.$$

Dually, if K is fixed and the hypothesis holds for all T's verifying $K \subseteq 4T$, then the assertion holds for any T.

There is no special magic to the number 4 in this proposition, and the same result holds if, instead, we assume in the hypothesis that it holds true for all $K \subseteq RT$, where $R > 2$ is some fixed constant. We use 4 only to simplify notation. Moreover, in the "dually" part, it is enough to assume that the condition holds for all $K \subseteq RT$ for some fixed $R > 1$, and if the body T is the Euclidean ball then $R > 1$ suffices in the first part as well.

This proposition is proved by an iterative scheme. We need two lemmas. The first one is based on a simple geometric iteration procedure.

LEMMA 4.4.6. *For any two symmetric convex bodies K, T in $\mathbb{R}^n$ and any integer $s \geqslant 1$,*

$$N(K,T) \leqslant N(K, 2^s T) \prod_{j=0}^{s-1} N(2^{1-j}K \cap 4T, T) \tag{4.4.5}$$

and

$$N(K,T) \leqslant N(K, 2^s T) \prod_{j=0}^{s-1} N\big(K, \mathrm{conv}\big(2^{j-1}T \cup \tfrac{1}{4}K\big)\big). \tag{4.4.6}$$

Proof. For (4.4.5) consider the following inequality

$$\begin{aligned} N(K, 2^j T) &\leqslant N(K, 2^{j+2}T \cap 2K) N(2K \cap 2^{j+2}T, 2^j T) \\ &\leqslant N(K, 2^{j+1}T) N(2K \cap 2^{j+2}T, 2^j T) \end{aligned}$$

which follows from Facts 4.1.7 and 4.1.9. Iterating this inequality from $j = 0$ to s, and dividing both entries in the second term by 2^j, gives (4.4.5). For (4.4.6), consider the following inequality which follows from Fact 4.1.7:

$$N(K, 2^j T) \leqslant N\Big(K, \mathrm{conv}(2^{j-1}T \cup \frac{1}{4}K)\Big) N\Big(\mathrm{conv}(2^{j-1}T \cup \frac{1}{4}K), 2^j T\Big).$$

Changing the convex hull in the second term to the Minkowski sum of the sets (which is bigger and thus harder to cover), using Fact 4.1.8, and multiplying both entries by 4, leads to

$$N(K, 2^j T) \leqslant N\Big(K, \operatorname{conv}(2^{j-1}T \cup \frac{1}{4}K)\Big) N(K, 2^{j+1}T).$$

Iterating the above argument yields (4.4.6). The proof of Lemma 4.4.6 is thus complete. □

For each factor in these products, the body that is being covered is included in 4 times the covering body, and so we will be able to use the assumption of the proposition. We will thus have a bound of the original covering number by a product of dual covering numbers. To collapse this product of many numbers to the product of just a few, we need the following super-multiplicativity inequalities for covering numbers.

LEMMA 4.4.7. *Let $A > a > 3B > 3b$. Then for any two symmetric convex bodies K and T,*

$$N(K \cap AT, aT) N(K \cap BT, bT) \leqslant N\Big(K \cap AT, \frac{b}{4}T\Big) \tag{4.4.7}$$

and

$$N\big(T, a \operatorname{conv}(K \cup \tfrac{1}{A}T)\big) N\big(T, b \operatorname{conv}(K \cup \tfrac{1}{B}T)\big) \leqslant N\big(T, \frac{b}{4}\operatorname{conv}(K \cup \tfrac{1}{A}T)\big). \tag{4.4.8}$$

Proof. For the first inequality, first notice that K enters the inequality only via its intersections with T-balls of radii $\leqslant A$, so we may as well assume that $K = K \cap AT$ to begin with. Denote $N_1 = N(K, aT)$ and $N_2 = N(K \cap BT, bT)$. Pick (by Fact 4.1.11) an aT-separated set $x_1, \ldots, x_{N_1}$ in K and a bT-separated set $y_1, \ldots, y_{N_2}$ in $K \cap BT$. Define a new set by $z_{i,j} = (x_i + y_j)/2$. All these points are in K, and there are $N_1 N_2$ of them. We shall show that, in addition, the $z_{i,j}$'s are $\frac{b}{2}T$-separated; this will, by Fact 4.1.11, imply that $N(K, \frac{b}{4}T) \geqslant N_1 N_2$, as required. To show the asserted separation, we denote by $\|\cdot\|$ the norm associated to the body T, and consider two cases. First, if we look at $\|z_{i,j} - z_{i,k}\|$, this is simply $\|y_j - y_k\|/2$ and it exceeds $b/2$. On the other hand if $k \neq i$, then by the triangle inequality $\|z_{i,j} - z_{k,l}\| \geqslant \|x_i - x_k\|/2 - \|y_j - y_l\|/2$, and using the fact that the y_i's are in BT we see that these quantities are greater than $a/2 - B$, which in turn exceeds $\frac{b}{2}$. This completes the proof of inequality (4.4.7).

For the second inequality in the lemma, we this time notice firstly that K enters the inequality only via the convex hull of its union with T-balls of radii $\geqslant \frac{1}{A}$, so we may as well assume that $K = K \cup \frac{1}{A}T$ to begin with. Denote $N_1 = N(T, aK)$ and $N_2 = N(T, b \operatorname{conv}(K \cup \frac{1}{B}B_2^n))$. Pick, by Fact 4.1.11, sets $\{x_1, \ldots, x_{N_1}\}$ and $\{y_1, \ldots, y_{N_2}\}$ in T which are respectively aK-separated and $b(\alpha K + \frac{1-\alpha}{B}T)$-separated, where $\alpha = \frac{a}{2a-b} \in (0,1)$ (note that $\alpha K + \frac{1-\alpha}{B}T \subseteq \operatorname{conv}(K \cup \frac{1}{B}T)$). Define $z_{i,j} = \frac{b}{2a}x_i + (1 - \frac{b}{2a})y_j$. All these points are in T, and there are $N_1 N_2$ of them. As above, it will be enough to show that the $z_{i,j}$'s are $\frac{b}{2}K$-separated, i.e., that whenever $(i,j) \neq (k,l)$, we have

$$z_{k,l} \notin z_{i,j} + \frac{b}{2}K.$$

When looking at $j = l$, this is the same as asking that $\frac{b}{2a}x_k \notin \frac{b}{2a}x_i + \frac{b}{2}K$, which follows from the separation of x_i and x_k. When looking at $j \neq l$ and noticing that

$x_i, x_k \in T$, we see that it suffices to show that

$$(1 - \tfrac{b}{2a})y_j \notin (1 - \tfrac{b}{2a})y_l + 2\frac{b}{2a}T + \frac{b}{2}K.$$

Under our hypotheses, the above follows from the separation of y_l and y_j. Indeed, $\frac{1}{2}(1-\frac{b}{2a})^{-1} = \alpha$, just by the definition of α. On the other hand, it is readily verified that the assumption $a > 3B > 2B + b$ implies $\frac{1}{a}(1 - \frac{b}{2a})^{-1} < \frac{1-\alpha}{B}$. The proof is thus complete. □

REMARK 4.4.8. With more careful arguing, any factor less than $1/2$ instead of $1/4$ can be obtained in the lemma (with then stronger conditions on a, B). Moreover, if we work all along with separated sets instead of covering numbers, then we may arrive at any factor less than 1: for a factor $1 - \varepsilon$ (which corresponds to $\frac{1}{2} - \frac{\varepsilon}{2}$ for covering numbers), we need the condition $a > 3(\frac{1-\varepsilon}{\varepsilon})B$. This can be used to improve slightly some of the constants in Theorem 4.4.4.

Proof of Proposition 4.4.5. Let us first assume that the centrally symmetric convex body T is fixed. The assumption in the proposition is that there exist two constants $c, C > 0$, such that for every L which is a centrally symmetric convex body with $L \subseteq 4T$, we have

$$N(L, T) \leqslant N(T^\circ, cL^\circ)^C. \tag{4.4.9}$$

Let K be any centrally symmetric convex body. Denote by s the smallest integer such that $N(K, 2^s T) = 1$. Inequality (4.4.5) of Lemma 4.4.6 tells us that

$$N(K, T) \leqslant \prod_{j=0}^{s-1} N(2^{1-j}K \cap 4T, T).$$

We use, for every j, inequality (4.4.9) for the body $L = 2^{1-j}K \cap 4T$ (which is clearly inside $4T$), to get that

$$N(K, T) \leqslant \prod_{j=0}^{s-1} N(T^\circ, c(2^{1-j}K \cap 4T)^\circ)^C,$$

and rewrite it as

$$N(K, T)^{1/C} \leqslant \prod_{j=0}^{s-1} N\left(T^\circ, c2^{j-1}\mathrm{conv}\big(K^\circ \cup \tfrac{1}{2^{j+1}}T^\circ\big)\right). \tag{4.4.10}$$

We would now like to collapse this product using (4.4.8) of Lemma 4.4.7 (with the body T° acting as T and K° as K).

However, an additional trick is required since for two neighboring factors in the products the condition of Lemma 4.4.7 does not hold, and so they cannot be "collapsed." For example, for two such factors in (4.4.10) one has $a = c2^{j-1}$ and $B = 2^j$, and so one cannot hope for $a > 3B$. We thus split the product into several parts, say l sub-products, for some constant l. We do this by grouping separately the factors corresponding to the j's with some fixed remainder modulo l. If two terms in the product are l terms apart, that is, correspond to some j and $j - l$, then we have $a = c2^{j-1}$ and $B = 2^{j-l+1}$, so as long as $c2^{j-1} > 32^{j-l+1}$, that is, $l > \log\frac{3}{c} + 2$, the two terms are collapsable. We will require a bit more, because we also lose a factor 4 in every use of (4.4.8), but these terms do not accumulate and are each time consumed in the condition of the next step.

More precisely, we denote $l = [\log \frac{12}{c}] + 3$, and rewrite the right hand side of (4.4.10) as (without loss of generality we may assume that $\log l$ divides $s+1$, otherwise we increase s so that it is the case)

$$\prod_{k=0}^{l-1} \left(\prod_{j=0}^{\frac{(s-(l-1))}{l}} N(T^\circ, c2^{(lj+k)-1}\mathrm{conv}(K^\circ \cup \frac{1}{2^{(lj+k)+1}} T^\circ)) \right).$$

We will bound from above each of the l products, using (4.4.8). To collapse the last term in the kth product with the one before last in the k-th product we need to check that $a = c2^{(s-l+1+k)}$ is greater that $3B = 3 \times 2^{(s-2l+1+k)}$, which clearly holds from the choice of l. We thus bound the k-th product by

$$N(T^\circ, \frac{c}{4} 2^{(s-2l+k)}\mathrm{conv}(K^\circ \cup \frac{1}{2^{(s-2l+k)+2}} T^\circ)) \times$$
$$\prod_{j=0}^{\frac{(s-(l-1))}{l}-2} N(T^\circ, c2^{(lj+k)-1}\mathrm{conv}(K^\circ \cup \frac{1}{2^{(lj+k)+1}} T^\circ)).$$

From here onward all the steps are the same; we just need to make sure at each stage j that

$$\frac{c}{4} 2^{(lj+k)-1} \geqslant 3 \times 2^{(lj-l+k)+1}. \tag{4.4.11}$$

As before, this can be rewritten as $2^{l-2} \geqslant \frac{12}{c}$, which clearly holds for our choice of l. We point out that the factors $1/4$ in $b/4$ do not accumulate, but enter into the quantity a of the next step.

Continuing this way with all the factors of these products, we arrive at

$$N(K,T)^{1/C} \leqslant \prod_{k=0}^{l-1} \left(N(T^\circ, \frac{c}{4} 2^{k-1}\mathrm{conv}(K^\circ \cup \frac{1}{2^{k+1}} T^\circ)) \right),$$

which clearly implies

$$N(K,T)^{1/C} \leqslant N\left(T^\circ, \frac{c}{8} K^\circ\right)^l.$$

This completes the proof of the reduction proposition in the case where T is a fixed body. We get the estimate $c' = \frac{c}{8}$ and $C' = Cl = C([\log \frac{12}{c}] + 3)$.

The proof of the second case, when one assumes the condition in the reduction proposition for a fixed body K and all T with $T \supseteq \frac{1}{4} K$, and deduce it (with different constants) for all T's, is fully analogous. One first uses the bound (4.4.6) instead of (4.4.5) of Lemma 4.4.6, then the assumption in the proposition, and then collapses the product using (4.4.7) instead of (4.4.8). □

4.4.3. Proof of the duality theorem

To complete the proof of Theorem 4.4.4, combine Lemma 4.4.1 with the reduction Proposition 4.4.5. The constants we arrive at are $\alpha = c' = \frac{1}{16 \times 8} = \frac{1}{128}$ and $\beta = C' = 10$.

4.4.4. Convexified Separation numbers

In this subsection we discuss a different, and related, notion of separation, which we call convex separation. We prove a duality theorem for the cardinality of maximal convexly separated sets, Theorem 4.4.10, and this proof is quite simple, using basically the Hahn-Banach theorem. We delay outlining the connection between the usual covering/separation and the convexified separation Part II since we need the notion of type which has not yet been introduced in this book. However, we remark here that for a wide class of bodies, in the case of bounded radius (which is, for duality purposes, sufficient in view of Proposition 4.4.5) the notion of maximal convexly separated sets and the usual covering numbers are equivalent. It will apply in particular for all ℓ_p balls ($1 < p < \infty$) as well as all uniformly convex and uniformly smooth bodies. This equivalence, together with Theorem 4.4.10 will imply a much more general duality theorem.

The convexified setting demystifies, in a sense, the connections between covering numbers and covering numbers of duals, enabling us to extract the one step where a duality connection between the geometry of a space and its dual appears. The connections (which we shall portray in Part II) between convexified and usual covering are geometrical considerations; while often delicate and involved, they are always set in a given normed space, and thus are conceptually simpler.

Recall that the separation number $M(K,T)$ is the maximal number of points in K such that the distance between each two, in the norm whose unit ball is T, is greater than 1. We now define a new number $\hat{M}(K,T)$, which can be thought of as the cardinality of a "maximal convexly separated set", and plays the central role in this section:

$$\hat{M}(K,T) = \max\{N : \exists x_1,\dots,x_N \in K \text{ s.t. } (x_j+T)\cap \operatorname{conv}\{x_i, i<j\} = \emptyset\}. \tag{4.4.12}$$

We shall refer to any sequence satisfying the condition (4.4.12) as *T-convexly separated.* Leaving out the convex hull operation "conv" leads to the usual T-separated set. Thus we clearly have $\hat{M}(K,T) \leqslant M(K,T)$. We emphasize that, as opposed to the usual notions of packing and covering, the *order* of the points is important here.

Remark 4.4.9. The above definition is very natural from the point of view of complexity theory and optimization. A standard device in constructing geometric algorithms is a "separation oracle": if $x \notin K$, the algorithm returns a functional efficiently separating x from K. It is arguable that the quantity $\hat{M}(K,\cdot)$ correctly describes the complexity of a set K with respect to many such algorithms.

While it is still an open problem whether the duality of entropy conjecture holds in full generality, the corresponding duality statement for convex separation is fairly straightforward. We have the next convexified duality theorem.

Theorem 4.4.10 (Artstein-Milman-Szarek-Tomczak). *For any pair of centrally symmetric convex bodies $K,T \subset \mathbb{R}^n$ we have*

$$\hat{M}(K,T) \leqslant \hat{M}(T^\circ, K^\circ/2)^2.$$

Proof of Theorem 4.4.10. Let $R := \sup\{\|x\|_T : x \in K\}$; i.e., R is the radius of K with respect to the gauge of T. We will show that

(i) $\hat{M}(T^\circ, K^\circ/2) \geqslant \hat{M}(K,T)/\lceil 4R \rceil$.
(ii) $\hat{M}(T^\circ, K^\circ/2) \geqslant \lceil 4R \rceil$.

Once the above are proved, Theorem 4.4.10 readily follows. To show (i), denote $N = \hat{M}(K,T)$ and let $x_1, \dots, x_N$ be a T-convexly separated sequence in K. Then, by (the elementary version of) the Hahn-Banach theorem, there exist separating functionals $y_1, \dots, y_N \in T^\circ$ such that

$$1 \leqslant i < j \leqslant N \Longrightarrow \langle y_j, x_j - x_i \rangle = \langle y_j, x_j \rangle - \langle y_j, x_i \rangle \geqslant 1, \tag{4.4.13}$$

a condition which is in fact equivalent to (x_j) being T-convexly separated. Now $x_j \in K \subseteq RT$ and $y_j \in T^\circ$ imply that $-R \leqslant \langle y_j, x_j \rangle \leqslant R$, and hence dividing $[-R, R]$ into $\lceil 4R \rceil$ subintervals of length $\leqslant 1/2$ we may deduce that one of these subintervals contains $M \geqslant N/\lceil 4R \rceil$ of the numbers $\langle y_j, x_j \rangle$. To simplify the notation, assume that this occurs for $j = 1, \dots, M$, that is

$$1 \leqslant i, j \leqslant M \Longrightarrow -1/2 \leqslant \langle y_i, x_i \rangle - \langle y_j, x_j \rangle \leqslant 1/2. \tag{4.4.14}$$

Combining (4.4.13) and (4.4.14) we obtain for $1 \leqslant i < j \leqslant M$

$$\langle y_i - y_j, x_i \rangle = \langle y_i, x_i \rangle - \langle y_j, x_j \rangle + \langle y_j, x_j \rangle - \langle y_j, x_i \rangle \geqslant -1/2 + 1 = 1/2,$$

which is again a condition of type (4.4.13) and thus shows that the sequence $y_M, \dots, y_1$ is $(K^\circ/2)$-convexly separated. Notice that the order has changed - indeed the above shows that y_i is separated from any convex combination $\sum_{j>i} \lambda_j y_j$ with $\sum_{j>i} \lambda_j = 1$ since

$$\Big\| y_i - \sum_{j>i} \lambda_j y_j \Big\|_{K^\circ} \geqslant \sum_{j>i} \lambda_j \langle y_i - y_j, x_i \rangle \geqslant 1/2.$$

Hence $\hat{M}(T^\circ, K^\circ/2) \geqslant M \geqslant N/\lceil 4R \rceil$, which is exactly the conclusion of (i).

To show (ii) we note that R is also the radius of T° with respect to the gauge of K°. Since we are in a finite-dimensional space, that radius is attained and so there is a segment $I := (-y, y) \subseteq T^\circ$ with $\|y\|_{K^\circ} = R$. This implies that $M(I, K^\circ/2) \geqslant \lceil 4R \rceil$. However, in dimension one, separated and convexly separated sets coincide; this allows to conclude that $\hat{M}(T^\circ, K^\circ/2) \geqslant \hat{M}(I, K^\circ/2) \geqslant \lceil 4R \rceil$, as required. □

REMARK 4.4.11. We stress again that under boundedness conditions there is a very wide class of bodies for which convex separation and usual separation/covering numbers are equivalent. This will follow from a fact called "Maurey's lemma" and will be discussed in Part II. It will imply a much more general duality of covering numbers relation.

As mentioned in the previous remark, for a wide class of bodies the duality theorem applies with universal constants. Without *any* assumption on the bodies one still has a duality relation with an extra factor which depends logarithmically on the dimension.

THEOREM 4.4.12. *Let $K, T \subset \mathbb{R}^n$ be centrally symmetric convex bodies. Then*

$$N(K,T) \leqslant N\Big(T^\circ, \frac{c}{\log(n+1)} K^\circ\Big)^{C \log(n+1) \log\log(n+2)}$$

where $1 > c > 0$ and $C > 1$ are universal.

4.5. Notes and remarks

Covering numbers

In this section we present fundamental results on covering numbers that play a crucial role in the development of the theory in the next chapters.

Theorem 4.1.20 is due to Milman and Pajor [**461**]. Lemma 4.1.22, Lemma 4.1.23 and Lemma 4.1.24 can be found in Milman's paper [**450**] and will be used in Chapter 8, in the proof of the Bourgain-Milman inequality and of the reverse Brunn-Minkowski inequality via isomorphic symmetrization.

Notions of weighted covering numbers and weighted separation numbers for convex sets were defined by Artstein-Avidan and Raz in [**33**] and studied in more depth by Artstein-Avidan and Slomka in [**34**]. Given a non-negative Borel regular measure μ on $\mathbb{R}^n$ and compact $K, T \subset \mathbb{R}^n$ such that T has non-empty interior, we say that μ is a covering measure of K by T if $\mu * \mathbf{1}_T \geqslant \mathbf{1}_K$ (where $\mathbf{1}_A$ denotes the indicator function of the set A). The *weighted* covering number of K by T is defined by

$$N^*(K,T) = \inf\left\{\int_{\mathbb{R}^n} d\mu \,:\, \mu * \mathbf{1}_T \geqslant \mathbf{1}_K\,, \mu \text{ non} - \text{negative Borel regular}\right\}.$$

Clearly, $N^*(K,T) \leqslant N(K,T)$. A weighted notion of separation is defined similarly: a measure μ is said to be T−separated if $\mu * \mathbf{1}_T \leqslant 1$, and the weighted separation number of K by T is

$$M^*(K,T) = \sup\left\{\int_K d\nu \,:\, \nu * \mathbf{1}_T \leqslant 1,\ \nu \text{ non} - \text{negative Borel regular}\right\}.$$

Under these definitions, which are motivated by similar weighted notions considered by L. Lovász in combinatorics [**396**], it was shown in [**34**] (based partially on [**33**]) that

$$M^*(K,T) = N^*(K,-T),$$

and that the weighted and the usual notions are not much different, in the centrally symmetric case:

$$N(K,T-T) \leqslant N^*(K,T) \leqslant N(K,T).$$

The weighted counterparts of covering numbers are in some sense "better behaved" than their integer-valued classical counterparts; for examples and applications, see [**34**], among them an application pertaining to Hadwiger's conjecture mentioned in Remark 4.1.16 and elaborated upon below.

Packings by balls

Let α_n denote the maximal volume of a symmetric ellipsoid $\mathcal{E}$ in $\mathbb{R}^n$ that does not contain integer points other than the origin. The question to estimate α_n is closely related to packings by balls. Recall that a family $P = \{x_i + rB_2^n : i \in I\}$ is a packing if the balls $x_i + rB_2^n$ have disjoint interiors. Then, the upper and the lower density of P are defined as follows: for every $R > 0$ we set $N(R) = \text{card}(\{i \in I : (x_i + rB_2^n) \cap (RB_2^n) \neq \emptyset\})$, and we define

$$\overline{\delta}(P) = \limsup_{R\to\infty} \frac{N(R)\kappa_n r^n}{\kappa_n R^n} \quad \text{and} \quad \underline{\delta}(P) = \liminf_{R\to\infty} \frac{N(R)\kappa_n r^n}{\kappa_n R^n}.$$

If $\overline{\delta}(P) = \underline{\delta}(P)$ then this common value is the density $\delta(P)$ of P.

A lattice packing is a packing of the form $P = \{x + rB_2^n : x \in \Lambda\}$, where Λ is a lattice in $\mathbb{R}^n$. One can check that, then, $\delta(P) = \frac{\kappa_n r^n}{\det \Lambda}$. Let δ_n denote the supremum of $\delta(P)$ over all lattice packings with balls of radius 1. Then, one can check that $\alpha_n = 2^n \delta_n$. Blichfeldt proved in [**77**] that $\delta_n \leqslant \frac{n+2}{2} 2^{-n/2}$. This was improved to $\delta_n \leqslant 2^{-0.5096n+o(n)}$ by Sidelnikov [**563**], and then to $2^{-0.5237n+o(n)}$ by Levenshtein [**378**]. As we mentioned in the text, the best known estimate is $\delta_n \leqslant 2^{-0.599n+o(n)}$, obtained by Kabatianski and

Levenshtein in [**320**]. For more information and a sketch of the proof of this last estimate, see [**193**].

The second question that we discuss in Remark 4.1.16 was asked by Hadwiger in [**296**]. It seems that it had been previously studied by Levi in [**379**], and independently, by Gohberg and Markus in [**252**]. The best known upper bound is due to Rogers, see [**525**]. For a comprehensive survey of this problem and known results see Brass, Moser and Pach [**115**]. For a comprehensive survey and more recent developments see Brass, Moser and Pach [**115**], Chapter 3 of the book by Bezdek [**71**] and the two papers, Bezdek and Litvak [**72**] and Gluskin and Litvak [**250**].

Sudakov's inequality and its dual

Sudakov's inequality [**574**] in its original form gives a lower bound for the expectation of the supremum of a Gaussian process; this form will be presented in Chapter 9. Theorem 4.2.1 is a direct application to the covering numbers of a convex body that follows from Sudakov's inequality once the geometric translation is done. The original proof of the dual statement, Theorem 4.2.2, is due to Pajor and Tomczak-Jaegermann [**486**]. The argument that allows one to pass from Sudakov's inequality to its dual and vice versa is due to Tomczak-Jaegermann [**592**] and is based on Lemma 4.2.5 which is closely related to the observation that $B_2^n \subset K + K^\circ$. The latter is already present in an earlier work of Firey [**208**]; he proved that $(K + K^\circ)/2$ contains B_2^n, where the polarity is with respect to any point in the interior of K. We follow the route to deduce Sudakov's inequality from its dual (although at present this is immediate from the duality of entropy theorem of Section 4.3). We first present Talagrand's proof of the dual Sudakov inequality for its elegance and beauty; his argument appears in [**374**] and in [**108**].

Entropy numbers and approximation numbers

Various entropy quantities have been defined as useful tools to measure the degree of compactness of operators. We refer to the books by Pietsch [**499**], Carl and Stephani [**140**], König [**358**] and Tomczak-Jaegermann [**593**] for their applications to operator theory.

The approximation numbers $a_n(u)$, the Gelfand numbers $c_n(u)$ and the Kolmogorov numbers $d_n(u)$ first appeared in the literature as diameters of sets - see the paper of Kolmogorov [**355**]. A detailed account of the development of this theory and its significance for approximation theory is given in the book of Pinkus [**501**]. Covering numbers $N(A, B)$ appear in the work of Kolmogorov [**356**] and Kolmogorov-Tikhomirov [**357**]. The introduction of entropy numbers $\{e_n(u)\}$ is usually attributed to Pietsch [**499**]; they had been considered earlier by Mityagin [**476**], Mityagin and Pelczynski [**477**], Lorentz [**395**], Triebel [**594**]. The first results on the comparison of the entropy numbers of a compact operator with its approximation, Gelfand and Kolmogorov numbers are due to Pietsch. Theorem 5.2 is due to Carl [**137**]. See also [**141**]. Schütt [**559**] has obtained very precise and useful estimates on the entropy numbers of diagonal operators between symmetric Banach spaces (see also the paper oF Gordon König and Schütt [**261**]).

Duality of entropy

The conjecture that a strong connection between $N(K, T)$ and $N(T^\circ, K^\circ)$ exists is due to Pietsch; it can be found in his book [**499**]. There, it was stated in the language of operator theory, as a precise connection between the sequences $e_k(u)$ and $e_k(u^*)$ of an operator u and its adjoint.

The special case where one of the spaces is Hilbertian, or equivalently, where one of the bodies is a Euclidean ball, was considered the main case as it has a very natural geometric meaning as well as many applications (see e.g. Li and Linde [**383**]). Partial results were obtained, mainly in the end of the 1980's, by Gordon, König and Schütt [**260**], König and Milman [**361**], Pajor and Tomczak-Jaegermann [**488**] and Bourgain, Pajor, Szarek and Tomczak-Jaegermann in [**112**]. Later on, V. Milman and Szarek continued this investigation in [**468**] and [**469**].

The problem was finally settled by the duality of entropy theorem (Theorem 4.4.4) due to Artstein-Avidan, Milman and Szarek [**30**] (see also [**28**] and [**29**]). In these papers some additional information on the constants involved is obtained, for example that one may take in Theorem 4.4.4 the constant β to be $2+\varepsilon$ for any $\varepsilon>0$ (and the dependence of α on ε is determined).

Theorem 4.4.10 from [**31**] is a simple adaptation of an argument from [**112**]. Theorem 4.4.12 is due to E. Milman [**437**].

Entropy extension

The "entropy extension" results that we describe below, allow us to estimate the covering numbers $N(L,tK)$ of two centrally symmetric convex bodies L and K in $\mathbb{R}^n$ provided that we have some information on the covering numbers of their projections onto some subspace of $\mathbb{R}^n$. A typical result of this type is the following.

PROPOSITION 4.5.1. *Let K be a centrally symmetric convex body in $\mathbb{R}^n$ and assume that $B_2^n\subseteq \rho K$ for some $\rho\geqslant 1$. Let W be a subspace of $\mathbb{R}^n$ with* $\dim W=m$ *and $P_{W^\perp}(K)\supseteq B_{W^\perp}$. Then, we have*

$$N(B_2^n,4K)\leqslant \overline{N}(B_2^n,2K)\leqslant (3\rho)^m .$$

The next theorem is due to Litvak, Milman, Pajor and Tomczak-Jaegermann [**388**] (see also [**389**]).

THEOREM 4.5.2. *Let K,L be symmetric convex bodies in $\mathbb{R}^n$ and assume that $L\subseteq RK$. Let W be a subspace of $\mathbb{R}^n$ with* $\dim W=m$ *and $0<r<t<R$. Then, we have*

$$N(L,tK)\leqslant 2^m\left(\frac{2R+t}{t-r}\right)^m N\left(P_{W^\perp}(L),\frac{r}{2}P_{W^\perp}(K)\right).$$

Similar in nature is a result of Vershynin and Rudelson from [**597**] which allows one to obtain estimates for the covering number $N(B_2^n,K)$ through information on the projections of B_2^n and K onto some subspace: If K is a symmetric convex body in $\mathbb{R}^n$ such that $K\supseteq \delta B_n$ and if $P_F(K)\supseteq B_F$ for some $F\in G_{n,k}$, $k\geqslant (1-\varepsilon)n$, then

$$N(B_2^n,4K)\leqslant (C/\delta)^{2\varepsilon n}.$$

For an application of entropy extension see the paper of Giannopoulos Stavrakakis Tsolomitis and Vritsiou [**245**].

Metric entropy of the Grassmann manifold

Let $G_{n,k}$ be the Grassmann manifold of the k-dimensional subspaces of $\mathbb{R}^n$, equipped with the metric $\sigma_q(E,F)=\sigma_q(P_E-P_F)$, where P_E is the orthogonal projection onto E and $\sigma_q(T)$ is the ℓ_q^n norm of the n-tuple of singular values $(s_1(T),\dots,s_n(T))$ of an operator $T\in L(\mathbb{R}^n)$. Estimates for the covering numbers $N(G_{n,k},\sigma_q,\varepsilon)$ were given by Szarek [**580**] in connection with his solution of the finite dimensional basis problem in [**581**]. For any integers $1\leqslant k\leqslant n$ such that $k\leqslant n-k$, for any $1\leqslant q\leqslant\infty$ and for every $\varepsilon>0$, one has

$$\left(\frac{c}{\varepsilon}\right)^{k(n-k)}\leqslant N(G_{n,k},\sigma_q,\varepsilon k^{1/q})\leqslant \left(\frac{C}{\varepsilon}\right)^{k(n-k)},$$

where $c,C>0$ are absolute constants. Related results and a new proof of this fact can be found in Pajor's paper [**485**].

Metric entropy of the Banach-Mazur compactum

The metric entropy $H_n(\varepsilon) := \log_2 N(\mathcal{B}_n, d, 1+\varepsilon)$ of the n-dimensional Banach-Mazur compactum $\mathcal{B}_n$ with respect to the Banach-Mazur distance d was studied by Bronstein, who proved in [**123**] that $H_n(\varepsilon) \sim \varepsilon^{\frac{1-n}{2}}$. In his result, for each fixed dimension n, this equivalence holds for small enough $\varepsilon > 0$, the range of which may depend on n. Pisier [**510**] provides asymptotic estimates as $n \to \infty$. More precisely, he proves that for every $\varepsilon > 0$ one has

$$0 < \liminf_{n\to\infty} \frac{\log\log N(\mathcal{B}_n, d, 1+\varepsilon)}{n} \leqslant \limsup_{n\to\infty} \frac{\log\log N(\mathcal{B}_n, d, 1+\varepsilon)}{n} < \infty.$$

CHAPTER 5

Almost Euclidean subspaces of finite dimensional normed spaces

The starting point for our exposition of the asymptotic theory of convex bodies is Dvoretzky theorem (Theorem 5.1.2). In its geometric formulation it states that every high-dimensional centrally symmetric convex body has central sections of high dimension which are almost ellipsoidal. The dependence of the dimension k of these sections on the dimension n of the body was clarified with Milman's proof of this fact. His argument established a precise quantitative form:

> For every n-dimensional normed space $X = (\mathbb{R}^n, \|\cdot\|)$ and every $\varepsilon \in (0,1)$ there exist an integer $k \geqslant c\varepsilon^2 \log n$ and a k-dimensional subspace F of X which satisfies $d(F, \ell_2^k) \leqslant 1+\varepsilon$.

We present the proof of the Dvoretzky-Milman theorem (Theorem 5.1.3) in Section 5.2. It exploits the concentration of measure phenomenon for the Euclidean sphere S^{n-1}, more precisely, the deviation inequality of Theorem 3.2.2 for Lipschitz functions $f : S^{n-1} \to \mathbb{R}$, which implies that the values of $\|\cdot\|$ on S^{n-1} concentrate near their average

$$M = \int_{S^{n-1}} \|x\|\, d\sigma(x).$$

Starting from this observation and denoting by $k_X(\varepsilon)$, for $\varepsilon > 0$, the largest integer $k \leqslant n$ for which we can find a subspace F of $\mathbb{R}^n$ with $\dim(F) = k$ such that for every $x \in S_F$,

$$(1+\varepsilon)^{-1} M \leqslant \|x\| \leqslant M(1+\varepsilon),$$

Milman showed (see Theorem 5.2.10) that if b is the least positive constant so that $\|x\| \leqslant b|x|$ for all $x \in \mathbb{R}^n$, then

$$k_X(\varepsilon) \geqslant c\varepsilon^2 [\log^{-1}(2/\varepsilon)] n (M/b)^2$$

for every $\varepsilon \in (0,1)$. Applying this estimate for a centrally symmetric convex body in John position one can conclude the proof of the Dvoretzky-Milman theorem.

A remarkable fact is that the dependence of k on n, M and b in Milman's proof is sharp in full generality. As we will see in Section 5.3, if we define the *critical dimension* $k(X)$ of X as the largest positive integer $k \leqslant n$ for which

$$\nu_{n,k}\Big(\Big\{F \in G_{n,k} : \frac{1}{2} M|x| \leqslant \|x\| \leqslant 2M|x| \text{ for all } x \in F\Big\}\Big) \geqslant 1 - \frac{k}{n+k}.$$

then we have the *asymptotic formula*

$$k(X) \simeq n(M/b)^2.$$

In Section 5.4 we consider the classical spaces ℓ_p^n, $1 \leqslant p \leqslant \infty$, and we determine the order of $k_{p,n} = k(\ell_p^n)$ as a function of p and n. In Section 5.5 we discuss the volume ratio theorem: recall from Section 2.4b that the *volume ratio* of a centrally symmetric convex body K in $\mathbb{R}^n$ is the quantity

$$\mathrm{vr}(K) = \inf\Big\{\Big(\frac{\mathrm{Vol}_n(K)}{\mathrm{Vol}_n(\mathcal{E})}\Big)^{1/n} : \mathcal{E} \subseteq K\Big\},$$

where the infimum is taken over all the ellipsoids $\mathcal{E}$ which are contained in K. Szarek and Tomczak-Jaegermann, generalizing previous work of Kashin, showed that if K is a centrally symmetric convex body in $\mathbb{R}^n$ such that $B_2^n \subseteq K$ and $\mathrm{Vol}_n(K) = \alpha^n \mathrm{Vol}_n(B_2^n)$ for some $\alpha > 1$ then, for every $1 \leqslant k \leqslant n$, a random subspace $E \in G_{n,k}$ satisfies with probability greater than $1 - e^{-n}$

$$B_2^n \cap E \subseteq K \cap E \subseteq (c\alpha)^{\frac{n}{n-k}} B_2^n \cap E,$$

where $c > 0$ is an absolute constant. In particular, $k(X_K) \simeq n$ up to a constant depending only on α.

The volume ratio theorem has a "global formulation". This was noted already by Kashin for the case of K being the cross-polytope. Under the same assumptions, one has that there exists $U \in O(n)$ satisfying

$$B_2^n \subseteq K \cap U(K) \subseteq c\alpha^2 B_2^n,$$

where $c > 0$ is an absolute constant. This is an example of a general principle asserting that local statements which describe the geometry of sections of a convex body K have their analogue in global statements which relate K to its orthogonal images. In Section 5.6 we give more examples illustrating the same principle: one of them is the global version of Dvoretzky theorem stating that for any $\varepsilon \in (0, 1/2)$ there exist an integer

$$t \leqslant \frac{c}{\varepsilon^2}\Big(\frac{b}{M}\Big)^2$$

and t orthogonal transformations $U_1, \dots, U_t \in O(n)$ such that

$$\frac{M}{1+\varepsilon} B_2^n \subseteq \frac{1}{t}\sum_{j=1}^{t} U_j^*(K^\circ) \subseteq (1+\varepsilon) M\, B_2^n.$$

In the next section (Section 5.7) we give a sample of results that indicate "phase transition" and "threshold" type behaviour; naturally, the statements are of isomorphic type and hold true up to some universal constants, but they are precise for any convex body.

Section 5.8 presents a small ball estimate of Klartag and Vershynin: For every norm $\|\cdot\|$ on $\mathbb{R}^n$ and every $0 < \varepsilon < \frac{1}{2}$ we have

$$\sigma(\{x \in S^{n-1} : \|x\| < \varepsilon M\}) < (c_1\varepsilon)^{c_2 d(K)} \leqslant (c_1\varepsilon)^{c_3 k(X_K)},$$

where $c_1, c_2, c_3 > 0$ are absolute constants, and

$$d(K) = \min\Big\{ -\log \sigma\Big(\Big\{x \in S^{n-1} : \|x\| \leqslant \frac{M(K)}{2}\Big\}\Big), n\Big\}.$$

An interesting feature of this result is that it demonstrates a difference between the lower and the upper inclusions in Dvoretzky theorem: the expected upper bound for the diameter of random sections of a convex body sometimes continues to hold in much larger dimensions than the critical dimension of the body. The proof is based on the B-theorem of Cordero-Erausquin, Fradelizi and Maurey that will be discussed in Part II of thisbook.

5.1. Dvoretzky type theorems

The starting point for Dvoretzky type theorems is one of the results of Dvoretzky and Rogers about the distribution of the contact points of a centrally symmetric convex body and its maximal volume ellipsoid which we already saw in Section 2.1d (Theorem 2.1.26).

PROPOSITION 5.1.1 (Dvoretzky-Rogers). *Let B_2^n be the maximal volume ellipsoid of the centrally symmetric convex body K in $\mathbb{R}^n$. There exist $k \simeq \sqrt{n}$ and orthonormal vectors $z_1, \ldots, z_k$ in $\mathbb{R}^n$ such that for all $a_1, \ldots, a_k \in \mathbb{R}$,*

$$\frac{1}{\sqrt{2}} \max_{i \leqslant k} |a_i| \leqslant \Big\| \sum_{i=1}^k a_i z_i \Big\| \leqslant \Big(\sum_{i=1}^k a_i^2 \Big)^{1/2}.$$

Proposition 5.1.1 asserts that for every centrally symmetric convex body K whose maximal volume ellipsoid is B_2^n, there exist $k \simeq \sqrt{n}$ and a k-dimensional subspace F of $\mathbb{R}^n$ such that $B_2^n \cap F \subseteq K \cap F \subseteq 2Q_F$, where Q_F denotes the unit cube in F for an appropriately chosen coordinate system. Grothendieck asked whether Q_F can be replaced by $B_2^n \cap F$ in this statement. He did not specify what the dependence of k on n might be, asking just that k should increase to infinity with n. A short time after, Dvoretzky answered Grothendieck's question in the affirmative.

THEOREM 5.1.2 (Dvoretzky). *Let $\varepsilon > 0$ and let k be a positive integer. There exists $N = N(k, \varepsilon)$ with the following property: Whenever X is a normed space of dimension $n \geqslant N$ we can find a k-dimensional subspace F of X with $d_{BM}(F, \ell_2^k) \leqslant 1 + \varepsilon$.*

Here d_{BM} denotes Banach-Mazur distance; see Definition 2.1.2 in Section 2.1.

Geometrically speaking, every high-dimensional centrally symmetric convex body has central sections of high dimension which are almost ellipsoidal. The dependence of $N(k, \varepsilon)$ on k and ε became a very important question, and Dvoretzky theorem took a much more precise quantitative form.

THEOREM 5.1.3 (Dvoretzky-Milman). *Let X be an n-dimensional normed space. For every $\varepsilon \in (0,1)$ there exist an integer $k \geqslant c\varepsilon^2 \log n$ and a k-dimensional subspace F of X which satisfies $d(F, \ell_2^k) \leqslant 1 + \varepsilon$.*

This means that Theorem 5.1.2 holds true with $N(k, \varepsilon) = \exp(c\varepsilon^{-2} k)$. Dvoretzky's original proof gave an estimate $N(k, \varepsilon) = \exp(c\varepsilon^{-2} k^2 \log k)$. Later, Milman established the estimate $N(k, \varepsilon) = \exp(c\varepsilon^{-2} |\log \varepsilon| k)$ with a different approach, which gave rise to the concentration of measure method discussed in Chapter 3. The logarithmic in ε term was removed by Gordon, and then by Schechtman.

The logarithmic dependence of k on n is best possible for small values of ε. We will see this in Section 5.4 by analyzing the example of ℓ_∞^n.

The right dependence of $N(k, \varepsilon)$ on ε for a fixed (even small) positive integer k is not clear. Using ideas from the theory of irregularities of distribution, Bourgain and Lindenstrauss have shown that the choice $N(k, \varepsilon) = c(k)\varepsilon^{-\frac{k-1}{2}} |\log \varepsilon|$ is possible for spaces X with a 1-symmetric basis. There are numerous connections of this question with other branches of mathematics (algebraic topology, number theory, harmonic analysis). We will discuss the best known estimate in Section 5.6 and provide a full proof in Part II.

5.2. Milman's proof

In this section we describe Milman's proof of Theorem 5.1.3. The parameter M of a normed space $X = (\mathbb{R}^n, \|\cdot\|)$, that is introduced in the next definition, will be very important for the proof.

5.2.1. The parameter M

DEFINITION 5.2.1. Let $X = (\mathbb{R}^n, \|\cdot\|)$ have unit ball K. We define

$$M = M(X_K) = \int_{S^{n-1}} \|x\| \, d\sigma(x),$$

the average of $\|\cdot\|$ on the sphere S^{n-1}. Clearly $M(X_K) = w(K^\circ)$. We also set $r(x) = \|x\|$ and denote by L_r the Lévy mean of r on S^{n-1}.

REMARKS 5.2.2. (i) The parameter M depends not only on the body K but also on the Euclidean structure we have chosen in $\mathbb{R}^n$. If we assume that $\frac{1}{a}|x| \leqslant \|x\| \leqslant b|x|$ and that $a, b > 0$ are the smallest constants for which this is true for all $x \in \mathbb{R}^n$, then we have the trivial bounds $\frac{1}{a} \leqslant M \leqslant b$.

(ii) We can broaden the class of bodies K for which the parameter $M = M(K)$ is defined and include bodies K that are star-shaped with respect to the origin: we set

$$M(K) = \int_{S^{n-1}} p_K(x) \, d\sigma(x),$$

where p_K is the Minkowski functional of K given by

$$p_K(x) := \inf\{t > 0 : x \in tK\}.$$

Clearly M is monotone decreasing with respect to inclusion, that is, if $K \subset K'$ then $\|x\|_K \geqslant \|x\|_{K'}$, and more generally $p_K(x) \geqslant p_{K'}(x)$, for every x and then $M(K) \geqslant M(K')$.

(iii) For a body $K \subset \mathbb{R}^n$, under the assumption that $\frac{1}{a}|x| \leqslant \|x\| \leqslant b|x|$ for the *best* such constants a, b, we have that there exists a point $x \in \partial K \cap \frac{1}{b}S^{n-1}$. Therefore, for $\theta = bx \in S^{n-1}$, we have that $h_K(\theta) = \|\theta\|_{K^\circ} = 1/b$, and this shows that K is included in a slab of width $2/b$, $S_b(\theta) := \{x : |\langle x, \theta\rangle| \leqslant 1/b\}$, and we have $M(K) \geqslant M(S_b)$. It is not hard to check that (say, for the slab $\{x : |x_1| \leqslant 1/b\}$) corresponding to $\theta = e_1 = (1, 0, \dots, 0)$ one has

$$M(S_b) = b \int_{S^{n-1}} |x_1| \, d\sigma(x) \simeq \frac{b}{\sqrt{n}}.$$

We thus get that, if $\frac{1}{a}|x| \leqslant \|x\| \leqslant b|x|$ with b best possible, then

$$\frac{b}{M} \leqslant c\sqrt{n}$$

where c is a universal constant. Note also that by the standard estimates (3.1.2) for the measure of a spherical cap, the median of the norm associated with $S_b(\theta)$ (which is $\|z\|_{S_b(\theta)} = b|\langle z, \theta\rangle|$) is of the order of $b/\sqrt{n}$. Since $K \subset S_b(\theta)$, by monotonicity we get that

$$L_r \geqslant c' b/\sqrt{n}.$$

(iv) For every $q > 0$ we define

$$M_q = M_q(X_K) = \left(\int_{S^{n-1}} \|x\|^q d\sigma(x) \right)^{1/q}.$$

In this notation $M = M_1$ and as a consequence of the Kahane-Khinchine type inequalities discussed in Section 3.5, one can check that $M_1 \leqslant M_2 \leqslant \sqrt{\pi/2}M_1$ independently of the dimension and the norm. Indeed, this is a consequence for example of Theorem 3.5.11 applied for the Gaussian measure together with changing integration on the sphere with integration of the homogeneous function with respect to the Gaussian measure, which amounts to multiplication by a constant which can be computed directly. In fact we will prove in Section 5.7.2 that, for every $1 \leqslant q \leqslant n$,

$$\max\left\{ M_1, c_1 \frac{b\sqrt{q}}{\sqrt{n}} \right\} \leqslant M_q \leqslant \max\left\{ 2M_1, c_2 \frac{b\sqrt{q}}{\sqrt{n}} \right\},$$

where $c_1, c_2 > 0$ are absolute constants.

(v) Let $g_1, \dots, g_n$ be independent standard Gaussian random variables on some probability space Ω and $\{e_1', \dots, e_n'\}$ be any orthonormal basis in $\mathbb{R}^n$. Integration in polar coordinates establishes the identity

$$\left(\int_\Omega \Big\| \sum_{i=1}^n g_i(\omega) e_i' \Big\|^2 d\omega \right)^{1/2} = \sqrt{n} M_2.$$

Using the symmetry of the g_i and the triangle inequality for $\|\cdot\|$ we get

$$\int_\Omega \Big\| \sum_{i=1}^k g_i(\omega) e_i' \Big\| d\omega \leqslant \int_\Omega \Big\| \sum_{i=1}^n g_i(\omega) e_i' \Big\| d\omega \tag{5.2.1}$$

for every $1 \leqslant k \leqslant n$. To prove this inequality, one uses the triangle inequality inductively. Indeed, for $k \geqslant 2$ we have

$$2\Big\| \sum_{i=1}^{k-1} g_i e_i' \Big\| \leqslant \Big\| \sum_{i=1}^{k-1} g_i e_i' + e_k' \Big\| + \Big\| \sum_{i=1}^{k-1} g_i e_i' - e_k' \Big\|,$$

and hence

$$\int_\Omega \Big\| \sum_{i=1}^{k-1} g_i(\omega) e_i' \Big\| d\omega \leqslant \int_\Omega \Big\| \sum_{i=1}^{k} g_i(\omega) e_i' \Big\| d\omega.$$

Then, (5.2.1) follows by induction. Combining (5.2.1) with the previous observations we have

$$M(F) \leqslant c\sqrt{n/k}M \tag{5.2.2}$$

for every k-dimensional subspace F of X_K.

A last useful fact is provided by the next lemma which asserts that L_r and M are always comparable.

LEMMA 5.2.3. *Let $K \subset \mathbb{R}^n$ and denote $r(x) = \|x\|$, and L_r its Lévy mean. We have*

$$\frac{1}{2} \leqslant \frac{M}{L_r} \leqslant c,$$

where $c > 0$ is an absolute constant. Moreover, we have the estimate which is sometimes more useful

$$\left|\frac{M}{L_r} - 1\right| \leqslant c' \frac{b}{\sqrt{n}M}$$

for universal c'.

Proof. Without loss of generality we may assume that $|x| \leqslant \|x\| \leqslant b|x|$, where b is best possible. We know that for every $\varepsilon > 0$

$$\sigma(\{x : |r(x) - L_r| \geqslant b\varepsilon\}) \leqslant 2\exp(-c_1\varepsilon^2 n).$$

We write

$$|M - L_r| \leqslant \int_{S^{n-1}} |r(x) - L_r|\, d\sigma(x) = \int_0^\infty \sigma(\{x : |r(x) - L_r| \geqslant t\})\, dt.$$

Set $b\varepsilon = t$ to get

$$|M - L_r| \leqslant \int_0^\infty 2\exp\Big(- c_1 t^2 \frac{n}{b^2}\Big) dt = c_3 \frac{b}{\sqrt{n}}.$$

Therefore,

$$\left|\frac{M}{L_r} - 1\right| \leqslant c_3 \frac{b}{\sqrt{n}L_r}.$$

From Remark 5.2.2 (iii) we know that

$$L_r \geqslant c_4 b/\sqrt{n}.$$

Putting the above two facts together we get that

$$M - L_r \leqslant c_3 \frac{b}{\sqrt{n}} \leqslant c_5 L_r.$$

In the other direction, observe that

$$M = \int_{S^{n-1}} \|x\| d\sigma(x) \geqslant \int_{\{x:\ \|x\| \geqslant L_r\}} \|x\|\, d\sigma(x) \geqslant \frac{1}{2} L_r.$$

Therefore, $M/L_r \geqslant 1/2$. □

5.2.2. Milman's argument

Milman's proof of Dvoretzky theorem is based on the concentration of measure phenomenon for the Euclidean sphere S^{n-1}. More precisely, we will exploit the deviation inequality of Theorem 3.2.2 for Lipschitz functions $f : S^{n-1} \to \mathbb{R}$, which we restate below.

PROPOSITION 5.2.4. *Let $f : S^{n-1} \to \mathbb{R}$ be a Lipschitz continuous function with constant b. Then, for every $\varepsilon > 0$,*

$$\sigma\left(\{x \in S^{n-1} : |f(x) - \mathbb{E}(f)| \geqslant b\varepsilon\}\right) \leqslant 4\exp(-c_1\varepsilon^2 n).$$

where $c_1 > 0$ is an absolute constant.

Proposition 5.2.4 asserts that the values of every *good function* f defined on S^{n-1} *concentrate* around its expectation (a similar statement holds true for the Lévy mean L_f of f): the measure of the set on which f takes values *close* to $\mathbb{E}(f)$ is practically equal to 1. The additional step to get to a theorem of the type mentioned above is to use this big measure on which f is almost constant and show that we can extract a *subspace* of high dimension, on whose sphere f is almost constant.

To this end we use a δ-net on the sphere of a k-dimensional subspace, namely a set $\mathcal{N} \subset S^{k-1}$ such that for every $y \in S^{k-1}$ there exists $x \in \mathcal{N}$ such that $|x-y| < \delta$. We use the following lemma, which is almost identical to Corollary 4.1.15 from Chapter 4, the only difference being that the covering set is chosen on the boundary of the ball.

LEMMA 5.2.5. *Let $\delta > 0$ and $k \in \mathbb{N}$. There is a δ-net $\mathcal{N}$ for S^{k-1} with cardinality $|\mathcal{N}| \leqslant \left(1+\frac{2}{\delta}\right)^k$.*

Proof. Let $\{x_i\}_{i=1}^N$ be a maximal δ-separated set in S^{k-1}, namely such that any two of its points have distance greater than or equal to δ. Then $\{x_i\}_{i=1}^N$ is a δ-net for S^{k-1}. Since the sets $x_i + \frac{1}{2}\delta B_2^k$ have disjoint interiors and are all included in $B_2^k + \frac{\delta}{2}B_2^k$, we get that

$$N\left(\tfrac{\delta}{2}\right)^k \mathrm{Vol}_k\left(B_2^k\right) \leqslant \left(1+\tfrac{\delta}{2}\right)^k \mathrm{Vol}_k(B_2^k),$$

which gives $N \leqslant (1+\frac{2}{\delta})^k$. □

Let $X = (\mathbb{R}^n, \|\cdot\|)$ be an n-dimensional normed space. The function $r : S^{n-1} \to \mathbb{R}$ defined by $r(x) = \|x\|$ is Lipschitz continuous with constant b which is the smallest positive real number such that $\|x\| \leqslant b|x|$ for all $x \in \mathbb{R}^n$. Observe that

$$\mathbb{E}(r) = \int_{S^{n-1}} \|x\|\, d\sigma(x) = M(X).$$

LEMMA 5.2.6. *Let $X = (\mathbb{R}^n, \|\cdot\|)$ be an n-dimensional normed space with $\|x\| \leqslant b|x|$ for all $x \in X$. Assume $m \leqslant \frac{1}{4}\exp(c_1\varepsilon^2 n/2)$ and $y_i \in S^{n-1}$, $i = 1, \ldots, m$. There exists $B \subset O(n)$ with $\nu_n(B) \geqslant 1 - \exp(-c_1\varepsilon^2 n/2)$ such that*

$$M - b\varepsilon \leqslant \|Uy_i\| \leqslant M + b\varepsilon$$

for every $U \in B$ and $i = 1, \ldots, m$.

Proof. Recall that if $x_0 \in S^{n-1}$ then, for every $A \subset S^{n-1}$,

$$\sigma(A) = \nu_n(\{U \in O(n) : Ux_0 \in A\}).$$

Consider the set

$$A = \{x \in S^{n-1} : M - b\varepsilon \leqslant \|x\| \leqslant M + b\varepsilon\}.$$

By Proposition 5.2.4,

$$\sigma(A) \geqslant 1 - 4e^{-c_1\varepsilon^2 n}.$$

It follows that if we define

$$B_i = \{U \in O(n) : M - b\varepsilon \leqslant \|Uy_i\| \leqslant M + b\varepsilon\}$$

for $i = 1, \ldots, m$, then

$$\nu_n(B_i) > 1 - 4e^{-c_1\varepsilon^2 n}.$$

Setting $B = \bigcap_{i=1}^m B_i$ we see that

$$\nu_n(B) \geqslant 1 - \sum_{i=1}^m \nu_n(B_i^c) \geqslant 1 - 4m\exp(-c_1\varepsilon^2 n),$$

and for every $U \in B$ and $1 \leqslant i \leqslant m$ we have

$$M - b\varepsilon \leqslant \|Uy_i\| \leqslant M + b\varepsilon.$$

If we assume that $m \leqslant \frac{1}{4}\exp(c_1\varepsilon^2 n/2)$, our claim follows. □

By taking the subset in Lemma 5.2.6 as a δ-net in a subspace of dimension k, we arrive at the following

PROPOSITION 5.2.7. *Let $X = (\mathbb{R}^n, \|\cdot\|)$ be an n-dimensional normed space with $\|x\| \leqslant b|x|$ for all $x \in X$. Let $\delta, \varepsilon \in (0, 1)$. If $(1 + 2/\delta)^k \leqslant \frac{1}{4} \exp(c_1 \varepsilon^2 n/2)$ then we may find $\Gamma \subset G_{n,k}$, with $\nu_{n,k}(\Gamma) \geqslant 1 - \exp(-c_1 \varepsilon^2 n/2)$, such that: for every $F \in \Gamma$ there exists a δ-net $\mathcal{N}_F$ for S_F such that*

$$M - b\varepsilon \leqslant \|x\| \leqslant M + b\varepsilon$$

for all $x \in \mathcal{N}_F$.

Proof. Let $\delta, \varepsilon \in (0, 1)$ and $k \in \mathbb{N}$ satisfy the condition in the statement of the proposition. We fix a subspace F_0 of $\mathbb{R}^n$ with $\dim(F_0) = k$. From Lemma 5.2.5 we can find a δ-net $\{y_1, \ldots, y_m\}$ for the unit sphere $S_{F_0} = S^{n-1} \cap F_0$ of F_0, with $m \leqslant (1 + 2/\delta)^k$. Consider the set $B \subset O(n)$ which is promised by Lemma 5.2.6. If $U \in B$ and $1 \leqslant i \leqslant m$ then

$$M - b\varepsilon \leqslant \|U y_i\| \leqslant M + b\varepsilon.$$

We set $F_U := U(F_0)$ and $x_i := U y_i$ for $i = 1, \ldots, m$. Since U is an orthogonal transformation, $\{x_1, \ldots, x_m\}$ is a δ-net for S_{F_U}, which satisfies

$$M - b\varepsilon \leqslant \|x_i\| \leqslant M - b\varepsilon.$$

We set $\Gamma = \{F_U : U \in B\}$ and observe that

$$\nu_{n,k}(\Gamma) = \nu_{n,k}(\{U(F_0) : U \in B\}) = \nu_n(B).$$

$\square$

Using now the fact that $\|\cdot\|$ is a norm, we will extend, for $F \in \Gamma$, the estimate of Proposition 5.2.7 from a δ-net $\mathcal{N}_F$ of S_F to the whole sphere S_F, by the method of successive approximation.

PROPOSITION 5.2.8. *Let $X = (\mathbb{R}^n, \|\cdot\|)$ be an n-dimensional normed space with $\|x\| \leqslant b|x|$ for all $x \in X$. Let $\delta, \varepsilon \in (0, 1)$. If $(1 + 2/\delta)^k \leqslant \frac{1}{4} \exp(c_1 \varepsilon^2 n/2)$ then we may find $\Gamma \subset G_{n,k}$, with $\nu_{n,k}(\Gamma) \geqslant 1 - \exp(-c_1 \varepsilon^2 n/2)$, such that for every $F \in \Gamma$ and for every $y \in S_F$ we have*

$$(5.2.3) \qquad \frac{1 - 2\delta}{1 - \delta} M - \frac{b\varepsilon}{1 - \delta} \leqslant \|y\| \leqslant \frac{M + b\varepsilon}{1 - \delta}.$$

Proof. The subset Γ will be the same as given by Proposition 5.2.7. Let $y \in S_F$. There exists $x_0 \in \mathcal{N}_F$ such that $|y - x_0| = \delta_1 < \delta$. Then, $\frac{y - x_0}{\delta_1} \in S_F$, and hence we can find $x_1 \in \mathcal{N}_F$ with $\left|\frac{y - x_0}{\delta_1} - x_1\right| = \delta_2 < \delta$. Then,

$$|y - x_0 - \delta_1 x_1| = \delta_1 \delta_2 < \delta^2.$$

Inductively, we find $x_0, \ldots, x_n \in \mathcal{N}_F$ and $\delta_1, \ldots, \delta_n$ such that

$$\left| y - \sum_{i=0}^{n} \Big(\prod_{j=0}^{i} \delta_j\Big) x_i \right| \leqslant \delta^{n+1},$$

where $\delta_0 = 1$, $\delta_i < \delta$ and so $\prod_{j=0}^{i} \delta_j \leqslant \delta^i$. Since $\delta < 1$,

$$y = \sum_{i=0}^{\infty} \Big(\prod_{j=0}^{i} \delta_j\Big) x_i.$$

We get

$$\|y\| = \Big\| \sum_{i=0}^{\infty} \Big(\prod_{j=0}^{i} \delta_j \Big) x_i \Big\| \leqslant \sum_{i=0}^{\infty} \delta^i \|x_i\| \leqslant (M + b\varepsilon) \sum_{i=0}^{\infty} \delta^i = \frac{M + b\varepsilon}{1 - \delta}.$$

On the other hand,

$$\begin{aligned} \|y\| &\geqslant \|x_0\| - \Big\| \sum_{i=1}^{\infty} \Big(\prod_{j=0}^{i} \delta_j \Big) x_i \Big\| \geqslant M - b\varepsilon - \frac{\delta}{1-\delta}(M + b\varepsilon) \\ &= \frac{1 - 2\delta}{1 - \delta} M - \frac{b\varepsilon}{1 - \delta}. \end{aligned}$$

It follows that (5.2.3) holds true for all $y \in S_F$. □

DEFINITION 5.2.9. Given $\varepsilon \in (0,1)$ we denote by $k_X(\varepsilon)$ the largest integer $k \leqslant n$ for which we can find a subspace F of $\mathbb{R}^n$ with $\dim(F) = k$ such that for every $x \in S_F$,

$$(1 + \varepsilon)^{-1} M \leqslant \|x\| \leqslant M(1 + \varepsilon). \tag{5.2.4}$$

With a suitable choice of δ, ε we obtain a first estimate on the dimension of "spherical" sections of $K = K_X = \{x : \|x\| \leqslant 1\}$.

THEOREM 5.2.10 (Milman). *Let $X = (\mathbb{R}^n, \|\cdot\|)$ and assume that $\|x\| \leqslant b|x|$ for all $x \in \mathbb{R}^n$. For any $\varepsilon \in (0,1)$ we have $k_X(\varepsilon) \geqslant c\varepsilon^2 [\log^{-1}(2/\varepsilon)] n (M/b)^2$. Equivalently, if*

$$k \leqslant k_X(\varepsilon) = c_2 \varepsilon^2 [\log^{-1}(2/\varepsilon)] n (M/b)^2,$$

then we can find a subspace F of $\mathbb{R}^n$ with $\dim(F) = k$ such that (5.2.4) is satisfied for every $x \in S_F$. Moreover, this holds for all F in a subset $A \subset G_{n,k}$ of measure at least $\nu_{n,k}(A) \geqslant 1 - \exp(-c_3 \varepsilon^2 k)$.

Proof. From Proposition 5.2.8, if $\zeta, \delta \in (0,1)$ and if $k \in \mathbb{N}$ satisfies $(1 + 2/\delta)^k \leqslant \frac{1}{4}\exp(c_1 \zeta^2 n/2)$, then for a random k-dimensional subspace F of $\mathbb{R}^n$ and for every $x \in S_F$ we have

$$\frac{1 - 2\delta}{1 - \delta} M - \frac{b\zeta}{1 - \delta} \leqslant \|x\| \leqslant \frac{M + b\zeta}{1 - \delta}.$$

It now suffices to choose $\delta, \zeta \in (0,1)$ so that

$$\frac{M + b\zeta}{1 - \delta} \leqslant M(1 + \varepsilon) \quad \text{and} \quad \frac{M}{1 + \varepsilon} \leqslant \frac{1 - 2\delta}{1 - \delta} M - \frac{b\zeta}{1 - \delta}.$$

Observe that $\zeta = \frac{M\delta}{b}$ and $\delta = \frac{\varepsilon}{6}$ satisfy our conditions.

Next, we compute the largest value of k for which

$$\Big(1 + \frac{12}{\varepsilon}\Big)^k \leqslant \frac{1}{4} \exp\Big(\frac{c_1}{72} \varepsilon^2 n \Big(\frac{M}{b}\Big)^2\Big).$$

We need

$$k \log \frac{C}{\varepsilon} \leqslant c' \varepsilon^2 n (M/b)^2,$$

and hence, the largest k we may get is of the order of $\varepsilon^2 [\log^{-1}(2/\varepsilon)] n (M/b)^2$. Adjusting the constant c_2 we may arrive at the desired probability estimate for this event. □

For the proof of Theorem 5.1.3 it remains to choose a position for the unit ball of X and to estimate the ratio M/b from below. The next lemma will be useful for our lower bound for M.

LEMMA 5.2.11. *For every $1 \leqslant m \leqslant n$, denoting $x = (x_1, \dots, x_n)$, we have that*

$$\int_{S^{n-1}} \max_{1 \leqslant j \leqslant m} |x_j| \, d\sigma(x) \geqslant c_4 \Big(\frac{\log m}{n} \Big)^{1/2},$$

where $c_4 > 0$ is an absolute constant.

Proof. Integration in polar coordinates shows that

$$\int_{\mathbb{R}^m} \max_{1 \leqslant j \leqslant m} |t_j| \, d\gamma_m(t) = \int_{\mathbb{R}^n} \max_{1 \leqslant j \leqslant m} |t_j| \, d\gamma_n(t) = \lambda_n \int_{S^{n-1}} \max_{1 \leqslant j \leqslant m} |x_j| \, d\sigma(x),$$

where $\lambda_n \simeq \sqrt{n}$. On the other hand, using the inequality

$$\frac{1}{\sqrt{2\pi}} \int_s^\infty e^{-t^2/2} dt \geqslant \frac{1}{\sqrt{2\pi}} \frac{s}{s^2+1} e^{-s^2/2}$$

which is valid for all $s > 0$, we may write

$$\begin{aligned} \gamma_m\Big(\Big\{ t : \max_{1 \leqslant j \leqslant m} |t_j| < s \Big\} \Big) &= (2\pi)^{-m/2} \int_{-s}^{s} \cdots \int_{-s}^{s} \exp\Big(-\frac{1}{2} \sum_{j=1}^m t_j^2 \Big) \, dt_1 \dots dt_m \\ &= \Big(\frac{1}{\sqrt{2\pi}} \int_{-s}^{s} e^{-t^2/2} dt \Big)^m \\ &\leqslant \Big(1 - \frac{2s}{(s^2+1)\sqrt{2\pi}} e^{-s^2/2} \Big)^m . \end{aligned}$$

If we choose $s = \sqrt{\log m}$, and m is large enough, we see that

$$\gamma_m\Big(\Big\{ t : \max_{j \leqslant m} |t_j| \geqslant \sqrt{\log m} \Big\} \Big) \geqslant \frac{1}{2}.$$

Then,

$$\begin{aligned} \int_{S^{n-1}} \max_{j \leqslant m} |x_j| \, d\sigma(x) &\simeq \frac{1}{\sqrt{n}} \int_{\mathbb{R}^m} \max_{j \leqslant m} |t_j| \, d\gamma_m(t) \\ &\geqslant \frac{\sqrt{\log m}}{\sqrt{n}} \gamma_m\Big(\Big\{ t : \max_{j \leqslant m} |t_j| \geqslant \sqrt{\log m} \Big\} \Big) \\ &\geqslant \frac{1}{2} \Big(\frac{\log m}{n} \Big)^{1/2}, \end{aligned}$$

and the result follows. □

THEOREM 5.2.12. *Let $X = (\mathbb{R}^n, \|\cdot\|)$ and assume that B_2^n is the maximal volume ellipsoid of its unit ball K. For every $\varepsilon \in (0,1)$ we can find a subspace F of X of dimension $k \geqslant c\varepsilon^2 [\log^{-1}(2/\varepsilon)] \log n$, with the following property: for every $x \in S_F$,*

$$(1+\varepsilon)^{-1} M \leqslant \|x\| \leqslant M(1+\varepsilon).$$

Proof. Since we may use Theorem 5.2.10, all we need to show is that in John position one has that

$$\frac{M}{b} \geqslant c\sqrt{\frac{\log n}{n}}.$$

By the Dvoretzky-Rogers lemma (Proposition 2.1.23) we can find an orthonormal basis $\{x_1, \dots, x_n\}$ with $\|x_i\| \geqslant 1/4$ for all $i = 1, \dots, n$. Note that

$$M = \int_{S^{n-1}} \Big\| \sum_{i=1}^n a_i x_i \Big\| \, d\sigma(a) = \int_{S^{n-1}} \int_{E_2^n} \Big\| \sum_{i=1}^n \varepsilon_i a_i x_i \Big\| \, d\varepsilon \, d\sigma(a).$$

We shall use the fact that for all $j = 1, \dots, n$,

$$\int_{E_2^n} \Big\| \sum_{i=1}^{n} \varepsilon_i y_i \Big\| \, d\varepsilon \geqslant \|y_j\|,$$

which can be proved in a similar way as (5.2.1).

Setting $y_i = a_i x_i$ we have

$$\int_{E_2^n} \Big\| \sum_{i=1}^{n} \varepsilon_i a_i x_i \Big\| \, d\varepsilon \geqslant \max_{1 \leqslant i \leqslant n} \|a_i x_i\|.$$

Thus, we get

$$M \geqslant \int_{S^{n-1}} \max_{1 \leqslant i \leqslant n} \|a_i x_i\| \, d\sigma(a).$$

Since for $i \leqslant n$ we have $\|x_i\| \geqslant 1/4$, we get

$$M \geqslant \frac{1}{4} \int_{S^{n-1}} \max_{1 \leqslant i \leqslant n} |a_i| \, d\sigma(a),$$

and Lemma 5.2.11 gives

$$M \geqslant c\sqrt{\frac{\log n}{n}}.$$

The proof is complete. □

We may now complete the proof of Theorem 5.1.3 with an extra term, which is logarithmic in ε, appearing in the estimate for $k_X(\varepsilon)$. As we will see in Section 9.3, this term can be removed.

Proof of Theorem 5.1.3 *with an extra logarithmic term.* Let $X = (\mathbb{R}^n, \|\cdot\|)$ be an n-dimensional normed space and let $\varepsilon \in (0,1)$. We choose $\delta = \varepsilon/4$ so that $(1+\delta)^2 \leqslant 1+\varepsilon$.

There exists $T \in GL_n$ such that B_2^n is the maximal volume ellipsoid of $T(K_X)$. We define $r(x) = \|x\|_{T(K_X)}$ and consider the expectation M of r on S^{n-1}. We can find $k \geqslant c(\delta)\log n = c(\varepsilon)\log n$ and a k-dimensional subspace F of $\mathbb{R}^n$, such that

$$\frac{M}{1+\delta}\,(B_2^n \cap F) \subset T(K_X) \cap F \subset (1+\delta)M\,(B_2^n \cap F).$$

If $F_1 = T^{-1}(F)$, we have $d(F_1, \ell_2^k) \leqslant (1+\delta)^2 \leqslant 1+\varepsilon$. □

Geometrically, Theorem 5.1.3 states the following: for every centrally symmetric convex body K in $\mathbb{R}^n$ and every $\varepsilon \in (0,1)$ there exist $k \geqslant c\varepsilon^2 \log n$, a subspace F of $\mathbb{R}^n$ with $\dim(F) = k$ and an ellipsoid $\mathcal{E}$ in F, such that

(5.2.5) $$\mathcal{E} \subset K \cap F \subset (1+\varepsilon)\mathcal{E}.$$

The dual formulation, which is obtained by applying this result to K° and then taking polars in F, reads as follows: for every centrally symmetric convex body K in $\mathbb{R}^n$ and every $\varepsilon \in (0,1)$ there exist $k \geqslant c\varepsilon^2 \log n$, a subspace F of $\mathbb{R}^n$ with $\dim(F) = k$ and an ellipsoid $\mathcal{E}$ in F, such that

(5.2.6) $$\mathcal{E} \subset P_F(K) \subset (1+\varepsilon)\mathcal{E}.$$

The next lemma shows that, by reducing the dimension k to $k/2$, we can actually replace the ellipsoid by a ball in (5.2.5) without making any assumption on the original position of K. Indeed, this is simply because every k-dimensional ellipsoid has a section of dimension $\lfloor k/2 \rfloor$ which is a ball.

LEMMA 5.2.13. *Let $\mathcal{E}$ be an ellipsoid in $\mathbb{R}^k$ with center at 0 and write $k = 2s$ or $k = 2s - 1$ for some $s \geqslant 1$. There is an s-dimensional subspace F of $\mathbb{R}^k$ such that $\mathcal{E} \cap F$ is a Euclidean ball in F.*

Sketch of the Proof. Since this is a simple linear algebra exercise, we only give an outline, and leave the reader with the precise computations. Assuming

$$\mathcal{E} = \left\{ x \in \mathbb{R}^k : \sum_{i=1}^{k} a_i^2 \langle x, e_i \rangle^2 \leqslant 1 \right\}$$

for some $0 < a_1 \leqslant a_2 \leqslant \cdots \leqslant a_s \leqslant \cdots \leqslant a_k$ and an orthonormal basis $\{e_i\}_{i \leqslant k}$ of $\mathbb{R}^k$, we define vectors v_i, $1 \leqslant i \leqslant s$, as follows: $v_s = e_s$, whereas, for every $1 \leqslant i \leqslant s-1$, if $a_{k+1-i} = a_s$, then we set $v_i = e_{k+1-i}$, otherwise we define

$$b_i^2 = \frac{a_s^2 - a_i^2}{a_{k+1-i}^2 - a_s^2} \qquad \text{and} \qquad v_i = \frac{e_i + b_i e_{k+1-i}}{\sqrt{b_i^2 + 1}}.$$

We shall take F to be the subspace spanned by the vectors v_i. One checks that $\{v_i : 1 \leqslant i \leqslant s\}$ is an orthonormal basis of F and for every $1 \leqslant i \leqslant s-1$ and every $x \in F$ we have

$$\sum_{i=1}^{s} a_s^2 \langle x, v_i \rangle^2 = \sum_{i=1}^{k} a_i^2 \langle x, e_i \rangle^2.$$

This shows that $\mathcal{E} \cap F$ is a ball of radius a_s^{-1} in F (a_s^{-1} being as large a radius as any s-dimensional Euclidean ball contained in $\mathcal{E}$ can have). □

5.3. The critical dimension $k(X)$

In the first part of this section we shall describe a result of Milman and Schechtman which implies that the lower bound for the dimension of Euclidean sections of a convex body given in Theorem 5.2.10 is in fact asymptotically an equality, in a sense which we now explain.

Let $X = (\mathbb{R}^n, \|\cdot\|)$ be an n-dimensional normed space. The proof of the Dvoretzky-Milman theorem, being probabilistic in nature, gives that a subspace F of X with $\dim(F) = \lfloor c\varepsilon^2 n(M/b)^2 \rfloor$ is $(1+\varepsilon)$-Euclidean with high probability. This leads to the definition of the following characteristics of X.

DEFINITION 5.3.1. *Let X be an n-dimensional normed space. We define $k(X)$ as the largest positive integer $k \leqslant n$ for which*

$$\nu_{n,k}\Big(\Big\{F \in G_{n,k} : \frac{1}{2}M|x| \leqslant \|x\| \leqslant 2M|x| \text{ for all } x \in F\Big\}\Big) \geqslant 1 - \frac{k}{n+k}.$$

We also define $\tilde{k}(X)$ as the largest positive integer $k \leqslant n$ for which

$$\nu_{n,k}\Big(\Big\{F \in G_{n,k} : \frac{1}{2}M|x| \leqslant \|x\| \leqslant 2M|x| \text{ for all } x \in F\Big\}\Big) \geqslant 1/2.$$

It will also be convenient to define $k_0(X)$ to be the largest positive $k \leq n$ for which there exists a subspace $F \in G_{n,k}$ with $\frac{1}{2}M|x| \leq \|x\| \leq 2M|x|$ for all $x \in F$. Thus $k(X) \leq \tilde{k}(X) \leq k_0(X)$. We see that $k(X)$ (or $\tilde{k}(X)$) is the largest possible dimension $k \leqslant n$ for which most of the k-dimensional subspaces of X are 4-Euclidean. (The fact that we define "most" by the formula $\frac{n}{n+k}$ is not intuitive, but will play a role below.) From Theorem 5.2.10 applied with $\varepsilon = 1/2$ we immediately get

THEOREM 5.3.2 (Dvoretzky-Milman). *Let $X = (\mathbb{R}^n, \|\cdot\|_K)$ and assume that $\|x\| \leqslant b|x|$ for all $x \in \mathbb{R}^n$. One has*

$$\tilde{k}(X) \geqslant cn(M/b)^2$$

for a universal c.

Proof. We use Theorem 5.2.10 with $\varepsilon = \frac{1}{2}$, say. It tells us that for some

$$k \geqslant c_1 n(M/b)^2,$$

we can find a subset $A \subset G_{n,k}$ of measure at least $\nu_{n,k}(A) \geqslant 1 - \exp(-c_2 k)$ such that for any subspace $F \in A$ and for any $x \in F$ one has

$$\frac{2}{3}M|x| \leqslant \|x\| \leqslant \frac{3}{2}M|x|.$$

The measure of A is certainly greater than $1/2$ if we assume that $n(M/b)^2 > C$, for some large enough absolute constant C. This assumption is reflected only in the constant in the statement of the theorem.

Note that if $n(M/b)^2 \geqslant C \log n$ then the same reasoning shows that $k(X) \geqslant cn(M/b)^2$. □

What is surprisingly simple, yet shows that in this estimate there is more than meets the eye, is the observation of Milman and Schechtman that an inverse inequality holds true.

THEOREM 5.3.3. *For every n-dimensional normed space X one has*

$$k(X) \leqslant 8n(M/b)^2.$$

Proof. We fix orthogonal subspaces $E^1, \ldots, E^t$ of dimension $\leqslant k(X)$ such that $\mathbb{R}^n = \sum_{i=1}^t E^i$. We may do this with an integer $t < \frac{n}{k} + 1$. By the definition of $k(X)$, all orthogonal images of each E^i are 4-Euclidean, except a portion $\frac{k}{n+k}$. We can thus find $U \in O(n)$ such that

$$\frac{1}{2}M|x| \leqslant \|x\| \leqslant 2M|x|, \qquad x \in U(E^i)$$

for all $i = 1, \ldots, t$, so long as $t \cdot \frac{k}{n+k} < 1$, which is precisely the case.

Having this decomposition of $\mathbb{R}^n$, every $x \in \mathbb{R}^n$ can be written in the form $x = \sum_{i=1}^t x_i$, where $x_i \in U(E^i)$. Since the x_i are orthogonal, we get

$$\|x\| \leqslant 2M \sum_{i=1}^{t} |x_i| \leqslant 2M\sqrt{t}|x|.$$

This means that $b \leqslant 2M\sqrt{t}$ and, since $t < (n+k)/k \leqslant 2n/k$, we see that $k(X) \leqslant 8n(M/b)^2$. □

The above shows that the following *asymptotic formula* holds true:

THEOREM 5.3.4 (Milman-Schechtman). *Let X be an n-dimensional normed space such that $M/b \geq c\sqrt{\frac{\log n}{n}}$. Then,*

$$k(X) \simeq n(M/b)^2.$$

The critical dimension $k(X)$ satisfies an interesting duality relation. Recall that, if $X = (\mathbb{R}^n, \|\cdot\|)$ is an n-dimensional normed space, then the dual norm is defined by

$$\|x\|^* = \sup\{|\langle x, y\rangle| : \|y\| \leqslant 1\}.$$

If $a^{-1}|x| \leqslant \|x\| \leqslant b|x|$ for all x, it is clear that $b^{-1}|x| \leqslant \|x\|^* \leqslant a|x|$ for all x. Then, if we define

$$\tilde{k}^*(X) = \tilde{k}(X^*) \quad \text{and} \quad M^*(X) = M(X^*),$$

we have, by Theorem 5.3.4, that

$$\tilde{k}^*(X) \geq n(M^*/a)^2.$$

On the other hand, a trivial application of the Cauchy-Schwarz inequality shows that

$$MM^* \geqslant \left(\int_{S^{n-1}} \sqrt{\|x\|^*}\sqrt{\|x\|}d\sigma(x)\right)^2 \geqslant \left(\int_{S^{n-1}} |\langle x, x\rangle|^{1/2} d\sigma(x)\right)^2 = 1.$$

This gives

$$\tilde{k}(X)\tilde{k}^*(X) \geqslant cn^2 \frac{(MM^*)^2}{(ab)^2} \geqslant \frac{cn^2}{(ab)^2}.$$

From John's theorem we can choose the position of the unit ball of X so that $ab \leqslant \sqrt{n}$. This immediately proves the next non-trivial fact found by Figiel, Lindenstrauss and Milman.

THEOREM 5.3.5 (Figiel-Lindenstrauss-Milman). *For every n-dimensional normed space X there exists a Euclidean structure for which one has*

$$\tilde{k}(X)\tilde{k}^*(X) \geqslant c_1 n,$$

where $c_1 > 0$ is an absolute constant.

This shows that for every pair (X, X^*) there exists a Euclidean structure such that at least one of the parameters $\tilde{k}(X)$ and $\tilde{k}^*(X)$ is greater than $c\sqrt{n}$.

An interesting application of the same line of reasoning is given by the next proposition.

PROPOSITION 5.3.6. *Let $X = (\mathbb{R}^n, \|\cdot\|)$ be an n-dimensional normed space, and let $a, b > 0$ be the smallest positive constants such that $a^{-1}|x| \leqslant \|x\| \leqslant b|x|$ for all $x \in \mathbb{R}^n$. There exists a subspace F of dimension*

$$\dim(F) \geqslant c_1 n \min\left\{(M/b)^2, (M^*/a)^2\right\}$$

such that

$$\|P_F : X \to X\| \leqslant c_2 MM^*,$$

where P_F is the orthogonal projection onto F and $c_1, c_2 > 0$ are absolute constants.

Proof. Applying Theorem 5.2.10 with $\varepsilon = \frac{1}{3}$ to both X and X^* we may find a subspace F of dimension $\dim(F) \geqslant c_1 n \min\left\{(M/b)^2, (M^*/a)^2\right\}$ so that

$$\frac{2M}{3}|x| \leqslant \|x\| \leqslant \frac{4M}{3}|x| \quad \text{and} \quad \frac{2M^*}{3}|x| \leqslant \|x\|^* \leqslant \frac{4M^*}{3}|x| \tag{5.3.1}$$

for all $x \in F$. Then, for every $y \in \mathbb{R}^n$ we have

$$|P_F(y)|^2 = \langle P_F(y), y\rangle \leqslant \|P_F(y)\|^* \|y\| \leqslant \frac{4M^*}{3}|P_F(y)|\,\|y\|,$$

and hence

$$|P_F(y)| \leqslant \frac{4M^*}{3}\|y\|.$$

Now, using (5.3.1) we can write

$$\|P_F(y)\| \leqslant \frac{4M}{3}|P_F(x)| \leqslant \frac{16}{9}MM^*\|y\|,$$

which completes the proof. □

It is surprising that Theorem 5.3.5 has an application which is of interest in combinatorics in general. Note that a polytope may have as few as $(n+1)$ vertices and $(n+1)$ facets, as does the simplex. If the polytope is centrally symmetric, however, then the one with fewest facets is the cube: it has $2n$ facets, but on the other hand it has 2^n vertices. Likewise, the cross polytope $B(\ell_1^n)$ has only $2n$ vertices, but has 2^n facets. It turns out that a Dvoretzky-type theorem such as Theorem 5.3.5 implies that, in general, a centrally symmetric polytope cannot have few facets and few vertices at the same time.

To demonstrate this we shall use the next proposition, which states that if P is a centrally symmetric polytope in $\mathbb{R}^n$, and if we denote by $f(P)$ the number of its facets, then

$$k_0(P) \leqslant \log f(P).$$

(In particular the same is true for $\tilde{k}(P)$.) This is because to approximate a ball in $\mathbb{R}^k$ one needs many facets, and any facet of a section of P must come from a unique facet of P.

Proposition 5.3.7. *If P is a polytope with m facets in $\mathbb{R}^k$ and if $B_2^k \subseteq P \subseteq aB_2^k$, then*

$$m \geqslant \exp(k/2a^2).$$

Proof. We write P in the form

$$P = \{x \in \mathbb{R}^k : \langle x, v_j\rangle \leqslant 1\ ,\ j \leqslant m\}.$$

Since $B_2^k \subset P$, we must have $|v_j| \leqslant 1$ for every $j = 1, \dots, m$. From the inclusion of P in aB_2^k it follows that for every $\theta \in S^{n-1}$ there exists $j \leqslant m$ for which $\langle \theta, v_j\rangle \geqslant 1/a$.

We set $u_j = v_j/|v_j|$, $j = 1, \dots, m$. Since $|v_j| \leqslant 1$,

$$\{\theta \in S^{k-1} : \langle \theta, v_j\rangle \geqslant 1/a\} \subseteq \{\theta \in S^{k-1} : \langle \theta, u_j\rangle \geqslant 1/a\},$$

and hence

$$S^{k-1} \subset \bigcup_{j=1}^{m} \{\theta \in S^{k-1} : \langle \theta, u_j\rangle \geqslant 1/a\}. \tag{5.3.2}$$

Each $\{\theta \in S^{k-1} : \langle \theta, u_j\rangle \geqslant 1/a\}$ is a cap in S^{k-1}, which is centered at u_j and has angular radius $2\arcsin(1/2a)$. Using the estimates from Remark 3.1.8 for the measure of a spherical cap we have that the normalized Lebesgue measure of each such cap is at most $\exp\left(-\frac{k}{2a^2}\right)$. We thus get that

$$1 = \sigma(S^{k-1}) \leqslant m\exp\left(-\frac{k}{2a^2}\right).$$

which completes the proof. □

If we let $v(P)$ stand for the number of vertices of P, then we have from standard convexity that $v(P) = f(P^\circ)$, so that we also have

$$k_0^*(P) \leqslant \log f(P^\circ) = \log v(P).$$

From Theorem 5.3.5 we have that P has a position $\tilde{P}$ such that $k(\tilde{P})k^*(\tilde{P}) \geqslant cn$, which proves the next combinatorial result, again of Figiel, Lindenstrauss and Milman.

THEOREM 5.3.8 (Figiel-Lindenstrauss-Milman). *Let P be a centrally symmetric polytope in $\mathbb{R}^n$. Then,*

$$(\log f(P)) \cdot (\log v(P)) \geqslant cn,$$

where $f(P)$ is the number of facets of P and $v(P)$ is the number of its vertices.

Note that the symmetry condition is very important, since, as we note above, the simplex has only $(n+1)$ vertices and only $(n+1)$ facets. While Dvoretzky's theorem in the form of Theorem 5.2.10 holds also for non-symmetric convex bodies (one may follow the proof and check that, if the norm is replaced by the gauge function of a convex body, all the estimates remain valid) the main point of difference is that in John's theorem for non-symmetric bodies one has n instead of $\sqrt{n}$, so that we only get $\tilde{k}(X)\tilde{k}^*(X) \geqslant c$, a triviality.

We close this section with a lemma showing that, for every $1 \leqslant k \leqslant n$, a random $F \in G_{n,k}$ satisfies $M(F) \leqslant CM(X)$, with high probability, for a universal constant $C > 0$ independent of dimension.

LEMMA 5.3.9. *There exist constants $c_1, c_2 > 0$ such that for any $n \in \mathbb{N}$, given $X = (\mathbb{R}^n, \|\cdot\|)$ and $1 \leqslant k < n$ we have*

$$\nu_{n,k}(\{F \in G_{n,k} : M(F) \geqslant c_1 t M(X)\}) \leqslant e^{-c_2 t^2 k}, \tag{5.3.3}$$

for all $t > 1$.

Proof. An application of Theorem 3.2.2 shows that

$$\sigma(\{\theta \in S^{n-1} : \|\theta\| > (1+\varepsilon)M(X)\}) \leqslant c_3 e^{-c_4\varepsilon^2 k(X)} \leqslant e^{-c_5\varepsilon^2},$$

for all $\varepsilon > 0$. This is equivalent to the fact that

$$\int_{S^{n-1}} \exp(u\|\theta\|/M(X))\, d\sigma(\theta) \leqslant e^{c_6 u^2}, \tag{5.3.4}$$

for all $u > 1$. Now, we consider independent random points $\theta_1, \dots, \theta_k$ which are uniformly distributed over the sphere. From (5.3.4), using independence and Hölder's inequality we get

$$\mathbb{E}\left(\exp\left(\frac{u}{kM(X)}\sum_{i=1}^k \|\theta_i\|\right)\right) \leqslant e^{c_6 u^2/k} \tag{5.3.5}$$

for all $u > 1$. Using Markov's inequality and choosing $u = u(t)$ in an optimal way we see that

$$\mathbb{P}\left(\sum_{i=1}^k \|\theta_i\| > c_7 t k M(X)\right) \leqslant e^{-t^2 k}, \tag{5.3.6}$$

for all $t > 1$, provided that $c_7 > 0$ is large enough.

We know that $F = \mathrm{span}\{\theta_1, \dots, \theta_k\}$ is uniformly distributed on $G_{n,k}$ almost surely. Then, we can write:

$$\nu_{n,k}(\{F \in G_{n,k} : M(F) > 2c_7tM(X)\}) \leqslant \frac{\mathbb{P}(\sum_{i=1}^k \|\theta_i\| > c_7tkM(X))}{\mathbb{P}(\sum_{i=1}^k \|\theta_i\| > c_7tkM(X) \mid M(F) > 2c_7tM(X))}.$$

As an upper bound for the numerator we use (5.3.6). For the denominator we may write:

$$\begin{aligned}
&\mathbb{P}\left(\sum_{i=1}^k \|\theta_i\| > c_7tkM(X) \mid M(F) > 2c_7tM(X)\right) \\
&\geqslant \mathbb{P}\left(\frac{1}{k}\sum_{i=1}^k \|\theta_i\| > \frac{M(F)}{2} \mid M(F) > 2c_7tM(X)\right) \\
&= \frac{1}{\mathbb{P}(\{F \in G_{n,k} : M(F) \geqslant 2c_7tM(X)\})}\int_{\{F\in G_{n,k}:M(F)>2c_7tM(X)\}} \mathbb{P}(A_F)\, d\nu_{n,k}(F) \\
&\geqslant \min_{F\in G_{n,k}} \mathbb{P}(A_F),
\end{aligned}$$

where A_F is the event:

$$A_F = \left\{\frac{1}{k}\sum_{i=1}^k \|\theta_i\| > \frac{M(F)}{2} \mid \mathrm{span}\{\theta_1, \dots, \theta_k\} = F\right\}.$$

In order to complete the proof it suffices to give a lower bound for $\mathbb{P}(A_F)$, uniformly in $F \in G_{n,k}$. To this end, fix $F \in G_{n,k}$. Note that if we condition on $F = \mathrm{span}\{\theta_1, \dots, \theta_k\}$ then each of the vectors θ_i is uniformly distributed in S_F. Thus, we may write:

$$\begin{aligned}
M(F) &= \mathbb{E}\left(\frac{1}{k}\sum_{i=1}^k \|\theta_i\| \mid \mathrm{span}\{\theta_1, \dots, \theta_k\} = F\right) \\
&\leqslant C\sqrt{k}M(F)\mathbb{P}(A_F) + \frac{M(F)}{2}(1 - \mathbb{P}(A_F)),
\end{aligned}$$

where we have used the fact that $\|\theta\| \leqslant C\sqrt{\dim F}M(F) = C\sqrt{k}M(F)$ for every $\theta \in S_F$. It follows that $\mathbb{P}(A_F) \geqslant c_8/\sqrt{k}$. Since F was arbitrary we obtain:

$$\nu_{n,k}(\{F \in G_{n,k} : M(F) \geqslant c_7tM(X)\}) \leqslant c_8^{-1}\sqrt{k}e^{-t^2k} < e^{-c_9t^2k},$$

for all $t > 1$. □

5.4. Euclidean subspaces of ℓ_p^n

In this section we consider the classical spaces ℓ_p^n, $1 \leqslant p \leqslant \infty$, and we determine the order of $k_{p,n} = k(\ell_p^n)$ as a function of p and n.

THEOREM 5.4.1. *If $1 \leqslant p \leqslant 2$ then $k_{p,n} \simeq n$.*

Proof. We know that $k_p \geqslant cn(M/b)^2$, where $b = \max\{\|x\|_p : x \in S^{n-1}\}$ and

$$M = M(\ell_p^n) = \int_{S^{n-1}} \|x\|_p d\sigma(x).$$

Since $1 \leqslant p \leqslant 2$, from Hölder's inequality we have

$$\|x\|_p \leqslant n^{\frac{1}{p}-\frac{1}{2}}|x|$$

for every $x \in \mathbb{R}^n$ with equality for $x = (\frac{1}{\sqrt{n}}, \ldots, \frac{1}{\sqrt{n}})$. It follows that $b = n^{\frac{1}{p}-\frac{1}{2}}$.

Using Hölder's inequality once more, we see that $\|x\|_1 \leqslant n^{1/q}\|x\|_p$ for every $x \in \mathbb{R}^n$, where q is the conjugate exponent of p. Therefore,

$$\begin{aligned} M(\ell_p^n) &\geqslant \int_{S^{n-1}} n^{-1/q}\|x\|_1 \, d\sigma(x) = n^{-1/q}\sum_{i=1}^n \int_{S^{n-1}} |x_i| \, d\sigma(x) \\ &= n^{1-\frac{1}{q}} \int_{S^{n-1}} |x_1| \, d\sigma(x). \end{aligned}$$

This last integral has been estimated already and we know that it is of the order of $\frac{1}{\sqrt{n}}$. Combining the above we get

$$k_{p,n} \geqslant cn(M(\ell_p^n)/b)^2 \geqslant c_1 n.$$

Since (obviously) $k_{p,n} \leqslant n$, the proof is complete. □

THEOREM 5.4.2. $k_{\infty,n} \simeq \log n$.

Proof. Note that from Theorem 5.3.2 we have $k_{\infty,n} \geqslant c\log n$. The fact that this estimate is optimal follows from the fact that the cube has only $2n$ facets, which means that any of its sections has at most $2n$ facets, and a polytope with not-many facets cannot be close to a ball. To make this precise, we use Proposition 5.3.7.

Assume that for some $k \in \mathbb{N}$ there exists a k-dimensional subspace of ℓ_∞^n such that $d(F, \ell_2^k) \leqslant 4$. The cube Q_n has $2n$ facets, which implies that $Q_n \cap F$ has $m \leqslant 2n$ facets and there exists an ellipsoid $\mathcal{E}$ in F such that $\mathcal{E} \subset Q_n \cap F \subset 4\mathcal{E}$. Applying a linear transformation, we find a polytope $P_1 = T(P) \subset \mathbb{R}^k$ with m facets, which satisfies

$$B_2^k \subset P_1 \subset 4B_2^k.$$

Proposition 5.3.7 shows that $2n \geqslant m \geqslant \exp(k/32)$, which gives

$$k \leqslant 32\log(2n).$$

It follows that $k_{\infty,n} \leqslant 32\log(2n)$, and hence $k_{\infty,n} \simeq \log n$. □

The case of $p \in (2, \infty)$ remains, and is more difficult.

THEOREM 5.4.3. *If* $2 < q < \infty$ *then* $c_1 n^{2/q} \leqslant k_{q,n} \leqslant c_2(q)n^{2/q}$, *where* $c_2(q) \simeq q$.

Proof. Since $q > 2$ we have $\|x\|_q \leqslant |x|$ for every $x \in \mathbb{R}^n$, and so, $b = 1$. On the other hand, from the computation in the proof of Theorem 5.4.1 we see that

$$M(\ell_q^n) \geqslant n^{\frac{1}{q}-\frac{1}{2}}.$$

This shows that

$$k_{q,n} \geqslant c_1 n(n^{\frac{1}{q}-\frac{1}{2}})^2 = c_1 n^{\frac{2}{q}}.$$

For the right hand side inequality, we assume that there exists a k-dimensional subspace of ℓ_q^n which is 4-isomorphic to ℓ_2^k. Equivalently, there exist $u_j = (u_{j1}, \ldots, u_{jn})$, $1 \leqslant j \leqslant k$, with the following property: for every choice of real numbers $a_1, \ldots, a_k$,

$$\Big(\sum_{j=1}^k |a_j|^2\Big)^{1/2} \leqslant \Big\|\sum_{j=1}^k a_j u_j\Big\|_q \leqslant 4\Big(\sum_{j=1}^k |a_j|^2\Big)^{1/2}.$$

In particular, for every choice of signs $\varepsilon_1,\dots,\varepsilon_k$ we have

$$k^{q/2}\leqslant\Big\|\sum_{j=1}^k\varepsilon_ju_j\Big\|_q^q=\sum_{i=1}^n\Big|\sum_{j=1}^k\varepsilon_ju_{ji}\Big|^q.$$

We integrate on E_2^k and apply Khintchine's inequality to get

$$k^{q/2}\leqslant\sum_{i=1}^n\int_{E_2^k}\Big|\sum_{j=1}^k\varepsilon_ju_{ji}\Big|^q d\varepsilon\leqslant B_q^q\sum_{i=1}^n\Big(\sum_{j=1}^k u_{ji}^2\Big)^{q/2},\tag{5.4.1}$$

where $B_q\simeq\sqrt{q}$. Now, observe that for every $1\leqslant i\leqslant n$, if we set $a_i=u_{ji}$ we get

$$\begin{aligned}\sum_{j=1}^k u_{ji}^2=\sum_{j=1}^k u_{ji}u_{ji}&\leqslant\Big(\sum_{l=1}^n\Big|\sum_{j=1}^k u_{ji}u_{jl}\Big|^q\Big)^{1/q}=\Big\|\sum_{j=1}^k u_{ji}u_j\Big\|_q\\&\leqslant 4\Big(\sum_{j=1}^k u_{ji}^2\Big)^{1/2}.\end{aligned}$$

It follows that

$$\sum_{j=1}^k u_{ji}^2\leqslant 4^2$$

and going back to (5.4.1) we see that

$$k_{q,n}^{q/2}\leqslant(4B_q)^q n.$$

In other words, $k_{q,n}\leqslant 16B_q^2n^{2/q}\simeq qn^{2/q}$. □

5.5. Volume ratio and Kashin's theorem

DEFINITION 5.5.1. Let K be a centrally symmetric convex body in $\mathbb{R}^n$. Recall that the *volume ratio* of K is the quantity

$$\mathrm{vr}(K)=\inf\Big\{\Big(\frac{\mathrm{Vol}_n(K)}{\mathrm{Vol}_n(\mathcal{E})}\Big)^{1/n}:\mathcal{E}\subseteq K\Big\},$$

where the infimum is taken over all the ellipsoids $\mathcal{E}$ which are contained in K. It is clear that volume ratio is invariant under invertible linear transformations of $\mathbb{R}^n$: $\mathrm{vr}(K)=\mathrm{vr}(T(K))$ for all $T\in GL_n$.

EXAMPLE 5.5.2. Let K be a centrally symmetric convex body in $\mathbb{R}^n$. If $\|\cdot\|$ is the corresponding norm, then

$$\begin{aligned}\int_{\mathbb{R}^n}e^{-\|x\|^p}dx&=\int_{\mathbb{R}^n}\int_{\|x\|}^\infty pt^{p-1}e^{-t^p}dtdx\\&=\int_0^\infty pt^{p-1}e^{-t^p}\mathrm{Vol}_n(\{x:\|x\|\leqslant t\})dt\\&=\mathrm{Vol}_n(K)\int_0^\infty pt^{n+p-1}e^{-t^p}dt\\&=\mathrm{Vol}_n(K)\Gamma\big(\tfrac{n}{p}+1\big).\end{aligned}$$

If we choose $K = B_p^n$, $1 \leqslant p < \infty$, we can compute the first integral in another way, simply noting that

$$\int_{\mathbb{R}^n} e^{-\|x\|_p^p} dx = \left(2\int_0^\infty e^{-t^p}\,dt\right)^n = [2\Gamma(1/p+1)]^n.$$

Therefore,

$$\mathrm{Vol}_n(B_p^n) = \frac{[2\Gamma(\frac{1}{p}+1)]^n}{\Gamma(\frac{n}{p}+1)}.$$

Observe that, if $1 \leqslant p \leqslant 2$ then the maximal volume ellipsoid of B_p^n is $n^{\frac{1}{2}-\frac{1}{p}}B_2^n$. It follows that

$$\mathrm{vr}(B_p^n) = \frac{2\Gamma(\frac{1}{p}+1)[\Gamma(\frac{n}{2}+1)]^{\frac{1}{n}}}{n^{\frac{1}{2}-\frac{1}{p}}[\Gamma(\frac{n}{p}+1)]^{\frac{1}{n}}\sqrt{\pi}} \leqslant C,$$

where $C > 0$ is an absolute constant. We say that the unit balls of ℓ_p^n, $1 \leqslant p \leqslant 2$ have *uniformly bounded volume ratio.*

We will show that if a body K has bounded volume ratio, then the space X_K has subspaces F of dimension k proportional to n (with any chosen proportion) such that their Banach-Mazur distance to ℓ_2^k is bounded by a function which depends only on k/n and on the volume ratio of K. In Section 5.4 we saw, using a direct computation, that in the case of ℓ_p^n with $1 \leqslant p < 2$ this is true *for some* $k \simeq n$. In this section we prove the following theorem, that was put forward by Szarek and Tomczak-Jaegermann, generalizing works of Kashin, and uses the notion of volume ratio.

THEOREM 5.5.3 (Kashin, Szarek-Tomczak). *Let K be a centrally symmetric convex body in $\mathbb{R}^n$ such that $B_2^n \subseteq K$ and $\mathrm{Vol}_n(K) = \alpha^n \mathrm{Vol}_n(B_2^n)$ for some $\alpha > 1$. For every $1 \leqslant k \leqslant n$, a random subspace $F \in G_{n,k}$ satisfies with probability greater than $1 - e^{-n}$:*

$$B_F \subseteq K \cap F \subseteq (c\alpha)^{\frac{n}{n-k}} B_F,$$

where $c > 0$ is an absolute constant.

Proof. Since $B_2^n \subseteq K$, we have $\|x\| \leqslant |x|$ for every $x \in \mathbb{R}^n$. Let $k \leqslant n$. We may write

$$\int_{G_{n,k}}\int_{S_F} \|x\|^{-n} d\sigma_F(x)\, d\nu_{n,k}(F) = \int_{S^{n-1}} \|x\|^{-n} d\sigma(x) = \frac{\mathrm{Vol}_n(K)}{\mathrm{Vol}_n(B_2^n)} = \alpha^n.$$

From Markov's inequality, the measure of the set of $F \in G_{n,k}$ which satisfy

$$\int_{S_F} \|x\|^{-n} d\sigma_F(x) \leqslant (e\alpha)^n$$

is greater than $1 - e^{-n}$. Denote this set by $A \subset G_{n,k}$. We claim that any $F \in A$ satisfies

$$B_F \subseteq K \cap F \subseteq (c\alpha)^{\frac{n}{n-k}} B_F.$$

Indeed, let F be such a subspace. Then, applying Markov's inequality again, we see that for every $r \in (0,1)$

$$\sigma_F(\{x \in S_F : \|x\| \geqslant r\}) \geqslant 1 - (er\alpha)^n. \tag{5.5.1}$$

Fix $x \in S_F$. From Remark 3.1.8 we know that the volume of the spherical cap around x of radius $r/2$ is at least $(r/6)^k$. If $(er\alpha)^n < (r/6)^k$ then there must be a nonempty intersection between this cap and the set in equation (5.5.1):

$$B(x, r/2) \cap \{y \in S_F : \|y\| \geqslant r\} \neq \emptyset.$$

Then, we may find $y \in S_F$ with $|x - y| \leqslant r/2$ and $\|y\| \geqslant r$. By the triangle inequality we get

$$\|x\| \geqslant \|y\| - \|x - y\| \geqslant r - |x - y| \geqslant r/2.$$

Since $x \in S_F$ was arbitrary, this shows that

$$B_F \subseteq K \cap F \subseteq \frac{2}{r} B_F.$$

It remains to choose an optimal r satisfying

$$e^n 6^k \alpha^n r^{n-k} < 1,$$

which suggests $r_{\max} = (6e\alpha)^{-\frac{n}{n-k}}$. This completes the proof. $\square$

Our next result is a "global formulation" of the volume ratio theorem.

THEOREM 5.5.4. *Let K be a centrally symmetric convex body in $\mathbb{R}^n$ such that $B_2^n \subseteq K$ and $\mathrm{Vol}_n(K) = \alpha^n \mathrm{Vol}_n(B_2^n)$ for some $\alpha > 1$. There exists $U \in O(n)$ with the property*

$$B_2^n \subset K \cap U(K) \subset c\alpha^2 B_2^n,$$

where $c > 0$ is an absolute constant.

Proof. Note that

$$\|x\|_{K \cap U(K)} = \max\{\|Ux\|, \|x\|\} \geqslant G_(x) := \frac{\|Ux\| + \|x\|}{2}$$

for all $U \in O(n)$ and $x \in \mathbb{R}^n$. Since $B_2^n \subset K \cap U(K)$ for every $U \in O(n)$, the theorem will follow if we find $U \in O(n)$ such that $G_U(\theta) \geqslant \frac{1}{c\alpha^2}$ for all $\theta \in S^{n-1}$. We have

$$\begin{aligned}
\int_{O(n)} \int_{S^{n-1}} \frac{1}{\|U\theta\|^n \|\theta\|^n} d\sigma(\theta) d\nu_n(U) &= \int_{S^{n-1}} \Big(\int_{O(n)} \frac{1}{\|U\theta\|^n} d\nu_n(U) \Big) \frac{1}{\|\theta\|^n} d\sigma(\theta) \\
&= \int_{S^{n-1}} \Big(\int_{S^{n-1}} \frac{1}{\|\phi\|^n} d\sigma(\phi) \Big) \frac{1}{\|\theta\|^n} d\sigma(\theta) \\
&= \Big(\int_{S^{n-1}} \frac{1}{\|\theta\|^n} d\sigma(\theta) \Big)^2 \\
&= \alpha^{2n}.
\end{aligned}$$

Therefore, we can find $U \in O(n)$ which satisfies

$$\int_{S^{n-1}} \left(\frac{2}{\|U\theta\| + \|\theta\|} \right)^{2n} d\sigma(\theta) \leqslant \int_{S^{n-1}} \frac{1}{\|U\theta\|^n \|\theta\|^n} d\sigma(\theta) \leqslant \alpha^{2n}.$$

Let $\theta \in S^{n-1}$ and set $G_U(\theta) = t$. If $\phi \in S^{n-1}$ and $|\theta - \phi| \leqslant t$, then the fact that G_U is a norm with Lipschitz constant 1 gives

$$G_U(\phi) \leqslant G_U(\theta) + G_U(\phi - \theta) \leqslant t + |\phi - \theta| \leqslant 2t.$$

On the other hand, $\sigma(B(\theta, t)) \geqslant (t/3)^n$, and hence

$$\left(\frac{t}{3}\right)^n \frac{1}{(2t)^{2n}} \leqslant \sigma(B(\theta, t)) \frac{1}{(2t)^{2n}} \leqslant \int_{S^{n-1}} \left(\frac{1}{G_U(\theta)} \right)^{2n} d\sigma(\theta) \leqslant \alpha^{2n}.$$

This implies that $t \geqslant 1/(c\alpha^2)$ for some absolute constant $c > 0$, which completes the proof. □

REMARK 5.5.5. The proofs of the two theorems are of the same nature. This is an example of a much more general principle: local statements (like Theorem 5.5.3) which describe the geometry of sections of a convex body K have their analogue in global statements (like Theorem 5.5.4) which concern operations involving K and its orthogonal images. While in many cases the local and the global statements do not naturally follow one from the other, and each requires a separate proof, in this case we show how Theorem 5.5.3 directly implies Theorem 5.5.4.

Second proof of Theorem 5.5.4. Let K be a centrally symmetric convex body which contains B_2^n and satisfies $\mathrm{Vol}_n(K)/\mathrm{Vol}_n(B_2^n) = \alpha^n$. The proof of Theorem 5.5.3 shows that the measure of subspaces F of $\mathbb{R}^n$ of dimension $\lfloor n/2 \rfloor + 1$ which satisfy

$$B_F \subset K \cap F \leqslant c\alpha^2 B_F$$

is greater than $1 - e^{-n}$, where $B_F := B_2^n \cap F$. Therefore, we may find F such that

$$\frac{1}{c\alpha^2}|x| \leqslant \|x\| \leqslant |x|$$

for all $x \in F$ and all $x \in F^\perp$ simultaneously.

Define $U = P_F - P_{F^\perp}$, where $P_F, P_{F^\perp}$ are the orthogonal projections onto $F, F^\perp$ respectively. Then $U \in O(n)$, and if we write $x = x_1 + x_2$ with $x_1 \in F$, $x_2 \in F^\perp$,

$$\begin{aligned} 2|x| \geqslant 2\|x\|_{K\cap U(K)} &\geqslant \|x\| + \|Ux\| \\ = \|x_1 + x_2\| + \|x_1 - x_2\| &\geqslant 2\max\{\|x_1\|, \|x_2\|\} \\ \geqslant \|x_1\| + \|x_2\| &\geqslant \frac{1}{c\alpha^2}(|x_1| + |x_2|) \\ \geqslant \frac{1}{c\alpha^2}|x|. \end{aligned}$$

It follows that $B_2^n \subset K \cap U(K) \subset (2c\alpha^2)B_2^n$. □

In the example of ℓ_1^n we get a very interesting application of Theorem 5.5.4 which also goes back to Kashin.

THEOREM 5.5.6. *There exist vectors $y_1, \ldots, y_{2n} \in S^{n-1}$ such that*

$$c\sqrt{n}|x| \leqslant \sum_{j=1}^{2n} |\langle x, y_j\rangle| \leqslant 2\sqrt{n}|x|$$

for every $x \in \mathbb{R}^n$, where $c > 0$ is an absolute constant.

Proof. The maximal volume ellipsoid of B_1^n is $n^{-1/2}B_2^n$, and its volume ratio is bounded by an absolute constant $C > 0$. From the proof of Theorem 5.5.4, we can find $U \in O(n)$ with the property

$$2\sqrt{n} \geqslant \|x\|_1 + \|Ux\|_1 \geqslant \frac{\sqrt{n}}{C_1^2}$$

for every $x \in S^{n-1}$, where $C_1 > 0$ is some (other) absolute constant. We set $y_i = e_i$ and $y_{n+i} = U^*(e_i)$, $i = 1, \dots, n$. Then,

$$2\sqrt{n}|x| \geqslant \sum_{j=1}^{2n} |\langle x, y_j \rangle| \geqslant \frac{\sqrt{n}}{C_1^2}|x|$$

for every $x \in \mathbb{R}^n$. □

There exists yet another type of proof for the spherical sections of bounded volume ratio bodies which goes by way of covering statements. Let us illustrate it briefly.

Third proof of Theorem 5.5.4. A bounded volume ratio body has an exponential covering by Euclidean balls, indeed if $B_2^n \subset K$ and $\frac{\mathrm{Vol}_n(K)}{\mathrm{Vol}_n(B_2^n)} \leqslant \alpha^n$ then

$$N(K, 2B_2^n) \leqslant \frac{\mathrm{Vol}_n(K + B_2^n)}{\mathrm{Vol}_n(B_2^n)} \leqslant \frac{\mathrm{Vol}_n(2K)}{\mathrm{Vol}_n(B_2^n)} \leqslant (2\alpha)^n.$$

Denote the centres of such a covering by $\{y_i\}_{i=1}^N$.

Pick a large R and consider the sphere RS^{n-1}. We shall estimate the probability over a random rotation U, that two of the caps $(y_i + 2B_2^n) \cap RS^{n-1}$ and $U(y_j + 2B_2^n) \cap RS^{n-1}$ will intersect. This is the same as the probability that $y_i \in Uy_j + 4B_2^n$, which in turn is the volume of the intersection $(y_j + 4B_2^n) \cap RS^{n-1}$ with respect to the total volume of RS^{n-1}. This volume is at most the normalized measure of a cap of radius $4/R$ on S^{n-1}, which in turn is at most (see the proof of Theorem 3.1.5)

$$\frac{\sqrt{n}}{2} \sin^{n-2}(4/R) \leqslant (cR)^{-n}.$$

By choosing R large enough, we can make sure that all N^2 such intersections will, with high probability, be empty. Indeed, for this probability to be, say, greater than $1 - e^{-n}$ we require

$$(2\alpha)^{2n}(cR)^{-n} < e^{-n}$$

which amounts to $R = 4e\alpha^2/c$. This is the same estimate (up to the different constants) as the one we had in Theorem 5.5.4. □

Remark 5.5.7. Let us emphasize that, while it seems that we have proven a more general statement, namely that, whenever $N(K, 2B_2^n)$ has an exponential bound and $B_2^n \subset K$, we get a result similar to Theorem 5.5.4, we always have that

$$\frac{\mathrm{Vol}_n(K)}{\mathrm{Vol}_n(B_2^n)} \leqslant 2^n N(K, 2B_2^n),$$

so that when such an exponential bound exists we are automatically in the bounded volume ratio situation.

5.6. Global form of the Dvoretzky-Milman theorem

In this section we give one more example illustrating the *principle* that local statements have their corresponding global equivalents and vice versa. The term "global" refers to properties of a convex body and its images under linear transformations in the space of full dimension while the term "local" refers to the structure of lower dimensional sections and projections of the body. It appears that if we are interested in the asymptotic behaviour of the relevant quantities, that is, as the dimension grows to infinity, then there is an exact parallelism between the two theories; the global (geometric) asymptotic theory and the local theory.

We have seen that if $X = (\mathbb{R}^n, \|\cdot\|)$ is an n-dimensional normed space then, for every $\varepsilon \in (0,1)$ there exists a subspace F of dimension $k \geqslant c\varepsilon^2 n(M/b)^2$ such that for all $x \in F$

$$\frac{M}{1+\varepsilon}|x| \leqslant \|x\| \leqslant (1+\varepsilon)M|x|,$$

where M is the expectation of $\|\cdot\|$ on S^{n-1} and b is the smallest positive constant for which $\|x\| \leqslant b|x|$ for all $x \in \mathbb{R}^n$. Moreover, the measure of the set of subspaces $F \in G_{n,k}$ for which this holds is close to 1. We will call this statement a *local form* of Dvoretzky-type theorem.

The next theorem represents the *global form* of the above estimate for k.

THEOREM 5.6.1 (Bourgain-Lindenstrauss-Milman). *Let $X = (\mathbb{R}^n, \|\cdot\|)$ be an n-dimensional normed space with unit ball K. Let $M = \int_{S^{n-1}} \|x\|\, d\sigma(x)$ and b the smallest positive constant for which $\|x\| \leqslant b|x|$ for all $x \in \mathbb{R}^n$. Then, for any $\varepsilon \in (0,1/2)$ there exist an integer*

$$t \leqslant \frac{c}{\varepsilon^2}\Big(\frac{b}{M}\Big)^2$$

and t orthogonal transformations $U_1, \ldots, U_t \in O(n)$ such that

$$\frac{M}{1+\varepsilon}|x| \leqslant \frac{1}{t}\sum_{j=1}^{t} \|U_j(x)\| \leqslant (1+\varepsilon)M|x| \qquad (x \in \mathbb{R}^n).$$

Equivalently,

$$\frac{M}{1+\varepsilon}\, B_2^n \subseteq \frac{1}{t}\sum_{j=1}^{t} U_j^*(K^\circ) \subseteq (1+\varepsilon)M\, B_2^n.$$

Moreover, this statement holds true for a random choice of $U_1, \ldots, U_t \in O(n)$ with probability greater than $1 - \exp(-c_5\varepsilon^2 tnM^2/b^2)$.

For the proof we will use the next version of Bernstein's inequality (see Section 3.5.1): if $\{g_j\}_{j\leqslant t}$ are independent random variables with mean 0 on some probability space $(\Omega, \mathcal{A}, \mathbb{P})$ and if $\|g_j\|_{\psi_2} \leqslant \alpha$ for all $1 \leqslant j \leqslant t$ then, for every $\varepsilon \in (0,1)$,

$$\mathbb{P}\Big(\Big|\sum_{j=1}^{t} g_j\Big| > \varepsilon t\Big) \leqslant 2\exp\Big(-\frac{\varepsilon^2 t}{8\alpha^2}\Big). \tag{5.6.1}$$

Recall that the ψ_2-norm of g is defined by

$$\|g\|_{\psi_2} := \inf\left\{t > 0 : \int_\Omega \exp\left((|f|/t)^2\right)\, d\mathbb{P} \leqslant 2\right\}.$$

Marcus and Pisier have proved that, for every norm $\|\cdot\|$ on $\mathbb{R}^n$ and every $x \in \mathbb{R}^n$, the map $g : O(n) \to \mathbb{R}$ with $U \mapsto \|U(x)\|$ has ψ_2-norm

$$\|g\|_{L^{\psi_2}(\nu_n)} \leqslant c\|g\|_{L^1(\nu_n)}, \tag{5.6.2}$$

where $c>0$ is an absolute constant (we will see a proof of this result in Chapter 9).

Proof of Theorem 5.6.1. Fix $t \in \mathbb{N}$ and $y \in S^{n-1}$. Let $0 < \varepsilon < 1/2$ and $\{x_i\}_{i \leqslant m}$ be an ε-net for S^{n-1} with cardinality $m \leqslant (4/\varepsilon)^n$. For every $i \leqslant m$ we consider the random variables $\{g_{ij}\}_{j \leqslant t}$ distributed as

$$g(U) = \|U(x)\| - M$$

on $(O(n), \nu_n)$, where

$$M = \int_{S^{n-1}} \|x\| \, d\sigma(x) = \int_{O(n)} \|U(y)\| \, d\nu_n(U).$$

Each g_{ij} has mean 0 and since $\|\cdot\|$ is Lipschitz with constant b we have

$$\begin{aligned}\nu_n(\{U : |g_{ij}(U)| \geqslant sM\}) &= \sigma\left(\{y \in S^{n-1} : |\|y\| - M| \geqslant sM\}\right) \\ &\leqslant 2e^{-s^2 n M^2/(2b^2)}\end{aligned}$$

for every $s > 0$, which implies

$$\|g_{ij}\|_{L^1(\nu_n)} \leqslant \frac{c_1 b}{\sqrt{n}}.$$

It follows from (5.6.2) that also $\|g_{ij}\|_{L^{\psi_2}(\nu_n)} \leqslant c_2 b/\sqrt{n}$. Applying (5.6.1) we get that for each $1 \leqslant i \leqslant m$

$$\mathbb{P}\left(\left\{(U_1, \ldots, U_t) : \left|t^{-1}\sum_{j=1}^{t} \|U_j(x_i)\| - M\right| \geqslant \varepsilon M\right\}\right) \leqslant 2\exp(-c_3\varepsilon^2 t n M^2/b^2),$$

where $c_3 > 0$ is an absolute constant.

We ask that $m\exp(-c_3\varepsilon^2 t n M^2/(2b^2)) < \exp(-c_5\varepsilon^2 n M^2 t/(2b^2))$ for some $c_5 < c_3$. This is satisfied if

$$t \geqslant c_4\varepsilon^{-2}\log(2/\varepsilon)\left(\frac{b}{M}\right)^2,$$

and then we have

$$\left|t^{-1}\sum_{j=1}^{t} \|U_j(x_i)\| - M\right| \leqslant \varepsilon M$$

for all $i = 1, \ldots, m$. The standard successive approximation argument shows that

$$\left|t^{-1}\sum_{j=1}^{t} \|U_j(x)\| - M\right| \leqslant \frac{3\varepsilon}{1-\varepsilon} M$$

for all $x \in S^{n-1}$. The term $\log(1/\varepsilon)$ in our estimate for t can be removed by a refinement of the argument (this was done by Schmuckenschläger). Moreover, this is satisfied with probability greater than $1 - \exp(-c_5\varepsilon^2 t n M^2/b^2)$ with respect to $(U_1, \ldots, U_t) \in O(n) \times \cdots \times O(n)$. $\square$

Looking at the proofs of the local and the global form of Dvoretzky theorem one realizes that, although they are independent of each other, they use the same tools. Milman and Schechtman obtained an "asymptotic formula" which relates them directly: Recall that $k(X)$ is the smallest integer $k \leqslant n$ with the property that

$$\nu_{n,k}\left(\left\{F \in G_{n,k} : \frac{1}{2}M|x| \leqslant \|x\| \leqslant 2M|x| \text{ for all } x \in F\right\}\right) \geqslant 1 - \frac{k}{n+k}.$$

Now, define $t(X)$ as the smallest integer t for which there exist orthogonal transformations $U_1, \dots, U_t \in O(n)$ such that

$$\frac{M}{2}|x| \leqslant \frac{1}{t}\sum_{i=1}^{t} \|U_i(x)\| \leqslant 2M|x| \text{ for all } x \in \mathbb{R}^n.$$

With these definitions we have:

THEOREM 5.6.2 (Milman-Schechtman). *There exist absolute constants* $0 < c, C$ *such that*

$$\frac{1}{C}n \leqslant k(X)t(X) \leqslant Cn$$

for every n-dimensional normed space X such that $(M/b)^2 \geq c\frac{\log n}{n}$.

The reader should not be too surprised at this point, since she has seen already in Theorem 5.3.4 that $k(X)$ has a formula, namely $k(X) \simeq n(M/b)^2$, and has also seen an upper bound for $t(X)$ by $C(b/M)^2$ by using Theorem 5.6.1 with $\varepsilon = 1/4$, say. The remaining job, namely bounding $t(X)$ from below, is based on the next lemma. Note that in the definition of $t(X)$ we do not require that a large measure of orthogonal transformations satisfies the inequality, but merely the existence of one such t-tuple of transformations.

LEMMA 5.6.3. *Let $\|\cdot\|$ be a norm on $\mathbb{R}^n$ and let $U_1, \dots, U_t \in O(n)$ such that the norm*

$$\|x\|' := \frac{1}{t}\sum_{i=1}^{t} \|U_i(x)\|$$

satisfies $\|x\|' \leqslant C|x|$ for all $x \in \mathbb{R}^n$ and some constant $C > 0$. Then,

$$\|x\| \leqslant C\sqrt{t}\,|x|$$

for all $x \in \mathbb{R}^n$.

Proof. We write K for the unit ball of the norm $\|\cdot\|$. Let $b = \max\{\|x\| : x \in S^{n-1}\}$ and consider $y \in S^{n-1}$ with $\|y\| = b$. Then,

$$|\langle x, y\rangle| \leqslant 1/b \text{ for all } x \in K.$$

We shall now prove that $b \leqslant C\sqrt{t}$. We define $x_i = U_i^*(y)$, $i = 1, \dots, t$, and for any choice of $\lambda_i > 0$ we set

$$A_i^\varepsilon := A_i^\varepsilon(\lambda_1, \dots, \lambda_t) = \Big\{x \in S^{n-1} : \varepsilon\langle x, x_i\rangle \geqslant \frac{\lambda_i}{b}\Big\}, \qquad 1 \leqslant i \leqslant t,\ \varepsilon = \pm 1.$$

Observe that if $\bigcap_{i=1}^{t}(A_i^+ \cup A_i^-) \neq \emptyset$ then $\sum_{i=1}^{t}\lambda_i \leqslant Ct$. To see this, assume that for some $z \in S^{n-1}$ we have

$$|\langle U_i(z), y\rangle| = |\langle z, x_i\rangle| \geqslant \lambda_i/b$$

for all $1 \leqslant i \leqslant t$. Then, $\|U_i(z)\| \geqslant \lambda_i$ for all i, and this gives

$$\sum_{i=1}^{t}\lambda_i \leqslant \sum_{i=1}^{t}\|U_i(z)\| = t\|z\|' \leqslant Ct.$$

Next, we choose $\varepsilon_i = \pm 1$ such that $\left|\sum_{i=1}^t \varepsilon_i x_i\right|$ is maximal among all choices of signs. By the parallelogram law, having chosen the maximal "diagonal",

$$\left|\sum_{i=1}^t \varepsilon_i x_i\right| \geqslant \Big(\sum_{i=1}^t |x_i|^2\Big)^{1/2} = \sqrt{t}.$$

Setting $z = (\sum_{i=1}^t \varepsilon_i x_i)/\left|\sum_{i=1}^t \varepsilon_i x_i\right|$ we have $z \in S^{n-1}$ and choosing $\lambda_i^* = b|\langle z, x_i\rangle|$ we obviously have $z \in \bigcap_{i=1}^t (A_i^+ \cup A_i^-)$. Therefore,

$$\sum_{i=1}^t \lambda_i^* = b\sum_{i=1}^t |\langle z, x_i\rangle| \leqslant Ct.$$

Observe that

$$\sum_{i=1}^t |\langle z, x_i\rangle| \geqslant \Big\langle z, \sum_{i-1}^t \varepsilon_i x_i\Big\rangle = \left|\sum_{i=1}^t \varepsilon_i x_i\right| \geqslant \sqrt{t}.$$

Combining the above we have $b\sqrt{t} \leqslant Ct$, and the lemma is proved. □

Proof of Theorem 5.6.2. Since we know that $k(X) \simeq n(\frac{M}{b})^2$ and that $t(X) \leqslant c(\frac{b}{M})^2$ we only need to explain that $t(X) \geqslant c'(\frac{b}{M})^2$. Indeed, by the definition of $t(X)$ there exist $U_i \in O(n)$ such that

$$\frac{M}{2}|x| \leqslant \frac{1}{t(X)}\sum_{i=1}^{t(X)} \|U_i(x)\| \leqslant 2M|x| \quad \text{for all } x \in \mathbb{R}^n,$$

which in particular means that for all x

$$\frac{1}{t(X)}\sum_{i=1}^{t(X)} \|U_i(x)\| \leqslant 2M|x|.$$

By the previous lemma this implies that

$$b \leqslant 2M\sqrt{t(X)},$$

where $b = \max\{\|x\| : x \in K\}$. □

5.7. Isomorphic phase transitions and thresholds

In this section we present some results that indicate "phase transition" and "threshold" type behaviour; such statements appear in statistical physics and combinatorics respectively. However, our results are isomorphic analogues of the phenomena that are observed in these two areas. They are true up to some universal factors, but they are applicable to any (convex) sets, without any special symmetries or structure.

The "simplest" phase transition result that we have already seen in Section 5.4 is the behaviour of the critical dimension of ℓ_p^n-spaces. We saw that

$$k_{q,n} = k(\ell_q^n) \simeq \begin{cases} n & 1 \leqslant q \leqslant 2 \\ qn^{2/q} & 2 \leqslant q \leqslant \log n \end{cases}$$

(and, of course $\ell_{\log n}^n \simeq \ell_\infty^n$). Here, $k(X)$ denotes the largest dimension k such that X contains a 2-isomorphic copy of ℓ_2^k.

A second example is given by Theorem 5.6.2. Let $X = (\mathbb{R}^n, \|\cdot\|)$ be an n-dimensional normed space with unit ball K, and let $M = \int_{S^{n-1}} \|x\|\, d\sigma(x)$ and b be the smallest positive constant for which $\|x\| \leqslant b|x|$ for all $x \in \mathbb{R}^n$. Define

$$t_0 = \left(\frac{b}{M}\right)^2$$

and for every $t \in \mathbb{N}$ and any $U_1, \ldots, U_t \in O(n)$ consider the space $X_t = X_t(U_1, \ldots, U_t)$ with norm

$$\|x\|_t = \frac{1}{t}\sum_{j=1}^{t} \|U_j(x)\|.$$

With this notation, Theorem 5.6.2 states that if $t \geqslant ct_0$ then a random choice of $U_1, \ldots, U_t \in O(n)$ satisfies with probability greater than $1 - \exp(-ctnM^2/b^2)$

$$\|x\|_t \simeq M|x| \qquad (x \in \mathbb{R}^n).$$

On the other hand, if $1 \leqslant t \leqslant t_0$ then Lemma 5.6.3 shows that

$$b(X) \simeq \sqrt{t}\, b(X_t)$$

for a random choice of $U_1, \ldots, U_t \in O(n)$. Therefore, we observe a phase transition of the function $t \mapsto b(X_t)$ at t_0: we have a stabilization after the critical value t_0, and the reason for this stabilization is the fact that the average norm $\|\cdot\|_t$ becomes (almost) Euclidean (with high probability) around the value t_0.

In the next subsections we will add a few more results and then comment on them.

5.7.1. Local example: diameter of random projections

The next proposition gives an upper bound for the diameter of a random k-dimensional projection of a centrally symmetric convex body.

PROPOSITION 5.7.1. *Let K be a centrally symmetric convex body in $\mathbb{R}^n$ with $K \subset aB_2^n$ and let $1 \leqslant k < n$. Then there exists a subset $\Gamma \subset G_{n,k}$ with measure greater than $1 - e^{-ck}$ such that the orthogonal projection of K onto any subspace $F \in \Gamma$ satisfies*

$$P_F(K) \subset C \max\{M^*(K), a\sqrt{k/n}\}\, B_2^n, \tag{5.7.1}$$

where $c > 0, C > 1$ are absolute constants.

REMARK 5.7.2. In fact we know a much more precise estimate, with constant C which is close to 1 as $k \to n$ and as k gets below $ck(X^*)$.

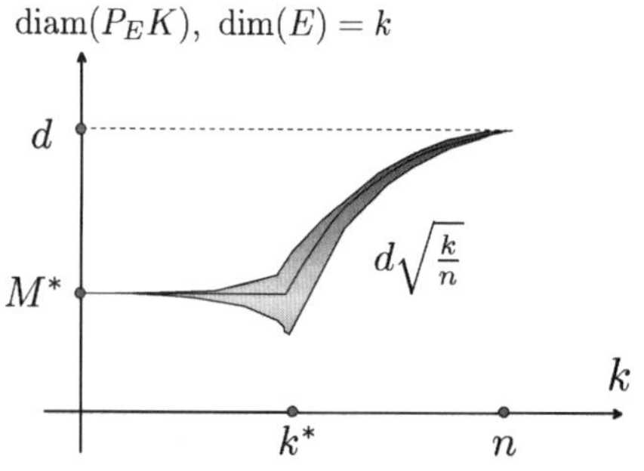

Proof. Note that the diameter of $P_F(K)$ is $\max_{u\in S^{n-1}\cap F} h_K(u)$ where h_K is the support function of K. The Lipschitz constant of h_K is bounded by a. We shall use a net argument. By successive approximation, as in the proof of Proposition 5.2.8, fixing F and a $\frac{1}{2}$-net $\mathcal{N}$ on $S_F = S^{n-1}\cap F$ we have

$$\max_{u\in S_F} h_K(u) \leqslant 2\max_{z\in\mathcal{N}} h_K(z).$$

Indeed; if $u_0 \in S_F$ then by successive approximation, there are $\delta_j \leqslant \frac{1}{2^j}$ and $(z_j) \subset \mathcal{N}$ such that $u_0 = z_0 + \sum_{j=1}^{\infty}\delta_j z_j$. Using the continuity, homogeneity and subadditivity of h_K we obtain:

$$|h_K(u_0)| \leqslant \sum_{j=0}^{\infty} 2^{-j}|h_K(z_j)| \leqslant 2\max_{z\in\mathcal{N}}|h_K(z)|.$$

Next fix some $\frac{1}{2}$-net $\mathcal{N}$ of S_{F_0} for some fixed F_0, with cardinality at most 5^k. For any $z\in\mathcal{N}$ we have by Proposition 5.2.4 that

$$\begin{aligned}\nu(\{U \in O(n) : h_K(Uz) \geqslant t + M^*(K)\}) \\ = \sigma\left(\left\{\theta \in S^{n-1} : h_K(\theta) \geqslant t + \int_{S^{n-1}} h_K(u)d\sigma(u)\right\}\right) \\ \leqslant \exp(-c_1t^2n/a^2).\end{aligned}$$

Therefore

$$\nu(\{U \in O(n) :\ \exists z \in \mathcal{N} \text{ s. t. } h_K(Uz) \geqslant t + M^*(K)\}) \leqslant 5^k e^{-c_1t^2n/a^2}.$$

Thus, for a measure of $U\in O(n)$ which is greater than $e^{\ln(5)k-c_1t^2n/a^2}$ we have

$$\max_{z\in\mathcal{N}}|h_K(Uz)| \leqslant t + M^*(K).$$

Choosing $t = c_2a\sqrt{k/n}$ for a suitable c_2 we get a measure of $B \subset G_{n,k}$ which is greater than $1-e^{-k}$ such that for every $F\in B$ we have

$$\max_{\theta\in S_F}|h_K(\theta)| \leqslant 2\left(c_2a\sqrt{k/n} + M^*(K)\right),$$

as needed. □

Note. In fact, one has that the reverse inequality

$$R(P_F(K)) \geqslant c\max\{M^*(K), \sqrt{k/n}R(K)\}$$

holds for most $F\in G_{n,k}$, where $R(A)$ is the minimal radius of a ball circumscribing A. To see this, first note that if $x\in K$ and $|x| = R(K)$ then, for most $F\in G_{n,k}$ we have $|P_F(x)| \geqslant c\sqrt{k/n}|x|$, and hence $R(P_F(K)) \geqslant c\sqrt{k/n}R(K)$. On the other hand, if $\sqrt{k/n}R(K) \leqslant c'w(K)$ for a small enough absolute constant $0 < c' < 1$ then we know that most k-dimensional projections of K are isomorphic Euclidean balls of radius $w(K)$. For any $1\leqslant k\leqslant n$ consider the parameter

$$R_k(K) = \int_{G_{n,k}} R(P_F(K))\,d\nu_{n,k}(F).$$

Proposition 5.7.1 and the preceding remark show that there exist absolute constants $c, C > 0$ such that for any n and any centrally symmetric convex body K in $\mathbb{R}^n$,

$$c\sqrt{k/n}\,R(K) \leqslant R_k(K) \leqslant C\sqrt{k/n}\,R(K)$$

if $k^* \leqslant k \leqslant n$ and

$$cw(K) \leqslant R_k(K) \leqslant Cw(K)$$

if $1 \leqslant k \leqslant k^*$. We observe the stabilization of the function $k \mapsto R_k(K)$ at the critical value k^*, and a unified form of behaviour for the whole family of convex bodies in high-dimensional spaces.

We emphasize again that the *only* reason for stabilization for small k is that a "random" projection $P_F(K)$ for $\dim(F) \lesssim \varepsilon^2 k^*$ includes a ball. That is, for a random subspace F of dimension $\lesssim \varepsilon^2 k^*$, the set $P_F(K)$ itself is almost a Euclidean ball, in the sense that

$$(1-\varepsilon)w(K)B_F \subseteq P_F(K) \subseteq (1+\varepsilon)w(K)\,B_F.$$

In fact, the same statement holds for general sets of points (non-convex) by considering their convex hull, and there the reason for stabilization of diameter of a random projection for small k is that the projection includes an ε-net of a Euclidean ball (of radius $w(K)/2$).

5.7.2. Global example: averages of norms on the sphere

The next example of phase transition has more complicated behaviour, with "constant" behaviour of a different nature. Let $X = (\mathbb{R}^n, \|\cdot\|)$ be an n-dimensional normed space and let K be its unit ball. As we described briefly in Section 5.2, one may extend the definition of $M(X)$ as follows. First, for every $q \neq 0$ we define

$$M_q := M_q(K) = \left(\int_{S^{n-1}} \|\theta\|^q d\sigma(\theta)\right)^{1/q}.$$

Litvak, Milman and Schechtman determined the behaviour of $M_q(X)$ as a function of q.

THEOREM 5.7.3 (Litvak-Milman-Schechtman). *Let K be a centrally symmetric convex body in $\mathbb{R}^n$ and let $\|\cdot\|$ be the corresponding norm on $\mathbb{R}^n$. We denote by b the smallest constant for which $\|x\| \leqslant b|x|$ holds true for every $x \in \mathbb{R}^n$. Then,*

$$\max\left\{M_1, c_1\frac{b\sqrt{q}}{\sqrt{n}}\right\} \leqslant M_q \leqslant \max\left\{2M_1, c_2\frac{b\sqrt{q}}{\sqrt{n}}\right\}$$

for all $q \in [1, n]$, where $c_1, c_2 > 0$ are absolute constants.

REMARK 5.7.4. Recall that in the non-degenerate case, i.e. $(M/b)^2 \geq \frac{\log n}{n}$, we have $k(X) \simeq n(M/b)^2$. Thus, Theorem 5.7.3 can be also formulated in the following way. We have:

(1) $M_q \simeq M$ for $1 \leqslant q \leqslant k(X)$.
(2) $M_q \simeq b\sqrt{q/n}$ for $k(X) \leqslant q \leqslant n$.
(3) $M_q \simeq b$ for $q \geqslant n$.

The change of behaviour of M_q occurs when $q \simeq n(M_1/b)^2$. This value is equivalent to $k(X)$. It is clear that $M_q \leqslant M_r$ if $q \leqslant r$. Since $M_q \leqslant b$ for all q and $M_n \simeq b$, it follows that $M_r \simeq b$ if $r \geqslant n$. In other words, we have a second change of behaviour of M_q at the point $q = n$. So, we see two (isomorphic) phase transitions: at $q = k(X)$ and at $q = n$.

Proof of Theorem 5.7.3. The function $\|x\| : S^{n-1} \to \mathbb{R}$ is Lipschitz continuous with constant b. Theorem 3.2.2 shows that

$$\sigma\big(\{x \in S^{n-1} : \big|\ \|x\| - M_1\ \big| > t\}\big) \leqslant 2\exp(-c_1 t^2 n/b^2)$$

for all $t > 0$. Then,

$$\begin{aligned}\int_{S^{n-1}} \big|\ \|x\| - M_1\ \big|^q d\sigma(x) &\leqslant 2q\int_0^\infty t^{q-1}\exp(-c_1t^2n/b^2)\,dt \\ &= \Big(\frac{b}{\sqrt{c_1 n}}\Big)^q 2q\int_0^\infty s^{q-1}\exp(-s^2)\,ds \\ &\leqslant \Big(C\frac{b\sqrt{q}}{\sqrt{n}}\Big)^q,\end{aligned}$$

for some absolute constant $C > 0$. The triangle inequality in $L^q(S^{n-1})$ implies that

$$M_q - M_1 \leqslant \|\ \|x\| - M_1\|_q \leqslant C\frac{b\sqrt{q}}{\sqrt{n}}. \tag{5.7.2}$$

In other words,

$$M_q \leqslant 2\max\Big\{M_1, C\frac{b\sqrt{q}}{\sqrt{n}}\Big\}.$$

For the left hand side inequality we observe that C is contained in a strip of width $1/b$: there exists $z \in S^{n-1}$ with $\|z\| = b$ such that $C \subset \{y : |\langle y, z\rangle| \leqslant 1/b\}$. It follows that for every $t > 0$

$$tC \subset \{y :\ |\langle y, z\rangle| \leqslant t/b\}.$$

This can be also written in the form

$$\{x \in S^{n-1} : \|x\| \geqslant t\} \supset B_t := \{x \in S^{n-1} : |\langle x, z\rangle| \geqslant t/b\}$$

for every $t > 0$. A simple computation shows that if $t \leqslant b/3$ then

$$\sigma(B_t) \geqslant \frac{c_2\sqrt{n}t}{b}\exp(-c_3nt^2/b^2) \tag{5.7.3}$$

where $c_2, c_3 > 0$ are absolute constants. To see this we write

$$\sigma(B_t) = \frac{2}{I_{n-2}}\int_a^{\pi/2}\cos^{n-2}\theta d\theta$$

where $I_n = \int_{-\pi/2}^{\pi/2}\cos^n\theta d\theta$ and $a = \arcsin(t/b)$. We know that $I_n \leqslant \sqrt{\frac{\pi}{2n}}$, therefore

$$\sigma(B_t) \geqslant c\sqrt{n}\int_a^\gamma \cos^{n-2}\theta d\theta \geqslant (\gamma - a)c\sqrt{n}\cos^{n-2}\gamma$$

where $\gamma = \arcsin(2t/b)$. Assuming that $t \leqslant b/3$ and substituting we get (5.7.3). It follows that

$$\begin{aligned}M_q &\geqslant t\big[\sigma(\{x \in S^{n-1} : \|x\| \geqslant t\})\big]^{1/q} \geqslant t\big[\sigma(B_t)\big]^{1/q} \\ &\geqslant t\left(\frac{c_2\sqrt{n}t}{b}\right)^{1/q}\exp(-c_3nt^2/(qb^2)).\end{aligned}$$

for all $t \leqslant b/3$. Choosing $t = b\sqrt{q}/(3\sqrt{n})$ we conclude the proof. $\square$

Next, for every $q \geqslant 1$ we define $t_q(X)$ as the smallest integer for which there exist $U_1, \ldots, U_t \in O(n)$ such that

$$\frac{M_q}{2}|x| \leqslant \left(\frac{1}{t}\sum_{i=1}^{t} \|U_i(x)\|^q\right)^{1/q} \leqslant 2M_q|x|$$

for all $x \in \mathbb{R}^n$. The parameters $t_q(X)$ were studied by Litvak, Milman and Schechtman who obtained a generalization of the formula

$$t(X) \simeq \frac{n}{k(X)} \simeq \left(\frac{b}{M(X)}\right)^2$$

of Theorem 5.6.2.

THEOREM 5.7.5 (Litvak-Milman-Schechtman). *There exist absolute constants* $c_1, c_2 > 0$ *such that, for every normed space* $X = (\mathbb{R}^n, \|\cdot\|)$*:*

(i) *If* $q > 2$ *then*

$$c_1 t(X)\left(\frac{M_1}{M_q}\right)^2 \leqslant t_q^{2/q}(X) \leqslant c_2 t(X)\left(\frac{M_1}{M_q}\right)^2.$$

(ii) *If* $1 \leqslant q \leqslant 2$ *then*

$$c_1 t(X) \leqslant t_q(X) \leqslant c_2 t(X).$$

REMARKS 5.7.6. The following are some non-precise mathematical statements which are on a more philosophical level, meant to point out some phenomena accompanying high-dimensional processes; the reader will notice their appearance several times in our presentation of the theory:

(1) The constant behaviour (at the "end" of a process) corresponds to stabilization for one reason (only): "maximal" symmetry is achieved.
(2) In addition, in the "global" process, i.e. when we study changes in the whole space, not in subspaces or projections, the "best" possibility, we are looking for, coincides approximately with a "random" selection. However, there is an amount of freedom in identifying the right notion of randomness which may change from problem to problem.
(3) More complicated processes with two (or more) phase transitions are superpositions of "elementary" ones.
(4) The constant behaviour (but at the "start" of a process) corresponds to *inertia*; in the behaviour of M_q, it is a *concentration phenomenon* and in the case of t_q (at $q = 2$) the reason behind it is a *convexification*: Khintchine's inequality transforms L_p-averages to $L_{p/2}$-averages which, for $1 \leqslant p \leqslant 2$, behave in the same way as L_1 because of the lack of convexity (i.e. the same as for the case $p = 2$).

5.7.3. Thresholds in problems of approximation

Let $x_i \in S^{n-1}$ and $I_i = [-x_i, x_i]$ be an interval. We approximate a Euclidean ball rB_2^n (of some radius $r \sim \frac{1}{\sqrt{n}}$) by $K_N = \frac{1}{N}\sum_1^N I_i$. Then, Kashin's theorem [**326**] states that

$$\inf_{\{x_i\}} d_G(K_N, B_2^n) = \begin{cases} \infty & N < n \\ \sqrt{n} & N = n \\ C(\lambda) & N = \lambda n, \lambda > 1. \end{cases}$$

Let $\lambda < 2$. Gluskin showed in [**249**] that

$$C(\lambda) \sim \min\left\{\sqrt{n}, \sqrt{\left(\log\frac{1}{\lambda-1}\right)\Big/(\lambda-1)}\right\}.$$

We see a sharp *threshold* at "n".

A similar picture may be observed for different convex bodies K in the problem of approximating B_2^n by averaging $U_i(K)$, $U_i \in O(n)$. And, changing the parameter of study from "decay of diameter of a generic average K_N" to "distance to the Euclidean ball" of this average, often changes the behaviour we observe from "phase transition" type to "threshold" type behaviour. It would be interesting to demonstrate similar changes from threshold type behaviour to phase transition type in problems of asymptotic combinatorics, which usually deals with thresholds.

5.8. Small ball estimates

In this section we introduce a parameter which is bounded from below by the critical dimension and which gives small ball probability estimates that are useful in many situations.

DEFINITION 5.8.1. Let K be a centrally symmetric convex body in $\mathbb{R}^n$. We define

$$d(K) = \min\left\{-\log\sigma\left(\left\{x \in S^{n-1} : \|x\| \leqslant \frac{M(K)}{2}\right\}\right), n\right\}.$$

The parameter d was introduced by Klartag and Vershynin in [**348**], where it was also observed that $d(K)$ is always larger than $k(X_K)$

PROPOSITION 5.8.2. *Let K be a centrally symmetric convex body in $\mathbb{R}^n$. Then,*

$$d(K) \geqslant ck(X_K),$$

where $c > 0$ is an absolute constant.

Proof. From Theorem 3.2.2 it follows that

$$\sigma(\{x \in S^{n-1} : |\,\|x\| - M(K)| > tM(K)\}) \leqslant \exp(-ct^2k(X_K))$$

for every $t > 0$. Choosing $t = 1/2$ we get the result. □

In [**348**], Klartag and Vershynin prove the following estimate as well.

THEOREM 5.8.3 (Klartag-Vershynin). *For every $0 < \varepsilon < \frac{1}{2}$ we have*

$$\sigma(\{x \in S^{n-1} : \|x\| < \varepsilon M\}) < (c_1\varepsilon)^{c_2 d(K)} \leqslant (c_1\varepsilon)^{c_3 k(X_K)},$$

where $c_1, c_2, c_3 > 0$ are absolute constants.

The proof is based on the B-theorem of Cordero-Erausquin, Fradelizi and Maurey. We will present the proof of Theorem 5.8.4 in Part II, but shall use it already here to derive the small ball estimate above.

THEOREM 5.8.4 (B-Theorem). *Let K be a centrally symmetric convex body in $\mathbb{R}^n$. Then, the function*

$$t \mapsto \gamma_n(e^t K)$$

is log-concave on $\mathbb{R}$.

Note. It is useful to also write the conclusion of Theorem 5.8.4 in the following form: if K is a centrally symmetric convex body in $\mathbb{R}^n$ then

$$\gamma_n(a^\lambda b^{1-\lambda}K) \geqslant \gamma_n(aK)^\lambda \gamma_n(bK)^{1-\lambda} \tag{5.8.1}$$

for all $a, b > 0$ and $\lambda \in (0,1)$. We also need the next lemma.

LEMMA 5.8.5. *If K is a star body in $\mathbb{R}^n$ then*

$$\tfrac{1}{2}\,\sigma\big(S^{n-1} \cap \tfrac{1}{2}K\big) \leqslant \gamma_n(\sqrt{n}K) \leqslant \sigma(S^{n-1} \cap 2K) + e^{-cn}.$$

Proof. Write σ_r for the rotationally invariant probability measure on rS^{n-1}. For the left-hand side inequality observe that, since K is star-shaped,

$$\begin{aligned}\gamma_n(\sqrt{n}K) &\geqslant \gamma_n(2\sqrt{n}B_2^n \cap \sqrt{n}K)\\ &\geqslant \gamma_n(2\sqrt{n}B_2^n)\sigma_{2\sqrt{n}}(2\sqrt{n}S^{n-1} \cap \sqrt{n}K)\\ &= \gamma_n(2\sqrt{n}B_2^n)\sigma\left(S^{n-1} \cap \tfrac{1}{2}K\right).\end{aligned}$$

From Markov's inequality we have

$$\gamma_n(\{x : |x| \geqslant 2\sqrt{n}\}) \leqslant \frac{1}{4n}\int_{\mathbb{R}^n} |x|^2 d\gamma_n(x) = \frac{1}{4},$$

and hence

$$\gamma_n(2\sqrt{n}B_2^n) = 1 - \gamma_n(\{x : |x| \geqslant 2\sqrt{n}\}) \geqslant \frac{3}{4}.$$

This shows that

$$\sigma\left(S^{n-1} \cap \tfrac{1}{2}K\right) \leqslant 2\gamma_n(\sqrt{n}K).$$

Next, observe that

$$\sqrt{n}K \subseteq \left(\tfrac{1}{2}\sqrt{n}B_2^n\right) \cup C\left(\tfrac{1}{2}\sqrt{n}S^{n-1} \cap \sqrt{n}K\right)$$

where, for $A \subseteq \frac{1}{2}\sqrt{n}S^{n-1}$, we write $C(A)$ for the positive cone generated by A. It follows that

$$\gamma_n(\sqrt{n}K) \leqslant \gamma_n\left(\tfrac{1}{2}\sqrt{n}B_2^n\right) + \sigma_{\frac{\sqrt{n}}{2}}\left(\tfrac{1}{2}\sqrt{n}S^{n-1} \cap \sqrt{n}K\right).$$

Now,

$$\sigma_{\frac{\sqrt{n}}{2}}\left(\tfrac{1}{2}\sqrt{n}S^{n-1} \cap \sqrt{n}K\right) = \sigma(S^{n-1} \cap 2K),$$

and a direct computation shows that

$$\gamma_n(\rho\sqrt{n}B_2^n) \leqslant \left(\frac{\rho\sqrt{n}}{\sqrt{2\pi}}\right)^n |B_2^n|,$$

for all $0 < \rho \leqslant 1$. It follows that

$$\gamma_n\left(\tfrac{1}{2}\sqrt{n}B_2^n\right) \leqslant e^{-cn},$$

for some absolute constant $c > 0$. □

Proof of Theorem 5.8.3. Let $m = \mathrm{med}(\|\cdot\|)$ denote the median of $\|\cdot\|$ with respect to the measure σ on the unit sphere S^{n-1}. Markov's inequality shows that

$$\frac{m}{2} \leqslant \int_{\{\theta : \|\theta\| \geqslant m\}} \|\theta\|\, d\sigma(\theta) \leqslant M.$$

Set $L = m\sqrt{n}K$. According to Lemma 5.8.5, we have

$$\gamma_n(2L) \geqslant \frac{1}{2}\sigma(S^{n-1} \cap mK) \geqslant \frac{1}{4}, \tag{5.8.2}$$

by the definition of the Lévy mean. On the other hand, using Lemma 5.8.5 again, we have

$$\begin{aligned}\gamma_n(\tfrac{1}{8}L) &\leqslant \sigma\left(S^{n-1}\cap \frac{m}{4}K\right)+e^{-cn} \\ &= \sigma\left(\left\{\theta\in S^{n-1}:\|\theta\|\leqslant \frac{m}{4}\right\}\right)+e^{-cn} \\ &\leqslant \sigma\left(\left\{\theta\in S^{n-1}:\|\theta\|\leqslant \frac{M}{2}\right\}\right)+e^{-cn} \\ &\leqslant 2e^{-c_1 d(K)},\end{aligned} \tag{5.8.3}$$

where $c_1>0$ is a suitable absolute constant (note that $d(K)\leqslant n$ by definition). We may assume that $0<\varepsilon<e^{-3}$. We apply the B-theorem for the body L, with $a=\varepsilon$, $b=2$ and $\lambda=3(\log\frac{1}{\varepsilon})^{-1}$. This gives

$$\gamma_n(\varepsilon L)^{\frac{3}{\log(1/\varepsilon)}}\gamma_n(2L)^{1-\frac{3}{\log(1/\varepsilon)}}\leqslant \gamma_n(\varepsilon^{\frac{3}{\log(1/\varepsilon)}}2^{1-\frac{1}{\log(1/\varepsilon)}}L)\leqslant \gamma_n(\tfrac{1}{8}L). \tag{5.8.4}$$

Combining (5.8.2), (5.8.3) and (5.8.4) we see that

$$\gamma_n(\varepsilon L)\leqslant 8e^{c_2 d(K)\log\varepsilon}\leqslant 8\varepsilon^{c_2 d(K)}\leqslant (c_3\varepsilon)^{c_2 d(K)}.$$

According to Lemma 5.8.5, and taking into account Lemma 5.2.3, we can transfer this estimate to the spherical measure, thus obtaining

$$\sigma\left(\{\theta\in S^{n-1}:\|\theta\|\leqslant \varepsilon M\}\right)\leqslant (c_4\varepsilon)^{c_5 d(K)}.$$

Adjusting the constants we get the theorem. □

Theorem 5.8.3 yields the following reverse Hölder inequalities.

COROLLARY 5.8.6. *Let K be a centrally symmetric convex body in $\mathbb{R}^n$. Then, for every $1\leqslant q<c_1 d(K)$,*

$$c_2 M\leqslant M_{-q}:=\left(\int_{S^{n-1}}\frac{1}{\|x\|^q}d\sigma(x)\right)^{-1/q}\leqslant c_3 M.$$

In other words, for every $0<q<c_1 d(K)$ we have

$$M_{-q}\simeq M.$$

Proof. The right hand side inequality follows easily from Hölder's inequality. For the left hand side inequality we use integration by parts and the small ball probability estimate. □

In particular, since $d(K)\geqslant ck(X_K)$, we always have the following.

THEOREM 5.8.7. *Let K be a centrally symmetric convex body in $\mathbb{R}^n$. Then, $M_q\simeq M_{-q}$ for all $1\leqslant q\leqslant k(X_K)$.*

Proof. From Theorem 5.7.3 we have $M_q\simeq M$ for all $q\leqslant k(X_K)$. From Theorem 5.8.6 we get $M_{-q}\simeq M$ for all $q\leqslant k(X_K)$. Combining these two facts we get the result. □

5.9. Dependence on ε

Let $\varepsilon \in (0,1)$. In Section 5.3 we saw that every n-dimensional normed space $X = (\mathbb{R}^n, \|\cdot\|)$ has a subspace F of dimension $k \geqslant c\varepsilon^2 n(M/b)^2$ such that $d(F, \ell_2^k) \leqslant 1+\varepsilon$, where $M = \int_{S^{n-1}} \|x\| d\sigma(x)$ and $b = \max\{\|x\| : x \in S^{n-1}\}$.

The next example is due to Figiel and shows that the dependence on ε cannot be improved in this statement. More precisely, we will build an n-dimensional normed space which is 2-isomorphic to ℓ_2^n, so that in particular $M/b \leqslant 1/2$, but which cannot have sections of dimension higher than $c\varepsilon^2 n$ which are $(1+\varepsilon)$-isomorphic to Euclidean because the existence of those would imply that some ℓ_p^n has "too good" (meaning too high-dimensional) Euclidean sections.

THEOREM 5.9.1. *Let $\varepsilon \in (0,1)$. If $n > \varepsilon^{-4}$, then there exists a 1-symmetric normed space X with $d(X, \ell_2^n) \leqslant 2$, which also has the following property: if F is a subspace of X and $d(F, \ell^{\dim(F)}) \leqslant 1+\varepsilon$, then $\dim(F) \leqslant c\varepsilon^2 n$, where $c > 0$ is an absolute constant.*

Proof. Let $\varepsilon \in (0,1)$ and $n > \varepsilon^{-4}$. We define $p \in (2,4)$ by the equation $n^{\frac{1}{p}-\frac{1}{2}} = 2\varepsilon$ and we consider the norm

$$\|x\| = |x| + \|x\|_p$$

on $\mathbb{R}^n$. Then clearly $|x| \leqslant \|x\| \leqslant 2|x|$ and in fact

$$(1 + n^{\frac{1}{p}-\frac{1}{2}})|x| \leqslant \|x\| \leqslant 2|x|.$$

Assume that for some subspace F of $\mathbb{R}^n$ and some $\alpha > 1$ we have

$$\alpha|x| \leqslant \|x\| \leqslant \alpha(1+\varepsilon)|x|$$

for all $x \in F$. Then, by the original bounds on the norm $\|\cdot\|$,

$$1 + \frac{\varepsilon}{2} \leqslant \frac{1 + n^{\frac{1}{p}-\frac{1}{2}}}{1+\varepsilon} \leqslant \alpha \leqslant 2,$$

which implies that, for all $x \in F$,

$$(\alpha - 1)|x| \leqslant \|x\|_p \leqslant ((1+\varepsilon)\alpha - 1)|x| = (\alpha - 1 + \varepsilon\alpha)|x|$$

for all $x \in F$. Since $\alpha \geqslant \frac{1+2\varepsilon}{1+\varepsilon}$, we see that

$$\varepsilon\alpha \leqslant 4(\alpha - 1).$$

Thus, setting $\beta = \alpha - 1$ we get

$$\beta|x| \leqslant \|x\|_p \leqslant 5\beta|x|$$

From the results of Section 5.4 it follows that

$$\dim(F) \leqslant Cn^{2/p} = C\left(n^{\frac{1}{p}-\frac{1}{2}}\right)^2 n = 4C\varepsilon^2 n$$

for some absolute constant $C > 0$. $\square$

However, Schechtman proved that the dependence on ε in the general estimate $k_X(\varepsilon) \geqslant c(\varepsilon)\log n$ can be improved from $c(\varepsilon) \geqslant c\varepsilon^2$ to $c(\varepsilon) \geqslant c\varepsilon\big(\log\frac{2}{\varepsilon}\big)^{-2}$:

THEOREM 5.9.2 (Schechtman). *There exists an absolute constant $c_0 > 0$ with the following property: for every $\varepsilon \in (0,1)$, for every $n \geqslant 1$ and for every n-dimensional normed space X, there exist*

$$k \geqslant c_0\varepsilon\Big(\log\frac{2}{\varepsilon}\Big)^{-2}\log n$$

and a k-dimensional subspace F of X such that $d(F,\ell_2^k) \leqslant 1+\varepsilon$.

The proof involves some techniques which we shall develop in later chapters, hence we defer it to Part II. Let us only outline the general idea. We distinguish two cases:

(1) If $c\varepsilon^2 n(M/b)^2 \geqslant c_0\varepsilon\big(\log\frac{2}{\varepsilon}\big)^{-2}\log n$ then the result follows from our known proof of Dvoretzky theorem.
(2) If this is not the case then we show that X has a subspace which is well isomorphic to ℓ_∞^m for some relatively large m, and then we apply Dvoretzky theorem for ℓ_∞^m to conclude the proof.

5.10. Notes and remarks

Dvoretzky type theorems

Theorem 5.1.2 appears in [**177**] and [**178**]. Dvoretzky notes in [**177**] that it answers a conjecture of Grothendieck from [**275**]: "in view of the obvious implications of results of this nature in the theory of Banach spaces and functional analysis in general, this result, or various weaker variations, have often been conjectured; however, the only explicit statement of this conjecture in print known to the author is in a recent paper by A. Grothendieck". The surveys of Lindenstrauss [**384**] and Milman [**456**] are devoted to the theorem and its applications.

For the numerous connections of Dvoretzky-type theorems with other branches of mathematics (algebraic topology, number theory, harmonic analysis) see the paper of Milman [**451**]. We shall explain some of them here in this section.

Milman's proof

Theorem 5.2.10 is due to Milman [**441**]. Note that his argument, as presented in Section 5.2, does not really use the symmetry of K, but of course the estimate for M/b is no longer valid for non-symmetric K. However, versions of a non-symmetric Dvoretzky-type theorem do exist (see e.g. Larman and Mani [**366**]). We also mention that the original Milman proof works over the complex field and over quaternions, a case which is sometimes of interest and does not follow formally from the theorem for real Banach spaces. Proofs of Dvoretzky theorem were also given later by Szankowski [**577**] and Figiel [**203**].

The critical dimension $k(X)$

The definition of the critical dimension (Definition 5.3.1) was given by Milman and Schechtman in [**466**]. It remains a curious open question whether the same result holds also if we change the quantity $1-\frac{k}{k+n}$ in the definition to a fixed number, say $1/2$. Theorem 5.3.5 and Theorem 5.3.8 are from the paper [**205**] of Figiel, Lindenstrauss and Milman.

Theorem 5.4.1 is also from [**205**]. The proof of the lower estimate in Theorem 5.4.3 was given by Bennett, Dor, Goodman, Johnson and Newman in [**63**]. The upper estimate, as well as Theorem 5.4.2, appear already in the original paper [**441**] of Milman.

Volume ratio and Kashin's theorem

Theorem 5.5.3 was put forward by Szarek and Tomczak-Jaegermann in [**579**] and [**586**], but is implicit in the work of Kashin [**326**], who focused in the case of ℓ_1^n but used a volume computation. A global version, analogous to that in Theorem 5.5.4, was also already considered in some form in the original Kashin paper. One more proof of the volume ratio theorem is given by Klartag in [**333**].

In the case of ℓ_1^{2n} the volume ratio theorem implies the surprising fact, known as Kashin decomposition, that one may find two orthogonal n-dimensional subspaces E and F of $\mathbb{R}^{2n}$ such that $\|x\|_1 \simeq \sqrt{n}|x|$ for all $x \in E$ and all $x \in F$. This result, which was motivated by questions in approximation theory, is counterintuitive at first sight because the Banach-Mazur distance between ℓ_1^{2n} and ℓ_2^{2n} is $\sqrt{2n}$. In fact the assertion holds for almost all (with respect to the Haar measure on $G_{2n,n}$) decompositions $E \oplus E^{\perp}$ of $\mathbb{R}^{2n}$. The articles of Anderson [**18**], Schechtman [**539**], Litvak, Pajor, Rudelson, Tomczak-Jaegermann and Vershynin [**393**] and [**394**] discuss the same phenomenon for different notions of randomness.

The problem of providing a construction of a Kashin decomposition of ℓ_1^{2n} is open. A relaxed question posed by Milman in [**449**] asks for a construction of Euclidean subspaces of ℓ_1^n of dimension proportional to n. One would like to construct a subspace $E \subset \ell_1^n$ with $\dim(E) = k \geqslant cn$ and $d(E, \ell_2^k) \leqslant C$, for all large enough n and some absolute constants $0 < c < 1 < C$. Rudin [**530**], in a work related to the Λ_p-problem, constructed Euclidean sections of ℓ_1^n of dimension $k \simeq \sqrt{n}$ using finite fields. Nemirovski [**484**] has proved that if S is the $5n$-dimensional simplex then there exists an explicit affine image T of an explicit n-dimensional section of S that satisfies $B_2^n \subseteq T \subseteq cB_2^n$ Ben-Tal and Nemirovski (see [**484**]) have proved for some absolute constant $c > 0$. Moreover, given $\varepsilon \in (0, 1)$, one can replace c by $1 + \varepsilon$ in the statement above if one starts with a simplex S of dimension $N \simeq n \log(2/\varepsilon)$. We do not know of a similar result for the cross-polytope (which is considered as a "centrally symmetric simplex").

Almost Euclidean subspaces of ℓ_1^n are of great interest to the computer science community, since they have found numerous applications: to high-dimensional nearest neighbor search (see e.g. Indyk [**317**]), to error-correcting codes over reals and compressive sensing (see e.g. Kashin and Temlyakov [**328**], Guruswami, Lee and Razborov [**290**] and Guruswami, Lee and Wigderson [**291**]) to vector quantization (see e.g. Lyubarskii and Vershynin [**400**]) and to other problems. There has been a renewed interest in the problem, see the articles of Artstein-Avidan and Milman [**25**], Szarek [**584**], Indyk [**318**], Lovett and Sodin [**399**], Guruswami, Lee and Wigderson [**291**], Guruswami, Lee and Razborov [**290**], Indyk and Szarek [**319**] and the references therein. Several explicit constructions were provided, using expanders, extractors and error-correcting codes. Some of them are trying to maximize the dimension and minimize the distortion (i.e. the distance to Euclidean space), others use limited randomness and aim at the optimal dimension and distortion that is achieved by the probabilistic method described in Section 5.5.

So far, using explicit constructions one can achieve either arbitrarily low distortion or arbitrarily high subspace dimension, but it is not known if one can have both. Allowing randomness that is sub-linear in n one can get subspace dimension close to the full dimension and constant distortion. One can also have distortion very close to 1, but at the cost of using super-linear randomness.

Global form of Dvoretzky type theorems – isomorphic phase transition

Bourgain, Lindenstrauss and Milman [**106**] proved Theorem 5.6.1, that is, the global form of the Dvoretzky-Milman theorem. See also Schmuckenschläger [**548**] for the dependence on ε. A proof of the Marcus-Pisier inequality is given in Chapter 9.

Theorem 5.6.2 is due to Milman and Schechtman, from the paper [**466**], as well as Lemma 5.6.3. For the precise statement in Remark 5.7.2 see Milman [**453**]. The proof of Theorem 5.7.5 can be found in the paper of Litvak, Milman and Schechtman [**390**]. For phase transition see Milman's [**457**] and [**458**].

We mention that a scheme for reducing the number of random operators involved, grouping the operators in some special way, was given in Artstein-Milman [**25**].

Small ball estimates

The parameter $d(K)$ was introduced by Klartag and Vershynin in [**348**]. The main question in the article concerned the diameter of proportional random sections of high-dimensional convex bodies. The results in [**348**] reveal a difference between the lower and the upper inclusions in Dvoretzky theorem: the expected upper bound for the diameter of random sections of a convex body sometimes continues to hold in much larger dimensions than the critical dimension of the body. An example is given by the cube; while $k(Q_n) \simeq \log n$, one has that $d(Q_n)$ is polynomial in n.

The B-theorem was proved in [**154**] by Cordero-Erausquin, Fradelizi and Maurey. It provides an affirmative answer to a question of Banaszczyk (see Latała [**368**]) and we shall elaborate on it in Part II. In the same article, the question is generalized as follows: one says that a probability measure μ and a convex body K in $\mathbb{R}^n$ satisfy the *B*-theorem if the function $t \mapsto \mu(e^t K)$ is log-concave. This problem is studied in $\mathbb{C}^n$ and it is shown by complex interpolation that the *B*-theorem holds true for a more general class of sets and measures. It is also shown that the concavity of $t \mapsto \mu(e^t K)$ is true for any unconditional log-concave probability measure μ provided that the convex body K is also unconditional.

Lemma 5.8.5 is from [**348**], Theorem 5.8.7 combines the result of Klartag and Vershynin with the one of Litvak, Milman and Schechtman, and plays a key role in Paouris' work on the negative moments $I_{-q}(\mu)$ of an isotropic log-concave probability measure μ that will be discussed in detail in Chapter 10. The result of Klartag and Vershynin heavily depends on the paper of Latała and Oleszkiewicz [**369**].

Essentially Euclidean convex bodies

As is clearly seen in this chapter, the asymptotic study of convex bodies revolves around properties which are of Euclidean type. For example, it is an immediate corollary from Theorem 5.2.10 that if a space has distance $R > 0$ to Euclidean space, then it has a $k = \lambda(\varepsilon, R)n$ dimensional subspace which is $(1+\varepsilon)$-isometric to Euclidean space.

It is natural to ask whether the theory's close link to Euclidean space is an artefact of the methods and our current knowledge, or that this family of mathematical results may only be true for Euclidean type spaces. It was observed by Bourgain and Tzafriri [**114**] that for ℓ_p^n, $p > 2$, and for subspaces of ℓ_∞^n a result of this type cannot hold.

It turns out that in fact the mathematics itself here is (essentially) Euclidean. Let us give one instance of such a result, which was proved in the paper [**391**] by Litvak, Milman and Tomczak-Jaegermann. The *only* spaces for which a theorem of the form mentioned above may hold are spaces whose unit balls are "$(\lambda, d(\varepsilon))$-essentially Euclidean", a property that is formally defined as follows: Given a function $d(\varepsilon) \geqslant 1$ and $0 < \lambda < 1$, a symmetric convex body K is called (λ, d)-essentially-Euclidean if there exists a $[\lambda n]$-dimensional subspace F such that for every $\varepsilon \in (0,1)$ there exists a further $[\varepsilon\lambda n]$-dimensional subspace $F \subset E$ such that $K \cap F$ is $d(\varepsilon)$-isomorphic to a Euclidean ball.

It is shown in [**391**] that if a given convex body K satisfies, for some given $\varepsilon_0 > 0, R \geqslant 2(1+\varepsilon_0)$ and $\lambda > 0$, that for any T which satisfies $d_G(K,T) \leqslant R$, there exists some $[\lambda n]$ dimensional subspace E with $d_G(K \cap E, T \cap E) \leqslant 1+\varepsilon_0$, then K must be (λ, d)-essentially Euclidean for some function $d(\varepsilon)$ which depends in an explicit way on R, ε_0 and λ.

This result, and similar results of the same nature given in [**391**], show that many of the constructions of the theory of this book cannot be extended beyond essentially Euclidean spaces.

Dependence on ε

Theorem 5.9.2 is due to Schechtman [**540**]. To find an ℓ_∞^m subspace, under suitable conditions, one uses the result of Alon and Milman [**10**]. A full proof will be presented in Part II. Theorem 5.9.1 has been observed by Figiel and can be found in [**541**].

It was observed by Milman in [**451**] that a problem of Knaster would imply a proof of Dvoretzky theorem with a good dependence on ε. Knaster's problem [**349**] is the following question: Is it true that for any continuous function $f : S^{n-1} \to \mathbb{R}^m$ and any choice of points $x_1, \dots, x_k \in S^{n-1}$, where $k = n - m + 1$, there exists $U \in SO(n)$ such that $f(Ux_1) = \cdots = f(Ux_k)$? The question was known to have a negative answer for $m > 2$ and in the case $m = 2$ for some values of n (see Makeev [**411**] and Chen [**147**] for more recent references). In the case $m = 1$ some partial positive results were known: besides the case $n = 2$ that can be easily checked, the answer is affirmative for $n = 3$ see Floyd [**214**] or when $x_1, \dots, x_n$ form an orthonormal basis (see Kakutani [**324**] and Yamabe-Yojobo [**609**]). Milman explains in [**451**] that assuming a positive answer to Knaster's problem for $m = 1$ and any n, we can get the estimate

$$k(X) \geqslant c(n, \varepsilon) \log n / \log(2/\varepsilon)$$

for every n-dimensional normed space X, where $c(n, \varepsilon) \to 1$ as $n \to \infty$ and $\varepsilon \to 0$. In fact the same is true (with a slightly different estimate) if the conjecture is correct for much smaller samples of points, $\{x_i\}_{i=1}^{k_0}$ where $k_0 \simeq n^\alpha$ for a fixed $\alpha > 0$.

Kashin and Szarek [**327**] showed that the answer to Knaster's problem is negative when $m = 1$, at least when n is sufficiently large. Their negative result applies also to k somewhat smaller than n, but does not exclude the possibility of a Knaster type result being true for $k \simeq n^\alpha$ for some $\alpha > 0$. In order to state their result it is useful to introduce the following notation. If $d, n \in \mathbb{N}$ with $d \leqslant n$, and if A is a subset of $\mathbb{R}^d$ and $U : \mathbb{R}^d \to \mathbb{R}^n$ is an isometry, then we say that U is a Knaster embedding of A into ℓ_∞^n if $\|U(x)\|_\infty$ is constant for $x \in A$.

THEOREM 5.10.1 (Kashin-Szarek). *For any k in $\mathbb{N}$ there exists $A = \{x_1, \dots, x_k\} \subset S^{k-1}$ such that if $n \leqslant k\lfloor \log(k/2) \rfloor / 32$ then there is no Knaster embedding U of A into ℓ_∞^n.*

This implies that $\|U(x_i)\|_\infty$ cannot be constant if $k = n$ and n is large enough. Thus, the answer to Knaster's problem is negative for large enough n, even if we restrict our attention to convex functions. A question posed in [**327**] leaves open the possibility to use ideas related to Knaster's problem for a proof of Dvoretzky theorem. It is conceivable that if $n \geqslant kd$ then for any $f : S^{n-1} \to \mathbb{R}$ and any $x_1, \dots, x_k \in S^{d-1}$ there exists an isometry $U : \mathbb{R}^d \to \mathbb{R}^n$ such that $f(Ux_1) = \cdots = f(Ux_k)$. This is true if $f(x) = \|x\|_\infty$.

The case $k = 2$ is the only one in which an optimal dependence in ε is known.

THEOREM 5.10.2. *For any convex body $K \subset \mathbb{R}^n$ there exists a 2-dimensional subspace E such that*

$$d_{BM}(K \cap E, B_2^2) \leqslant 1 + \frac{2\pi^2}{3(n+1)^2}$$

The proof, which was noted by M. Gromov, is simple; it can be found in [**451**] and makes use of a non-trivial generalization of the Borsuk-Ulam theorem.

Algebraic form of Dvoretzky-type theorem

The next theorem of Dol'nikov and Karasev [**175**], which answers in the affirmative a conjecture of Gromov and Milman (see [**451**] and [**456**]) resembles Theorem 5.1.2 of Dvoretzky on almost spherical sections of convex bodies. Here one considers polynomials instead of convex bodies, and unlike Theorem 5.1.3 and its relatives, it gives exact "roundness" rather than isomorphic one.

THEOREM 5.10.3 (Dol'nikov and Karasev). *Let $d \in \mathbb{N}$ be even and let $k \in \mathbb{N}$. There exists $n(d, k) \in \mathbb{N}$ such that for any homogeneous polynomial f of degree d on $\mathbb{R}^n$, where $n > n(d, k)$, there exists a subspace $V \in G_{n,k}$ such that $f|_V$ is proportional to the $(d/2)^{th}$ power of the standard quadratic form*

$$Q(x_1, \dots, x_n) = x_1^2 + x_2^2 + \cdots + x_n^2.$$

One should divide the statement here into two cases: one is when the degree d is odd, and then the polynomial is actually identically 0 on some k dimensional subspace; in this case the existence of such a subspace was proved very long ago by Birch [**74**]. In the case of even degree d it was conjectured in [**451**] that the dimension of the space $n(d,k)$ may be not too large, of the order k^d, but this remains open. The above theorem uses a topological Borsuk-Ulam type theorem; as a consequence, it does not provide any reasonable estimate on n.

The algebraic form of Dvoretzky theorem has its origin in the observation that ℓ_p^n for even $p = 2k$ allows isometric embeddings of ℓ_2^d for $d \simeq n^{1/p}$. In [**451**] the connection with Waring-type theorems was emphasized. It was realized by Hurwitz that the presentation of a power of a quadratic form as a sum of powers of linear forms was essential for the problem of Waring. He has proved it for $k = 2, 4$ and the general case was proved by Hilbert [**303**] and simplified by Schmidt [**546**]. Let us mention a special case of this curious embedding, of ℓ_2^4 in ℓ_4^{12}, which follows from the following identity of Lucas [**401**] and Hurwitz [**307**]:

$$6\left(\sum_{i=1}^{4} x_i^2\right)^2 = \sum_{1\leqslant i<j\leqslant 4} (x_i + x_j)^4 + \sum_{1\leqslant i<j\leqslant 4} (x_i - x_j)^4.$$

In the paper of H. König [**359**] several explicit constructions of embeddings are given, for instance when $n = 4^m$ then ℓ_2^n embeds isometrically into $\ell_2^{n(n+2)/2}$ over $\mathbb{R}$, while over $\mathbb{C}$ when n is an odd prime power, ℓ_2^n embeds into $\ell_4^{n^2+n}$ and if $q = n-1$ is a prime power, ℓ_2^n embeds isometrically into $\ell_4^{n^2+1}$. The latter implies also an isometric embedding of ℓ_2^m into $\ell_2^{(3/4)m^2+3}$ over $\mathbb{R}$, when $m = 2q-2$ for q a prime power, which in turn yields cubature formulae of degree 5 on S^{m-1} with $(3/2)m^2+6$ points, see [**359**] for details.

CHAPTER 6

The ℓ-position and the Rademacher projection

In this chapter we discuss upper bounds for the parameter $M(K)M^*(K)$, or equivalently, the product of the mean width of K and the mean width of its polar. This parameter will play an important role in the next chapter. To get a decent bound one needs to minimize this parameter over all positions of the convex body. We shall show that this quantity can be bounded from above by a parameter of the space $(X, \|\cdot\|_K)$ which is called its K-convexity constant, and which in turn can be bounded from above, for X of dimension n, by $c \log[d_{BM}(X, \ell_2^n)+1] \leqslant c' \log n$ for universal c, c'. This estimate for the K-convexity constant is due to Pisier and as we will see it is one of the fundamental facts in the asymptotic theory.

Let us briefly describe the circle of ideas that will form this chapter. We start with an n-dimensional normed space X and consider any norm α on $L(\ell_2^n, X)$, the space of linear operators $u : \ell_2^n \to X$. The dual norm is defined on $L(X, \ell_2^n)$ by

$$\alpha^*(v) = \sup\{\mathrm{tr}(vu) : \alpha(u) \leq 1\}.$$

As we will see in Section 6.4, for any norm α on $L(\ell_2^n, X)$, there exists $u : \ell_2^n \to X$ such that $\alpha(u) = 1$ and $\alpha^*(u^{-1}) = n$. In Section 6.4 we also introduce the ℓ-norm on $L(\ell_2^n, X)$. This was defined by Figiel and Tomczak-Jaegermann as follows:

$$\ell(u) = \left(\mathbb{E}\Big\|\sum_{i=1}^n g_i u(e_i)\Big\|^2\right)^{1/2},$$

where $\{g_1, \ldots, g_n\}$ are independent standard Gaussian random variables on some probability space, and $\{e_1, \ldots, e_n\}$ is the standard orthonormal basis of $\mathbb{R}^n$. A straightforward computation gives

$$\ell(u) \simeq \sqrt{n}\, w(u^*(K^\circ)),$$

where K is the unit ball of X. This formula connects the ℓ-norm to the mean width.

It is more instructive to replace the Gaussians by the Rademacher functions $r_i : E_2^n \to \{-1, 1\}$ defined by $r_i(\epsilon) = \epsilon_i$, where $E_2^n = \{-1, 1\}^n$ is viewed as a probability space with the uniform measure μ_n. An inequality of Maurey and Pisier (see the notes and remarks section) shows that

$$\ell(u) \simeq \left(\int_{E_2^n} \|\sum_{i=1}^n r_i(\epsilon) u(e_i)\|^2 d\mu_n(\epsilon)\right)^{1/2}$$

up to a $\sqrt{\log n}$-term. Consider the Walsh functions $w_A(\epsilon) = \prod_{i\in A} r_i(\epsilon)$, where $A \subseteq \{1, \ldots, n\}$. It is not hard to see that every function $f : E_2^n \to X$ is uniquely represented in the form

$$f(\epsilon) = \sum_A w_A(\epsilon) x_A,$$

for some $x_A \in X$. The space of all functions $f : E_2^n \to X$ becomes a Banach space with the norm

$$\|f\|_{L_2(X)} = \left(\int_{E_2^n} \|f(\epsilon)\|^2 d\mu_n(\epsilon) \right)^{1/2}$$

The Rademacher projection $Rad_n : L_2(X) \to L_2(X)$ is the operator sending $f = \sum w_A x_A$ to the function $Rad_n(f) := \sum_{i=1}^n r_i x_{\{i\}}$. Denote by $K_r(X)$ the norm of this projection. Pisier's inequality, which is presented in Section 6.3, asserts that

$$K_r(X) \leq c \log[d_{BM}(X, \ell_2^n) + 1],$$

where $c > 0$ is an absolute constant. A result of Figiel and Tomczak-Jaegermann (see Section 6.4) shows the relevance of this estimate to the study of the ℓ-norm: using trace duality they showed that there exists $u : \ell_2^n \to X$ such that

$$\ell(u)\ell((u^{-1})^*) \leq nK_r(X).$$

Combining the above with John's theorem we see in Section 6.5 that there exists a linear image $\tilde{K}$ of the unit ball K of X so that

$$M(\tilde{K})M^*(\tilde{K}) \leqslant c \log n,$$

where $c > 0$ is an absolute constant.

The K-convexity constant of X is defined similarly as the norm of some projection operator on a space of X valued functions defined on $(\mathbb{R}^n, \gamma_n)$, onto a subspace spanned by linear combinations of vectors in X with coefficients being polynomials of degree one in each variable. To this end we shall in the first section describe these polynomials and their properties, the function spaces involved and their corresponding norms, and the relevant projection operators. In Section 6.2 we describe the upper bound for the norm of this projection.

Finally, in Section 6.6, we compare the K-convexity constant with $K_r(X)$, show that they are equivalent, and discuss the optimality of the upper bound for the norms of the two projection operators.

It is worthwhile to mention that the study of K-convexity, Rademacher projection, and of the spaces $L_2(\Omega, X)$ and $L_2(\Omega, X^*)$ gave rise in the 1970's and 80's to the extremely powerful and beautiful type-cotype theory mostly developed by Maurey and Pisier. We shall outline some of these developments in Part II of this book.

6.1. Hermite polynomials

We shall work in $\mathbb{R}^n$, endowed with the Euclidean distance and the n-dimensional Gaussian measure γ_n. The properties of this distribution were discussed in Chapter 3, Section 3.1.1. In particular its invariance under orthogonal transformations was mentioned, which in particular means that for every measurable function $f : \mathbb{R} \to \mathbb{R}^+$ and for every $t = (t_1, \ldots, t_n) \in \mathbb{R}^n$,

$$\int_{\mathbb{R}^n} f(\langle t, x \rangle) d\gamma_n(x) = \int_{\mathbb{R}} f(\|t\|_2 x_1) d\gamma_1(x_1).$$

More generally, for all integers $n, m \geqslant 1$, for every $t = (t_1, \dots, t_m) \in \mathbb{R}^m$ and for every function $f : \mathbb{R}^n \to \mathbb{R}^+$,

$$\int_{(\mathbb{R}^n)^m} f\Big(\sum_{i=1}^m t_i x^i\Big) d\gamma_n(x^1) \cdots d\gamma_n(x^m) = \int_{\mathbb{R}^n} f\left(\Big(\sum_{i=1}^m t_i^2\Big)^{1/2} x^1\right) d\gamma_n(x^1). \tag{6.1.1}$$

We shall use this fact soon. We start by defining the Hermite polynomials in one real variable.

Definition 6.1.1 (Hermite polynomials, $n = 1$). The sequence $(h_k)_{k \geqslant 0}$ of the Hermite polynomials in one real variable is generated by the formula

$$\exp\left(tx - \frac{t^2}{2}\right) = \sum_{k=0}^\infty \frac{t^k}{k!} h_k(x) \tag{6.1.2}$$

holding for all $t \in \mathbb{R}$.

As a consequence we get several equivalent formulae for the k-th Hermite polynomial. It is given by

$$h_k(x) = \frac{d^k}{dt^k}\Big|_{t=0} \exp\left(tx - \frac{t^2}{2}\right).$$

Equivalently

$$h_k(x) := (-1)^k e^{x^2/2} \frac{d^k}{dx^k}\big(e^{-x^2/2}\big). \tag{6.1.3}$$

Next we define the sequence of the Hermite polynomials in n real variables.

Definition 6.1.2 (Hermite polynomials, $n \geqslant 1$). The sequence (H_α) of the Hermite polynomials in n real variables, with $\alpha = (\alpha_1, \dots, \alpha_n) \in \mathbb{N}^n$ is given by

$$H_\alpha(x_1, \dots, x_n) := \prod_{i=1}^n h_{\alpha_i}(x_i).$$

Under the notation

$$\alpha! := \prod_{i=1}^n \alpha_i! \quad \text{and} \quad t^\alpha := \prod_{i=1}^n t_i^{\alpha_i}$$

for every $t = (t_1, \dots, t_n) \in \mathbb{R}^n$, we easily arrive at an analogous formula to that in (6.1.2) generating the Hermite polynomials on $\mathbb{R}^n$:

$$\exp\left(\langle t, x\rangle - \frac{1}{2}\sum_{i=1}^n t_i^2\right) = \sum_{\alpha \in \mathbb{N}^n} \frac{t^\alpha}{\alpha!} H_\alpha(x), \qquad x = (x_1, \dots, x_n). \tag{6.1.4}$$

In the next proposition we state some simple facts about Hermite polynomials which can be verified by induction, differentiation, and using the above equivalent definitions.

Proposition 6.1.3. *The sequence $(h_k)_{k \geqslant 0}$ of the Hermite polynomials in one real variable satisfies the recursive relation $h_0 \equiv 1$,*

$$h_k(x) = \Big(x - \frac{d}{dx}\Big)(h_{k-1}(x)) = \Big(x - \frac{d}{dx}\Big)^k (h_0(x)), \ k = 1, 2, \dots \tag{6.1.5}$$

For every $k \geqslant 0$, the k-th Hermite polynomial is a polynomial of degree k with leading coefficient 1, in particular it is integrable in $L_p(\gamma_1)$ for every $p > 0$ and

$$\|h_k\|_{L_2(\gamma_1)} = \sqrt{k!}, \quad k \geqslant 0. \tag{6.1.6}$$

In addition, the sequence $(h_k)_{k\geqslant 0}$ is orthogonal in $L_2(\gamma_1)$ and spans the dense subspace of all polynomials.

COROLLARY 6.1.4. *The sequence of the Hermite polynomials in n variables is orthogonal in $L_2(\mathbb{R}^n, \gamma_n)$ and spans the dense subspace of all polynomial functions on $\mathbb{R}^n$. Moreover, for every $\alpha = (\alpha_1, \ldots, \alpha_n) \in \mathbb{N}^n$,*

$$\|H_\alpha\|^2_{L_2(\mathbb{R}^n,\gamma_n)} = \alpha! = \prod_{i=1}^n \alpha_i!. \tag{6.1.7}$$

DEFINITION 6.1.5. Let $k \geqslant 0$. We define the orthogonal projection $Q_k : L_2(\mathbb{R}^n, \gamma_n) \to H_k$ where H_k is the (closed) span in $L_2(\mathbb{R}^n, \gamma_n)$ of the Hermite polynomials of degree exactly k, namely of

$$\Big\{H_\alpha \mid \alpha \in \mathbb{N}^n, \sum_{i=1}^n \alpha_i = k\Big\}.$$

In particular

$$Q_0(f) = \int_{\mathbb{R}^n} f d\gamma_n \quad \text{for every } f \in L_2(\mathbb{R}^n, \gamma_n).$$

PROPOSITION 6.1.6. *The sequence of orthogonal projections $\{Q_k \mid k \geqslant 0\}$ on $L_2(\mathbb{R}^n, \gamma_n)$ satisfies*

$$\mathrm{Id} = \sum_{k\geqslant 0} Q_k \tag{6.1.8}$$

and for every $\varepsilon \in [-1, 1]$, the operator

$$T_\varepsilon := \sum_{k\geqslant 0} \varepsilon^k Q_k \tag{6.1.9}$$

is a positive contraction on $L_2(\mathbb{R}^n, \gamma_n)$.

Proof. To prove (6.1.8) and (6.1.9), it suffices to work with polynomials in the variables $x_1, x_2, \ldots, x_n$, or with exponential functions of the form $x \mapsto e^{\langle t,x\rangle}$, where $t \in \mathbb{R}^n$, since the linear span of either of these two families of functions is dense in $L_2(\mathbb{R}^n, \gamma_n)$.

Let $-1 \leqslant \varepsilon \leqslant 1$. We introduce the operator $T(\varepsilon)$ defined on the dense subspace S of all polynomials by the formula

$$T(\varepsilon)f(x) := \int_{\mathbb{R}^n} f\big(\varepsilon x + (1-\varepsilon^2)^{1/2} y\big) d\gamma_n(y) \text{ for every polynomial } f. \tag{6.1.10}$$

Clearly $T(\varepsilon)$ is linear and maps polynomials into polynomials. Moreover, by Cauchy-Schwarz inequality and then (6.1.1) we have

$$\|T(\varepsilon)f\|_{L_2} \leqslant \left(\int_{\mathbb{R}^n}\int_{\mathbb{R}^n} |f\big(\varepsilon x + (1-\varepsilon^2)^{1/2} y\big)|^2 d\gamma_n(y) d\gamma_n(x)\right)^{1/2} = \|f\|_{L_2}.$$

This means that for an arbitrary $f \in L_2(\mathbb{R}^n, \gamma_n)$, (6.1.10) makes sense for almost every $x \in \mathbb{R}^n$. Therefore, $T(\varepsilon)$ extends to a linear contraction on L_2 and it is positive in the sense that $T(\varepsilon)f \geqslant 0$ for all $f \geqslant 0$. In the case $\varepsilon = 1$, it is the identity operator.

To complete the proof, we have to show that

$$T(\varepsilon) \equiv T_\varepsilon = \sum_{k\geqslant 0} \varepsilon^k Q_k.$$

To do so, it suffices to check that the two operators coincide on the (dense) family of functions

$$\phi_t(x) := \exp\Big(\langle t, x\rangle - \frac{1}{2}\sum_{i\geqslant 1} t_i^2\Big).$$

It is easy to check that $T(\varepsilon)\phi_t = \phi_{\varepsilon t}$. On the other hand, setting $|\alpha| := \sum_{i=1}^n \alpha_i$ for every $\alpha \in A$, and taking the generating formula (6.1.4) into account, we see that

$$\left(\sum_{k\geqslant 0} \varepsilon^k Q_k\right)\phi_t = \sum_{\alpha\in\mathbb{N}^n} \varepsilon^{|\alpha|}\frac{t^\alpha}{\alpha!}H_\alpha,$$

hence by (6.1.4) again,

$$\left(\sum_{k\geqslant 0} \varepsilon^k Q_k\right)\phi_t = \phi_{\varepsilon t}.$$

This completes the proof. □

We next introduce a normed space of functions with values in X.

DEFINITION 6.1.7. Let $L_2(\Omega, \mathbb{P})$ be the space of L_2-integrable real functions defined on some probability space $(\Omega, \mathbb{P})$. For every Banach space X, with norm $\|\cdot\|_X$, we can define the space $L_2(\Omega, \mathbb{P}) \otimes X$ whose elements are functions of the form

$$\omega \in \Omega \mapsto \left(\sum_{i\in I} f_i \otimes x_i\right)(\omega) := \sum_{i\in I} f_i(\omega)\cdot x_i, \tag{6.1.11}$$

where the index set I is finite, the x_i are vectors in X and the f_i are functions in $L_2(\Omega, \mathbb{P})$. We equip this linear space with the norm

$$\|\sum_{i\in I} f_i \otimes x_i\|_{L_2(\Omega,\mathbb{P})\otimes X} := \left(\int_\Omega \|\sum_{i\in I} f_i(\omega)\cdot x_i\|_X^2 d\mathbb{P}(\omega)\right)^{1/2}, \tag{6.1.12}$$

and we write $L_2(\Omega, \mathbb{P}; X)$ for the completion of $L_2(\Omega, \mathbb{P}) \otimes X$ with respect to this norm. Below we shall write, for convenience, $L_2(\Omega, \mathbb{P}; X) = L_2(X)$.

The following lemma asserts that $L_2(\Omega, \mathbb{P}; X^*)$ acts on $L_2(\Omega, \mathbb{P}; X)$ as a dual space.

LEMMA 6.1.8. *Let g be any function in $L_2(\Omega, \mathbb{P}; X^*)$. Then g defines a linear functional on $L_2(\Omega, \mathbb{P}; X)$ by*

$$f \in L_2(\Omega, \mathbb{P}; X) \mapsto \langle g, f\rangle := \int_\Omega g(\omega)\big(f(\omega)\big)d\mathbb{P}(\omega)$$

In addition, by the Cauchy-Schwarz inequality,

$$\Big|\int_\Omega g(\omega)\big(f(\omega)\big)d\mathbb{P}(\omega)\Big| \leqslant \|f\|_{L_2(X)}\|g\|_{L_2(X^*)}, \tag{6.1.13}$$

which means the linear functional $\tilde{g}$ *defined by* g *satisfies* $\|\tilde{g}\|_{(L_2(X))^*} \leqslant \|g\|_{L_2(X^*)}$. *The reverse inequality also holds, since*

$$\|\tilde{g}\|_{L_2(X^*)} = \sup\left\{\int_\Omega g(\omega)\big(f(\omega)\big)d\mathbb{P}(\omega) \mid \|f\|_{L_2(X)} \leqslant 1\right\}. \tag{6.1.14}$$

Similarly, for every $f \in L_2(\Omega, \mathbb{P}; X)$,

$$\|f\|_{L_2(X)} = \sup\left\{\int_\Omega h(\omega)\big(f(\omega)\big)d\mathbb{P}(\omega) \mid \|h\|_{L_2(X^*)} \leqslant 1\right\}. \tag{6.1.15}$$

Proof. To see why (6.1.14) holds, suppose first that g is a step function, namely that there exist finite sequences $\{A_i\}$, $\{a_i\}$ and $\{x_i^*\}$ of measurable subsets of Ω, of real scalars and of vectors in X^*, respectively, such that $g = \sum_i a_i \mathbf{1}_{A_i} \otimes x_i^*$. Then, we can also write g as

$$g = \sum_{j=1}^s \mathbf{1}_{B_j} \otimes y_j^*$$

where $\{B_j\}_{j=1}^s$ is a finite sequence of pairwise disjoint measurable subsets of Ω (which have non-zero measure) and $y_j^* \in X^*$. Obviously,

$$\begin{aligned}
\|g\|_{L_2(X^*)} &= \left(\sum_{j=1}^s \mathbb{P}(B_j)\|y_j^*\|_{X^*}^2\right)^{1/2} = \left|\left(\sqrt{\mathbb{P}(B_j)}\|y_j^*\|_{X^*}\right)_{j=1}^s\right| \\
&= \left\langle \left(\sqrt{\mathbb{P}(B_j)}\|y_j^*\|_{X^*}\right)_{j=1}^s, (t_j)_{j=1}^s \right\rangle = \sum_{j=1}^s t_j \sqrt{\mathbb{P}(B_j)}\|y_j^*\|_{X^*}
\end{aligned}$$

for some vector $(t_j)_{j=1}^s \in \mathbb{R}^s$ with $|(t_1, \dots, t_s)| = 1$. But then, for every $j \leqslant s$ and for every $\varepsilon > 0$, we can find $x_j \in X$ with $\|x_j\|_X = 1$ such that $y_j^*(x_j) \geqslant \|y_j^*\|_{X^*} - \varepsilon$, and thus, if we set

$$f_\varepsilon := \sum_{j=1}^s \frac{t_j}{\sqrt{\mathbb{P}(B_j)}} \mathbf{1}_{B_j} \otimes x_j \in L_2(\Omega, \mathbb{P}) \otimes X,$$

we will have that $\|f_\varepsilon\|_{L_2(X)} = |(t_1, \dots, t_s)| = 1$ and

$$\begin{aligned}
\int_\Omega g(\omega)\big(f_\varepsilon(\omega)\big)d\mathbb{P}(\omega) &= \sum_{j=1}^s t_j \sqrt{\mathbb{P}(B_j)} y_j^*(x_j) \\
&\geqslant \sum_{j=1}^s t_j \sqrt{\mathbb{P}(B_j)}\|y_j\|_{X^*} - \varepsilon \sum_{j=1}^s t_j \sqrt{\mathbb{P}(B_j)} \\
&\geqslant \|g\|_{L_2(X^*)} - \varepsilon,
\end{aligned}$$

as required. Now, using the density of the step functions in $L_2(\Omega, \mathbb{P}; X^*)$ as well as (6.1.13), we can deduce (6.1.14) for every function $g \in L_2(\Omega, \mathbb{P}; X^*)$. The second assertion, namely (6.1.15), can be verified in the same way and we omit the proof. $\square$

6.2. Pisier's inequality

A natural way to define a linear operator from $L_2(\Omega, \mathbb{P}) \otimes X$ to itself combining an operator T on $L_2(\Omega, \mathbb{P})$ and an operator V on X is to set

$$(T \otimes V)\left(\sum_{i\in I} f_i \otimes x_i\right) = \sum_{i\in I} T(f_i) \otimes V(x_i). \tag{6.2.1}$$

Our aim is to give upper bounds for the norm of the Gaussian projection $Q_1 \otimes \mathrm{Id}_X : L_2(\mathbb{R}^n, \gamma_n) \otimes X \to L_2(\mathbb{R}^n, \gamma_n) \otimes X$, where X is an n-dimensional normed space. It is not hard to check that $Q_1 \otimes \mathrm{Id}_X$ is always bounded with norm $\leqslant n$, no matter what the space X is. A second upper bound may be given in terms of the Banach-Mazur distance from X to Euclidean space.

PROPOSITION 6.2.1. *Let $T : L_2(\mathbb{R}^n, \gamma_n) \to L_2(\mathbb{R}^n, \gamma_n)$ be a bounded operator and let X be an n-dimensional normed space. We consider the operator $T \otimes \mathrm{Id}_X$ defined on $L_2(\mathbb{R}^n, \gamma_n) \otimes X$. If T is positive then $T \otimes \mathrm{Id}_X$ is extended to a bounded operator on $L_2(X)$, and $\|T \otimes \mathrm{Id}_X\| = \|T\|$. In general,*

$$\|T \otimes \mathrm{Id}_X\| \leqslant d(X, \ell_2^n)\|T\|.$$

Proof. Assume first that $X = \ell_2^n$, and let $\{e_i\}_{i=1}^n$ be an orthonormal basis of X. Given a bounded operator $T : L_2(\mathbb{R}^n, \gamma_n) \to L_2(\mathbb{R}^n, \gamma_n)$ and any $F = \sum_{j\in J} f_j \otimes x_j \in L_2(X)$ we have

$$\langle e_i, (T \otimes \mathrm{Id}_X)(F)\rangle = T(g_i),$$

where $g_i = \langle e_i, F\rangle$. It follows that

$$\|F(\omega)\|_X = \Big(\sum_{i=1}^n |g_i(\omega)|^2\Big)^{1/2}$$

and

$$\|(T \otimes \mathrm{Id}_X)(F)(\omega)\|_X = \Big(\sum_{i=1}^n |T(g_i)(\omega)|^2\Big)^{1/2}.$$

Then,

$$\begin{aligned}\|(T \otimes \mathrm{Id}_X)(F)\|_{L_2(X)} &= \Big(\sum_{i=1}^n \|T(g_i)\|^2_{L_2(\mathbb{R}^n,\gamma_n)}\Big)^{1/2}\\ &\leqslant \|T\|\Big(\sum_{i=1}^n \|g_i\|^2_{L_2(\mathbb{R}^n,\gamma_n)}\Big)^{1/2} = \|T\|\,\|F\|_{L_2(X)}.\end{aligned}$$

This shows that $\|T \otimes \mathrm{Id}_X\| \leqslant \|T\|$.

For the general case, we can pass through ℓ_2^n at the cost of a $d(X, \ell_2^n)$ term (see also the proof of Proposition 6.3.6 (c)). □

DEFINITION 6.2.2 (K-convexity constant). Let X be an n-dimensional normed space. We define the K-convexity constant of X as

$$K(X) := \|Q_1 \otimes \mathrm{Id}_X\|_{L_2(\mathbb{R}^n,\gamma_n;X)\to L_2(\mathbb{R}^n,\gamma_n;X)}. \tag{6.2.2}$$

REMARK 6.2.3. One can check that $K(X^*) = K(X)$ for every finite dimensional normed space X. Lemma 6.1.8 allows us to give a simple proof of this fact: recall that the subspace S of all polynomials in the variables $x_1, \ldots, x_n$, is dense in

$L_2(\mathbb{R}^n, \gamma_n)$, and it is exactly the span of the sequence $\{H_\alpha \mid \alpha \in A\}$ of the Hermite polynomials. By orthogonality of the Hermite polynomials, we have that

$$(6.2.3)\quad \langle (Q_1 \otimes \mathrm{Id}_{X^*})g, f\rangle = \sum_{|\alpha|=1} \int_\Omega x^*_{\alpha,g}(x_{\alpha,f}) H_\alpha^2(\omega) d\mathbb{P}(\omega) = \langle g, (Q_1 \otimes \mathrm{Id}_X) f\rangle$$

for every function $g = \sum_{\alpha\in A} H_\alpha \otimes x^*_{\alpha,g} \in S \otimes X^*$ and every function $f = \sum_{\alpha\in A} H_\alpha \otimes x_{\alpha,f} \in S \otimes X$ (note that only a finite number of the vectors $x_{\alpha,f}$ and $x^*_{\alpha,g}$ are non-zero). It follows that

$$\begin{aligned}
&\|(Q_1 \otimes \mathrm{Id}_{X^*})g\|_{L_2(X^*)} \\
&= \sup\{\langle (Q_1 \otimes \mathrm{Id}_{X^*})g, f\rangle \mid f \in S \otimes X, \|f\|_{L_2(X)} \leqslant 1\} \\
&= \sup\{\langle g, (Q_1 \otimes \mathrm{Id}_X) f\rangle \mid f \in S \otimes X, \|f\|_{L_2(X)} \leqslant 1\} \\
&\leqslant \sup\{\|g\|_{L_2(X^*)} \|(Q_1 \otimes \mathrm{Id}_X) f\|_{L_2(X)} \mid f \in S \otimes X, \|f\|_{L_2(X)} \leqslant 1\} \\
&= K(X)\|g\|_{L_2(X^*)},
\end{aligned}$$

which shows that X^* is K-convex and that $K(X^*) \leqslant K(X)$. The reverse inequality follows from applying (6.1.15) and (6.2.3) in a similar way for every $f \in S \otimes X$.

The main result of this section is Pisier's estimate for $K(X)$.

THEOREM 6.2.4 (Pisier). *There exists an absolute constant C such that, for every n-dimensional normed space X, we have*

$$(6.2.4)\qquad K(X) \leqslant C \log[d(X, \ell_2^n) + 1].$$

For the proof we need the next

LEMMA 6.2.5. *Let X be a Banach space and let $(x_n)_{n\geqslant 0}$ be a sequence of vectors in X with only finitely many non-zero terms. Suppose that*

$$(6.2.5)\qquad \max_{-1\leqslant \varepsilon \leqslant 1} \left\| \sum_{n=0}^\infty \varepsilon^n x_n \right\| \leqslant 1.$$

If $D = \max_{n\geqslant 0} \|x_n\|$, then we have

$$(6.2.6)\qquad \|x_1\| \leqslant c \log(D+1)$$

where c is an absolute constant.

We will need a classical inequality of Bernstein, which gives a bound for $\|Q'\|_\infty$ in terms of $\|Q\|_\infty$ when Q is a trigonometric polynomial of degree n, and a corollary for polynomials with Banach-space valued coefficients.

LEMMA 6.2.6 (Bernstein). *If Q is a trigonometric polynomial of degree n, then*

$$(6.2.7)\qquad \|Q'\|_\infty \leqslant 2n\|Q\|_\infty.$$

If $P(t) = \sum_{0\leqslant k\leqslant n} z_k t^k$ is a polynomial with complex coefficients, then $P(t)$ satisfies the inequality

$$(6.2.8)\qquad |z_1| = |P'(0)| \leqslant 4n \sup_{|t|\leqslant \frac{1}{2}} |P(t)|.$$

Finally, if $P(t) = \sum_{0\leqslant k\leqslant n} x_k t^k$ where $x_k \in X$ then

$$(6.2.9)\qquad \|x_1\| \leqslant 4n \sup_{|t|\leqslant \frac{1}{2}} \|P(t)\|.$$

Proof of Lemma 6.2.5. We apply inequality (6.2.9) to the polynomial

$$P_n(t) := \sum_{0\leqslant k\leqslant n} x_k t^k, \quad n \geqslant 1.$$

Using the triangle inequality, we have for every $n \geqslant 1$ that

$$\begin{aligned}\|x_1\| &\leqslant 4n\left[\sup_{|\varepsilon|\leqslant\frac{1}{2}}\Big\|\sum_{k=0}^{\infty}\varepsilon^k x_k\Big\| + \sum_{k>n}2^{-k}D\right]\\ &\leqslant 4n\big(1+D2^{-n}\big).\end{aligned}$$

Choosing $n = \lceil \log_2 D\rceil$, we obtain the result. □

Proof of theorem 6.2.4. We will use the positive contractions T_ε from Proposition 6.1.6, as well as Lemma 6.2.5. Recall that the subspace S of all polynomials is exactly the linear span of $\{H_\alpha \,|\, \alpha \in A\}$ and is dense in $L_2(\mathbb{R}^n, \gamma_n)$.

For any $f \in S \otimes X$ with $\|f\|_{L_2(X)} \leqslant 1$, consider the sequence

$$\{(Q_k \otimes \mathrm{Id}_X)(f) \,|\, k \geqslant 0\}$$

of vectors in $L_2(X)$, which has only finitely many non-zero terms. We recall that $(T_\varepsilon \otimes \mathrm{Id}_X)(f) = \sum_{k\geqslant 0}\varepsilon^k(Q_k \otimes \mathrm{Id}_X)(f)$, therefore, by Proposition 6.1.6

$$\max_{-1\leqslant\varepsilon\leqslant 1}\Big\|\sum_{k\geqslant 0}\varepsilon^k(Q_k\otimes\mathrm{Id}_X)(f)\Big\|_{L_2(X)} \leqslant \Big(\max_{-1\leqslant\varepsilon\leqslant 1}\|T_\varepsilon\|_{L_2(X)\to L_2(X)}\Big)\cdot\|f\|_{L_2(X)} \leqslant 1.$$

In addition, by Proposition 6.2.1, for every $k \geqslant 0$,

$$\|(Q_k\otimes\mathrm{Id}_X)(f)\|_{L_2(X)} \leqslant \big(d(X,\ell_2^n)\|Q_k\|_{L_2\to L_2}\big)\cdot\|f\|_{L_2(X)} \leqslant d(X,\ell_2^n),$$

so we may apply Lemma 6.2.5 with $D = d(X,\ell_2^n)$; we obtain that

$$\|(Q_1\otimes\mathrm{Id}_X)(f)\|_{L_2(X)} \leqslant C\log[d(X,\ell_2^n)+1]$$

for every $f \in S\otimes X$ with $\|f\|_{L_2(X)} \leqslant 1$. Now, the desired bound for $K(X) = \|Q_1\otimes\mathrm{Id}_X\|_{L_2(X)\to L_2(X)}$ follows from the density of $S\otimes X$ in $L_2(X)$. □

6.3. The Rademacher projection

Before applying the bounds on the K convexity constant, we would like to illustrate an alternative route in which the Gaussian measure is replaced by the uniform measure on the discrete cube. The same ideas work very well and one gets another projection constant, called the Rademacher constant, which in turn can be used to bound MM^*.

DEFINITION 6.3.1. For every $n \in \mathbb{N}$ we consider the discrete cube $E_2^n = \{-1,1\}^n$ and the Rademacher functions $r_i : E_2^n \to \{-1,1\}$ defined by

$$r_i(\epsilon) = r_i(\epsilon_1,\ldots,\epsilon_i,\ldots,\epsilon_n) = \epsilon_i$$

as in Section 3.5a. If $\emptyset \neq A \subseteq [n] := \{1,\ldots,n\}$, we define $w_A : E_2^n \to \{-1,1\}$ by

$$w_A(\epsilon) = \prod_{i\in A} r_i(\epsilon).$$

We also agree that $w_\emptyset \equiv 1$. The functions w_A, $A \subseteq \{1,\ldots,n\}$, are the Walsh functions. Note that $r_i = w_{\{i\}}$ and that $w_A w_B = w_{A\triangle B}$ where $A\triangle B = (A\setminus B)\cup(B\setminus A)$

We view E_2^n as a probability space with the uniform measure μ_n. If $f : E_2^n \to \mathbb{R}$, then

$$\int_{E_2^n} f(\epsilon) d\mu_n(\epsilon) = \frac{1}{2^n} \sum_{\epsilon \in E_2^n} f(\epsilon).$$

More generally, if X is a Banach space and $f : E_2^n \to X$, we define

$$\int_{E_2^n} f(\epsilon) d\mu_n(\epsilon) = \frac{1}{2^n} \sum_{\epsilon \in E_2^n} f(\epsilon) \in X.$$

The space of all functions $f : E_2^n \to X$ becomes a Banach space with any of the equivalent (see Theorem 3.5.2 in Section 3.5) norms

$$\|f\|_{L_q(E_2^n;X)} := \left(\int_{E_2^n} \|f(\epsilon)\|_X^q d\mu_n(\epsilon) \right)^{1/q},$$

where $q \in [1, +\infty)$. It is easy to check, and is captured in the following proposition, that the Walsh functions form a complete orthonormal basis of $L_2(E_2^n; \mathbb{R})$, and so every function $f : E_2^n \to X$ can be expressed as sum of elements of the set $\{w_A \mid A \subseteq [n]\} \otimes X$.

PROPOSITION 6.3.2. *The Walsh functions satisfy the orthogonality relations*

$$\sum_{\epsilon} w_A(\epsilon) w_B(\epsilon) = 2^n \delta_{AB} \tag{6.3.1}$$

and

$$\sum_{A} w_A(\epsilon) w_A(\zeta) = 2^n \delta_{\epsilon\zeta}, \tag{6.3.2}$$

where $\delta_{xy} = 1$ *if* $x = y$ *and* $\delta_{xy} = 0$ *if* $x \neq y$. *Moreover, every* $f : E_2^n \to X$ *is uniquely represented in the form*

$$f(\epsilon) = \sum_{A} w_A(\epsilon) x_A,$$

for some $x_A \in X$.

Proof. The orthogonality relations are simple - the first follows since $w_A w_B = w_{A \triangle B}$ and clearly $\sum_\epsilon w_{A \triangle B}(\epsilon) = 0$ unless $A \triangle B = \emptyset$. For the second, one only needs to notice that if we fix an index j such that $\epsilon_j \neq \zeta_j$ then for any $A \subset \{1, \dots, n\}$ with $j \notin A$ we have $w_A(\epsilon) w_A(\zeta) = -w_{\tilde{A}}(\epsilon) w_{\tilde{A}}(\zeta)$, where $\tilde{A} = A \cup \{j\}$. To show that any $f : E_2^n \to X$ is uniquely representable, given such f we define for $A \subseteq \{1, \dots, n\}$ the vector

$$x_A = \int_{E_2^n} w_A(\epsilon) f(\epsilon) d\mu_n(\epsilon).$$

Then, using (6.3.2), we see that for every $\epsilon \in E_2^n$

$$\begin{aligned} \sum_A w_A(\epsilon) x_A &= \sum_A w_A(\epsilon) \left(\int_{E_2^n} w_A(\zeta) f(\zeta) d\mu_n(\zeta) \right) \\ &= \int_{E_2^n} f(\zeta) \left(\sum_A w_A(\epsilon) w_A(\zeta) \right) d\mu_n(\zeta) = f(\epsilon). \end{aligned}$$

To prove uniqueness, assume that $f(\epsilon) = \sum_A w_A(\epsilon) y_A$ for some $y_A \in X$. Then,

$$x_A = \int_{E_2^n} w_A(\zeta) f(\zeta) d\mu_n(\zeta) = \int_{E_2^n} \left(\sum_B w_B(\zeta) y_B \right) w_A(\zeta) d\mu_n(\zeta)$$
$$= \frac{1}{2^n} \sum_B y_B \left(\sum_{\zeta \in E_2^n} w_B(\zeta) w_A(\zeta) \right) = y_A,$$

where we have used (6.3.1) this time. □

DEFINITION 6.3.3. The Rademacher projection of $f : E_2^n \to X$ is the function $Rad_n f : E_2^n \to X$ defined by

$$Rad_n f(\epsilon) = \sum_{i=1}^n r_i(\epsilon) x_{\{i\}},$$

where $f = \sum w_A x_A$. We consider the linear operator $f \mapsto Rad_n(f)$ on $L_2(E_2^n; X)$, and if this operator is bounded, then we set

$$K_r(X) := \|Rad_n\|_{L_2(E_2^n;X) \to L_2(E_2^n;X)}. \tag{6.3.3}$$

As with the K-convexity constant of a Banach space X, in the case where X is finite dimensional we wish to examine how good an estimate we can give for $K_r(X)$. In this case, it is simpler to begin by writing $Rad_n f$ as the convolution of f with the sum of the Rademacher functions, where the convolution of a function $f : E_2^n \to X$ and a function $g : E_2^n \to \mathbb{R}$ is defined by

$$(f * g)(\epsilon) = \int_{E_2^n} f(\epsilon\zeta) g(\zeta) d\mu_n(\zeta). \tag{6.3.4}$$

(Here $\epsilon\zeta$ is pointwise product.)

LEMMA 6.3.4. *Define $g_r : E_2^n \to \mathbb{R}$ by $g_r(\epsilon) = \sum_{i=1}^n r_i(\epsilon)$. Then, for every $f : E_2^n \to X$ we have*

$$Rad_n f = f * g_r.$$

Proof. Direct computation shows that

$$(f * g_r)(\epsilon) = \int_{E_2^n} \left(\sum_A w_A(\epsilon\zeta) x_A \right) \left(\sum_{i=1}^n r_i(\zeta) \right) d\mu_n(\zeta)$$
$$= \sum_{i=1}^n \sum_A x_A w_A(\epsilon) \left(\int_{E_2^n} w_A(\zeta) r_i(\zeta) d\mu_n(\zeta) \right)$$
$$= \sum_{i=1}^n r_i(\epsilon) x_{\{i\}} = Rad_n f(\epsilon),$$

as claimed. □

The following proposition is an analogue to Lemma 6.1.8. Here it is much easier to show that $L_2(E_2^n; X^*)$ is the dual space of $L_2(E_2^n; X)$ since the latter can be naturally identified with an ℓ_2 sum of 2^n copies of X.

PROPOSITION 6.3.5. *Let X be a Banach space and X^* be its dual. For every $f : E_2^n \to X$ and $\phi : E_2^n \to X^*$ the following hold true:*

(*a*) *The linear functional corresponding to* ϕ *which is defined on* $L_2(E_2^n;X)$ *by*

$$\langle \phi, f\rangle := \int_{E_2^n} [\phi(\epsilon)](f(\epsilon))d\mu_n(\epsilon) \tag{6.3.5}$$

is bounded and has norm equal to $\|\phi\|_{L_2(E_2^n;X^*)}$.

(*b*) *There exists* $\psi : E_2^n \to X^*$ *such that* $\|\psi\|_{L_2(E_2^n;X^*)} = 1$ *and*

$$\|f\|_{L_2(E_2^n;X)} = \langle \psi, f\rangle = \int_{E_2^n} [\psi(\epsilon)](f(\epsilon))d\mu_n(\epsilon).$$

(*c*) *Let* H *be a Hilbert space and* $h : E_2^n \to H$ *given by* $h = \sum_A w_A x_A$ *for some set* $\{x_A\}_{A\subset\{1,\dots n\}} \subset H$. *Then,*

$$\|h\|^2_{L_2(E_2^n;H)} = \sum_A \|x_A\|_H^2.$$

Proof. For the first two points we repeat the proof given in Lemma 6.1.8; note that all functions now, in either $L_2(E_2^n;X)$ or $L_2(E_2^n;X^*)$, can be thought of as step functions. To prove (*c*), observe that by (6.3.1)

$$\begin{aligned}\int_{E_2^n} \|h(\epsilon)\|_H^2 d\mu_n(\epsilon) &= \int_{E_2^n} \Big\|\sum_A w_A(\epsilon)x_A\Big\|_H^2 d\mu_n(\epsilon)\\ &= \int_{E_2^n} \Big\langle \sum_A w_A(\epsilon)x_A, \sum_B w_B(\epsilon)x_B\Big\rangle_H d\mu_n(\epsilon)\\ &= \sum_A\sum_B \langle x_A, x_B\rangle_H \int_{E_2^n} w_A(\epsilon)w_B(\epsilon)d\mu_n(\epsilon)\\ &= \sum_A \langle x_A, x_A\rangle_H = \sum_A \|x_A\|_H^2,\end{aligned}$$

whence we are done. □

In the following proposition we summarize basic facts about convolutions of functions on E_2^n.

PROPOSITION 6.3.6. *Let* X *be a Banach space,* $f : E_2^n \to X$, $g : E_2^n \to \mathbb{R}$, *and let* $g(\epsilon) = \sum_A w_A(\epsilon)c_A$ *be the representation of* g. *Then,*

(*a*) $\|f * g\|_{L_2(E_2^n;X)} \leqslant \|f\|_{L_2(E_2^n;X)}\|g\|_{L_1(E_2^n;\mathbb{R})}$.

(*b*) *If* $X = H$ *is a Hilbert space, then*

$$\|f * g\|_{L_2(E_2^n;H)} \leqslant \|f\|_{L_2(E_2^n;H)} \max_A |c_A|.$$

(*c*) *If* $d(X,H)$ *is the Banach-Mazur distance of* X *to a Hilbert space* H, *then*

$$\|f * g\|_{L_2(E_2^n;X)} \leqslant \|f\|_{L_2(E_2^n;X)} \max_A |c_A| d(X,H).$$

Proof. (*a*) Since $f * g : E_2^n \to X$, by Proposition 6.3.5(*b*) we can find $\phi : E_2^n \to X^*$ such that $\|\phi\|_{L_2(E_2^n;X^*)} = 1$ and

$$\begin{aligned}
\|f * g\|_{L_2(E_2^n;X)} &= \int_{E_2^n} [\phi(\epsilon)]\big((f * g)(\epsilon)\big) d\mu_n(\epsilon) \\
&= \int_{E_2^n} g(\zeta) \int_{E_2^n} [\phi(\epsilon)]\big(f(\epsilon\zeta)\big) d\mu_n(\epsilon) d\mu_n(\zeta) \\
&\leqslant \int_{E_2^n} |g(\zeta)| \int_{E_2^n} \|\phi(\epsilon)\|_{X^*} \|f(\epsilon\zeta)\|_X d\mu_n(\epsilon) d\mu_n(\zeta) \\
&\leqslant \int_{E_2^n} |g(\zeta)| \left(\int_{E_2^n} \|\phi(\epsilon)\|_{X^*}^2 d\epsilon\right)^{1/2} \left(\int_{E_2^n} \|f(\epsilon\zeta)\|_X^2 d\mu_n(\epsilon)\right)^{1/2} d\mu_n(\zeta) \\
&= \|f\|_{L_2(E_2^n;X)} \|g\|_{L_1(E_2^n;\mathbb{R})}.
\end{aligned}$$

(*b*) Note that if $f = \sum_A w_A x_A$ and $g = \sum_A c_A w_A$ are the representations of f and g respectively, then $f * g = \sum_A c_A x_A w_A$. This follows by a direct computation:

$$\begin{aligned}
(f * g)(\epsilon) &= \int_{E_2^n} f(\epsilon\zeta)\Big(\sum_A w_A(\zeta) c_A\Big) d\mu_n(\zeta) = \sum_A \Big(\int_{E_2^n} f(\epsilon\zeta) w_A(\zeta) d\mu_n(\zeta)\Big) c_A \\
&= \sum_A \Big(\int_{E_2^n} f(\epsilon\zeta) w_A(\epsilon\zeta) d\mu_n(\zeta)\Big) c_A w_A(\epsilon) = \sum_A x_A c_A w_A(\epsilon).
\end{aligned}$$

From Proposition 6.3.5(*c*), we have

$$\begin{aligned}
\|f * g\|_{L_2(E_2^n;H)} &= \Big(\sum_A \|c_A x_A\|_H^2\Big)^{1/2} \\
&\leqslant \max_A |c_A| \Big(\sum_A \|x_A\|_H^2\Big)^{1/2} = \Big(\max_A |c_A|\Big) \|f\|_{L_2(E_2^n;H)}.
\end{aligned}$$

(*c*) Given $\varepsilon > 0$, there exists an isomorphism $T : X \to H$ such that $\|T\|\|T^{-1}\| \leqslant (1 + \varepsilon) d(X, H)$. Since $f * g = \sum_A c_A x_A w_A$, we have

$$T[(f * g)(\epsilon)] = \sum_A w_A(\epsilon) c_A T(x_A).$$

Then, by the definition of the norms and by Proposition 6.3.5(*c*),

$$\begin{aligned}
\|f * g\|_{L_2(E_2^n;X)} &= \|T^{-1} \circ T \circ (f * g)\|_{L_2(E_2^n;X)} \\
&\leqslant \|T^{-1}\|\|T \circ (f * g)\|_{L_2(E_2^n;H)} \\
&= \|T^{-1}\| \Big(\sum_A \|c_A T(x_A)\|_H^2\Big)^{1/2} \\
&\leqslant \max_A |c_A| \, \|T^{-1}\|\|T \circ f\|_{L_2(E_2^n;H)} \\
&\leqslant \max_A |c_A| \, \|T^{-1}\|\|T\|\|f\|_{L_2(E_2^n;X)} \\
&\leqslant \|f\|_{L_2(E_2^n;X)} \max_A |c_A| (1 + \varepsilon) d(X, H).
\end{aligned}$$

This completes the proof. □

REMARK 6.3.7. Lemma 6.3.4 shows that $Rad_n f = f * g_r$, where $g_r = \sum_{i=1}^n r_i$. By the orthogonality of the Rademacher functions we have that $\|g\|_{L_1(E_2^n;\mathbb{R})} \leqslant \|g\|_{L_2(E_2^n;\mathbb{R})} = \sqrt{n}$, therefore Proposition 6.3.6(a) gives

$$\|Rad_n f\|_{L_2(E_2^n;X)} \leqslant \sqrt{n}\|f\|_{L_2(E_2^n;X)},$$

which is a first bound on the norm of this projection. Part (c) of the same proposition gives at least as good an estimate when X is isomorphic to a Hilbert space H and $d(X,H) \leqslant \sqrt{n}$ (as is the case with n-dimensional normed spaces by John theorem): since for the function g_r we have $\max_A |c_A| = 1$, we see that

$$\|Rad_n f\|_{L_2(E_2^n;X)\to L_2(E_2^n;H)} \leqslant d(X,H). \tag{6.3.6}$$

Finally, if $X = H$ is a Hilbert space, then part (b) of the proposition shows that $\|Rad_n f\|_{L_2(E_2^n;H)\to L_2(E_2^n;H)} = 1$.

Inequality (6.3.6) is analogous to the bound Proposition 6.2.1 gives us for the K-convexity constant of an n-dimensional normed space. As with that bound, we can significantly improve (6.3.6) to depend only logarithmically on the Banach-Mazur distance between X and ℓ_2^n.

THEOREM 6.3.8 (Pisier). *There exists an absolute constant C such that, for every n-dimensional normed space X, we have*

$$\|Rad_n\|_{L_2(E_2^n;X)\to L_2(E_2^n;X)} \leqslant C\log[d(X,\ell_2^n)+1]. \tag{6.3.7}$$

In particular, for every n-dimensional normed space X,

$$\|Rad_n\|_{L_2(E_2^n;X)\to L_2(E_2^n;X)} \leqslant C\log(n+1).$$

For the proof we build a special signed measure using Lemma 6.2.6.

PROPOSITION 6.3.9. *Let $l \in \mathbb{N}$, $l \geqslant 2$. There is a signed measure μ on $[-\frac{1}{2},\frac{1}{2}]$ with total variation $\|\mu\| \leqslant 4l$, which satisfies the following*

$$\int_{-1/2}^{1/2} t\,d\mu(t) = 1, \qquad \int_{-1/2}^{1/2} t^k d\mu(t) = 0, \qquad k = 0,2,\ldots,l. \tag{6.3.8}$$

Proof. We denote by $\mathcal{P}_l$ the space of all polynomials with real coefficients that have degree less than or equal to l, and we define a functional $F : \mathcal{P}_l \to \mathbb{R}$ by

$$F(p) = p'(0).$$

Then inequality (6.2.8) of Lemma 6.2.6 shows that

$$|F(p)| = |p'(0)| \leqslant 4l \max_{t\in[-1/2,1/2]} |p(t)|, \qquad p \in \mathcal{P}_l.$$

By the Hahn-Banach theorem, we can extend F to some $\tilde{F} \in \big(C[-1/2,1/2]\big)^*$ with $\|\tilde{F}\| \leqslant 4l$. By the Riesz representation theorem, there exists a signed measure μ on $[-1/2,1/2]$ with $\|\mu\| \leqslant 4l$, which satisfies

$$\int_{-1/2}^{1/2} t^k d\mu(t) = \tilde{F}(t^k) = (t^k)'|_{t=0}, \qquad k = 0,1,\ldots,l,$$

or in other words satisfies (6.3.8). $\square$

Proof of Theorem 6.3.8. Let $2 \leqslant l \leqslant n$ to be determined later in the proof and consider the measure μ ensured by Proposition 6.3.9. Define

$$g(\epsilon) = \int_{-1/2}^{1/2} \prod_{i=1}^{n} \big(1 + tr_i(\epsilon)\big) d\mu(t).$$

Since $\prod_{i=1}^{n}\big(1 + tr_i(\epsilon)\big) = \sum_{k=0}^{n} t^k \sum_{|A|=k} w_A(\epsilon)$, we have

$$\begin{aligned} g(\epsilon) &= \int_{-1/2}^{1/2} \sum_{k=0}^{n} t^k \sum_{|A|=k} w_A(\epsilon) d\mu(t) \\ &= \sum_{k=0}^{n} \left(\left(\int_{-1/2}^{1/2} t^k d\mu(t) \right) \sum_{|A|=k} w_A(\epsilon) \right) \\ &= \sum_{i=1}^{n} r_i(\epsilon) + \sum_{k=l+1}^{n} \left(\left(\int_{-1/2}^{1/2} t^k d\mu(t) \right) \sum_{|A|=k} w_A(\epsilon) \right), \end{aligned}$$

where we have used the properties of μ.

We set $g_1 = g_r = \sum_{i=1}^{n} r_i$, $g_2 = \sum_{k=l+1}^{n} \big[\big(\int_{-1/2}^{1/2} t^k d\mu(t) \big) \sum_{|A|=k} w_A \big]$, so that $g = g_1 + g_2$. We also have that

$$\begin{aligned} \|g\|_{L_1(\mathbb{R})} &= \int_{E_2^n} \left| \int_{-1/2}^{1/2} \prod_{i=1}^{n} \big(1 + tr_i(\epsilon)\big) d\mu(t) \right| d\mu_n(\epsilon) \\ &\leqslant \int_{E_2^n} \left(\int_{-1/2}^{1/2} \prod_{i=1}^{n} \big(1 + tr_i(\epsilon)\big) d\mu^+(t) + \int_{-1/2}^{1/2} \prod_{i=1}^{n} \big(1 + tr_i(\epsilon)\big) d\mu^-(t) \right) d\mu_n(\epsilon) \\ &= \int_{-1/2}^{1/2} \left(\int_{E_2^n} \prod_{i=1}^{n} \big(1 + tr_i(\epsilon)\big) d\mu_n(\epsilon) \right) d|\mu|(t). \end{aligned}$$

Since for every $t \in [-1/2, 1/2]$

$$\begin{aligned} \int_{E_2^n} \prod_{i=1}^{n} \big(1 + tr_i(\epsilon)\big) d\mu_n(\epsilon) &= \sum_{k=0}^{n} t^k \left(\sum_{|A|=k} \int_{E_2^n} w_A(\epsilon) d\mu_n(\epsilon) \right) \\ &= \int_{E_2^n} w_{\emptyset}(\epsilon) d\mu_n(\epsilon) = 1, \end{aligned}$$

it follows that

$$\|g\|_{L_1(E_2^n;\mathbb{R})} \leqslant \int_{-1/2}^{1/2} d|\mu| = \|\mu\| \leqslant 4l$$

and hence by Proposition 6.3.6(a)

$$\|f * g\|_{L_2(E_2^n;X)} \leqslant \|f\|_{L_2(E_2^n;X)} \|g\|_{L_1(E_2^n;\mathbb{R})} \leqslant 4l \|f\|_{L_2(E_2^n;X)}. \tag{6.3.9}$$

Next, observe that if $g_2 = \sum_A w_A c_A^{g_2} = \sum_{k=l+1}^{n} \sum_{|A|=k} w_A c_A^{g_2}$, then

$$c_A^{g_2} = \int_{-1/2}^{1/2} t^k d\mu(t) \ \text{ where } k = |A|,$$

so by Proposition 6.3.6(c) we have that

$$\begin{aligned} \|f * g_2\|_{L_2(E_2^n;X)} &\leqslant \|f\|_{L_2(E_2^n;X)} \max_A |c_A^{g_2}|\, d(X,H) \\ &= \|f\|_{L_2(E_2^n;X)} \max_{l<k\leqslant n} \left| \int_{-1/2}^{1/2} t^k \, d\mu(t)\right| d(X,H) \\ &\leqslant \|f\|_{L_2(E_2^n;X)} \frac{1}{2^{l+1}} \|\mu\|\, d(X,H) \\ &\leqslant \|f\|_{L_2(E_2^n;X)} \frac{4l}{2^{l+1}}\, d(X,H). \end{aligned} \tag{6.3.10}$$

Combining (6.3.9) and (6.3.10), we conclude that

$$\begin{aligned} \|Rad_n f\|_{L_2(E_2^n;X)} &= \|f * (g - g_2)\|_{L_2(E_2^n;X)} \\ &= \|(f * g) - (f * g_2)\|_{L_2(E_2^n;X)} \\ &\leqslant \|(f * g)\|_{L_2(E_2^n;X)} + \|(f * g_2)\|_{L_2(E_2^n;X)} \\ &\leqslant \|f\|_{L_2(E_2^n;X)} \left(4l + 4l\frac{d(X,H)}{2^{l+1}}\right). \end{aligned}$$

This holds true for every $l \in \mathbb{N}$, $l \geqslant 2$, and if we choose it so that $2^{l-1} \leqslant d(X,H) + 1 < 2^l$, we will obtain

$$\|Rad_n f\|_{L_2(E_2^n;X)} \leqslant 6l\|f\|_{L_2(E_2^n;X)} \leqslant c\log[d(X,H)+1]\|f\|_{L_2(E_2^n;X)}$$

for some absolute constant c, as required. □

REMARK 6.3.10. The K-convexity constant and the Rademacher constant of a finite dimensional normed space are equivalent. More precisely, in Section 6.6 we show that if X is an n-dimensional normed space, then

$$\frac{2}{\pi} K_r(X) \leqslant K(X) \leqslant K_r(X).$$

6.4. The ℓ-norm

6.4.1. Trace duality

DEFINITION 6.4.1. Let α be a norm on $L(\mathbb{R}^n)$. We can always define a dual norm on $L(\mathbb{R}^n)$ by setting

$$\alpha^*(v) := \sup\{\mathrm{tr}(vu) : \alpha(u) \leqslant 1\}.$$

We note in passing that in fact there is a notion, commonly used in operator theory, of an *operator norm*, which is a norm α on the space of operators (say, in finite dimension n, on $L(\mathbb{R}^n)$) satisfying two additional properties: that $\alpha(u) = \|u\|_{\mathrm{op}}$ for rank one operators u, and that $\alpha(u_1u_2u_3) \leqslant \|u_1\|_{\mathrm{op}}\alpha(u_2)\|u_3\|_{\mathrm{op}}$. The usual definition of "operator norms" involves considering the category of all Banach spaces (or normed spaces of a given dimension) and operator norms between them satisfy the above conditions for all operators u_1, u_2 and u_3 acting between any four spaces; i.e. $u_i : E_{i-1} \to E_i$. One may check that if α is an operator norm then so is α^*. We say that α and α^* are in trace duality. The next lemma of Lewis applies to any pair of trace dual norms.

LEMMA 6.4.2. *For any norm α on $L(\mathbb{R}^n)$ there exists $u \in L(\mathbb{R}^n)$ such that $\alpha(u) = 1$ and $\alpha^*(u^{-1}) = n$.*

Proof. We first observe that $n \leqslant \alpha(v)\alpha^*(v^{-1})$ for every $v \in GL_n$, since

$$\frac{n}{\alpha(v)} = \frac{1}{\alpha(v)}\mathrm{tr}(\mathrm{Id}) = \mathrm{tr}\left(v^{-1}\frac{v}{\alpha(v)}\right) \leqslant \alpha^*(v^{-1}).$$

Choose $u \in GL_n$ with $\alpha(u) \leqslant 1$ such that

$$|\det u| = \max\{|\det v| : v \in GL_n, \alpha(v) \leqslant 1\}.$$

Existence of such a u follows by continuity, and it is clear that $\alpha(u) = 1$. Let $v \in L(\mathbb{R}^n)$ and $\varepsilon > 0$ be small enough. Then, $(u+\varepsilon v)/\alpha(u+\varepsilon v)$ has α-norm equal to 1. Thus,

$$|\det(u+\varepsilon v)| \leqslant (\alpha(u+\varepsilon v))^n \, |\det u|.$$

It follows that

$$|\det[u^{-1}(u+\varepsilon v)]|^{\frac{1}{n}} \leqslant \alpha(u+\varepsilon v),$$

and, since α is a norm,

$$|\det(\mathrm{Id}+\varepsilon u^{-1}v)|^{\frac{1}{n}} \leqslant \alpha(u) + \varepsilon\alpha(v) = 1 + \varepsilon\alpha(v).$$

This means that

$$\frac{[\det(\mathrm{Id}+\varepsilon u^{-1}v)]^{\frac{1}{n}} - 1}{\varepsilon} \leqslant \alpha(v),$$

and letting $\varepsilon \to 0^+$ we obtain

$$\frac{\mathrm{tr}(u^{-1}v)}{n} \leqslant \alpha(v).$$

Since v was arbitrary, we conclude that $\alpha^*(u^{-1}) \leqslant n$. □

6.4.2. The ℓ norm of an operator, and trace duality

Let $\{e_i\}_{i=1}^n$ be the standard orthonormal basis of $\mathbb{R}^n$ and let $X = (\mathbb{R}^n, \|\cdot\|)$ be an n-dimensional normed space. We will apply the trace duality of the previous subsection to the ℓ-norm on $L(\ell_2^n, X)$ which was defined by Figiel and Tomczak-Jaegermann as follows. For every $u : \ell_2^n \to X$ we define

$$\ell(u) = \left(\int_{\mathbb{R}^n} \Big\|\sum_{i=1}^n x_i u(e_i)\Big\|^2 d\gamma_n(x)\right)^{1/2} = \left(\mathbb{E}\Big\|\sum_{i=1}^n g_i u(e_i)\Big\|^2\right)^{1/2},$$

where $g_1, \ldots, g_n$ are independent standard Gaussian random variables on some probability space.

PROPOSITION 6.4.3. *ℓ is a norm on $L(\ell_2^n, X)$.*

Proof. This is a direct verification. In fact, it has properties of an operator norm, though it is defined only on operators from ℓ_2^n. □

REMARK 6.4.4. If $u : \ell_2^n \to X$ is an isomorphism, then $(u^{-1})^* : \ell_2^n \to X^*$. We can then write

$$\ell((u^{-1})^*) = \left(\int_{\mathbb{R}^n} \Big\|\sum_{i=1}^n x_i (u^{-1})^*(e_i)\Big\|_*^2 d\gamma_n(x)\right)^{1/2},$$

where $\|\cdot\|_*$ is the norm of X^*.

THEOREM 6.4.5 (Figiel-Tomczak). *Let $X = (\mathbb{R}^n, \|\cdot\|)$ be an n-dimensional normed space. There exists $u : \ell_2^n \to X$ such that*

$$\ell(u)\ell\big((u^{-1})^*\big) \leqslant nK(X).$$

Proof. By Lemma 6.4.2 we can find an isomorphism $u : \ell_2^n \to X$ such that $\ell(u)\ell^*(u^{-1}) = n$, where

$$\ell^*(u^{-1}) = \sup\{\mathrm{tr}(u^{-1}v) \mid \ell(v) \leqslant 1\};$$

we will show that this u satisfies the conclusion of the theorem.

Recall that S is the subspace of all polynomials in the variables $x_1, \dots, x_n$ and H_{e_i} is the Hermite polynomial of degree one corresponding to the index e_i so that $H_{e_i}(x) = x_i$, and define $f \in S \otimes X^*$ by

$$f(x) = \sum_{i=1}^n x_i\big((u^{-1})^*(e_i)\big) = \sum_{i=1}^n \big(H_{e_i} \otimes \big((u^{-1})^*(e_i)\big)\big)(x).$$

Then $\ell\left((u^{-1})^*\right) = \|f\|_{L_2(X^*)}$, and given any $\varepsilon > 0$, by Lemma 6.1.8 we can find a function

$$g_\varepsilon = \sum_{\alpha \in \mathbb{N}^n} H_\alpha \otimes z_{\alpha, g_\varepsilon} \in S \otimes X$$

with the properties that $\|g_\varepsilon\|_{L_2(X)} = 1$ and

$$\begin{aligned}\ell((u^{-1})^*) - \varepsilon \leqslant \langle f, g_\varepsilon\rangle &= \int_{\mathbb{R}^n} \Big\langle \sum_{i=1}^n x_i\big((u^{-1})^*(e_i)\big), \sum_{\alpha\in\mathbb{N}^n} H_\alpha(x) z_{\alpha,g_\varepsilon}\Big\rangle d\gamma_n(x) \\ &= \sum_{i=1}^n \sum_{\alpha\in\mathbb{N}^n} \left(\int_{\mathbb{R}^n} x_i H_\alpha(x) d\gamma_n(x)\right) \langle (u^{-1})^*(e_i), z_{\alpha,g_\varepsilon}\rangle.\end{aligned}$$

By orthogonality of the Hermite polynomials, and since $\|H_{e_i}\|_{L_2(\gamma_n)} = 1$ for every $i = 1, \dots, n$, we get that

$$\ell((u^{-1})^*) - \varepsilon \leqslant \sum_{i=1}^n \langle (u^{-1})^*(e_i), z_{e_i, g_\varepsilon}\rangle.$$

We now define $v_\varepsilon : \ell_2^n \to X$ setting $v_\varepsilon(e_i) = z_{e_i, g_\varepsilon}$. Then,

$$\begin{aligned}\ell((u^{-1})^*) - \varepsilon &\leqslant \sum_{i=1}^n \langle (u^{-1})^*(e_i), v_\varepsilon(e_i)\rangle \\ &= \sum_{i=1}^n \langle e_i, (u^{-1}v_\varepsilon)(e_i)\rangle = \mathrm{tr}(u^{-1}v_\varepsilon) \leqslant \ell^*(u^{-1})\ell(v_\varepsilon),\end{aligned}$$

where by definition

$$\begin{aligned}\ell(v_\varepsilon) &= \left(\int_{\mathbb{R}^n} \Big\|\sum_{i=1}^n x_i v_\varepsilon(e_i)\Big\|^2 d\gamma_n(x)\right)^{1/2} = \left(\int_{\mathbb{R}^n} \Big\|\sum_{i=1}^n x_i z_{e_i,g_\varepsilon}\Big\|^2 d\gamma_n(x)\right)^{1/2} \\ &= \|(Q_1 \otimes \mathrm{Id}_X) g_\varepsilon\|_{L_2(X)} \leqslant K(X) \cdot \|g_\varepsilon\|_{L_2(X)} = K(X).\end{aligned}$$

This means that for every $\varepsilon > 0$,

$$\ell((u^{-1})^*) - \varepsilon \leqslant K(X)\ell^*(u^{-1}),$$

and hence,

$$\ell(u)\ell\big((u^{-1})^*\big) \leqslant \ell(u)\ell^*(u^{-1})K(X) = nK(X)$$

as required. $\square$

REMARK 6.4.6. The alert reader will have noticed that in fact the bound above corresponds to a bound for the product $M(K)M^*(K)$ where K is a linear image (determined by u) of the unit ball of X. We discuss this in detail in Section 6.5.

6.4.3. Trace duality for ℓ_r

We may, again, follow the route which was detailed in Section 6.3. To this end we define the ℓ_r-norm on $L(\mathbb{R}^n)$ using the Rademacher functions instead of gaussian variables. If $\{e_i\}_{i=1}^n$ is the standard orthonormal basis of $\mathbb{R}^n$, and if $X = (\mathbb{R}^n, \|\cdot\|)$ is an n-dimensional normed space, we define

$$\ell_r(u) = \left(\int_{E_2^n} \|\sum_{i=1}^n r_i(\epsilon)u(e_i)\|^2 d\mu_n(\epsilon)\right)^{1/2} \tag{6.4.1}$$

for every $u : \ell_2^n \to X$.

It is again standard to check that

PROPOSITION 6.4.7. *ℓ_r is a norm on $L(\ell_2^n, X)$.*

We can apply trace duality to this norm and get an exact analogue to Theorem 6.4.5. Indeed, whenever we have an isomorphism $u : \ell_2^n \to X$, then $(u^{-1})^* : \ell_2^n \to X^*$ and we can write

$$\ell_r((u^{-1})^*) = \left(\int_{E_2^n} \|\sum_{i=1}^n r_i(\epsilon)(u^{-1})^*(e_i)\|_*^2 d\mu_n(\epsilon)\right)^{1/2},$$

where $\|\cdot\|_*$ is the norm of X^*. We prove

THEOREM 6.4.8. *Let $X = (\mathbb{R}^n, \|\cdot\|)$ be an n-dimensional normed space. There exists $u : \ell_2^n \to X$ such that*

$$\ell_r(u)\ell_r((u^{-1})^*) \leqslant nK_r(X).$$

Proof. Since ℓ_r is a norm on $L(\mathbb{R}^n)$, by Lemma 6.4.2 we can find an isomorphism $u : \ell_2^n \to X$ such that $\ell_r(u)\ell_r^*(u^{-1}) = n$, where

$$\ell_r^*(u^{-1}) = \sup\{\mathrm{tr}(u^{-1}v) : \ell_r(v) \leqslant 1\}.$$

We define $f : E_2^n \to X^*$ by

$$f(\epsilon) = \sum_{i=1}^n r_i(\epsilon)((u^{-1})^*(e_i)).$$

Proposition 6.3.5 shows that for some $\phi : E_2^n \to X$ with $\|\phi\|_{L_2(E_2^n;X)} = 1$ and representation $\phi(\epsilon) = \sum_A w_A(\epsilon)x_A$, we have

$$\begin{aligned}\ell_r((u^{-1})^*) &= \int_{E_2^n} \langle f(\epsilon), \phi(\epsilon)\rangle d\mu_n(\epsilon) = \int_{E_2^n} [f(\epsilon)](\phi(\epsilon))d\mu_n(\epsilon) \\ &= \int_{E_2^n} \langle \sum_{i=1}^n r_i(\epsilon)((u^{-1})^*(e_i)), \sum_A w_A(\epsilon)x_A\rangle d\mu_n(\epsilon) \\ &= \sum_{i=1}^n \sum_A \left(\int_{E_2^n} r_i(\epsilon)w_A(\epsilon)d\mu_n(\epsilon)\right) \langle (u^{-1})^*(e_i), x_A\rangle.\end{aligned}$$

Then, by the orthonormality of the Walsh functions, we get that

$$\ell_r\big((u^{-1})^*\big) = \sum_{i=1}^{n} \langle (u^{-1})^*(e_i), x_{\{i\}}\rangle.$$

We now define $v : \ell_2^n \to X$, setting $v(e_i) = x_{\{i\}}$. Then,

$$\begin{aligned}\ell_r\big((u^{-1})^*\big) &= \sum_{i=1}^{n} \langle (u^{-1})^*(e_i), v(e_i)\rangle \\ &= \sum_{i=1}^{n} \langle e_i, (u^{-1}v)(e_i)\rangle = \mathrm{tr}(u^{-1}v) \leqslant \ell_r^*(u^{-1})\ell_r(v).\end{aligned}$$

But by the definition of $\ell_r(v)$, we have

$$\begin{aligned}\ell_r(v) &= \Big(\int_{E_2^n} \Big\|\sum_{i=1}^{n} r_i(\epsilon)v(e_i)\Big\|^2 d\mu_n(\epsilon)\Big)^{1/2} = \Big(\int_{E_2^n} \Big\|\sum_{i=1}^{n} r_i(\epsilon)x_{\{i\}}\Big\|^2 d\mu_n(\epsilon)\Big)^{1/2} \\ &= \|Rad_n(\phi)\|_{L_2(E_2^n;X)} \leqslant K_r(X)\cdot \|\phi\|_{L_2(E_2^n;X)} = K_r(X).\end{aligned}$$

This means that

$$\ell_r\big((u^{-1})^*\big) \leqslant \ell_r^*(u^{-1})K_r(X),$$

and hence,

$$\ell_r(u)\ell_r\big((u^{-1})^*\big) \leqslant \ell_r(u)\ell_r^*(u^{-1})K_r(X) = nK_r(X).$$

□

By translating the quantities $\ell_r(u)$ and $\ell_r\big((u^{-1})^*\big)$ in the language of convex bodies, the previous theorem takes the following form.

COROLLARY 6.4.9. *Let K be a centrally symmetric convex body in $\mathbb{R}^n$. There exists a linear image K_1 of K such that*

$$\Big(\int_{E_2^n} \Big\|\sum_{i=1}^{n} r_i(\epsilon)e_i\Big\|_{K_1}^2 d\mu_n(\epsilon)\Big)^{1/2} \Big(\int_{E_2^n} \Big\|\sum_{i=1}^{n} r_i(\epsilon)e_i\Big\|_{K_1^\circ}^2 d\mu_n(\epsilon)\Big)^{1/2} \leqslant nK_r(X),$$

where $X = \big(\mathbb{R}^n, \|\cdot\|_K\big)$. □

6.5. The MM^*-estimate

In this section we translate Theorem 6.4.5 to the language of convex bodies, and derive the MM^* estimate promised in the introduction. Our starting point is the next formula.

$$M(K) = \int_{S^{n-1}} \|u\| d\sigma(u) = c_n \frac{1}{\sqrt{n}} \Big(\int_{\mathbb{R}^n} \|x\|^2 d\gamma_n(x)\Big)^{1/2} \tag{6.5.1}$$

where c_n is bounded between two absolute constants.

Indeed, this was mentioned in Remark 5.2.2 right after the definition of M and the more general $M_q = \big(\int_{S^{n-1}} \|u\|^q d\sigma(u)\big)^{1/q}$. We have that

$$M_2 = \frac{1}{\sqrt{n}} \Big(\int_{\mathbb{R}^n} \|x\|^2 d\gamma_n(x)\Big)^{1/2}$$

and then $M_1 \leqslant M_2 \leqslant \sqrt{2}M_1$.

THEOREM 6.5.1 (MM^*-estimate). *Let $X = (\mathbb{R}^n, \|\cdot\|)$ be an n-dimensional normed space and T be its unit ball. There exists a linear image T_1 of T such that*

$$M(T_1)M^*(T_1) \leqslant C\,K(X),$$

where $C > 0$ is an absolute constant.

Proof. By Theorem 6.4.5, there exists $u : \ell_2^n \to X$ with $\ell(u)\ell\big((u^{-1})^*\big) \leqslant nK(X)$. We set $T_1 = u^{-1}(T)$. Note that

$$\left(\int_{\mathbb{R}^n} \|u(x)\|_T^2 d\gamma_n(x)\right)^{1/2} = \left(\int_{\mathbb{R}^n} \|x\|_{T_1}^2 d\gamma_n(x)\right)^{1/2}$$

and

$$\left(\int_{\mathbb{R}^n} \big\|\big((u^{-1})^*(x)\big)\big\|_{T^\circ}^2 d\gamma_n(x)\right)^{1/2} = \left(\int_{\mathbb{R}^n} \|x\|_{T_1^\circ}^2 d\gamma_n(x)\right)^{1/2}.$$

Then, (6.5.1) shows that

$$M(T_1)M^*(T_1) \leqslant \frac{C}{n}\ell(u)\ell\big((u^{-1})^*\big) \leqslant C\,K(X)$$

as claimed. □

The next theorem, which is a direct consequence of Theorem 6.5.1, may be viewed as a *reverse Urysohn inequality.*

THEOREM 6.5.2. *Let $X = (\mathbb{R}^n, \|\cdot\|)$ be an n-dimensional normed space and T be its unit ball. There exists a linear image T_1 of T with volume $\mathrm{Vol}_n(T_1) = 1$ and mean width*

$$w(T_1) \leqslant c\sqrt{n}K(X), \tag{6.5.2}$$

where $c > 0$ is an absolute constant.

(Note that the volume normalization together with Urysohn's inequality ensures $w(K) \geqslant c\sqrt{n}$.)

Proof. Recall that $w(C) = 2M(C^\circ)$ for every centrally symmetric convex body C in $\mathbb{R}^n$. Then, Theorem 6.5.1 states that we can find $A \in GL_n$ such that $T_1 = A(T)$ satisfies

$$w(T_1)w(T_1^\circ) \leqslant c_1\,K(X).$$

We may clearly assume that $\mathrm{Vol}_n(T_1) = 1$. Then we observe that, from (1.5.3),

$$w(T_1^\circ) \geqslant \left(\frac{\kappa_n}{\mathrm{Vol}_n(T_1)}\right)^{1/n} \geqslant \frac{c_2}{\sqrt{n}}.$$

It follows that

$$\frac{c_2}{\sqrt{n}}w(T_1) \leqslant w(T_1)w(T_1^\circ) \leqslant K(X),$$

and this proves (6.5.2). □

Combining Pisier's inequality for $K(X)$ with John's theorem, we arrive at a very good estimate for the minimal mean width of an n-dimensional centrally symmetric convex body.

COROLLARY 6.5.3. *Every centrally symmetric convex body K in $\mathbb{R}^n$ has a linear image K_1 with volume 1 and mean width $w(K_1) \leqslant c\sqrt{n}\log n$.*

One can also show that the symmetry of K is not necessary in this last statement.

THEOREM 6.5.4. *Let K be a convex body in $\mathbb{R}^n$. There is an affine image K_1 of K with volume 1, such that*

$$w(K_1) \leqslant c\sqrt{n}\log n.$$

Proof. We may assume that the origin is in K. Consider the difference body $K - K$ of K. There is a linear transformation T of $\mathbb{R}^n$ such that $\operatorname{Vol}_n(T(K-K)) = 1$ and

$$w(T(K-K)) \leqslant c\sqrt{n}\log n.$$

Observe that $T(K - K) = TK - TK$, and

$$w(TK - TK) = \int_{S^{n-1}} h_{TK-TK}(u)\,d\sigma(u) = \int_{S^{n-1}} [h_{TK}(u) + h_{-TK}(u)]\,d\sigma(u)$$
$$= \int_{S^{n-1}} [h_{TK}(u) + h_{TK}(-u)]\,d\sigma(u) = 2w(TK).$$

On the other hand, by the Rogers-Shephard inequality (Theorem 1.5.2) we have

$$\operatorname{Vol}_n(TK) \geqslant 4^{-n}\operatorname{Vol}_n(TK - TK) = 4^{-n}.$$

This shows that $K_1 = c_1 TK$ has volume 1 for some $c_1 \leqslant 4$. From the above,

$$w(K_1) \leqslant 4w(TK) \leqslant 2c\sqrt{n}\log n.$$

$\square$

6.6. Equivalence of the two projections

In this section we will see that the K-convexity constant and the Rademacher constant of a finite dimensional normed space are equivalent up to an absolute constant.

THEOREM 6.6.1. *Let X be an n-dimensional normed space. Then, we have*

$$\frac{2}{\pi}K_r(X) \leqslant K(X) \leqslant K_r(X).$$

The proof is based on the next two lemmas.

LEMMA 6.6.2. *Let $\|\cdot\|$ be a norm on $\mathbb{R}^n$. If g_i are independent standard Gaussian random variables on some probability space Ω and r_i are the Rademacher functions on E_2^n, then*

$$\left(\int_{E_2^n} \left\|\sum_{i=1}^n r_i(\epsilon)e_i\right\|^2 d\mu_n(\epsilon)\right)^{1/2} \leqslant \sqrt{\frac{\pi}{2}}\left(\int_{\Omega}\left\|\sum_{i=1}^n g_i(\omega)e_i\right\|^2 d\omega\right)^{1/2}.$$

Proof. We note that

$$\int_{\Omega}|g_i(\omega)|d\omega = \frac{2}{\sqrt{2\pi}}\int_0^{\infty} te^{-\frac{t^2}{2}}dt = \sqrt{\frac{2}{\pi}}.$$

It follows that

$$\begin{aligned}\left(\int_{E_2^n}\Big\|\sum_{i=1}^n r_i(\epsilon)e_i\Big\|^2 d\epsilon\right)^{1/2} &= \frac{1}{\sqrt{\frac{2}{\pi}}}\left(\int_{E_2^n}\Big\|\sum_{i=1}^n r_i(\epsilon)\left(\int_\Omega |g_i(\omega)|d\omega\right)e_i\Big\|^2 d\epsilon\right)^{1/2}\\ &= \sqrt{\frac{\pi}{2}}\left(\int_{E_2^n}\Big\|\int_\Omega\sum_{i=1}^n r_i(\epsilon)|g_i(\omega)|e_i d\omega\Big\|^2 d\epsilon\right)^{1/2}\\ &\leqslant \sqrt{\frac{\pi}{2}}\left(\int_{E_2^n}\int_\Omega\Big\|\sum_{i=1}^n r_i(\epsilon)|g_i(\omega)|e_i\Big\|^2 d\omega d\epsilon\right)^{1/2}\\ &= \sqrt{\frac{\pi}{2}}\left(\int_\Omega\Big\|\sum_{i=1}^n g_i(\omega)e_i\Big\|^2 d\omega\right)^{1/2},\end{aligned}$$

if we take into account the fact that the variables g_i on Ω and $r_i|g_i|$ on $E_2^n\times\Omega$ have the same distribution. □

LEMMA 6.6.3. *Let (Ω_1,μ_1) and (Ω_2,μ_2) be two probability spaces. Let $f_i,g_i\in L_2(\Omega_1,\mu_1)$ and $\alpha_i,\beta_i\in L_2(\Omega_2,\mu_2)$, $i=1,\dots,n$. Let X be a Banach space and assume that there exist $C_1,C_2>0$ such that*

$$\left(\int_{\Omega_2}\Big\|\sum_{i=1}^n\alpha_i(\omega)x_i\Big\|^2 d\mu_2(\omega)\right)^{1/2}\leqslant C_1\left(\int_{\Omega_1}\Big\|\sum_{i=1}^n f_i(\omega)x_i\Big\|^2 d\mu_1(\omega)\right)^{1/2}$$

and

$$\left(\int_{\Omega_2}\Big\|\sum_{i=1}^n\beta_i(\omega)x_i^*\Big\|_*^2 d\mu_2(\omega)\right)^{1/2}\leqslant C_2\left(\int_{\Omega_1}\Big\|\sum_{i=1}^n g_i(\omega)x_i^*\Big\|_*^2 d\mu_1(\omega)\right)^{1/2}$$

for all $x_1,\dots,x_n\in X$ and $x_1^*,\dots,x_n^*\in X^*$. Let $T:L_2(\Omega_1,\mu_1;X)\to L_2(\Omega_1,\mu_1;X)$ and $S:L_2(\Omega_2,\mu_2;X)\to L_2(\Omega_2,\mu_2;X)$ be given by

$$T(\phi)=\sum_{i=1}^n f_i\int_{\Omega_1}g_i(\omega)\phi(\omega)\,d\mu_1(\omega)\quad\text{and}\quad S(\phi)=\sum_{i=1}^n\alpha_i\int_{\Omega_2}\beta_i(\omega)\phi(\omega)\,d\mu_2(\omega).$$

Then,

$$\|S\|\leqslant C_1C_2\|T\|.$$

Proof. A simple plugging in of the definitions gives that the adjoint operator T^*, just like in Lemma 6.1.8, is an operator $T^*:L_2(\Omega_1,\mu_1;X^*)\to L_2(\Omega_1,\mu_1;X^*)$ and is given by

$$T^*(\psi)=\sum_{i=1}^n g_i\int_{\Omega_1}f_i(\omega)\psi(\omega)\,d\mu_1(\omega).$$

We fix $\phi\in L_2(\Omega_2,\mu_2;X)$ and set

$$w=\sum_{i=1}^n f_i\int_{\Omega_2}\beta_i(\omega)\phi(\omega)\,d\mu_2(\omega)\in L_2(\Omega_1,\mu_1;X).$$

There exists a norming functional $\psi \in L_2(\Omega_1, \mu_1; X^*)$ such that $\|\psi\|_{L_2(X^*)} = 1$ and $\int_{\Omega_1} \langle \psi(\omega), w(\omega)\rangle \, d\mu_1(\omega) = \|w\|_{L_2(X)}$. Then,

$$\begin{aligned}
\|S(\phi)\| &= \left(\int_{\Omega_2} \left\| \sum_{i=1}^n \alpha_i(\omega) \int_{\Omega_2} \beta_i(\omega')\phi(\omega')\, d\mu_2(\omega') \right\|^2 d\mu_2(\omega) \right)^{1/2} \\
&\leqslant C_1 \|w\|_{L_2(\Omega_1,\mu_1;X)} = C_1 \int_{\Omega_1} \langle \psi(\omega'), w(\omega')\rangle \, d\mu_1(\omega') \\
&= C_1 \int_{\Omega_2} \int_{\Omega_1} \sum_{i=1}^n f_i(\omega')\beta_i(\omega) \langle \psi(\omega'), \phi(\omega)\rangle d\mu_1(\omega')\, d\mu_2(\omega) \\
&= C_1 \int_{\Omega_2} \Big\langle \sum_{i=1}^n \beta_i(\omega) \int_{\Omega_1} f_i(\omega')\psi(\omega')\, d\mu_1(\omega'), \phi(\omega) \Big\rangle \, d\mu_2(\omega) \\
&\leqslant C_1 \|\phi\|_{L_2(\Omega_2,\mu_2;X)} \left(\int_{\Omega_2} \left\| \sum_{i=1}^n \beta_i(\omega) \int_{\Omega_1} f_i(\omega')\psi(\omega')\, d\mu_1(\omega') \right\|^2 d\mu_2(\omega) \right)^{1/2} \\
&\leqslant C_1 C_2 \|\phi\|_{L_2(\Omega_2,\mu_2;X)} \|T^*(\psi)\| \leqslant C_1 C_2 \|\phi\|_{L_2(\Omega_2,\mu_2;X)} \|T\|,
\end{aligned}$$

where the last inequality follows since ψ is of norm 1 and since $\|T\| = \|T^*\|$. The proof of the lemma is complete. □

Proof of Theorem 6.6.1. We shall apply the previous lemma for the spaces $(\Omega_1, \mu_1) = (\mathbb{R}^n, \gamma_n)$ and $(\Omega_2, \mu_2) = (E_2^n, \mu_n)$. Lemma 6.6.2 shows that the conditions of Lemma 6.6.3 are satisfied with $C_1 = C_2 = \sqrt{\pi/2}$. We let $T = Q_1 \otimes \mathrm{Id}_X$ and $S = Rad_n$ and get from Lemma 6.6.3 that

$$\|Rad_n\| \leqslant \frac{\pi}{2} \|Q_1 \otimes \mathrm{Id}_X\|.$$

This shows the left hand side inequality.

For the right hand side inequality we shall use that Bernoulli random variables (represented here by the r_i's) converge, in average, to Gaussians. For every $i = 1, \dots, n$ and $m \geqslant 1$ we define

$$\delta_{i,m} = \frac{1}{\sqrt{m}} \sum_{j=0}^{m-1} r_{im+j},$$

where (r_s) is the sequence of Rademacher functions on $[0, 1]$. By the central limit theorem the joint distribution of $(\delta_{1,m}, \dots, \delta_{n,m})$ converges to that of $(g_1, \dots, g_n)$ as $m \to \infty$, where g_i are independent standard Gaussian random variables on some probability space (Ω, μ). It follows that

$$\mathbb{E}\, F(\delta_{1,m}, \dots, \delta_{n,m}) \to \mathbb{E}\, F(g_1, \dots, g_n)$$

as $m \to \infty$, for every continuous function $F : \mathbb{R}^n \to \mathbb{R}$ which satisfies e.g. the condition $F(u)\exp(-\|u\|_1) \to 0$ as $\|u\|_1 \to \infty$.

Let X be an n-dimensional normed space and let $x_1, \dots, x_n \in X$. Consider the function

$$F(u_1, \dots, u_n) = \left\| \sum_{i=1}^n u_i x_i \right\|^2.$$

Given $\phi \in L_2(\Omega, \mu; X)$ we use Lemma 6.6.3 (more precisely, the argument that we used for its proof) to write

$$\text{(6.6.1)} \qquad \|(Q_1 \otimes \mathrm{Id}_X)(\phi)\| = \left(\int_\Omega \left\| \sum_{i=1}^n g_i \int_\Omega g_i \phi \, d\mu \right\|^2 d\mu \right)^{1/2}$$

$$\leqslant \|\phi\|_{L_2(X)} \lim_{m\to\infty} \left(\mathbb{E} \left\| \sum_{i=1}^n \delta_{i,m} \, \mathbb{E}\left(\delta_{i,m}\psi\right) \right\|^2 \right)^{1/2}$$

for some $\psi \in L_2(X^*)$ with $\|\psi\|_{L_2(X^*)} = 1$.

Now, for every $1 \leqslant i \leqslant n$ and $0 \leqslant j \leqslant m-1$ we define

$$w(im+j) = \mathbb{E}\left(r_{im+j}\psi\right)$$

and consider the permutation σ_j of $\{im, im+1, \ldots, (i+1)m-1\}$ which is defined by

$$\sigma_j(im+k) = im + (k+j) \bmod m \qquad (k = 0, \ldots, m-1).$$

Using the convexity of $\|\cdot\|_{L_2(X^*)}$ we get

$$\left(\mathbb{E} \left\| \sum_{i=1}^n \delta_{i,m} \, \mathbb{E}\left(\delta_{i,m}\psi\right) \right\|^2 \right)^{1/2} = \left(\mathbb{E} \left\| \sum_{i=1}^n \sum_{k=0}^{m-1} \sum_{j=0}^{m-1} \frac{1}{\sqrt{m}} r_{im+k} w(im+j) \right\|^2 \right)^{1/2}$$

$$\leqslant \frac{1}{m} \sum_{j=0}^{m-1} \left(\mathbb{E} \left\| \sum_{i=1}^n \sum_{k=0}^{m-1} r_{im+k} w(\sigma_j(im+k)) \right\|^2 \right)^{1/2}$$

$$= \left(\mathbb{E} \left\| \sum_{s=1}^{nm} r_s w(s) \right\|^2 \right)^{1/2} = \|Rad_{nm}(\psi)\|_{L_2(X^*)}$$

$$\leqslant K_r(X^*) \|\psi\|_{L_2(X^*)} = K_r(X).$$

Letting $m \to \infty$ and taking into account (6.6.1) we conclude that

$$\|(Q_1 \otimes \mathrm{Id}_X)(\phi)\| \leqslant K_r(X) \, \|\phi\|_{L_2(X)}.$$

It follows that $K(X) = \|(Q_1 \otimes \mathrm{Id}_X)(\phi)\| \leqslant K_r(X)$. □

6.7. Bourgain's example

Let X be an n-dimensional normed space. In Section 6.3 we saw that the Rademacher constant $K_r(X)$ is bounded by $C \log[d(X, \ell_2^n) + 1] \leqslant C \log(n+1)$. In the case of a space X with an unconditional basis, one can improve this bound to $K_r(X) \leqslant C\sqrt{\log(n+1)}$. However, Bourgain showed that, in general, the bound $O(\log n)$ is best possible. His example is an L_D^∞-space, where D is a random set of Walsh functions on $E_2^n = \{-1, 1\}^n$.

THEOREM 6.7.1 (Bourgain). *For every $n \geqslant 1$ we can find a subset D of $\{w_A : A \subseteq \{1, \ldots, n\}\}$ and a function $\phi \in L_D^\infty(E_2^n)$ such that:*

(i) $\|\phi\|_\infty \leqslant C$,
(ii) $\widehat{\phi}(k) = \frac{1}{\sqrt{n}}$ *for all $k = 1, \ldots, n$,*

(iii) $\log|D| \leqslant C\sqrt{n}$,

where $C > 0$ is an absolute constant.

Proof. We define $\chi : E_2^n \to \mathbb{C}$ by

$$\chi(\epsilon) = \prod_{k=1}^{n}\Big(1 + \frac{i}{\sqrt{n}}\epsilon_k\Big).$$

Then,

$$\|\chi\|_\infty = \Big(1 + \frac{1}{n}\Big)^{n/2} \leqslant 3.$$

We fix a positive integer m (which will be later chosen of order $\log n$). For every $S \subseteq \{1, \ldots, n\}$ with $|S| = k > m$ we consider a $0-1$ valued random variable ξ_S with $\mathbb{E}(\xi_S) = \frac{2^k}{k!}$. We also assume that the random variables ξ_S are independent.

We write

$$\beta(\epsilon) := \operatorname{Im}\chi(\epsilon) = \sum_S \frac{1}{n^{|S|/2}}\nu_S w_S(\epsilon),$$

where $\nu_S \in \{0, 1, -1\}$. Next, we consider the random set

$$D(\omega) = \{S \subseteq \{1, \ldots, n\} : |S| \leqslant m \text{ or } \xi_S(\omega) = 1\}$$

and the random function

$$\psi_\omega := \sum_{k=0}^{m}\sum_{|S|=k}\frac{1}{n^{k/2}}\nu_S w_S + \sum_{k=m+1}^{n}\sum_{|S|=k}\xi_S(\omega)\frac{k!}{2^k}\frac{1}{n^{k/2}}\nu_S w_S.$$

Note that all non-zero Walsh coefficients $\widehat{\psi_\omega}$ of ψ_ω correspond to $S \in D(\omega)$.

Choose $m \simeq \log n$. We first compute

$$\mathbb{E}(|D(\omega)|) = \sum_{k=0}^{m}\binom{n}{k} + \sum_{k=m+1}^{n}\binom{n}{k}\frac{2^k}{k!} \leqslant 2n^m + 4^{3\sqrt{n}} \leqslant 4^{4\sqrt{n}}. \tag{6.7.1}$$

Then, we write

$$\begin{aligned}\int\|\psi_\omega\|_\infty\, d\omega &\leqslant \int\Big\|\psi_\omega - \int\psi_\omega d\omega\Big\|_\infty d\omega + \|\beta\|_\infty \\ &\leqslant 3 + \sum_{k=m+1}^{n}\Big\|\sum_{|S|=k}\big(\xi_S(\omega) - \mathbb{E}(\xi_S)\big)\frac{k!}{2^k}\frac{1}{n^{k/2}}\nu_S w_S\Big\|_\infty d\omega \\ &\leqslant 3 + 4\sum_{k=m+1}^{n}\frac{k!}{2^k}\frac{1}{n^{k/2}}\iint\Big\|\sum_{|S|=k}\delta_S\xi_S(\omega)\nu_S w_S\Big\|_\infty d\omega\, d\delta,\end{aligned}$$

where $(\delta_S)_{|S|>m}$ is an independent set of Walsh functions. Note also that if $f : E_2^n \to \mathbb{R}$ then $\|f\|_\infty \leqslant 2\|f\|_n$.

Using Khintchine's inequality, we get

$$\int\Big\|\sum_{|S|=k}\delta_S\xi_S(\omega)\nu_S w_S\Big\|_n d\delta \leqslant \sqrt{n}\Big(\sum_{|S|=k}\xi_S(\omega)\Big)^{1/2},$$

and hence

$$\mathbb{E}_\omega \int \left\| \sum_{|S|=k} \delta_S \xi_S(\omega) \nu_S w_S \right\|_n d\delta \leqslant \sqrt{n} \binom{n}{k}^{1/2} \left(\frac{2^k}{k!} \right)^{1/2}.$$

It follows that

$$\int \|\psi_\omega\|_\infty \, d\omega \leqslant 3 + 8\sqrt{n} \sum_{k=m+1}^{n} \left(\frac{k!}{2^k} \frac{1}{n^k} \binom{n}{k} \right)^{1/2} \leqslant 3 + 8\sqrt{n} 2^{-m/2}. \tag{6.7.2}$$

From (6.7.1) and (6.7.2) we see that there exists ω such that $\|\psi_\omega\|_\infty \leqslant 8$ and $\log|D(\omega)| \leqslant 5\sqrt{n}$. We define $\phi = \psi_\omega$ and $D = D(\omega)$. Then $\phi \in L_D^\infty$ and satisfies our claim. □

Observe that if we set $X = L_D^\infty$ and consider the function $F(x) = \phi_x$, then

$$Rad_n F(x)(\epsilon) = \sum \widehat{\phi}(k) x_k \epsilon_k.$$

From Theorem 6.7.1 (ii) and (iii) it follows that

$$\|Rad_n F\| = \sqrt{n} \simeq \log(\dim(X)).$$

This proves the following.

THEOREM 6.7.2 (Bourgain). *There exists a sequence $N_k \to \infty$ such that for every k we can find a normed space X_k with $\dim(X_k) = N_k$ and $K_r(X_k) \geqslant c \log N_k$, where $c > 0$ is an absolute constant.* □

6.8. Notes and remarks

The notion of K-convexity appears in the paper of Maurey and Pisier [**421**]. The original definition concerned the Rademacher constant $K_r(X)$. The class of K-convex Banach spaces, the ones for which $K(X) < \infty$, was characterized by Pisier in [**506**] as follows:

THEOREM 6.8.1 (Pisier). *A Banach space X is not K-convex if and only if for every $n \geqslant 1$ and every $\varepsilon > 0$ there exists an n-dimensional subspace F_n of X such that $d_{BM}(F_n, \ell_1^n) \leqslant 1 + \varepsilon$.*

While Theorem 6.8.1 is formulated for an infinite dimensional space X, in our setting of finite dimensional normed spaces it should be seen as a statement on a family of finite dimensional spaces $\{X_n\}$ with $\dim(X_n) = n \to \infty$. Then the result may be reformulated for this family as follows: If $K(X_n)$ is not uniformly bounded then for every k and $\varepsilon > 0$ there exist X_n and a k-dimensional subspace $F \subset X_n$ such that $d_{BM}(F_n, \ell_1^k) \leqslant 1 + \varepsilon$.

We mention the fact that Lemma 6.6.2 has a reverse direction which is highly nontrivial, called the Maurey-Pisier Lemma. For general norms, it includes a factor logarithmic in dimension. The equivalence of the two sides in the inequality of Lemma 6.6.2 (even if only up to a logarithmic factor) is quite surprising geometrically - since the integral on one side is compared with, on the other side, the average on a relatively small and highly structured set of points - the vertices of a cube. It will be discussed in detail in Part II of this book, where a whole chapter will be devoted to type-cotype theory; a proof of Theorem 6.8.1 will be provided there as well. The reader is also referred to [**507**] for more details and a beautiful proof of Lemma 6.6.2 based on a method due to Kwapien.

Hermite polynomials

Basic properties of Hermite polynomials are discussed in detail in the book [**19**] of Andrews, Askey and Roy. Proposition 6.1.6 introduces a classical property of the Ornstein-Uhlenbeck semigroup that will be studied in more detail in Part II of this book.

Pisier's inequality

The first proof of Pisier's inequality (Theorem 6.2.4) appeared in [**505**] (see also a remark in [**504**]). Pisier arrived at this theorem in a few steps. In his article [**502**], where the first answer on the question to estimate the norm of the Rademacher projection was given, he proved that $K(X) \leqslant c(d_{BM}(X,\ell_2^n))^{2/3}$. Let us note that all the deep consequences of Theorem 6.2.4 that we present in the next two chapters would also follow, in their essential part, from this first estimate. This makes one expect that new interesting consequences may appear in the future, making use of the remarkable Pisier's logarithmic result in its full strength. Our exposition follows the one from Pisier's book [**508**] (where a simpler argument is used, employing Lemma 6.2.5 from [**110**]).

Let X be an n-dimensional normed space. In Section 6.3 we saw that the Rademacher constant $K_r(X)$ is bounded by $C\log[d(X,\ell_2^n)+1] \leqslant C\log(n+1)$. In the case of a space X with an unconditional basis, one can improve this bound to $K_r(X) \leqslant C\sqrt{\log(n+1)}$. In Section 6.7 we present a result of Bourgain who showed in [**97**] that, in general, the bound $O(\log n)$ is best possible.

The Rademacher projection

Our exposition in this section follows the appendix of Bourgain and Milman's [**110**] and the book of Milman and Schechtman [**464**].

The Rademacher projection can be thought of as a linearization procedure - given 2^n points in a Banach space X, $\{x_i\}$, one labels them according to the vertices of the cube to attain a function $f:\{-1,1\}^n \to X$, and then $Rad_n f$ can be encoded by n vectors $\{y_i\}_{i=1}^n$, the coefficients of r_i; this point of view was used in [**111**] (see also [**507**]) where Bourgain, Milman and Wolfson introduce a notion of type for metric spaces and prove an analogue of Theorem 6.8.1 for metric spaces. They also showed that if the n-dimensional Hamming cube can be C-Lipschitz embedded into some Banach space then this space contains a k-dimensional subspace which is 2-isomorphic to ℓ_1^k where k increases to infinity when $n\to\infty$. A connected but slightly different notion, equivalent to metric type, was considered earlier by P. Enflo in [**188**] (see also [**189**]).

The ℓ-norm

Lemma 6.4.2, in the form it is written, is due to Lewis. The ℓ-norm was introduced by Figiel and Tomczak-Jaegermann in [**206**] and Theorem 6.4.5 appears in the same paper.

The MM^*-estimate

Although Bourgain's example shows that one cannot expect in general a better dependence on the dimension for the norm of the Rademacher projection, it is unknown whether the MM^*-estimate is optimal. It remains an open question to check if there exists an absolute constant $c>0$ such that for every centrally symmetric convex body K in $\mathbb{R}^n$ one can find a position K_1 of K with $M(K_1)M^*(K_1) \leqslant c\sqrt{\log n}$ (note that the minimum of MM^* over all positions of the cube is of the order of $\sqrt{\log n}$). If true, the $\sqrt{\log n}$-bound would make many things "exact". In particular, this would imply a sharp reverse Urysohn inequality since Theorem 6.5.4 would take the following form: *every convex body K in $\mathbb{R}^n$ has an* *affine image* K_1 *of volume* 1 *such that*

$$w(K_1) \leqslant c\sqrt{n}\sqrt{\log n}.$$

It is interesting to observe that although the minimal mean width position of a convex body admits a nice isotropic geometric characterization (Theorem 2.2.4), this extremal position has not been exploited in this direction. A similar situation appears with Petty's characterization of the minimal surface area position (Theorem 2.3.1): Ball's sharp reverse isoperimetric inequality made use of John position. For non-symmetric estimates on MM^*, see Rudelson [**528**].

Equivalence of the two projections

The equivalence of the Rademacher and Gaussian projection were known to Pisier from a very early stage of the theory, see e.g. [**503**]. Lemma 6.6.2 which gives the estimate $K_r(X) \leqslant (\pi/2)K(X)$ is standard; together with the reverse inequality $K(X) \leqslant K_r(X)$ it appears in the paper of Figiel and Tomczak-Jaegermann [**206**].

CHAPTER 7

Proportional Theory

7.1. Introduction

Let $X = (\mathbb{R}^n, \|\cdot\|)$ be an n-dimensional normed space. Milman's estimate for Dvoretzky theorem (Theorem 5.2.10), describes the family of k-dimensional subspaces of X in the case $k \leqslant k(X) \simeq n(M/b)^2$, where $M = \int_{S^{n-1}} \|x\| \, d\sigma(x)$ and $b = \max\{\|x\| : x \in S^{n-1}\}$: most of them are c-Euclidean for some absolute constant $c > 0$. It may happen that $k(X)$ is proportional to n, but the example of the cube (with $k(X) \simeq \log n$, see Theorem 5.4.2) shows that "proportional Euclidean structure" cannot be expected in full generality. However, surprisingly enough, there is non-trivial information about subspaces of dimension λn, even when the "proportion" $\lambda \in (0,1)$ is very close to 1. The first step in this direction is Milman's M^*-estimate.

THEOREM 7.1.1 (Milman). *There exist two functions $f, h : (0,1) \to \mathbb{R}^+$ such that for every $\lambda \in (0,1)$ and every n-dimensional normed space $X = (\mathbb{R}^n, \|\cdot\|)$, the set $B \subset G_{n,\lfloor \lambda n \rfloor}$ of subspaces $F \in G_{n,\lfloor \lambda n \rfloor}$ satisfying*

$$\frac{f(\lambda)}{M^*}|x| \leqslant \|x\|$$

for all $x \in F$, has measure at least $1 - \exp(-h(\lambda)n)$ where $M^* = \int_{S^{n-1}} \|x\|^* d\sigma(x)$.

In a more geometric language, Theorem 7.1.1 says that for every centrally symmetric convex body K in $\mathbb{R}^n$ and every $\lambda \in (0,1)$, a random $\lfloor \lambda n \rfloor$-dimensional section $K \cap F$ of K satisfies the inclusion

$$K \cap F \subseteq \frac{M^*(K)}{f(\lambda)} B_2^n \cap F.$$

In other words, the diameter of a random "proportional section" of a high-dimensional centrally symmetric convex body K is controlled by the mean width $w(K) = M^*(K)$ of the body; a random section of K does not feel the diameter $R(K)$ but the radius $M^*(K)$ which is, roughly speaking, the level r at which half of the supporting hyperplanes of rB_2^n intersect the body K.

If we apply Theorem 7.1.1 to the polar body K° of K and then consider the polar bodies of $K^\circ \cap F$ and $B_2^n \cap F$ for a random subspace F, we obtain the following equivalent statement: for any $\lambda \in (0,1)$, with high probability on subspaces $F \in G_{n,\lfloor \lambda n \rfloor}$ we have

$$P_F(K) \supseteq \frac{f(\lambda)}{M(K)} B_2^n \cap F. \tag{7.1.1}$$

In Sections 7.2 and 7.3 we present several proofs of the M^*-estimate; based on these, we will be able to say more about the best possible function f for which the theorem holds true and about the corresponding estimate for the probability of subspaces in which this occurs. In Section 7.4 we shall show an important application of the M^* estimate, Milman's quotient of a subspace theorem. It states that for any normed space $X = (\mathbb{R}^n, \|\cdot\|_K)$ and any proportion $\lambda \in (0,1)$ we may find subspaces $F \subseteq E \subset \mathbb{R}^n$ with $\dim F = \lambda n$ such that $P_F(K \cap E)$ is $c(\lambda)$ isomorphic to an ellipsoid. That is, a quotient space of a subspace of X, of proportional dimension, is isomorphically Euclidean. In Section 7.5 we discuss some improvement of the M^* estimate (which is of course an estimate for the outer-radius of a random section) and complement it by a lower bound for the outer-radius of sections of K, which holds for *all* subspaces. We also compare "best" sections with "random" ones of slightly lower dimension. In Section 7.6 we provide a linear relation between the outer-radius of a section of K and the outer-radius of a section of K°. This relies on the so called "distance lemma" which is in itself a useful tool.

7.2. First proofs of the M^*-estimate

In this section we present three proofs of Theorem 7.1.1. The first two are due to V. Milman and they are the first chronologically as well.

7.2.1. Approach through volume

The first proof of Theorem 7.1.1 was based on the volume ratio theorem (Theorem 5.5.3) and Urysohn's inequality (Theorem 1.5.11).

THEOREM 7.2.1. *Theorem* 7.1.1 *holds with* $f(\lambda) = C^{-1/(1-\lambda)}$ *and* $h(\lambda) = 1$.

Proof. Let K denote the unit ball of X, and denote $M^* = M^*(X) = M^*(K)$. It is clear that we can work with the body $\frac{1}{M^*}K$ instead of K, and hence without loss of generality we may assume that $M^* = 1$. Consider the body $T = \mathrm{conv}(K \cup B_2^n)$. Since $T \subseteq K + B_2^n$, we have

$$M^*(T) \leqslant M^*(K + B_2^n) = M^*(K) + M^*(B_2^n) = 2.$$

From Urysohn's inequality, Theorem 1.5.11, we have

$$\left(\frac{\mathrm{Vol}_n(T)}{\mathrm{Vol}_n(B_2^n)}\right)^{1/n} \leqslant M^*(T) \leqslant 2.$$

Since $T \supseteq B_2^n$, the volume ratio theorem, Theorem 5.5.3, shows that, for every $\lambda \in (0,1)$, a subspace $F \in G_{n,\lfloor \lambda n \rfloor}$ satisfies

$$\frac{1}{C^{1/(1-\lambda)}}|x| \leqslant \|x\|_T$$

for all $x \in F$, with probability at least $1 - e^{-n}$. Here $C > 0$ is an absolute constant coming from Theorem 5.5.3. Finally, note that $\|x\|_T \leqslant \|x\|$ because $T \supseteq K$. This completes the proof. □

Note. An inspection of the proof of Theorem 5.5.3 shows that one can actually have that, for every $\lambda \in (0,1)$ and $t > 1$, a subspace $F \in G_{n,\lfloor \lambda n \rfloor}$ satisfies

$$\frac{1}{(Ct)^{1/(1-\lambda)}}|x| \leqslant \|x\|_T$$

for all $x \in F$, with probability at least $1 - t^{-n}$.

7.2.2. Approach through the isoperimetric inequality

Milman's second proof of Theorem 7.1.1 is based on the isoperimetric inequality for S^{n-1} and establishes linear dependence on $(1-\lambda)$. Recall that for every Borel subset A of S^{n-1} and any $t > 0$, the t-neighborhood of A is the set $A_t = \{x : \mathrm{dist}(x, A) < t\}$ where the distance is with respect to the geodesic metric. We will use the following lemma.

LEMMA 7.2.2. *There exists an absolute constant $c > 0$ and a function $\gamma(\lambda) \geqslant c(1-\lambda)$ such that for any Borel set $A \subseteq S^{n-1}$ with $\sigma(A) \geqslant 1/2$, for every $\lambda \in (1/2, 1)$, the measure of subspaces F of dimension $k = \lfloor \lambda n \rfloor$ which satisfy*

$$F \cap S^{n-1} \subseteq A_{\frac{\pi}{2}-\gamma},$$

is greater than $1 - \exp(-c(1-\lambda)n)$.

Proof. Let $I_k = \int_0^{\pi/2} \sin^k t\, dt$. By the spherical isoperimetric inequality, for any $0 < \varepsilon < \frac{\pi}{2}$ and any $x \in S^{n-1}$ we have

$$\begin{aligned}\sigma(A_{\frac{\pi}{2}-\varepsilon}) &\geqslant \sigma(B(x, \pi - \varepsilon)) = 1 - \sigma(B(x, \varepsilon))\\ &= 1 - \frac{1}{2I_{n-2}} \int_0^{\varepsilon} \sin^{n-2} t\, dt \geqslant 1 - c_1(\varepsilon)\sqrt{n} \sin^{n-2} \varepsilon.\end{aligned}$$

Double integration over $G_{n,k}$ gives

$$\int_{G_{n,k}} \sigma_k(A_{\frac{\pi}{2}-\varepsilon} \cap F) d\nu_{n,k}(F) = \sigma(A_{\frac{\pi}{2}-\varepsilon}) \geqslant 1 - c_1(\varepsilon)\sqrt{n} \sin^{n-2} \varepsilon.$$

So, by Markov's inequality, with probability greater than $1 - \sin^{n-\sqrt{\lambda n}} \varepsilon$ on subspaces $F \in G_{n,k}$,

$$\sigma_k(A_{\frac{\pi}{2}-\varepsilon} \cap F) \geqslant 1 - c_2(\varepsilon)\sqrt{n} \sin^{\sqrt{\lambda n}} \varepsilon,$$

where $c_2(\varepsilon) > 0$ is a constant depending on ε. On the other hand, setting $s = k^{-3/2}$, for every $x \in S^{n-1} \cap F$ and any $0 < \delta < \frac{\pi}{2}$ we have

$$\begin{aligned}\sigma_F\left(B(x, \delta)\right) &= \frac{1}{2I_{n-2}} \int_0^{\delta} \sin^{k-2} t\, dt \geqslant \frac{1}{2I_{k-2}} \int_{(1-s)\delta}^{\delta} \sin^{k-2} t\, dt\\ &\geqslant c_3\sqrt{k}\delta s \sin^{k-2}((1-s)\delta) \geqslant \frac{c_4}{k} \sin^{k-1} \delta,\end{aligned}$$

where $c_4 > 0$ is an absolute constant and σ_F is the rotationally invariant probability measure on $S^{n-1} \cap F$. This implies that if

$$\frac{c_4}{k} \sin^{k-1} \delta + 1 - c_2(\varepsilon)\sqrt{n} \sin^{\sqrt{\lambda n}} \varepsilon > 1, \tag{7.2.1}$$

then for any $x \in S^{n-1}$

$$A_{\frac{\pi}{2}-\varepsilon} \cap B(x, \delta) \neq \emptyset,$$

and hence $x \in A_{\frac{\pi}{2}-\varepsilon+\delta}$, which implies

$$F \cap S^{n-1} \subseteq A_{\frac{\pi}{2}-\gamma},$$

where $\gamma = \varepsilon - \delta$.

We proceed to analyze the condition (7.2.1). Rearranging and setting $\lambda = k/n$ we require that

$$\sin^{k-1} \delta \geqslant c_5(\varepsilon)\lambda n^{3/2} \sin^{\sqrt{\lambda n}} \varepsilon,$$

or equivalently (taking n-th roots, and swallowing one term λ into the constant)

$$\sin\varepsilon \leqslant \alpha_n \sin^{\sqrt{\lambda}}\delta = \alpha_n \sin^{\sqrt{\lambda}}(\varepsilon-\gamma), \tag{7.2.2}$$

where $\alpha_n \leqslant (c_4/n)^{\frac{3}{2n}}$. Fix $\varepsilon = \pi/4$. Then, (7.2.2) takes the form

$$2^{\frac{\sqrt{\lambda}-1}{2\sqrt{\lambda}}}\alpha_n^{-\frac{1}{\sqrt{\lambda}}} \leqslant \cos\gamma(1-\tan\gamma).$$

Taking logarithms we see that we need

$$\frac{\sqrt{\lambda}-1}{2\sqrt{\lambda}}\log 2 + \frac{1}{\sqrt{\lambda}}|\log\alpha_n| \leqslant \log(\cos\gamma) + \log(1-\tan\gamma).$$

Note that $\log(\cos\gamma) \simeq -\gamma^2$ and $\log(1-\tan\gamma) \simeq -\gamma$ for small γ, and hence

$$\log(\cos\gamma) + \log(1-\tan\gamma) \geqslant c_6\gamma.$$

Also,

$$|\log\alpha_n| \leqslant \frac{c_7\log n}{n}.$$

Therefore, we may choose $\gamma > 0$

$$\gamma \geqslant c_8\Big[\frac{1-\sqrt{\lambda}}{2\sqrt{\lambda}} - \frac{\log n}{\sqrt{\lambda}n}\Big]$$

so that (7.2.2) will be satisfied. Note that as $\lambda \geqslant \frac{1}{2}$ then the condition can be simplified to $\gamma \geqslant c(1-\lambda)$ where $c > 0$ is an absolute constant.

The argument shows that a subspace $F \in G_{n,k}$ satisfies the above with probability greater than $1-(\sin\varepsilon)^{(1-\sqrt{\lambda})n}$, where $\varepsilon = \pi/4$. Therefore, this probability is greater than $1-e^{-c(1-\lambda)n}$. □

THEOREM 7.2.3. *Theorem* 7.1.1 *holds with* $f(\lambda) = C(1-\lambda)$ *and probability estimate* $1-e^{-c(1-\lambda)n}$.

Proof. Consider the set

$$A = \{y \in S^{n-1} : \|y\|_* \leqslant 2M^*\}.$$

From Markov's inequality we have $\sigma(A) \geqslant 1/2$.

We may assume that $\lambda \in (1/2, 1)$. From Lemma 7.2.2 we know that a random subspace F of dimension $k = \lfloor\lambda n\rfloor$ satisfies with probability greater than $1-e^{-c(1-\lambda)n}$

$$F \cap S^{n-1} \subseteq A_{\frac{\pi}{2}-\gamma},$$

where $\gamma \geqslant c(1-\lambda)$ and $c > 0$ is an absolute constant. Consider any F with this property and let $x \in S^{n-1}\cap F$. Since $\mathrm{dist}(x, A) \leqslant \frac{\pi}{2}-\gamma$, there exists $y \in A$ such that

$$\sin\gamma \leqslant |\langle x, y\rangle| \leqslant \|y\|_*\|x\| \leqslant 2M^*\|x\|,$$

and since $\sin\gamma \geqslant \frac{2}{\pi}\gamma \geqslant \frac{2c}{\pi}(1-\lambda)$, the theorem follows. □

7.2.3. Approach through covering

Next, we give a third proof of Theorem 7.1.1 in its equivalent dual form (7.1.1); the argument below is due to Litvak, Milman and Pajor and employs the *Johnson-Lindenstrauss lemma.*

LEMMA 7.2.4 (Johnson-Lindenstrauss). *There exist absolute constants $c_i > 0$, $i = 1,2,3$, such that if $\sqrt{c_1/k} < \varepsilon < 1$ and $N < \exp(c_2\varepsilon^2 k)$ then for every $\{y_1,\dots,y_N\} \subset S^{n-1}$ and every k-dimensional subspace F of $\mathbb{R}^n$ there exists a set $B \subseteq O(n)$ of measure $\nu_n(B) \geqslant 1-\exp(-c_3\varepsilon^2 k)$ such that for every $U \in B$ we have*

$$(1-\varepsilon)\sqrt{k/n} \leqslant |P_F U(y_j)| \leqslant (1+\varepsilon)\sqrt{k/n}$$

for all $1 \leqslant j \leqslant N$.

Proof. Consider the function $f(x) = |P_F(x)|^2$ on S^{n-1}. It is easy to check that f is a Lipschitz continuous function with constant 2, and $\mathbb{E}(f) = k/n$. By the spherical isoperimetric inequality (see Theorem 3.2.2) it follows that, for every $\varepsilon \in (0,1)$,

$$\sigma\left(\left\{x \in S^{n-1} : \left||P_F(x)|^2 - \frac{k}{n}\right| \geqslant \varepsilon\frac{k}{n}\right\}\right) \leqslant \sqrt{\pi/2}\exp(-\varepsilon^2 k/8). \tag{7.2.3}$$

We define

$$B_\varepsilon = \left\{U \in O(n) : \frac{(1-\varepsilon)k}{n} \leqslant |P_F U(y_j)|^2 \leqslant \frac{(1+\varepsilon)k}{n} \text{ for all } j = 1,\dots,N\right\}.$$

If $\varepsilon > \sqrt{c_1/k}$ then (7.2.3) shows that

$$\begin{aligned}1-\nu_n(B_\varepsilon) &\leqslant N\sigma\left(\left\{x \in S^{n-1} : \left||P_F(x)|^2 - \frac{k}{n}\right| \geqslant \varepsilon\frac{k}{n}\right\}\right)\\ &\leqslant N\sqrt{\pi/2}\exp(-\varepsilon^2 k/8)\\ &\leqslant \exp(-c_3\varepsilon^2 k)\end{aligned}$$

provided that the absolute constant c_2 is chosen small enough. □

LEMMA 7.2.5. *Let T and K be two centrally symmetric convex bodies in $\mathbb{R}^m$. Assume that there exist $0 < \alpha < 1$ and $x_i \in \alpha T$ such that $T \subseteq \bigcup(x_i + K)$. Then,*

$$T \subseteq \frac{1}{1-\alpha}K.$$

Proof. Let s be the smallest positive number for which $\|x\|_K \leqslant s\|x\|_T$ for all $x \in \mathbb{R}^m$. Assume that $y \in T$. There exists $x_i \in \alpha T$ such that $y - x_i \in K$. Then

$$\|x_i\|_K \leqslant s\|x_i\|_T \leqslant \alpha s,$$

and hence

$$\|y\|_K \leqslant \|y-x_i\|_K + \|x_i\|_K \leqslant 1 + \alpha s.$$

Since $y \in T$ was arbitrary, we have $s \leqslant 1+\alpha s$. This shows that $s \leqslant 1/(1-\alpha)$. □

We will also use the dual Sudakov inequality, that is, Theorem 4.2.2 from Chapter 4, which says that if K is a centrally symmetric convex body in $\mathbb{R}^n$ then, for every $t > 0$,

$$N(B_2^n, tK) \leqslant \exp\left(cn(M/t)^2\right),$$

where $c > 0$ is an absolute constant.

THEOREM 7.2.6. *Theorem 7.1.1 holds with $f(\lambda) = C(1-\lambda)^2$ and probability estimate $1 - e^{-c(1-\lambda)^2 k}$, for some absolute constants c, C.*

Proof. Let K be a centrally symmetric convex body in $\mathbb{R}^n$; let $\lambda \in (0,1)$ and $k = \lfloor \lambda n \rfloor$. Without loss of generality we may assume that $\lambda > 1/9$.

Let $t > 0$ be a constant which will be suitably chosen. By the dual Sudakov inequality we can find $N \leqslant \exp\left(cn(M/t)^2\right)$ and $x_1, \ldots, x_N \in B_2^n$ such that $B_2^n \subseteq \bigcup_{i \leqslant N}(x_i + tK)$. If $\sqrt{c_1/k} < \varepsilon < 1$ and

(7.2.4) $$c_2 \varepsilon^2 k \leqslant cn(M/t)^2$$

where $c_1, c_2 > 0$ are the constants in Lemma 7.2.4, then there exists $B \subset G_{n,k}$ with measure at least $1 - e^{-c_3 \varepsilon^2 k}$ such that for any $F \in B$ and all $1 \leqslant i \leqslant N$ we have

$$|P_F(x_i)| \leqslant (1+\varepsilon)\sqrt{\lambda}.$$

Then,

$$B_F := B_2^n \cap F \subseteq \bigcup_{i=1}^{N}(y_i + tP_F(K))$$

where $y_i = P_F(x_i)$ and $|y_i| \leqslant (1+\varepsilon)\sqrt{\lambda}$ for all $1 \leqslant i \leqslant N$. If

(7.2.5) $$\alpha := (1+\varepsilon)\sqrt{\lambda} < 1,$$

we may apply Lemma 7.2.5 to conclude that

(7.2.6) $$B_F \subseteq \frac{t}{1-\alpha}P_F(K).$$

It remains to choose ε and t. We set $\varepsilon = \frac{1-\sqrt{\lambda}}{2\sqrt{\lambda}}$. Since $\lambda > 1/9$ we have $\sqrt{c_1/k} < \varepsilon < 1$ (the left hand side inequality holds true if we assume that n is large enough). Also, $\alpha = \frac{1+\sqrt{\lambda}}{2} < 1$; that is, (7.2.5) is satisfied. Finally, (7.2.4) is satisfied if we choose

$$t = \sqrt{\frac{4c}{c_2}}\frac{M}{1-\sqrt{\lambda}} = \frac{c_4 M}{1-\sqrt{\lambda}}.$$

Going back to (7.2.6) we get

$$B_F \subseteq \frac{2c_4 M}{(1-\sqrt{\lambda})^2}P_F(K) \subseteq \frac{c_5 M}{(1-\lambda)^2}P_F(K).$$

$\square$

7.3. Proofs with the optimal dependence

In this section another proof of Theorem 7.1.1. As we will see, one can have $f(\lambda) \geqslant c\sqrt{1-\lambda}$, where $c > 0$ is an absolute constant. This optimal dependence in the M^*-estimate was first obtained by Pajor and Tomczak-Jaegermann. In fact, Gordon showed that one may even choose c to be almost 1, but this we shall only state in this section, and defer the proof to Chapter 9.

THEOREM 7.3.1 (Pajor-Tomczak). *There exist absolute constants $c, c' > 0$ such that for every $\lambda \in (0,1)$ and every n-dimensional normed space $X = (\mathbb{R}^n, \|\cdot\|)$, the set $B \subset G_{n,\lfloor \lambda n \rfloor}$ of subspaces $F \in G_{n,\lfloor \lambda n \rfloor}$ satisfying*

$$\frac{c\sqrt{1-\lambda}}{M^*}|x| \leqslant \|x\|$$

for all $x \in F$, has measure at least $1 - \exp(-c'(1-\lambda)n)$.

REMARK 7.3.2. It is useful to give an equivalent formulation of the M^*-estimate in terms of the Gelfand numbers and the ℓ-norm. Let $X = (\mathbb{R}^n, \|\cdot\|)$ be an n-dimensional normed space with unit ball K, and let $u : \ell_2^n \to X^*$ be an isomorphism. Note that $c_k(u^*) \leqslant r$ is equivalent to the existence of a subspace $F \in G_{n,n-k+1}$ such that $|u^*(x)| \leqslant r\|x\|$ for all $x \in F$, which is in turn equivalent to

$$(u^{-1})^*(K) \cap F' \subseteq rB_{F'}$$

where $F' = (u^*)^{-1}(F)$. Then, Theorem 7.3.1 shows that

$$c_k(u^*) \leqslant c\sqrt{n/k}M^*(u^*(K)).$$

On the other hand,

$$\begin{aligned}\ell(u) = \mathbb{E}\left\|\sum_{i=1}^n g_i u(e_i)\right\|_{K^\circ} &= \mathbb{E}\left\|\sum_{i=1}^n g_i e_i\right\|_{u^{-1}(K^\circ)} = \mathbb{E}\left\|\sum_{i=1}^n g_i e_i\right\|_{(u^*(K))^\circ}\\ &\simeq \sqrt{n}M^*(u^*(K)).\end{aligned}$$

Setting $Y = X^*$ we see that Theorem 7.3.1 can be formulated in the following (in fact, equivalent) way:

THEOREM 7.3.3. *There exists an absolute constant $C > 0$ such that for any $n \geqslant 1$, for any n-dimensional normed space space Y and for any operator $u : \ell_2^n \to Y$,*

$$\sup_{1\leqslant k\leqslant n}\left(\sqrt{k}c_k(u^*)\right) \leqslant C\ell(u). \tag{7.3.1}$$

We shall present an alternative proof of Theorem 7.3.1 that was given later by Milman. The first main ingredient is the upper bound for the diameter of a random k-dimensional projection of a centrally symmetric convex body that we discussed in Section 5.7.1, and which appeared as Proposition 5.7.1.

LEMMA 7.3.4. *Let L be a symmetric convex body in $\mathbb{R}^n$ with $L \subset dB_2^n$ and $1 \leqslant k < n$. Then with probability greater that $1 - e^{-c_0 k}$ a random $F \in G_{n,k}$ satisfies*

$$P_F(L) \subseteq C\left(M^*(L) + d\sqrt{k/n}\right)(B_2^n \cap F), \tag{7.3.2}$$

where $C, c_0 > 0$ are absolute constants.

We will also use the fact that for any $u \in S^{n-1}$

$$\nu_{n,k}\left(\left\{F \in G_{n,k} : |P_F(u)| \leqslant c_1\sqrt{k/n}\right\}\right) \leqslant e^{-c_2 k}, \tag{7.3.3}$$

where $c_1, c_2 > 0$ are absolute constants, which follows easily from concentration and was already mentioned in (7.2.3).

Proof of Theorem 7.3.1. We will show that if K is a centrally symmetric convex body in $\mathbb{R}^n$ then for all $1 \leqslant k < n$ the measure of $F \in G_{n,n-k}$ which satisfy

$$\mathrm{diam}(K \cap F) \leqslant C\sqrt{n/k}M^*(K),$$

is greater than $1 - e^{-ck}$.

Let $t = \gamma\sqrt{n/k}M^*(K)$, where $\gamma > 0$ is an absolute constant which will be suitably chosen. By Sudakov's inequality (Theorem 4.2.1) there exists a t-net $\mathcal{N}$ of K (with respect to the Euclidean metric) of cardinality

$$|\mathcal{N}| \leqslant \exp(cn(M^*(K)/t)^2) = e^{c_1 k},$$

where $c_1 = c/\gamma^2$.

Recall that for any fixed $y \in \mathbb{R}^n$ and for a random projection onto a k-dimensional subspace F one has

$$|P_F(y)| \geqslant c\sqrt{k/n}\,|y| \tag{7.3.4}$$

with probability greater than $1 - e^{-k}$. Applying the union bound over $\mathcal{N}$ we get

$$|P_F(y)| \geqslant c\sqrt{k/n}\,|y| \tag{7.3.5}$$

for all $y \in \mathcal{N}$ with probability greater than $1 - |\mathcal{N}|e^{-k} > 1 - e^{-c_2 k}$, where $c_2 = 1 - c/\gamma^2 > c_2/2$ if $\gamma > 0$ is large enough.

Note also that, by Lemma 7.3.4, with probability at least $1 - e^{-c_0 k}$ with respect to $F \in G_{n,k}$, we have

$$\mathrm{diam}(P_F(tB_2^n \cap 2K)) \leqslant C(M^*(K) + t\sqrt{k/n}) = C(1+\gamma)M^*(K). \tag{7.3.6}$$

Let $E \in G_{n,n-k}$ such that $F := E^\perp$ satisfies both (7.3.5) and (7.3.6). For every $x \in K \cap E$ there exists $y_x \in \mathcal{N}$ with $|x - y_x| \leqslant t$. Since $x \in \ker(P_{E^\perp})$, we have $|P_{E^\perp}(y_x)| = |P_{E^\perp}(x - y_x)|$. Moreover, since K is convex and centrally symmetric, we also have $x - y_x \in 2K$, and hence $x - y_x \in tB_2^n \cap 2K$. Taking into account (7.3.6) we conclude that for any $x \in K \cap E$ there exists $y_x \in \mathcal{N}$ such that

$$|P_{E^\perp}(y_x)| = |P_{E^\perp}(x - y_x)| \leqslant \mathrm{diam}(P(tB_2^n \cap 2K)) \leqslant C(1+\gamma)M^*(K). \tag{7.3.7}$$

Then, (7.3.5) implies that

$$|y_x| \leqslant c^{-1}\sqrt{n/k}|P_{E^\perp}(y_x)| \leqslant C_1(1+\gamma)\sqrt{n/k}M^*(K).$$

Hence, for any $x \in K \cap E$,

$$|x| \leqslant |y_x| + |x - y_x| \leqslant C_1(1+\gamma)\sqrt{n/k}M^*(K) + t \leqslant C_2\sqrt{n/k}M^*(K),$$

by our choice of t. □

7.3.1. Gordon's formulation of the M^*-estimate

Many of the applications of Milman's M^* estimate actually use a fixed proportion $\lambda < 1$ and for these results the delicate study of the asymptotic behaviour of $f(\lambda)$ as $\lambda \to 1$ is not important. In fact, in such cases the very first proof given above as Theorem 7.2.1 provides an extremely good probability estimate. However, it turned out that for some very beautiful applications which we will discuss later in this chapter, not only is the $c\sqrt{1-\lambda}$ behaviour of f as $\lambda \to 1$ important, but even the value of the constant c in Theorem 7.3.1 needs to be taken very close to 1. The previous approaches do not yield this result, for which we need Gordon's sharp formulation of the M^*-estimate, given in Theorem 7.3.5 below. At the same time we would like to mention that the estimate in Theorem 7.3.1 (with some constant $c > 0$) plays an important role in operator theory.

The constant a_k which appears in the theorem below is defined (for $1 \leqslant k \leqslant n$) by

$$a_k := \frac{\sqrt{2}\Gamma\left(\frac{k+1}{2}\right)}{\Gamma\left(\frac{k}{2}\right)} = \sqrt{k}\Big(1 - \frac{1}{4k} + O(k^{-2})\Big); \tag{7.3.8}$$

the second equation follows from Stirling's formula.

THEOREM 7.3.5 (Gordon). *Let K be a convex body in $\mathbb{R}^n$ with $0 \in \mathrm{int}(K)$. For any $0 < \gamma < 1$ and $1 \leqslant k < n$ there exists $B \subset G_{n,n-k}$ with measure at least $1 - \frac{7}{2}\exp\Big(-\frac{1}{18}(1-\gamma)^2 a_k^2\Big)$ such that for any $F \in B$ and for all $x \in F$,*

$$\frac{\gamma a_k}{a_n M^*(K)}|x| \leqslant \|x\|. \tag{7.3.9}$$

The proof is postponed and will appear in Chapter 9, as it is based on Gordon's min-max principle of Slepian type for Gaussian processes, which will be dicussed there. Taking into account the estimate for a_k in (7.3.8) we immediately see that for every convex body K in $\mathbb{R}^n$ with $0 \in \mathrm{int}(K)$ and for every $1 \leqslant k < n$ there exists a subspace $F \in G_{n,n-k}$ such that

$$|x| \leqslant C\sqrt{n/k}M^*(K)\|x\|_K \tag{7.3.10}$$

for all $x \in F$, where $C > 0$ is an absolute constant which may be chosen very close to 1, assuming k and n are sufficiently large.

7.4. Milman's quotient of subspace theorem

Milman's quotient of subspace theorem states that by performing two operations on an n-dimensional space X, taking first a subspace and then a quotient of it, we can always arrive at a new space of dimension proportional to n which is (independently of n) close to Euclidean. In order to interpret this in the language of convex bodies, observe that if K is the unit ball of $X = (\mathbb{R}^n, \|\cdot\|)$ and if $G \subseteq E \subseteq X$ then E/G is isometrically isomorphic to the subspace $F := E \cap G^{\perp}$ equipped with the norm induced by $P_F(K \cap E)$.

We write $QS(X)$ for the class of all quotient spaces of subspaces of X; a space $Y \in QS(X)$ is of the form E/G where $G \subset E \subset X$. It is useful to note that $QS(X)$ is the same as the class $SQ(X)$ of all subspaces of quotient spaces of X. Indeed, if $F \subset E \subset \mathbb{R}^n$ one sees that $P_F(K \cap E) = (P_{F+E^{\perp}}(K)) \cap F$. This implies the following very useful property: if $Y \in QS(X)$ then every subspace or quotient space of Y also belongs to $QS(X)$ and $QS(Y) \subseteq QS(X)$. Thus, every iteration of the operation of choosing a quotient of a subspace leads to an element of $QS(X)$.

The precise statement of the quotient of subspace theorem is the following.

THEOREM 7.4.1 (Milman). *Let X be an n-dimensional normed space. For every $1 \leqslant k < n$ there exists $Y \in QS(X)$ with $\dim(Y) = n - k$ and*

$$d(Y, \ell_2^{n-k}) \leqslant C\frac{n}{k}\log\Big(\frac{Cn}{k}\Big), \tag{7.4.1}$$

where $C > 0$ is an absolute constant.

Geometrically, the quotient of subspace theorem asserts that for every centrally symmetric convex body K in $\mathbb{R}^n$ and any $\alpha \in (0,1)$ we can find subspaces $F \subseteq E$ with $\dim(F) \geqslant \alpha n$ and an ellipsoid $\mathcal{E}$ in F such that

$$\mathcal{E} \subset P_F(K \cap E) \subset c(1-\alpha)^{-1} |\log(c(1-\alpha))| \mathcal{E}.$$

The proof of Theorem 7.4.1 is based on the M^*-estimate and an iteration procedure which makes essential use of the ℓ-position. We also need the next two *iteration lemmas.*

LEMMA 7.4.2. *Let $f : [0,1) \to \mathbb{R}^+$ be a bounded function. Assume that there exist $\gamma > 0$, $C \geqslant 1$ and $\theta \in (0,1)$ such that*

$$f(\delta^2) \leqslant C(1-\delta)^{-\gamma} f(\delta)^\theta$$

for all $\delta \in (0,1)$. Then,

$$f(\delta) \leqslant C(\theta)(1-\delta)^{-\frac{\gamma}{1-\theta}}$$

for all $\delta \in (0,1)$, where $C(\theta) = 2^{\frac{\gamma}{(1-\theta)^2}} C^{\frac{1}{1-\theta}}$.

Proof. We first consider the particular case where $\gamma = 0$ and $C = 1$. Let $\delta \in [0,1)$. We need to show that $f(\delta) \leqslant 1$. Note that for any $k \geqslant 1$ we can write δ in the form $\delta = t^{2^k}$ for some $t \in [0,1)$. Next, using the assumption we write

$$f(t^{2^k}) \leqslant f(t^{2^{k-1}})^\theta \leqslant \cdots \leqslant f(t)^{\theta^k}.$$

Since f is bounded by some $M > 0$, we see that

$$f(\delta) = f(t^{2^k}) \leqslant M^{\theta^k}$$

for all $k \geqslant 1$. On observing that $\lim_{k\to\infty} M^{\theta^k} = 1$ we conclude that $f(\delta) \leqslant 1$, as needed.

We reduce the general case to the previous one by considering the function

$$\varphi(\delta) = C(\theta)^{-1}(1-\delta)^{\frac{\gamma}{1-\theta}} f(\delta).$$

Since $1 - \delta^2 \leqslant 2(1-\delta)$ we check that

$$\varphi(\delta^2) \leqslant C(\theta)^{-1}[2(1-\delta)]^{\frac{\gamma}{1-\theta}} f(\delta^2) \leqslant \varphi(\delta)^\theta.$$

It remains to observe that φ is bounded, as $\gamma > 0$. The first part of the proof thus shows that $\varphi(\delta) \leqslant 1$ for all $\delta \in [0,1)$, which proves the lemma. $\square$

LEMMA 7.4.3. *Let $f : [0,1) \to \mathbb{R}^+$ be a bounded function such that $f(t) \geqslant 1$ for all $0 \leqslant t < 1$. Assume that there exists $\gamma > 0$ such that*

$$f(\delta^2) \leqslant \frac{\gamma}{1-\delta} \log(e^2 f(\delta)) \tag{7.4.2}$$

for all $\delta \in [0,1)$. Then,

$$f(\delta) \leqslant \frac{32e\gamma}{1-\delta} \log |e\gamma(1-\delta)^{-1}|$$

for all $\delta \in [0,1)$.

Proof. We first observe that, for every $0 < \theta < \frac{1}{2}$ and for all $y > e^2$,

$$y^\theta \geqslant e\theta \log y, \tag{7.4.3}$$

With equality, if we fix $y \geqslant e^2$, for $\theta = (\log y)^{-1} < \frac{1}{2}$.

We rewrite the hypothesis of the lemma in the following equivalent form: for all $\delta \in [0,1)$ and all $0 < \theta < \frac{1}{2}$,

$$f(\delta^2) \leqslant \frac{\gamma}{e\theta(1-\delta)}\,(e^2 f(\delta))^{\theta}.$$

Then, Lemma 7.4.2 shows that

$$f(\delta) \leqslant \frac{A(\theta)}{(1-\delta)^{\frac{1}{1-\theta}}}, \tag{7.4.4}$$

where

$$A(\theta) = \Big(2^{\frac{1}{1-\theta}}\frac{e^{2\theta}\gamma}{e\theta}\Big)^{\frac{1}{1-\theta}}.$$

Using the elementary inequality $\frac{1}{1-\theta} \leqslant 1+2\theta$ for $\theta \in (0,\frac{1}{2})$, we see that

$$A(\theta) \leqslant B(\theta) := 16\Big(\frac{e^{2\theta}\gamma}{e\theta}\Big)\Big(\frac{\gamma}{\theta}\Big)^{2\theta}.$$

Since $(1/\theta)^{2\theta} \leqslant e^{2/e} \leqslant e$ for all $\theta \in (0,\frac{1}{2})$, we check that $B(\theta) \leqslant \frac{16\gamma}{\theta}(e\gamma)^{2\theta}$, and hence (7.4.4) takes the simpler form

$$f(\delta) \leqslant \frac{16\gamma}{(1-\delta)\theta}\left[\frac{e^2\gamma^2}{(1-\delta)^2}\right]^{\theta}.$$

Using the equality case in (7.4.2), with $\theta = 1/\log(y)$ and $y = \frac{e^2\gamma^2}{(1-\delta)^2}$ we get that

$$f(\delta) \leqslant \frac{16e\gamma}{1-\delta}\log\left[\frac{e^2\gamma^2}{(1-\delta)^2}\right].$$

This completes the proof. □

Proof of Theorem 7.4.1. Denote by $QS(X)$ the class of all quotient spaces of a subspace of X, and consider the function $f : [0,1) \to \mathbb{R}^+$ defined by

$$f(\delta) = \inf\{d(F,\ell_2^k) : F \in QS(X), \mathrm{dim}F \geqslant \delta n\}.$$

Let us prove that

$$f(\delta^2) \leqslant c(1-\delta)^{-1}\log(e^2 f(\delta)).$$

This will complete the proof since we may then use Lemma 7.4.3 to conclude that

$$f(\delta) \leqslant c'(1-\delta)^{-1}|\log(c''(1-\delta))|$$

for all $\delta \in [0,1)$.

Note that what we should prove is that for any normed space X of dimension k there exists a subspace $F \in QS(X)$ of dimension $m \geqslant \delta k$ such that $d(F,\ell_2^m) \leqslant c(1-\delta)^{-1}\log[d(X,\ell_2^k)+1]$.

The proof of this fact is a double application of the M^*-estimate. We may assume that $K = B_X$ is in the ℓ-position: then, by Theorem 6.5.2 we have $M(K)M^*(K) \leqslant c\log[d(X,\ell_2^n)+1]$. Let $\sqrt{\delta} = \lambda \in (0,1)$. We shall show that there exist a subspace E of X with $\dim(E) \geqslant \lambda k$ and a subspace F of E^* with $\dim(F) = m \geqslant \lambda^2 k$, such that $d(F,\ell_2^m) \leqslant c(1-\delta)^{-1}\log[d(X,\ell_2^n)+1]$.

By Theorem 7.3.1, a random λk-dimensional subspace E of X satisfies

$$\frac{c_1\sqrt{1-\lambda}}{M^*(K)}|x| \leqslant \|x\| \leqslant b|x| \qquad (x \in E) \tag{7.4.5}$$

Moreover, since (7.4.5) holds with probability greater than $1-\exp(-c\lambda k)$ over $E \in G_{k,\lambda k}$, we may assume that $M(K\cap E) \leqslant c_2 M(K)$. Indeed, this follows from

Lemma 5.3.9. Therefore, repeating the same argument for $(K\cap E)^\circ$, we may find a subspace F of E with $\dim(F)=m\geqslant \lambda^2 k$ and

$$\frac{c_3\sqrt{1-\lambda}}{M(K)}|x|\leqslant \frac{c_1\sqrt{1-\lambda}}{M^*((K\cap E)^\circ)}|x|\leqslant \|x\|_{(K\cap E)^\circ}\leqslant \frac{M^*(K)}{c_1\sqrt{1-\lambda}}|x|$$

for every $x\in F$. Since K is in the ℓ-position, we obtain, letting $B_2^n\cap F=B_F$,

$$d(P_E(K^\circ)\cap F,B_F)\leqslant c_4(1-\lambda)^{-1}M(K)M^*(K)\leqslant c_5(1-\lambda)^{-1}\log[d(X,\ell_2^k)+1].$$

Taking polars in F we conclude that

$$d(P_F(K\cap E),B_F)\leqslant \frac{c_5}{1-\lambda}\log[d(X,\ell_2^n)+1]\leqslant \frac{2c_5}{1-\delta}\log(e^2 d(X,\ell_2^k)).$$

as needed. □

7.5. Asymptotic formulas for random sections

We begin with a very simple improvement of the M^*-estimate, in which we first intersect the body with a big ball, then use the M^*-estimate to produce a section with small diameter; the "trick" is that if this diameter is smaller than the radius of the ball with which we intersected to begin with, we know that also the corresponding section of the original body had this diameter. More precisely:

LEMMA 7.5.1. *Let $\lambda\in(0,1)$ and $\varepsilon\in(0,1)$. For every $n\geqslant n_0=n_0(\varepsilon,\lambda)$ and every centrally symmetric convex body K in $\mathbb{R}^n$ we can find a set $B\subseteq G_{n,k}$, where $k=\lfloor\lambda n\rfloor$, of measure $\nu_{n,k}(B)\geqslant 1-4\exp(-c\varepsilon^2(n-k))$, such that*

$$K\cap F\subseteq rB_2^n\cap F$$

for every $F\in B$, where r is the solution of the equation

$$M^*(K\cap rB_2^n)=(1-\varepsilon)\sqrt{1-\lambda}\,r.$$

Proof. As in Gordon's form of the M^*-estimate, Theorem 7.3.5, let $a_s=\sqrt{2}\Gamma(\frac{s+1}{2})/\Gamma(\frac{s}{2})$. One can check that $\frac{a_{n-k}}{a_n}\geqslant(\frac{n-k-1}{n})^{1/2}$, and this implies that for $n\geqslant n_0(\varepsilon,\lambda)$ we have

$$\frac{(1-\frac{\varepsilon}{2})a_{n-k}}{a_n(1-\varepsilon)\sqrt{1-\lambda}}\geqslant 1+\frac{\varepsilon}{2}.$$

Suppose that r satisfies the equation $M^*(K\cap rB_2^n)=(1-\varepsilon)\sqrt{1-\lambda}\,r$. Then Theorem 7.3.5 shows that

$$\|x\|_{K\cap rB_2^n}\geqslant \frac{(1-\frac{\varepsilon}{2})a_{n-k}}{a_nM^*(K\cap rB_2^n)}|x|\geqslant \frac{1+\frac{\varepsilon}{2}}{r}|x|\qquad (x\in F)$$

for all F in a subset $B\subset G_{n,k}$ of measure $\nu_{n,k}(B)\geqslant 1-\frac{7}{2}\exp(-\frac{1}{72}a_{n-k}^2\varepsilon^2)$. Since $\|x\|_{K\cap rB_2^n}=\max\left\{\|x\|_K,\frac{1}{r}|x|\right\}$, this shows that actually

$$\|x\|_K\geqslant \frac{1}{r}|x|$$

for every $F\in B$ and for all $x\in F$, and this completes the proof since $a_{n-k}^2\simeq n-k$. □

Next we would like to show that the upper estimates given by Lemma 7.5.1 can be complemented by lower estimates for every proportion $\lambda\in(0,1)$. The bound turns out to hold *for every* $\lfloor\lambda n\rfloor$-dimensional section of K. More precisely, denote for $A\subset\mathbb{R}^n$ its outer-radius by $R(A)=\inf\{r: A\subset rB_2^n\}$. Then:

THEOREM 7.5.2 (Giannopoulos-Milman-Tsolomitis). *Let* $\lambda \in (0,1)$ *and* $\delta > 0$ *satisfy* $(1+\delta)\sqrt{\frac{2(1-\lambda)}{2-\lambda}} < 1$. *If* $n \geqslant n_1 = n_1(\lambda, \delta)$, *then for every centrally symmetric convex body* $K \subset \mathbb{R}^n$ *and for every* $F \in G_{n,\lfloor \lambda n \rfloor}$ *we have*

$$R(K \cap F) \geqslant \frac{1}{3}\delta\sqrt{1-\lambda}r$$

where r *is the solution of the equation*

$$M^*(K \cap rB_2^n) = (1+\delta)\sqrt{\frac{2(1-\lambda)}{2-\lambda}}r. \tag{7.5.1}$$

To prove Theorem 7.5.2 we need some preparations. The main tool in the proof of Theorem 7.5.2 is an isoperimetric theorem of Gromov, which we next quote. We do not provide its proof in this book. Recall that the t-extension of a subset $A \subset S^{n-1}$ is defined by $A_t = \{x \in S^{n-1} : \rho(x, A) < t\}$.

THEOREM 7.5.3 (Gromov). *Assume that* $k < n$ *are positive integers,* n *is even and* $n - k = 2^m - 1$ *for some positive integer* m. *For every odd continuous function* $f : S^{k-1} \to S^{n-1}$ *and every* $t > 0$,

$$\sigma_n\left((f(S^{k-1}))_t\right) \geqslant \sigma_{n,k}(t),$$

where $\sigma_{n,k}(t)$ *is the measure of the* t*-extension of* S^{k-1} *in* S^{n-1}.

It is not known if Theorem 7.5.3 holds true for all $k < n$. However, by embedding into a higher dimensional sphere, and noting that for every $t > 0$, for every symmetric Borel set $A \subseteq S^{n-1} \subset S^{m-1}$ and every $m \geqslant n$, one has $\sigma_n(A_t) \geqslant \sigma_m(A_t)$, Vershynin showed that Theorem 7.5.3 implies the following estimate for general n (again, we shall not, in this text, prove this implication):

PROPOSITION 7.5.4. *Assume that* $k < n$ *are positive integers. For every odd continuous function* $f : S^{k-1} \to S^{n-1}$ *and every* $t > 0$,

$$\sigma_n\left((f(S^{k-1}))_t\right) \geqslant \sigma_{2n-k,k-2}(t),$$

where $\sigma_{m,k}(t)$ *is the measure of the* t*-extension of* S^{k-1} *in* S^{m-1}.

As a consequence, we shall show that if a certain body includes a ball of radius $a < 1$, and when it is projected onto some k dimensional F, the projection includes a much larger ball, of radius $b > 1$, then K must include a significant part of S^{n-1}. The extremal case essentially being the convex hull of a ball of radius a with a k-dimensional ball of radius b. The proof of this fact, due to Vershynin, makes essential use of Proposition 7.5.4.

THEOREM 7.5.5. *Let* K *be a centrally symmetric convex body in* $\mathbb{R}^n$ *and assume that for some* $a < 1 < b$ *and some* $F \in G_{n,k}$, $k > 2$ *we have*

$$aB_2^n \subseteq K \text{ and } bB_2^n \cap F \subseteq P_F(K).$$

Then,

$$\sigma_n(K \cap S^{n-1}) \geqslant \sigma_{2n-k,k-2}(\theta),$$

where $\theta = \arcsin(a) - \arcsin(a/b)$.

Proof. Since $bB_2^n \cap F \subseteq P_F(K)$, there exists an odd continuous function $g : bS_F \to K$ such that $|g(bx)| \geqslant b$ for all $x \in S_F$. Consider the function $f : S_F \to S^{n-1}$ defined by $f(x) = g(bx)/|g(bx)|$. We may clearly identify S_F with S^{k-1} and then apply

Proposition 7.5.4 to get $\sigma(A_\phi) \geqslant \sigma_{2n-k,k-2}(\phi)$ for every $\phi > 0$, where $A = f(S_F)$. Next, using the fact that $|g(bx)| \geqslant b > a$ and simple trigonometry, observe that

$$K \cap S^{n-1} \supseteq \mathrm{conv}\{\pm g(bx), aB_2^n\} \cap S^{n-1} \supseteq B(f(x), \theta) \cap S^{n-1}$$

for every $x \in S_F$, where $\theta = \arcsin(a) - \arcsin(a/b)$. Finally, use Proposition 7.5.4. □

To use Theorem 7.5.5 one needs a way to estimate $\sigma_{n,k}(\theta)$. For us the following simple bound will be enough.

LEMMA 7.5.6. *Let $\delta \in (0,1)$ and let $k = \lambda n$ for some positive integer $k < n$. If $n \geqslant \frac{4\lambda}{(1-\lambda)\delta^2}$ and*

$$\sin^2\theta > (1+\delta)(1-\lambda),$$

then, $\sigma_{n,k}(\theta) > 1/2$.

Proof. Observe that $\sigma_{n,k}(\theta) = \mathrm{Prob}(Y_n \leqslant \sin^2\theta)$, where Y_n is a random variable with distribution $\mathrm{Beta}(\frac{(1-\lambda)n}{2}, \frac{\lambda n}{2})$. Since

$$\mathbb{E}(Y_n) = 1-\lambda \text{ and } \mathrm{Var}(Y_n) = \frac{2\lambda(1-\lambda)}{n+2},$$

a simple application of Chebyshev's inequality shows that

$$\mathrm{Prob}(Y_n > (1-\lambda)+t) \leqslant \frac{\mathrm{Var}(Y_n)}{t^2} \leqslant \frac{2\lambda(1-\lambda)}{(n+2)t^2}$$

for every $t > 0$. Choosing $t = \delta(1-\lambda)$ we get the lemma. □

Theorem 7.5.5 implies that if a body K contains the Euclidean unit ball then a condition of the form $M(K) > g(\lambda)$ implies that the inradius of *every* $\lceil \lambda n \rceil$-dimensional projection $P_F(K)$ of K cannot be big. More precisely, denote the in-radius of a set $A \subset \mathbb{R}^m$ by $r(A) = \sup\{r : rB_2^m \subset A\}$. Then, we have:

PROPOSITION 7.5.7. *Let $\lambda \in (0,1)$ and let K be a centrally symmetric convex body in $\mathbb{R}^n$ such that $B_2^n \subseteq K$. If $M(K) > \beta(\lambda) = \sqrt{\frac{2(1-\lambda)}{2-\lambda}}$ and $n \geqslant C(M(K) - \beta(\lambda))^{-2}$, then*

$$r(P_F(K)) \leqslant \frac{3}{M(K) - \beta(\lambda)}$$

for every $F \in G_{n,\lfloor \lambda n \rfloor}$.

Proof. Let $k = \lfloor \lambda n \rfloor$ and let L be the Lévy mean of $\|\cdot\|$ on S^{n-1}. As in the proof of Lemma 5.2.3 we have $|M - L| \leqslant \frac{c_1}{\sqrt{n}}$ for some absolute $c_1 > 0$ (since we have assumed that the in-radius of K is greater than 1).

Consider $F \in G_{n,k}$ such that $r := r(P_F(K))$ is maximal. If $(M - \frac{c_1}{\sqrt{n}})r \leqslant 1$ then there is nothing to prove since $\frac{c_1}{\sqrt{n}} \leqslant \frac{2M+\beta}{3}$ if $n \geqslant C(M-\beta)^{-2}$.

Otherwise, since $(M - \frac{c_1}{\sqrt{n}})K \supseteq (M - \frac{c_1}{\sqrt{n}})B_2^n$ we can apply Theorem 7.5.5 to the body $(M - \frac{c_1}{\sqrt{n}})K$. It follows that

$$\sigma_n\left(\left(M - \frac{c_1}{\sqrt{n}}\right)K \cap S^{n-1}\right) \geqslant \sigma_{2n-k,k-2}(\theta)$$

where $\theta = \arcsin\left(M - \frac{c_1}{\sqrt{n}}\right) - \arcsin(1/r)$. On the other hand,

$$\sigma_n\big((M - c_1/\sqrt{n})K \cap S^{n-1}\big) \leqslant \sigma_n(LK \cap S^{n-1}) \leqslant 1/2.$$

We set $\lambda_0 = \frac{k-2}{2n-k}$ and $\delta_0 = \frac{M-\beta(\lambda)}{\beta(\lambda)}$. From Lemma 7.5.6 it follows that (for $n \geqslant n_0(\lambda_0, \delta_0) \geqslant C(M-\beta(\lambda))^{-2}$) we must have

$$\sin\theta \leqslant \sqrt{(1+\delta_0)\frac{2(n-k-1)}{2n-k}} < \sqrt{1+\delta_0}\beta(\lambda). \tag{7.5.2}$$

Observe that

$$\begin{aligned}\sin\theta &= \frac{(M-c_1/\sqrt{n})}{r}\sqrt{r^2-1} - \frac{1}{r}\sqrt{1-(M-c_1/\sqrt{n})^2}\\ &= \frac{M-c_1/\sqrt{n}}{r}\left(\sqrt{r^2-1} - \sqrt{(M-c_1/\sqrt{n})^{-2}-1}\right)\\ &\geqslant \frac{M-c_1/\sqrt{n}}{r}\frac{r^2-(M-c_1/\sqrt{n})^{-2}}{r+(M-c_1/\sqrt{n})^{-1}} = \left(M-\frac{c_1}{\sqrt{n}}\right) - \frac{1}{r}.\end{aligned}$$

Then, (7.5.2) gives

$$r\left(\left(M-\frac{c_1}{\sqrt{n}}\right) - \sqrt{1+\delta_0}\beta(\lambda)\right) \leqslant 1.$$

Finally, under the assumption $n \geqslant C(M-\beta(\lambda))^{-2}$ it is easily checked that $\frac{c_1}{\sqrt{n}} + \sqrt{1+\delta_0}\beta(\lambda) \leqslant \left(1+\frac{2\delta_0}{3}\right)\beta(\lambda)$. Combined with the definition of δ_0, this proves the proposition. □

By duality, Proposition 7.5.7 implies

PROPOSITION 7.5.8. *Let $\lambda \in (0,1)$ and let K be a centrally symmetric convex body in $\mathbb{R}^n$ such that $K \subseteq B_2^n$. If $M^*(K) > \beta(\lambda) = \sqrt{\frac{2(1-\lambda)}{2-\lambda}}$ and $n \geqslant C(M^*(K)-\beta(\lambda))^{-2}$, then*

$$R(K\cap F) \geqslant \frac{M^*(K)-\beta(\lambda)}{3}$$

for every $F \in G_{n,\lfloor \lambda n\rfloor}$.

Finally, this proposition implies Theorem 7.5.2

Proof of Theorem 7.5.2. Given a centrally symmetric convex body K consider $K' = \frac{1}{r}K \cap B_2^n$ where r is the solution of the equation in the statement of the Theorem. Then K' satisfies the conditions of Proposition 7.5.8 and $M^*(K') - \beta(\lambda) = \delta\beta(\lambda) \geqslant \delta\sqrt{1-\lambda}$. We thus apply Proposition 7.5.8 and get $R(K\cap F) = rR(K'\cap F) \geqslant \frac{\delta}{3}\sqrt{1-\lambda}$ as needed. □

So far in this section we have seen the following two facts about the outer-radius of the sections of a centrally symmetric convex body K in $\mathbb{R}^n$: Let $\lambda \in (0,1)$.

(i) If r_1 is the solution of the equation

$$M^*(K\cap rB_2^n) = (1-\varepsilon)\sqrt{1-\lambda}\, r \tag{7.5.3}$$

then *a random* $F \in G_{n,\lfloor\lambda n\rfloor}$ satisfies

$$R(K\cap F) \leqslant r_1.$$

(ii) If r_2 is the solution of the equation

$$M^*(K\cap rB_2^n) = (1+\delta)\sqrt{\frac{2(1-\lambda)}{2-\lambda}}r, \tag{7.5.4}$$

then

$$R(K\cap F)\geqslant \frac{1}{3}\delta\sqrt{1-\lambda}r_2$$

for every $F\in G_{n,\lfloor\lambda n\rfloor}$. Note that the functions of λ in the equations (7.5.3) and (7.5.4) are comparable (in fact, they would almost coincide if we would have been allowed to apply Gromov's isoperimetric theorem for *every pair* of n and k).

In the remainder of the section we shall compare the smallest outer-radius of a $\lfloor\lambda n\rfloor$-dimensional section of a convex body:

$$a(\lambda,K)=\min\left\{R(K\cap F):\ F\in G_{n,\lfloor\lambda n\rfloor}\right\}$$

with the "typical" one:

$$b(\lambda,K)=\min\left\{r>0:R(K\cap F)\leqslant r:\ \text{with probability }\geqslant 1/2\text{ in }G_{n,\lfloor\lambda n\rfloor}\right\}.$$

It is clear that $a(\lambda,K)\leqslant b(\lambda,K)$ for all λ and K. However, combining (7.5.3) and (7.5.4) we have

THEOREM 7.5.9 (Giannopoulos-Milman-Tsolomitis). *There exists* $n_0\in\mathbb{N}$ *and a universal* $C>1$ *such that*

$$\frac{1}{C}b(0.8,K)\leqslant a(0.9,K) \tag{7.5.5}$$

for every $n\geqslant n_0$ *and every centrally symmetric convex body* K *in* $\mathbb{R}^n$.

The choice of 0.9 and 0.8 was arbitrary. In fact the theorem applies for any $0<\mu<1$ instead of 0.9 and then some proportion $<\frac{\mu}{2-\mu}$ instead of 0.9. One may also get explicit bounds for the dependence of C on these proportions as they reach their respective limits (see the notes and remarks at the end of this chapter).

Proof. Let $\varepsilon\in(0,1)$ be a constant that will be suitably chosen. Let K be a centrally symmetric convex body in $\mathbb{R}^n$ and let r_1 be the solution of the equation

$$M^*(K\cap rB_2^n)=(1-\varepsilon)\sqrt{1-0.8}\,r.$$

If n is large enough, then from (7.5.3) we have

$$b(0.8,K)\leqslant r_1. \tag{7.5.6}$$

We choose

$$\varepsilon=0.01$$

and then, one can check that

$$(1-\varepsilon)\sqrt{1-0.8}\geqslant(1+\varepsilon)\sqrt{\frac{2(1-0.9)}{2-0.9}}.$$

It follows that if r_2 is the solution of the equation

$$M^*(K\cap rB_2^n)=(1+\varepsilon)\sqrt{\frac{2(1-0.9)}{2-0.9}}\,r$$

then $r_1\leqslant r_2$. Now, from (7.5.4) we get

$$\frac{1}{3}\varepsilon\sqrt{1-0.9}\,r_2\leqslant a(0.9,K).$$

Combining this with (7.5.6) we complete the proof of (7.5.5). $\square$

7.6. Linear duality relations

In this section we describe some unexpected linear duality relation between a convex body K and its polar K°. It is clear from the definition of the polar body that if K is "large" in some sense, then its polar body is "small". A fundamental result in this direction is the fact that

$$s(K)^{1/n} := \left(\mathrm{Vol}_n(K)\mathrm{Vol}_n(K^\circ)\right)^{1/n} \simeq \mathrm{Vol}_n(B_2^n)^{2/n}$$

for every centrally symmetric convex body in $\mathbb{R}^n$. This is a consequence of the Blaschke-Santaló inequality, which we saw in Chapter 1, Section 1.5.4, and its reverse (the Bourgain-Milman inequality) which will be discussed in detail in the next chapter. Here, we describe some duality relations that follow from the M^*-estimate.

DEFINITION 7.6.1. Let $X = (\mathbb{R}^n, \|\cdot\|)$ be an n-dimensional normed space. For every $r > 0$ we define

$$t(X,r) := \max\left\{k \leqslant n : \text{there exists } F \in G_{n,k} : \frac{1}{r}|x| \leqslant \|x\| \text{ for every } x \in F\right\}.$$

Equivalently, we may write $K \cap F \subset rB_2^n$.

We present a proof of the following result of Milman.

THEOREM 7.6.2 (Milman). *Let $\kappa \in (0,1)$. For every $n \geqslant n_0(\kappa)$ and every n-dimensional normed space $X = (\mathbb{R}^n, \|\cdot\|)$ and every $r > 0$,*

$$t(X,r) + t\left(X^*, \frac{4}{\kappa r}\right) \geqslant (1-\kappa)n. \tag{7.6.1}$$

It is useful to compare this statement with Theorem 5.3.5. Recall that $k(X)$ (the *critical dimension* of X) was defined as the largest positive integer $k \leqslant n$ for which

$$\nu_{n,k}\left(\left\{E_k \in G_{n,k} : \frac{1}{2}M|x| \leqslant \|x\| \leqslant 2M|x| \ \text{ for all } x \in E_k\right\}\right) \geqslant 1 - \frac{k}{n+k}.$$

In Section 5.4 we saw that for every n-dimensional normed space X one has $k(X) \simeq n(M/b)^2$. From the inequality $k(x) \geq cn(M/b)^2$ it follows that (Theorem 5.3.5)

$$k(X)k(X^*) \geqslant cn$$

for some absolute constant $c > 0$. In **[205]** it is proved that this inequality is optimal; there exist n-dimensional spaces with $k(X) \simeq k(X^*) \simeq \sqrt{n}$. Observe that the definition of $k(X)$ requires that a subspace is c-Euclidean in contrast with the definition of $t(X)$ which requires much less. This explains why one could expect a much stronger duality relation.

The main ingredient in the proof of Theorem 7.6.2 is the next *distance lemma.* Recall that the geometric distance between two centrally symmetric convex bodies K, T in $\mathbb{R}^n$ is defined by

$$d_G(K,T) = \inf\{ab : \frac{1}{b}T \subseteq K \subseteq aT\}.$$

LEMMA 7.6.3 (distance lemma). *Let $\|\cdot\|$ be a norm on $\mathbb{R}^n$ such that $a^{-1}|x| \leqslant \|x\| \leqslant b|x|$ for all $x \in \mathbb{R}^n$. If*

$$\left(\frac{\|x\|}{b}\right)^2 + \left(\frac{\|x\|^*}{a}\right)^2 = t > 1 \tag{7.6.2}$$

for some $x \in S^{n-1}$, then

$$d_G(K, B_2^n) \leqslant \frac{1}{t-1},$$

where K is the unit ball of X.

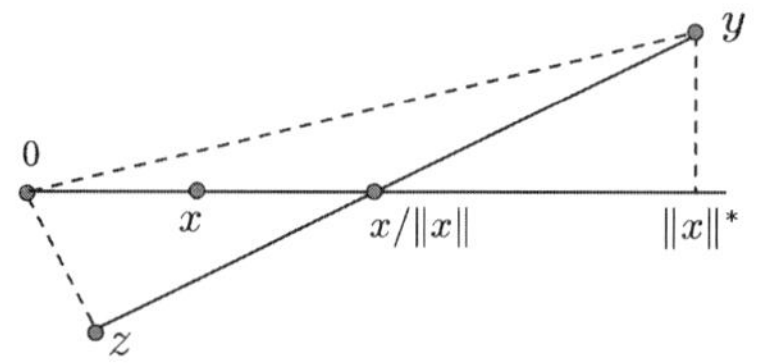

Proof. We assume that $x \in S^{n-1}$ satisfies (7.6.2) and we choose $y \in K$ such that $\|x\|^* = \langle y, x\rangle$. Note that $|y| \leqslant a$. We work in the plane $E = \text{span}\{x, y\}$. If $z = (1+\lambda)\frac{x}{\|x\|} - \lambda y$ (for some $\lambda > 0$) is the point of the smallest Euclidean norm on the line joining $\frac{x}{\|x\|}$ and y, we have

$$\|z\| \geqslant (1+\lambda) - \lambda\|y\| \geqslant 1$$

because $y \in K$, and hence $|z| \geqslant 1/b$.

Observe that the triangles $\triangle(0, z, x/\|x\|)$ and $\triangle(y, \|x\|^* x, x/\|x\|)$ are similar. Therefore,

$$\frac{\|x\|}{b} \leqslant \frac{|z|}{1/\|x\|} \leqslant \frac{\sqrt{a^2 - (\|x\|^*)^2}}{\sqrt{a^2 - (\|x\|^*)^2 + (\|x\|^* - \|x\|^{-1})^2}},$$

which implies

$$\frac{1}{a^2b^2} \leqslant \frac{1 - (\|x\|^*/a)^2}{\|x\|^2a^2 - 2\|x\|\,\|x\|^* + 1}.$$

This in turn can be written in the form

$$\frac{1}{a^2b^2} \leqslant \frac{2\|x\|\,\|x\|^*}{a^2b^2} + 1 - \left[\left(\frac{\|x\|}{b}\right)^2 + \left(\frac{\|x\|^*}{a}\right)^2\right] \leqslant \frac{1}{ab}t + 1 - t. \tag{7.6.3}$$

Setting $u = \frac{1}{ab}$ we get

$$u^2 - tu + t - 1 = (u-1)(u-(t-1)) \leqslant 0.$$

Since $u \leqslant 1$ we must have $u \geqslant t-1$, which shows that

$$d_G(K, B_2^n) \leqslant ab = \frac{1}{u} \leqslant \frac{1}{t-1},$$

as claimed. □

REMARK 7.6.4. The following variants of Lemma 7.6.3 will be used in the proof of Theorem 7.6.2. Let L and L^* denote the Lévy means of $\|\cdot\|$ and $\|\cdot\|^*$ on S^{n-1}.

If

$$\left(\frac{L}{b}\right)^2+\left(\frac{L^*}{a}\right)^2=t>1$$

then

$$d_G(K,B_2^n)\leqslant\frac{1}{t-1}.$$

This follows immediately from Lemma 7.6.3 since we can always find $x\in S^{n-1}$ with $\|x\|\geqslant L$ and $\|x\|^*\geqslant L^*$. On the other hand, from the proof of Lemma 5.2.3 we know that

$$|M-L|\leqslant\frac{c_1b}{\sqrt{n}}\quad\text{and}\quad|M^*-L^*|\leqslant\frac{c_1a}{\sqrt{n}}$$

where $c_1>0$ is an absolute constant. Taking into account that $M/b\leqslant 1$ and $M^*/a\leqslant 1$ we see that

$$\left(\frac{L}{b}\right)^2+\left(\frac{L^*}{a}\right)^2\geqslant\left(\frac{M}{b}-\frac{c_1}{\sqrt{n}}\right)^2+\left(\frac{M^*}{a}-\frac{c_1}{\sqrt{n}}\right)^2\geqslant\left(\frac{M}{b}\right)^2+\left(\frac{M^*}{a}\right)^2-\frac{2c_1}{\sqrt{n}}.$$

Therefore, if we assume that

$$\left(\frac{M}{b}\right)^2+\left(\frac{M^*}{a}\right)^2=t>1+\frac{2c_1}{\sqrt{n}}$$

then we can still conclude that $d_G(K,B_2^n)\leqslant\frac{1}{t-(1+2c_1/\sqrt{n})}$.

It is useful to include here yet another form of Lemma 7.6.3:

LEMMA 7.6.5 (distance lemma). *Let $\|\cdot\|$ be a norm on $\mathbb{R}^n$ such that $a^{-1}|x|\leqslant\|x\|\leqslant b|x|$ for all $x\in\mathbb{R}^n$. If there exists $x\in S^{n-1}$ such that*

$$1-\left(\frac{\|x\|}{b}\right)^2\leqslant c\left(\frac{\|x\|^*}{a}\right)^2$$

for some $0<c<1$, then

$$d_G(K,B_2^n)\leqslant\frac{\sqrt{1-c}}{1-\sqrt{c}}\frac{1}{\sqrt{t-1}}.\tag{7.6.4}$$

Proof. The proof is very similar to the previous one. Going back to (7.6.3) we set $t_1=\|x\|^*/a$ and $t_2=\|x\|/b$. Then, $t_1^2+t_2^2=t$ and there exists $c_1\leqslant c$ such that $1-t_2^2=c_1t_1^2$, therefore

$$t-1=(1-c_1)t_1^2.\tag{7.6.5}$$

We may rewrite (7.6.3) in the form

$$u^2\leqslant 2ut_1t_2+1-t\leqslant 2ut_1+1-t_1^2-t_2^2,\tag{7.6.6}$$

where $u=\frac{1}{ab}$. From (7.6.6) we get

$$(u-t_1)^2\leqslant 1-t_2^2=c_1t_1^2.\tag{7.6.7}$$

Now, we distinguish two cases. If $u\geqslant\frac{\sqrt{t-1}}{\sqrt{1-c_1}}$ then

$$ab=\frac{1}{u}\leqslant\frac{\sqrt{1-c_1}}{\sqrt{t-1}}\leqslant\frac{1}{\sqrt{t-1}}.$$

If $u\leqslant\frac{\sqrt{t-1}}{\sqrt{1-c_1}}$ then we use (7.6.7): we have

$$u\geqslant t_1-\sqrt{c_1}t_1=(1-\sqrt{c_1})t_1,$$

and hence

$$ab = \frac{1}{u} \leqslant \frac{1}{(1-\sqrt{c_1})t_1} = \frac{\sqrt{1-c_1}}{1-\sqrt{c_1}}\frac{1}{\sqrt{t-1}} \leqslant \frac{\sqrt{1-c}}{1-\sqrt{c}}\frac{1}{\sqrt{t-1}},$$

where we have used (7.6.5) and the fact that the function $c \mapsto \frac{\sqrt{1-c}}{1-\sqrt{c}}$ is increasing on $(0,1)$. This second bound covers the first one, so the proof of (7.6.4) is complete. $\square$

The second main ingredient in the proof of Theorem 7.6.2 is Lemma 7.5.1 and its dual lemma, which is simply using Lemma 7.5.1 for the body K° and then taking duals on both sides. This gives

LEMMA 7.6.6. *Let $\mu \in (0,1)$ and $\varepsilon \in (0,1)$. For every $n \geqslant n_0 = n_0(\varepsilon, \lambda)$ and every centrally symmetric convex body K in $\mathbb{R}^n$ we can find a set $B \subseteq G_{n,k}$, where $k = \lfloor \mu n \rfloor$, of measure $\nu_{n,k}(B) \geqslant 1 - 4\exp(-c\varepsilon^2(n-k))$ such that*

$$\rho B_2^n \cap F \subseteq P_F(K)$$

for every $F \in B$, where ρ is the solution of the equation

$$M(\mathrm{conv}(K \cup \rho B_2^n)) = (1-\varepsilon)\frac{\sqrt{1-\mu}}{\rho}.$$

Proof of Theorem 7.6.2. Let $\kappa \in (0,1)$ and let $n \geqslant n_0(\kappa)$. We fix $\lambda, \mu \in (0,1)$ with $\lambda + \mu = 1 - \kappa$, and let $\varepsilon = \kappa/8$. Find $r, \rho > 0$ such that

$$M^*(K \cap rB_2^n) = (1-\varepsilon)\sqrt{1-\lambda}\, r \quad \text{and} \quad M(\mathrm{conv}(K \cup \rho B_2^n)) = (1-\varepsilon)\frac{\sqrt{1-\mu}}{\rho}.$$

We consider the body

$$T = \mathrm{conv}((K \cap rB_2^n) \cup \rho B_2^n).$$

Then, $T \supseteq K \cap rB_2^n$ and $T \subseteq \mathrm{conv}(K \cup \rho B_2^n)$, which implies

$$M^*(T) \geqslant M^*(K \cap rB_2^n) \quad \text{and} \quad M(T) \geqslant M(\mathrm{conv}(K \cup \rho B_2^n)).$$

Since $\rho B_2^n \subseteq T \subseteq rB_2^n$ and

$$\begin{aligned}(M(T)\rho)^2 + (M^*(T)/r)^2 &\geqslant (1-\varepsilon)^2[2 - (\lambda+\mu)] = (1-\varepsilon)^2(1+\kappa)\\ &\geqslant 1 + \frac{\kappa}{4} + \frac{2c_1}{\sqrt{n}},\end{aligned}$$

we may apply Lemma 7.6.3 (or more precisely Remark 7.6.4) to get

$$\frac{r}{\rho} = d_G(T, B_2^n) \leqslant \frac{4}{\kappa}.$$

Now we may apply Lemma 7.5.1 to find a subspace F_1 of dimension $k_1 = [\lambda n]$ such that $K \cap F_1 \subseteq T \cap F_1 \subseteq rB_{F_1}$, and Lemma 7.6.6 to find a subspace F_2 of dimension $k_2 = [\mu n]$ such that $(r\kappa/4)B_{F_2} \subseteq \rho B_{F_2} \subseteq P_{F_2}(T) \subseteq P_{F_2}(K)$. This shows that

$$t(X, r) + t\big(X^*, \frac{4}{\kappa r}\big) \geqslant k_1 + k_2 \geqslant (1-\kappa)n,$$

as claimed. $\square$

7.7. Notes and remarks

First proofs of the M^*-estimate

The M^*-estimate was presented for the first time by Milman at the Laurent Schwartz Colloquium in 1983. This proof was using Urysohn's inequality and is given in Theorem 7.2.1 (see [**444**]). The next proof, namely that of Theorem 7.2.3 is given in [**445**]. The third proof given, of Theorem 7.2.6, is from Litvak, Milman and Pajor [**387**].

Proofs with the optimal dependence

The argument of Pajor and Tomczak appears in [**487**]. The proof of Theorem 7.3.1 that we present is from [**453**]. Note that the Pajor-Tomczak proof is very close in spirit to that of a similar result of Garnaev and Gluskin [**221**] for the concrete space ℓ_1^n. Milman's argument is very close to Makovoz's proof of the Garnaev-Gluskin theorem in [**412**]. Theorem 7.3.1 (with some constant $c > 0$) plays and important role in operator theory, see for example [**593**] and [**509**].

Gordon's theorem appears in [**258**] and will be studied in Chapter 9. It has the following variant, which asserts only the existence of F, but is with $\gamma = 1$:

THEOREM 7.7.1. *Let K be a convex body in $\mathbb{R}^n$ with $0 \in \mathrm{int}(K)$. Then, for any $1 \leqslant k < n$ there exists $F \in G_{n,n-k}$ such that*

$$|x| \leqslant \frac{a_n M^*}{a_k} \|x\|_K$$

for all $x \in F$.

Milman's quotient of subspace theorem

Theorem 7.4.1 is from [**445**]. Milman used the M^*-estimate and introduced the idea of iteration. Our presentation here incorporates the optimal dependence in the M^*-estimate; the formulation and the proof of the iteration Lemma 7.4.3 come from Pisier [**509**]. However, similar iteration lemmas can be found in [**444**] or [**447**] (see also the article of Dilworth and Szarek [**172**]).

With these tools one sees that for every centrally symmetric convex body K in $\mathbb{R}^n$ and any $\alpha \in (0,1)$ one can find subspaces $G \subset E$ with $\dim(G) \geqslant \alpha n$ and an ellipsoid $\mathcal{E}$ in G such that

$$\mathcal{E} \subset P_G(K \cap E) \subset c(1-\alpha)^{-1} |\log(c(1-\alpha))| \mathcal{E}.$$

It is not known whether this "almost linear" dependence on $1-\alpha$ is optimal.

Asymptotic formulas for random sections

The question to obtain upper and lower bounds for the diameter of random $\lfloor \lambda n \rfloor$-dimensional central sections of a centrally symmetric convex body through the solutions of equations of the form $M^*(K \cap rB_2^n) = h(\lambda) r$ was studied by Giannopoulos and Milman in [**229**] and [**231**]. Theorem 7.5.2 is due to Giannopoulos, Milman and Tsolomitis [**237**] and gives better estimates.

The main tool for this improvement is Gromov's theorem on the isoperimetry of waists which appeared in [**270**]. It was first observed by Vershynin that it could be used in this context. Vershynin's results are from [**597**]. The asymptotic behaviour of $\sigma_{n,k}(\theta)$ has been determined by Artstein in [**22**]: Let $\lambda \in (0,1)$. Then, the following estimates hold as $n \to \infty$.

(1) If $\sin^2\theta > 1-\lambda$, then

$$\sigma_{n,k}(\theta) \simeq 1 - \frac{1}{\sqrt{n\pi}} \frac{\sqrt{\lambda(1-\lambda)}}{\sin^2\theta - (1-\lambda)} e^{\frac{n}{2} u(\lambda,\theta)}.$$

(2) If $\sin^2\theta < 1-\lambda$, then

$$\sigma_{n,k}(\theta) \simeq \frac{1}{\sqrt{n\pi}} \frac{\sqrt{\lambda(1-\lambda)}}{(1-\lambda)-\sin^2\theta} e^{\frac{n}{2}u(\lambda,\theta)},$$

where

$$u(\lambda,\theta) = (1-\lambda)\ln\frac{(1-\lambda)}{\sin^2\theta} + \lambda\ln\frac{\lambda}{\cos^2\theta}.$$

In particular, there exists a critical value $\theta(\lambda) = \arcsin(\sqrt{1-\lambda})$ such that if $k \geqslant \lambda n$ and $\theta > \theta(\lambda)$ then $\sigma_{n,k}(\theta) \to 1$ as $n \to \infty$.

Minimal and typical diameter of proportional sections

The precise statement (see [**237**]) of the general case of Theorem 7.5.9 is as follows:

THEOREM 7.7.2. *Let $0 < \mu < 1$ and $0 < s < \frac{1}{2-\mu}$. There exists $n_0 = n_0(\mu, s)$ such that*

$$\left(\frac{c\mu(1-s(2-\mu))}{1-s\mu}\sqrt{1-\mu}\right) b(s\mu, K) \leqslant a(\mu, K)$$

for every $n \geqslant n_0$ and every centrally symmetric convex body K in $\mathbb{R}^n$.

Assuming that Gromov's theorem (Theorem 7.5.3) holds without any restriction on n and k one would be able to prove the following version of Theorem 7.5.2: *Let $\lambda \in (0,1)$ and $\varepsilon > 0$ satisfy $(1+\varepsilon)\sqrt{1-\lambda} < 1$ and let $n \geqslant n_1(\lambda,\varepsilon) \simeq \frac{1}{(1-\lambda)\varepsilon^2}$. If K is a centrally symmetric convex body in $\mathbb{R}^n$, and if r_2 is the solution of the equation*

$$M^*(K \cap rB_2^n) = (1+\varepsilon)\sqrt{1-\lambda}r,$$

then

$$R(K \cap F) \geqslant \frac{1}{2}\varepsilon\sqrt{1-\lambda}r_2$$

for every $F \in G_{n,\lfloor \lambda n \rfloor}$.

This would give a very precise "asymptotic formula" for the diameter of random $\lfloor \lambda n \rfloor$-dimensional sections of any high-dimensional centrally symmetric convex body. Solving the single "asymptotic equation" $M^*(K \cap rB_2^n) \asymp \sqrt{1-\lambda}\, r$ would give both an upper and a lower bound (up to a constant depending on λ) for the radius of a random $K \cap F$, $F \in G_{n,\lfloor \lambda n \rfloor}$.

We would also have the next improved version of Theorem 7.7.2: *Let $\mu, s \in (0,1)$. There exists $n_0 = n_0(\mu, s)$ such that*

$$b(s\mu, K) \leqslant \frac{c(1-s\mu)}{(1-s)\mu\sqrt{1-\mu}} a(\mu, K)$$

for every $n \geqslant n_0$ and every centrally symmetric convex body K in $\mathbb{R}^n$.

Results similar to those given in this section were obtained by Litvak, Pajor and Tomczak-Jaegermann in [**392**]. They prove

THEOREM 7.7.3. *Let $1 \leqslant k < m < n$, $a > 0$, and let K be a convex body in $\mathbb{R}^n$ such that there exists a subspace $E \in G_{n,n-k}$ with $R(K \cap E) \leqslant 2a$. Then there is a subset $B \subset G_{n,n-m}$ of measure at least $1 - 2\exp(-m/2)$ such that for all $F \in B$*

$$R(K \cap F) \leqslant a\left(C\sqrt{n/m}\right)^{\frac{m}{m-k}}.$$

The proof uses Gaussian random matrices and covering estimates.

Random sections and intersection of random rotations

Let K be a centrally symmetric convex body in $\mathbb{R}^n$ and let $t,k \geqslant 2$ be two integers. We define the minimal circumradius of an intersection of t rotations of K by

$$r_t(K) = \min\{\rho > 0 : \exists\, U_1,\dots,U_t \in O(n) \text{ s.t. } U_1(K)\cap\cdots\cap U_t(K) \subseteq \rho B_2^n\},$$

and the "upper radius" of a random $\lceil n/k\rceil$-dimensional central section of K by

$$R_k(K) = \min\Big\{\rho > 0 : \nu_{n,\lceil \frac{n}{k}\rceil}\Big(F\in G_{n,\lceil \frac{n}{k}\rceil} : R(K\cap F)\leqslant \rho\Big) \geqslant 1-\frac{1}{k+1}\Big\}.$$

It was proved by Milman in [**455**] that if $\rho > R_k(K)$ then starting from a set of pairwise orthogonal $\lceil n/k\rceil$-dimensional subspaces $F_1,\dots,F_k$ which satisfy $K\cap F_i \subseteq \rho(B_2^n\cap F_i)$ and $F_1\oplus\cdots\oplus F_k = \mathbb{R}^n$, one can build rotations $U_1,\dots,U_t\in O(n)$, for some $t\leqslant 2k$, such that

$$\frac{1}{t}\sum_{j=1}^t U_j(K^\circ) \supseteq \frac{1}{\rho\sqrt{k}}\,B_2^n.$$

Dualizing this fact, one has that $U_1^*(K)\cap\cdots\cap U_t^*(K)\subseteq \rho\sqrt{k}B_2^n$, which shows that

$$r_{2k}(K)\leqslant \sqrt{k}R_k(K).$$

The next theorem (see Giannopoulos and Milman [**230**] and Giannopoulos, Milman and Tsolomitis [**237**]) provides a converse inequality.

THEOREM 7.7.4. *There exist $c_1,c_2>0$ such that for every integer $t\geqslant 2$ and every $n\geqslant 2(t+1)$, the inequality*

$$R_{c_1t}(K)\leqslant c_2\sqrt{t}r_t(K)$$

holds true for every centrally symmetric convex body K in $\mathbb{R}^n$.

The proof uses Borsuk's theorem and the following *conditional M-estimate* from [**237**]: for any $\lambda\in(0,1)$, if r is the solution of the equation $M^*(K\cap rB_2^n) = \sqrt{\frac{3+\lambda}{4}}r$, then $R(K^\circ\cap F)\leqslant \frac{2(\sqrt{2}+1)}{(1-\lambda)^{1/2}r}$ for a random $F\in G_{n,\lfloor\lambda n\rfloor}$.

Linear duality relations

The results of this section, and in particular the distance lemmas, are from Milman's [**454**]; see also [**455**]. The actual result is:

THEOREM 7.7.5. *Let $\kappa\in(0,1)$. For every n-dimensional normed space $X=(\mathbb{R}^n,\|\cdot\|)$ and every $r>0$,*

$$t(X,r)+t\left(X^*,\frac{1}{\kappa r}\right)\geqslant (1-\kappa)n - o(n).$$

and follows from applying the variant (Theorem 7.7.1) of Gordon's result with $\varepsilon=0$ and without the probability estimate (just existence).

Coordinate theory

We fix an orthonormal basis $\{e_1,\dots,e_n\}$ of $\mathbb{R}^n$ and for every non-empty $\sigma\subseteq\{1,\dots,n\}$ we consider the coordinate subspace $\mathbb{R}^\sigma=\text{span}\{e_j : j\in\sigma\}$. A coordinate version of the M^*-estimate was established by Giannopoulos and Milman in [**228**]: If K is an ellipsoid in $\mathbb{R}^n$, then for every $\lambda\in(0,1)$ we can find $\sigma\subseteq\{1,\dots,n\}$ of cardinality $|\sigma|\geqslant(1-\lambda)n$ such that

$$P_{\mathbb{R}^\sigma}(K)\supseteq \frac{[\lambda/\log(1/\lambda)]^{1/2}}{M(K)}B_2^n\cap\mathbb{R}^\sigma.$$

This observation (which has its origin in two papers of Giannopoulos [**222**], [**223**]) has consequences for the question of the maximal Banach-Mazur distance to the cube and the proportional Dvoretzky-Rogers factorization theorem that will be discussed in Part II. The proof is based on an isomorphic version of the Sauer-Shelah lemma from combinatorics, which was proved by Szarek and Talagrand [**585**] and is close in spirit to the theory of restricted invertibility of operators which was developed by Bourgain and Tzafriri [**113**]. As the example of the cube shows, one cannot have a coordinate M^*-estimate for an arbitrary convex body. Under assumptions which guarantee the existence of "large ellipsoids" of any proportional dimension inside the body, one can use the above ellipsoidal result and obtain coordinate M^*-estimates. This is done in [**228**] for bodies whose volume ratio or cotype-2 constant is well-bounded.

Rudelson and Vershynin [**529**] obtained another class of coordinate results. Assume that K is a centrally symmetric convex body in $\mathbb{R}^n$ such that the norm $\|\cdot\|$ induced by K satisfies the conditions $\|x\| \leqslant |x|$ for all x and $M = M(K) \geqslant \delta$ for some positive constant $\delta > 0$. Then, there exist two positive numbers s and t with $c\delta \leqslant t \leqslant 1$ and $st \geqslant \delta / \log^{3/2}(2/\delta)$ and a subset σ of $\{1, \ldots, n\}$ with cardinality $|\sigma| \geqslant s^2 n$, such that

$$\Big\| \sum_{i \in \sigma} a_i e_i \Big\| \geqslant \frac{ct}{\sqrt{n}} \sum_{i \in \sigma} |a_i|$$

for all choices of reals a_i, $i \in \sigma$. From this statement one can recover Elton's theorem about spaces which contain large dimensional copies of ℓ_1's [**187**] in an optimal form. Note that the space $X = (\mathbb{R}^n, \|\cdot\|)$ satisfies $k(X) \simeq n(M/b)^2 \geqslant \delta n$. In other words, the result concerns spaces which have Euclidean subspaces of some dimension proportional to n (depending on δ). The estimate of Rudelson and Vershynin shows that

$$K \cap \mathbb{R}^\sigma \subseteq c(\delta)\sqrt{n} B_1^\sigma .$$

This may be viewed as a coordinate version of the M^*-estimate for this class of bodies. The formulation is dual to the one in [**228**]: one now considers sections instead of projections. To feel the analogy even more, we state the following "condition-free" version of the result in Rudelson and Vershynin [**529**]: Let K be a symmetric convex body in $\mathbb{R}^n$ with $B_2^n \subseteq K$. There exists a subset σ of $\{1, \ldots, n\}$ with cardinality $|\sigma| \geqslant cf(M)n$, such that

$$M \cdot (K \cap \mathbb{R}^\sigma) \subseteq \sqrt{|\sigma|} B_1^\sigma ,$$

where $f(x) = x \log^{-3/2}(2/x)$. Compare with the M^*-estimate: one has sections of the body inside an appropriate ℓ_1-ball on coordinate subspaces (this is weaker, but the example of ℓ_1^n shows that it is natural). Also, the parameter $1/M^*$ is replaced by M (which is stronger). However, the estimates hold for some proportional dimensions and not for any proportion.

CHAPTER 8

M-position and the reverse Brunn–Minkowski inequality

8.1. Introduction

One of the deepest results in asymptotic geometric analysis is the existence of an M-ellipsoid, namely that for every convex body K there exists an ellipsoid of the same volume which can replace it, in many computations, up to universal constants. This result was discovered by V. Milman and leads to many conclusions, among them the reverse Santaló inequality and the reverse Brunn-Minkowski inequality mentioned in the title of the chapter, though things were not discovered exactly in this order.

Let us begin by describing the reverse Santaló inequality. We recall that the volume product, sometimes also called the Mahler product, is defined by

$$s(K) := \mathrm{Vol}_n(K)\mathrm{Vol}_n(K^\circ).$$

We have already encountered it in Chapter 1, Section 1.5.4, where we proved the classical Blaschke-Santaló inequality which states that, given a centrally symmetric convex body K in $\mathbb{R}^n$, the volume product $s(K)$ is less than or equal to the volume product $s(B_2^n)$, and equality holds if and only if K is an ellipsoid. In the opposite direction, a well-known conjecture of Mahler states that $s(K) \geqslant 4^n/n!$ for every centrally symmetric convex body K (i.e., the cube is a minimizer for $s(K)$ among centrally symmetric convex bodies) and that $s(K) \geqslant (n+1)^{n+1}/(n!)^2$ in the not necessarily symmetric case, meaning that in this case the simplex is a minimizer. This has been verified for some classes of bodies, e.g. zonoids and 1-unconditional bodies. The first main result in this chapter is an isomorphic version of the reverse Santaló inequality, also called the Bourgain-Milman inequality, which states that there exists an absolute constant $c > 0$ such that

$$\left(\frac{s(K)}{s(B_2^n)}\right)^{1/n} \geqslant c$$

for every centrally symmetric convex body K in $\mathbb{R}^n$ (from which it can then easily be inferred, using the Rogers-Shephard inequality, that $s(K)^{1/n} \geqslant c' s(B_2^n)^{1/n}$ for every convex body K which contains the origin in its interior, with $c' = c/4$). The inequality was first proved in 1985 and answers the question of Mahler in the asymptotic sense: for every centered convex body K in $\mathbb{R}^n$, the affine invariant $s(K)^{1/n}$ is of the order of $1/n$. A few other proofs have appeared, the most recent of which give the best known lower bounds for the constant c in the symmetric case and exploit tools from quite diverse areas: Kuperberg has shown that in the centrally symmetric case we have $c \geqslant 1/2$, and his proof uses tools from differential geometry, while Nazarov's proof uses multivariable complex analysis and leads to the bound $c \geqslant \pi^2/32$.

The original proof of the reverse Santaló inequality employed a dimension descending procedure which was based on Milman's quotient of subspace theorem, thus depended on the MM^*-estimate, Theorem 6.5.1 (together with Pisier's inequality, Theorem 6.2.4). We present it in Section 8.2. Later, Milman offered a second approach, which introduced an "isomorphic symmetrization" technique. This is a symmetrization scheme which is in many ways different from the classical symmetrizations. In each step, none of the natural parameters of the body is being preserved, but the ones which are of interest remain under control. The MM^*-estimate is again crucial for the proof. We describe this second proof in Section 8.3.

The existence of an "M-ellipsoid" associated with any centrally symmetric convex body K in $\mathbb{R}^n$ was proved by Milman: there exists an absolute constant $C > 0$ such that for any centrally symmetric convex body K in $\mathbb{R}^n$ we can find an origin symmetric ellipsoid $\mathcal{E}_K$ satisfying $\mathrm{Vol}_n(K) = \mathrm{Vol}_n(\mathcal{E}_K)$ and

$$\begin{aligned} &\frac{1}{C}\mathrm{Vol}_n(\mathcal{E}_K + T)^{1/n} \leqslant \mathrm{Vol}_n(K + T)^{1/n} \leqslant C\,\mathrm{Vol}_n(\mathcal{E}_K + T)^{1/n}, \\ &\frac{1}{C}\mathrm{Vol}_n(\mathcal{E}_K^\circ + T)^{1/n} \leqslant \mathrm{Vol}_n(K^\circ + T)^{1/n} \leqslant C\,\mathrm{Vol}_n(\mathcal{E}_K^\circ + T)^{1/n}, \end{aligned} \tag{8.1.1}$$

for every convex body T in $\mathbb{R}^n$. In fact (see Remark 8.4.2) in addition one has

$$\mathrm{Vol}_n(K \cap \mathcal{E}_K) \geqslant C^{-n}\mathrm{Vol}_n(K) \qquad \text{and} \qquad \mathrm{Vol}_n(K^\circ \cap \mathcal{E}_K^\circ) \geqslant C^{-n}\mathrm{Vol}_n(K^\circ),$$

and even more - for every convex body $T \subset \mathbb{R}^n$

$$\mathrm{Vol}_n(\mathcal{E}_K \cap T)^{1/n} \simeq \mathrm{Vol}_n(K \cap T)^{1/n} \quad \text{and} \quad \mathrm{Vol}_n(\mathcal{E}_K^\circ \cap T)^{1/n} \simeq \mathrm{Vol}_n(K^\circ \cap T)^{1/n}.$$

These inequalities mean that, essentially, the convex body can be replaced by its M-ellipsoid in many volume computations, and is a very strong tool. Sometimes one defines the quantity

$$M(K,T) = \left(\frac{\mathrm{Vol}_n(K+T)}{\mathrm{Vol}_n(K\cap T)}\,\frac{\mathrm{Vol}_n(K^\circ + T^\circ)}{\mathrm{Vol}_n(K^\circ \cap T^\circ)}\right)^{1/n}.$$

In this notation, the result regarding the existence of M-ellipsoids reads:

There exists an absolute constant $C > 0$ with the following property: for every $n \in \mathbb{N}$ and every centrally symmetric convex body K in $\mathbb{R}^n$ we can find an ellipsoid $\mathcal{E}$ in $\mathbb{R}^n$ such that

$$M(K,\mathcal{E}) \leqslant C.$$

The existence of M-ellipsoids can be equivalently established by introducing the M-position of a convex body. To any given centered convex body K in $\mathbb{R}^n$ we can apply a linear transformation and find a position $\tilde{K} = u_K(K)$ of volume $\mathrm{Vol}_n(\tilde{K}) = \mathrm{Vol}_n(K)$ such that (8.1.1) is satisfied with $\mathcal{E}_K$ a multiple of B_2^n. This is the so-called M-position of K. Note at this point that "being in M-position" is not well defined, since there is a constant C involved. To be more precise, let us use the following definition.

DEFINITION 8.1.1. We say that a centrally symmetric convex body $K \subset \mathbb{R}^n$ is in M-position with constant C if letting rB_2^n have the same volume as K (namely,

$r^n = \frac{\mathrm{Vol}(K)}{\mathrm{Vol}(B_2^n)}$) we have

$$\begin{aligned}(8.1.2) \qquad \frac{1}{C}\mathrm{Vol}_n(rB_2^n + T)^{1/n} &\leqslant \mathrm{Vol}_n(K+T)^{1/n} \leqslant C\,\mathrm{Vol}_n(rB_2^n+T)^{1/n},\\ \frac{1}{C}\mathrm{Vol}_n(r^{-1}B_2^n + T)^{1/n} &\leqslant \mathrm{Vol}_n(K^\circ+T)^{1/n} \leqslant C\,\mathrm{Vol}_n(r^{-1}B_2^n+T)^{1/n},\end{aligned}$$

for every convex body T in $\mathbb{R}^n$.

Thus, Milman's theorem asserts that there exists a universal constant C such that for any n and any centrally symmetric convex body K in $\mathbb{R}^n$ there exists $A \in SL_n$ such that AK is in M-position with constant C.

Since for two Euclidean balls (of any radii) the Brunn-Minkowski inequality is an equality, it follows from the existence of M-ellipsoids that for every pair of convex bodies K_1 and K_2 in $\mathbb{R}^n$ and for all $t_1, t_2 > 0$, their M-positions $\tilde{K}_1$ and $\tilde{K}_2$ satisfy

$$(8.1.3) \qquad \mathrm{Vol}_n(t_1\tilde{K}_1 + t_2\tilde{K}_2)^{1/n} \leqslant c'\left(t_1\mathrm{Vol}_n(\tilde{K}_1)^{1/n} + t_2\mathrm{Vol}_n(\tilde{K}_2)^{1/n}\right),$$

where $c' > 0$ is an absolute constant, and that (8.1.3) remains true if we replace $\tilde{K}_1$ or $\tilde{K}_2$ (or both) by their polars. This statement is Milman's reverse Brunn-Minkowski inequality.

Another way to define the M-position of a convex body is through covering numbers. As follows from Milman's proof, there exists an absolute constant $\beta > 0$ such that every centered convex body K in $\mathbb{R}^n$ has a linear image $\tilde{K}$ which satisfies $\mathrm{Vol}_n(\tilde{K}) = \mathrm{Vol}_n(B_2^n)$ and

$$(8.1.4) \qquad \max\{N(\tilde{K}, B_2^n), N(B_2^n, \tilde{K}), N(\tilde{K}^\circ, B_2^n), N(B_2^n, \tilde{K}^\circ)\} \leqslant \exp(\beta n).$$

It turns out that for a body to satisfy these inequalities, is equivalent to being in M-position in the sense defined above, and so these terms are sometimes mixed and one might say that a convex body K which satisfies (8.1.4) is in M-position "with constant β". To clarify this let us prove that, up to constants, these two properties imply each other.

LEMMA 8.1.2. *If a centrally symmetric convex body $K \subset \mathbb{R}^n$ with* $\mathrm{Vol}_n(K) = \mathrm{Vol}_n(B_2^n)$ *is in M-position with constant C then it satisfies* (8.1.4) *with constant* $\beta = \ln(3C)$. *Similarly, if a centrally symmetric convex body K with* $\mathrm{Vol}_n(K) = \mathrm{Vol}_n(B_2^n)$ *satisfies* (8.1.4) *with constant β then it is in M position with constant* $c_2\beta$.

Proof. For the first implication, since K is in M position, its M-ellipsoid is B_2^n, and thus using the standard volume estimate for covering numbers (Theorem 4.1.13) we get

$$\begin{aligned} N(K, B_2^n) &\leqslant \frac{\mathrm{Vol}_n(K + B_2^n/2)}{\mathrm{Vol}_n(B_2^n/2)} \leqslant C^n\frac{\mathrm{Vol}_n((3/2)B_2^n)}{\mathrm{Vol}_n(B_2^n/2)} = (3C)^n\\ N(K^\circ, B_2^n) &\leqslant \frac{\mathrm{Vol}_n(K^\circ + B_2^n/2)}{\mathrm{Vol}_n(B_2^n/2)} \leqslant C^n\frac{\mathrm{Vol}_n((3/2)B_2^n)}{\mathrm{Vol}_n(B_2^n/2)} = (3C)^n. \end{aligned}$$

We could stop here, using duality of entropy numbers, but actually showing the other two inequalities is also very simple

$$N(B_2^n, K) \leqslant \frac{\mathrm{Vol}_n(B_2^n + K/2)}{\mathrm{Vol}_n(K/2)} \leqslant C^n \frac{\mathrm{Vol}_n((3/2)K)}{\mathrm{Vol}_n(K/2)} = (3C)^n$$
$$N(B_2^n, K^\circ) \leqslant \frac{\mathrm{Vol}_n(B_2^n + K^\circ/2)}{\mathrm{Vol}_n(K^\circ/2)} \leqslant C^n \frac{\mathrm{Vol}_n((3/2)K^\circ)}{\mathrm{Vol}_n(K^\circ/2)} = (3C)^n.$$

For the second implication, assume that K satisfies (8.1.4); then, there exist $\{x_i\}_{i=1}^N$ for $N \leqslant \exp(\beta n)$ such that $K \subset \bigcup_{i=1}^N (x_i + B_2^n)$. Therefore for every T

$$K + T \subset \bigcup_{i=1}^{N} [(x_i + B_2^n) + T].$$

In particular, $\mathrm{Vol}_n(K+T) \leqslant (\exp(\beta))^n \mathrm{Vol}_n(B_2^n + T)$. Since $N(K^\circ, B_2^n) \leqslant \exp(\beta n)$ as well, we get similarly that

$$\mathrm{Vol}_n(K^\circ + T) \leqslant (\exp(\beta))^n \mathrm{Vol}_n(B_2^n + T).$$

As for the reverse inequality, we simply exchange the roles of K and B_2^n. By (8.1.4) there exist $\{y_i\}_{i=1}^N$ for $N = \exp(\beta n)$ such that $B_2^n \subset \bigcup_{i=1}^N (x_i + K)$. Therefore for every T

$$B_2^n + T \subset \bigcup_{i=1}^{N} [(x_i + K) + T)].$$

In particular

$$\mathrm{Vol}_n(B_2^n + T) \leqslant (\exp(\beta))^n \mathrm{Vol}_n(K + T).$$

The same proof works for K° of course. □

In fact, the following stronger form of one of the implications is true as well.

LEMMA 8.1.3. *For every $c > 0$ there exists $C(c) > 0$ such that the following holds. Let $n \in \mathbb{N}$ and let $K \subset \mathbb{R}^n$ be a centrally symmetric convex body and let $r = \left(\frac{\mathrm{Vol}_n(K)}{\mathrm{Vol}_n(B_2^n)}\right)^{1/n}$. Assume that $N(K, rB_2^n) \leqslant \exp(cn)$. Then K is in M position with constant $C(c)$.*

Proof. Assume without loss of generality that $r = 1$. Note that by duality of entropy numbers, Theorem 4.4.4, we have that for some $c_1 = c_1(c)$, $N(B_2^n, K^\circ) \leqslant \exp(c_1 n)$. If we show that $N(K^\circ, B_2^n) \leqslant \exp(c_3 n)$ for some $c_3(c)$, and use duality again, then by Lemma 8.1.2 we are done. We use the volume bound for covering numbers (Theorem 4.1.13) and Theorem 1.5.10 to get

$$N(K^\circ, B_2^n) \leqslant \frac{\mathrm{Vol}_n(K^\circ + B_2^n/2)}{\mathrm{Vol}_n(B_2^n/2)} \leqslant 2^n \frac{\mathrm{Vol}_n(B_2^n)}{\mathrm{Vol}_n((K^\circ + B_2^n/2)^\circ)}.$$

Note that $K^\circ + B_2^n/2 \subset \mathrm{conv}(2K^\circ, B_2^n)$ so that $(K^\circ + B_2^n/2)^\circ \supset (K/2) \cap B_2^n \supset (K \cap B_2^n)/2$ and we get

$$N(K^\circ, B_2^n) \leqslant 4^n \frac{\mathrm{Vol}_n(B_2^n)}{\mathrm{Vol}_n(K \cap B_2^n)}.$$

However, we know that $\mathrm{Vol}_n(K \cap B_2^n) N(K, B_2^n) \geqslant \mathrm{Vol}_n(K)$ and thus we get (using that $\mathrm{Vol}_n(K) = \mathrm{Vol}_n(B_2^n)$) that

$$N(K^\circ, B_2^n) \leqslant 4^n N(K, B_2^n) \leqslant \exp((c_1 + \ln(4))n)$$

as needed. □

Pisier has proposed a different approach to these results, which allows one to find a whole family of special M-ellipsoids satisfying stronger entropy estimates. The precise statement is as follows. For every $0 < \alpha < 2$ and every centrally symmetric convex body K in $\mathbb{R}^n$, there exists a linear image $\tilde{K}$ of K which satisfies $\mathrm{Vol}_n(\tilde{K}) = \mathrm{Vol}_n(B_2^n)$ and

$$(8.1.5)\quad \max\{N(\tilde{K}, tB_2^n), N(B_2^n, t\tilde{K}), N(\tilde{K}^\circ, tB_2^n), N(B_2^n, t\tilde{K}^\circ)\} \leqslant \exp\left(\frac{c(\alpha)n}{t^\alpha}\right)$$

for every $t \geqslant 1$, where $c(\alpha)$ is a constant depending only on α, with $c(\alpha) = O\big((2-\alpha)^{-1}\big)$ as $\alpha \to 2$. We then say that $\tilde{K}$ is in M-position of order α (or α-regular M-position). We describe Pisier's approach in the last part of this chapter.

8.2. The Bourgain-Milman inequality

In this section we describe a proof of the reverse Santaló inequality; it is based on a dimension descending procedure which exploits the quotient of subspace theorem. It will be convenient for us to introduce the normalized version of the volume product.

DEFINITION 8.2.1. If K is a centrally symmetric convex body in $\mathbb{R}^n$ we set

$$\overline{s}(K) = \left(\frac{\mathrm{Vol}_n(K)\mathrm{Vol}_n(K^\circ)}{\mathrm{Vol}_n(B_2^n)^2}\right)^{1/n}.$$

THEOREM 8.2.2 (Bourgain-Milman). *There exists an absolute constant $c > 0$ such that for every centrally symmetric convex body K in $\mathbb{R}^n$,*

$$\overline{s}(K) \geqslant c.$$

Proof. For every positive integer N we define

$$\alpha_N := \inf\{\overline{s}(K) : 1 \leqslant n \leqslant N, K \text{ is a centrally symmetric convex body in } \mathbb{R}^n\}.$$

Let $n \leqslant N$ and consider a centrally symmetric convex body K in $\mathbb{R}^n$. Applying the quotient of subspace theorem (Theorem 7.4.1) we may find subspaces $F \subset E \subset \mathbb{R}^n$ such that $\dim E = m$ and $\dim F = k \geqslant n/2$ and an ellipsoid $\mathcal{E}$ in F so that

$$\mathcal{E} \subseteq P_F(K \cap E) \subseteq C_0\mathcal{E},$$

where $C_0 \geqslant 1$ is an absolute constant. Then, we easily check that

$$(8.2.1)\qquad C_0^{-k}s(\mathcal{E}) \leqslant s(P_F(K\cap E)) \leqslant C_0^k s(\mathcal{E}).$$

From the affine invariance of $s(\cdot)$ it is clear that $s(\mathcal{E}) = s(B_2^k) = \kappa_k^2$.

We apply Lemma 1.5.6 to get

$$\mathrm{Vol}_n(K) \geqslant 2^{-n}\mathrm{Vol}_m(K\cap E)\mathrm{Vol}_{n-m}(P_{E^\perp}(K)),$$

and then apply Lemma 1.5.6 again so that we have

$$(8.2.2)\quad \mathrm{Vol}_n(K) \geqslant 2^{-2n}\mathrm{Vol}_k(P_F(K\cap E))\mathrm{Vol}_{m-k}(K\cap E\cap F^\perp)\mathrm{Vol}_{n-m}(P_{E^\perp}(K)).$$

Working in the same way with K° we get

$$(8.2.3)\quad \mathrm{Vol}_n(K^\circ) \geqslant 2^{-2n}\mathrm{Vol}_k(P_E(K^\circ)\cap F)\mathrm{Vol}_{m-k}(P_{F^\perp}P_E(K^\circ))\mathrm{Vol}_{n-m}(K^\circ\cap E^\perp).$$

We define

$$K_1 = P_{E^\perp}(K) \quad\text{and}\quad K_2 = K\cap E\cap F^\perp.$$

We also set $d_1 = \dim E^{\perp}$ and $d_2 = \dim(E \cap F^{\perp})$. Then, $d_1 + d_2 = n - k$. Multiplying (8.2.2) and (8.2.3) we get

$$s(K) \geqslant 2^{-4n} s(P_F(K \cap E)) s(K_1) s(K_2). \tag{8.2.4}$$

By the definition of α_N we have

$$s(K_1) \geqslant \alpha_N^{d_1} \kappa_{d_1}^2 \quad \text{and} \quad s(K_2) \geqslant \alpha_N^{d_2} \kappa_{d_2}^2.$$

Using also (8.2.1) we finally get

$$s(K) \geqslant 2^{-4n} C_0^{-k} \kappa_k^2 \kappa_{d_1}^2 \kappa_{d_2}^2 \alpha_N^{d_1+d_2}.$$

It follows that

$$\overline{s}(K) \geqslant \alpha_N^{\frac{n-k}{n}} C_0^{-\frac{k}{n}} \left(\frac{\kappa_k^2 \kappa_{d_1}^2 \kappa_{d_2}^2}{16^n \kappa_n^2} \right)^{\frac{1}{n}}. \tag{8.2.5}$$

Now, observe that if we apply the right hand side inequality of (1.5.1) for the Euclidean unit ball we get $\kappa_{n-k} = \kappa_{d_1+d_2} \leqslant \kappa_{d_1} \kappa_{d_2}$ and $\kappa_n \leqslant \kappa_{n-k} \kappa_k$. Therefore,

$$\kappa_k \kappa_{d_1} \kappa_{d_2} \geqslant \kappa_k \kappa_{n-k} \geqslant \kappa_n.$$

Going back to (8.2.5), and using the fact that $\alpha_N \leqslant \alpha_1 = 1$ and $\frac{n-k}{n} \leqslant \frac{1}{2}$, we see that

$$\overline{s}(K) \geqslant \frac{\sqrt{\alpha_N}}{16 C_0}. \tag{8.2.6}$$

Since K was arbitrary, we conclude that $\alpha_N \geqslant \sqrt{\alpha_N}/(16C_0)$, and hence

$$\alpha_N \geqslant c := (16 C_0)^{-2}.$$

This is true for every positive integer N; thus, the theorem is proved. □

We conclude this section with a duality of entropy result of König and Milman [**361**] which follows from the Bourgain-Milman inequality.

THEOREM 8.2.3 (König-Milman). *For every pair of centrally symmetric convex bodies K and T in $\mathbb{R}^n$ one has*

$$C^{-n} N(T^{\circ}, K^{\circ}) \leqslant N(K, T) \leqslant C^n N(T^{\circ}, K^{\circ}),$$

where $C > 0$ is an absolute constant.

Proof. In the following line of inequalities we use the inclusion $T^{\circ} \subseteq (K \cap T)^{\circ}$, Fact 4.1.9, and Theorem 4.1.13, together with Theorem 8.2.2.

$$\begin{aligned} N(T^{\circ}, K^{\circ}) &\leqslant N((K \cap T)^{\circ}, K^{\circ}) \leqslant \frac{\mathrm{Vol}_n(2(K \cap T)^{\circ} + K^{\circ})}{\mathrm{Vol}_n(K^{\circ})} \\ &\leqslant \frac{\mathrm{Vol}_n(3(K \cap T)^{\circ})}{\mathrm{Vol}_n(K^{\circ})} = 3^n \frac{\mathrm{Vol}_n((K \cap T)^{\circ})}{\mathrm{Vol}_n(K^{\circ})} \\ &\leqslant (3c)^n \frac{\mathrm{Vol}_n(K)}{\mathrm{Vol}_n(K \cap T)} = (6c)^n \frac{\mathrm{Vol}_n(K)}{\mathrm{Vol}_n(2(K \cap T))} \\ &\leqslant (6c)^n N(K, 2(K \cap T)) \leqslant (6c)^n N(K, T), \end{aligned}$$

where we have used the constant c from Theorem 8.2.2. The right hand side inequality is proved in the same way (replacing K and T with T° and K° respectively). □

8.3. Isomorphic symmetrization

In this section we introduce Milman's isomorphic symmetrization technique and give an alternative proof of the reverse Santaló inequality, Theorem 8.2.2.

The idea of the proof is as follows. For every n-dimensional normed space $X=(\mathbb{R}^n,\|\cdot\|)$ we define $d_X:=d_{BM}(X,\ell_2^n)$, the Banach Mazur distance between X and Euclidean space. We start with an arbitrary centrally symmetric convex body K in $\mathbb{R}^n$ and set $b(K)=\max\{\|x\|_K : x\in B_2^n\}$. We know by Theorem 6.5.1 from Chapter 6 (together with Theorem 6.2.4) that there exists $T\in SL_n$ such that the body $\tilde{K}=T(K)$ has the additional properties $M(\tilde{K})=1$ and $M(\tilde{K}^\circ)\leqslant c\log(d_{X_K}+1)$. Since $s(K)=s(\tilde{K})$ we may assume to begin with that K satisfies these two inequalities. Since $M(K)=1$, we have $1\leqslant b(K)\leqslant c_0\sqrt{n}$, see Remark 5.2.2 in Chapter 5.

We shall define an iterative procedure. At each step we shall replace K by a new body K_1 which will have more or less the same volume as K, while the Banach-Mazur distance to Euclidean, $d_{X_{K_1}}$, will be much smaller than d_{X_K}. At the same time, we will be able to control the volume of K_1° so that the quantities $s(K_1)^{1/n}$ and $s(K)^{1/n}$ will be of the same order. This procedure is called "isomorphic symmetrization". After a finite number of steps, we shall reach an isomorphic Euclidean ball.

To this end we shall need the following proposition, assuring us that indeed at every step the volume parameter is controlled.

PROPOSITION 8.3.1. *Let K be a centrally symmetric convex body in $\mathbb{R}^n$ with $b(K)\leqslant c_0\sqrt{n}$ and $M(K)=1$, and let $\alpha>1$. Set*

$$\lambda^*=M^*(K)\alpha \quad \lambda_*=M(K)\alpha=\alpha,$$

and define

$$K_1=\mathrm{conv}\left[(K\cap\lambda^*B_2^n)\cup\frac{1}{\lambda_*}B_2^n\right].$$

Then

$$\mathrm{Vol}_n(K_1)\geqslant\mathrm{Vol}_n(K)\exp\big(-cn/\alpha^2\big)$$

and, as long as, say, $\alpha<n^{1/3}$ we also have

$$\mathrm{Vol}_n(K_1)\leqslant\mathrm{Vol}_n(K)\ \exp\big(cn/\alpha^2\big).$$

Proof. We shall use the following simple observation about covering numbers (which is simply the case $P=\{0\}$ in Lemma 4.1.17 from Chapter 4):

$$\mathrm{Vol}_n(K)\leqslant\sum_{i=1}^{N}\mathrm{Vol}_n(K\cap(x_i+sB_2^n))\leqslant N(K,sB_2^n)\,\mathrm{Vol}_n(K\cap sB_2^n). \tag{8.3.1}$$

By the definition of K_1 and (8.3.1) we have

$$\mathrm{Vol}_n(K_1)\geqslant\mathrm{Vol}_n(K\cap\lambda^*B_2^n)\geqslant N(K,\lambda^*B_2^n)^{-1}\mathrm{Vol}_n(K).$$

Using Sudakov's inequality (Theorem 4.2.1) we get

$$\mathrm{Vol}_n(K_1)\geqslant\mathrm{Vol}_n(K)\ \exp\Big(-cn\Big(\frac{M(K^\circ)}{\lambda^*}\Big)^2\Big)=\mathrm{Vol}_n(K)\ \exp\big(-cn/\alpha^2\big),$$

where $c>0$ is an absolute constant.

For the second inequality, bounding the volume of K_1 from below, we shall use Lemma 4.1.23 from Chapter 4 which states that given a convex body $L\subset\mathbb{R}^n$ and a centrally symmetric convex body $K \subset \mathbb{R}^n$ such that $L \subseteq bK$ for some $b \geqslant 1$ we have

$$\mathrm{Vol}_n\Big(\mathrm{conv}\big(K \cup L\big)\Big) \leqslant 3ebn\, N(L, K)\mathrm{Vol}_n(K).$$

We shall use this lemma with $L = \frac{1}{\alpha}B_2^n$ and $b = \frac{b(K)}{\alpha} \leqslant b(K)$ (observe that $\lambda_* = \alpha > 1$). By the definition of K_1 and the dual Sudakov inequality (Theorem 4.2.2) we see that

$$\begin{aligned}\mathrm{Vol}_n(K_1) &\leqslant \mathrm{Vol}_n\Big(\mathrm{conv}\Big(K \cup \frac{1}{\lambda_*}B_2^n\Big)\Big)\\ &= \mathrm{Vol}_n\Big(\mathrm{conv}\,\Big(K \cup \frac{1}{\alpha}B_2^n\Big)\Big)\\ &\leqslant 3en\frac{b(K)}{\alpha}N(B_2^n, \alpha K)\,\mathrm{Vol}_n(K)\\ &\leqslant 3enb(K)\mathrm{Vol}_n(K)\,\exp\Big(cn\Big(\frac{M(K)}{\alpha}\Big)^2\Big)\\ &= 3enb(K)\,\mathrm{Vol}_n(K)\,\exp\big(cn/\alpha^2\big).\end{aligned}$$

The term $3enb(K)$ can be absorbed by the exponential term since we have assumed $\alpha \leqslant n^{1/3}$, and hence, given also that $b(K) \leqslant c_0\sqrt{n}$, we have $2enb(K) \leqslant \exp\big(c'n/\alpha^2\big)$ for some absolute constant c'. Thus, the proof of the lemma is complete. $\square$

Combining the two parts of the proposition we get that under the condition of the proposition

$$\exp\big(-cn/\alpha^2\big) \leqslant \frac{\mathrm{Vol}_n(K)}{\mathrm{Vol}_n(K_1)} \leqslant \exp\big(cn/\alpha^2\big). \tag{8.3.2}$$

Note that the polar body K_1° of K_1 is the set

$$K_1^\circ = \mathrm{conv}\,\Big(K^\circ \cup \Big(\frac{1}{\lambda^*}B_2^n\Big)\Big) \cap \lambda_* B_2^n.$$

Similar to Proposition 8.3.1 we can show

PROPOSITION 8.3.2. *Under the same assumptions as in Proposition* 8.3.1 *we have*

$$\exp\big(-cn/\alpha^2\big) \leqslant \frac{\mathrm{Vol}_n(K^\circ)}{\mathrm{Vol}_n(K_1^\circ)} \leqslant \exp\big(cn/\alpha^2\big). \tag{8.3.3}$$

Proof. By the definition of K_1° we have

$$\mathrm{Vol}_n(K_1^\circ) \geqslant \mathrm{Vol}_n(K^\circ \cap \lambda_* B_2^n) \geqslant N(K^\circ, \lambda_* B_2^n)^{-1}\mathrm{Vol}_n(K^\circ).$$

Here we have used (8.3.1) again. Using Sudakov's inequality, we get

$$\mathrm{Vol}_n(K_1^\circ) \geqslant \mathrm{Vol}_n(K^\circ)\,\exp\Big(-cn\Big(\frac{M(K)}{\lambda_*}\Big)^2\Big) = \mathrm{Vol}_n(K^\circ)\,\exp\big(-cn/\alpha^2\big).$$

On the other hand, we can use Lemma 4.1.23 again, this time for K° (observe that $\lambda^* = M(K)M^*(K)\alpha \geqslant 1$) and the dual Sudakov inequality, to get

$$\begin{aligned}\mathrm{Vol}_n(K_1^\circ) &\leqslant \mathrm{Vol}_n\Big(\mathrm{conv}\Big(K^\circ \cup \frac{1}{\lambda^*}B_2^n\Big)\Big)\\ &\leqslant 3en\frac{b(K^\circ)}{\lambda^*}N(B_2^n, \lambda^* K^\circ)\,\mathrm{Vol}_n(K^\circ)\\ &\leqslant 3enb(K^\circ)\mathrm{Vol}_n(K^\circ)\,\exp\Big(cn\Big(\frac{M^*(K)}{\lambda^*}\Big)^2\Big)\\ &= 3enb(K^\circ)\mathrm{Vol}_n(K^\circ)\,\exp\big(cn/\alpha^2\big).\end{aligned}$$

As before, the term $3enb(K^\circ)$ can be omitted by adjusting the constant c of the exponential term, since $M^*(K) \leqslant c'\log(d_{X_K}+1) \leqslant c'\log(n+1)$ implies that $b(K^\circ) \leqslant c''\sqrt{n}\log n$. □

REMARK 8.3.3. Introducing K_1 we have managed to improve the distance $d_{X_{K_1}}$. Since $\alpha > 1$, we have $1 \leqslant M(K)M(K^\circ)\alpha^2 = \lambda^*\lambda_*$ and, from the definition of K_1,

$$\lambda_*^{-1}B_2^n \subseteq K_1 \subseteq \lambda^* B_2^n.$$

Therefore,

$$d_{X_{K_1}} \leqslant \lambda^*\lambda_* = M(K)M(K^\circ)\alpha^2 \preceq \alpha^2\log(d_{X_K}+1).$$

Choosing $\alpha \simeq \log n$ and using the fact that $d_{X_K} \leqslant \sqrt{n}$ by John's theorem, we get

$$d_{X_{K_1}} \leqslant c(\log n)^3 \tag{8.3.4}$$

and

$$\exp\big(-cn/\alpha^2\big) \leqslant \frac{s(K)}{s(K_1)} \leqslant \exp\big(cn/\alpha^2\big). \tag{8.3.5}$$

Now, we can repeat this step.

Second proof of the reverse Santaló inequality. We shall define an inductive procedure. Consider the sequence α_j with

$$\alpha_1 = \log n, \quad \alpha_2 = \log(\log n) = \log^{(2)} n, \dots, \alpha_t = \log^{(t)} n$$

where t is the smallest positive integer with the property

$$\alpha_t = \log^{(t)} n < 2.$$

First Step: We start with an arbitrary centrally symmetric convex body K in $\mathbb{R}^n$. Applying a suitable linear transformation, we may assume that $M(K) = 1$ and $M(K^\circ) \preceq \log(d_{X_K}+1)$ (during all this procedure, $\preceq \Theta$ means $\leqslant c\cdot\Theta$ where $c > 0$ is an absolute constant that may be chosen in advance). We choose $\alpha_1 = \log n$ and set

$$\lambda_1^* = M(K^\circ)\alpha_1 \text{ and } \lambda_{1,*} = M(K)\alpha_1.$$

Then we define

$$K_1 = \mathrm{conv}\Big((K \cap \lambda_1^* B_2^n) \cup \frac{1}{\lambda_{1,*}}B_2^n\Big).$$

We saw that

$$d_{X_{K_1}} \preceq (\log n)^3$$

and

$$\exp\big(-cn/(\log n)^2\big) \leqslant \frac{s(K)}{s(K_1)} \leqslant \exp\big(cn/(\log n)^2\big).$$

Second Step: At this step we start with the body K_1, which we may assume, by applying a suitable linear transformation, that it satisfies the additional assumptions that $M(K_1) = 1$ and $M(K_1^\circ) \preceq \log(d_{X_{K_1}} + 1)$. We choose $\alpha_2 = \log^{(2)} n$ and we set

$$\lambda_2^* = M(K_1^\circ)\alpha_2 \text{ and } \lambda_{2,*} = M(K_1)\alpha_2.$$

Then we define

$$K_2 = \mathrm{conv}\Big((K_1 \cap \lambda_2^* B_2^n) \cup \frac{1}{\lambda_{2,*}} B_2^n\Big).$$

As before, we see that $d_{X_{K_2}} \leqslant M(K_1)M(K_1^\circ)\alpha_2^2 \preceq (\log^{(2)} n)^3$ and

$$\exp\big(-cn/(\log^{(2)} n)^2\big) \leqslant \frac{s(K_1)}{s(K_2)} \leqslant \exp\big(cn/(\log^{(2)} n)^2\big),$$

and hence

$$\exp\Big(-cn\sum_{j=1}^{2} \alpha_j^{-2}\Big) \leqslant \frac{s(K)}{s(K_2)} \leqslant \exp\Big(cn\sum_{j=1}^{2} \alpha_j^{-2}\Big).$$

Continuing in the same way, after t steps we arrive at a centrally symmetric convex body K_t satisfying

$$d_{X_{K_t}} \preceq (\log^{(t)} n)^3 \leqslant c_1.$$

In fact, the construction shows that there exists $r > 0$ such that

$$rB_2^n \subseteq K_t \subseteq c_1 r B_2^n.$$

By duality,

$$\frac{1}{c_1 r} B_2^n \subseteq K_t^\circ \subseteq \frac{1}{r} B_2^n$$

and from the definition of $s(K_t)$ we see that

$$\frac{1}{c_1^n} \cdot s(B_2^n) \leqslant s(K_t) \leqslant c_1^n \cdot s(B_2^n).$$

We also have

$$\exp\Big(-cn\sum_{j=1}^{t} \alpha_j^{-2}\Big) \leqslant \frac{s(K)}{s(K_t)} \leqslant \exp\Big(cn\sum_{j=1}^{t} \alpha_j^{-2}\Big).$$

These two inequalities imply that

$$C^{-n}\exp\Big(-cn\sum_{j=1}^{t} \alpha_j^{-2}\Big)\, s(B_2^n) \leqslant s(K) \leqslant C^n \exp\Big(cn\sum_{j=1}^{t} \alpha_j^{-2}\Big)\, s(B_2^n).$$

We also observe that, for every $1 \leqslant j < t$,

$$\begin{aligned}\alpha_{j+1} = \log \alpha_j &\Rightarrow \alpha_j = \exp(\alpha_{j+1})\\ &\Rightarrow \alpha_1 = \exp(\alpha_2) = \cdots = \exp(\cdots \exp(\alpha_t)),\end{aligned}$$

which gives

$$\sum_{j=1}^{t} \alpha_j^{-2} = \frac{1}{\alpha_t^2} + \frac{1}{\exp(\alpha_t^2)} + \cdots + \frac{1}{\exp(\cdots \exp(\alpha_t^2))} < C_1.$$

It follows that

$$\frac{1}{C_2^n} s(B_2^n) \leqslant s(K) \leqslant C_2^n\, s(B_2^n)$$

for some absolute constant $C_2 > 0$. □

Note that this proof of the reverse Santaló inequality gives also an upper bound for $s(K)$. Clearly, this upper bound is not as good as the sharp bound given by the Blaschke-Santaló inequality of Theorem 1.5.10.

8.4. Milman's reverse Brunn-Minkowski inequality

In this section we prove the reverse Brunn-Minkowski inequality. For any centrally symmetric convex body K in $\mathbb{R}^n$ we first establish the existence of an M-ellipsoid $\mathcal{E}_K$ for K; this is done with the method of isomorphic symmetrization, which was introduced in the last proof of the reverse Santaló inequality. Our aim is to prove the following

THEOREM 8.4.1 (Milman). *For every centrally symmetric convex body K in $\mathbb{R}^n$ there exists an ellipsoid $\mathcal{E}_K$ satisfying*

$$\mathrm{Vol}_n(\mathcal{E}_K) = \mathrm{Vol}_n(K),$$

and such that for any centrally symmetric convex body P in $\mathbb{R}^n$,

$$C^{-n}\mathrm{Vol}_n(\mathcal{E}_K + P) \leqslant \mathrm{Vol}_n(K + P) \leqslant C^n\mathrm{Vol}_n(\mathcal{E}_K + P)$$

$$C^{-n}\mathrm{Vol}_n((\mathcal{E}_K)^\circ + P) \leqslant \mathrm{Vol}_n(K^\circ + P) \leqslant C^n\mathrm{Vol}_n((\mathcal{E}_K)^\circ + P)$$

where $C > 0$ is an absolute constant.

For the proof we need two results from Chapter 4 regarding covering numbers, Lemma 4.1.17 which states that for convex centrally symmetric K and P

$$N(K, sB_2^n) \geqslant \frac{\mathrm{Vol}_n(K + P)}{\mathrm{Vol}_n((sB_2^n \cap K) + P)}, \tag{8.4.1}$$

and Lemma 4.1.24 which states that if $B_2^n \subseteq sbK$ for some $s > 0$ and some $b \geqslant 1$ then

$$\mathrm{Vol}_n\Big(\mathrm{conv}\left(K \cup \frac{1}{s}B_2^n\right) + P\Big) \leqslant 2ebnN(B_2^n, sK)\,\mathrm{Vol}_n(K + P). \tag{8.4.2}$$

Proof of Theorem 8.4.1. We may assume that $\mathrm{Vol}_n(K) = \mathrm{Vol}_n(B_2^n)$. We set $K_0 := K$ and consider $T \in SL_n$ such that $K_0' = T(K_0)$ has the following properties:

(1) $1 \leqslant b(K_0') \leqslant c\sqrt{n}\,\log n$.
(2) $1 \leqslant M(K_0') \leqslant c\log n$.
(3) $M(K_0')\,M((K_0')^\circ) \preceq \log d_{X_{K_0}} \preceq \log n$.

For $\alpha_1 = \log n$ we set

$$\lambda_1^* = M((K_0')^\circ)\alpha_1 \text{ and } \lambda_{1,*} = M(K_0')\alpha_1$$

and consider the body

$$K_1 = \mathrm{conv}\Big((K_0' \cap \lambda_1^* B_2^n) \cup \frac{1}{\lambda_{1,*}}B_2^n\Big).$$

Then,

$$d_{X_{K_1}} \leqslant \lambda_1^*\lambda_{1,*} = M(K_\circ')\,M((K_\circ')^\circ)\alpha_1^2 \preceq (\log n)^3.$$

For every centrally symmetric convex body P, using (8.4.2) and the dual Sudakov inequality (Theorem 4.2.2), we have

$$\begin{aligned}\mathrm{Vol}_n(K_1+P) &\leqslant \mathrm{Vol}_n\Big(\mathrm{conv}\Big(K_0'\cup\frac{1}{\lambda_{1,*}}B_2^n\Big)+P\Big)\\ &\leqslant 2ebn\,N(B_2^n,\lambda_{1,*}K_0')\,\mathrm{Vol}_n(K_0'+P)\\ &\leqslant 2ebn\,\exp\Big(cn\Big(\frac{M(K_0')}{\lambda_1^*}\Big)^2\Big)\,\mathrm{Vol}_n(K_0'+P)\\ &= 2ebn\,\exp\big(cn/\alpha_1^2\big)\,\mathrm{Vol}_n(K_0'+P)\\ &\leqslant \exp\big(c'n/\alpha_1^2\big)\,\mathrm{Vol}_n(K_0'+P),\end{aligned}$$

since the term $2ebn$ can be absorbed by the exponential term (by changing the value of the constant c) because $b\leqslant c\sqrt{n}\log n$ and $\alpha_1\leqslant\log n$.

Conversely, using (8.4.1) and Sudakov's inequality, we see that for every centrally symmetric convex body P in $\mathbb{R}^n$ we also have

$$\begin{aligned}\mathrm{Vol}_n(K_1+P) &= \mathrm{Vol}_n\Big(\mathrm{conv}\Big((K_0'\cap\lambda_1^*B_2^n)\cup\frac{1}{\lambda_{1,*}}B_2^n\Big)+P\Big)\\ &\geqslant \mathrm{Vol}_n((K_0'\cap\lambda_1^*B_2^n)+P)\geqslant N(K_0',\lambda_1^*B_2^n)^{-1}\,\mathrm{Vol}_n(K_0'+P)\\ &\geqslant \mathrm{Vol}_n(K_0'+P)\,\exp\Big(-cn\Big(\frac{M((K_0')^\circ)}{\lambda_1^*}\Big)^2\Big)\\ &= \mathrm{Vol}_n(K_0'+P)\,\exp\big(-cn/\alpha_1^2\big).\end{aligned}$$

Combining the two inequalities we have

$$\exp\big(-cn/\alpha_1^2\big)\,\mathrm{Vol}_n(K_0'+P)\leqslant\mathrm{Vol}_n(K_1+P)\leqslant\exp\big(cn/\alpha_1^2\big)\,\mathrm{Vol}_n(K_0'+P)$$

for every centrally symmetric convex body P in $\mathbb{R}^n$ (and it is easily checked that the same holds even if P is not a body, but merely a centrally symmetric, compact convex set in $\mathbb{R}^n$). In particular, setting $P=\{0\}$ we see that

$$\exp\big(-cn/\alpha_1^2\big)\,\mathrm{Vol}_n(K_0')\leqslant\mathrm{Vol}_n(K_1)\leqslant\exp\big(cn/\alpha_1^2\big)\,\mathrm{Vol}_n(K_0').$$

In other words, $\mathrm{Vol}_n(K_0)^{1/n}=\mathrm{Vol}_n(K_0')^{1/n}\simeq\mathrm{Vol}_n(K_1)^{1/n}$.

We continue with an inductive procedure, similar to the one in the proof of the reverse Santaló inequality. Actually, our choice for α_j will be exactly the same. For every j we get a centrally symmetric convex body K_j with the following property: for every centrally symmetric convex body (or compact set) P in $\mathbb{R}^n$,

$$\begin{aligned}\text{(8.4.3)}\quad \exp(-cn\alpha_j^{-2})\mathrm{Vol}_n(K_{j-1}'+P) &\leqslant \mathrm{Vol}_n(K_j+P)\\ &\leqslant \exp(cn\alpha_j^{-2})\mathrm{Vol}_n(K_{j-1}'+P),\end{aligned}$$

where $K_j'=T_j(K_j)$ is defined so that it satisfies $\mathrm{Vol}_n(K_j')=\mathrm{Vol}_n(K_j)$ and $M(K_j')M((K_j')^\circ)\preceq\log d_{X_{K_j}}$. We end this procedure at the first step t for which $d_{X_{K_t}}\leqslant 8$. Since the sum $\sum_{j=1}^t\alpha_j^{-2}$ is (independently of t) bounded by a constant c_1, setting $P=\{0\}$ and using all the inequalities (8.4.3) we see that

$$\mathrm{Vol}_n(K_t)^{1/n}\simeq\mathrm{Vol}_n(K_0)^{1/n}$$

(the same holds true for every K_j). By the above it is also evident that at step t not only will we have $d_{X_{K_t}}\leqslant 8$, but also we will have found positive numbers α, b

with $b/\alpha \leqslant 8$ such that

$$\alpha B_2^n \subseteq K_t \subseteq b B_2^n.$$

We define $r > 0$ by the equation

$$\mathrm{Vol}_n(rB_2^n) = \mathrm{Vol}_n(K_t).$$

From the original assumption that $\mathrm{Vol}_n(K_0) = \mathrm{Vol}_n(B_2^n)$ and from our volume estimates, it follows that $\alpha, r, b \simeq 1$.

From the inclusion $K_t \subseteq bB_2^n$ we see that, for every centrally symmetric convex body $P \subseteq \mathbb{R}^n$,

$$\begin{aligned}\mathrm{Vol}_n(K_t + P)^{1/n} &\leqslant \mathrm{Vol}_n(bB_2^n + P)^{1/n} \leqslant \left(\frac{b}{r}\right) \mathrm{Vol}_n(rB_2^n + P)^{1/n} \\ &\leqslant 8\mathrm{Vol}_n(rB_2^n + P)^{1/n}\end{aligned}$$

using the fact that $1 \leqslant b/r \leqslant b/\alpha \leqslant 8$. Similarly

$$\begin{aligned}\mathrm{Vol}_n(K_t + P)^{1/n} &\geqslant \mathrm{Vol}_n(\alpha B_2^n + P)^{1/n} \geqslant \left(\frac{\alpha}{r}\right) \mathrm{Vol}_n(rB_2^n + P)^{1/n} \\ &\geqslant \frac{1}{8}\mathrm{Vol}_n(rB_2^n + P)^{1/n}\end{aligned}$$

using the fact that $1 \geqslant \alpha/r \geqslant \alpha/b \geqslant 1/8$.

Setting $\mathcal{E}_t = rB_2^n$ we find an ellipsoid with $\mathrm{Vol}_n(\mathcal{E}_t) = \mathrm{Vol}_n(K_t)$ and such that, for every centrally symmetric convex body $P \subseteq \mathbb{R}^n$,

$$8^{-n}\mathrm{Vol}_n(\mathcal{E}_t + P) \leqslant \mathrm{Vol}_n(K_t + P) \leqslant 8^n \mathrm{Vol}_n(\mathcal{E}_t + P).$$

Now, for every P, we see that

$$\begin{aligned}\mathrm{Vol}_n(K'_{t-1} + P)^{1/n} &\leqslant \exp\left(c/\alpha_t^2\right)\mathrm{Vol}_n(K_t + P)^{1/n} \\ &\leqslant 8 \cdot \exp\left(c/\alpha_t^2\right)\mathrm{Vol}_n(\mathcal{E}_t + P)^{1/n},\end{aligned}$$

and

$$\begin{aligned}\mathrm{Vol}_n(K'_{t-1} + P)^{1/n} &\geqslant \exp\left(-c/\alpha_t^2\right)\mathrm{Vol}_n(K_t + P)^{1/n} \\ &\geqslant \frac{1}{8} \cdot \exp\left(-c/\alpha_t^2\right)\mathrm{Vol}_n(\mathcal{E}_t + P)^{1/n},\end{aligned}$$

where $K'_{t-1} = T_{t-1}(K_{t-1})$. Considering the ellipsoid $\mathcal{E}_{t-1} = T_{t-1}^{-1}(\mathcal{E}_t)$ and using the fact that $|\det T_{t-1}| = 1$, we may write

$$\begin{aligned}\mathrm{Vol}_n(K_{t-1} + P)^{1/n} &= \mathrm{Vol}_n\big(T_{t-1}\big(K_{t-1} + T_{t-1}^{-1}(P)\big)\big)^{1/n} \\ &\leqslant 8 \cdot \exp\left(c/\alpha_t^2\right)\mathrm{Vol}_n\big(T_{t-1}\big(\mathcal{E}_t + T_{t-1}^{-1}(P)\big)\big)^{1/n} \\ &= 8 \cdot \exp\left(c/\alpha_t^2\right)\mathrm{Vol}_n(\mathcal{E}_{t-1} + P)^{1/n}.\end{aligned}$$

Note also that

$$\mathrm{Vol}_n(\mathcal{E}_{t-1}) = \mathrm{Vol}_n(T_{t-1}^{-1}(\mathcal{E}_t)) = \mathrm{Vol}_n(\mathcal{E}_t), \tag{8.4.4}$$

which means that the volume of the ellipsoid has not changed. Thus, for the body K_{t-1}, as well, we have found an ellipsoid $\mathcal{E}_{t-1}$ with the required properties.

Following the steps of the previous procedure backwards and taking into account the fact that the volumes of the ellipsoids that we consider at each step are all equal, we finally arrive back at the first step and we get an inequality of the form

$$\begin{aligned}\frac{1}{8}\exp\Big(-c\sum_{j=1}^{t}\alpha_j^{-2}\Big)\,\mathrm{Vol}_n(\mathcal{E}_0+P)^{1/n} &\leqslant \mathrm{Vol}_n(K_0+P)^{1/n}\\ &\leqslant 8\exp\Big(c\sum_{j=1}^{t}\alpha_j^{-2}\Big)\,\mathrm{Vol}_n(\mathcal{E}_0+P)^{1/n}.\end{aligned}$$

Since the sum $\sum_{j=1}^{t}\alpha_j^{-2}$ is uniformly bounded, we get

$$\frac{1}{C}\mathrm{Vol}_n(\mathcal{E}_0+P)^{1/n}\leqslant \mathrm{Vol}_n(K_0+P)^{1/n}\leqslant C\mathrm{Vol}_n(\mathcal{E}_0+P)^{1/n}$$

for every centrally symmetric convex body P in $\mathbb{R}^n$, where $\mathcal{E}_0$ is an ellipsoid with

$$\mathrm{Vol}_n(\mathcal{E}_0)^{1/n}=\mathrm{Vol}_n(K_t)^{1/n}\simeq \mathrm{Vol}_n(K_0)^{1/n}.$$

Considering a simple (homothetical) transformation of $\mathcal{E}_0$ and slightly adjusting the absolute constant C, we see now that with the body $K=K_0$ we can associate an ellipsoid $\mathcal{E}_K$ which satisfies both conclusions in the first inequality of the theorem.

The second line of inequalities can either be proved directly, using similar methods, or instead we can use the same reasoning as in Lemma 8.1.2 together with duality for covering numbers, to get the desired result. More precisely, since we already know

$$N(\mathcal{E}_K,K)\leqslant \frac{\mathrm{Vol}_n(2\mathcal{E}_K+K)}{\mathrm{Vol}_n(K)}\leqslant (3C)^n$$

then by duality for covering numbers, Theorem 4.4.4 we have that

$$N(K^\circ,(\mathcal{E}_K)^\circ)\leqslant C_1^n$$

and thus for any P

$$\mathrm{Vol}_n(K^\circ+P)\leqslant N(K^\circ,(\mathcal{E}_K)^\circ)\mathrm{Vol}_n((\mathcal{E}_K)^\circ+P)\leqslant C_1^n\mathrm{Vol}_n((\mathcal{E}_K)^\circ+P).$$

Similarly we work, as in Lemma 8.1.2, with the lower bound. □

REMARK 8.4.2. It is useful to notice that for a body in M-position (say, with constant C) and with, say, $\mathrm{Vol}_n(K)=\mathrm{Vol}_n(B_2^n)$ one has, by definition,

$$\mathrm{Vol}_n(K+B_2^n)\leqslant (2C)^n\mathrm{Vol}_n(K)$$

but also one has, by Lemma 4.1.17, that

$$\mathrm{Vol}_n(K\cap B_2^n)\geqslant \frac{\mathrm{Vol}_n(K)}{N(K,B_2^n)}\geqslant C_1^{-n}\mathrm{Vol}_n(K)$$

where we have used Lemma 8.1.2. Similarly for the polars

$$\mathrm{Vol}_n(K^\circ+B_2^n)\leqslant C_2^n\mathrm{Vol}_n(K^\circ)\qquad\text{and}\qquad \mathrm{Vol}_n(K^\circ\cap B_2^n)\geqslant C_3^{-n}\mathrm{Vol}_n(K^\circ).$$

We are now ready to prove the reverse Brunn-Minkowski inequality.

THEOREM 8.4.3. *For every centrally symmetric convex body K in $\mathbb{R}^n$ there exists $U_K\in SL_n$ such that: if K_1,K_2 are centrally symmetric convex bodies in $\mathbb{R}^n$ then, for every $t>0$,*

$$\mathrm{Vol}_n(U_{K_1}(K_1)+tU_{K_2}(K_2))^{1/n}\leqslant C\left(\mathrm{Vol}_n(K_1)^{1/n}+t\mathrm{Vol}_n(K_2)^{1/n}\right),$$

where $C>0$ is a numerical constant which is independent of the dimension or of the choice of the bodies $K_i, i=1,2$.

Proof. Applying Theorem 8.4.1 for each of the bodies K_i, $i = 1, 2$ we can find ellipsoids $\mathcal{E}_{K_i}$ with
$$\mathrm{Vol}_n(\mathcal{E}_{K_i}) = \mathrm{Vol}_n(K_i)$$
such that, for every centrally symmetric convex body P in $\mathbb{R}^n$,

(8.4.5) $$\mathrm{Vol}_n(K_i + P)^{1/n} \leqslant C \mathrm{Vol}_n(\mathcal{E}_{K_i} + P)^{1/n}.$$

We define $U_{K_i} \in SL_n$ such that
$$U_{K_i}(\mathcal{E}_{K_i}) = r_i B_2^n, \qquad i = 1, 2.$$

Then, by (8.4.5) we obtain
$$\mathrm{Vol}_n(U_{K_i}(K_i) + U_{K_i}(P))^{1/n} \leqslant C \mathrm{Vol}_n(U_{K_i}(\mathcal{E}_{K_i}) + U_{K_i}(P))^{1/n},$$
and hence
$$\mathrm{Vol}_n(U_{K_i}(K_i) + \tilde{P})^{1/n} \leqslant C \mathrm{Vol}_n(r_i B_2^n + \tilde{P})^{1/n}$$
for every centrally symmetric convex body $\tilde{P}$ in $\mathbb{R}^n$, for $i = 1$ or 2. Applying this, first with $i = 1$ and $\tilde{P} = t U_{K_2}(K_2)$, and then with $i = 2$ and $\tilde{P} = r_1 B_2^n$, we see that
$$\mathrm{Vol}_n(U_{K_1}(K_1) + t U_{K_2}(K_2))^{1/n} \leqslant C \mathrm{Vol}_n(r_1 B_2^n + t U_{K_2}(K_2))^{1/n}$$
$$\leqslant C^2 \mathrm{Vol}_n(r_1 B_2^n + t r_2 B_2^n)^{1/n}$$
$$= C^2 \left(\mathrm{Vol}_n(r_1 B_2^n)^{1/n} + t \mathrm{Vol}_n(r_2 B_2^n)^{1/n} \right)$$
$$= C^2 \left(\mathrm{Vol}_n(U_{K_1}(\mathcal{E}_{K_1}))^{1/n} + t \mathrm{Vol}_n(U_{K_2}(\mathcal{E}_{K_2}))^{1/n} \right)$$
$$= C^2 \left(\mathrm{Vol}_n(\mathcal{E}_{K_1})^{1/n} + t \mathrm{Vol}_n(\mathcal{E}_{K_2})^{1/n} \right)$$
$$= C^2 \left(\mathrm{Vol}_n(K_1)^{1/n} + t \mathrm{Vol}_n(K_2)^{1/n} \right),$$
which proves the theorem. $\square$

8.5. Extension to the non-symmetric case

In this section we indicate how one can extend the main results of this chapter to the not necessarily symmetric setting. The fact that the lower bound for $s(K)$ for symmetric convex bodies implies a lower bound for $s(K)$ for convex bodies which include the origin is straightforward:

THEOREM 8.5.1. *There exists a universal constant $c_1 > 0$ such that for any $n \in \mathbb{N}$ and any convex body $K \subset \mathbb{R}^n$ which includes 0 we have*
$$\mathrm{Vol}_n(K) \mathrm{Vol}_n(K^\circ) \geqslant c_1^n \kappa_n^2.$$

Proof. Consider $T = K - K$, and note that $T^\circ \subset K^\circ$. By Theorem 8.2.2 we know that $\mathrm{Vol}_n(T) \mathrm{Vol}_n(T^\circ) \geqslant c^n \kappa_n^2$ and by Rogers-Shephard inequality (Theorem 1.5.2) we know that $\mathrm{Vol}_n(K) \geqslant 4^{-n} \mathrm{Vol}_n(T)$ so that
$$\mathrm{Vol}_n(K) \mathrm{Vol}_n(K^\circ) \geqslant 4^{-n} \mathrm{Vol}_n(T) \mathrm{Vol}_n(T^\circ) \geqslant (c/4)^n \kappa_n^2.$$
$\square$

Regarding the extension to the non-symmetric case of the other results shown in this chapter, a main tool, introduced for this purpose by Milman and Pajor, is

Theorem 4.1.20 which was discussed and proven in Chapter 4. We recall that it states that if K and L are two convex bodies in $\mathbb{R}^n$ with the same barycenter then,

$$\mathrm{Vol}_n(K)\,\mathrm{Vol}_n(L) \leqslant \mathrm{Vol}_n(K+L)\,\mathrm{Vol}_n(K\cap(-L)). \tag{8.5.1}$$

In particular, if K has its barycentre at the origin then

$$\mathrm{Vol}_n(K\cap(-K)) \geqslant 2^{-n}\mathrm{Vol}_n(K). \tag{8.5.2}$$

Theorem 8.2.3 can be extended to the non-symmetric case:

THEOREM 8.5.2. *Let K and T be two convex bodies in $\mathbb{R}^n$, with barycenter at the origin. Then,*

$$C^{-n}N(T^\circ,K^\circ) \leqslant N(K,T) \leqslant C^nN(T^\circ,K^\circ),$$

where $C>0$ is an absolute constant.

Proof. We write

$$\begin{aligned} N(\mathrm{conv}(K\cup(-K)),T\cap(-T)) &\leqslant N(K-K,T\cap(-T))\\ &\leqslant N(K-K,K)N(K,T)N(T,T\cap(-T))\\ &\leqslant N(K-K,K\cap(-K))N(K,T)N(T-T,T\cap(-T))\\ &\leqslant 9^n\frac{\mathrm{Vol}_n(K-K)}{\mathrm{Vol}_n(K\cap(-K))}\,\frac{\mathrm{Vol}_n(T-T)}{\mathrm{Vol}_n(T\cap(-T))}N(K,T)\\ &\leqslant C_1^n\,N(K,T)\end{aligned}$$

where $C_1 = 9\cdot 8^2$, using (8.5.2) twice in the last step.

From Theorem 8.2.3 we know that

$$N(\mathrm{conv}(T^\circ\cup(-T)^\circ),K^\circ\cap(-K)^\circ) \leqslant C_2^nN(\mathrm{conv}(K\cup(-K)),T\cap(-T)).$$

It follows that

$$N(T^\circ,K^\circ) \leqslant N(\mathrm{conv}(T^\circ\cup(-T)^\circ),K^\circ\cap(-K)^\circ) \leqslant C^n\,N(K,T)$$

with $C=C_1C_2$.

For the reverse inequality we work in a similar way. We need to control the ratios $\mathrm{Vol}_n(K^\circ-K^\circ)/\mathrm{Vol}_n(K^\circ\cap(-K)^\circ)$ and $\mathrm{Vol}_n(T^\circ-T^\circ)/\mathrm{Vol}_n(T^\circ\cap(-T)^\circ)$, which is easily done if we use the Blaschke-Santaló and the Bourgain-Milman inequality; for example,

$$\begin{aligned}\frac{\mathrm{Vol}_n(K^\circ-K^\circ)}{\mathrm{Vol}_n(K^\circ\cap(-K)^\circ)} &\leqslant 2^n\frac{\mathrm{Vol}_n(\mathrm{conv}(K^\circ\cup(-K)^\circ)}{\mathrm{Vol}_n(K^\circ\cap(-K)^\circ)} \leqslant \frac{2^n}{c^n}\frac{\mathrm{Vol}_n(K\cup(-K))}{\mathrm{Vol}_n(K\cap(-K))}\\ &\leqslant \frac{16^n}{c^n},\end{aligned}$$

where $c>0$ is the constant in Theorem 8.2.2. □

To extend the notion of M-position to the non-symmetric case it is convenient to use the covering numbers formulation. We shall make use of the following simple corollary of Lemma 8.1.3.

LEMMA 8.5.3. *Let $K\subset\mathbb{R}^n$ be a convex body with barycenter at the origin. Assume $K-K$ is in M-position with constant C. Then $K\cap(-K)$ is in M-position with constant $C_1(C)$ depending only on C.*

Proof. Denote by $r_1, r_2 > 0$ the constants such that $\mathrm{Vol}_n(K-K) = \mathrm{Vol}_n(r_1 B_2^n)$ and $\mathrm{Vol}_n(K\cap(-K)) = \mathrm{Vol}_n(r_2 B_2^n)$. By (8.5.2) and the Rogers Shephard inequality we know that

$$r_1^n = \frac{\mathrm{Vol}_n(K-K)}{\kappa_n} \leqslant 8^n \frac{\mathrm{Vol}_n(K\cap(-K))}{\kappa_n} = (8r_2)^n$$

so that $r_2 \leqslant r_1 \leqslant 8r_2$. By Lemma 8.1.2, since $K-K$ is in M-position, for some $c_1 = c_1(C)$ we have $N(K-K, r_1 B_2^n) \leqslant \exp(c_1 n)$. Therefore

$$N(K\cap(-K), r_2 B_2^n) \leqslant N(K-K, r_1 B_2^n) N(r_1 B_2^n, \frac{r_1}{8} B_2^n) \leqslant \exp((c_1(C)+c_2)n).$$

By Lemma 8.1.3 we have that $K\cap(-K)$ is also in M position, with some other constant depending only on $c_1(C)+c_2$. □

THEOREM 8.5.4. *Let $K \subset \mathbb{R}^n$ be a convex body with $\mathrm{Vol}_n(K) = \mathrm{Vol}_n(B_2^n)$ and barycenter at the origin. Assume that $K-K$ is in M-position with some constant C. Then*

$$\max\{N(K, B_2^n), N(K^\circ, B_2^n), N(B_2^n, K), N(B_2^n, K^\circ)\} \leqslant \exp(cn)$$

for some $c > 0$ depending only on C.

Proof. Clearly the M-ellipsoid of $K-K$ is rB_2^n with $1 < r \leqslant 4$ by Rogers-Shephard inequality. Therefore $N(K, B_2^n) \leqslant N(K-K, rB_2^n) N(rB_2^n, B_2^n) \leqslant \exp(cn)$ by Lemma 8.1.2. Similarly $N(B_2^n, K^\circ) \leqslant N(B_2^n, (K-K)^\circ) \leqslant \exp(cn)$ by the same reasoning. Since $K\cap(-K)$ is also in M position, by (8.5.2) we have that its M-ellipsoid in $r'B_2^n$ with $1/2 \leqslant r' < 1$ and thus $N(B_2^n, K) \leqslant N(B_2^n, r'B_2^n) N(r'B_2^n, K\cap(-K)) \leqslant \exp(c'n)$ for c' depending only on C. Similarly $N(K^\circ, B_2^n) \leqslant N((K\cap(-K))^\circ, B_2^n) \leqslant \exp(c'n)$. □

8.6. Applications of the M-position

In this section we present a few first applications of the M-position. The next result shows that random projections of a body in M-position have bounded volume ratio:

THEOREM 8.6.1. *Let K be a centrally symmetric convex body in $\mathbb{R}^n$. Assume that $\mathrm{Vol}_n(K) = \mathrm{Vol}_n(B_2^n)$ and $N(K, B_2^n) \leqslant \exp(\beta n)$ for some constant $\beta > 0$. Then, for any $1 \leqslant k < n$, a random orthogonal projection $P_E(K)$ of K onto a k-dimensional subspace E of $\mathbb{R}^n$ has volume ratio bounded by a constant $C(\beta, n/(n-k))$.*

Proof. Observing that

$$\mathrm{Vol}_n(B_2^n) = \mathrm{Vol}_n(K) \leqslant N(K, B_2^n)\,\mathrm{Vol}_n(K\cap B_2^n) \leqslant e^{\beta n}\mathrm{Vol}_n(K\cap B_2^n),$$

and using the Blaschke-Santaló inequality we get

$$\mathrm{Vol}_n(\mathrm{conv}(K^\circ \cup B_2^n))^{1/n} \leqslant C\mathrm{Vol}_n(B_2^n)^{1/n},$$

where C depends only on β. In other words, $W = \mathrm{conv}(K^\circ \cup B_2^n)$ has bounded volume ratio, hence we may apply Theorem 5.5.3 and deduce that

$$K^\circ \cap E \subseteq W \cap E \subseteq C(\beta)^{\frac{n}{n-k}} B_E$$

for a random $E \in G_{n,k}$ (as usual, $B_E = B_2^n \cap E$). By duality, this implies that

$$P_E(K) \supseteq rB_E,$$

where $r = C(\beta)^{-\frac{n}{n-k}}$. Since

$$\begin{aligned}\mathrm{Vol}_k(P_E(K)) &\leqslant N(P_E(K), B_E)\mathrm{Vol}_k(B_E) \leqslant N(K, B_2^n)\mathrm{Vol}_k(B_E)\\ &\leqslant \exp(\beta n)\mathrm{Vol}_k(B_E),\end{aligned}$$

this shows that

$$\big(\mathrm{Vol}_k(P_E(K))/\mathrm{Vol}_k(rB_E)\big)^{1/k} \leqslant C(\beta, k/n) = \exp(\beta n/k)C(\beta)^{\frac{n}{n-k}},$$

whence we are done. □

Our next application will be based on the following lemma.

PROPOSITION 8.6.2. *Let T be a centrally symmetric convex body in $\mathbb{R}^n$. If $R(T\cap E) \leqslant r$ for all E in a subset of $G_{n,n/2}$ of measure greater than $1/2$ then there exists $U \in O(n)$ such that $R(T \cap U(T)) \leqslant \sqrt{2}\, r$.*

Proof. We use a standard argument (see Pisier's book [**509**] or Milman [**455**]). From the assumption we know that there exists $E \in G_{n,n/2}$ such that

$$\|y\|_T \geqslant \frac{1}{r}|y| \tag{8.6.1}$$

for all $y \in E$ and all $y \in E^{\perp}$. We write $P_1 = P_E$ and $P_2 = P_{E^{\perp}}$. Then, we write $I = P_1 + P_2$ and we define $U = P_1 - P_2 \in O(n)$. Let $x \in \mathbb{R}^n$. We write $x = x_1 + x_2$, where $x_1 = P_1(x)$ and $x_2 = P_2(x)$. Then,

$$\begin{aligned}\|x_1 + x_2\|_T + \|x_1 - x_2\|_T &\geqslant 2\max\{\|x_1\|_T, \|x_2\|_T\} \geqslant \frac{2}{r}\max\{|x_1|, |x_2|\}\\ &\geqslant \frac{\sqrt{2}}{r}\sqrt{|x_1|^2 + |x_2|^2} = \frac{\sqrt{2}}{r}|x|.\end{aligned}$$

This means that

$$\|x\|_T + \|x\|_{U^{-1}(T)} \geqslant \frac{\sqrt{2}}{r}|x|,$$

or equivalently, since $U = U^*$,

$$2\mathrm{conv}(T^\circ \cup U(T^\circ)) \supseteq T^\circ + U(T^\circ) \supseteq \frac{\sqrt{2}}{r}B_2^n. \tag{8.6.2}$$

Taking polars we conclude the proof. □

In the proof of Theorem 8.6.1 we saw that if K is a centrally symmetric convex body in $\mathbb{R}^n$ $\mathrm{Vol}_n(K) = \mathrm{Vol}_n(B_2^n)$ and $N(K, B_2^n) \leqslant \exp(\beta n)$ for some constant $\beta > 0$, then

$$R(K^\circ \cap E) \leqslant C(\beta)$$

for a random $E \in G_{n,n/2}$. Then, Proposition 8.6.2 shows that there exists $U \in O(n)$ such that

$$K + U(K) \supseteq c(\beta)B_2^n.$$

On the other hand, the reverse Brunn-Minkowski inequality applies to K and $U(K)$; it is not hard to check that

$$\mathrm{Vol}_n(K + U(K))^{1/n} \leqslant C_1(\beta)\mathrm{Vol}_n(B_2^n)^{1/n}.$$

This proves the next

THEOREM 8.6.3. *Let K be a centrally symmetric convex body in $\mathbb{R}^n$ with $\mathrm{Vol}_n(K) = \mathrm{Vol}_n(B_2^n)$ and $N(K, B_2^n) \leqslant \exp(\beta n)$ for some constant $\beta > 0$. We may find $U \in O(n)$ so that the volume ratio of $K + U(K)$ is bounded by a constant depending only on β.*

REMARK 8.6.4. In Chapter 5, we saw (Theorem 5.5.4) that if T is a centrally symmetric convex body in $\mathbb{R}^n$ such that $\mathrm{vr}(T) = \alpha$ then there exists $U \in O(n)$ with the property

$$d_G(T \cap U(T), B_2^n) \leqslant c\alpha^2$$

where $c > 0$ is an absolute constant. Combining this fact with Theorem 8.6.3 we see that two simple operations can transform a convex body to an isomorphic ellipsoid.

THEOREM 8.6.5 (Milman). *Let K be a centrally symmetric convex body in $\mathbb{R}^n$ which is in M-position with constant β. We may find $U, V \in O(n)$ so that if $T = \mathrm{conv}(K \cup U(K))$ and $D = T \cap V(T)$ then*

$$d_G(D, B_2^n) \leqslant C(\beta).$$

8.7. α-regular M-position: Pisier's approach

In this section we present the approach of Pisier to the existence of M-ellipsoids, from which amounts a stronger form of M-position, the so-called α-regular M-position. Throughout this section we follow very closely the presentation of these results in [**508**] and in [**509**].

A geometric formulation of Pisier's main result is the following.

THEOREM 8.7.1 (Pisier). *For every $\alpha > 1/2$ there exists a constant $C = C(\alpha) > 0$ with the following property: for every $n \in \mathbb{N}$ and for every centrally symmetric convex body K in $\mathbb{R}^n$ one may find a position $\tilde{K}$ of K such that for every $1 \leqslant k < n$ one may find $F, H \in G_{n,n-k}$ satisfying*

$$\tilde{K} \cap F \subseteq C(n/k)^{\alpha} B_F \quad \text{and} \quad \tilde{K}^{\circ} \cap H \subseteq C(n/k)^{\alpha} B_F. \tag{8.7.1}$$

(as usual, B_F denotes $B_2^n \cap F$). Moreover, the constant $C(\alpha)$ is $O((\alpha - 1/2)^{-1/2})$ as $\alpha \to 1/2$.

To present Pisier's approach, and the fact that it is gives a strengthening of the original M-position, it is convenient to use the language of the Gelfand and Kolmogorov numbers of operators which we introduced in Chapter 4. Recall that if $u : \ell_2^n \to X$ is a linear operator then, for every $k \geqslant 1$,

$$c_k(u) = \inf\{\|u|_F\| \;:\; F \subset X \;,\; \mathrm{codim}(F) < k\}$$

and

$$d_k(u) = \inf\{\|Q_E u\| : \; E \subset Y, \; \dim(E) < k\}$$

where $Q_E : X \to X/E$ is the quotient map. Recall also that the M^*-estimate can be formulated as follows (Theorem 7.3.3): there exists an absolute constant $C_1 > 0$ such that for any Banach space X, for any $n \geqslant 1$ and for any operator $u : \ell_2^n \to X$,

$$\sup_{k \geqslant 1} \left(\sqrt{k} c_k(u^*)\right) \leqslant C_1 \ell(u). \tag{8.7.2}$$

Note that if $u \equiv \mathrm{Id} : \ell_2^n \to X = (\mathbb{R}^n, \|\cdot\|)$ where $\|\cdot\|$ is the norm induced by the centrally symmetric convex body $\tilde{K}$, then $c_k(u^{-1}) \leqslant r$ is equivalent to the existence of a subspace $F \in G_{n,n-k+1}$ such that $|x| \leqslant r\|x\|$ for all $x \in F$, which is in turn equivalent to

$$\tilde{K} \cap F \subseteq rB_F.$$

On the other hand, $d_k(u) \leqslant r$ is equivalent to the existence of a subspace $E \in G_{n,k-1}$ such that $\|x\|_{X/E} \leqslant r|x|$ for all $x \in E$, which is in turn equivalent to $P_{E^\perp}(\tilde{K}) \supseteq r^{-1}B_E$, and hence to

$$\tilde{K}^\circ \cap H \subseteq rB_H,$$

where $H = E^\perp$. In view of these observations, Theorem 8.7.1 is equivalent to the next statement (we simply set $\tilde{K} = u^{-1}(K)$ in order to pass from Theorem 8.7.2 to Theorem 8.7.1).

THEOREM 8.7.2 (Pisier). *For every $\alpha > 1/2$ there exists a constant $C = C(\alpha) > 0$ with the following property: for every $n \in \mathbb{N}$ and for every n-dimensional, real or complex, normed space X there exists an isomorphism $u : \ell_2^n \to X$ such that*

$$d_k(u) \leqslant C(n/k)^\alpha \quad \text{and} \quad c_k(u^{-1}) \leqslant C(n/k)^\alpha \tag{8.7.3}$$

for every $k = 1, \ldots, n$. Moreover, the constant $C(\alpha)$ is $O((\alpha - 1/2)^{-1/2})$ as $\alpha \to 1/2$.

Recall that by Carl's Theorem 4.3.4, for every $\alpha > 0$ there exists a constant ρ_α such that

$$\sup_{k\leqslant n} k^\alpha e_k(u) \leqslant \rho_\alpha \sup_{k\leqslant n} k^\alpha c_k(u),$$

and

$$\sup_{k\leqslant n} k^\alpha e_k(u) \leqslant \rho_\alpha \sup_{k\leqslant n} k^\alpha d_k(u).$$

Taking into account the fact that we have $c_k(u) = d_k(u^*)$ and $d_k(u) = c_k(u^*)$ we get as a corollary from Theorem 8.7.2 the following

THEOREM 8.7.3 (Pisier). *For every $\alpha > 1/2$ there exists a constant $C_2 = C_2(\alpha) > 0$ with the following property: for every $n \in \mathbb{N}$ and every n-dimensional normed space X there exists an isomorphism $u : \ell_2^n \to X$ such that*

$$\max\{e_k(u), e_k(u^*), e_k(u^{-1}), e_k((u^{-1})^*)\} \leqslant C_2(n/k)^\alpha$$

for every $k = 1, \ldots, n$. Moreover, $C_2(\alpha) = O((\alpha - 1/2)^{-1/2})$ as $\alpha \to 1/2$. □

The geometric reformulation of the theorem is the following.

THEOREM 8.7.4 (Pisier). *For every $p < 2$ there exists a constant A_p such that for every symmetric convex body K in $\mathbb{R}^n$ we can find an ellipsoid $\mathcal{E}$ in $\mathbb{R}^n$ which satisfies*

$$\max\{N(K, t\mathcal{E}), N(\mathcal{E}, tK), N(K^\circ, t\mathcal{E}^\circ), N(\mathcal{E}^\circ, tK^\circ)\} \leqslant \exp(A_p n/t^p)$$

for all $t \geqslant 1$. □

DEFINITION 8.7.5. Let $\alpha > 1/2$. We say that a symmetric convex body K is *α-regular* if

$$\max\{N(K, tB_2^n), N(B_2^n, tK), N(K^\circ, tB_2^n), N(B_2^n, tK^\circ)\} \leqslant \exp(A_p n/t^p)$$

for every $t \geqslant 1$, where $p = 1/\alpha$. Theorem 8.7.4 shows that every symmetric convex body has a linear image which is α-regular.

We will present the proof of Theorem 8.7.2 in Subsection 8.7.2; it is based on complex interpolation, which is introduced in the next subsection. Before this, let us show a few important consequences of the theorem.

A strong form of the reverse Brunn-Minkowski inequality can be proved for α-regular bodies.

THEOREM 8.7.6. *Let $\alpha > 1/2$ and let $K_1, \ldots, K_m$ be multiples of α-regular symmetric convex bodies in $\mathbb{R}^n$. Then,*

$$\mathrm{Vol}_n(K_1 + \cdots + K_m)^{1/n} \leqslant b(\alpha) m^{\alpha} \sum_{j=1}^{m} \mathrm{Vol}_n(K_j)^{1/n}$$

where $b(\alpha)$ is a positive constant depending only on α.

Proof. From the definition of an α-regular body, for every $t \geqslant 1$ we have

$$\max\{N(K_j, tr_j B_2^n), N(r_j B_2^n, tK_j)\} \leqslant \exp(A_p n/t^p)$$

where $p = 1/\alpha$ and $r_j > 0$. Therefore,

$$\begin{aligned} N(K_1 + \cdots + K_m, t(r_1 B_2^n + \cdots + r_m B_2^n)) &\leqslant \prod_{j=1}^{m} N(K_j, tr_j B_2^n) \\ &\leqslant \exp(A_p nm/t^p). \end{aligned}$$

It follows that

$$\begin{aligned} \mathrm{Vol}_n(K_1 + \cdots + K_m)^{1/n} &\leqslant t \exp(A_p m/t^p) \mathrm{Vol}_n(r_1 B_2^n + \cdots + r_m B_2^n)^{1/n} \\ &= t \exp(A_p m/t^p) \left(\mathrm{Vol}_n(r_1 B_2^n)^{1/n} + \cdots + \mathrm{Vol}_n(r_m B_2^n)^{1/n} \right). \end{aligned}$$

From $N(r_j B_2^n, K_j) \leqslant \exp(A_p n)$ we see that

$$\mathrm{Vol}_n(r_j B_2^n)^{1/n} \leqslant \exp(A_p) \mathrm{Vol}_n(K_j)^{1/n}$$

and hence

$$\begin{aligned} \mathrm{Vol}_n(K_1 + \cdots + K_m)^{1/n} &\leqslant t \exp(A_p + A_p m/t^p) \\ &\quad \times \left(\mathrm{Vol}_n(K_1)^{1/n} + \cdots + \mathrm{Vol}_n(K_m)^{1/n} \right). \end{aligned}$$

Choosing $t = m^{\alpha} = m^{1/p}$ we get the result. □

A dual statement can be proved in a similar way.

THEOREM 8.7.7. *Let $\alpha > 1/2$ and let $K_1, \ldots, K_m$ be multiples of α-regular symmetric convex bodies in $\mathbb{R}^n$. Then,*

$$\mathrm{Vol}_n(K_1 \cap \cdots \cap K_m)^{1/n} \geqslant \delta(\alpha) m^{-1-\alpha} \min\left\{ \mathrm{Vol}_n(K_1)^{1/n}, \ldots, \mathrm{Vol}_n(K_m)^{1/n} \right\}$$

where $\delta(\alpha)$ is a positive constant depending only on α.

8.7.1. Complex interpolation

We say that two Banach spaces X_0 and X_1 are *compatible* if there is a Hausdorff topological vector space Z such that X_0 and X_1 are subspaces of Z (in this case we will sometimes write $\overline{X} = (X_0, X_1)$ and call $\overline{X}$ a Banach couple). Then, we can define their sum $X_0 + X_1$ and their intersection $X_0 \cap X_1$; they can be equipped with the norms

$$\|x\|_{X_0+X_1} = \inf\{\|x_0\|_{X_0} + \|x_1\|_{X_1} : x = x_0 + x_1, x_0 \in X_0, x_1 \in X_1\}$$

and

$$\|x\|_{X_0\cap X_1} = \max\{\|x\|_{X_0}, \|x\|_{X_1}\}.$$

It can be checked (see [**66**, Lemma 2.3.1]) that $X_0 \cap X_1$ and $X_0 + X_1$ are Banach spaces. We are interested in the finite dimensional case, and hence this is trivially satisfied.

Next, we associate with each Banach couple $\overline{X}$ an interpolation functor using the theory of vector valued analytic functions. First, we define a Banach space $\mathcal{F} = \mathcal{F}(X_0, X_1)$ of analytic functions in the following way. We consider the strip $S = \{z \in \mathbb{C} : 0 < \mathrm{Re} z < 1\}$, its closure $\overline{S} = \{z \in \mathbb{C} : 0 \leqslant \mathrm{Re} z \leqslant 1\}$, and the space $\mathcal{F}$ of all functions $f : \overline{S} \to X_0 + X_1$ which are bounded and continuous on $\overline{S}$, analytic on S and also have the property that the functions $t \mapsto f(it)$ and $t \mapsto f(1 + it)$ have values in X_0 and X_1 respectively, they are continuous and tend to 0 as $|t| \to \infty$.

Clearly, $\mathcal{F}(X_0, X_1)$ is a vector space. We define a norm on $\mathcal{F}(X_0, X_1)$ by

$$\|f\|_{\mathcal{F}} = \max\left(\sup_{t\in\mathbb{R}} \|f(it)\|_{X_0}, \sup_{t\in\mathbb{R}} \|f(1 + it)\|_{X_1}\right).$$

One can check that $\mathcal{F}(X_0, X_1)$ equipped with this norm becomes a Banach space. To see this, let $\{f_n\}$ be a sequence in $\mathcal{F}(X_0, X_1)$ with $\sum_{n=1}^{\infty} \|f_n\|_{\mathcal{F}} < \infty$. From the definition of $\|\cdot\|_{X_0+X_1}$ we check that for every $z \in \overline{S}$

$$\|f_n(z)\|_{X_0+X_1} \leqslant \max\left(\sup_{t\in\mathbb{R}} \|f_n(it)\|_{X_0}, \sup_{t\in\mathbb{R}} \|f_n(1 + it)\|_{X_1}\right) = \|f_n\|_{\mathcal{F}}.$$

Since $X_0 + X_1$ is a Banach space, $\sum_{n=1}^{\infty} f_n$ converges uniformly on S to a function f with values in $X_0 + X_1$. It follows that f is analytic on S and bounded and continuous on $\overline{S}$. Since $\|f_n(j + it)\|_{X_j} \leqslant \|f_n\|_{\mathcal{F}}$, $j = 0, 1$, a similar argument shows that $\sum_{n=1}^{\infty} f_n(j + it)$ converges uniformly to some function which coincides with $f(j + it)$. Thus, $f \in \mathcal{F}(X_0, X_1)$ and $\sum_{n=1}^{\infty} f_n$ converges to f in $\mathcal{F}(X_0, X_1)$.

Now, we can define the functor C_θ which sends every $\theta \in (0, 1)$ to the interpolation space $C_\theta(X_0, X_1) = X_\theta := [X_0, X_1]_\theta$, defined as the space of all $x \in X_0 + X_1$ for which there exists $f \in \mathcal{F}$ such that $x = f(\theta)$. On the space X_θ we define the norm

$$\|x\|_\theta = \inf\{\|f\|_{\mathcal{F}} : f \in \mathcal{F}(X_0, X_1), f(\theta) = x\}.$$

One can check (see [**66**, Theorem 4.1.2]) that X_θ is a Banach space and an intermediate space between $X_0 \cap X_1$ and $X_0 + X_1$. Also, the functor C_θ is exact with exponent θ; this means that if $\overline{Y} = (Y_0, Y_1)$ is another Banach couple and if T is an operator which is bounded from X_0 to Y_0 and from X_1 to Y_1 then T maps X_θ into Y_θ and

$$\|T : X_\theta \to Y_\theta\| \leqslant \|T : X_0 \to Y_0\|^{1-\theta} \|T : X_1 \to Y_1\|^\theta.$$

In what follows we need a very special case of this assertion. We assume that X_0 and X_1 are both $\mathbb{C}^n$ equipped with two different norms (then, X_θ is also $\mathbb{C}^n$ for all θ).

Claim. For every $\theta \in (0,1)$ and for any $x \in X_\theta$ one has

$$\|x\|_{X_\theta} \leqslant \|x\|_{X_0}^{1-\theta} \|x\|_{X_1}^{\theta}. \tag{8.7.4}$$

Proof. We set $\alpha_0 = \|x\|_{X_0}$, $\alpha_1 = \|x\|_{X_1}$ and define

$$f(z) = \alpha_0^{1-\theta}\alpha_1^{\theta}(\alpha_0^{1-z}\alpha_1^{z})^{-1}x.$$

Then $f \in \mathcal{F}(X_0, X_1)$, $\|f\|_{\mathcal{F}} = \alpha_0^{1-\theta}\alpha_1^{\theta}$ and $f(\theta) = x$. The result follows from the definition of $\|\cdot\|_\theta$. □

We will also need the dual inequality. For every $\theta \in (0,1)$ and every linear functional $u \in (X_0 \cap X_1)^*$ we define

$$\|u\|_\theta^* = \sup\{|u(x)| : x \in X_0 \cap X_1, \|x\|_{X_\theta} \leqslant 1\}. \tag{8.7.5}$$

Then, we have the following.

Claim. For every $u \in (X_0 \cap X_1)^*$,

$$\|u\|_\theta^* \leqslant \left(\|u\|_0^*\right)^{1-\theta} \left(\|u\|_1^*\right)^{\theta}. \tag{8.7.6}$$

The proof of this claim is a consequence of the *three lines lemma*: If $g : \overline{S} \to \mathbb{C}$ is a bounded continuous function, which is holomorphic on S, then, for every $\theta \in (0,1)$ and every $t \in \mathbb{R}$ we have

$$|g(\theta)| \leqslant \left(\sup_t |g(it)|\right)^{1-\theta} \left(\sup_t |g(1+it)|\right)^{\theta}. \tag{8.7.7}$$

Let $x \in X_\theta$ with $\|x\|_{X_\theta} \leqslant 1$ and let $f \in \mathcal{F}(X_0, X_1)$ with $f(\theta) = x$. Applying the three lines lemma to the function

$$g(z) = u(f(z))$$

we get

$$|u(x)| = |u(f(\theta))| = |g(\theta)| \leqslant \left(\|u\|_0^*\right)^{1-\theta} \left(\|u\|_1^*\right)^{\theta},$$

which proves (8.7.6). □

The main result of this subsection is the next theorem, which will replace Pisier's inequality for the K-convexity constant for our goals.

THEOREM 8.7.8. *Let (X_0, X_1) be a couple of compatible complex Banach spaces. If X_1 is a Hilbert space then for every $\theta \in (0,1)$ we have*

$$K(X_\theta) \leqslant \phi(\theta)$$

where $\phi(\theta)$ is a constant depending only on θ. Moreover, we have $\phi(\theta) = O(1/\theta)$ as $\theta \to 0$.

For the proof of Theorem 8.7.8 we need two lemmas. The first one is a more or less direct consequence of (8.7.4) and (8.7.6).

LEMMA 8.7.9. *Let Y be a complex Banach space and let $T : X_0 + X_1 \to Y$ and $R : Y \to X_0 \cap X_1$ be bounded $\mathbb{C}$-linear operators. Then, $T_\theta := T\,|_{X_\theta} : X_\theta \to Y$ and $R_\theta := R : Y \to X_\theta$ are bounded operators and satisfy*

$$c_k(R_\theta) \leqslant \|R_0\|^{1-\theta}(c_k(R_1))^\theta$$

and

$$d_k(T_\theta) \leqslant \|T_0\|^{1-\theta}(d_k(T_1))^\theta$$

for all $k \geqslant 1$.

Proof. The first inequality is an immediate consequence of (8.7.4). For the second inequality we may assume for simplicity that the operators T_θ, T_1 are compact and that $X_0 \cap X_1$ is dense in both X_0 and X_1 (these conditions are trivially satisfied in our case). Using the equalities $d_k(T_\theta) = c_k(T_\theta^*)$ and $d_k(T_1) = c_k(T_1^*)$ we get the second inequality from (8.7.6). □

The second lemma is elementary. Given $z \in \mathbb{C}$ we denote by $\mathrm{Arg}(z)$ the argument of z which satisfies $-\pi < \mathrm{Arg}(z) \leqslant \pi$.

LEMMA 8.7.10. *Let $D_0 = [-1, 1]$, $D_1 = \{z \in \mathbb{C} : |z| = 1\}$ and*

$$D_\theta = \left\{ z \in \mathbb{C} : \left|\mathrm{Arg}\left(\frac{z+1}{z-1}\right)\right| = \frac{\theta\pi}{2} \right\}, \quad \theta \in (0, 1).$$

Then, for every $w \in D_\theta$ there exists $G : \overline{S} \to \mathbb{C}$ which is holomorphic on S and satisfies: $G(\theta) = w$ and $G(it) \in D_0$, $G(1 + it) \in D_1$ for all $t \in \mathbb{R}$.

Proof. Observe that D_θ is the set of the points z in the complex plane for which the angle $\widehat{(-1)z1}$ is equal to $\pi - \frac{\pi\theta}{2}$.

Let $w \in D_\theta$. If $\mathrm{Im}(w) > 0$ then there exists a real number s such that $w = i\tan(\frac{\pi\theta}{4} + is)$. We define

$$G(z) = i\tan\left(\frac{\pi}{4}z + is\right).$$

If $\mathrm{Im}(w) < 0$ then there exists a real number s such that $w = -i\tan(\frac{\pi\theta}{4} + is)$. In this case we define

$$G(z) = -i\tan\left(\frac{\pi}{4}z + is\right).$$

In both cases we verify that G is holomorphic on a neighborhood of S and satisfies $G(\theta) = w$ and $G(it) \in D_0$, $G(1 + it) \in D_1$ for all $t \in \mathbb{R}$. □

Proof of Theorem 8.7.8. Recall from Chapter 6 that, for every $k \geqslant 0$, Q_k is the orthogonal projection $Q_k : L_2(\mathbb{R}^n, \gamma_n) \to H_k$ where H_k is the (closed) span in $L_2(\mathbb{R}^n, \gamma_n)$ of the Hermite polynomials of degree exactly k. According to Proposition 6.1.6, we have $\mathrm{Id} = \sum_{k\geqslant 0} Q_k$ and for every $\varepsilon \in [-1, 1]$, the operator $T_\varepsilon := \sum_{k\geqslant 0} \varepsilon^k Q_k$ is a positive contraction on $L_2 := L_2(\mathbb{R}^n, \gamma_n)$.

For every $z \in \mathbb{C}$ with $|z| \leqslant 1$ we define

$$T(z) = \sum_{k\geqslant 0} z^k Q_k.$$

Note that $T(z)$ is a norm-1 operator on L_2. Let $0 < \theta < 1$. We will find $\delta = \delta(\theta) > 0$ such that, for every $z \in \mathbb{C}$ with $|z| \leqslant \delta$,

$$\|T(z) \otimes I_{X_\theta}\|_{L_2\otimes X_\theta \to L_2\otimes X_\theta} \leqslant 1.$$

We set $T_\theta(z) = T(z) \otimes I_{X_\theta}$ and $Q_{1,\theta} = Q_1 \otimes I_{X_\theta}$. From Cauchy's formula we see that

$$Q_1 = \frac{1}{\delta}\int_0^{2\pi} e^{-is}T(\delta e^{is})\frac{ds}{2\pi},$$

and similarly,

$$Q_{1,\theta} = \frac{1}{\delta}\int_0^{2\pi} e^{-is}T_\theta(\delta e^{is})\frac{ds}{2\pi}.$$

From Jensen's inequality,

$$K(X_\theta) = \|Q_{1,\theta}\|_{L_2\otimes X_\theta\to L_2\otimes X_\theta} \leqslant \frac{1}{\delta}\sup_{|z|=\delta}\|T_\theta(z)\|_{L_2\otimes X_\theta\to L_2\otimes X_\theta}. \tag{8.7.8}$$

We proceed with the choice of δ.

From Proposition 6.2.1 we know that for every $z \in D_0 = [-1, 1]$ one has

$$\|T(z) \otimes I_{X_0}\|_{L_2\otimes X_0\to L_2\otimes X_0} \leqslant 1. \tag{8.7.9}$$

Also, since X_1 is a Hilbert space, for every $z \in D_1$ we have

$$\|T(z) \otimes I_{X_1}\|_{L_2\otimes X_1\to L_2\otimes X_1} \leqslant 1. \tag{8.7.10}$$

Using Lemma 8.7.10 we will show that for every $z \in D_\theta$ we have

$$\|T(z) \otimes I_{X_\theta}\|_{L_2\otimes X_\theta\to L_2\otimes X_\theta} \leqslant 1. \tag{8.7.11}$$

From [**66**, Theorem 5.1.2] we know that the spaces $(L_2\otimes X_0, L_2\otimes X_1)_\theta$ and $L_2\otimes X_\theta$ are isometrically isomorphic. This implies that for every $\phi \in L_2\otimes X_\theta$ with $\|\phi\| < 1$, there exists $f \in \mathcal{F} := \mathcal{F}(L_2 \otimes X_0, L_2 \otimes X_1)$ such that $\|f\|_{\mathcal{F}} < 1$ and $f(\theta) = \phi$. We fix $z \in D_\theta$ and consider the function G of Lemma 8.7.10. Recall that $G(\theta) = z$.

We define $W : \overline{S} \to (L_2 \otimes X_0) + (L_2 \otimes X_1)$ with

$$W(u) = T(G(u))f(u).$$

Then, W is holomorphic on S and $W(\theta) = T(z)\phi$.

For simplicity we write $Y_0 = L_2 \otimes X_0$ and $Y_1 = L_2 \otimes X_1$. Then,

$$\|T(z)\phi\|_{(Y_0,Y_1)_\theta} \leqslant \max\{\sup_t \|W(it)\|_{Y_0}, \sup_t \|W(1+it)\|_{Y_1}\}.$$

From (8.7.9), (8.7.10) and Lemma 8.7.10 it follows that

$$\|T(z)\phi\|_{(Y_0,Y_1)_\theta} \leqslant 1.$$

Since $(Y_0, Y_1)_\theta = L_2 \otimes X_\theta$, we conclude that for every $z \in D_\theta$

$$\|T(z)\|_{L_2\otimes X_\theta\to L_2\otimes X_\theta} \leqslant 1. \tag{8.7.12}$$

Note that $T(z_1z_2) = T(z_1)T(z_2)$. Thus, (8.7.12) holds for every z inside the domain with boundary D_θ. In particular, it holds true for every $z \in \mathbb{C}$ with $|z| \leqslant \delta(\theta) := \tan(\pi\theta/4)$. From (8.7.8) we get

$$K(X_\theta) \leqslant (\tan(\pi\theta/4))^{-1}.$$

This proves the theorem. □

REMARK 8.7.11. Pisier's inequality (Theorem 6.2.4) is a consequence of Theorem 8.7.8. We may assume that $X = \mathbb{C}^n$ (the real case follows). We may also assume that the identity operator $I : \ell_2^n \to X$ satisfies $\|I\| = d(X, \ell_2^n)$ and $\|I^{-1}\| = 1$. Consider the interpolation space $X_\theta = (X, \ell_2^n)_\theta$. From Lemma 8.7.9 with $k = 1$, we have

$$d(X, X_\theta) \leqslant [d(X, \ell_2^n)]^\theta.$$

From Theorem 8.7.8 we have $K(X_\theta) \leqslant \phi(\theta)$. This implies that

$$K(X) \leqslant d(X, X_\theta)K(X_\theta) \leqslant \phi(\theta)[d(X, \ell_2^n)]^\theta \leqslant \frac{c}{\theta}[d(X, \ell_2^n)]^\theta.$$

Choosing $\theta \simeq 1/\log[d(X, \ell_2^n) + 1]$ we get

$$K(X) \leqslant c_1 \log[d(X, \ell_2^n) + 1].$$

8.7.2. Proof of Theorem 8.7.2

We can now prove Theorem 8.7.2: we fix $\alpha > 1/2$ and we show that there exists a constant $C = C(\alpha) > 0$ such that for every n-dimensional, real or complex, normed space X there exists an isomorphism $u : \ell_2^n \to X$ such that

$$d_k(u) \leqslant C(n/k)^\alpha \quad \text{and} \quad c_k(u^{-1}) \leqslant C(n/k)^\alpha$$

for every $k \geqslant 1$.

REMARK 8.7.12. If we assume that the K-convexity constant of X is $O(1)$ then the result is clearly true. Because then, there exists an isomorphism $u : \ell_2^n \to X$ such that

$$\ell(u) \leqslant \sqrt{n} \quad \text{and} \quad \ell((u^{-1})^*) \leqslant K(X)\sqrt{n}$$

and then the M^*-estimate (8.7.2) gives

$$d_k(u) = c_k(u^*) \leqslant C_1(n/k)^{1/2} \quad \text{and} \quad c_k(u^{-1}) \leqslant C_1 K(X)(n/k)^{1/2}.$$

Proof of Theorem 8.7.2. We first consider the complex case. We fix $\alpha > 1/2$ and we denote by C the smallest positive constant for which the theorem holds true. Our aim is to show that C is bounded by a constant that depends only on α.

Let $X = (\mathbb{C}^n, \|\cdot\|)$. By the definition of C we may find an isomorphism $u : \ell_2^n \to X$ such that

$$d_k(u) \leqslant C(n/k)^\alpha \quad \text{and} \quad c_k(u^{-1}) \leqslant C(n/k)^\alpha$$

for every $k \geqslant 1$. We may also assume that u is the identity operator.

We define $\theta = (\alpha - 1/2)/\alpha \in (0, 1)$ and we consider the interpolation space $X_\theta = (X, \ell_2^n)_\theta$ which corresponds to the pair (X, ℓ_2^n). We write $u_\theta : X_\theta \to X$ for the identity operator mapping X_θ to X. From the MM^*-estimate we know that there exists an isomorphism $v : \ell_2^n \to X_\theta$ such that

$$\ell(v) = \sqrt{n} \quad \text{and} \quad \ell((v^{-1})^*) \leqslant K(X_\theta)\sqrt{n}.$$

Then the M^*-estimate (8.7.2) and Theorem 8.7.8 show that

$$d_k(v) = c_k(v^*) \leqslant C_1(n/k)^{1/2} \quad \text{and} \quad c_k(v^{-1}) \leqslant C_1\phi(\theta)(n/k)^{1/2}$$

for every $k \geqslant 1$.

For $\theta = 0, 1$ we have $u_0 = I : X \to X$ and $u_1 = u : \ell_2^n \to X$. Then, Lemma 8.7.9 shows that

$$d_k(u_\theta) \leqslant \|u_0\|^{1-\theta} d_k(u_1)^\theta = d_k(u)^\theta$$

and

$$c_k(u_\theta^{-1}) \leqslant \|u_0^{-1}\|^{1-\theta} c_k(u_1^{-1})^\theta = c_k(u^{-1})^\theta.$$

From the definition of C it follows that

$$d_k(u_\theta) \leqslant (C(n/k)^\alpha)^\theta \quad \text{and} \quad c_k(u_\theta^{-1}) \leqslant (C(n/k)^\alpha)^\theta$$

for every $k \geqslant 1$.

We consider the operator $w = u_\theta v : \ell_2^n \to X$. From (4.3.2) we know that the approximation numbers are submultiplicative; in particular,

$$d_{2k-1}(w) \leqslant d_k(u_\theta)d_k(v) \leqslant C_1 C^\theta (n/k)^{\alpha\theta+1/2} = C_1 C^\theta (n/k)^\alpha$$

and

$$c_{2k-1}(w^{-1}) \leqslant c_k(u_\theta^{-1})c_k(v^{-1}) \leqslant C_1\phi(\theta)C^\theta(n/k)^\alpha.$$

If we define $r = 2^\alpha C_1 C^\theta$ then these inequalities give

$$d_k(w) \leqslant r(n/k)^\alpha \quad \text{and} \quad c_k(w^{-1}) \leqslant r\phi(\theta)(n/k)^\alpha$$

for every $k \geqslant 1$. Setting $w_1 = \phi(\theta)^{1/2}w$ we get

$$d_k(w_1) \leqslant r\phi(\theta)^{1/2}(n/k)^\alpha \quad \text{and} \quad c_k(w_1^{-1}) \leqslant r\phi(\theta)^{1/2}(n/k)^\alpha$$

for every $k \geqslant 1$. By the definition of C,

$$C \leqslant r\phi(\theta)^{1/2} = C_1 C^\theta 2^\alpha \phi(\theta)^{1/2},$$

and hence

$$C \leqslant C(\alpha) = \left(C_1 2^\alpha \phi(\theta)^{1/2}\right)^{\frac{1}{1-\theta}},$$

where $\theta = (\alpha - 1/2)/\alpha$. Note that from our estimate for the behaviour of $\phi(\theta)$ as $\theta \to 0$ it follows that $C(\alpha) = O((\alpha - 1/2)^{-1})$ as $\alpha \to 1/2$.

Now, we can pass to the real case with the next argument. Let X be an n-dimensional real normed space. There exists a complex n-dimensional normed space $\tilde{X}$ such that X is isometrically embedded into a real n-dimensional subspace of $\tilde{X}$ (viewed as a real $2n$-dimensional space) and such that there exists an $\mathbb{R}$-linear projection $P : \tilde{X} \to X$ of norm 1. For example, we can choose $\tilde{X}$ to be the space of all $\mathbb{R}$-linear operators from $\mathbb{C}$ to X equipped with the operator norm. We may also assume that $X \subseteq \tilde{X}$ if we identify X with its image through this embedding.

By the complex form of the theorem, there exists an isomorphism $u_1 : \ell_2^n(\mathbb{C}) \to \tilde{X}$ such that

$$d_k(u_1) \leqslant C(n/k)^\alpha \quad \text{and} \quad c_k(u_1^{-1}) \leqslant C(n/k)^\alpha \tag{8.7.13}$$

for every $k = 1, \ldots, n$. We set $H = u_1^{-1}(X)$. Then, we can identify H with $\ell_2^n(\mathbb{R})$. If we define $u = Pu_1|_H : H \to X$, then u is an $\mathbb{R}$-linear isomorphism and we have

$$d_{2k-1}(u) \leqslant d_k(u_1) \quad \text{and} \quad c_{2k-1}(u^{-1}) \leqslant c_k(u_1^{-1}) \tag{8.7.14}$$

where the approximation numbers are understood in the real sense for u, u^{-1} and in the complex sense for u_1, u_1^{-1}. It follows that u satisfies the conclusion of the theorem with constant $C' = 2^\alpha C$. The constant 2^α is added in order to pass from d_{2k-1}, c_{2k-1} to d_k, c_k. □

8.8. Notes and remarks

The Bourgain-Milman inequality

Theorem 8.2.2 is from Bourgain and Milman [**110**] and provides an asymptotic affirmative answer to Mahler's conjecture. We present a simplified version of the argument which can be found in Pisier [**509**].

Mahler's conjecture appears in [**409**] and [**410**] in connection with some questions from the geometry of numbers. The conjecture has been verified for some classes of bodies:

(1) For the class of 1-unconditional convex bodies by Saint-Raymond [**531**]; the equality cases were clarified by Meyer [**433**] and Reisner [**519**].
(2) For the class of zonoids; this was proved by Reisner (see [**517**] and [**518**]). A short proof of Mahler's conjecture for zonoids appears in [**263**].

Kuperberg's proof of the reverse Santaló inequality appears in [**363**] (see [**362**] for an earlier attempt that provided a weaker estimate with an elementary argument). The proof of Nazarov can be found in [**482**].

Theorem 8.2.3 was proved by König and Milman in [**361**].

Isomorphic symmetrization

V. Milman introduced the method of isomorphic symmetrization in [**450**]. The proof of the reverse Santaló inequality that we present in this section comes from that paper.

The reverse Brunn-Minkowski inequality

The reverse Brunn-Minkowski inequality was proved by Milman in [**448**]. The proof of Theorem 8.4.1 with the method of isomorphic symmetrization that we present in the text is from [**450**].

Extension to the non-symmetric case

The non-symmetric case of the reverse Santaló inequality already appeared in the paper of Bourgain and Milman, as a simple consequence of the symmetric case. The complete result regarding the reverse Brunn-Minkowski inequality was shown in the paper of Milman and Pajor [**461**].

Pisier's approach

Theorem 8.7.2 is from Pisier's paper [**508**]. See also his book [**509**, Chapter 7]. We follow closely, with his permission, the presentation of these results given there.

Facts from complex interpolation that are used in the proof can be found in the book of Bergh and Lofstrom [**66**]. Note that the existence of α-regular M-ellipsoids, established by Pisier in the centrally symmetric case, does not have any known extension to the not-necessarily symmetric setting. This is in analogy with the fact that we do not have, in this case, full knowledge on the norm of the Rademacher projection and the optimal form of the MM^* estimate is still unknown.

Essential uniqueness of M-ellipsoid

Since the existence of M-ellipsoids is established in an isomorphic way, we have not provided a geometric characterization and uniqueness for the M-ellipsoid of a convex body. However, surprisingly enough, in some essential way the M-ellipsoids of a given body are unique as shown by Milman and Pajor in [**462**]. In order to describe their result we need a formal definition of essential uniqueness.

Definition. Let $c : (0,1) \to \mathbb{R}^+$. We say that two ellipsoids $\mathcal{E}_1$ and $\mathcal{E}_2$ are *essentially c-equivalent* if for every $\varepsilon \in (0,1)$ there exists a subspace F of $\mathbb{R}^n$ with $\dim(F) \geqslant (1-\varepsilon)n$ such that

$$\frac{1}{c(\varepsilon)}\mathcal{E}_2 \cap F \subseteq \mathcal{E}_1 \cap F \subseteq c(\varepsilon)\mathcal{E}_2 \cap F.$$

With this definition, we have:

Theorem. *Let $\beta > 0$. There exists a function $c : (0,1) \to \mathbb{R}^+$, depending only on β, with the following property: if K is a convex body in $\mathbb{R}^n$ and if $\mathcal{E}_1, \mathcal{E}_2$ are two ellipsoids with $\mathrm{Vol}_n(K) = \mathrm{Vol}_n(\mathcal{E}_1) = \mathrm{Vol}_n(\mathcal{E}_2)$ that satisfy*

$$\max\{N(K,\mathcal{E}_i), N(\mathcal{E}_i, K), N(K^\circ, \mathcal{E}_i^\circ), N(\mathcal{E}_i^\circ, K^\circ\} \leqslant \exp(\beta n), \tag{8.8.1}$$

then $\mathcal{E}_1$ and $\mathcal{E}_2$ are essentially c-equivalent.

Given the value of β, the argument in [**462**] provides the estimate $c(\varepsilon) \leqslant \exp(\beta/\varepsilon)$ for the function c.

CHAPTER 9

Gaussian approach

The geometric study of random processes, and especially of Gaussian processes, has strong connections with asymptotic geometric analysis, so it is useful to introduce some of its methods and main results. This is the purpose of this chapter. Besides the applications which we mention next, and the fact that they offer an alternative point of view on some basic theorems that we have already established, the tools that we present in this chapter will appear again in Part II of this book. We stress, however, that as this book is not a book on Gaussian processes, we limit the discussion to the specific topics most relevant to our theory.

This chapter introduces a "Gaussian approach" to some of the main results that we presented in previous chapters, including sharp versions of the Dvoretzky-Milman theorem and of the M^*-estimate:

(i) Gordon's Theorem 9.2.6, which is proved in Section 9.2, asserts that for every $\varepsilon \in (0,1)$, for every $k > 1$, for every $n \geqslant \exp(c\varepsilon^{-2}k)$, and for any centrally symmetric convex body K in $\mathbb{R}^n$, we may find a k-dimensional subspace F of $\mathbb{R}^n$ such that

$$d(K \cap F, B_2^k) \leqslant \frac{1+\varepsilon}{1-\varepsilon}.$$

In other words, $k_{X_K}(\varepsilon) \geqslant c\varepsilon^2 \log n$.

(ii) Gordon's Theorem 9.3.7, which is proved in Section 9.3, asserts that for every convex body K in $\mathbb{R}^n$ with $0 \in \text{int}(K)$ and for every $1 \leqslant k < n$ there exists a subspace $F \in G_{n,n-k}$ such that

$$|x| \leqslant a_{n,k}\sqrt{n/k}\, w(K)\|x\|_K$$

for all $x \in F$, where $a_{n,k} \to 1$ as $k \to \infty$.

The proof of these results is based on comparison principles for Gaussian processes, due to Gordon, which extend a theorem of Slepian.

We start with the classical bounds of Sudakov and Dudley for the expectation $\mathbb{E}\sup_{t\in T} Z_t$ of the supremum of a Gaussian process $\mathcal{Z} = (Z_t)_{t\in T}$ in terms of the metric entropy $N_\varepsilon(\mathcal{Z})$ of $\mathcal{Z}$ (see Section 9.1 for the definitions). The main result of Section 9.1 states that there exist absolute constants $c_1, c_2 > 0$ so that, for every Gaussian process $\mathcal{Z} = (Z_t)_{t\in T}$,

$$c_1 \sup_{\varepsilon>0}\left(\varepsilon\sqrt{\log N_\varepsilon(\mathcal{Z})}\right) \leqslant \mathbb{E}\sup_{t\in T} Z_t \leqslant c_2 \int_0^\infty \sqrt{\log N_\varepsilon(\mathcal{Z})}\, d\varepsilon.$$

The lower bound is Sudakov inequality and the upper bound is Dudley's inequality. A direct consequence of the left hand side inequality is Sudakov inequality for the covering numbers $\overline{N}(K, \varepsilon B_2^n)$ of a convex body (that was obtained in a different way in Chapter 4). The proof of the Dudley-Sudakov inequality is based on Slepian's comparison theorem. A number of more delicate min-max principles, due to Gordon, that extend Slepian's lemma are proved in Section 9.5. These are applied in Sections 9.2 and 9.3 to yield alternative proofs of the Dvoretzky-Milman theorem and of the M^*-estimate respectively.

Finally, in Section 9.4 we discuss two results on the norm of random operators. The first one, Chevet's inequality, concerns the norm of the random $n\times m$ matrix $G=\sum_{i=1}^n\sum_{j=1}^m g_{ij}x_i^*\otimes y_j$ where X and Y are normed spaces, $x_i^*\in X^*$, $y_j\in Y$, and g_{ij} are independent standard Gaussian random variables, when viewed as an operator $G:X\to Y$. The proof is based again on Slepian's lemma. Next, we discuss an inequality of Marcus and Pisier which allows us to replace random Gaussian operators by random orthogonal ones (this fact has already been used in Chapter 5, for the proof of the global form of Dvoretzky theorem; see Theorem 5.6.1): if X is a normed space and $\{x_{ij}\}_{i,j\leqslant m}\subset X$ then for any $p>0$ one has

$$\left(\int_{O(m)}\left\|\sum_{i,j}U_{ij}x_{ij}\right\|^p d\nu_m(U)\right)^{1/p}\simeq\frac{1}{\sqrt{m}}\left(\mathbb{E}\left\|\sum_{i,j}g_{ij}x_{ij}\right\|^p\right)^{1/p},$$

up to a constant depending on p. Both Chevet inequality and Marcus-Pisier inequality will be used again in Part II.

9.1. Dudley, and another look at Sudakov

DEFINITION 9.1.1. Let $(\Omega,\mathcal{A},\mathbb{P})$ be a probability space, and let $\mathcal{Z}=(Z_t)_{t\in T}$ be a family of real random variables on Ω, indexed by a set T. We say that $\mathcal{Z}$ is a *Gaussian process* if every finite linear combination $a_1Z_{t_1}+\cdots+a_mZ_{t_m}$ of Z_t is a normal random variable with mean zero.

EXAMPLES 9.1.2. (i) Let $g_1,\dots,g_N$ be independent random variables with $g_i\sim N(0,1)$. Then, $\mathcal{Z}=\{g_1,\dots,g_N\}$ is a Gaussian process, since for every choice of real numbers $a_1,\dots,a_N$ we have

$$a_1g_1+\cdots+a_Ng_N\sim N(0,(a_1^2+\cdots+a_N^2)^{1/2}).$$

(ii) Let $g_1,\dots,g_N$ be independent random variables with $g_i\sim N(0,1)$ and let $T\subseteq\mathbb{R}^N$. We write $\{e_1,\dots,e_N\}$ for the standard orthonormal basis of $\mathbb{R}^N$. For every $t\in T$ we define a random variable Z_t setting

$$Z_t=\left\langle t,\sum_{i=1}^N g_ie_i\right\rangle=\sum_{i=1}^N t_ig_i.$$

Each Z_t is a normal random variable with mean 0, and we easily check that

$$a_1Z_{t_1}+\cdots+a_kZ_{t_k}=Z_{a_1t_1+\cdots+a_kt_k}$$

for every $k\in\mathbb{N}$ and any $a_1,\dots,a_k\in\mathbb{R}$, $t_1,\dots,t_k\in T$. So, $\mathcal{Z}=(Z_t)_{t\in T}$ is a Gaussian process.

We can always view a Gaussian process $\mathcal{Z}=(Z_t)_{t\in T}$ as a subset of $L^2(\Omega,\mathcal{A},\mathbb{P})$. We write B for the open unit ball of $L^2(\Omega,\mathcal{A},\mathbb{P})$ and for every $\varepsilon>0$ we define the *ε-entropy number of $\mathcal{Z}$* by

$$N_\varepsilon(\mathcal{Z})=\min\left\{N\in\mathbb{N}:\text{ there exist } t_1,\dots,t_N\in T\text{ s.t. }\mathcal{Z}\subseteq\bigcup_{i=1}^N(Z_{t_i}+\varepsilon B)\right\}.$$

If the latter set is empty, we set $N_\varepsilon(\mathcal{Z})=\infty$.

The distance $\|Z_t - Z_s\|_2$ of Z_t, Z_s in $L^2(\Omega, \mathcal{A}, \mathbb{P})$ induces a distance on the index set T in a natural way. For every $t, s \in T$ we define

$$d(t,s) = \|Z_t - Z_s\|_2.$$

In this way, the ε-entropy numbers of $\mathcal{Z}$ correspond to the *ε-entropy numbers* of (T, d): we have

$$N_\varepsilon(\mathcal{Z}) = N(T, d, \varepsilon),$$

where

$$N(T,d,\varepsilon) := \min\left\{ N \in \mathbb{N} : \text{ there exist } t_1, \ldots, t_N \in T \text{ s.t. } T \subseteq \bigcup_{i=1}^{N} B(t_i, \varepsilon) \right\}$$

and $B(t, \varepsilon) = \{s \in T : d(t,s) < \varepsilon\}$.

We will be interested in sharp bounds for *the expectation of* $\sup_{t\in T} Z_t$

$$\mathbb{E} \sup_{t\in T} Z_t := \sup\left\{ \mathbb{E}\sup_{t\in F} Z_t : F \subseteq T, |F| < +\infty \right\}$$

in terms of the geometry of (T, d).

The most important example for us is example (ii) above, namely when $K \subset \mathbb{R}^n$ is a convex body and for every $t \in K$

$$Z_t = \left\langle t, \sum_{i=1}^{N} g_i e_i \right\rangle,$$

for $g_1, \ldots, g_N$ independent random variables with $g_i \sim N(0,1)$. In this case, one easily checks that the induced distance is the Euclidean norm $d(s,t) = |s-t|$ and thus

$$\overline{N}(K, \varepsilon B_2^n) = N_\varepsilon(\{Z_t : t \in K\}).$$

Furthermore, the number $\mathbb{E}\sup_{t\in K} Z_t$ has a concrete geometric interpretation - it is no other than the mean width of K multiplied by a suitable constant.

LEMMA 9.1.3. *For every convex body K in $\mathbb{R}^n$,*

$$\mathbb{E} \sup_{t\in K} \left\langle t, \sum_{i=1}^{n} g_i e_i \right\rangle = c_n w(K),$$

where c_n is a constant depending on n, with $c_n \simeq \sqrt{n}$.

Proof. A computation that we have already seen in the proof of Theorem 4.2.2 shows that

$$\begin{aligned}
\mathbb{E}\sup_{t\in K}\langle t, \sum_{i=1}^{n} g_i e_i\rangle &= \frac{1}{(2\pi)^{n/2}} \int_{\mathbb{R}^n} \sup_{t\in K} \langle t, x\rangle e^{-|x|^2/2} dx \\
&= \frac{n\mathrm{Vol}_n(B_2^n)}{(2\pi)^{n/2}} \int_{S^{n-1}} \int_0^\infty \max_{t\in K}\langle t, \theta\rangle u^n e^{-u^2/2} du\, d\sigma(\theta) \\
&= \frac{n}{2^{n/2}\Gamma\left(\frac{n}{2}+1\right)} \int_0^\infty u^n e^{-u^2/2} du \cdot \int_{S^{n-1}} \max_{t\in K}\langle t, \theta\rangle\, d\sigma(\theta) \\
&= \frac{\sqrt{2}\,\Gamma\left(\frac{n+1}{2}\right)}{\Gamma\left(\frac{n}{2}+1\right)} \cdot w(K) \\
&= c_n w(K) \simeq \sqrt{n} w(K),
\end{aligned}$$

if we use polar coordinates and the fact that $\mathrm{Vol}_n(B_2^n) = \pi^{n/2}/\Gamma\left(\frac{n}{2}+1\right)$. □

In the following subsections we will discuss the two classical bounds for $\mathbb{E}\sup_{t\in T} Z_t$ of Sudakov and Dudley, which can be summarized in the next theorem.

THEOREM 9.1.4 (Sudakov-Dudley). *There exist absolute constants* $c_1, c_2 > 0$ *with the following property: for every Gaussian process* $\mathcal{Z} = (Z_t)_{t\in T}$,

$$c_1 \sup_{\varepsilon>0} \left(\varepsilon\sqrt{\log N_\varepsilon(\mathcal{Z})}\right) \leqslant \mathbb{E}\sup_{t\in T} Z_t \leqslant c_2 \int_0^\infty \sqrt{\log N_\varepsilon(\mathcal{Z})}d\varepsilon.$$

The lower bound is the inequality of Sudakov and the upper bound is the inequality of Dudley. The proof of both inequalities is based on Slepian's lemma (Theorem 9.1.6 below), which will reduce the problem to the simpler example of independent gaussian random variables, which we next discuss.

Let us consider the Gaussian process $\mathcal{Z} = \{g_1, \ldots, g_N\}$ of Example 9.1.2 (i). The independence of g_i and a simple computation show that

$$\|g_i - g_j\|_2^2 = \mathbb{E}(g_i - g_j)^2 = \mathbb{E}(g_i^2) + \mathbb{E}(g_j^2) - 2\mathbb{E}(g_i)\,\mathbb{E}(g_j) = 2$$

if $i \neq j$, so $N_\varepsilon(\mathcal{Z}) = 1$ if $\varepsilon > \sqrt{2}$ and $N_\varepsilon(\mathcal{Z}) = N$ if $0 < \varepsilon \leqslant \sqrt{2}$ (this is like covering the points $\{e_1, \ldots, e_N\}$ in $\mathbb{R}^N$ by ε-balls, with centers forced to be at these points). Therefore,

$$\sup_{\varepsilon>0} \left(\varepsilon\sqrt{\log N_\varepsilon(\mathcal{Z})}\right) = \sqrt{2}\sqrt{\log N} = \int_0^\infty \sqrt{\log N_\varepsilon(\mathcal{Z})}d\varepsilon.$$

The next proposition shows that the assertion of Theorem 9.1.4 holds true in this case and in particular the two bounds are sharp (as they are the same). This is a standard and well known fact in probability theory.

PROPOSITION 9.1.5. *Let* $g_1, \ldots, g_N$, $N \geqslant 2$, *be independent random variables with* $g_i \sim N(0,1)$. *Then,*

$$c_1\sqrt{\log N} \leqslant \mathbb{E}\max_{i\leqslant N} g_i \leqslant c_2\sqrt{\log N},$$

where $c_1, c_2 > 0$ *are absolute constants.*

Proof. Let $q \geqslant 1$. By Hölder's inequality,

$$\mathbb{E}\max_{i\leqslant N} g_i \leqslant \mathbb{E}\max_{i\leqslant N} |g_i| \leqslant \mathbb{E}\left[\Big(\sum_{i\leqslant N} |g_i|^q\Big)^{1/q}\right] \leqslant \Big(\sum_{i\leqslant N} \mathbb{E}|g_i|^q\Big)^{1/q}.$$

One can compute

$$\begin{aligned}\mathbb{E}|g|^q &= \frac{2}{\sqrt{2\pi}}\int_0^\infty x^q e^{-x^2/2}dx = \frac{2}{\sqrt{2\pi}}\int_0^\infty (2y)^{\frac{q-1}{2}} e^{-y}dy \\ &= \frac{2^{q/2}}{\sqrt{\pi}}\Gamma\Big(\frac{q+1}{2}\Big).\end{aligned}$$

Therefore,

$$\mathbb{E}\max_{i\leqslant N} g_i \leqslant \left(N\frac{2^{q/2}}{\sqrt{\pi}}\Gamma\Big(\frac{q+1}{2}\Big)\right)^{1/q} \leqslant c\sqrt{q}N^{1/q}.$$

If we choose $q \sim \log N$, we see that

$$\mathbb{E}\max_{i\leqslant N} g_i \leqslant c_2\sqrt{\log N},$$

where $c_2 > 0$ is an absolute constant.

For the reverse inequality we first show that there exist an absolute constant $\alpha > 0$ and some $n_0 \in \mathbb{N}$ so that, if $g \sim N(0,1)$ and if $N \geqslant n_0$, then

$$\mathbb{P}(g \geqslant \alpha\sqrt{\log N}) \geqslant \frac{1}{N}.$$

To see this, note that

$$\begin{aligned}\mathbb{P}(g \geqslant \alpha\sqrt{\log N}) &= \frac{1}{\sqrt{2\pi}}\int_{\alpha\sqrt{\log N}}^{\infty} e^{-x^2/2}dx \\ &\geqslant \frac{1}{\sqrt{2\pi}}\int_{\alpha\sqrt{\log N}}^{2\alpha\sqrt{\log N}} e^{-x^2/2}dx \\ &= \frac{\alpha\sqrt{\log N}}{\sqrt{2\pi}}e^{-2\alpha^2\log N} \\ &= \frac{\alpha\sqrt{\log N}}{\sqrt{2\pi}}N^{-2\alpha^2} \geqslant \frac{1}{N}\end{aligned}$$

if, for example, $\alpha = 1/2$ and n_0 is suitably chosen. Then, for every $N \geqslant n_0$ we have

$$\mathbb{P}\left(\max_{i\leqslant N} g_i \leqslant \frac{1}{2}\sqrt{\log N}\right) = \left[\mathbb{P}\left(g \leqslant \frac{1}{2}\sqrt{\log N}\right)\right]^N \leqslant \left(1-\frac{1}{N}\right)^N \leqslant \frac{1}{e},$$

and Markov's inequality gives

$$\mathbb{E}\max_{i\leqslant N} g_i \geqslant \frac{1}{2}\sqrt{\log N}\cdot\mathbb{P}\left(\max_{i\leqslant N} g_i \geqslant \frac{1}{2}\sqrt{\log N}\right) \geqslant \frac{1}{2}\left(1-\frac{1}{e}\right)\sqrt{\log N}$$

if $N \geqslant n_0$. It is now clear that if we choose a suitable absolute constant $c_1 > 0$, we will obtain the estimate

$$c_1\sqrt{\log N} \leqslant \mathbb{E}\max_{i\leqslant N} g_i$$

for every $N \geqslant 2$. □

Let us now state Slepian's lemma, Slepian Lemma whose proof will be given in Section 9.5.

THEOREM 9.1.6 (Slepian). *Let $(\Omega, \mathcal{A}, \mathbb{P})$ be a probability space and let $X = (X_1, \dots, X_n)$, $Y = (Y_1, \dots, Y_n)$ be two n-tuples of Gaussian random variables, defined on Ω, with mean value equal to 0. Assume that*

$$\mathbb{E}(X_i^2) = \mathbb{E}(Y_i^2) \tag{9.1.1}$$

$$\mathbb{E}(X_iX_j) \geqslant \mathbb{E}(Y_iY_j) \tag{9.1.2}$$

for every $i, j = 1, \dots, n$. Then, for all $t_1, \dots, t_n \in \mathbb{R}$,

$$\mathbb{P}\left(\bigcup_{j=1}^{n}\{X_j > t_j\}\right) \leqslant \mathbb{P}\left(\bigcup_{j=1}^{n}\{Y_j > t_j\}\right).$$

In particular,

$$\mathbb{E}\max_{j\leqslant n} X_j \leqslant \mathbb{E}\max_{j\leqslant n} Y_j.$$

We often need a variant of Theorem 9.1.6 in which the assumptions concern the L_2-norms $\|X_i - X_j\|_2$ and $\|Y_i - Y_j\|_2$, without assuming that X_i and Y_i have the same L_2-norm. We give a simple, complete and independent proof of the relevant variant of Slepian's lemma, with an extra factor 2 which is not really needed but will not bother us in any of the applications.

THEOREM 9.1.7 (Sudakov). *Let $(\Omega, \mathcal{A}, \mathbb{P})$ be a probability space and let $X = (X_1, \ldots, X_n)$, $Y = (Y_1, \ldots, Y_n)$ be two n-tuples of Gaussian random variables with mean zero, defined on Ω. We assume that*

$$\|X_i - X_j\|_2 \leqslant \|Y_i - Y_j\|_2 \tag{9.1.3}$$

for every $i, j = 1, \ldots, n$. Then,

$$\mathbb{E} \max_{i \leqslant n} X_i \leqslant \mathbb{E} \max_{i \leqslant n} Y_i.$$

REMARK 9.1.8. The proof of Theorem 9.1.7 is given in Section 9.5 with an extra constant 2 in the left hand side of the inequality, which suffices for all our applications.

Let us also mention that Fernique proved

THEOREM 9.1.9 (Fernique). *Under the assumptions of Theorem 9.1.7, if $F : \mathbb{R}^+ \to \mathbb{R}$ is a non-negative convex increasing function then*

$$\mathbb{E}\, F\left(\max_{i,j} |X_i - X_j|\right) \leqslant \mathbb{E}\, F\left(\max_{i,j} |Y_i - Y_j|\right).$$

We next show how we may apply Slepian's lemma to obtain a proof of Sudakov's inequality.

THEOREM 9.1.10 (Sudakov inequality). *There exists an absolute constant $c_1 > 0$ with the following property: for every Gaussian process $\mathcal{Z} = (Z_t)_{t \in T}$,*

$$c_1 \sup_{\varepsilon > 0} \left(\varepsilon \sqrt{\log N_\varepsilon(\mathcal{Z})}\right) \leqslant \mathbb{E} \sup_{t \in T} Z_t.$$

The proof will be based on the left hand side inequality of the next lemma (the right hand side inequality will be used in the proof of Dudley's inequality).

LEMMA 9.1.11. *Let $Z = (Z_1, \ldots, Z_n)$ be a Gaussian process. We set*

$$A = \min_{i \neq j} \|Z_i - Z_j\|_2 \text{ and } B = \max_{i \neq j} \|Z_i - Z_j\|_2.$$

Then,

$$\frac{c_1}{2\sqrt{2}} A \sqrt{\log n} \leqslant \mathbb{E} \max_{i \leqslant n} Z_i \leqslant c_2 \sqrt{2} B \sqrt{\log n},$$

where c_1, c_2 are the constants from Proposition 9.1.5.

Proof. We consider independent standard Gaussian random variables $g_1, \ldots, g_n$ (independent from Z) and we set

$$X_i = \frac{g_i}{\sqrt{2}} \min_{i \neq j} \|Z_i - Z_j\|_2 = \frac{A g_i}{\sqrt{2}}$$

and

$$Y_i = \frac{g_i}{\sqrt{2}} \max_{i \neq j} \|Z_i - Z_j\|_2 = \frac{B g_i}{\sqrt{2}}.$$

Then, for every $i \neq j$ we have

$$\|X_i - X_j\|_2 = \min_{i \neq j} \|Z_i - Z_j\|_2 \leqslant \max_{i \neq j} \|Z_i - Z_j\|_2 = \|Y_i - Y_j\|_2.$$

From Theorem 9.1.7,

$$\frac{1}{2}\mathbb{E}\max_{i\leqslant n} X_i \leqslant \mathbb{E}\max_{i\leqslant n} Z_i \leqslant 2\mathbb{E}\max_{i\leqslant n} Y_i.$$

Proposition 9.1.5 implies that

$$\frac{1}{2}\mathbb{E}\max_{i\leqslant n} X_i = \frac{1}{2\sqrt{2}}\min_{i\neq j}\|Z_i - Z_j\|_2 \cdot \mathbb{E}\max_{i\leqslant n} g_i \geqslant \frac{c_1 A}{2\sqrt{2}}\sqrt{\log n}$$

and

$$2\mathbb{E}\max_{i\leqslant n} Y_i = \frac{2}{\sqrt{2}}\max_{i\neq j}\|Z_i - Z_j\|_2 \cdot \mathbb{E}\max_{i\leqslant n} g_i \leqslant c_2\sqrt{2}B\sqrt{\log n},$$

which proves the result. □

Proof of Theorem 9.1.10. We fix $\varepsilon > 0$ and assume that $N_\varepsilon(\mathcal{Z})$ is finite. Consider a subset $\{Z_1, \dots, Z_n\}$ of $\mathcal{Z}$ which is maximal with respect to the condition

$$\|Z_i - Z_j\|_2 \geqslant \varepsilon$$

for all $i \neq j$. Then,

$$\mathcal{Z} \subseteq \bigcup_{i=1}^{n}(Z_i + \varepsilon B),$$

and hence $N_\varepsilon(\mathcal{Z}) \leqslant n$. From Lemma 9.1.11 we get

$$\frac{c_1}{2\sqrt{2}}\varepsilon\sqrt{\log n} \leqslant \mathbb{E}\max_{i\leqslant n} Z_i \leqslant \mathbb{E}\sup_{t\in T} Z_t,$$

so

$$\varepsilon\sqrt{\log N_\varepsilon(\mathcal{Z})} \leqslant \frac{2\sqrt{2}}{c_1}\mathbb{E}\sup_{t\in T} Z_t.$$

The argument shows that if $N_\varepsilon(\mathcal{Z}) = \infty$ then both sides of this inequality are infinite. □

REMARK 9.1.12. Let K be a convex body in $\mathbb{R}^n$. Recall that the covering number $\overline{N}(K, \varepsilon B_2^n)$ of K is defined by

$$\overline{N}(K, \varepsilon B_2^n) := \min\Big\{N : \text{there exist}\, t_1, \dots, t_N \in K \text{ s.t. } K \subseteq \bigcup_{i=1}^{N}(t_i + \varepsilon B_2^n)\Big\},$$

where B_2^n is the Euclidean unit ball. We have seen that if $g_1, \dots, g_n$ are independent random variables with $g_i \sim N(0,1)$ and if $\{e_1, \dots, e_n\}$ is the standard orthonormal basis of $\mathbb{R}^n$, then the family $\mathcal{Z} = \{Z_t \mid t \in K\}$ defined by

$$Z_t = \Big\langle t, \sum_{i=1}^{n} g_i e_i\Big\rangle = \sum_{i=1}^{n} t_i g_i \tag{9.1.4}$$

is a Gaussian process and

$$\overline{N}(K, \varepsilon B_2^n) = N_\varepsilon(\mathcal{Z}). \tag{9.1.5}$$

Sudakov's inequality implies that

$$\log \overline{N}(K, \varepsilon B_2^n) \leqslant \frac{c}{\varepsilon^2}\left(\mathbb{E}\sup_{t\in K}\Big\langle t, \sum_{i=1}^{n} g_i e_i\Big\rangle\right)^2. \tag{9.1.6}$$

A direct consequence of Lemma 9.1.3 and Theorem 9.1.10 is Sudakov's inequality for the covering numbers of a convex body (which was proved in Section 4.2): There exists an absolute constant $C > 0$ with the following property: if K is a convex body in $\mathbb{R}^n$, then

$$\overline{N}(K, \varepsilon B_2^n) \leqslant \exp\left(\frac{Cnw^2(K)}{\varepsilon^2}\right) \tag{9.1.7}$$

for every $\varepsilon > 0$.

Let us now pass to the proof of Dudley's inequality; this is the right hand side inequality in Theorem 9.1.4.

THEOREM 9.1.13 (Dudley). *There exists an absolute constant $c_2 > 0$ with the following property: for every Gaussian process $\mathcal{Z} = (Z_t)_{t\in T}$,*

$$\mathbb{E} \sup_{t\in T} Z_t \leqslant c_2 \int_0^\infty \sqrt{\log N_\varepsilon(\mathcal{Z})}d\varepsilon.$$

We shall follow the presentation of the proof of this result given by M. Talagrand.

Proof. Let F be a finite subset of T and let $\delta > 1$. We define

$$D = \max_{i,j\in F} \|Z_i - Z_j\|_2,$$

and, for every $n \geqslant 0$, we set

$$\varepsilon_n = \delta\frac{D}{2^n} \quad \text{and} \quad N_n = N_{\varepsilon_n}(\mathcal{Z}_F),$$

where $\mathcal{Z}_F = \{Z_t : t \in F\}$.

By the definition of N_n, we can find $F_n \subseteq F$ with $|F_n| = N_n$, such that

$$\mathcal{Z}_F \subseteq \bigcup_{t\in F_n} (Z_t + \varepsilon_n B).$$

In particular, we may choose $F_0 = \{t_0\}$ for any $t_0 \in F$.

Let $s \in F$. For every $n \geqslant 0$ there exists $t_n(s) \in F_n$ such that

$$\|Z_s - Z_{t_n(s)}\|_2 < \varepsilon_n.$$

Then, for every $s \in F$ and every $n \geqslant 1$ we have

$$\|Z_{t_n(s)} - Z_{t_{n-1}(s)}\|_2 \leqslant \|Z_{t_n(s)} - Z_s\|_2 + \|Z_s - Z_{t_{n-1}(s)}\|_2 < 2\varepsilon_{n-1}, \tag{9.1.8}$$

and, since F is finite,

$$Z_s = Z_{t_0} + \sum_{n=1}^\infty (Z_{t_n(s)} - Z_{t_{n-1}(s)}).$$

(Note that the last sum is finite: for large enough n, $Z_{t_n(s)} = Z_s$). It follows that

$$\max_{s\in F} Z_s \leqslant Z_{t_0} + \sum_{n=1}^\infty \max_{s\in F}(Z_{t_n(s)} - Z_{t_{n-1}(s)}), \tag{9.1.9}$$

and hence

$$\mathbb{E}\left(\max_{s\in F} Z_s\right) \leqslant \sum_{n=1}^\infty \mathbb{E}\left(\max_{s\in F}(Z_{t_n(s)} - Z_{t_{n-1}(s)})\right). \tag{9.1.10}$$

From the right hand side inequality of Lemma 9.1.11 we get

$$\mathbb{E}\left(\max_{s\in F}(Z_{t_n(s)} - Z_{t_{n-1}(s)})\right) \leqslant 4\sqrt{2}c_2\varepsilon_{n-1}\sqrt{\log(N_nN_{n-1})} \leqslant C\varepsilon_{n-1}\sqrt{\log N_n},$$

because (N_n) is increasing, and the set $\{Z_{t_n(s)} - Z_{t_{n-1}(s)} : s \in F\}$ has cardinality less than or equal to $|F_n| \cdot |F_{n-1}| = N_n N_{n-1}$ and diameter (in $L^2(\Omega)$) bounded by $4\varepsilon_{n-1}$, because of (9.1.8).

Going back to (9.1.10), we obtain

$$\begin{aligned}\mathbb{E}\max_{s\in F} Z_s &\leqslant C\sum_{n=1}^{\infty}\varepsilon_{n-1}\sqrt{\log N_{\varepsilon_n}(\mathcal{Z}_F)}\\ &\leqslant 4C\sum_{n=1}^{\infty}\int_{\varepsilon_{n+1}}^{\varepsilon_n}\sqrt{\log N_\varepsilon(\mathcal{Z}_F)}d\varepsilon\\ &\leqslant 4C\int_0^{\infty}\sqrt{\log N_\varepsilon(\mathcal{Z}_F)}d\varepsilon.\end{aligned}$$

We used the fact that $\varepsilon \mapsto N_\varepsilon(\mathcal{Z}_F)$ is decreasing, and the fact that $\varepsilon_{n-1} = 4(\varepsilon_n - \varepsilon_{n+1})$.

To complete the proof, we observe that for every $\varepsilon > 0$ and $F \subseteq T$,

$$N_\varepsilon(\mathcal{Z}_F) \leqslant N_{\varepsilon/2}(\mathcal{Z}).$$

Indeed, if $\mathcal{Z} \subseteq \bigcup_{j=1}^N (Z_{t_j} + (\varepsilon/2)B)$, we consider the set

$$A_F = \{j \leqslant N \mid [Z_{t_j} + (\varepsilon/2)B] \cap \mathcal{Z}_F \neq \emptyset\},$$

for every $j \in A_F$ we choose $s_j \in F$ with $\|Z_{s_j} - Z_{t_j}\|_2 \leqslant \varepsilon/2$, and, using the triangle inequality, we easily check that $\mathcal{Z}_F \subseteq \bigcup_{j\in A_F}(Z_{s_j} + \varepsilon B)$.

Thus, we conclude that

$$\mathbb{E}\max_{s\in F} Z_s \leqslant 4C\int_0^{\infty}\sqrt{\log N_{\varepsilon/2}(\mathcal{Z})}d\varepsilon = 8C\int_0^{\infty}\sqrt{\log N_\varepsilon(\mathcal{Z})}d\varepsilon,$$

and the proof of the theorem is complete, by the definition of $\mathbb{E}\sup_{t\in T} Z_t$. □

REMARK 9.1.14 (application to covering numbers). Let K be a convex body in $\mathbb{R}^n$. We define the k-th entropy number $e_k(K)$ as follows:

$$e_k(K) = e_k(K, B_2^n) = \min\{t > 0 : N(K, tB_2^n) \leqslant 2^{k-1}\}.$$

This definition is in complete analogy with the one given in Section 4.3: if K is centrally symmetric then $e_k(u) = e_k(\mathrm{Id})$, where $\mathrm{Id} : \ell_2^n \to (\mathbb{R}^n, \|\cdot\|_K)$ is the identity operator. Then, Lemma 9.1.3 and Theorem 9.1.13 imply the following general upper bound for the mean width of K:

$$M^*(K) = w(K) \leqslant \frac{C}{\sqrt{n}}\sum_{k=1}^{\infty}\frac{1}{\sqrt{k}}e_k(K, B_2^n), \tag{9.1.11}$$

where $C > 0$ is an absolute constant. For the proof of (9.1.11) we only have to consider the Gaussian process $\mathcal{Z} = (Z_t)_{t\in K}$ of Remark 9.1.12 and then compare the right hand side of (9.1.11) with Dudley's integral

$$\int_0^{\infty}\sqrt{\log N_\varepsilon(\mathcal{Z})}d\varepsilon = \int_0^{\infty}\sqrt{\log \overline{N}(K, \varepsilon B_2^n)}d\varepsilon.$$

9.2. Gaussian proof of Dvoretzky theorem

In this section we present a "Gaussian proof" of Dvoretzky theorem which is due to Gordon. It is based on a min-max principle for Gaussian processes, whose original formulation is the following.

THEOREM 9.2.1 (Gordon). *Let X_{ij} and Y_{ij} $(1 \leqslant i \leqslant n, 1 \leqslant j \leqslant m)$ be Gaussian random variables with mean zero such that*

(i) $\mathbb{E}(X_{ij}^2) = \mathbb{E}(Y_{ij}^2)$ *for all* i, j.
(ii) $\mathbb{E}(X_{ij}X_{ik}) \leqslant \mathbb{E}(Y_{ij}Y_{ik})$ *for all* $1 \leqslant i \leqslant n$ *and* $1 \leqslant j, k \leqslant m$.
(iii) $\mathbb{E}(X_{ij}X_{lk}) \geqslant \mathbb{E}(Y_{ij}Y_{lk})$ *for all* $1 \leqslant i \neq l \leqslant n$ *and* $1 \leqslant j, k \leqslant m$.

Then, for any choice of reals (t_{ij}) *one has*

$$\mathbb{P}\Big(\bigcap_{i=1}^{n}\bigcup_{j=1}^{m}\{Y_{ij} \geqslant t_{ij}\}\Big) \leqslant \mathbb{P}\Big(\bigcap_{i=1}^{n}\bigcup_{j=1}^{m}\{X_{ij} \geqslant t_{ij}\}\Big). \tag{9.2.1}$$

In particular,

$$\mathbb{E}\,\min_i \max_j Y_{ij} \leqslant \mathbb{E}\,\min_i \max_j X_{ij}.$$

The proof of Theorem 9.2.1 will be given in Section 9.5. What we need is the next application.

PROPOSITION 9.2.2. *Let $T \subset \mathbb{R}^n$ and $S \subset \mathbb{R}^m$ be finite sets. We set $R = \max\{|v| : v \in S\}$ and for every $u = (u_1, \ldots, u_n) \in T$ and $v = (v_1, \ldots, v_m) \in S$ we define the Gaussian random variables*

$$X_{u,v} = \sum_{i=1}^{n}\sum_{j=1}^{m} u_i v_j g_{ij} + R|u|g$$

$$Y_{u,v} = |u|\sum_{j=1}^{m} v_j g_j + R\sum_{i=1}^{n} u_i h_i,$$

where g, g_j, h_i, g_{ij} are independent standard Gaussian random variables on some probability space $(\Omega, \mathcal{A}, \mathbb{P})$. Then, for any choice of reals $\{t_{u,v}\}$ we have

$$\begin{aligned}\mathbb{P}\Big(\bigcap_{u\in T}\bigcup_{v\in S}\{Y_{u,v} \geqslant t_{u,v}\}\Big) &\leqslant \mathbb{P}\Big(\bigcap_{u\in T}\bigcup_{v\in S}\{X_{u,v} \geqslant t_{u,v}\}\Big)\\ &\leqslant \mathbb{P}\Big(\bigcup_{u,v}\{X_{u,v} \geqslant t_{u,v}\}\Big) \leqslant \mathbb{P}\Big(\bigcup_{u,v}\{Y_{u,v} \geqslant t_{u,v}\}\Big).\end{aligned} \tag{9.2.2}$$

In particular,

$$\mathbb{E}(\min_u \max_v Y_{u,v}) \leqslant \mathbb{E}(\min_u \max_v X_{u,v}) \leqslant \mathbb{E}(\max_{u,v} X_{u,v}) \leqslant \mathbb{E}(\max_{u,v} Y_{u,v}). \tag{9.2.3}$$

Proof. We first check that for all $u, u' \in T$ and all $v, v' \in S$

$$\mathbb{E}(X_{u,v}X_{u',v'}) - \mathbb{E}(Y_{u,v}Y_{u',v'}) = (R^2 - \langle v, v'\rangle)(|u||u'| - \langle u, u'\rangle) \geqslant 0, \tag{9.2.4}$$

and hence $\mathbb{E}(X_{u,v}X_{u',v'}) \geqslant \mathbb{E}(Y_{u,v}Y_{u',v'})$.

The first inequality in (9.2.2) is immediate from Theorem 9.2.1 (note that, because of (9.2.4), all the assumptions of that theorem are satisfied). Using (9.2.4) and Theorem 9.2.1 for the processes $\{X_{u,v} : (u,v) \in T \times S\}$ and $\{Y_{u,v} : (u,v) \in T \times S\}$ with $n = 1$, we see that

$$\mathbb{P}\Big(\bigcup_{u,v}\{X_{u,v} \geqslant t_{u,v}\}\Big) \leqslant \mathbb{P}\Big(\bigcup_{u,v}\{Y_{u,v} \geqslant t_{u,v}\}\Big).$$

The second inequality in (9.2.2) is obvious.

Finally, using the integration by parts formula

$$\mathbb{E}(X) = \int_0^\infty \mathbb{P}(X > t)dt - \int_0^\infty \mathbb{P}(X < -t)dt, \tag{9.2.5}$$

we get (9.2.3). □

DEFINITION 9.2.3. In what follows, we consider a centrally symmetric convex body K in $\mathbb{R}^n$. Given $y_1, \ldots, y_m \in \mathbb{R}^n$ we set

$$\varepsilon_2(y_1, \ldots, y_m) = \sup\Big\{\Big(\sum_{j=1}^m \langle y_j, x\rangle^2\Big)^{1/2} : x \in K^\circ\Big\},$$

where K° is the polar body of K.

We also set

$$a_k = \mathbb{E}\left[\Big(\sum_{i=1}^k g_i^2\Big)^{1/2}\right] = \sqrt{2}\Gamma((k+1)/2)/\Gamma(k/2).$$

From this formula we easily check that $a_k a_{k+1} = k$ for all k. In order to estimate a_k observe that

$$a_k^2 \leqslant \mathbb{E}\Big(\sum_{j=1}^k g_j^2\Big) = k,$$

and hence $a_k\sqrt{k+1} \geqslant a_k a_{k+1} = k$. It follows that

$$a_k = \sqrt{k}\left(1 - \frac{1}{4k} + \frac{1}{32k^2} + O(k^{-3})\right).$$

The main technical statement is the following.

THEOREM 9.2.4. *Let K be a centrally symmetric convex body in $\mathbb{R}^n$. Let $y_1, \ldots, y_m \in \mathbb{R}^n$ and let $\{e_1, \ldots, e_k\}$ be the standard orthonormal basis of ℓ_2^k. If $G_\omega : \ell_2^k \to \ell_2^n$ is the random Gaussian operator*

$$G_\omega = \sum_{i=1}^k \sum_{j=1}^m g_{ij}(\omega) e_i \otimes y_j$$

where g_{ij} are independent standard Gaussian random variables, then

$$\begin{aligned} \mathbb{E}\Big\|\sum_{j=1}^m g_j y_j\Big\| - a_k\varepsilon_2(y_1, \ldots, y_m) &\leqslant \mathbb{E}\min_{|u|=1}\|G_\omega(u)\| \\ &\leqslant \mathbb{E}\max_{|u|=1}\|G_\omega(u)\| \leqslant \mathbb{E}\Big\|\sum_{j=1}^m g_j y_j\Big\| + a_k\varepsilon_2(y_1, \ldots, y_m). \end{aligned} \tag{9.2.6}$$

Proof. We set $T = S^{k-1}$ and

$$S = \{(\langle y_1, v\rangle, \dots, \langle y_m, v\rangle) : v \in \mathrm{bd}(K^\circ)\} \subset \mathbb{R}^m.$$

Note that

$$R = \max\{|z| : z \in S\} = \varepsilon_2(y_1, \dots, y_m).$$

Both sets are compact, and hence, by an approximation argument we may apply Proposition 9.2.2 to get

$$\mathbb{E}(\min_u \max_v Y_{u,v}) \leqslant \mathbb{E}(\min_u \max_v X_{u,v}) \leqslant \mathbb{E}(\max_{u,v} X_{u,v}) \leqslant \mathbb{E}(\max_{u,v} Y_{u,v}). \tag{9.2.7}$$

for the processes

$$X_{u,v} = \sum_{i=1}^{k}\sum_{j=1}^{m} u_i\langle y_j, v\rangle g_{ij} + \varepsilon_2(y_1, \dots, y_m) g$$

$$Y_{u,v} = \sum_{j=1}^{m} \langle y_j, v\rangle g_j + \varepsilon_2(y_1, \dots, y_m)\sum_{i=1}^{k} u_i h_i,$$

where $u \in S^{k-1}$ and $v \in \mathrm{bd}(K^\circ)$. Note that

$$\sum_{i=1}^{k}\sum_{j=1}^{m} u_i\langle y_j, v\rangle g_{ij}(\omega) = \langle G_\omega(u), v\rangle.$$

Then, in order to obtain the assertion of Theorem 9.2.4 we only have to observe that

$$\min_u \max_v Y_{u,v} = \Big\|\sum_{j=1}^{m} g_j y_j\Big\| - \varepsilon_2(y_1, \dots, y_m)\Big(\sum_{i=1}^{k} h_i^2\Big)^{1/2}$$

$$\max_u \max_v Y_{u,v} = \Big\|\sum_{j=1}^{m} g_j y_j\Big\| + \varepsilon_2(y_1, \dots, y_m)\Big(\sum_{i=1}^{k} h_i^2\Big)^{1/2}$$

$$\min_u \max_v X_{u,v} = \min_u \|G_\omega(u)\| + \varepsilon_2(y_1, \dots, y_m) g$$

$$\max_u \max_v X_{u,v} = \max_u \|G_\omega(u)\| + \varepsilon_2(y_1, \dots, y_m) g,$$

and then substitute into (9.2.6) and compute the expectations. □

THEOREM 9.2.5. *Let K be a centrally symmetric convex body in $\mathbb{R}^n$. Let $y_1, \dots, y_m \in \mathbb{R}^n$ and let $\{e_1, \dots, e_k\}$ be the standard orthonormal basis of ℓ_2^k. Assume that*

$$\mathbb{E}\Big\|\sum_{j=1}^{m} g_j y_j\Big\| > a_k \varepsilon_2(y_1, \dots, y_m),$$

where $g_1, \dots, g_m$ are independent standard Gaussian random variables on some probability space $(\Omega, \mathcal{A}, \mathbb{P})$. Then, there exists a k-dimensional subspace F of $\mathbb{R}^n$ such that

$$d(K \cap F, B_2^k) \leqslant D := \frac{\mathbb{E}\Big\|\sum_{j=1}^{m} g_j y_j\Big\| + a_k \varepsilon_2(y_1, \dots, y_m)}{\mathbb{E}\Big\|\sum_{j=1}^{m} g_j y_j\Big\| - a_k \varepsilon_2(y_1, \dots, y_m)}.$$

Proof. For any $\omega \in \Omega$ we define $F(\omega) = \max\{\|G_\omega(x)\| : x \in S^{k-1}\}$ and $f(\omega) = \min\{\|G_\omega(x)\| : x \in S^{k-1}\}$. Theorem 9.2.4 shows that

$$\mathbb{E}\Big\|\sum_{j=1}^{m} g_j y_j\Big\| - a_k\varepsilon_2(y_1,\dots,y_m) \leqslant \mathbb{E}(f(\omega)) \tag{9.2.8}$$

$$\leqslant \mathbb{E}(F(\omega)) \leqslant \mathbb{E}\Big\|\sum_{j=1}^{m} g_j y_j\Big\| + a_k\varepsilon_2(y_1,\dots,y_m).$$

Therefore, we may find $\omega_0 \in \Omega$ so that $F(\omega_0) \leqslant Df(\omega_0)$. Since

$$f(\omega_0)|x| \leqslant \|G_{\omega_0}(x)\| \leqslant F(\omega_0)|x|$$

for all $x \in \ell_2^k$, this means that if $F = \mathrm{span}\{G_{\omega_0}(x) : x \in \ell_2^k\}$ and if $\mathcal{E} = \frac{1}{F(\omega_0)}G_{\omega_0}(B_2^k)$, then $\mathcal{E}$ is an ellipsoid in F satisfying

$$\mathcal{E} \subseteq K \cap F \subseteq D \cdot \mathcal{E}.$$

The theorem follows. □

We can now give Gordon's proof of Dvoretzky theorem.

THEOREM 9.2.6 (Gordon). *There exists an absolute constant $c > 0$ with the following property: for every $\varepsilon \in (0,1)$, for every $k > 1$, for every $n \geqslant \exp(c\varepsilon^{-2}k)$, and for any centrally symmetric convex body K in $\mathbb{R}^n$, we may find a k-dimensional subspace F of $\mathbb{R}^n$ such that*

$$d(K \cap F, B_2^k) \leqslant \frac{1+\varepsilon}{1-\varepsilon}.$$

Proof. From Theorem 2.1.24 we may assume that $B_2^n \subseteq K$ and that there exist $y_1,\dots,y_n \in \mathbb{R}^n$ and an orthonormal basis $\{z_j\}_{j\leqslant n}$ of $\mathbb{R}^n$ with the next three properties:

(1) Each y_j is a contact point of K and B_2^n, that is,
$\|y_j\| = \|y_j\|^* = |y_j| = 1$ for all $j = 1,\dots,n$.

(2) Each y_i can be written in the form $y_i = \sum_{j=1}^{i} a_{ij}z_j$, where $\sum_{j=1}^{i-1} a_{ij}^2 = 1 - a_{ii}^2 \leqslant \frac{i-1}{n}$.

(3) For any $1 \leqslant m \leqslant n$ and for any choice of real numbers $\{t_i\}_{i\leqslant m}$,

$$\Big\|\sum_{i=1}^{m} t_i y_i\Big\| \leqslant \sqrt{2 + \frac{m(m-1)}{n}}\Big(\sum_{i=1}^{m} t_i^2\Big)^{1/2}.$$

To see this, we write

$$\begin{aligned}\Big\|\sum_{i=1}^m t_iy_i\Big\|^2 &\leqslant \Big|\sum_{i=1}^m t_iy_i\Big|^2 = \sum_{j=1}^m\Big(\sum_{i=j}^m t_ia_{ij}\Big)^2 \leqslant \sum_{j=1}^m\left[2t_j^2a_{jj}^2 + 2\left(\sum_{i=j+1}^m t_ia_{ij}\right)^2\right]\\
&\leqslant 2\sum_{j=1}^m\left[t_j^2a_{jj}^2 + \left(\sum_{i=j+1}^m t_i^2\right)\left(\sum_{k=j+1}^m a_{kj}^2\right)\right]\\
&= 2\sum_{i=1}^m\left[a_{ii}^2 + \sum_{k=1}^m\sum_{j=1}^{\min(i-1,m-1)} a_{kj}^2\right]t_i^2\\
&\leqslant 2\sum_{i=1}^m\left(1+\sum_{k=1}^m\frac{k-1}{n}\right)t_i^2 = \left[2+\frac{m(m-1)}{n}\right]\sum_{i=1}^m t_i^2.\end{aligned}$$

From the third condition we see that, for every $x \in K^\circ$ and every $t = (t_1,\ldots,t_m) \in S^{m-1}$,

$$\sum_{j=1}^m t_j\langle y_j, x\rangle = \Big\langle\sum_{j=1}^m t_jy_j, x\Big\rangle \leqslant \Big\|\sum_{i=1}^m t_iy_i\Big\| \leqslant \sqrt{2+\frac{m(m-1)}{n}},$$

and hence

$$\left(\sum_{j=1}^m\langle y_j, x\rangle^2\right)^{1/2} \leqslant \sqrt{2+\frac{m(m-1)}{n}},$$

which implies

$$\varepsilon_2(y_1,\ldots,y_m) \leqslant \sqrt{2+\frac{m(m-1)}{n}} \leqslant \sqrt{2+\frac{m^2}{n}}.$$

From the first condition we see that

$$\begin{aligned}\mathbb{E}\Big\|\sum_{j=1}^m g_jy_j\Big\| &\geqslant \mathbb{E}\Big(\max_{1\leqslant k\leqslant m}\Big\langle\sum_j g_jy_j, y_k\Big\rangle\Big)\\
&\geqslant \mathbb{E}\Big(\max_{1\leqslant k\leqslant m}\Big(g_k - \sum_{j\neq k}|\langle y_j, y_k\rangle g_j|\Big)\Big).\end{aligned}$$

On the other hand, for all $1 \leqslant i < k \leqslant m$ we have

$$\langle y_i, y_k\rangle \leqslant \sum_{j=1}^i|a_{ij}a_{kj}| \leqslant \Big(\sum_{j=1}^{k-1}a_{kj}^2\Big)^{1/2} \leqslant \Big(\sum_{j=1}^{m-1}a_{kj}^2\Big)^{1/2} \leqslant \sqrt{\frac{m-1}{n}},$$

which shows that

$$\sum_{j\neq k}|\langle y_j, y_k\rangle g_j| \leqslant \Big(\sum_{j=1}^m g_j^2\Big)^{1/2}\frac{m-1}{\sqrt{n}}.$$

It follows that

$$\begin{aligned}\mathbb{E}\Big\|\sum_{j=1}^m g_jy_j\Big\| &\geqslant \mathbb{E}\Big(\max_{1\leqslant k\leqslant m} g_k - \Big(\sum_{j=1}^m g_j^2\Big)^{1/2}\frac{m-1}{\sqrt{n}}\Big)\\
&\geqslant c_1\sqrt{\log m} - \sqrt{m^3/n},\end{aligned}$$

where $c_1 > 0$ is an absolute constant. We choose $m = \lfloor n^{1/3} \rfloor$ and use Theorem 9.2.5. Using also the fact that $a_k \leqslant \sqrt{k}$ we see that there exists $F \in G_{n,k}$ such that $d(K \cap F, B_2^k) \leqslant \frac{1+\varepsilon}{1-\varepsilon}$, provided that $n \geqslant \exp(c\varepsilon^{-2}k)$ for some absolute constant $c > 0$. □

Note that the dependence on ε in Theorem 9.2.6 is exactly the one in Theorem 5.1.3: for every n-dimensional normed space X and for every $\varepsilon \in (0,1)$ there exist an integer $k \geqslant c\varepsilon^2 \log n$ and a k-dimensional subspace F of X which satisfies $d(F, \ell_2^k) \leqslant 1 + \varepsilon$.

9.3. Gaussian proof of the M^*-estimate

Gordon gave a proof of the M^*-estimate based on his min-max principle. The starting point of his argument is the following variant of Theorem 9.1.7.

THEOREM 9.3.1 (Gordon). *Let X_{ij} and Y_{ij} $(1 \leqslant i \leqslant n, 1 \leqslant j \leqslant m)$ be Gaussian random variables with mean zero such that*

(i) $\mathbb{E}|X_{ij} - X_{ik}|^2 \leqslant \mathbb{E}|Y_{ij} - Y_{ik}|^2$ *for all* $1 \leqslant i \leqslant n$ *and* $1 \leqslant j, k \leqslant m$.
(ii) $\mathbb{E}|X_{ij} - X_{lk}|^2 \geqslant \mathbb{E}|Y_{ij} - Y_{lk}|^2$ *for all* $1 \leqslant i \neq l \leqslant n$ *and* $1 \leqslant j, k \leqslant m$.

Then,

$$\mathbb{E} \min_i \max_j X_{ij} \leqslant \mathbb{E} \min_i \max_j Y_{ij}.$$

REMARK 9.3.2. Note that the special case $n = 1$ of Theorem 9.3.1 reduces to Theorem 9.1.7.

THEOREM 9.3.3. *Let $\{g_{ij}\}, \{h_i\}, \{g_j\}$ and g $(1 \leqslant i \leqslant n, 1 \leqslant j \leqslant k)$ be independent standard Gaussian random variables on some probability space $(\Omega, \mathcal{A}, \mathbb{P})$. Consider the random Gaussian operator $G : \mathbb{R}^n \to \mathbb{R}^k$ defined by*

$$G(\omega) = \sum_{i=1}^n \sum_{j=1}^k g_{ij}(\omega) e_i \otimes e_j',$$

where $\{e_i\}_{i=1}^n, \{e_j'\}_{j=1}^k$ are the standard orthonormal basis vectors in $\mathbb{R}^n$ and $\mathbb{R}^k$ respectively. Then, for every closed subset S of $\mathbb{R}^n$ one has

$$\begin{aligned}\mathbb{E}\left(\min_{x \in S} |G(\omega)(x)|\right) &\geqslant \mathbb{E}\Big(\min_{x \in S} \Big\{|x| \Big(\sum_{j=1}^k g_j^2\Big)^{1/2} + \sum_{i=1}^n h_i x_i\Big\}\Big) \\ &\geqslant a_k \min_{x \in S} |x| - \mathbb{E}\left(\max_{x \in S} \sum_{i=1}^n h_i x_i\right)\end{aligned}$$

and

$$\begin{aligned}\mathbb{E}\left(\max_{x \in S} |G(\omega)(x)|\right) &\leqslant \mathbb{E}\Big(\max_{x \in S} \Big\{|x| \Big(\sum_{j=1}^k g_j^2\Big)^{1/2} + \sum_{i=1}^n h_i x_i\Big\}\Big) \\ &\leqslant a_k \max_{x \in S} |x| + \mathbb{E}\left(\max_{x \in S} \sum_{i=1}^n h_i x_i\right).\end{aligned}$$

Proof. We define two Gaussian processes indexed by $x \in S$ and $y \in S^{k-1}$ as follows:

$$X_{x,y} = |x| \sum_{j=1}^{k} g_j y_j + \sum_{i=1}^{n} h_i x_i$$

and

$$Y_{x,y} = \langle G(\omega)(x), y \rangle = \sum_{i=1}^{n} \sum_{j=1}^{k} g_{ij} x_i y_j.$$

We observe that for all $x, x' \in S$ and $y, y' \in S^{k-1}$

$$\begin{aligned}
&\mathbb{E}\,|X_{x,y} - X_{x',y'}|^2 - \mathbb{E}\,|Y_{x,y} - Y_{x',y'}|^2 \\
&= |x|^2 + |x'|^2 - 2|x|\,|x'|\langle y, y'\rangle - 2\langle x, x'\rangle(1 - \langle y, y'\rangle) \\
&\geqslant |x|^2 + |x'|^2 - 2|x|\,|x'|\langle y, y'\rangle - 2|x|\,|x'|(1 - \langle y, y'\rangle) \\
&\geqslant 0.
\end{aligned}$$

Note that we have equality if $x = x'$. Applying Theorem 9.3.1 we get

$$\mathbb{E} \min_{x \in S} \max_{y \in S^{k-1}} X_{x,y} \leqslant \mathbb{E} \min_{x \in S} \max_{y \in S^{k-1}} Y_{x,y}.$$

From Slepian's inequality we also have

$$\mathbb{E} \max_{x,y} X_{x,y} \geqslant \mathbb{E} \max_{x,y} Y_{x,y}.$$

Combining these two inequalities we get the theorem. □

The next corollary of Theorem 9.3.3 will be convenient for applications.

PROPOSITION 9.3.4. *Let S be a closed subset of S^{n-1}. In the notation of the previous theorem, set*

$$s = \mathbb{E}\left(\max_{x \in S} \sum_{i=1}^{n} h_i x_i \right).$$

If $1 \leqslant k < n$ and $a_k > s$ then there exists an operator $T : \ell_2^n \to \ell_2^k$ such that

$$\frac{|Tx|}{|Ty|} \leqslant \frac{a_k + s}{a_k - s}$$

for all $x, y \in S$.

Proof. The first result of Theorem 9.3.3 shows that

$$\mathbb{E}\left(\min_{x \in S} |G(\omega)(x)| \right) \geqslant a_k - s > 0.$$

The second one shows that

$$\mathbb{E}\left(\max_{x \in S} |G(\omega)(x)| \right) \leqslant a_k + s.$$

Therefore, there exists $\omega \in \Omega$ such that the operator $T = G(\omega)$ satisfies our claim. □

Our second corollary of Theorem 9.3.3 will be used for the proof of the M^*-estimate.

PROPOSITION 9.3.5. *Let $1 \leqslant k < n$ and consider the random operator $G : \mathbb{R}^n \to \mathbb{R}^k$ defined by*

$$G(\omega) = \sum_{i=1}^{n} \sum_{j=1}^{k} g_{ij}(\omega) e_i \otimes e_j'.$$

If S is a closed subset of S^{n-1} such that

$$a_k > s = \mathbb{E}\left(\max_{x \in S} \sum_{i=1}^{n} x_i h_i\right),$$

then

$$\mathbb{E} \min_{x \in S} |G(\omega)(x)| > 0$$

which is in turn equivalent to the fact that there exists a subspace $F \in G_{n,n-k}$ such that $F \cap S = \emptyset$.

Proof. The first assertion of the proposition is a direct consequence of the inequality

$$\mathbb{E}\left(\min_{x \in S} |G(\omega)(x)|\right) \geqslant a_k \min_{x \in S} |x| - \mathbb{E}\left(\max_{x \in S} \sum_{i=1}^{n} h_i x_i\right)$$

of Theorem 9.3.3 since $|x| = 1$ for all $x \in S$.

Assume that $\mathbb{E}\, \min_{x \in S} |G(\omega)(x)| > 0$. Then, we choose $\omega_0 \in \Omega$ such that $\min_{x \in S} |G(\omega_0)(x)| > 0$ and we set $F = G(\omega_0)^{-1}(0)$. It is clear that $\dim(F) = n - k$ and that $F \cap S = \emptyset$. Conversely, since every $F \in G_{n,n-k}$ can be written in the form $G(\omega)^{-1}(0)$, the compactness of S implies that if $G(\omega)^{-1}(0) \cap S = \emptyset$ for some ω then $\min_{x \in S} |G(\omega)(x)| > 0$ on a set $A \subset \Omega$ of positive measure, which shows that $\mathbb{E}\, \min_{x \in S} |G(\omega)(x)| > 0$. □

THEOREM 9.3.6. *Let K be a convex body in $\mathbb{R}^n$ with $0 \in \mathrm{int}(K)$. For every $r > 0$ set*

$$K_r = r^{-1} K \cap B_2^n.$$

Let $1 \leqslant k < n$ and set

$$r_0 := \inf\left\{r > 0 : a_k > \mathbb{E}\left\|\sum_{i=1}^{n} h_i e_i\right\|_{K_r^\circ}\right\} \leqslant a_k^{-1} \mathbb{E}\left\|\sum_{i=1}^{n} h_i e_i\right\|_{K^\circ}.$$

Then, there exists $F \in G_{n,n-k}$ such that

$$|x| \leqslant r_0 \|x\|_K \tag{9.3.1}$$

for all $x \in F$.

Proof. We first observe that

$$s_r := \mathbb{E}\left(\max_{x \in K_r} \sum_{i=1}^{n} x_i h_i\right) = \mathbb{E}\left\|\sum_{i=1}^{n} h_i e_i\right\|_{K_r^\circ}.$$

If $a_k > s_r$ then Proposition 9.3.5 shows that there exists $F \in G_{n,n-k}$ such that $F \cap (S^{n-1} \cap K_r) = \emptyset$, or equivalently,

$$|x| < r\|x\|_K$$

for all $x \in F$. Taking the infimum over all such r we get (9.3.1).

It remains to check that $r_0 \leqslant a_k^{-1}\mathbb{E}\,\|\sum_{i=1}^n h_ie_i\|_{K^\circ}$. We just note that $K_r \subseteq r^{-1}K$, which shows that

$$\Big\|\sum_{i=1}^n h_ie_i\Big\|_{K_r^\circ} \leqslant r^{-1}\mathbb{E}\Big\|\sum_{i=1}^n h_ie_i\Big\|_{K^\circ}$$

for every $r>0$. By the definition of r_0 we have

$$a_k = \mathbb{E}\Big\|\sum_{i=1}^n h_ie_i\Big\|_{K_{r_0}^\circ} \leqslant r_0^{-1}\mathbb{E}\Big\|\sum_{i=1}^n h_ie_i\Big\|_{K^\circ},$$

which proves our claim. □

Taking into account the estimate for a_k in Definition 9.2.3 we obtain a sharp form of the M^*-estimate.

THEOREM 9.3.7 (Gordon's form of M^*-estimate). *Let K be a convex body in $\mathbb{R}^n$ with $0\in \mathrm{int}(K)$. For every $1\leqslant k<n$ there exists a subspace $F\in G_{n,n-k}$ such that*

$$|x| \leqslant a_k^{-1}\mathbb{E}\Big\|\sum_{i=1}^n g_ie_i\Big\|_{K^\circ}\|x\|_k \leqslant C\sqrt{n/k}\,w(K)\|x\|_K$$

for all $x\in F$, where $C>0$ is an absolute constant.

Proof. Recall that

$$w(K) \simeq \frac{1}{\sqrt{n}}\mathbb{E}\Big\|\sum_{i=1}^n g_ie_i\Big\|_{K^\circ}.$$

Then, the theorem follows immediately from Theorem 9.3.6. □

A modification of the above arguments leads to a similar result for a random subspace $F\in G_{n,n-k}$, which we next quote without detailing the proof, see Gordon [**258**] for the details. This was used in some of the applications of the M^*-estimate that we presented in Chapter 7.

THEOREM 9.3.8. *For every convex body K in $\mathbb{R}^n$ with $0\in\mathrm{int}(K)$ and for any $0<\gamma<1$ and $1\leqslant k<n$, one can find $\mathcal{A}\subseteq G_{n,n-k}$ with*

$$\nu_{n,n-k}(\mathcal{A}) \geqslant 1-\frac{7}{2}\exp\Big(-\tfrac{1}{18}(1-\gamma)^2a_k^2\Big) \tag{9.3.2}$$

such that, for every $E\in\mathcal{A}$ and for all $x\in E$,

$$\frac{\gamma a_k}{a_n w(K)}\,|x| \leqslant \|x\|.$$

9.4. Random orthogonal factorizations

We first introduce some notation. As in the previous section, given a normed space X and an n-tuple of vectors $x_1,\dots,x_n\in X$ we set

$$\varepsilon_2(x_1,\dots,x_n) = \sup\left\{\Big(\sum_{j=1}^n |x^*(x_j)|^2\Big)^{1/2} : \|x^*\|\leqslant 1\right\}.$$

One can check that if $T:\ell_2^n\to X$ is the operator defined by $T=\sum_{j=1}^n e_j\otimes x_j$ then $\|T\| = \varepsilon_2(x_1,\dots,x_n)$.

Chevet's inequality estimates the expectation of the operator norm of a Gaussian matrix.

THEOREM 9.4.1 (Chevet's inequality). *Let X and Y be two normed spaces and let $\{x_i^*\}_{i=1}^n \subset X^*$ and $\{y_j\}_{j=1}^m \subset Y$. Then,*

$$\frac{A}{2} \leqslant \mathbb{E}\left\|\sum_{i=1}^n \sum_{j=1}^m g_{ij} x_i^* \otimes y_j\right\| \leqslant \sqrt{2}A,$$

where

$$A := \varepsilon_2(x_1^*, \ldots, x_n^*)\mathbb{E}\left\|\sum_{j=1}^m g_j y_j\right\| + \varepsilon_2(y_1, \ldots, y_m)\mathbb{E}\left\|\sum_{i=1}^n g_i x_i^*\right\|.$$

Proof. We will use Theorem 9.1.7 to obtain the right hand side inequality with the slightly worse constant $2\sqrt{2}$ instead of $\sqrt{2}$. This is the part of the theorem that we use in this book and the precise value of the constant is not important for our needs.

We define $D = \{(x, y^*) : \|x\| \leqslant 1, \|y^*\| \leqslant 1\} \subset X \times Y^*$. For every $z = (x, y^*) \in D$ we define

$$X_z = \sum_{i,j} g_{ij} \langle x, x_i^* \rangle \langle y_j, y^* \rangle$$

and

$$Y_z = \sqrt{2}\left(\varepsilon_2(x_1^*, \ldots, x_n^*) \sum_{j=1}^m h_j \langle y_j, y^* \rangle + \varepsilon_2(y_1, \ldots, y_m) \sum_{i=1}^n g_i \langle x, x_i^* \rangle\right),$$

where g_{ij}, g_i and h_j are independent standard Gaussian random variables. We easily check that $\{X_z\}$ and $\{Y_z\}$ satisfy the assumptions of Theorem 9.1.7, and hence

$$\mathbb{E}\left\|\sum_{i,j} g_{ij} x_i^* \otimes y_j\right\| = \mathbb{E}\sup_z X_z \leqslant 2\mathbb{E}\sup_z Y_z = 2\sqrt{2}A,$$

as claimed. $\square$

Note. Comparison of Gaussian and Rademacher averages implies that we also have

$$\mathbb{E}\left\|\sum_{i,j} \epsilon_{ij} x_i^* \otimes y_j\right\| \leqslant \sqrt{\pi/2}\,\mathbb{E}\left\|\sum_{i,j} g_{ij} x_i^* \otimes y_j\right\| \leqslant \sqrt{\pi}A,$$

where $\{\epsilon_{ij}\}$ are independent Bernoulli random variables.

Next, we discuss an inequality of Marcus and Pisier which allows us to replace random Gaussian operators by random orthogonal ones. We consider the orthogonal group $O(m)$ equipped with the Haar probability measure ν_m. Recall that if $U, V \in O(m)$ are fixed and $G = \frac{1}{\sqrt{m}}(g_{ij})$ denotes a random $m \times m$ Gaussian matrix (where g_{ij} are independent standard Gaussian random variables on some probability space (Ω, P)) then the matrix UGV has the same distribution as G and also the joint distribution of UG with respect to $\nu_m \times P$ on $O(m) \times \Omega$ is the same as the distribution of G and the joint distribution of $U|G|$ (where $|G| = (GG^*)^{1/2}$) is again the same as that of G. To see this, note that for any $\omega \in \Omega$ one may find $V(\omega) \in O(m)$ so that $|G(\omega)| = V(\omega)G(\omega)$ and that, by the invariance of ν_m, for any ω one has that the distribution of $UV(\omega)G(\omega)$ is the same as that of $UG(\omega)$; since this is true for every $\omega \in \Omega$, we get that $U|G| \sim UG \sim G$.

THEOREM 9.4.2 (Marcus-Pisier). *Let X be a normed space and let $\{x_{ij}\}_{i,j\leqslant m} \subset X$. For any $p > 0$ one has*

$$\frac{c}{\sqrt{m}}\left(\mathbb{E}\left\|\sum_{i,j} g_{ij}x_{ij}\right\|^p\right)^{1/p} \leqslant \left(\int_{O(m)}\left\|\sum_{i,j} U_{ij}x_{ij}\right\|^p d\nu_m(U)\right)^{1/p} \leqslant \frac{C}{\sqrt{m}}\left(\mathbb{E}\left\|\sum_{i,j} g_{ij}x_{ij}\right\|^p\right)^{1/p},$$

where $c, C > 0$ are absolute constants.

Proof. We prove only the right hand side, which is the more difficult one, and the one we later on use. For any $m \times m$ matrix $A = (a_{ij})$ we set $\mathrm{tr}(AX) = \sum_{i,j} a_{ij}x_{ij}$. With this notation we want to show that

$$\left(\int_{O(m)} \|\mathrm{tr}(UX)\|^p \, d\nu_m(U)\right)^{1/p} \leqslant C\left(\mathbb{E}\,\|\mathrm{tr}(GX)\|^p\right)^{1/p}.$$

We first show that $\mathbb{E}\,|G| = \delta_m I$ for some $\delta_m \geqslant \delta > 0$. To see this, we observe that for every fixed $U \in O(m)$ we have $U|G|U^{-1} = |UGU^{-1}|$ and UGU^{-1} has the same distribution as G, therefore, $U(\mathbb{E}|G|)U^{-1} = \mathbb{E}|UGU^{-1}| = \mathbb{E}|G|$. Since $\mathbb{E}|G|$ commutes with all $U \in O(m)$, it must be a multiple of the identity matrix. In order to estimate δ_m we use Theorem 9.4.1. We have

$$\mathbb{E}\|G : \ell_2^m \to \ell_2^m\| = \frac{1}{\sqrt{m}}\mathbb{E}\left\|\sum_{i,j} g_{ij}e_i \otimes e_j\right\| \leqslant \frac{2\sqrt{2}}{\sqrt{m}}\mathbb{E}\left\|\sum_j g_je_j\right\| = 2\sqrt{2}$$

and using the fact that $(\mathbb{E}\|G\|^2)^{1/2} \simeq \mathbb{E}\|G\|$ and $(\mathbb{E}(\mathrm{tr}|G|)^2)^{1/2} \simeq \mathbb{E}(\mathrm{tr}|G|)$ we get

$$m = \mathbb{E}\,\mathrm{tr}(GG^*) \leqslant (\mathbb{E}(\mathrm{tr}|G|)^2)^{1/2}(\mathbb{E}\|G\|^2)^{1/2} \leqslant C\mathbb{E}(\mathrm{tr}|G|),$$

which implies that

$$\delta_m = \frac{\mathrm{tr}(\mathbb{E}|G|)}{m} = \frac{\mathbb{E}(\mathrm{tr}|G|)}{m} \geqslant \frac{1}{C} = \delta > 0.$$

Now, we can write

$$\begin{aligned}\left(\int_{O(m)} \|\mathrm{tr}(UX)\|^p \, d\nu_m(U)\right)^{1/p} &\leqslant \frac{1}{\delta_m}\left(\int_{O(m)} \|\mathrm{tr}(U(\mathbb{E}|G|)X)\|^p \, d\nu_m(U)\right)^{1/p} \\ &\leqslant \frac{1}{\delta_m}\left(\int_{O(m)} \|\mathrm{tr}(U|G|X)\|^p \, d\nu_m(U)\right)^{1/p} \\ &\leqslant \frac{1}{\delta}\left(\mathbb{E}\,\|\mathrm{tr}(GX)\|^p\right)^{1/p},\end{aligned}$$

using the fact that $U|G|$ has the same distribution as G. $\square$

Combining the above we see that if $x_i^* \in X^*$, $y_j \in Y$, $1 \leqslant i,j \leqslant n$, then

$$\left(\int_{O(n)} \left\| \sum_{i,j} U_{ij} x_i^* \otimes y_j \right\|^2 d\nu_n(U) \right)^{1/2} \leqslant \frac{CA}{\sqrt{n}},$$

where

$$A := \varepsilon_2(x_1^*, \dots, x_n^*) \mathbb{E} \left\| \sum_{j=1}^n g_j y_j \right\| + \varepsilon_2(y_1, \dots, y_n) \mathbb{E} \left\| \sum_{i=1}^n g_i x_i^* \right\|.$$

9.5. Comparison principles for Gaussian processes

In this last section we provide the proofs of most of the comparison principles that we used in this chapter. Recall that the standard Gaussian measure on $\mathbb{R}^n$ is the Borel probability measure γ_n defined by

$$\gamma_n(B) = \frac{1}{(2\pi)^{n/2}} \int_B \exp(-|x|^2) dx$$

for every Borel subset B of $\mathbb{R}^n$.

Let $(\Omega, \mathcal{A}, \mathbb{P})$ be a probability space. A random variable $N : \Omega \to \mathbb{R}^n$ is a *standard Gaussian random variable* on $\mathbb{R}^n$ if $\mathbb{P}(N \in B) = \gamma_n(B)$ for every Borel subset B of $\mathbb{R}^n$. We say that a random variable $X : \Omega \to \mathbb{R}$ is normally distributed on $\mathbb{R}$ if $X = \sigma N + m$ for some standard Gaussian random variable N on $\mathbb{R}$ and some $\sigma \geqslant 0$ and $m \in \mathbb{R}$. We write μ for the distribution $\mathrm{dist}(X)$ of X (i.e. the probability measure which is defined on $\mathbb{R}$ by $\mu(B) = \mathbb{P}(X \in B)$). Then, we have the following:

(1) The expectation and the variance of X are given by

$$\mathbb{E}(X) = m \quad \text{and} \quad \mathrm{Var}(X) = \sigma^2.$$

(2) The characteristic function of X is given by

$$\widehat{\mu}(-t) = \mathbb{E}(e^{itX}) = e^{imt - \frac{1}{2}\sigma^2 t^2}$$

for every $t \in \mathbb{R}$.

We say that X is a $N(m, \sigma^2)$ random variable. If $\sigma = 0$ then $\mu = \delta_m$ (point mass at m), while if $\sigma > 0$ we have

$$d\mu(x) = \frac{1}{\sigma\sqrt{2\pi}} \exp\left(-\frac{(x-m)^2}{2\sigma^2} \right) dx,$$

which implies that

$$\mathbb{E}(f(X)) = \frac{1}{\sigma\sqrt{2\pi}} \int_{-\infty}^{\infty} f(x) e^{-(x-m)^2/2\sigma^2} dx$$

for every $f \in L_1(\mu)$ and every Borel measurable function $f : \mathbb{R} \to [0, \infty)$.

PROPOSITION 9.5.1. *Let $X = (X_1, \dots, X_n) : \Omega \to \mathbb{R}^n$ be a random vector. The following are equivalent:*

1. *There exist an $n \times n$ matrix A and a vector $m \in \mathbb{R}^n$ such that* $\mathrm{dist}(X) = \mathrm{dist}(AN + m)$, *where N is a standard normal random variable.*
2. *For every choice of $t_1, \dots, t_n \in \mathbb{R}$, the random variable $Y = \sum_{i=1}^n t_i X_i$ is normally distributed on $\mathbb{R}$.*

3. *There exist* $a \in \mathbb{R}^n$ *and a positive semidefinite quadratic form* Q *on* $\mathbb{R}^n$ *such that* $\mathbb{E}(e^{i\langle y,X\rangle}) = e^{i\langle a,y\rangle - \frac{1}{2}Q(y)}$ *for every* $y \in \mathbb{R}^n$.

Suppose that $X = (X_1, \ldots, X_n)$ satisfies the three equivalent conditions in Proposition 9.5.1. Consider the covariance matrix $\Gamma = (\gamma_{ij})$ of X, where

$$\gamma_{ij} = \mathbb{E}\big([X_i - \mathbb{E}(X_i)]\,[X_j - \mathbb{E}(X_j)]\big), \qquad i,j = 1, \ldots, n.$$

Then, under the notation of Proposition 9.5.1, we have:

(1) $\mathbb{E}(X) = m$, $\mathbb{E}(Y) = \langle t, m\rangle$ and $\mathrm{Var}(Y) = \|A^*t\|_2^2$ for any $t = (t_1, \ldots, t_n)$.
(2) $AA^* = \Gamma$, $a = m$ and $Q(y) = \langle \Gamma y, y\rangle = \mathrm{Var}(\langle y, X\rangle)$ for every $y \in \mathbb{R}^n$.

Fourier inversion formula. *If* $f, \widehat{f} \in L^1(\mathbb{R}^n)$, *then*

$$f(z) = \frac{1}{(2\pi)^n} \int_{\mathbb{R}^n} \widehat{f}(y) e^{i\langle y,z\rangle} dy \tag{9.5.1}$$

almost everywhere on $\mathbb{R}^n$. *Moreover,* (9.5.1) *holds true at every* $z \in \mathbb{R}^n$ *that is a point of continuity of* f. □

Assume that the random vector $X = (X_1, \ldots, X_n)$ is normally distributed on $\mathbb{R}^n$, with $\mathbb{E}(X_i) = 0$, $i = 1, \ldots, n$. If the matrix Γ is invertible then the inversion formula implies the following (we use $\mathcal{L}(x)$ to denote the distribution law of the random variable X):

PROPOSITION 9.5.2. *If* $\mathcal{L}(X) = \mathcal{L}(AN)$, *then* X *has a density which is given by*

$$g(z) = (2\pi)^{-n} \int_{\mathbb{R}^n} \exp(i\langle y, z\rangle - \langle \Gamma y, y\rangle/2) dy,$$

where $\Gamma = AA^*$ *is the covariance matrix of* X.

Slepian's lemma (Theorem 9.1.6) asserts that if $X = (X_1, \ldots, X_n)$, $Y = (Y_1, \ldots, Y_n)$ are two n-tuples of Gaussian random variables with mean value equal to 0, and if

$$\mathbb{E}(X_i^2) = \mathbb{E}(Y_i^2) \tag{9.5.2}$$

$$\mathbb{E}(X_iX_j) \geqslant \mathbb{E}(Y_iY_j) \tag{9.5.3}$$

for every $i, j = 1, \ldots, n$ then, for all $t_1, \ldots, t_n \in \mathbb{R}$,

$$\mathbb{P}\left(\bigcup_{j=1}^n \{X_j > t_j\}\right) \leqslant \mathbb{P}\left(\bigcup_{j=1}^n \{Y_j > t_j\}\right).$$

In particular,

$$\mathbb{E} \max_{j\leqslant n} X_j \leqslant \mathbb{E} \max_{j\leqslant n} Y_j.$$

Proof of Theorem 9.1.6. It is enough to show that for any $t_1, \ldots, t_n \in \mathbb{R}$

$$\mathbb{P}\big(\cup \{X_j > t_j\}\big) \leqslant \mathbb{P}\big(\cup \{Y_j > t_j\}\big).$$

For any positive definite matrix $\Gamma = (\gamma_{ij})$ we write $Z = (Z_1, \ldots, Z_n)$ for the centered Gaussian random vector with covariance matrix Γ, and $g(z_1, \ldots, z_n; \Gamma)$ for its density. The main observation is that if $j \neq k$ then

$$\frac{\partial g}{\partial \gamma_{jk}} = \frac{\partial^2 g}{\partial z_j \partial z_k}.$$

This follows if we use Fourier inversion formula to write

$$g(z_1,\dots,z_n;\Gamma)=\frac{1}{(2\pi)^n}\int_{\mathbb{R}^n}\exp(i\langle x,z\rangle-\tfrac{1}{2}\langle\Gamma x,x\rangle)\,dx$$

and then differentiate under the integral sign. We fix $t_1,\dots,t_n$ and define

$$Q(Z,\Gamma)=\mathbb{P}\big(\cap\{Z_j\leqslant t_j\}\big)=\int_{-\infty}^{t_1}\cdots\int_{-\infty}^{t_n}g(z_1,\dots,z_n;\Gamma)\,dz.$$

Then,

$$\frac{\partial Q(Z,\Gamma)}{\partial\gamma_{jk}}=\int_{-\infty}^{t_1}\cdots\int_{-\infty}^{t_n}\frac{\partial^2}{\partial z_j\partial z_k}g(z_1,\dots,z_n;\Gamma)\,dz,$$

an integral of $g(z_1,\dots,t_j,\dots,t_k,\dots,z_n;\Gamma)$ with respect to the remaining $(n-2)$-variables. From this expression we see that

$$\frac{\partial Q(Z,\Gamma)}{\partial\gamma_{jk}}\geqslant 0. \tag{9.5.4}$$

We denote by Γ_X and Γ_Y the covariance matrices of X and Y. We may assume that both are positive definite. If we set $\Gamma(\theta)=\theta\Gamma_X+(1-\theta)\Gamma_Y$, $\theta\in[0,1]$, then (9.5.2), (9.5.3) and (9.5.4) show that the function $q(\theta)=1-Q(Z,\Gamma(\theta))$ is decreasing (check that $q'\leqslant 0$). In particular, $q(0)\geqslant q(1)$, or equivalently $Q(Z,\Gamma_X)\leqslant Q(Z,\Gamma_Y)$, and the result follows. □

Theorem 9.1.7 asserts that under the assumption

$$\|X_i-X_j\|_2\leqslant\|Y_i-Y_j\|_2 \tag{9.5.5}$$

for every $i,j=1,\dots,n$, we have

$$\mathbb{E}\max_{i\leqslant n}X_i\leqslant\mathbb{E}\max_{i\leqslant n}Y_i.$$

(or, as we have proven it, with a factor 2 on the left hand side).

Proof of Theorem 9.1.7. If we set $X_i'=X_i-X_1$ and $Y_i'=Y_i-Y_1$, then the assumption (9.5.5) is still valid and

$$\mathbb{E}\max_{i\leqslant n}X_i'=\mathbb{E}\max_{i\leqslant n}X_i\quad,\quad\mathbb{E}\max_{i\leqslant n}Y_i'=\mathbb{E}\max_{i\leqslant n}Y_i.$$

Thus, we can make the additional assumption that $X_1=Y_1=0$. Then, from (9.5.5) we get

$$\|X_i\|_2\leqslant\|Y_i\|_2 \tag{9.5.6}$$

for all $i=1,\dots,n$. We consider a standard Gaussian random variable g, independent from X_i and Y_i, on Ω, and we set

$$\begin{aligned}C&=\max_{i\leqslant n}\|Y_i\|_2\\ \tilde{X}_i&=X_i+C\cdot g\\ \tilde{Y}_i&=Y_i+\big(C^2-\|Y_i\|_2^2+\|X_i\|_2^2\big)^{1/2}g=Y_i+b_i\cdot g.\end{aligned}$$

Note that $C^2-\|Y_i\|_2^2+\|X_i\|_2^2\geqslant 0$, and hence b_i are well defined. From (9.5.6) it follows that

$$b_i\leqslant C. \tag{9.5.7}$$

By the definition of $\tilde{X}_i$ and $\tilde{Y}_i$ we have

$$\|\tilde{X}_i-\tilde{X}_j\|_2=\|X_i-X_j\|_2$$

and

$$\|\tilde{Y}_i - \tilde{Y}_j\|_2 = \|(Y_i - Y_j) + g(b_i - b_j)\|_2 \tag{9.5.8}$$
$$= \left(\|Y_i - Y_j\|_2^2 + |b_i - b_j|^2\right)^{1/2} \geqslant \|Y_i - Y_j\|_2.$$

This shows that

$$\|\tilde{X}_i - \tilde{X}_j\|_2 \leqslant \|\tilde{Y}_i - \tilde{Y}_j\|_2 \tag{9.5.9}$$

for every $i, j = 1, \ldots, n$. Moreover,

$$\|\tilde{X}_i\|_2^2 = \|X_i\|_2^2 + C^2 = \|Y_i\|_2^2 + b_i^2 = \|\tilde{Y}_i\|_2^2. \tag{9.5.10}$$

From (9.5.9) and (9.5.10) it is clear that the n-tuples $\tilde{X} = (\tilde{X}_1, \ldots, \tilde{X}_n)$ and $\tilde{Y} = (\tilde{Y}_1, \ldots, \tilde{Y}_n)$ satisfy the assumptions of Theorem 9.1.6. Then,

$$\mathbb{E} \max_{i \leqslant n} \tilde{X}_i \leqslant \mathbb{E} \max_{i \leqslant n} \tilde{Y}_i. \tag{9.5.11}$$

We observe that

$$\mathbb{E} \max_{i \leqslant n} \tilde{X}_i = \mathbb{E}\left(\max_{i \leqslant n} X_i + C \cdot g\right) = \mathbb{E} \max_{i \leqslant n} X_i \tag{9.5.12}$$

and, by (9.5.7),

$$\mathbb{E} \max_{i \leqslant n} \tilde{Y}_i \leqslant \mathbb{E} \max_{i \leqslant n} (Y_i + b_i \cdot g^+) \leqslant \mathbb{E} \max_{i \leqslant n} Y_i + C \cdot \mathbb{E} g^+.$$

On the other hand, direct computation shows that $\mathbb{E}(Y_i^+) = \|Y_i\|_2 \, \mathbb{E}(g^+)$, and hence

$$C = \max_{i \leqslant n} \|Y_i\|_2 = \max_{i \leqslant n} \frac{\mathbb{E}(Y_i^+)}{\mathbb{E}(g^+)} \leqslant \frac{\mathbb{E}\left(\max_{i \leqslant n} Y_i^+\right)}{\mathbb{E}(g^+)} = \frac{\mathbb{E}\left(\max_{i \leqslant n} Y_i\right)}{\mathbb{E}(g^+)}$$

because $\max_{i \leqslant n} Y_i^+ = \max_{i \leqslant n} Y_i$ by $Y_1 \equiv 0$. Therefore,

$$C\, \mathbb{E}(g^+) \leqslant \mathbb{E} \max_{i \leqslant n} Y_i,$$

or equivalently,

$$\mathbb{E} \max_{i \leqslant n} \tilde{Y}_i \leqslant 2 \mathbb{E} \max_{i \leqslant n} Y_i. \tag{9.5.13}$$

The result follows from (9.5.11), (9.5.12) and (9.5.13). □

Next, we discuss Gordon's min-max principle. We will deduce Theorem 9.2.1 from a more general comparison principle. The proof of the next theorem is due to Kahane.

THEOREM 9.5.3. *Let X_j and Y_j $(1 \leqslant j \leqslant n)$ be Gaussian random variables with mean zero. Assume that for some $A, B \subseteq \{1, \ldots, n\} \times \{1, \ldots, n\}$ we have*

(i) $\mathbb{E}(X_i X_j) \leqslant \mathbb{E}(Y_i Y_j)$ *for all* $(i, j) \in A$.
(ii) $\mathbb{E}(X_i X_j) \geqslant \mathbb{E}(Y_i Y_j)$ *for all* $(i, j) \in B$.
(iii) $\mathbb{E}(X_i X_j) = \mathbb{E}(Y_i Y_j)$ *for all* $(i, j) \notin A \cup B$.

Let $f : \mathbb{R}^n \to \mathbb{R}$ be a function whose second derivatives in the sense of distributions satisfy $D_{ij}(f) \geqslant 0$ if $(i, j) \in A$ and $D_{ij}(f) \leqslant 0$ if $(i, j) \in B$. Then,

$$\mathbb{E}\, f(X_1, \ldots, X_n) \leqslant \mathbb{E}\, f(Y_1, \ldots, Y_n).$$

Proof. We may assume that $X = (X_1, \dots, X_n)$ and $Y = (Y_1, \dots, Y_n)$ are independent. For every $0 \leqslant t \leqslant 1$ set

$$X(t) = \sqrt{1-t}X + \sqrt{t}Y$$

and

$$\varphi(t) = \mathbb{E}\, f(X(t)).$$

Differentiating we see that

$$\varphi'(t) = \sum_{i=1}^{n} \mathbb{E}\,[(D_i f)(X(t))X_i'(t)].$$

We fix t and i, and we show that the expectation in the right hand side is non-negative. We have

$$\mathbb{E}\,(X_j(t)X_i'(t)) = \frac{1}{2}(Y_jY_i - X_jX_i),$$

and using the assumptions of the theorem we see that

$$X_j(t) = \alpha_j X_i'(t) + W_j,$$

where W_j is orthogonal to $X_i'(t)$, and $\alpha_j \geqslant 0$ if $(i,j) \in A$, $\alpha_j \leqslant 0$ if $(i,j) \in B$, and $\alpha_j = 0$ if $(i,j) \notin A \cup B$. We view $\mathbb{E}\,[(D_i f)(X(t))X_i'(t)]$ as a function of the α_j's for $(i,j) \in A \cup B$. This is an increasing function of α_j when $(i,j) \in A$, and a decreasing function of α_j when $(i,j) \in B$. Since

$$\mathbb{E}\,[(D_i f)(W)X_i'(t)] = \mathbb{E}\,((D_i f)(W))\,\mathbb{E}\,(X_i'(t)) = 0,$$

we finally get

$$\sum_{i=1}^{n} \mathbb{E}\,[(D_i f)(X(t))X_i'(t)] \geqslant 0.$$

Then, $\varphi(0) \leqslant \varphi(1)$, which is precisely the assertion of the theorem. □

Proof of Theorem 9.2.1. We set $N = nm$ and for every $1 \leqslant s \leqslant N$ we denote by $i(s)$ and $j(s)$ the unique $1 \leqslant i(s) \leqslant n$ and $1 \leqslant j(s) \leqslant m$ for which $s = m(i(s) - 1) + j(s)$. Then, for each $1 \leqslant s \leqslant N$ we set $X_s = X_{i(s),j(s)}$ and $Y_s = Y_{i(s),j(s)}$. We define $A, B \subseteq \{1, \dots, N\} \times \{1, \dots, N\}$ by

$$A = \{(s,t) : i(s) = i(t)\}$$
$$B = \{(s,t) : i(s) \neq i(t)\},$$

and observe that the hypotheses of Theorem 9.5.3 are satisfied. Next, given $t_{ij} \in \mathbb{R}$, we consider the set

$$C = \bigcup_{i=1}^{n} \bigcap_{\{1 \leqslant s \leqslant N : i(s) = i\}} \{x \in \mathbb{R}^N : x_s > t_{i,j(s)}\}$$

and apply Theorem 9.5.3 to the function $f = \mathbf{1}_C$ to obtain the result. □

REMARK 9.5.4. One can also give a proof of Slepian's lemma by applying Theorem 9.5.3 with $A = \{(i,j) : i \neq j\}$, $B = \emptyset$ and the function

$$f(x_1, \dots, x_n) = \prod_{i=1}^{n} \mathbf{1}_{\{-\infty, t_i\}}(x_i).$$

to obtain (9.2.1). Then, integration by parts shows that

$$\mathbb{E}\,\max_{i \leqslant n} X_i \leqslant \mathbb{E}\,\max_{i \leqslant n} Y_i.$$

9.6. Notes and remarks

Gaussian processes: Sudakov and Dudley inequalities

Theorem 9.1.4 combines the bounds of Sudakov and Dudley. Dudley proved the upper bound in [**176**] and conjectured the lower bound that was later proved by Sudakov in [**574**]. Theorem 9.1.6 was proved by Slepian in [**564**] (see also Sudakov [**574**], Badrikian and Chevet [**36**] and Fernique [**199**]).

The exact constants in Proposition 9.1.5, namely the exact asymptotics for C_N (which satisfies, by the proposition, $c_1 \leqslant C_N \leqslant c_2$) such that

$$\mathbb{E} \max_{i \leqslant N} g_i = C_N \sqrt{\log N},$$

were provided by Cramér in his book [**156**].

For the exact conclusion in Slepian's lemma one needs the assumption $\mathbb{E}(X_i^2) = \mathbb{E}(Y_i^2)$, which is not assumed in Fernique's result. A result which connects the two was given by Pinelis, who wrote many papers about the following question: Assume ξ and η are two random variables which satisfy that for a sufficiently large family of functions (say, convex) $F : [0, \infty) \to [0, \infty)$ we have $\mathbb{E}(F(\xi)) \leqslant \mathbb{E}(F(\eta))$ (comparison of "generalized moments"). Then, can we show that $\mathbb{P}(\xi > t) \leq C\mathbb{P}(\eta > t)$ for all $t > 0$ (comparison of tails up to some constant C)?

First he addressed this problem in [**500**], where he considered the class of even functions with convex second derivative, and showed that the inequality for moments implies comparison of tails with constant $C = 2e^3/9$ (which is moreover best possible). He used this result to prove Eaton's conjecture: If $X = a_1\varepsilon_1 + \cdots + a_n\varepsilon_n$ with $a_1^2 + \cdots + a_n^2 = 1$ is a normalized linear combination of independent Rademacher variables ε_k then we have $\mathbb{P}(X > t) \leq C\mathbb{P}(Z > t) \sim Ce^{-t^2}/t$, where Z is a standard Gaussian variable. That is, one recovers the missing term $1/t$ which is usually "missed" when we only compare the Laplace transforms:

$$Ee^{\lambda X} \leq Ee^{\lambda Z} = e^{\lambda^2/2}.$$

A different proof of Eaton's conjecture was proposed by Bobkov, Götze and Houdré in [**86**]. For convex functions Pinelis showed that if for every convex function $F : [0, \infty) \to [0, \infty)$ we have $\mathbb{E}(F(\xi)) \leqslant \mathbb{E}(F(\eta))$, Then $\mathbb{P}(\xi > t) \leqslant \frac{e^2}{2}\mathbb{P}(\eta > t)$ for every t.

Gordon's min-max principle

Gordon removed the logarithmic in ε term from the estimate for $k_X(\varepsilon)$ in the Dvoretzky-Milman theorem. Theorem 9.2.6 was initially proved in [**255**], and a second proof was given in [**257**] which works equally well in the not necessarily symmetric case. For a different proof of the same fact see Schechtman [**537**].

The main tool in Gordon's argument is his min-max principle for Gaussian processes extending Slepian's lemma (see Theorem 9.2.1 and Proposition 9.2.2). A simple proof of these results was given by Kahane [**323**]; in Section 9.5 we present Kahane's result (Theorem 9.5.3) and deduce from this both Slepian's lemma and Gordon's min-max theorem.

Theorem 9.3.6, Gordon's sharp version of the M^*-estimate, comes from [**258**]. It is based on Theorem 9.3.1 which is a variant of the results in [**255**]; generalizations in various directions can be found in Gordon [**256**].

Chevet inequality

The proof of Theorem 9.4.1 which we provided comes from the book of Tomczak-Jaegermann [**593**] and follows an argument of Gordon from [**255**]. As we will see in Part II it is a very useful tool for questions regarding factorization of operators and Banach-Mazur distance estimates. Chevet's inequality was brought to the attention of the experts in our field by G. Pisier, and its first use in the theory was in two papers, one by Benyamini and Gordon in [**64**] and the second by Davis, Milman and Tomczak-Jaegermann in [**167**]. For a more advanced application of this method see the paper of Bourgain and Milman [**109**].

Marcus-Pisier inequality

Theorem 9.4.2 is due to Marcus and Pisier [**414**]. The right hand side inequality is highly non-trivial, while the left hand side is relatively simple and was observed by Davis and Garling (see Szarek [**582**], where the proof of both sides is reproduced, and Wagner [**603**]).

CHAPTER 10

Volume distribution in convex bodies

In this chapter we discuss new discoveries and a number of major open problems on the distribution of volume in high-dimensional convex bodies. The picture which arises is novel and non-trivial. A natural framework for this study is the *isotropic position* of a convex body that was briefly introduced in Chapter 2. We recall (see also Definition 10.1.1 below) that a convex body $K \subset \mathbb{R}^n$ is called isotropic if $\mathrm{Vol}_n(K) = 1$, its barycenter is at the origin and its inertia matrix is a multiple of the identity, that is, there exists a constant $L_K > 0$ such that

$$\int_K \langle x, \theta \rangle^2 dx = L_K^2$$

for every $\theta \in S^{n-1}$. The number L_K is then called the isotropic constant of K.

We will see in Section 10.1 that the affine class of any convex body K contains a unique, up to orthogonal transformations, isotropic convex body; this is the isotropic position of K.

The central theme in this chapter is the *hyperplane conjecture* (or *slicing problem*). The question is if there exists an absolute constant $c > 0$ such that $\max_{\theta \in S^{n-1}} \mathrm{Vol}_{n-1}(K \cap \theta^{\perp}) \geqslant c$ for every convex body K of volume 1 in $\mathbb{R}^n$ that has barycenter at the origin (formulated as Conjecture 10.1.17 below). We will show that an affirmative answer to this question is equivalent to the following statement (formulated as Conjecture 10.1.7 below):

There exists an absolute constant $C > 0$ such that

$$L_n := \max\{L_K : K \text{ is an isotropic convex body in } \mathbb{R}^n\} \leqslant C.$$

The notion of the isotropic constant and the conjecture can be reintroduced in the more general setting of finite log-concave measures, in a way that is equivalent to the above when we consider uniform measures on convex bodies. We say that a finite log-concave measure μ on $\mathbb{R}^n$ is isotropic if μ is a probability measure, its barycenter is at the origin and the covariance matrix $\mathrm{Cov}(\mu)$ of μ is the identity matrix. We give the precise definitions in Section 10.2, where we again see, as in the case of bodies, that any finite log-concave measure has an affine image that is isotropic. Furthermore, we describe K. Ball's theorem that, in fact, for some absolute $c > 1$,

$$L_n \leqslant \sup\{L_\mu : \mu \text{ is isotropic on } \mathbb{R}^n\} \leqslant cL_n.$$

Next, we present the best known upper bounds for L_n. Around 1985-6 (published in 1990), Bourgain obtained the upper bound $L_n \leqslant c\sqrt[4]{n}\log n$ and, in 2006, this estimate was improved by Klartag to $L_n \leqslant c\sqrt[4]{n}$. Bourgain 's argument is presented in Section 10.3 and Klartag's approach via the logarithmic Laplace transform is explained in Section 10.5. Actually, Klartag obtained a solution to an isomorphic version of the hyperplane conjecture, the "isomoprphic slicing problem", by showing that, for every convex body K in $\mathbb{R}^n$ and any $\varepsilon \in (0,1)$, one can find a centered convex body $T \subset \mathbb{R}^n$ and a point $x \in \mathbb{R}^n$ such that $(1+\varepsilon)^{-1}T \subseteq K + x \subseteq (1+\varepsilon)T$ and $L_T \leqslant C/\sqrt{\varepsilon}$ for some absolute constant $C > 0$. We also present an application of the proof of this theorem, which gives a new proof of the reverse Santaló inequality.

An additional essential ingredient in Klartag's proof of the bound $L_n \leqslant c\sqrt[4]{n}$, which is a beautiful and important result on its own right, is the following very useful deviation inequality of Paouris: if μ is an isotropic log-concave probability measure on $\mathbb{R}^n$ then

$$\mu(\{x \in \mathbb{R}^n : |x| \geqslant ct\sqrt{n}\}) \leqslant \exp\left(-t\sqrt{n}\right)$$

for every $t \geqslant 1$, where $c > 0$ is an absolute constant. The proof is presented in Section 10.4 along with the basic theory of the L_q-centroid bodies of an isotropic log-concave measure.

We have included some of the more technical computations in a short appendix to this chapter, which is Section 10.6.

Let us briefly discuss a number of other challenging conjectures and important results about isotropic log-concave measures. The first one is the *central limit problem*, that asks if the 1-dimensional marginals of high-dimensional isotropic log-concave measures μ are approximately Gaussian with high probability. It is generally known through results of Sudakov that, if μ is an isotropic probability measure on $\mathbb{R}^n$ that satisfies the following *thin shell condition*,

$$\mu\left(\left|\frac{|x|}{\sqrt{n}} - 1\right| \geqslant \varepsilon\right) \leqslant \varepsilon$$

for some $\varepsilon \in (0,1)$, then, for all directions θ in a subset A of S^{n-1} with $\sigma(A) \geqslant 1 - \exp(-c_1\sqrt{n})$, one has

$$|\mu(\{x : \langle x, \theta\rangle \leqslant t\}) - \Phi(t)| \leqslant c_2(\varepsilon + n^{-\alpha}) \qquad \text{for all } t \in \mathbb{R},$$

where $\Phi(t)$ is the standard Gaussian distribution function and $c_1, c_2, \alpha > 0$ are absolute constants. Thus, the central limit problem is reduced to the question whether every isotropic log-concave measure μ in $\mathbb{R}^n$ satisfies such a thin shell condition with $\varepsilon = \varepsilon_n$ tending to 0 as n tends to infinity. An affirmative answer to the problem was given by Klartag who obtained power-type estimates verifying the thin-shell condition; he showed that if μ is an isotropic log-concave measure on $\mathbb{R}^n$ then

$$\mathbb{E}\Big(\frac{|x|^2}{n} - 1\Big)^2 \leqslant \frac{C}{n^\alpha}$$

with some $\alpha \simeq 1/5$, and, as a consequence, that the density f_θ of $x \mapsto \langle x, \theta\rangle$ with respect to μ satisfies

$$\int_{-\infty}^{\infty} |f_\theta(t) - \gamma(t)|\, dt \leqslant \frac{1}{n^\kappa}$$

and

$$\sup_{|t| \leqslant n^\kappa} \left|\frac{f_\theta(t)}{\gamma(t)} - 1\right| \leqslant \frac{1}{n^\kappa},$$

for all θ in a subset A of S^{n-1} with measure $\sigma(A) \geqslant 1 - c_1 \exp(-c_2\sqrt{n})$, where γ is the density of a standard Gaussian random variable, and c_1, c_2, κ are absolute constants. Although some sharper estimates were obtained afterwards, the following quantitative conjecture remains open:

There exists an absolute constant $C > 0$ such that, for any $n \geqslant 1$ and any isotropic log-concave measure μ on $\mathbb{R}^n$, one has

$$\sigma_\mu^2 := \int_{\mathbb{R}^n} \big(|x| - \sqrt{n}\big)^2 d\mu(x) \leqslant C^2.$$

A third conjecture concerns the *Cheeger constant* Is_μ of an isotropic log-concave measure μ: this is defined as the best constant $\kappa \geqslant 0$ such that

$$\mu^+(A) \geqslant \kappa \min\{\mu(A), 1 - \mu(A)\}$$

for every Borel subset A of $\mathbb{R}^n$, where $\mu^+(A)$ is the Minkowski content of A. The *Kannan-Lovász-Simonovits conjecture* asks if there exists an absolute constant $c > 0$ such that

$$\mathrm{Is}_n := \inf\{\mathrm{Is}_\mu : \mu \text{ is isotropic log-concave measure on } \mathbb{R}^n\} \geqslant c.$$

Another way to formulate this conjecture is to ask for a *Poincaré inequality* to be satisfied by every isotropic log-concave measure μ on $\mathbb{R}^n$ with a constant $c > 0$ that is independent of the measure or the dimension n; more precisely, the KLS-conjecture is equivalent to asking if there exists an absolute constant $c > 0$ such that

$$c \int_{\mathbb{R}^n} \varphi^2 d\mu \leqslant \int_{\mathbb{R}^n} |\nabla \varphi|^2 d\mu$$

for every isotropic log-concave measure μ on $\mathbb{R}^n$ and for every smooth function φ with $\int_{\mathbb{R}^n} \varphi \, d\mu = 0$. We will come back to these questions in Part II and we will see that precise quantitative relations exist between any two of the aforementioned problems, the hyperplane conjecture, the thin-shell conjecture and the KLS-conjecture. For a detailed discussion of this area we refer to the book of Brazitikos, Giannopoulos, Valettas and Vritsiou [**116**].

10.1. Isotropic position

DEFINITION 10.1.1. A convex body K in $\mathbb{R}^n$ is called *isotropic* if it has volume $\mathrm{Vol}_n(K) = 1$, it is centered (i.e. its barycenter is at the origin), and there is a constant $\alpha > 0$ such that

$$\int_K \langle x, y\rangle^2 dx = \alpha^2 |y|^2 \tag{10.1.1}$$

for all $y \in \mathbb{R}^n$. Note that if K satisfies the isotropic condition (10.1.1) then

$$\int_K |x|^2 dx = \sum_{i=1}^n \int_K \langle x, e_i\rangle^2 dx = n\alpha^2.$$

Also, it is easily checked that if K is an isotropic convex body in $\mathbb{R}^n$ then $U(K)$ is also isotropic for every $U \in O(n)$.

REMARK 10.1.2. Following the reasoning of Lemma 2.1.13 one can easily check that the *isotropic condition* (10.1.1) is equivalent to each one of the following statements:

(1) For every $i,j=1,\dots,n$,

$$\int_K x_ix_j dx = \alpha^2\delta_{ij}, \tag{10.1.2}$$

where $x_j=\langle x,e_j\rangle$ are the coordinates of x with respect to some orthonormal basis $\{e_1,\dots,e_n\}$ of $\mathbb{R}^n$.

(2) For every $T\in L(\mathbb{R}^n)$,

$$\int_K \langle x,Tx\rangle dx = \alpha^2(\mathrm{tr}T). \tag{10.1.3}$$

10.1.1. Existence and uniqueness

The next proposition shows that every centered convex body has a linear image which satisfies the isotropic condition.

PROPOSITION 10.1.3. *Let K be a centered convex body in $\mathbb{R}^n$. There exists $T\in GL_n$ such that $T(K)$ is isotropic.*

Proof. The operator $M\in L(\mathbb{R}^n)$ defined by $M(y)=\int_K\langle x,y\rangle x dx$ is symmetric and positive definite; therefore, it has a symmetric and positive definite square root S. Consider the linear image $\tilde{K}=S^{-1}(K)$ of K. Then, for every $y\in\mathbb{R}^n$ we have

$$\begin{aligned}\int_{\tilde{K}}\langle x,y\rangle^2dx &= |\mathrm{det}S|^{-1}\int_K\langle S^{-1}x,y\rangle^2dx\\ &= |\mathrm{det}S|^{-1}\int_K\langle x,S^{-1}y\rangle^2dx\\ &= |\mathrm{det}S|^{-1}\Big\langle\int_K\langle x,S^{-1}y\rangle xdx,S^{-1}y\Big\rangle\\ &= |\mathrm{det}S|^{-1}\langle MS^{-1}y,S^{-1}y\rangle = |\mathrm{det}S|^{-1}|y|^2.\end{aligned} \tag{10.1.4}$$

Normalizing the volume of $\tilde{K}$ we get the result. □

Proposition 10.1.3 shows that every centered convex body K in $\mathbb{R}^n$ has a *position* $\tilde{K}$ which is isotropic. We say that $\tilde{K}$ is an *isotropic position* of K. The next theorem shows that the isotropic position of a convex body is uniquely determined up to orthogonal transformations, and arises as a solution of a minimization problem.

THEOREM 10.1.4. *Let K be a centered convex body of volume 1 in $\mathbb{R}^n$. Define*

$$B(K)=\inf\Big\{\int_{TK}|x|^2dx : T\in SL_n\Big\}. \tag{10.1.5}$$

Then, a position K_1 of K, of volume 1, is isotropic if and only if

$$\int_{K_1}|x|^2dx = B(K). \tag{10.1.6}$$

Furthermore, if K_1 and K_2 are isotropic positions of K then there exists $U\in O(n)$ such that $K_2=U(K_1)$.

Proof. Fix an isotropic position K_1 of K. Remark 10.1.2 shows that there exists $\alpha>0$ such that

$$\int_{K_1}\langle x,Tx\rangle dx=\alpha^2(\mathrm{tr}T)$$

for every $T \in L(\mathbb{R}^n)$. Then, for every $T \in SL_n$ we have

$$(10.1.7) \qquad \int_{TK_1} |x|^2 dx = \int_{K_1} |Tx|^2 dx = \int_{K_1} \langle x, T^*Tx\rangle dx$$
$$= \alpha^2 \mathrm{tr}(T^*T) \geqslant n\alpha^2 = \int_{K_1} |x|^2 dx,$$

where we have used the arithmetic-geometric means inequality in the form

$$\mathrm{tr}(T^*T) \geqslant n[\det(T^*T)]^{1/n}.$$

This shows that K_1 satisfies (10.1.6). In particular, the infimum in (10.1.5) is a minimum.

Note also that if we have equality in (10.1.7) then $T^*T = \mathrm{Id}$, and hence $T \in O(n)$. This shows that any other position $\tilde{K}$ of K which satisfies (10.1.6) is an orthogonal image of K_1, therefore it is isotropic.

Finally, if K_2 is some other isotropic position of K then the first part of the proof shows that K_2 satisfies (10.1.6). By the previous step, we must have $K_2 = U(K_1)$ for some $U \in O(n)$. □

REMARK 10.1.5. An alternative way to prove the above fact, that K is isotropic if K is a solution of the aforementioned minimization problem, is by using the simple variational argument of Sections 2.3 and 2.4. Let $T \in L(\mathbb{R}^n)$. For small $\varepsilon > 0$, $\mathrm{Id} + \varepsilon T$ is invertible, and hence $(\mathrm{Id} + \varepsilon T)/[\det(\mathrm{Id} + \varepsilon T)]^{1/n}$ preserves volumes. Consequently,

$$\int_K |x|^2 dx \leqslant \int_K \frac{|x + \varepsilon Tx|^2}{[\det(\mathrm{Id} + \varepsilon T)]^{2/n}} dx.$$

Note that $|x+\varepsilon Tx|^2 = |x|^2 + 2\varepsilon\langle x, Tx\rangle + O(\varepsilon^2)$ and $[\det(\mathrm{Id}+\varepsilon T)]^{2/n} = 1 + 2\varepsilon\frac{\mathrm{tr}(T)}{n} + O(\varepsilon^2)$. Therefore, letting $\varepsilon \to 0^+$, we see that

$$(10.1.8) \qquad \frac{\mathrm{tr}T}{n} \int_K |x|^2 dx \leqslant \int_K \langle x, Tx\rangle dx.$$

Since T was arbitrary, the same inequality holds with $-T$ instead of T, therefore

$$(10.1.9) \qquad \frac{\mathrm{tr}(T)}{n} \int_K |x|^2 dx = \int_K \langle x, Tx\rangle dx$$

for all $T \in L(\mathbb{R}^n)$. This condition implies that K is isotropic.

The preceding discussion shows that the following definition for the isotropic constant of a general convex body makes sense.

DEFINITION 10.1.6. *Let $K \subset \mathbb{R}^n$ be a convex body. Its isotropic constant L_K is defined by*

$$L_K^2 = \frac{1}{n} \min\left\{ \frac{1}{\mathrm{Vol}_n(T\tilde{K})^{1+\frac{2}{n}}} \int_{T\tilde{K}} |x|^2 dx \;\middle|\; T \in GL_n \right\},$$

where $\tilde{K} = K - \mathrm{bar}(K)$ is the centered translate of K.

Note that L_K depends only on the affine class of K. It will be useful to denote throughout this chapter

$$\overline{K} = \frac{1}{\mathrm{Vol}_n(K)^{1/n}} K.$$

Note also that if K is isotropic then for all $\theta \in S^{n-1}$ we have

$$\int_K \langle x, \theta \rangle^2 dx = L_K^2.$$

The constant L_K is called *the isotropic constant* of K. The first main problem that we discuss in this chapter is the so-called *hyperplane conjecture* (see the end of the next subsection for an explanation of the name).

CONJECTURE 10.1.7 (isotropic constant problem). *There exists an absolute constant $C > 0$ such that for any $n \in \mathbb{N}$ and any convex body $K \subset \mathbb{R}^n$ we have*

$$L_K \leqslant C.$$

Equivalently, if K is an isotropic convex body in $\mathbb{R}^n$, then

$$\int_K \langle x, \theta \rangle^2 dx \leqslant C^2$$

for every $\theta \in S^{n-1}$.

While this conjecture remains open, it is interesting to note that no single example of a convex body K with $L_K > 1$ is known. A reverse estimate, bounding L_K from below, is available and quite simple.

PROPOSITION 10.1.8. *For every isotropic convex body K in $\mathbb{R}^n$*

$$L_K \geqslant L_{B_2^n} \geqslant c,$$

where $c > 0$ is an absolute constant.

Proof. If $r_n = \kappa_n^{-1/n}$, then $\mathrm{Vol}_n(r_n B_2^n) = 1$ and $r_n B_2^n$ is isotropic. Let K be an isotropic convex body. Observe that $|x| > r_n$ on $K \setminus r_n B_2^n$ and $|x| \leqslant r_n$ on $r_n B_2^n \setminus K$. It follows that

$$\begin{aligned} nL_K^2 &= \int_K |x|^2 dx = \int_{K \cap r_n B_2^n} |x|^2 dx + \int_{K \setminus r_n B_2^n} |x|^2 dx \\ &\geqslant \int_{K \cap r_n B_2^n} |x|^2 dx + \int_{r_n B_2^n \setminus K} |x|^2 dx = \int_{r_n B_2^n} |x|^2 dx = nL_{B_2^n}^2. \end{aligned}$$

A simple computation shows that

$$L_{B_2^n}^2 = \frac{1}{n} \int_{r_n B_2^n} |x|^2 dx = \frac{1}{n} \frac{n\kappa_n}{n+2} r_n^{n+2} = \frac{\kappa_n^{-2/n}}{n+2} \geqslant c^2,$$

where $c > 0$ is an absolute constant, therefore $L_K \geqslant L_{B_2^n} \geqslant c$. $\square$

In the symmetric case, a first upper bound for the isotropic constant follows from John's Theorem.

PROPOSITION 10.1.9. *Let K be a centrally symmetric convex body in $\mathbb{R}^n$ Then $L_K \leqslant c d_{BM}(K, B_2^n)$. In particular, by John's theorem $L_K \leqslant c\sqrt{n}$.*

Proof. Since L_K is an affine invariant, we may assume that K satisfies $\mathrm{Vol}_n(K) = 1$ and that $rB_2^n \subseteq K$ is the "distance ellipsoid" for K, that is, $K \subset drB_2^n$ for $d = d_{BM}(K, B_2^n)$. Since $rB_2^n \subset K$ and $\mathrm{Vol}_n(K) = 1$ we have $r \leqslant c\sqrt{n}$. Theorem 10.1.4 shows that

$$nL_K^2 \leqslant \int_K |x|^2 dx \leqslant (rd)^2.$$

It follows that

$$L_K \leqslant \frac{rd}{\sqrt{n}} \leqslant cd.$$

□

10.1.2. Moments of inertia and maximal hyperplane sections

Let K be a centered convex body of volume 1 in $\mathbb{R}^n$. Let $M(K) = (m_{ij})_{i,j=1}^n$ be the matrix of inertia of K, which is defined by $m_{ij} = \int_K x_i x_j dx$. As we saw in the proof of Proposition 10.1.3, $M(K)$ has a symmetric and positive definite square root S. Consider the ellipsoid $\mathcal{E}_B(K) := S^{-1}(B_2^n)$; then

$$\|y\|_{\mathcal{E}_B(K)}^2 = |Sy|^2 = \langle Sy, Sy\rangle = \langle My, y\rangle = \int_K \langle x, y\rangle^2 dx.$$

This is called the *Binet ellipsoid* of K. Closely related is the Legendre ellipsoid of a convex body, which is the unique ellipsoid centered at the centre of mass of K and having the same moments of inertia (with respect to lines passing through this center of mass) as K. It is straightforward to check that this is precisely (a multiple of) the polar of the Binet ellipsoid.

REMARK 10.1.10. Observe that K is in isotropic position if and only if $\mathcal{E}_B(K) = L_K^{-1}B_2^n$.

The next proposition shows that the volume of $\mathcal{E}_B(K)$ is invariant under the action of SL_n.

PROPOSITION 10.1.11. *Let K be a centered convex body of volume 1 in $\mathbb{R}^n$. Then,*

$$\mathrm{Vol}_n(\mathcal{E}_B(K)) = \kappa_n L_K^{-n}.$$

Proof. If K is an isotropic convex body in $\mathbb{R}^n$ then $\mathcal{E}_B(K) = L_K^{-1}B_2^n$, and hence $\mathrm{Vol}_n(\mathcal{E}_B(K)) = \kappa_n L_K^{-n}$. It is easily checked that if $T \in SL_n$ then $M_{T(K)} = TM_KT^*$, and hence $|\det M_K| = |\det M_{T(K)}|$; furthermore, by definition we have $\mathcal{E}_B(T(K)) = S^{-1}(B_2^n)$ where $S^2 = M_{T(K)}$. It follows that

$$\mathrm{Vol}_n(\mathcal{E}_B(TK)) = \kappa_n|\det M_{T(K)}|^{-1/2} = \kappa_n|\det M_K|^{-1/2} = \mathrm{Vol}_n(\mathcal{E}_B(K))$$

for every $T \in SL_n$. □

COROLLARY 10.1.12. *Let K be a centered convex body of volume 1 in $\mathbb{R}^n$. There exists $\theta \in S^{n-1}$ such that*

$$\int_K \langle x, \theta\rangle^2 dx \leqslant L_K^2.$$

Proof. Note that by integration in polar coordinates

$$L_K^{-n} = \frac{\mathrm{Vol}_n(\mathcal{E}_B(K))}{\kappa_n} = \int_{S^{n-1}} \|\theta\|_{\mathcal{E}_B(K)}^{-n} d\sigma(\theta).$$

It follows that $\min_{\theta\in S^{n-1}} \|\theta\|_{\mathcal{E}_B(K)} \leqslant L_K$. □

In the rest of the subsection we discuss the relation of the moments of inertia of a centered convex body with the volume of its hyperplane sections that pass through the origin. In the isotropic case, the conclusion is the following (which already gives a hint about the name "hyperplane conjecture").

THEOREM 10.1.13. *Let K be an isotropic convex body in $\mathbb{R}^n$. For every $\theta \in S^{n-1}$ we have*

$$\frac{c_1}{L_K} \leqslant \mathrm{Vol}_{n-1}(K \cap \theta^{\perp}) \leqslant \frac{c_2}{L_K}, \tag{10.1.10}$$

where $c_1, c_2 > 0$ are absolute constants.

For the proof of the theorem we need the next proposition which explains that hyperplane sections through the center of mass are, up to an absolute constant, maximal. We postpone its proof, which is simple but technical, to Section 10.6 which contains several technical results. Note, however, that in the centrally symmetric case there is nothing to prove since the inequality holds with 1 instead of e thanks to Brunn's concavity principle.

PROPOSITION 10.1.14. *Let K be a centered convex body of volume 1 in $\mathbb{R}^n$. Let $\theta \in S^{n-1}$ and consider the function $f(t) = f_{K,\theta}(t) = \mathrm{Vol}_{n-1}(K \cap \{x : \langle x, \theta\rangle = t\})$, $t \in \mathbb{R}$. Then,*

$$\|f\|_\infty \leqslant ef(0) = e\mathrm{Vol}_{n-1}(K \cap \theta^{\perp}).$$

Proof of Theorem 10.1.13. Let $f := f_{K,\theta}$. To prove the left hand side of (10.1.10), we set $\beta = \int_0^{+\infty} f(t)dt$ and define

$$g(t) = \|f\|_\infty \chi_{[0,\beta/\|f\|_\infty]}(t).$$

Since $g \geqslant f$ on the support of g, we have

$$\int_0^s f(t)dt \leqslant \int_0^s g(t)dt$$

for every $0 \leqslant s \leqslant \beta/\|f\|_\infty$. The integrals of f and g on $[0, +\infty)$ are both equal to β. So,

$$\int_s^\infty g(t)dt \leqslant \int_s^\infty f(t)dt$$

for every $s \geqslant 0$. It follows that

$$\begin{aligned}
\int_0^\infty t^2 f(t)dt &= \int_0^\infty \int_0^t 2sf(t)dsdt = \int_0^\infty 2s\left(\int_s^\infty f(t)dt\right)ds \\
&\geqslant \int_0^\infty 2s\left(\int_s^\infty g(t)dt\right)ds = \int_0^\infty t^2 g(t)dt \\
&= \int_0^{\beta/\|f\|_\infty} t^2\|f\|_\infty dt = \frac{\beta^3}{3\|f\|_\infty^2}.
\end{aligned}$$

In the same way, setting $\alpha = \int_{-\infty}^0 f(t)dt$, we see that

$$\int_{-\infty}^0 t^2 f(t)dt \geqslant \frac{\alpha^3}{3\|f\|_\infty^2}.$$

It follows that

$$\int_K \langle x, \theta\rangle^2 dx = \int_{-\infty}^\infty t^2 f(t)\, dt \geqslant \frac{\beta^3 + \alpha^3}{3\|f\|_\infty^2},$$

and since $\alpha + \beta = \mathrm{Vol}_n(K) = 1$, we get

$$\left(\int_K \langle x, \theta\rangle^2 dx\right)^{1/2} \geqslant \frac{1}{2\sqrt{3}}\frac{1}{\|f\|_\infty} \geqslant \frac{1}{2\sqrt{3}}\frac{1}{ef(0)},$$

with the last inequality following from Proposition 10.1.14.

To prove the right hand side inequality of (10.1.10), we distinguish two cases. Assume first that there exists $s > 0$ such that $f(s) = f(0)/2$. Then,

$$\beta = \int_0^\infty f(t)dt \geqslant \int_0^s f(t)dt \geqslant sf(s) = sf(0)/2,$$

because, since f is log-concave, we easily see that $f(t) \geqslant f(0)^{1-t/s} f(s)^{t/s} \geqslant f(s)$ on $[0, s]$. On the other hand, if $t > s$, then

$$f(s) \geqslant [f(0)]^{1-\frac{s}{t}} [f(t)]^{\frac{s}{t}},$$

which implies that $f(t) \leqslant f(0)2^{-t/s}$. We now write

$$\begin{aligned}
\int_0^\infty t^2 f(t)dt &= \int_0^s t^2 f(t)dt + \int_s^\infty t^2 f(t)dt \\
&\leqslant \|f\|_\infty \int_0^s t^2 dt + \int_s^\infty t^2 f(0)2^{-t/s} dt \\
&\leqslant f(0)\left(e\frac{s^3}{3} + s^3 \int_1^\infty u^2 2^{-u} du\right) \\
&\leqslant c_0 f(0) s^3 \\
&\leqslant c_0 f(0) \left(\frac{2\beta}{f(0)}\right)^3 \\
&\leqslant c_1/[f(0)]^2,
\end{aligned}$$

taking into account the fact that $\beta < 1$.

Now, assume that, for every $s > 0$ on the support of f, we have $f(s) > f(0)/2$. Then, the role of s is played by $s_0 = \sup\{s > 0 : f(s) > 0\}$. We have $\beta \geqslant f(0)s_0/2$ and

$$\int_0^\infty t^2 f(t)dt = \int_0^{s_0} t^2 f(t)dt \leqslant ef(0)s_0^3/3;$$

combining the two inequalities we get the same estimate as before, without using the fact that $\log f$ is concave.

We work in the same way on $(-\infty, 0]$. It follows that

$$\int_K \langle x, \theta\rangle^2 dx = \int_0^\infty t^2 f(t)dt + \int_{-\infty}^0 t^2 f(t)dt \leqslant \left(\frac{c_2}{f(0)}\right)^2,$$

which completes the proof. □

REMARK 10.1.15. It is useful to note that the inradius $r(K)$ and the outer-radius $R(K)$ of an isotropic convex body K in $\mathbb{R}^n$ satisfy the bounds

$$c_1 L_K \leqslant r(K) \leqslant R(K) \leqslant c_2 n L_K, \tag{10.1.11}$$

where $c_1, c_2 > 0$ are absolute constants. The following simple argument proves the right hand side inequality: given $\theta \in S^{n-1}$, one knows that

$$\mathrm{Vol}_{n-1}(K \cap \theta^\perp) \simeq \frac{1}{L_K}. \tag{10.1.12}$$

Let $x_\theta \in K$ such that $\langle x_\theta, \theta\rangle = h_K(\theta)$ and consider the cone

$$C(\theta) = \mathrm{conv}(K \cap \theta^\perp, x_\theta).$$

Then $C(\theta) \subseteq K$, and hence

$$1 = \mathrm{Vol}_n(K) \geqslant \mathrm{Vol}_n(C(\theta)) = \frac{\mathrm{Vol}_{n-1}(K \cap \theta^\perp) \cdot h_K(\theta)}{n}.$$

It follows that $h_K(\theta) \leqslant c_2 n L_K$.

For the left hand side inequality we make use of Grünbaum's lemma (which is Proposition 1.5.16 in Chapter 1) which implies that $e^{-1} \leqslant \|f_{K,\theta}\|_\infty h_K(\theta)$ and using Proposition 10.1.14 we get that

$$e^{-1} \leqslant e\mathrm{Vol}_{n-1}(K \cap \theta^\perp) h_K(\theta).$$

Now, using (10.1.12) we see that $h_K(\theta) \geqslant c_1 L_K$, and since θ was arbitrary, this gives $r(K) \geqslant c_1 L_K$. In the symmetric case one actually has the bound $r(K) \geqslant L_K$, because $|\langle x, \theta\rangle| \leqslant h_K(\theta)$, and hence

$$h_K(\theta) \geqslant \left(\int_K \langle x, \theta\rangle^2 dx\right)^{1/2} = L_K$$

for every $\theta \in S^{n-1}$.

Remark 10.1.16. We can now easily prove the bound $L_K \leqslant c\sqrt{n}$ for every convex body K in $\mathbb{R}^n$, and not merely the symmetric ones. Indeed, assume that K is in isotropic position and write $r(K)B_2^n \subseteq K$ and $r(K) \geqslant c_1 L_K$, we get

$$\kappa_n (c_1 L_K)^n \leqslant \kappa_n (r(K))^n = \mathrm{Vol}_n(r(K)B_2^n) \leqslant \mathrm{Vol}_n(K) = 1.$$

It follows that $L_K \leqslant c_1^{-1} \kappa_n^{-1/n} \leqslant c_1' \sqrt{n}$.

Moreover, Theorem 10.1.13 reveals a close connection between the isotropic constant problem and the *slicing problem* or *hyperplane conjecture*:

Conjecture 10.1.17 (The slicing problem). There exists an absolute constant $c > 0$ with the following property: if K is a convex body in $\mathbb{R}^n$ with volume 1 and barycenter at the origin, there exists $\theta \in S^{n-1}$ such that

$$\mathrm{Vol}_{n-1}(K \cap \theta^\perp) \geqslant c. \tag{10.1.13}$$

It is not hard to see that the two conjectures are equivalent. Assume that the slicing problem has an affirmative answer. If K is isotropic, Theorem 10.1.13 shows that *all* sections $K \cap \theta^\perp$ have volume bounded by c_2/L_K. Since (10.1.13) must be true for at least one $\theta \in S^{n-1}$, we get $L_K \leqslant c_2/c$.

Conversely, if there exists an absolute bound C for the isotropic constant, then the slicing conjecture follows. An easy way to see this is through the Binet ellipsoid of inertia. Assume that K is a centered convex body of volume 1 in $\mathbb{R}^n$. According to Corollary 10.1.12, there exists $\theta \in S^{n-1}$ such that

$$\int_K \langle x, \theta\rangle^2 dx \leqslant L_K^2 \leqslant C^2.$$

Now, the proof of Theorem 10.1.13 shows that

$$\mathrm{Vol}_{n-1}(K \cap \theta^\perp) \geqslant c := \frac{1}{2\sqrt{3}eC}.$$

10.2. Isotropic log-concave measures

DEFINITION 10.2.1. Let μ be a Borel probability measure on $\mathbb{R}^n$ which is absolutely continuous with respect to Lebesgue measure. We shall say it is *isotropic* if it is centered and satisfies the isotropic condition

$$\int_{\mathbb{R}^n} \langle x, \theta \rangle^2 \, d\mu(x) = 1$$

for all $\theta \in S^{n-1}$. Similarly, we shall say that a centered log-concave function $f : \mathbb{R}^n \to [0, \infty)$ is isotropic if $\int f = 1$ and the measure $d\mu = f(x)dx$ is isotropic.

As in Remark 10.1.2, we easily check that a centered measure μ as above is isotropic if and only if for any $T \in L(\mathbb{R}^n)$ one has

$$\int_{\mathbb{R}^n} \langle x, Tx \rangle \, d\mu(x) = \mathrm{tr}(T),$$

or equivalently if $\int_{\mathbb{R}^n} x_i x_j \, d\mu(x) = \delta_{ij}$ for all $i, j = 1, \ldots, n$.

Note that if μ is isotropic, then

$$\int_{\mathbb{R}^n} |x|^2 \, d\mu(x) = n,$$

and more generally,

$$\int_{\mathbb{R}^n} |Tx|^2 d\mu(x) = \|T\|_{\mathrm{HS}}^2$$

for any $T \in L(\mathbb{R}^n)$.

Following the proof of Proposition 10.1.3, we can check that every non-degenerate absolutely continuous probability measure μ has an isotropic image $\nu = \mu \circ S$, where $S : \mathbb{R}^n \to \mathbb{R}^n$ is an affine map. Similarly, every log-concave $f : \mathbb{R}^n \to [0, \infty)$ with $0 < \int f < \infty$ has an isotropic image: there exist an affine isomorphism $S : \mathbb{R}^n \to \mathbb{R}^n$ and a positive number a such that $af \circ S$ is isotropic.

REMARK 10.2.2. It is useful to compare the definition of an isotropic convex body (Definition 10.1.1) with the definition of an isotropic log-concave measure. Note that a convex body K of volume 1 in $\mathbb{R}^n$ being isotropic implies that the covariance matrix of the measure $\mathbf{1}_K dx$ is $L_K \,\mathrm{Id}$. So, we see that a convex body K of volume 1 is isotropic if and only if the function $f_K := L_K^n \mathbf{1}_{\frac{1}{L_K} K}$ is an isotropic log-concave function.

DEFINITION 10.2.3 (General definition of the isotropic constant). Let f be a log-concave function with finite, positive integral. We define its *inertia* – or *covariance* – matrix $\mathrm{Cov}(f)$ as the matrix with entries

$$[\mathrm{Cov}(f)]_{ij} := \frac{\int_{\mathbb{R}^n} x_i x_j f(x)\, dx}{\int_{\mathbb{R}^n} f(x)\, dx} - \frac{\int_{\mathbb{R}^n} x_i f(x)\, dx}{\int_{\mathbb{R}^n} f(x)\, dx} \frac{\int_{\mathbb{R}^n} x_j f(x)\, dx}{\int_{\mathbb{R}^n} f(x)\, dx}.$$

Note that if f is isotropic then $\mathrm{Cov}(f)$ is the identity matrix. If f is the density of a measure μ we denote this matrix also by $\mathrm{Cov}(\mu)$. The *isotropic constant* is given by

$$L_f := \left(\frac{\sup_{x \in \mathbb{R}^n} f(x)}{\int_{\mathbb{R}^n} f(x) dx} \right)^{\frac{1}{n}} [\det \mathrm{Cov}(f)]^{\frac{1}{2n}}. \tag{10.2.1}$$

(and given a log-concave measure μ with density f_μ we let $L_\mu := L_{f_\mu}$).

Remark 10.2.4. It is worthwhile to note that (say, for a probability measure) $L_f \simeq \exp((\int f\ln f)/n)[\det \mathrm{Cov}(f)]^{\frac{1}{2n}}$. With this point of view, the slicing problem becomes a question about the comparison between two ways of measuring the "size" of a log-concave measure: the entropic way and the covariance-structure way. If these two ways are equivalent, this would mean that the isotropic constant is bounded.

With the above definition it is easy to check that the isotropic constant L_μ is an affine invariant; we have $L_\mu = L_{\lambda\mu\circ A}$, $L_f = L_{\lambda f\circ A}$ for every invertible affine transformation A of $\mathbb{R}^n$ and every positive number λ.

One can also prove a characterization of the isotropic constant which is completely analogous to the one in Theorem 10.1.4: if $f:\mathbb{R}^n\to[0,\infty)$ is a log-concave density, then

$$nL_f^2 = \inf_{\substack{T\in SL_n\\ y\in\mathbb{R}^n}}\Big(\sup_{x\in\mathbb{R}^n} f(x)\Big)^{2/n}\int_{\mathbb{R}^n}|Tx+y|^2 f(x)\,dx.$$

We leave this as a straightforward exercise.

The next proposition shows that the isotropic constants of all log-concave measures are uniformly bounded from below by a constant $c>0$ which is independent of the dimension. Its proof is analogous to the proof of Proposition 10.1.8.

Proposition 10.2.5. *Let $f:\mathbb{R}^n\to[0,\infty)$ be an isotropic log-concave function. Then*

$$L_f = \|f\|_\infty^{1/n} \geqslant c,$$

where $c>0$ is an absolute constant.

Proof. Since f is isotropic, we may write

$$\begin{aligned} n = \int |x|^2 f(x)\,dx &= \int_{\mathbb{R}^n}\left(\int_0^{|x|^2}\mathbf{1}\,dt\right)f(x)\,dx \\ &= \int_0^\infty\int_{\mathbb{R}^n}\mathbf{1}_{\{x:|x|^2\geqslant t\}}(x)f(x)\,dx\,dt \\ &= \int_0^\infty\int_{\mathbb{R}^n\setminus\sqrt{t}B_2^n} f(x)\,dx\,dt \\ &= \int_0^\infty\left(1-\int_{\sqrt{t}B_2^n} f(x)dx\right)dt \\ &\geqslant \int_0^{(\kappa_n\|f\|_\infty)^{-2/n}}[1-\kappa_n\|f\|_\infty t^{n/2}]\,dt \\ &= (\kappa_n\|f\|_\infty)^{-2/n}\frac{n}{n+2}. \end{aligned}$$

Since ${\kappa_n}^{-1/n}\simeq\sqrt{n}$, we get $\|f\|_\infty^{1/n}\geqslant c$ for some absolute constant $c>0$. □

The hyperplane conjecture for log-concave measures can now be stated as follows:

Conjecture 10.2.6. *Let $f:\mathbb{R}^n\to[0,\infty)$ be an isotropic log-concave density. Then*

$$\|f\|_\infty^{1/n}\leqslant C,$$

where $C>0$ is an absolute constant.

10.2.1. Convex bodies associated with log-concave functions

In this section we associate a family of sets $K_p(f)$ with any given log-concave function f; we prove that these are convex bodies and we describe some of their basic properties. The bodies $K_p(f)$ were introduced by K. Ball who also established their convexity. They will play an important role in the sequel as they allow us to study properties of log-concave measures through those of convex bodies and vice versa.

DEFINITION 10.2.7 (Ball). Let $f : \mathbb{R}^n \to [0,\infty)$ be a measurable function such that $f(0) > 0$. For any $p > 0$ we define the set $K_p(f)$ as follows:

$$K_p(f) = \left\{ x \in \mathbb{R}^n : \int_0^\infty r^{p-1} f(rx)\, dr \geqslant \frac{f(0)}{p} \right\}.$$

If f_μ is the density of a Borel probability measure μ and $f_\mu(0) > 0$, then we define

$$K_p(\mu) := K_p(f_\mu).$$

Note that from the definition it follows that the radial function of $K_p(f)$ is given by

$$\rho_{K_p(f)}(x) = \left(\frac{1}{f(0)} \int_0^\infty p r^{p-1} f(rx)\, dr \right)^{1/p}. \tag{10.2.2}$$

It is easy to check that if f is given by $f(x) = h(\|x\|_K)$ for some $h : [0,\infty) \to [0,\infty)$ then (when well defined) $K_p(f)$ will be a multiple of K. In particular, we will have use for the following special case:

LEMMA 10.2.8. *Let K be a convex body in $\mathbb{R}^n$ with $0 \in K$. Then, we have $K_p(\mathbf{1}_K) = K$ for all $p > 0$.*

Proof. For every $\theta \in S^{n-1}$ we have

$$\begin{aligned}
\rho^p_{K_p(\mathbf{1}_K)}(\theta) &= \frac{1}{\mathbf{1}_K(0)} \int_0^{+\infty} p r^{p-1}\, \mathbf{1}_K(r\theta)\, dr \\
&= \int_0^{\rho_K(\theta)} p r^{p-1}\, dr = \rho^p_K(\theta).
\end{aligned}$$

It follows that $K_p(\mathbf{1}_K) = K$. □

The next proposition describes some basic properties of the sets $K_p(f)$.

PROPOSITION 10.2.9. *Let $f, g : \mathbb{R}^n \to [0,\infty)$ be two integrable functions with $f(0) = g(0) > 0$, and set*

$$m = \inf\left\{ \frac{f(x)}{g(x)} : g(x) > 0 \right\} \quad \text{and} \quad M^{-1} = \inf\left\{ \frac{g(x)}{f(x)} : f(x) > 0 \right\}.$$

Then, for every $p > 0$ we have the following:

(i) $0 \in K_p(f)$.
(ii) *$K_p(f)$ is a star-shaped set.*
(iii) *$K_p(f)$ is symmetric if f is even.*
(iv) $m^{1/p} K_p(g) \subseteq K_p(f) \subseteq M^{1/p} K_p(g)$.

(v) *For any* $\theta \in S^{n-1}$ *we have*

$$\int_{K_{n+1}(f)} \langle x, \theta \rangle \, dx = \frac{1}{f(0)} \int_{\mathbb{R}^n} \langle x, \theta \rangle f(x) \, dx.$$

In particular, f is centered if and only if $K_{n+1}(f)$ is centered.

(vi) *For any* $\theta \in S^{n-1}$ *and* $p > 0$ *we have*

$$\int_{K_{n+p}(f)} |\langle x, \theta \rangle|^p \, dx = \frac{1}{f(0)} \int_{\mathbb{R}^n} |\langle x, \theta \rangle|^p f(x) \, dx.$$

(vii) *If* $p > -n$ *and V is a star-shaped body with gauge function* $\| \cdot \|_V$ *then*

$$\int_{K_{n+p}(f)} \|x\|_V^p \, dx = \frac{1}{f(0)} \int_{\mathbb{R}^n} \|x\|_V^p f(x) \, dx. \tag{10.2.3}$$

Proof. The first three statements (i), (ii) and (iii) can be easily checked. To prove (iv) we compare the radial functions of $K_p(f)$ and $K_p(g)$. We have

$$\begin{aligned}\rho_{K_p(f)}^p(x) &= \frac{1}{f(0)} \int_0^\infty p r^{p-1} f(rx) \, dr \leqslant \frac{M}{g(0)} \int_0^\infty p r^{p-1} g(rx) \, dr \\ &= (M^{1/p} \rho_{K_p(g)}(x))^p,\end{aligned}$$

and similarly $(m^{1/p} \rho_{K_p(g)}(x))^p \leqslant \rho_{K_p(f)}^p(x)$.

(v) Integrating in polar coordinates we see that, for any $\theta \in S^{n-1}$,

$$\begin{aligned}\int_{K_{n+1}(f)} \langle x, \theta \rangle dx &= n\kappa_n \int_{S^{n-1}} \langle \phi, \theta \rangle \int_0^{\rho_{K_{n+1}(f)}(\phi)} r^n dr d\sigma(\phi) \\ &= \frac{n\kappa_n}{f(0)} \int_{S^{n-1}} \langle \phi, \theta \rangle \int_0^\infty r^n f(r\phi) dr d\sigma(\phi) \\ &= \frac{1}{f(0)} \int_{\mathbb{R}^n} \langle x, \theta \rangle f(x) dx.\end{aligned}$$

(vi) The same argument shows that, for every $p > -n$ and $\theta \in S^{n-1}$,

$$\begin{aligned}\int_{K_{n+p}(f)} |\langle x, \theta \rangle|^p dx &= n\kappa_n \int_{S^{n-1}} |\langle \phi, \theta \rangle|^p \int_0^{\rho_{K_{n+p}(f)}(\phi)} r^{n+p-1} dr d\sigma(\phi) \\ &= \frac{n\kappa_n}{f(0)} \int_{S^{n-1}} |\langle \phi, \theta \rangle|^p \int_0^\infty r^{n+p-1} f(r\phi) dr d\sigma(\phi) \\ &= \frac{1}{f(0)} \int_{\mathbb{R}^n} |\langle x, \theta \rangle|^p f(x) dx.\end{aligned}$$

Working in the same manner we check (vii). □

Assuming the log-concavity of f we can prove that the sets $K_p(f)$, $p > 0$, are convex. To this end, we use Theorem 1.4.6 which appeared in Chapter 1.

THEOREM 10.2.10 (Ball). *Let $f : \mathbb{R}^n \to [0, \infty)$ be a log-concave function such that $f(0) > 0$. For every $p > 0$, $K_p(f)$ is a convex set.*

Proof. Let $p > 0$. Let $x, y \in K_p(f)$; then,

$$p \int_0^\infty r^{p-1} f(rx) \, dr \geqslant f(0) \quad \text{and} \quad p \int_0^\infty r^{p-1} f(ry) \, dr \geqslant f(0).$$

Let $\lambda, \mu > 0$ with $\lambda + \mu = 1$. We set $\gamma = 1/p$ and define $w, g, h : \mathbb{R}^+ \to \mathbb{R}^+$ by

$$w(r) = f(r^\gamma x), \qquad g(s) = f(s^\gamma y), \qquad h(t) = f(t^\gamma(\lambda x + \mu y)).$$

Since f is log-concave, for every pair $(r, s) \in \mathbb{R}^+ \times \mathbb{R}^+$ we have

$$\begin{aligned} h(M_{-\gamma}^\lambda(r,s)) &= f\Big(\frac{1}{\lambda r^{-\gamma} + \mu s^{-\gamma}}(\lambda x + \mu y)\Big) \\ &= f\Big(\frac{\lambda s^\gamma}{\lambda s^\gamma + \mu r^\gamma} r^\gamma x + \frac{\mu r^\gamma}{\lambda s^\gamma + \mu r^\gamma} s^\gamma y\Big) \\ &\geqslant w(r)^{\frac{\lambda s^\gamma}{\lambda s^\gamma + \mu r^\gamma}} g(s)^{\frac{\mu r^\gamma}{\lambda s^\gamma + \mu r^\gamma}}. \end{aligned}$$

Using Theorem 1.4.6 we get

$$\Big(\int_0^\infty f(r^\gamma(\lambda x + \mu y))dr\Big)^{-\gamma} \leqslant \lambda \Big(\int_0^\infty f(r^\gamma x)dr\Big)^{-\gamma} + \mu \Big(\int_0^\infty f(r^\gamma y)dr\Big)^{-\gamma}.$$

The change of variables $t = r^\gamma$ shows that

$$\begin{aligned} &\Big(p\int_0^\infty r^{p-1} f(r(\lambda x + \mu y))dr\Big)^{-1/p} \\ &\qquad \leqslant \lambda \Big(p\int_0^\infty r^{p-1} f(rx)dr\Big)^{-1/p} + \mu \Big(p\int_0^\infty r^{p-1} f(ry)dr\Big)^{-1/p} \\ &\qquad \leqslant \lambda(f(0))^{-1/p} + \mu(f(0))^{-1/p} = (f(0))^{-1/p}. \end{aligned}$$

This shows that

$$p\int_0^\infty r^{p-1} f(r(\lambda x + \mu y))dr \geqslant f(0),$$

and hence $\lambda x + \mu y \in K_p(f)$. □

To show that $K_n(f)$ is indeed a convex body, namely is compact and with non-empty interior, one simply computes its volume to see that it is non-zero and finite:

LEMMA 10.2.11. *For every measurable function $f : \mathbb{R}^n \to [0, \infty)$ such that $f(0) > 0$ we have*

$$\mathrm{Vol}_n(K_n(f)) = \frac{1}{f(0)} \int_{\mathbb{R}^n} f(x)dx.$$

In particular, if f is log-concave and such that $0 < \int_{\mathbb{R}^n} f < \infty$, then $K_n(f)$ is a convex body.

Proof. We can write

$$\begin{aligned} \mathrm{Vol}_n(K_n(f)) &= \int_{K_n(f)} \mathbf{1}\, dx \\ &= n\kappa_n \int_{S^{n-1}} \int_0^{\rho_{K_n(f)}(\phi)} r^{n-1} dr d\sigma(\phi) \\ &= \frac{n\kappa_n}{f(0)} \int_{S^{n-1}} \int_0^\infty r^{n-1} f(r\phi) dr d\sigma(\phi) \\ &= \frac{1}{f(0)} \int_{\mathbb{R}^n} f(x)dx \end{aligned}$$

using (10.2.2) and integration in polar coordinates. □

The fact that all of the convex sets $K_p(f)$, $p > 0$, are indeed convex bodies, namely they are compact and have non-empty interior, whenever the log-concave function f has finite, positive integral, is a consequence of the next proposition. The corollary following it gives an approximate formula for the volume of $K_{n+p}(f)$ when $p > 0$. We show the proofs for both of them in the appendix of this chapter, Section 10.6.

PROPOSITION 10.2.12. *Let $f : \mathbb{R}^n \to [0, \infty)$ be a log-concave function.*

(1) *If $0 < p \leqslant q$, then*

$$\frac{\Gamma(p+1)^{\frac{1}{p}}}{\Gamma(q+1)^{\frac{1}{q}}} K_q(f) \subseteq K_p(f) \subseteq \left(\frac{\|f\|_\infty}{f(0)}\right)^{\frac{1}{p}-\frac{1}{q}} K_q(f). \tag{10.2.4}$$

(2) *If f has its barycenter at the origin then, for every $0 < p \leqslant q$,*

$$\frac{\Gamma(p+1)^{\frac{1}{p}}}{\Gamma(q+1)^{\frac{1}{q}}} K_q(f) \subseteq K_p(f) \subseteq e^{\frac{n}{p}-\frac{n}{q}} K_q(f). \tag{10.2.5}$$

COROLLARY 10.2.13. *Let $f : \mathbb{R}^n \to [0, \infty)$ be a log-concave function with barycenter at the origin and $\int f = 1$. Then, for every $p > 0$ we have*

$$e^{-1} \leqslant f(0)^{\frac{1}{n}+\frac{1}{p}} \mathrm{Vol}_n(K_{n+p}(f))^{\frac{1}{n}+\frac{1}{p}} \leqslant e\frac{n+p}{n}, \tag{10.2.6}$$

while for $-n < p < 0$ we have

$$e^{-1} \leqslant f(0)^{\frac{1}{-p}-\frac{1}{n}} \mathrm{Vol}_n(K_{n+p}(f))^{\frac{1}{-p}-\frac{1}{n}} \leqslant e. \tag{10.2.7}$$

Our first application of the bodies $K_p(f)$ is the fact that a universal bound on the isotropic constant of convex bodies implies a bound for the isotropic constant of log-concave functions. We start with the symmetric case.

PROPOSITION 10.2.14 (Ball). *Let $f : \mathbb{R}^n \to [0, \infty)$ be an even log-concave function with finite, positive integral. Then, the body $T = K_{n+2}(f)$ is a centrally symmetric convex body with*

$$c_1 L_f \leqslant L_T \leqslant c_2 L_f,$$

where $c_1, c_2 > 0$ are absolute constants. Furthermore, if f is isotropic, then $\overline{T} = \frac{1}{\mathrm{Vol}_n(T)^{1/n}} T$ is an isotropic convex body.

Proof. Since f is even and log-concave, T is a centrally symmetric convex body; we also have $f(x) \leqslant f(0)$ for all $x \in \mathbb{R}^n$. Hence, $f(0) > 0$. From Proposition 10.2.9 (vi)

$$\int_T \langle x, \theta\rangle^2 \, dx = \frac{1}{f(0)} \int_{\mathbb{R}^n} \langle x, \theta\rangle^2 f(x)\, dx,$$

and more generally

$$\int_T \langle x, \theta\rangle\langle x, \phi\rangle \, dx = \frac{1}{f(0)} \int_{\mathbb{R}^n} \langle x, \theta\rangle\langle x, \phi\rangle f(x)\, dx$$

for all $\theta, \phi \in S^{n-1}$. It follows that

$$\mathrm{Vol}_n(T)\mathrm{Cov}(\mathbf{1}_T) = \frac{\int f}{f(0)}\mathrm{Cov}(f).$$

By the definition of the isotropic constant we obtain

$$L_T = L_{\mathbf{1}_T} = \frac{1}{\mathrm{Vol}_n(T)^{\frac{1}{2}+\frac{1}{n}}}\left(\frac{1}{f(0)}\int f\right)^{\frac{1}{2}+\frac{1}{n}} L_f.$$

On the other hand, applying Corollary 10.2.13 with $p=2$ we see that

$$\mathrm{Vol}_n(T)^{\frac{1}{2}+\frac{1}{n}} = \mathrm{Vol}_n(K_{n+2}(f))^{\frac{1}{2}+\frac{1}{n}} \simeq \left(\frac{1}{f(0)}\int_{\mathbb{R}^n} f(x)\,dx\right)^{\frac{1}{2}+\frac{1}{n}}.$$

This shows that $L_T \simeq L_f$. Finally, note that if f is isotropic then

$$\int_{\overline{T}} \langle x,\theta\rangle^2\,dx = \frac{1}{\mathrm{Vol}_n(T)^{1+\frac{2}{n}}}\int_T \langle x,\theta\rangle^2\,dx = \frac{1}{f(0)\mathrm{Vol}_n(T)^{1+\frac{2}{n}}}$$

for every $\theta\in S^{n-1}$, which shows that $\overline{T}$ is in isotropic position. $\square$

The next proposition shows that we can further reduce our study of the behaviour of the isotropic constant to the class of symmetric convex bodies.

Proposition 10.2.15. *For every convex body K we can find a centrally symmetric convex body T with the property that*

$$L_K \leqslant cL_T,$$

where $c>0$ is an absolute constant.

Proof. Without loss of generality, we may assume that K has volume 1 and barycenter at the origin. We define a function f supported on $K-K$ as follows:

$$f(x) = (\mathbf{1}_K * \mathbf{1}_{-K})(x) = \int_{\mathbb{R}^n} \mathbf{1}_K(y)\mathbf{1}_{-K}(x-y)\,dy = \mathrm{Vol}_n(K\cap(x+K)).$$

Given that for every x

$$\mathrm{Vol}_n(K\cap(x+K)) = \mathrm{Vol}_n(-x+(K\cap(x+K))) = \mathrm{Vol}_n((-x+K)\cap K),$$

we have that f is even. The fact that f is log-concave follows from the Brunn-Minkowski inequality (see e.g. the proof of Theorem 1.5.2). In addition, it is easy to check that $\int_{\mathbb{R}^n} f = 1$ and that $f(x)\leqslant f(0) = \mathrm{Vol}_n(K) = 1$. Therefore,

$$L_f = [\det \mathrm{Cov}(f)]^{\frac{1}{2n}}.$$

Next, since one easily checks that for any h and g with barycenter at 0 and total mass 1 one has

$$\mathrm{Cov}(h*g) = \mathrm{Cov}(h) + \mathrm{Cov}(g),$$

we get that

$$\mathrm{Cov}(f) = \mathrm{Cov}(K) + \mathrm{Cov}(-K).$$

As these are positive definite matrices it follows (Lemma 2.1.5) that

$$[\det \mathrm{Cov}(f)]^{1/n} \geqslant [\det \mathrm{Cov}(K)]^{1/n} + [\det \mathrm{Cov}(-K)]^{1/n} = 2[\det \mathrm{Cov}(K)]^{1/n},$$

and hence

$$L_K = [\det \mathrm{Cov}(K)]^{\frac{1}{2n}} \leqslant \frac{1}{\sqrt{2}}[\det \mathrm{Cov}(f)]^{\frac{1}{2n}} = \frac{1}{\sqrt{2}}L_f.$$

It is easy now to check that the body $T := K_{n+2}(f)$ has the desired properties: T is centrally symmetric because f is even, and in addition $L_T \simeq L_f \geqslant L_K$. $\square$

Assuming that the function $f:\mathbb{R}^n\to[0,\infty)$ is centered, but not necessarily even, we prefer to work with the centered body $K_{n+1}(f)$ instead of $K_{n+2}(f)$.

PROPOSITION 10.2.16. *Let $f : \mathbb{R}^n \to [0,\infty)$ be a centered log-concave function with finite, positive integral. Then, $T = K_{n+1}(f)$ is a centered convex body in $\mathbb{R}^n$ with*

$$c_1 L_f \leqslant L_T \leqslant c_2 L_f,$$

where $c_1, c_2 > 0$ are absolute constants.

Proof. Note that, by Theorem 10.6.2, f being centered implies $f(0) > 0$; thus, $K_{n+1}(f)$ is well-defined and by Proposition 10.2.9 (v) and Theorem 10.2.10 we know that it is a centered convex body. Without loss of generality we may assume that f is log-concave with $\int f = 1$, otherwise we work with $f_1 = \frac{f}{\int f}$ using the fact that $K_{n+1}(\lambda f) = K_{n+1}(f)$ and $L_{\lambda f} = L_f$ for any $\lambda > 0$. By Proposition 10.2.9 we have

$$\int_T |\langle x, \theta\rangle| \, dx = \frac{1}{f(0)} \int |\langle x, \theta\rangle| f(x) \, dx.$$

Borell's lemma (or, more precisely, its consequence given as Theorem 3.5.11) implies that for every $y \in \mathbb{R}^n$

$$\begin{aligned}\left(\frac{1}{\mathrm{Vol}_n(T)} \int_T \langle x, y\rangle^2 \, dx\right)^{1/2} &\simeq \frac{1}{\mathrm{Vol}_n(T)} \int_T |\langle x, y\rangle| \, dx \\ &= \frac{1}{f(0)\mathrm{Vol}_n(T)} \int |\langle x, y\rangle| f(x) \, dx \\ &\simeq \frac{1}{f(0)\mathrm{Vol}_n(T)} \left(\int \langle x, y\rangle^2 f(x) \, dx\right)^{1/2},\end{aligned}$$

which, combined with the fact that T and f are both centered, implies that there exist absolute constants $c_1, c_2 > 0$ such that as positive definite matrices

$$c_2 \mathrm{Cov}(\mathbf{1}_T) \leqslant (\mathrm{Vol}_n(T) f(0))^{-2} \mathrm{Cov}(f) \leqslant c_1 \mathrm{Cov}(\mathbf{1}_T).$$

Therefore

$$[\det \mathrm{Cov}(\mathbf{1}_T)]^{1/n} \simeq (\mathrm{Vol}_n(T) f(0))^{-2} [\det \mathrm{Cov}(f)]^{1/n}. \tag{10.2.8}$$

From the definition of the isotropic constant it follows that

$$\begin{aligned}L_T &= \frac{1}{\mathrm{Vol}_n(T)^{1/n}} [\det \mathrm{Cov}(T)]^{\frac{1}{2n}} \\ &\simeq \mathrm{Vol}_n(T)^{-1/n} (f(0)\mathrm{Vol}_n(T))^{-1} [\det \mathrm{Cov}(f)]^{\frac{1}{2n}} \\ &\simeq (f(0)\mathrm{Vol}_n(T))^{-1-\frac{1}{n}} L_f,\end{aligned}$$

where we have also used the fact that one has $\|f\|_\infty^{1/n} \simeq f(0)^{1/n}$ by Theorem 10.6.2. Finally, applying Proposition 10.2.13 with $p = 1$ we get that

$$e^{-1} \leqslant (f(0)\mathrm{Vol}_n(T))^{1+\frac{1}{n}} \leqslant e \frac{n+1}{n} \leqslant 2e. \tag{10.2.9}$$

This completes the proof. □

10.3. Bourgain's upper bound for the isotropic constant

In this section we give a proof of Bourgain's upper bound $L_n \leqslant C\sqrt[4]{n}\log n$, which is based on the ψ_1-behaviour of linear functionals on convex bodies. The subsequent argument is based on Bourgain's approach but it is more elementary, since it involves simpler entropy considerations (it is a variation of an argument of S. Dar). We start with the tools that we will use in the proof.

The first tool regards the maximum of a finite number of ψ_1 random variables. It is a consequence of Proposition 3.5.8 with $\alpha = 1$:

PROPOSITION 10.3.1. *Let K be an isotropic convex body in $\mathbb{R}^n$, and let $N \geqslant 2$ and $\theta_1, \dots, \theta_N \in S^{n-1}$. Then*

$$\int_K \max_{1 \leqslant i \leqslant N} |\langle x, \theta_i\rangle| \, dx \leqslant C L_K (\log N),$$

where $C > 0$ is an absolute constant.

Proof. Indeed, by Remark 3.5.12 (and the Khinchine type inequality in Theorem 3.5.11) the random variables $\langle \cdot, \theta\rangle$ on K satisfy the ψ_1-estimate

$$\|\langle \cdot, \theta\rangle\|_{\psi_1} \leqslant c\|\langle \cdot, \theta\rangle\|_2 = cL_K$$

for all $\theta \in S^{n-1}$, where $c > 0$ is universal, so one may apply Proposition 3.5.8. $\square$

The next tool is sometimes called a Dudley-Fernique decomposition. It is one of the tools in proving Dudley's theorem regarding entropy which we have discussed in Chapter 9. However, here it is given in a straightforward geometric way, and so we include the simple and direct proof.

THEOREM 10.3.2. *Let K be a convex body in $\mathbb{R}^n$, with $0 \in K$ and $K \subset RB_2^n$. There exist $Z_j \subseteq (3R/2^j)B_2^n$, $j \in \mathbb{N}$, with cardinality*

$$\log |Z_j| \leqslant cn \left(\frac{2^j w(K)}{R}\right)^2,$$

which satisfy the following: for every $x \in K$ and any $m \in \mathbb{N}$ we can find $z_j \in Z_j$, $j = 1, \dots, m$, and $w_m \in (R/2^m)B_2^n$ such that $x = z_1 + \cdots + z_m + w_m$.

Proof. We shall use elementary properties of covering numbers and Sudakov's bound of Theorem 4.2.1. For every $j \in \mathbb{N}$ we may find a subset N_j of K with cardinality

$$|N_j| = N(K, (R/2^j)B_2^n)$$

such that

$$K \subseteq \bigcup_{y \in N_j} (y + (R/2^j)B_2^n).$$

Theorem 4.2.1 shows that

$$\log |N_j| \leqslant cn \left(\frac{2^j w(K)}{R}\right)^2. \tag{10.3.1}$$

We set $N_0 = \{0\}$ and

$$W_j = N_j - N_{j-1} = \{y - y' : y \in N_j, y' \in N_{j-1}\}$$

for every $j \geqslant 1$. We define $Z_j = W_j \cap (3R/2^j)B_2^n$. Thus $\log|Z_j| \leqslant \log|W_j| \leqslant c'n\left(\frac{2^j w(K)}{R}\right)^2$. We need to show that for every $x \in K$ and any $m \in \mathbb{N}$ we can find $z_j \in W_j \cap (3R/2^j)B_2^n$, $j = 1, \dots, m$, and $w_m \in (R/2^m)B_2^n$ such that

$$x = z_1 + \cdots + z_m + w_m.$$

Given such x, by the definition of N_j, we can find $y_j \in N_j$, $j = 1, \dots, m$, such that

$$|x - y_j| \leqslant \frac{R}{2^j}.$$

We write

$$x = (y_1 - 0) + (y_2 - y_1) + \cdots + (y_m - y_{m-1}) + (x - y_m).$$

We set $y_0 = 0$ and $w_m = x - y_m$, $z_j = y_j - y_{j-1}$ for $j = 1, \dots, m$. Then, $|w_m| = |x - y_m| \leqslant R/2^m$, and $z_j \in N_j - N_{j-1} = W_j$. Also,

$$|z_j| \leqslant |x - y_j| + |x - y_{j-1}| \leqslant \frac{R}{2^j} + \frac{R}{2^{j-1}} = \frac{3R}{2^j}.$$

Finally, $x = z_1 + \cdots + z_m + w_m$ as claimed. □

We are now ready to give Dar's version of the proof of Bourgain's bound.

THEOREM 10.3.3. *If K is an isotropic convex body in $\mathbb{R}^n$ then*

$$L_K \leqslant c\sqrt[4]{n}\log n,$$

where $c > 0$ is an absolute constant.

Proof. By Theorem 6.5.2 there exists a symmetric and positive definite $T \in SL_n$ such that

$$w(TK) \leqslant c\sqrt{n}\log n.$$

We write

$$nL_K^2 = \int_K |x|^2 dx \leqslant \frac{\mathrm{tr}T}{n}\int_K |x|^2 = \int_K \langle x, Tx\rangle dx.$$

Therefore,

$$nL_K^2 \leqslant \int_K \max_{y \in TK} |\langle y, x\rangle| dx.$$

If $TK \subset RB_2^n$, we can use Theorem 10.3.2 to find $Z_j \subset (3R/2^j)B_2^n$ such that

$$\log|Z_j| \leqslant cn\left(\frac{w(TK)2^j}{R}\right)^2, \tag{10.3.2}$$

and so that for every $m \in \mathbb{N}$, every $y \in TK$ can be written in the form $y = z_1 + \cdots + z_m + w_m$ with $z_j \in Z_j$ and $w_m \in (R/2^m)B_2^n$. This implies that

$$\begin{aligned}\max_{y \in TK} |\langle y, x\rangle| &\leqslant \sum_{j=1}^m \max_{z \in Z_j} |\langle z, x\rangle| + \max_{w \in (R/2^m)B_2^n} |\langle w, x\rangle| \\ &\leqslant \sum_{j=1}^m \frac{3R}{2^j} \max_{z \in Z_j} |\langle \overline{z}, x\rangle| + \frac{R}{2^m}|x|,\end{aligned}$$

where $\overline{z}$ denotes the unit vector in the direction of z. Noting that (by Cauchy-Schwarz inequality) $\int_K |x|dx \leqslant \sqrt{n}L_K$ and using the above, we see that

$$\begin{aligned} nL_K^2 &\leqslant \sum_{j=1}^{m} \frac{3R}{2^j} \int_K \max_{z\in Z_j} |\langle \overline{z}, x\rangle| dx + \frac{R}{2^m}\int_K |x|dx \\ &\leqslant \sum_{j=1}^{m} \frac{3R}{2^j} \int_K \max_{z\in Z_j} |\langle \overline{z}, x\rangle| dx + \frac{R}{2^m}\sqrt{n}L_K. \end{aligned}$$

From Proposition 10.3.1 and (10.3.2) we get

$$nL_K^2 \leqslant \sum_{j=1}^{m} \frac{3R}{2^j} c_1 nL_K \left(\frac{w(TK)2^j}{R}\right)^2 + \frac{R}{2^m}\sqrt{n}L_K. \tag{10.3.3}$$

The sum on the right hand side is bounded by

$$c_2 L_K n w^2(TK)\frac{2^m}{R}.$$

Solving the equation

$$\frac{nw^2(TK)2^s}{R} = \frac{R\sqrt{n}}{2^s}$$

(where s here can be non-integer), we see that the optimal (integer) value of m satisfies the "equation"

$$\frac{R}{2^m} \simeq \sqrt[4]{n}w(TK).$$

Going back to (10.3.3), we obtain

$$nL_K^2 \leqslant c_3\sqrt{n}\sqrt[4]{n}w(TK)L_K.$$

Since $w(TK) \leqslant c_4\sqrt{n}\log n$, we get the result. □

The same proof, using the more general estimate for the maximum of ψ_α random variables (see Proposition 3.5.8), gives a better estimate for L_K under the assumption that the body K is "ψ_α in all directions" for some $\alpha \in [1, 2)$. More precisely,

THEOREM 10.3.4. *Let K be an isotropic convex body in $\mathbb{R}^n$. Assume that there exist $1 \leqslant \alpha < 2$ and $B_\alpha > 0$ such that K satisfies the ψ_α-estimate*

$$\|\langle \cdot, \theta\rangle\|_{\psi_\alpha} \leqslant B_\alpha \|\langle \cdot, \theta\rangle\|_2 = B_\alpha L_K$$

for every $\theta \in S^{n-1}$. Then

$$L_K \leqslant CB_\alpha^{\frac{\alpha}{2}}(2-\alpha)^{-\frac{\alpha}{2}} n^{\frac{1}{2}-\frac{\alpha}{4}}\log n,$$

where $C > 0$ is an absolute constant. □

We remark that Bourgain proved that if a body satisfies a uniform ψ_2 estimate in all directions then L_K is bounded by a function of its ψ_2-constant (see notes and remarks at the end of the chapter).

10.4. Paouris' deviation inequality

In this section we prove the following theorem.

THEOREM 10.4.1 (Paouris). *Let $n \in \mathbb{N}$ and let μ be an isotropic log-concave probability measure on $\mathbb{R}^n$. Then for every $t \geqslant 1$,*

$$\mu(\{x \in \mathbb{R}^n : |x| \geqslant ct\sqrt{n}\}) \leqslant \exp\left(-t\sqrt{n}\right), \tag{10.4.1}$$

where $c > 0$ is an absolute constant.

In the particular case of unconditional isotropic convex bodies, the inequality of Paouris had been previously proved by Bobkov and Nazarov. Their work, in turn, has its origin in the work of Schechtman, Zinn and Schmuckenschläger on the volume of the intersection of two L_p^n-balls.

10.4.1. Reduction to the behaviour of moments and Alesker's deviation estimate

We begin by describing a reduction of the problem to the behaviour of the moments of the function $x \mapsto |x|$. Given a log-concave probability measure μ on $\mathbb{R}^n$, for $q > -n$, $q \neq 0$ we define

$$I_q(\mu) = \left(\int_{\mathbb{R}^n} |x|^q d\mu(x)\right)^{1/q}.$$

We let $I_q(K) = I_q(\mathbf{1}_K dx)$ when K is a convex body of volume 1 in $\mathbb{R}^n$.

Clearly, since μ is a probability measure, we have that $I_q(\mu)$ is increasing in q. One may also compare $I_q(\mu)$ with $I_p(\mu)$ in the opposite direction as follows

LEMMA 10.4.2. *For every $y \in \mathbb{R}^n$ and every $p, q \geqslant 1$ we have*

$$\left(\int_{\mathbb{R}^n} |\langle \cdot, y\rangle|^{pq} d\mu(x)\right)^{1/pq} \leqslant c_1 q \left(\int_{\mathbb{R}^n} |\langle \cdot, y\rangle|^{p} d\mu(x)\right)^{1/p},$$

and

$$I_{pq}(\mu) \leqslant c_1 q I_p(\mu).$$

where $c_1 > 0$ is an absolute constant.

Proof. Indeed we have seen in Chapter 3, Theorem 3.5.11 that this Khintchine type inequality is simply a consequence of Borell's lemma for log-concave measures. □

The first step towards understanding the behaviour of the moments $I_q(K)$ was made by Alesker. He showed

THEOREM 10.4.3 (Alesker). *Let K be an isotropic convex body in $\mathbb{R}^n$. For all $q \geqslant 1$ we have*

$$I_q(K) \leqslant c\sqrt{q} I_2(K),$$

and letting $f(x) = |x|$, we have

$$\|f\|_{\psi_2} \leqslant c\sqrt{n} L_K,$$

where $c > 0$ is an absolute constant.

For the proof of Alesker's theorem we first note the following simple formula:

LEMMA 10.4.4. *Let K be a convex body of volume 1 in $\mathbb{R}^n$. For every $q \geqslant 1$,*

$$\left(\int_{S^{n-1}}\int_K |\langle x,\theta\rangle|^q dx d\sigma(\theta)\right)^{1/q} \simeq \sqrt{\frac{q}{q+n}}I_q(K).$$

Proof. For every $q \geqslant 1$ and $x \in \mathbb{R}^n$, we check that

$$\left(\int_{S^{n-1}} |\langle x,\theta\rangle|^q d\sigma(\theta)\right)^{1/q} \simeq \frac{\sqrt{q}}{\sqrt{q+n}}|x|. \tag{10.4.2}$$

To see this, using polar coordinates we first see that

$$\begin{aligned}\int_{B_2^n} |\langle x,y\rangle|^q dy &= n\omega_n \int_0^1 r^{n+q-1}dr \int_{S^{n-1}} |\langle x,\theta\rangle|^q d\sigma(\theta)\\ &= \frac{n\omega_n}{n+q}\int_{S^{n-1}} |\langle x,\theta\rangle|^q d\sigma(\theta).\end{aligned}$$

But we can also write the left hand side as

$$\begin{aligned}\int_{B_2^n} |\langle x,y\rangle|^q dy &= |x|^q \int_{B_2^n} \left|\left\langle \frac{x}{|x|},y\right\rangle\right|^q dy\\ &= |x|^q\int_{B_2^n} |\langle e_1,y\rangle|^q dy\\ &= 2\omega_{n-1}|x|^q\int_0^1 t^q(1-t^2)^{(n-1)/2}dt\\ &= \omega_{n-1}|x|^q\frac{\Gamma\left(\frac{q+1}{2}\right)\Gamma\left(\frac{n+1}{2}\right)}{\Gamma\left(\frac{n+q+2}{2}\right)}.\end{aligned}$$

Comparing the two expressions and using Stirling's formula we get (10.4.2). A simple application of Fubini's theorem gives the result. □

Proof of Theorem 10.4.3. By Lemma 3.5.5, the first assertion implies the second. Thus we concentrate on proving that for every $q > 1$

$$\left(\int_K |x|^q dx\right)^{1/q} \leqslant c_1\sqrt{q}\sqrt{n}L_K \tag{10.4.3}$$

for some absolute constant $c_1 > 0$. We know by Lemma 10.4.2 that for every $\theta \in S^{n-1}$

$$\int_K |\langle x,\theta\rangle|^q dx \leqslant c_2^q q^q L_K^q.$$

Integrating on the sphere we get

$$\int_{S^{n-1}}\int_K |\langle x,\theta\rangle|^q dx\, d\sigma(\theta) \leqslant c_2^q q^q L_K^q.$$

Taking into account Lemma 10.4.4, we see that

$$\left(\int_K |x|^q dx\right)^{1/q} \leqslant c_3 q\sqrt{\frac{n+q}{q}}L_K \leqslant c_4\sqrt{q}\sqrt{n}L_K,$$

provided that $q \leqslant n$. On the other hand, if $q > n$, using the fact that $K \subset cnL_K B_2^n$ (see Remark 10.1.15), we get

$$\left(\int_K |x|^q dx\right)^{1/q} \leqslant cnL_K \leqslant c\sqrt{q}\sqrt{n}L_K.$$

Combining the above we see that (10.4.3) holds true for all $q > 1$. □

Using Lemma 3.5.7 which is about the tails of ψ_2 functions we get

COROLLARY 10.4.5. *There exists an absolute constant $c > 0$ such that: if K is an isotropic convex body in $\mathbb{R}^n$ then*

$$\mathrm{Vol}_n(\{x \in K : |x| \geqslant c\sqrt{n}L_K s\}) \leqslant 2\exp(-s^2)$$

for every $s > 0$.

Paouris proved a very strong statement regarding the moments $I_q(\mu)$ for isotropic μ, from which his main theorem, Theorem 10.4.1 above, follows. In this subsection we state it, show how it implies Theorem 10.4.1 and another consequence of it. In the remainder of Section 10.4 we shall deal with proving the following theorem.

THEOREM 10.4.6 (Paouris). *Let μ be an isotropic log-concave measure on $\mathbb{R}^n$. For any integer $2 \leqslant q \leqslant c_1\sqrt{n}$ we have*

$$I_{-q}(\mu) \simeq I_q(\mu).$$

In particular, all moments $I_q(\mu)$ in this range are equivalent and so $I_q(\mu) \leqslant c_2 I_2(\mu) = c_2\sqrt{n}$ for $2 \leqslant q \leqslant \sqrt{n}$.

Proof of Theorem 10.4.1 *using Theorem* 10.4.6 Let μ be an isotropic log-concave probability measure in $\mathbb{R}^n$ and let $q \geqslant 2$. From Markov's inequality we have

$$\mu(\{x : |x| \geqslant e^3 I_q(\mu)\}) \leqslant e^{-3q}.$$

Then, from Borell's lemma for log-concave measures (Lemma 3.5.10) we get

$$\begin{aligned} \mu(\{x : |x| \geqslant e^3 I_q(\mu)t\}) &\leqslant (1 - e^{-3q}) \left(\frac{e^{-3q}}{1 - e^{-3q}}\right)^{(t+1)/2} \\ &\leqslant e^{-qt} \end{aligned} \tag{10.4.4}$$

for every $t \geqslant 1$. Choosing $q_0 = \sqrt{n}$, and using the fact that $I_{q_0}(\mu) \leqslant c_1\sqrt{n}$, we get

$$\mu(\{x : |x| \geqslant c_1 e^3\sqrt{n}t\}) \leqslant \exp(-\sqrt{n}t)$$

for every $t \geqslant 1$. This proves the theorem. □

To end this subsection we give yet another consequence of Theorem 10.4.6 which is a small ball probability estimate for an isotropic log-concave measure on $\mathbb{R}^n$.

THEOREM 10.4.7. *Let μ be an isotropic log-concave measure on $\mathbb{R}^n$. Then, for any $0 < \varepsilon < \varepsilon_0$ we have*

$$\mu(\{x : |x| \leqslant \varepsilon\sqrt{n}\}) \leqslant \varepsilon^{c\sqrt{n}},$$

where $\varepsilon_0, c > 0$ are absolute constants.

Proof. Let $2 \leqslant k \leqslant \sqrt{n}$. Then we can write

$$\begin{aligned} \mu(\{x : |x| \leqslant \varepsilon I_2(\mu)\}) &\leqslant \mu(\{x : |x| \leqslant c_1\varepsilon I_{-k}(\mu)\}) \\ &\leqslant (c_1\varepsilon)^k \leqslant \varepsilon^{k/2}, \end{aligned}$$

for all $0 < \varepsilon < c_1^{-2}$. Choosing $k \simeq \sqrt{n}$ we get the result. □

10.4.2. The L_q-centroid bodies

One of the main tools in the proof of Theorem 10.4.6 is the family of the L_q centroid bodies associated with a convex body or, more generally, with a log-concave function.

DEFINITION 10.4.8. Let $q \geqslant 1$ and let $f : \mathbb{R}^n \to [0, \infty)$ be a log-concave function with $\int f = 1$. We define the *L_q-centroid body* $Z_q(f)$ to be the symmetric convex body with support function given by

$$h_{Z_q(f)}(y) := \left(\int_{\mathbb{R}^n} |\langle x, y\rangle|^q f(x)\, dx \right)^{1/q}.$$

By Hölder's inequality we have that $Z_1(f) \subseteq Z_p(f) \subseteq Z_q(f)$ for every $1 \leqslant p \leqslant q < \infty$; in the case of convex bodies (that is, $f = \mathbf{1}_K$, in which case we sometimes write $Z_q(\mathbf{1}_K) = Z_q(K)$) we also see that $Z_q(K) \subseteq \mathrm{conv}(K, -K)$ for every $q \geqslant 1$, since

$$h_{Z_q(K)}(y) = \left(\int_K |\langle x, y\rangle|^q dx \right)^{1/q} \leqslant \max\left\{ h_K(y), h_K(-y) \right\}$$

for all $y \in \mathbb{R}^n$.

Our next task is to estimate the volume of $Z_q(f)$. We start with a reverse of the trivial inclusion $Z_q(K) \subset \mathrm{conv}(K, -K)$.

LEMMA 10.4.9. *Let K be a centered convex body of volume 1 in $\mathbb{R}^n$. Then, for every $\theta \in S^{n-1}$ and every $q \geqslant 1$,*

$$\int_K |\langle x, \theta\rangle|^q dx \geqslant \frac{\Gamma(q+1)\Gamma(n)}{2e\Gamma(q+n+1)} \max\left\{ h_K^q(\theta), h_K^q(-\theta) \right\}.$$

Proof. Consider the function $f_\theta(t) = \mathrm{Vol}_{n-1}(K \cap \{\langle x, \theta\rangle = t\})$. Brunn's principle implies that $f_\theta^{1/(n-1)}$ is concave. It follows that

$$f_\theta(t) \geqslant \left(1 - \frac{t}{h_K(\theta)}\right)^{n-1} f_\theta(0)$$

for all $t \in [0, h_K(\theta)]$. Therefore,

$$\begin{aligned}
\int_K |\langle x, \theta\rangle|^q dx &= \int_0^{h_K(\theta)} t^q f_\theta(t) dt + \int_0^{h_K(-\theta)} t^q f_{-\theta}(t) dt \\
&\geqslant \int_0^{h_K(\theta)} t^q \left(1 - \frac{t}{h_K(\theta)}\right)^{n-1} f_\theta(0) dt \\
&\quad + \int_0^{h_K(-\theta)} t^q \left(1 - \frac{t}{h_K(-\theta)}\right)^{n-1} f_\theta(0) dt \\
&= f_\theta(0) \left(h_K^{q+1}(\theta) + h_K^{q+1}(-\theta) \right) \int_0^1 s^q (1-s)^{n-1} ds \\
&= \frac{\Gamma(q+1)\Gamma(n)}{\Gamma(q+n+1)} f_\theta(0) \left(h_K^{q+1}(\theta) + h_K^{q+1}(-\theta) \right) \\
&\geqslant \frac{\Gamma(q+1)\Gamma(n)}{2\Gamma(q+n+1)} f_\theta(0) \left(h_K(\theta) + h_K(-\theta) \right) \cdot \max\left\{ h_K^q(\theta), h_K^q(-\theta) \right\}.
\end{aligned}$$

Since K has its center of mass at the origin, we have $\|f_\theta\|_\infty \leqslant e f_\theta(0)$, and hence

$$1 = \mathrm{Vol}_n(K) = \int_{-h_K(-\theta)}^{h_K(\theta)} f_\theta(t)dt \leqslant e\,(h_K(\theta) + h_K(-\theta))\, f_\theta(0).$$

This completes the proof. □

COROLLARY 10.4.10. *Let $n \in \mathbb{N}$ and $K \subset \mathbb{R}^n$ be a centered convex body with $\mathrm{Vol}_n(K) = 1$. Then*

$$c \leqslant \mathrm{Vol}_n(Z_n(K))^{1/n} \leqslant \mathrm{Vol}_n(K-K)^{1/n} \leqslant 4 \tag{10.4.5}$$

for some absolute constant $c > 0$.

Proof. Indeed, by Lemma 10.4.9 for $q = n$ we get that

$$cK \subseteq Z_n(K) \subseteq \mathrm{conv}(K, -K)$$

and using the Rogers-Shephard inequality (Theorem 1.5.2) we are done. □

This corollary has its counterpart in the log-concave world. We shall show

THEOREM 10.4.11. *Let f be a centered log-concave density in $\mathbb{R}^n$. Then,*

$$\frac{c_1}{f(0)^{1/n}} \leqslant \mathrm{Vol}_n(Z_n(f))^{1/n} \leqslant \frac{c_2}{f(0)^{1/n}}, \tag{10.4.6}$$

where $c_1, c_2 > 0$ are universal constants. Similarly, if μ is a centered log-concave probability measure on $\mathbb{R}^n$, then

$$\mathrm{Vol}_n(Z_n(\mu))^{1/n} \simeq \frac{1}{\|\mu\|_\infty^{1/n}}.$$

To this end we shall need to make use of the bodies $K_p(f)$ which were introduced in Definition 10.2.7. We start as follows (recall that we use $\overline{K}$ to denote the normalized copy of K, namely $\overline{K} = \frac{1}{\mathrm{Vol}_n(K)^{1/n}} K)$:

PROPOSITION 10.4.12. *Let $f : \mathbb{R}^n \to [0,\infty)$ be centered and log-concave with $\int f = 1$. For every $p \geqslant 1$,*

$$Z_p(\overline{K_{n+p}}(f))\mathrm{Vol}_n(K_{n+p}(f))^{\frac{1}{p}+\frac{1}{n}} f(0)^{1/p} = Z_p(f).$$

Proof. Let $p \geqslant 1$. From Proposition 10.2.9 (vi) we know that

$$\int_{K_{n+p}(f)} |\langle x,\theta\rangle|^p\, dx = \frac{1}{f(0)} \int_{\mathbb{R}^n} |\langle x,\theta\rangle|^p\, f(x)\, dx$$

for all $\theta \in S^{n-1}$. Since

$$\int_{K_{n+p}(f)} |\langle x,\theta\rangle|^p\, dx = \mathrm{Vol}_n(K_{n+p}(f))^{1+\frac{p}{n}} \int_{\overline{K_{n+p}}(f)} |\langle x,\theta\rangle|^p\, dx,$$

the result follows. □

With Proposition 10.4.12 at hand, we may use Proposition 10.2.12 and Corollary 10.2.13 to show

THEOREM 10.4.13. *Let f be a centered log-concave density in $\mathbb{R}^n$. Then, for every $1 \leqslant q \leqslant n$, one has*

$$c_1 f(0)^{1/n} Z_q(f) \subseteq Z_q(\overline{K_{n+1}}(f)) \subseteq c_2 f(0)^{1/n} Z_q(f) \tag{10.4.7}$$

where $c_1, c_2 > 0$ are absolute constants.

Proof. We may assume that q is a positive integer. By Proposition 10.4.12 and Corollary 10.2.13 we see that

$$\frac{1}{e}Z_q(\overline{K_{n+q}}(f)) \subseteq f(0)^{1/n}Z_q(f) \subseteq e\frac{n+q}{n}Z_q(\overline{K_{n+q}}(f)) \subseteq 2eZ_q(\overline{K_{n+q}}(f))$$

(given that $q \leqslant n$), so it suffices to compare $Z_q(\overline{K_{n+q}}(f))$ and $Z_q(\overline{K_{n+1}}(f))$. Using the inclusions in Proposition 10.2.12, we write

$$\begin{aligned} h_{Z_q(\overline{K_{n+q}}(f))}(\theta) &= \frac{1}{\mathrm{Vol}_n(K_{n+q}(f))^{\frac{1}{q}+\frac{1}{n}}}\left(\int_{K_{n+q}(f)} |\langle x,\theta\rangle|^q dx\right)^{1/q} \\ &\leqslant \frac{1}{\mathrm{Vol}_n(K_{n+q}(f))^{\frac{1}{q}+\frac{1}{n}}}\left(\int_{\frac{\Gamma(n+q+1)^{1/n+q}}{\Gamma(n+2)^{1/n+1}}K_{n+1}(f)} |\langle x,\theta\rangle|^q dx\right)^{1/q} \\ &= \left(\frac{|K_{n+1}(f)|}{\mathrm{Vol}_n(K_{n+q}(f))}\right)^{\frac{1}{q}+\frac{1}{n}}\left(\frac{\Gamma(n+q+1)^{\frac{1}{n+q}}}{\Gamma(n+2)^{\frac{1}{n+1}}}\right)^{1+\frac{n}{q}} h_{Z_q(\overline{K_{n+1}}(f))}(\theta) \\ &\leqslant \left(e^{\frac{n^2}{n+1}-\frac{n^2}{n+q}}\right)^{\frac{1}{q}+\frac{1}{n}}\left(\frac{\Gamma(n+q+1)^{\frac{1}{n+q}}}{\Gamma(n+2)^{\frac{1}{n+1}}}\right)^{1+\frac{n}{q}} h_{Z_q(\overline{K_{n+1}}(f))}(\theta) \end{aligned}$$

for every $\theta \in S^{n-1}$. Since

$$\begin{aligned} \left(\frac{\Gamma(n+q+1)^{\frac{1}{n+q}}}{\Gamma(n+2)^{\frac{1}{n+1}}}\right)^{1+\frac{n}{q}} &= \frac{(\Gamma(n+q+1))^{\frac{1}{q}}}{(\Gamma(n+2))^{\frac{1}{q}+\frac{q-1}{q(n+1)}}} \\ &\leqslant \left(\frac{\Gamma(n+q+1)}{\Gamma(n+2)}\right)^{\frac{1}{q}}\left(\frac{c_0 e}{n+1}\right)^{\frac{q-1}{q}} = (c_0 e)^{\frac{q-1}{q}}\prod_{i=1}^{q}\left(\frac{n+i}{n+1}\right)^{1/q} \leqslant 2c_0' e \end{aligned}$$

and

$$\left(e^{\frac{n^2}{n+1}-\frac{n^2}{n+q}}\right)^{\frac{1}{q}+\frac{1}{n}} \leqslant e^2,$$

we get that $Z_q(\overline{K_{n+q}}(f)) \subseteq c_1 Z_q(\overline{K_{n+1}}(f))$. In the same way we establish an analogous inverse inclusion, and this completes the proof. $\square$

Theorem 10.4.11 is a direct consequence of the previous theorem.

Proof of Theorem 10.4.11. Recall that, since f is centered, the body $K_{n+1}(f)$ is also centered. We know then (recall (10.4.5)) that

$$\mathrm{Vol}_n(Z_n(\overline{K_{n+1}}(f)))^{1/n} \simeq 1$$

and hence by the previous theorem

$$f(0)^{1/n}\mathrm{Vol}_n(Z_n(f))^{1/n} \simeq \mathrm{Vol}_n(Z_n(\overline{K_{n+1}}(f)))^{1/n} \simeq 1.$$

For the second claim, we merely observe that, if f is the log-concave density of μ, then

$$\mathrm{Vol}_n(Z_n(\mu))^{1/n} = \mathrm{Vol}_n(Z_n(f))^{1/n} \simeq \frac{1}{f(0)^{1/n}}.$$

But since $\mathrm{bar}(\mu) = 0$, Theorem 10.6.2 shows that

$$e^{-n}\|\mu\|_\infty \leqslant f(0) \leqslant \|\mu\|_\infty,$$

and the claim follows. $\square$

10.4.3. Two basic formulas

Let us explain how L_q centroid bodies will help us prove Theorem 10.4.6. Recall the parameter $w_q(K) = M_q(K^\circ)$ which was defined and studied in Chapter 5, Section 5.7.2, given by

$$w_q(K) = \left(\int_{S^{n-1}} h_K^q(x) d\sigma(x) \right)^{1/q}.$$

It turns out that one may produce the following precise formulas for $I_q(\mu)$ and $I_{-q}(\mu)$:

LEMMA 10.4.14. *Let μ be an n-dimensional log-concave probability measure. For every $q \geqslant 1$ we have that*

$$I_q(\mu) \simeq \sqrt{\frac{q+n}{q}} w_q(Z_q(\mu)).$$

LEMMA 10.4.15. *Let μ be an n-dimensional log-concave probability measure with barycenter at the origin. For every integer $1 \leqslant q < n$ we have*

$$I_{-q}(\mu) \simeq \sqrt{\frac{n}{q}} w_{-q}(Z_q(\mu)). \tag{10.4.8}$$

If one bears in mind Theorem 5.8.7, which states that for a convex body K, up to the parameter $k^*(K)$ we have $w_q(K) \simeq w_{-q}(K)$, then the path to proving Theorem 10.4.6 becomes clear: all we need to do (after proving the two lemmas above) is to prove that there is some $q^* \geqslant \sqrt{n}$ for which $k^*(Z_q(\mu)) \geqslant q$. Indeed, this will be shown in Section 10.4.4. But first, let us prove the two lemmas above.

Proof of Lemma 10.4.14. We use again the fact we used in Lemma 10.4.4, that for every $x \in \mathbb{R}^n$,

$$\left(\int_{S^{n-1}} |\langle x, \theta \rangle|^q d\sigma(\theta) \right)^{1/q} = a_{n,q} \frac{\sqrt{q}}{\sqrt{q+n}} |x|,$$

where $a_{n,q} \simeq 1$. Since

$$w_q(Z_q(\mu)) = \left(\int_{S^{n-1}} \int_{\mathbb{R}^n} |\langle x, \theta \rangle|^q d\mu(x) d\sigma(\theta) \right)^{1/q},$$

the lemma follows. $\square$

Although the formula for $I_{-q}(\mu)$ is of the same nature, its proof is much more involved and uses marginals of measures quite strongly. We thus need some preparation. Recall that, given an integrable function $f : \mathbb{R}^n \to [0, \infty)$, an integer $k \in [1, n]$ and a subspace $F \in G_{n,k}$, we define the *marginal* $\pi_F(f) : F \to [0, \infty)$ of f with respect to F by

$$\pi_F(f)(x) := \int_{x+F^\perp} f(y) dy. \tag{10.4.9}$$

Note that in the case $k = n$ we have $F = \mathbb{R}^n$ and $\pi_F f = f$. More generally, for every $\mu \in \mathcal{P}_n$ we can define the marginal of μ with respect to the k-dimensional subspace F setting

$$\pi_F(\mu)(A) := \mu(P_F^{-1}(A))$$

for all Borel subsets of F. Note that when μ has a (log-concave) density f_μ, then these two definitions coincide. We recall some basic properties of marginals, all of which are easy to prove and apply also to the slightly more general case of marginals of measures. Here $f:\mathbb{R}^n\to[0,\infty)$ is an integrable function, $1\leqslant k\leqslant n$ and $F\in G_{n,k}$.

(1) If f is even then the same holds true for $\pi_F(f)$.
(2) We have
$$\int_F \pi_F(f)(x)\,dx=\int_{\mathbb{R}^n} f(x)\,dx.$$
(3) For any measurable function $g:F\to\mathbb{R}$ we have
$$\int_{\mathbb{R}^n} g(P_Fx)f(x)\,dx=\int_F g(x)\pi_F(f)(x)\,dx.$$
(4) For every $\theta\in S_F$,
$$\int_F\langle x,\theta\rangle\pi_F(f)(x)dx=\int_{\mathbb{R}^n}\langle x,\theta\rangle f(x)dx. \tag{10.4.10}$$
In particular, if f is centered then, for every $F\in G_{n,k}$, $\pi_F(f)$ is centered.
(5) For every $p>0$ and $\theta\in S_F$,
$$\int_{\mathbb{R}^n}|\langle x,\theta\rangle|^p f(x)dx=\int_F|\langle x,\theta\rangle|^p\pi_F(f)(x)dx.$$
In particular, if f is isotropic, the same holds true for $\pi_F(f)$.
(6) If f is log-concave, then $\pi_F(f)$ is log-concave.

The next simple but crucial proposition relates projections of the centroid bodies to centroid bodies of the corresponding marginals.

PROPOSITION 10.4.16. *Let $f:\mathbb{R}^n\to[0,\infty)$ be a density on $\mathbb{R}^n$. For every $1\leqslant k\leqslant n$ and for every $F\in G_{n,k}$ and $q\geqslant 1$, one has*
$$P_F(Z_q(f))=Z_q(\pi_F(f)). \tag{10.4.11}$$

Proof. For every $q\geqslant 1$ and $\theta\in S_F$, we easily check that
$$\int_{\mathbb{R}^n}|\langle x,\theta\rangle|^q f(x)dx=\int_F|\langle x,\theta\rangle|^q\pi_F(f)(x)dx,$$
using the fact that $\langle x,\theta\rangle=\langle P_F(x),\theta\rangle$ for all $x\in\mathbb{R}^n$. □

The proof of Lemma 10.4.15 is based on two basic integral formulas, together with a comparison between $(\mathrm{Vol}_k(P_Ff))^{-1}$ and $\pi_F(f)(0)$.

We formulate the integral formulas as two separate propositions because they are of independent interest.

PROPOSITION 10.4.17. *Let f be a log-concave density on $\mathbb{R}^n$ and let $1\leqslant k<n$ be a positive integer. Then,*
$$I_{-k}(f)=c_{n,k}\left(\int_{G_{n,k}}\pi_F(f)(0)d\nu_{n,k}(F)\right)^{-1/k}, \tag{10.4.12}$$
where
$$c_{n,k}=\left(\frac{(n-k)\kappa_{n-k}}{n\kappa_n}\right)^{1/k}\simeq\sqrt{n}.$$

Proof. Let $1 \leqslant k < n$. Then, we have

$$\begin{aligned}
&\int_{G_{n,k}} \pi_F(f)(0)\, d\nu_{n,k}(F) \\
&= \int_{G_{n,n-k}} \pi_{E^\perp}(f)(0)\, d\nu_{n,n-k}(E) \\
&= \int_{G_{n,n-k}} \int_E f(y)\, dy\, d\nu_{n,n-k}(E) \\
&= \int_{G_{n,n-k}} (n-k)\kappa_{n-k} \int_{S_E} \int_0^\infty r^{n-k-1} f(r\theta)\, dr\, d\sigma_E(\theta)\, d\nu_{n,n-k}(E) \\
&= \frac{(n-k)\kappa_{n-k}}{n\kappa_n} n\kappa_n \int_{S^{n-1}} \int_0^\infty r^{n-k-1} f(r\theta)\, dr\, d\sigma(\theta) \\
&= \frac{(n-k)\kappa_{n-k}}{n\kappa_n} \int_{\mathbb{R}^n} |x|^{-k} f(x)\, dx = \frac{(n-k)\kappa_{n-k}}{n\kappa_n} I_{-k}^{-k}(f).
\end{aligned}$$

It follows that

$$I_{-k}(f) = \left(\frac{(n-k)\kappa_{n-k}}{n\kappa_n}\right)^{1/k} \left(\int_{G_{n,k}} \pi_F(f)(0)\, d\nu_{n,k}(F)\right)^{-1/k}.$$

One can check $c_{n,k} = \left(\frac{(n-k)\kappa_{n-k}}{n\kappa_n}\right)^{1/k} \simeq \sqrt{n}$ and the proof is complete. □

PROPOSITION 10.4.18. *Let C be a symmetric convex body in $\mathbb{R}^n$ and let $1 \leqslant k < n$ be a positive integer. Then,*

$$w_{-k}(C) \simeq \sqrt{k} \left(\int_{G_{n,k}} \mathrm{Vol}_k(P_F(C))^{-1} d\nu_{n,k}(F)\right)^{-1/k}.$$

Proof. Using the Blaschke-Santaló and the reverse Santaló inequality (Theorem 1.5.10 from Chapter 1 and the Bourgain Milman Theorem 8.2.2 from Chapter 8), we write

$$\begin{aligned}
w_{-k}^{-1}(C) &= \left(\int_{S^{n-1}} \frac{1}{h_C^k(\theta)} d\sigma(\theta)\right)^{1/k} \\
&= \left(\int_{G_{n,k}} \int_{S_F} \frac{1}{\|\theta\|_{(P_F C)^\circ}^k} d\sigma(\theta) d\nu_{n,k}(F)\right)^{1/k} \\
&= \left(\int_{G_{n,k}} \frac{\mathrm{Vol}_k(P_F(C))^\circ}{\mathrm{Vol}_k(B_2^k)} d\nu_{n,k}(F)\right)^{1/k} \\
&\simeq \left(\int_{G_{n,k}} \frac{\mathrm{Vol}_k(B_2^k)}{\mathrm{Vol}_k(P_F(C))} d\nu_{n,k}(F)\right)^{1/k},
\end{aligned}$$

and the result follows. □

Thus, to prove Lemma 10.4.15 all we need is to show that

$$\frac{1}{\mathrm{Vol}_k(P_F(Z_k(f)))^{1/k}} \simeq \pi_F(f)(0)^{1/k}.$$

This follows from Theorem 10.4.11 which states that for a centered log-concave density f on $\mathbb{R}^n$

$$\frac{c_1}{f(0)^{1/n}} \leqslant \mathrm{Vol}_n(Z_n(f))^{1/n} \leqslant \frac{c_2}{f(0)^{1/n}}, \tag{10.4.13}$$

where $c_1, c_2 > 0$ are universal constants.

Proof of Lemma 10.4.15. Let f be a centered log-concave density on $\mathbb{R}^n$. Then, for every $F \in G_{n,k}$, the function $\pi_F(f)$ is a centered log-concave density on F. So, we may apply (10.4.13) for $\pi_F(f)$. It follows that

$$\frac{c_1}{\pi_F(f)(0)^{1/k}} \leqslant \mathrm{Vol}_k(Z_k(\pi_F(f)))^{1/k} \leqslant \frac{c_2}{\pi_F(f)(0)^{1/k}}.$$

This fact, combined with Proposition 10.4.17 and Proposition 10.4.18, completes the proof of Lemma 10.4.15. $\square$

10.4.4. The parameter q_*

DEFINITION 10.4.19. Let μ be a centered log-concave probability measure in $\mathbb{R}^n$. We define

$$q_*(\mu) = \max\left\{q \geqslant 1 : c_0 n\left(\frac{w(Z_q(\mu))}{R(Z_q(\mu))}\right)^2 = q\right\},$$

where $c_0 > 0$ is an absolute constant such that $c_0 n\left(\frac{w(A)}{R(A)}\right)^2 \geqslant 1$ for every symmetric convex body A in $\mathbb{R}^n$. Note that the equation clearly has at least one solution since both sides are continuous functions of q and for $q = n$ the left hand side is bounded by $c_0 n$. Recall that Theorem 5.7.3, applied to the polar body $Z_q^\circ(\mu)$ of $Z_q(\mu)$, states that if $1 \leqslant q \leqslant C_1 k_*(Z_q(\mu))$ then $w_q(Z_q(\mu)) \leqslant C_2 w(Z_q(\mu))$. In particular, this is true for $q = q_*(\mu)$.

PROPOSITION 10.4.20. *For every centered log-concave probability measure μ in $\mathbb{R}^n$ we have*

$$q_*(\mu) \geqslant c\sqrt{n}\left(\frac{w(Z_2(\mu))}{R(Z_2(\mu))}\right)$$

for some absolute constant $c > 0$. In particular, for an isotropic μ

$$q_*(\mu) \geqslant c\sqrt{n}.$$

Proof. Let $q_* := q_*(\mu)$. We write

$$\begin{aligned} w(Z_{q_*}(\mu)) &\geqslant c_1 w_{q_*}(Z_{q_*}(\mu)) = c_1 a_{n,q_*}\sqrt{\frac{q_*}{n+q_*}} I_{q_*}(\mu) \\ &\geqslant c_2 a_{n,q_*}\sqrt{\frac{q_*}{n+q_*}} I_2(\mu) \\ &= c_2 a_{n,q_*}\sqrt{\frac{q_*}{n+q_*}}\,\sqrt{n} w_2(Z_2(\mu)). \end{aligned}$$

(Note that if $1 \leqslant q_* \leqslant 2$ then we still have $I_{q_*}(\mu) \simeq I_2(\mu)$). In other words,

$$w(Z_{q_*}(\mu)) \geqslant c_2\sqrt{q_*} w(Z_2(\mu)). \tag{10.4.14}$$

Since μ is a ψ_1-measure with an absolute constant c, by Theorem 3.5.11 we have

$$R(Z_{q_*}(\mu)) \leqslant c q_* R(Z_2(\mu)). \tag{10.4.15}$$

Using the definition of q_* and the inequalities (10.4.14) and (10.4.15), we write

$$q_* = n\left(\frac{w(Z_{q_*}(\mu))}{R(Z_{q_*}(\mu))}\right)^2 \geqslant c_3 n\frac{q_*}{q_*^2}\left(\frac{w(Z_2(\mu))}{R(Z_2(\mu))}\right)^2.$$

The first part of the proposition follows. Observe that μ is isotropic if and only if $Z_2(\mu)$ is a ball, in which case we get the bound $\sqrt{n}$. □

Finally, we have all the ingredients to prove Theorem 10.4.6.

Proof of Theorem 10.4.6. Let μ be a log-concave isotropic probability measure. By Proposition 10.4.20 we know that $q_*(\mu) \geqslant c\sqrt{n}$. Using Theorem 5.8.7 together with the definition of $q_* \simeq k^*(Z_{q_*}(\mu))$, we see that

$$w_{q_*}(Z_{q_*}(\mu)) \simeq w_{-q_*}(Z_{q_*}(\mu)).$$

Combining this fact with Lemma 10.4.14 and Lemma 10.4.15 we get that

$$I_{q_*}(\mu) \simeq \sqrt{\frac{q_*+n}{q_*}}w_{q_*}(Z_{q_*}(\mu)) \simeq \sqrt{\frac{n}{q_*}}w_{-q_*}(Z_{q_*}(\mu)) \simeq I_{-q_*}(\mu).$$

The proof is thus complete. □

10.5. The isomorphic slicing problem

10.5.1. Klartag's convex perturbations

In this section we describe Klartag's affirmative answer to the isomorphic slicing problem (note that for the purposes of Section 10.5.3 we state and prove the estimates for the isotropic constant of the approximating bodies T in more detail than is needed for the purposes of the isomorphic slicing problem itself).

THEOREM 10.5.1 (Klartag). *Let K be a convex body in $\mathbb{R}^n$. For every $\varepsilon \in (0,1)$ there exists a centered convex body $T \subset \mathbb{R}^n$ and a point $x \in \mathbb{R}^n$ such that*

$$\frac{1}{1+\varepsilon}T \subseteq K + x \subseteq (1+\varepsilon)T \tag{10.5.1}$$

and

$$L_T \leqslant \frac{C}{\sqrt{\varepsilon n s(K-K)^{1/n}}}, \tag{10.5.2}$$

where $s(K-K) = \mathrm{Vol}_n(K-K)\mathrm{Vol}_n((K-K)^\circ)$ and $C > 0$ is an absolute constant. In particular,

$$L_T \leqslant \frac{C'}{\sqrt{\varepsilon}}$$

for some absolute constant $C' > 0$.

The idea of the proof is to exchange K with a body T which will be $K_{n+1}(g)$ for some function g on (the centered copy of) K, which is not much different from $\mathbf{1}_K$. This function will be chosen from a family of so called "tilts" of $\mathbf{1}_K$, which are functions proportional to $e^{\langle x,\xi\rangle}\mathbf{1}_K$. First we state a lemma allowing us to compare the isotropic constant of f and of $K_{n+1}(g)$ when g is the centered translation of f and f has bounded variation on K.

LEMMA 10.5.2. *Let K be a convex body in $\mathbb{R}^n$ and let $f : K \to (0,\infty)$ be a log-concave function such that*

$$\sup_{x\in K} f(x) \leqslant m^{n+1} \inf_{x\in K} f(x)$$

for some $m > 1$. We write $x_0 = \mathrm{bar}(f)$ and denote the centered translation of f by g, namely $g(x) := f(x+x_0)$ for all $x \in K - x_0$. Then the body $T := K_{n+1}(g)$ is centered,

$$L_f = L_g \simeq L_T$$

and

$$\frac{1}{m}T \subseteq K - x_0 \subseteq mT. \tag{10.5.3}$$

Proof. It is easy to check that f seen as a function on all of $\mathbb{R}^n$ (where we set $f(x) = 0$ for $x \notin K$) is log-concave as well, and that $x_0 = \mathrm{bar}(f) \in K$. The facts that T is centered and that $L_T \simeq L_g = L_f$ follow easily from the results of Section 10.2 (and in particular from Proposition 10.2.16), so it remains to prove (10.5.3). Since $K_{n+1}(\lambda g) = K_{n+1}(g)$ for every $\lambda > 0$, we may assume without loss of generality that $g(0) = \mathbf{1}_{K-x_0}(0) = 1$, in which case

$$\inf\left\{\frac{g(x)}{\mathbf{1}_{K-x_0}(x)} : \mathbf{1}_{K-x_0}(x) > 0\right\} = \inf_{x\in K-x_0} g(x) = \inf_{y\in K} f(y) \geqslant m^{-(n+1)}$$

and

$$\inf\left\{\frac{\mathbf{1}_{K-x_0}(x)}{g(x)} : g(x) > 0\right\} = \left(\sup_{y\in K} f(y)\right)^{-1} \geqslant m^{-(n+1)}.$$

But then, by Lemma 10.2.8 and Proposition 10.2.9 (iv) we obtain

$$\frac{1}{m}K_{n+1}(g) \subseteq K_{n+1}\big(\mathbf{1}_{K-x_0}\big) = K - x_0 \subseteq mK_{n+1}(g),$$

which completes the proof. □

The way to choose f is "probabilistic" in nature. We consider the uniform measure on K, which we denote by $\mu = \mathbf{1}_K dx$. Together with it we shall consider a family of measures $\{\mu_\xi\}_{\xi\in\mathbb{R}^n}$ (one may call them "tilts" of μ) which will be probability measures with density proportional to $e^{\langle x,\xi\rangle}\mathbf{1}_K(x)$. The properties of these measures are closely related to the logarithmic Laplace transform of the measure μ. Let us recall the definition the logarithmic Laplace transform of a general Borel measure μ on $\mathbb{R}^n$ which is given by

$$\Lambda_\mu(\xi) = \log\left(\frac{1}{\mu(\mathbb{R}^n)}\int_{\mathbb{R}^n} e^{\langle\xi,x\rangle}d\mu(x)\right). \tag{10.5.4}$$

The properties relevant to our discussion are summarized in the following proposition.

PROPOSITION 10.5.3. *If $\mu = \mu_K$ is Lebesgue measure on some convex body K in $\mathbb{R}^n$, then*

$$(\nabla\Lambda_\mu)(\mathbb{R}^n) = \mathrm{int}(K). \tag{10.5.5}$$

If μ_ξ is the probability measure on $\mathbb{R}^n$ with density proportional to $e^{\langle\xi,x\rangle}\mathbf{1}_K(x)$, then

$$bar(\mu_\xi) = \nabla\Lambda_\mu(\xi) \tag{10.5.6}$$

and

$$(\text{Hess}\Lambda_\mu)(\xi) = \text{Cov}(\mu_\xi). \tag{10.5.7}$$

Moreover, the map $\nabla\Lambda_\mu$ *transports the measure* ν *with density* $\det(\text{Hess}\Lambda_\mu)(\xi)$ *to* μ. *Equivalently, for every continuous non-negative function* $\phi : \mathbb{R}^n \to \mathbb{R}$,

$$\int_K \phi(x)\,dx = \int_{\mathbb{R}^n} \phi(\nabla\Lambda_\mu(\xi)) \det(\text{Hess}(\Lambda_\mu)(\xi))\,d\xi = \int \phi(\nabla\Lambda_\mu(\xi))d\nu(\xi). \tag{10.5.8}$$

Proof. Let $F = \Lambda_\mu$, that is

$$F(x) = \log\left(\frac{1}{\text{Vol}_n(K)}\int_K e^{\langle x,y\rangle}\,dy\right). \tag{10.5.9}$$

Observe that F is a C^2-smooth, strictly convex function. Smoothness is clear, as we are integrating a smooth function on a compact set. The strict convexity follows from Cauchy-Schwarz inequality. Differentiating under the integral sign we get:

$$\nabla F(\xi) = \frac{\int_K ye^{\langle \xi,y\rangle}\,dy}{\int_K e^{\langle \xi,z\rangle}\,dz} = \int y\,d\mu_\xi(y) = \text{bar}(\mu_\xi). \tag{10.5.10}$$

Since μ_ξ is supported on the compact, convex set K we obtain that $\nabla F(\xi) = \text{bar}(\mu_\xi) \in K$ for all $\xi \in \mathbb{R}^n$. This shows that $\nabla F(\mathbb{R}^n) \subseteq K$.

To compute the Hessian we differentiate twice to get:

$$\begin{aligned}\frac{\partial^2 F(\xi)}{\partial\xi_j\partial\xi_i} &= \frac{\int_K x_ix_je^{\langle\xi,x\rangle}\,dx - \int_K x_ie^{\langle\xi,x\rangle}\,dx\int_K x_je^{\langle\xi,x\rangle}\,dx}{\left(\int_K e^{\langle\xi,x\rangle}\,dx\right)^2} \\ &= \int x_ix_j\,d\mu_\xi(x) - \int x_i\,d\mu_\xi(x)\int x_j\,d\mu_\xi(x) \\ &= \text{Cov}(\mu_\xi)_{ij}.\end{aligned} \tag{10.5.11}$$

Next we prove that $\nabla F(\mathbb{R}^n) = \text{int}(K)$. Let $z \in \text{bd}(K)$ be an exposed point of K. Then there exist $u \in \mathbb{R}^n$ and $t \in \mathbb{R}$ such that $\langle u, z\rangle = t$ and for any $x \in K$, $x \neq z$ we have $\langle x, z\rangle < t$. Consider the probability measure μ_{ru} for large $r > 0$. Its density $d\mu_{ru}(x)$ is proportional to $e^{r\langle u,x\rangle}\mathbf{1}_K(x)$ and it attains its unique maximum at z. Moreover, it is straightforward to verify that as $r \to \infty$

$$\mu_{ru} \xrightarrow{w^*} \delta_z, \tag{10.5.12}$$

where δ_z is the Dirac measure at z. Therefore, by (10.5.10) we obtain

$$\nabla F(ru) \to_{r\to\infty} z, \tag{10.5.13}$$

and so $z \in \overline{\nabla F(\mathbb{R}^n)}$.

Since z was an arbitrary exposed point of K, using a theorem of Straszewicz (see [**556**]) stating that $K = \overline{\text{conv}(\exp(K))}$ and since $\overline{\nabla F(\mathbb{R}^n)}$ is convex, we get that $K \subseteq \overline{\nabla F(\mathbb{R}^n)}$. Moreover, $\nabla F(\mathbb{R}^n)$ is open and combining with the fact $\nabla F(\mathbb{R}^n) \subseteq K \subseteq \overline{\nabla F(\mathbb{R}^n)}$ we conclude that $\nabla F(\mathbb{R}^n)$ is the interior of K. For the last part note that since F is strictly convex, ∇F is one-to-one. So, for any continuous function $g : \mathbb{R}^n \to \mathbb{R}$, changing variables $y = \nabla F(\xi)$ we get

$$\int_{\nabla F(\mathbb{R}^n)} g(y)\,dy = \int_{\mathbb{R}^n} g(\nabla F(\xi)) \det(\text{Hess}F)(\xi)\,d\xi = \int_{\mathbb{R}^n} g(\nabla F(\xi))\,d\nu(\xi). \tag{10.5.14}$$

This completes the proof of the proposition. □

Proof of Theorem 10.5.1. We may clearly assume without loss of generality that K is centered and that $\mathrm{Vol}_n(K)=1$, because L_T and $s(K-K)$ are affine invariants.

We denote again by $\mu=\mu_K$ the Lebesgue measure restricted on K, and

$$d\nu(\xi)=\det(\mathrm{Hess}\,\Lambda_\mu)(\xi)\equiv\det\mathrm{Cov}(\mu_\xi)d\xi$$

as in Proposition 10.5.3. Using (10.5.8) with $\phi=1$ we get that

$$\nu(\mathbb{R}^n)=\int_{\mathbb{R}^n}\mathbf{1}\det(\mathrm{Hess}\,\Lambda_\mu)(\xi)\,d\xi=\int_K\mathbf{1}\,dx=\mathrm{Vol}_n(K)=1.$$

Thus, for every $\varepsilon>0$ we may write

$$\begin{aligned}\mathrm{Vol}_n(\varepsilon n(K-K)^\circ)&\min_{\xi\in\varepsilon n(K-K)^\circ}\det\mathrm{Cov}(\mu_\xi)\leqslant\\ &\leqslant\int_{\varepsilon n(K-K)^\circ}\det\mathrm{Cov}(\mu_\xi)\,d\xi=\nu(\varepsilon n(K-K)^\circ)\leqslant 1,\end{aligned}$$

which shows that there exists $\xi_0\in\varepsilon n(K-K)^\circ$ such that

$$\begin{aligned}\det\mathrm{Cov}(\mu_{\xi_0})&=\min_{\xi\in\varepsilon n(K-K)^\circ}\det\mathrm{Cov}(\mu_\xi)\leqslant\mathrm{Vol}_n(\varepsilon n(K-K)^\circ)^{-1}\\ &\leqslant\left(\frac{4}{\varepsilon ns(K-K)^{1/n}}\right)^n,\end{aligned}$$

where we have used the Rogers-Shephard inequality, Theorem 1.5.2, in the last inequality. From the definition of μ_{ξ_0} and of the isotropic constant we have that

$$L_{\mu_{\xi_0}}=\left(\frac{\sup_{x\in K}e^{\langle\xi_0,x\rangle}}{\int_K e^{\langle\xi_0,x\rangle}dx}\right)^{\frac{1}{n}}[\det\mathrm{Cov}(\mu_{\xi_0})]^{\frac{1}{2n}}.$$

Since $\xi_0\in\varepsilon n(K-K)^\circ$ and $K\cup(-K)\subset K-K$, we know that $|\langle\xi_0,x\rangle|\leqslant\varepsilon n$ for all $x\in K$, therefore

$$\sup_{x\in K}e^{\langle\xi_0,x\rangle}\leqslant e^{\varepsilon n}\qquad\text{and}\qquad\sup_{x\in K}e^{\langle\xi_0,x\rangle}\geqslant e^{-\varepsilon n}.$$

On the other hand, since K is centered, from Jensen's inequality we have that

$$\int_K e^{\langle\xi_0,x\rangle}dx\geqslant e^{\left(\int_K\langle\xi_0,x\rangle\,dx\right)}=1.$$

Combining all these we get

$$L_{\mu_{\xi_0}}\leqslant\frac{2e^\varepsilon}{\sqrt{\varepsilon ns(K-K)^{1/n}}}.\tag{10.5.15}$$

Finally, we note that the function $f_{\xi_0}(x)=e^{\langle\xi_0,x\rangle}\mathbf{1}_K(x)$ (which is proportional to the density of μ_{ξ_0}) is obviously log-concave and satisfies

$$\sup_{x\in\mathrm{supp}(f_{\xi_0})}f_{\xi_0}(x)\leqslant e^{2\varepsilon n}\inf_{x\in\mathrm{supp}(f_{\xi_0})}f_{\xi_0}(x).$$

Therefore, applying Lemma 10.5.2, we can find a centered convex body T_{ξ_0} in $\mathbb{R}^n$ such that

$$L_{T_{\xi_0}}\simeq L_{f_{\xi_0}}=L_{\mu_{\xi_0}}\leqslant\frac{2e^\varepsilon}{\sqrt{\varepsilon ns(K-K)^{1/n}}}$$

and

$$\frac{1}{e^{2\varepsilon}}T_{\xi_0}\subseteq K-b_{\xi_0}\subseteq e^{2\varepsilon}T_{\xi_0}$$

where b_{ξ_0} is the barycenter of f_{ξ_0}. Given that $e^{2\varepsilon} \leqslant 1 + c\varepsilon$ when $\varepsilon \in (0,1)$, the result follows. □

10.5.2. Klartag's upper bound for the isotropic constant

Using Theorem 10.5.1 and Paouris' distributional inequality, Klartag was also able to slightly improve Bourgain's upper bound for the isotropic constant:

THEOREM 10.5.4 (Klartag). *Let K be a convex body in $\mathbb{R}^n$. Then*

$$L_K \leqslant Cn^{1/4}, \tag{10.5.16}$$

where $C > 0$ is an absolute constant.

Theorem 10.5.4 will follow from Theorem 10.5.1 combined with the following lemma.

LEMMA 10.5.5. *Let K, T be two convex bodies in $\mathbb{R}^n$ and $t \geqslant 1$. Suppose that*

$$\frac{1}{1+\frac{t}{\sqrt{n}}}(T+y) \subseteq K + x \subseteq \left(1+\frac{t}{\sqrt{n}}\right)(T+y) \tag{10.5.17}$$

for some $x, y \in \mathbb{R}^n$. Then

$$L_K \leqslant ctL_T,$$

where $c > 0$ is an absolute constant.

Proof. We may assume that $t < \sqrt{n}$, otherwise the conclusion of the lemma is trivial since by Remark 10.1.16 and Proposition 10.1.8 we have $L_K < c\sqrt{n} \leqslant c'\sqrt{n}L_K$. Note that (10.5.17) continues to hold (with possibly different $x, y \in \mathbb{R}^n$) if we translate either K or T or if we apply an invertible linear transformation to both of them, thus we may assume that T is in isotropic position. Then by Paouris's Theorem 10.4.1 we have that

$$\mathrm{Vol}_n(T \setminus Ct\sqrt{n}L_T B_2^n) \leqslant \exp(-4t\sqrt{n})$$

for some absolute constant $C \geqslant 1$. We set

$$K_1 = \left(1+\frac{t}{\sqrt{n}}\right)^{-1}(K+x) - y,$$

and by (10.5.17) we have $K_1 \subseteq T$, and hence

$$\mathrm{Vol}_n(K_1 \setminus Ct\sqrt{n}L_T B_2^n) \leqslant \exp(-4t\sqrt{n}). \tag{10.5.18}$$

By (10.5.17) we also see that

$$\begin{aligned} \mathrm{Vol}_n(K_1) &= \left(1+\frac{t}{\sqrt{n}}\right)^{-n}\mathrm{Vol}_n(K) \\ &\geqslant \left(1+\frac{t}{\sqrt{n}}\right)^{-2n}\mathrm{Vol}_n(T) > e^{-2t\sqrt{n}}, \end{aligned} \tag{10.5.19}$$

which combined with (10.5.18) gives

$$\mathrm{Vol}_n(K_1 \cap (Ct\sqrt{n}L_T B_2^n)) > \frac{\mathrm{Vol}_n(K)}{2}.$$

Therefore the median of the Euclidean norm on K_1, with respect to the uniform measure on K_1, is not larger than $Ct\sqrt{n}L_T$. Since K_1 is convex, and hence the uniform measure on K_1 is a log-concave probability measure, using Theorem 10.6.4 we obtain

$$\left(\frac{1}{\mathrm{Vol}_n(K_1)}\int_{K_1}|x|^2\,dx\right)^{1/2} \leqslant C't\sqrt{n}L_T, \tag{10.5.20}$$

for some absolute constant $C'>0$. Recall that by Theorem 10.1.4

$$\sqrt{n}L_K=\sqrt{n}L_{K_1}=\min\left\{\left(\frac{1}{\mathrm{Vol}_n(S(K_0))^{1+\frac{2}{n}}}\int_{S(K_0)}|x|^2\,dx\right)^{1/2}\,\Big|\, S\in GL_n\right\},$$

where K_0 is the centered translate of K_1, that is,

$$K_0=K_1-\mathrm{bar}(K_1)=K_1-\frac{1}{\mathrm{Vol}_n(K_1)}\int_{K_1}xdx.$$

It is also not hard to check that

$$\begin{aligned}\frac{1}{\mathrm{Vol}_n(K_0)^{1+\frac{2}{n}}}\int_{K_0}|x|^2\,dx &= \frac{1}{\mathrm{Vol}_n(K_1)^{1+\frac{2}{n}}}\int_{K_1}|x|^2\,dx-\frac{1}{\mathrm{Vol}_n(K_1)^{\frac{2}{n}}}|\mathrm{bar}(K_1)|^2\\ &\leqslant \frac{1}{\mathrm{Vol}_n(K_1)^{1+\frac{2}{n}}}\int_{K_1}|x|^2\,dx,\end{aligned}$$

and thus

$$\sqrt{n}L_K\leqslant\left(\frac{1}{\mathrm{Vol}_n(K_1)^{1+\frac{2}{n}}}\int_{K_1}|x|^2\,dx\right)^{1/2}\leqslant\frac{C't\sqrt{n}L_T}{\mathrm{Vol}_n(K_1)^{1/n}}\leqslant C''t\sqrt{n}L_T$$

by (10.5.19) and (10.5.20). This proves the lemma. □

Proof of Theorem 10.5.4. Let K be a convex body in $\mathbb{R}^n$. According to Theorem 10.5.1, given $\varepsilon\in(0,1)$ we can find a centered convex body $T=T_\varepsilon$ such that

$$L_T\leqslant\frac{C}{\sqrt{\varepsilon}}$$

and

$$\frac{1}{(1+\varepsilon)}T\subseteq K+x\subseteq(1+\varepsilon)T$$

for some $x=x_\varepsilon\in\mathbb{R}^n$. If we choose $\varepsilon=\frac{1}{\sqrt{n}}$, Lemma 10.5.5 shows that

$$L_K\leqslant cL_T\leqslant c\frac{C}{\sqrt{\varepsilon}}=C'\sqrt[4]{n},$$

which was the assertion of the theorem. □

10.5.3. Proof of the reverse Santaló inequality via the isotropic position

In Section 10.5.1 we proved (10.5.1) and (10.5.2) of Theorem 10.5.1 without using the reverse Santaló inequality. In this section we show how one can give an alternative proof of the reverse Santaló inequality (Theorem 8.2.2), which is different than the one we saw in Chapter 8 in that it does not use the MM^* estimate and Pisier's bound for the norm of the Rademacher projection. The proof is divided into two steps, where in the first step, using some simple entropy estimates for isotropic convex bodies, one proves a variant of the reverse Santaló inequality which involves the isotropic constant, and in the second step the isotropic constant is removed.

THEOREM 10.5.6. *Let K be an isotropic convex body in $\mathbb{R}^n$. For every $t>0$ we have*

$$N(K,tB_2^n)\leqslant 2\exp\left(\frac{4(n+1)I_1(K)}{t}\right)\leqslant 2\exp\left(\frac{cn^{3/2}L_K}{t}\right).$$

Proof. Consider the gauge function of K given by $p_K(x)=\inf\{\lambda>0: x\in\lambda K\}$. We define a Borel probability measure on $\mathbb{R}^n$ by

$$\mu(A)=\frac{1}{c_K}\int_A e^{-p_K(x)}dx,$$

where $c_K=\int_{\mathbb{R}^n}\exp(-p_K(x))dx$. Let $\{x_1,\dots,x_N\}$ be a maximal t-separated set in K, so that

$$i\neq j\Longrightarrow |x_i-x_j|\geqslant t$$

and $K\subseteq\bigcup_{i\leqslant N}(x_i+tB_2^n)$. (In particular, $N\geqslant N(K,tB_2^n)$.)

We choose $b>0$ so that $\mu(bB_2^n)\geqslant 1/2$. If we set $y_i=(2b/t)x_i$, then

$$\begin{aligned}\mu(y_i+bB_2^n)&=\frac{1}{c_K}\int_{bB_2^n}e^{-p_K(x+y_i)}dx\geqslant\frac{1}{c_K}\int_{bB_2^n}e^{-p_K(x)}e^{-p_K(y_i)}dx\\&=e^{-p_K(y_i)}\frac{1}{c_K}\int_{bB_2^n}e^{-p_K(x)}dx=e^{-\frac{2b}{t}p_K(x_i)}\mu(bB_2^n)\\&\geqslant e^{-2b/t}\mu(bB_2^n).\end{aligned}$$

The balls $y_i+bB_2^n$ have disjoint interiors, therefore

$$Ne^{-2b/t}\mu(bB_2^n)\leqslant\sum_{i=1}^N\mu(y_i+bB_2^n)=\mu\left(\bigcup_{i=1}^N(y_i+bB_2^n)\right)\leqslant 1.$$

It follows that

$$N(K,tB_2^n)\leqslant e^{2b/t}(\mu(bB_2^n))^{-1}\leqslant 2e^{2b/t}.$$

What remains is to estimate b from above. We first compute the constant

$$\begin{aligned}c_K&=\int_{\mathbb{R}^n}e^{-p_K(x)}dx=\int_{\mathbb{R}^n}\int_{p_K(x)}^\infty e^{-s}dsdx\\&=\int_0^\infty e^{-s}\mathrm{Vol}_n(\{x: p_K(x)\leqslant s\})ds\\&=\int_0^\infty\mathrm{Vol}_n(sK)\,e^{-s}ds=\int_0^\infty s^ne^{-s}ds=n!.\end{aligned}$$

It follows that

$$\begin{aligned}J:=\int_{\mathbb{R}^n}|x|d\mu(x)&=\frac{1}{c_K}\int_{\mathbb{R}^n}|x|\int_{p_K(x)}^\infty e^{-s}dsdx\\&=\frac{1}{n!}\int_0^\infty s^{n+1}e^{-s}ds\cdot\int_K|x|dx=(n+1)I_1(K).\end{aligned}$$

From Markov's inequality, $\mu(\{x\in\mathbb{R}^n: |x|>2J\})\leqslant 1/2$, which shows that $\mu(2JB_2^n)\geqslant 1/2$. If we choose $b=2J$, we get

$$N(K,tB_2^n)\leqslant 2\exp(4J/t)\leqslant 2\exp(4(n+1)I_1(K)/t)\leqslant 2\exp(cn^{3/2}L_K/t),$$

since $I_1\leqslant I_2=\sqrt{n}L_K$. $\square$

We use the entropy estimates we have for isotropic convex bodies in their dual form. The following Lemma follows from Theorem 10.5.6 together with duality for covering numbers (Theorem 4.4.4). We point out that in fact it can be shown to easily follow from considerations as in Theorem 4.2.6 which do not need the full strength of the duality theorem.

LEMMA 10.5.7. *Let K be an isotropic convex body in $\mathbb{R}^n$. Then, for every $t > 0$ one has*

$$\log N(B_2^n, tK^\circ) \leqslant \log N(B_2^n, t(K-K)^\circ) \leqslant \frac{c_2 n^{3/2} L_K}{t},$$

where $c_2 > 0$ is an absolute constant.

We next state the first step of the proof, namely a reverse Santaló inequality with lower bound depending on L_K.

PROPOSITION 10.5.8. *Let K be a convex body which contains 0 in its interior. Then,*

$$\text{Vol}_n(K-K)^{1/n}\text{Vol}_n(n(K-K)^\circ)^{1/n} \geqslant \frac{c_1}{L_K}, \tag{10.5.21}$$

where $c_1 > 0$ is an absolute constant. In particular

$$\text{Vol}_n(K)^{1/n}\text{Vol}_n(nK^\circ)^{1/n} \geqslant \frac{c_2}{L_K}, \tag{10.5.22}$$

where $c_2 > 0$ is an absolute constant.

Proof. We may assume that $\text{Vol}_n(K) = 1$. Since $\text{Vol}_n(K-K)\text{Vol}_n((K-K)^\circ)$ is an affine invariant, we may assume without loss of generality that K is in isotropic position. Then, by Lemma 10.5.7

$$N\big(B_2^n, \sqrt{n}L_K(K-K)^\circ\big) \leqslant C^n$$

for some absolute constant $C \geqslant 1$, which means that we can find $x_1, \dots, x_N$ with $N \leqslant C^n$ such that

$$B_2^n \subseteq \bigcup_{1\leqslant i\leqslant N} \big(x_i + \sqrt{n}L_K(K-K)^\circ\big).$$

It follows that

$$\text{Vol}_n(B_2^n) \leqslant C^n\text{Vol}_n(\sqrt{n}L_K(K-K)^\circ)),$$

and hence

$$\text{Vol}_n((K-K)^\circ)^{1/n} \geqslant \frac{\text{Vol}_n(B_2^n)^{1/n}}{C\sqrt{n}L_K} \geqslant \frac{c_1}{nL_K},$$

where $c_1 > 0$ is an absolute constant. Recalling also that by Brunn-Minkowski $\text{Vol}_n(K) = 1$ implies that $\text{Vol}_n(K-K)^{1/n}\text{Vol}_n((K-K)^\circ)^{1/n} \geqslant 2\text{Vol}_n((K-K)^\circ)^{1/n}$, we conclude the proof of (10.5.21). The second equation follows from the first using the Rogers-Shephard inequality (Theorem 1.5.2). □

In the second step of the proof we combine Proposition 10.5.8 with Theorem 10.5.1 to remove the isotropic constant L_K from the lower bound for $s(K)^{1/n}$ and thus re-prove the reverse Santaló inequality.

Proof of Theorem 8.2.2. Recall that as in the derivation of (10.5.22) from (10.5.21), it suffices to show the reverse Santaló inequality for symmetric bodies. Given then a symmetric body K in $\mathbb{R}^n$, we apply Klartag's method of convex perturbations, Theorem 10.5.1, to find a convex body $T \subset \mathbb{R}^n$ satisfying (10.5.1) and (10.5.2) of Theorem 10.5.1 with $\varepsilon = 1/2$, namely

$$\frac{2}{3}T \subseteq K + x \subseteq \frac{3}{2}T \tag{10.5.23}$$

for some $x \in \mathbb{R}^n$, and

$$L_T \leqslant \frac{c_0}{\sqrt{ns(K)^{1/n}}}$$

for some absolute constant c_0. Proposition 10.5.8 shows that

$$\mathrm{Vol}_n(T-T)^{1/n}\mathrm{Vol}_n(n(T-T)^\circ)^{1/n} \geqslant \frac{c_1}{L_T},$$

where $c_1 > 0$ is an absolute constant. Observe that by (10.5.23) we get

$$\frac{2}{3}(T-T) \subseteq K-K = 2K \subseteq \frac{3}{2}(T-T),$$

and thus $K^\circ \supseteq \frac{4}{3}(T-T)^\circ$. Combining the above, we obtain

$$\begin{aligned} ns(K)^{1/n} &= \mathrm{Vol}_n(nK^\circ)^{1/n}\mathrm{Vol}_n(K)^{1/n} \\ &\geqslant \frac{4}{9}\mathrm{Vol}_n(n(T-T)^\circ)^{1/n}\mathrm{Vol}_n(T-T)^{1/n} \geqslant \frac{c_1'}{L_T} \\ &\geqslant c_2\sqrt{ns(K)^{1/n}}. \end{aligned}$$

It follows that $s(K)^{1/n} \geqslant c_3/n$ with $c_3 = c_2^2$, which completes the proof. □

10.6. A few technical inequalities for log-concave functions

We consider this section as a short technical appendix to the main parts of the chapter. In it we collect some basic inequalities about log-concave functions and log-concave probability measures that were used in this chapter, The first two lemmas establish two basic properties of log-concave functions. We first show that every integrable log-concave function $f : \mathbb{R}^n \to [0,\infty)$ has finite moments of all orders; this is a consequence of the fact that $f(x)$ decays exponentially as $|x| \to \infty$.

LEMMA 10.6.1. *Let $f : \mathbb{R}^n \to [0,\infty)$ be a log-concave function with finite, positive integral. Then, there exist constants $A, B > 0$ such that $f(x) \leqslant Ae^{-B|x|}$ for all $x \in \mathbb{R}^n$. In particular, f has finite moments of all orders.*

Proof. Since $\int f > 0$, we can find $t \in (0,1)$ such that the set $C := \{x : f(x) > t\}$ has positive Lebesgue measure. Note that C is convex because f is log-concave; it follows that C is a convex set of positive volume, hence it has non-empty interior. Let $x_0 \in C$ and $r > 0$ such that $x_0 + rB_2^n \subseteq C$. Working with $f_1(\cdot) = f(\cdot + x_0)$ if needed, we may assume that $rB_2^n \subseteq C$.

We set $K = \{x : f(x) > t/e\}$. Then, Markov's inequality and the monotonicity of volume show that $0 < \mathrm{Vol}_n(K) < \infty$. Using the fact that K is convex, it has finite volume and contains rB_2^n, we see that it is bounded. So, we can find $R > 0$ such that $K \subset \frac{R}{2}B_2^n$. Then, for every x with $|x| > R$ we have $R\frac{x}{|x|} \notin K$, and hence $f(Rx/|x|) \leqslant t/e$, while $r\frac{x}{|x|} \in C$, which shows that $f(rx/|x|) \geqslant t$. Moreover, we may write

$$\frac{Rx}{|x|} = \frac{|x|-R}{|x|-r}\frac{rx}{|x|} + \frac{R-r}{|x|-r}x.$$

Since f is log-concave, we get

$$\frac{t}{e} \geqslant f\left(R\frac{x}{|x|}\right) \geqslant f\left(r\frac{x}{|x|}\right)^{\frac{|x|-R}{|x|-r}} f(x)^{\frac{R-r}{|x|-r}} \geqslant t^{\frac{|x|-R}{|x|-r}} f(x)^{\frac{R-r}{|x|-r}}.$$

It follows that

$$f(x) \leqslant te^{-\frac{|x|-r}{R-r}} < e^{-|x|/R}$$

for every $x \in \mathbb{R}^n$ with $|x| > R$. On the other hand, for every $x \in RB_2^n$ and for every $y \in \frac{x}{2} + \frac{r}{2}B_2^n$ we have by the log-concavity of f that

$$f(y) \geqslant \sqrt{f(x)f(2y-x)} \geqslant \sqrt{t}\,\sqrt{f(x)};$$

this combined with the integrability of f shows that there exists $M > 0$ such that $f(x) \leqslant M$ for every $x \in RB_2^n$. So, we can clearly find two constants $A, B > 0$, which depend on f, so that $f(x) \leqslant Ae^{-B|x|}$ for every $x \in \mathbb{R}^n$. □

The second result, which is due to Fradelizi [**215**], shows that the value of a centered log-concave function at the origin is comparable to its maximum (up to a constant depending on the dimension). Here, we present a different proof. Note that, if f is assumed even, then $f(0) = \|f\|_\infty$.

THEOREM 10.6.2. *Let $f : \mathbb{R}^n \to [0, \infty)$ be a centered log-concave function. Then,*

$$f(0) \leqslant \|f\|_\infty \leqslant e^n f(0).$$

Proof. We may assume that f is continuously differentiable and that $\int_{\mathbb{R}^n} f(y)dy = 1$. From Jensen's inequality we have

$$\log f(0) = \log f\left(\int_{\mathbb{R}^n} yf(y)dy\right) \geqslant \int_{\mathbb{R}^n} f(y)\log f(y)dy. \tag{10.6.1}$$

Let $x \in \mathbb{R}^n$. Using the fact that f is log-concave we have that

$$-\log f(x) \geqslant -\log f(y) + \langle x - y, \nabla(-\log f)(y)\rangle. \tag{10.6.2}$$

Multiplying both terms of the last inequality by $f(y)$, and then integrating with respect to y, we get

$$\begin{aligned} -f(x)\log f(x) &\geqslant -\int_{\mathbb{R}^n} f(y)\log f(y)dy + \int_{\mathbb{R}^n} \langle x-y, -\nabla f(y)\rangle\, dy \\ &\geqslant -\int_{\mathbb{R}^n} f(y)\log f(y)dy - n, \end{aligned} \tag{10.6.3}$$

where the last inequality follows if we integrate by parts (and since $f(y)$ decays exponentially as $|y| \to \infty$). Combining (10.6.1) and (10.6.3) we get

$$\log f(0) \geqslant \int_{\mathbb{R}^n} f(y)\log f(y)dx \geqslant \log f(x) - n,$$

for every $x \in \mathbb{R}^n$. Taking the supremum over all x we get the result. □

As an application we get a proof which was postponed from Section 10.1.2 of

Proposition 10.1.14. *Let K be a centered convex body of volume 1 in $\mathbb{R}^n$. Let $\theta \in S^{n-1}$ and consider the function $f(t) = f_{K,\theta}(t) = \mathrm{Vol}_{n-1}(K \cap \{\langle x, \theta\rangle = t\})$, $t \in \mathbb{R}$. Then,*

$$\|f\|_\infty \leqslant ef(0) = e\mathrm{Vol}_{n-1}(K \cap \theta^\perp).$$

Proof. By the Brunn-Minkowski inequality, f is a log-concave function on $\mathbb{R}$. Since K is centered, we easily check that f is centered as well. Thus, we may apply Theorem 10.6.2. □

The next result provides a small ball estimate for log-concave probability measures.

THEOREM 10.6.3 (Latała). *Let μ be a log-concave probability measure on $\mathbb{R}^n$. For any norm $\|\cdot\|$ on $\mathbb{R}^n$ and any $0\leqslant t\leqslant 1$ one has*

$$\mu(\{x:\|x\|\leqslant t\mathbb{E}_\mu(\|x\|)\})\leqslant Ct, \tag{10.6.4}$$

where $C>0$ is an absolute constant.

Proof. Let K be the unit ball of $(\mathbb{R}^n,\|\cdot\|)$. We define

$$\alpha=\inf\Big\{s>0:\mu(sK)\geqslant\frac{2}{3}\Big\}.$$

Then $\mu(\alpha K)\leqslant 2/3$ and applying Borell's lemma we see that

$$\begin{aligned}\mathbb{E}_\mu(\|x\|)&\leqslant\alpha+\alpha\int_1^\infty[1-\mu(s\alpha K)]\,ds\\ &\leqslant\alpha+\alpha\int_1^\infty\mu(\alpha K)\left(\frac{1-\mu(\alpha K)}{\mu(\alpha K)}\right)^{\frac{s+1}{2}}ds\\ &\leqslant\alpha+\frac{2\alpha}{3\sqrt{2}}\int_1^\infty 2^{-s/2}\,ds\leqslant c\alpha\end{aligned}$$

where $c>0$ is an absolute constant. We also have

$$1-\mu(3\alpha K)\leqslant\mu(\alpha K)\left(\frac{1-\mu(\alpha K)}{\mu(\alpha K)}\right)^2\leqslant\frac{1}{6},$$

which implies

$$\mu(\{x:\alpha\leqslant\|x\|<3\alpha\})\geqslant\frac{1}{6}.$$

We fix $k\in\mathbb{N}$ and for any $u\geqslant\frac{\alpha}{2k}$ we define

$$T(u)=\Big\{x:u-\frac{\alpha}{2k}\leqslant\|x\|<u+\frac{\alpha}{2k}\Big\}.$$

Since

$$\{x:\alpha\leqslant\|x\|<3\alpha\}=\bigcup_{s=1}^{2k}T\Big(\alpha+\frac{(2s-1)\alpha}{2k}\Big),$$

we may find $u_0>\alpha$ so that $\mu(T(u_0))\geqslant\frac{1}{12k}$. Note that

$$\lambda T(u)+(1-\lambda)\frac{\alpha}{2k}K\subseteq T(\lambda u) \tag{10.6.5}$$

for every $0<\lambda<1$, by the triangle inequality.

Claim. For every $k\in\mathbb{N}$ we have

$$\mu\left(\frac{\alpha}{2k}K\right)\leqslant\frac{48}{k}.$$

To see this, assume the contrary and then observe that for any $0<\lambda<\frac{1}{2}$ the log-concavity of μ and (10.6.5) imply

$$\mu(T(\lambda u_0))\geqslant\mu(T(u_0))^\lambda\mu\Big(\frac{\alpha}{2k}K\Big)^{1-\lambda}\geqslant\Big(\frac{1}{12k}\Big)^\lambda\Big(\frac{48}{k}\Big)^{1-\lambda}>\frac{2}{k}.$$

This shows that $\mu(T(u))>\frac{2}{k}$ for all $u<u_0/2$ (and hence, for all $u\leqslant\alpha/2$). We get a contradiction if we observe that the sets $T(s\alpha/k)$, $1\leqslant s\leqslant k/2$ are disjoint, they have measure greater than $\frac{2}{k}$, and they are also disjoint from $\frac{\alpha}{2k}K$.

Now, let $0 < t \leqslant 1/2$. We may find $k \in \mathbb{N}$ so that $\frac{1}{4k} \leqslant t \leqslant \frac{1}{2k}$, and then

$$\mu(\{x : \|x\| \leqslant t\alpha\}) \leqslant \mu\Big(\Big\{x : \|x\| \leqslant \frac{\alpha}{2k}\Big\}\Big) = \mu\Big(\frac{\alpha}{2k}K\Big) \leqslant \frac{48}{k} \leqslant 192t.$$

Since $\mathbb{E}_\mu(\|x\|) \leqslant c\alpha$ the theorem follows. □

A consequence of Theorem 10.6.3 is the next Kahane-Khintchine inequality for negative exponents.

THEOREM 10.6.4. *Let μ be a log-concave probability measure on $\mathbb{R}^n$. For any norm $\|\cdot\|$ on $\mathbb{R}^n$ and any $-1 < q < 0$ one has*

$$\mathbb{E}_\mu(\|x\|) \leqslant \frac{C}{1+q}\big(\mathbb{E}_\mu(\|x\|^q)\big)^{1/q}, \tag{10.6.6}$$

where $C > 0$ is an absolute constant.

Proof. We may assume that $\mathbb{E}_\mu(\|x\|) = 1$. We set $p = -q$ and using integration by parts and Theorem 10.6.3 we write

$$\begin{aligned}\mathbb{E}_\mu(\|x\|^q) &= \mathbb{E}_\mu\Big(\frac{1}{\|x\|^p}\Big) = p\int_0^\infty t^{p-1}\mu\Big(\Big\{\frac{1}{\|x\|} \geqslant t\Big\}\Big)\\ &\leqslant 1 + C_1 p\int_1^\infty t^{p-2}dt = 1 + \frac{C_1 p}{1-p} \leqslant \frac{1+C_1 p}{1-p} \leqslant \frac{e^{C_1 p}}{1-p}.\end{aligned}$$

It follows that

$$\big(\mathbb{E}_\mu(\|x\|^q)\big)^{1/q} \geqslant \Big(\frac{1-p}{e^{C_1 p}}\Big)^{1/p} = e^{-C_1}(1-p)\,(1-p)^{1/p-1} \geqslant \frac{1-p}{C}$$

because $(1-p)^{1-1/p}$ is bounded on $(0,1)$. □

We shall need two more technical inequalities for log-concave functions. These are then used for the proof of the inclusion relations and the volume estimates for the bodies $K_p(f)$. Note that by Lemma 10.6.1 all integrals appearing below are finite.

LEMMA 10.6.5. *Let $f : [0,\infty) \to [0,\infty)$ be a log-concave function with $f(0) > 0$. Then, the function*

$$G(p) := \Big(\frac{1}{f(0)\Gamma(p)}\int_0^\infty f(x)x^{p-1}\,dx\Big)^{1/p}$$

is a decreasing function of p on $(0,\infty)$.

Proof. Without loss of generality we may assume that $f(0) = 1$, otherwise we work with the log-concave function $f_1 = \frac{1}{f(0)}f$. Let $p > 0$. Applying the change of variables $y = cx$ we see that for every $c > 0$ one has

$$\int_0^\infty e^{-cx}x^{p-1}dx = \frac{1}{c^p}\int_0^\infty e^{-x}x^{p-1}dx = \frac{\Gamma(p)}{c^p}.$$

Thus, if we choose $c_p = \frac{1}{G(p)}$ we have

$$\int_0^\infty e^{-c_p x}x^{p-1}dx = \int_0^\infty f(x)x^{p-1}dx. \tag{10.6.7}$$

In particular, it cannot be that $e^{-c_p x} < f(x)$ for every $x \in (0, +\infty)$, hence the set $\{x > 0 : e^{-c_p x} \geqslant f(x)\}$ is non-empty and

$$x_0 := \inf\{x > 0 : e^{-c_p x} \geqslant f(x)\} \in [0, +\infty).$$

Obviously then

$$e^{-c_p x} < f(x) \text{ for every } 0 < x < x_0, \tag{10.6.8}$$

whereas for $x > x_0$ we can find $y \in [x_0, x) \cap \{y' > 0 : e^{-c_p y'} \geqslant f(y')\}$ and we can write

$$e^{-c_p y} \geqslant f(y) \geqslant f(x)^{\frac{y}{x}} f(0)^{1-\frac{y}{x}} = f(x)^{\frac{y}{x}} \tag{10.6.9}$$

by the log-concavity of f, which means that $f(x) \leqslant (e^{-c_p y})^{\frac{x}{y}} = e^{-c_p x}$. It follows that

$$\int_x^\infty f(t) t^{p-1} dt \leqslant \int_x^\infty e^{-c_p t} t^{p-1} dt \tag{10.6.10}$$

for every $x > x_0$. On the other hand, by (10.6.8) we see that for $x \leqslant x_0$

$$\int_0^x f(t) t^{p-1} dt \geqslant \int_0^x e^{-c_p t} t^{p-1} dt,$$

and thus (10.6.7) implies (10.6.10) for every $x \leqslant x_0$ as well.

Let us consider $q > p$. Then, by Fubini's theorem and by (10.6.7)

$$\begin{aligned}
\int_0^\infty f(x) x^{q-1} dx &= \int_0^\infty f(x) x^{p-1} \int_0^x (q-p) t^{q-p-1} dt dx \\
&= \int_0^\infty (q-p) t^{q-p-1} \int_t^\infty f(x) x^{p-1} dx dt \\
&\leqslant \int_0^\infty (q-p) t^{q-p-1} \int_t^\infty e^{-c_p x} x^{p-1} dx dt \\
&= \int_0^\infty e^{-c_p x} x^{q-1} dx = \frac{\Gamma(q)}{c_p^q}.
\end{aligned}$$

We conclude that

$$G(q) = \left(\frac{1}{\Gamma(q)} \int_0^\infty f(x) x^{q-1}\, dx\right)^{1/q} \leqslant \frac{1}{c_p} = G(p),$$

which was our claim. $\square$

The inequality that follows goes in the opposite direction.

LEMMA 10.6.6. *Let $f : [0, \infty) \to [0, \infty)$ be a log-concave function. Then,*

$$F(p) := \left(\frac{p}{\|f\|_\infty} \int_0^\infty x^{p-1} f(x)\, dx\right)^{1/p}$$

is an increasing function of p on $(0, \infty)$.

Proof. Without loss of generality we may assume that $\|f\|_\infty = 1$. Then, for any $0 < p < q$ and $\beta > 0$, we may write

$$\begin{aligned}\frac{F(q)^q}{q} = \int_0^\infty x^{q-1} f(x)\, dx &= \int_0^\beta x^{q-1} f(x)\, dx + \int_\beta^\infty x^{q-1} f(x)\, dx \\ &\geqslant \int_0^\beta x^{q-1} f(x)\, dx + \beta^{q-p} \int_\beta^\infty x^{p-1} f(x)\, dx \\ &= \beta^{q-p} \frac{F(p)^p}{p} - \beta^q \int_0^1 (x^{p-1} - x^{q-1}) f(\beta x)\, dx \\ &\geqslant \beta^{q-p} \frac{F(p)^p}{p} - \beta^q \left(\frac{1}{p} - \frac{1}{q}\right).\end{aligned}$$

The choice $\beta = F(p)$ minimizes the right hand side and gives the result. □

Finally, we are ready to prove Proposition 10.2.12 and its Corollary 10.2.13.

Proof of Proposition 10.2.12. Observe that (ii) is a direct consequence of (i) if we use Theorem 10.6.2: we know that if $\mathrm{bar}(f) = 0$ then $f(0) \leqslant \|f\|_\infty \leqslant e^n f(0)$.

For the right inclusion in (10.2.4) we use Lemma 10.6.6: for every $x \neq 0$ the function

$$F(p) := \left(\frac{p}{\|f\|_\infty} \int_0^\infty r^{p-1} f(rx)\, dr\right)^{1/p}$$

is an increasing function of p on $(0, \infty)$. Then,

$$\begin{aligned}\rho_{K_q(f)}(x) &= \left(\frac{q}{f(0)} \int_0^\infty r^{q-1} f(rx)\, dr\right)^{1/q} \\ &= \left(\frac{\|f\|_\infty}{f(0)}\right)^{1/q} \left(\frac{q}{\|f\|_\infty} \int_0^\infty r^{q-1} f(rx)\, dr\right)^{1/q} \\ &= \left(\frac{\|f\|_\infty}{f(0)}\right)^{1/q} F(q) \geqslant \left(\frac{\|f\|_\infty}{f(0)}\right)^{1/q} F(p) \\ &= \left(\frac{\|f\|_\infty}{f(0)}\right)^{1/q-1/p} \left(\frac{\|f\|_\infty}{f(0)}\right)^{1/p} F(p) \\ &= \left(\frac{\|f\|_\infty}{f(0)}\right)^{1/q-1/p} \rho_{K_p(f)}(x).\end{aligned}$$

For the left inclusion in (10.2.4) we use Lemma 10.6.5: for every $x \neq 0$ the function

$$G(p) := \left(\frac{1}{f(0)\Gamma(p)} \int_0^\infty r^{p-1} f(rx)\, dr\right)^{1/p}$$

is a decreasing function of p on $[1,\infty)$. Then,

$$\begin{aligned}\rho_{K_q(f)}(x) &= \left(\frac{q}{f(0)}\int_0^\infty r^{q-1}f(rx)\,dr\right)^{1/q}\\ &= \Gamma(q+1)^{1/q}\left(\frac{1}{f(0)\Gamma(q)}\int_0^\infty r^{q-1}f(rx)\,dr\right)^{1/q}\\ &= \Gamma(q+1)^{1/q}G(q)\\ &\leqslant \frac{\Gamma(q+1)^{1/q}}{\Gamma(p+1)^{1/p}}\Gamma(p+1)^{1/p}G(p)\\ &= \frac{\Gamma(q+1)^{1/q}}{\Gamma(p+1)^{1/p}}\rho_{K_p(f)}(x),\end{aligned}$$

and the proof is complete. □

Proof of Corollary 10.2.13. To prove (10.2.6), we first use (10.2.5) to get

$$e^{\frac{n^2}{n+p}-n}\mathrm{Vol}_n(K_n(f)) \leqslant \mathrm{Vol}_n(K_{n+p}(f)) \leqslant \frac{\Gamma(n+p+1)^{\frac{n}{n+p}}}{\Gamma(n+1)}\mathrm{Vol}_n(K_n(f)).$$

Thus, combining this inequality with Lemma 10.2.11 we see that, for every $p>0$,

$$\frac{e^{-\frac{np}{n+p}}}{f(0)} \leqslant \mathrm{Vol}_n(K_{n+p}(f)) \leqslant (\Gamma(n+p+1))^{\frac{n}{n+p}}\frac{1}{n!f(0)},$$

and hence

$$\frac{1}{e} \leqslant f(0)^{\frac{1}{n}+\frac{1}{p}}\mathrm{Vol}_n(K_{n+p}(f))^{\frac{1}{n}+\frac{1}{p}} \leqslant \frac{(\Gamma(n+p+1))^{\frac{1}{p}}}{(n!)^{\frac{n+p}{np}}}.$$

Using the bounds

$$\frac{(\Gamma(n+p+1))^{\frac{1}{p}}}{(n!)^{\frac{n+p}{np}}} \leqslant (n+p)\frac{(n!)^{\frac{1}{p}}}{(n!)^{\frac{n+p}{np}}} = \frac{n+p}{(n!)^{\frac{1}{n}}} \leqslant e\frac{n+p}{n},$$

we conclude the proof of (10.2.6). We verify (10.2.7) in a similar manner, using also the inequality

$$\frac{\Gamma(q+1)^{\frac{1}{q}}}{\Gamma(p+1)^{\frac{1}{p}}} \leqslant e^{\frac{q}{p}-1}$$

that holds true for all $0<p\leqslant q$. □

10.7. Notes and remarks

Isotropic position and hyperplane conjecture

The hyperplane conjecture appears for the first time in the work of Bourgain [**98**] on high-dimensional maximal functions associated with arbitrary convex bodies. The conjecture was stated in this form in the article of V. Milman and Pajor [**459**] and in the PhD Thesis of K. Ball [**37**]. We mention that the authors of this book have different opinions regarding the validity of the conjecture.

The duality between the Binet and the Legendre ellipsoid is computed in [**459**], where it is mentioned that this fact goes back to classical mechanics, and that the authors learned it from a footnote in the paper of Fritz John [**312**].

Bourgain's article [**98**] concerned high-dimensional maximal functions associated with arbitrary convex bodies. He was interested in bounds for the L_p-norm of the maximal function

$$M_K f(x) = \sup\left\{\frac{1}{\mathrm{Vol}_n(tK)}\int_{tK}|f(x+y)|\,dy \mid t>0\right\}$$

of $f\in L^1_{\mathrm{loc}}(\mathbb{R}^n)$, where K is a centrally symmetric convex body in $\mathbb{R}^n$. Let $C_p(K)$ denote the best constant such that $\|M_K f\|_p \leqslant C_p(K)\|f\|_p$ is satisfied. Bourgain showed that there exists an absolute constant $C>0$ (independent of n and K) such that

$$\|M_K\|_{L_2(\mathbb{R}^n)\to L_2(\mathbb{R}^n)} \leqslant C.$$

Earlier, Stein had proved in [**570**] that if $K=B_2^n$ is the Euclidean unit ball then $C_p(B_2^n)$ is bounded independently of the dimension for all $p>1$. By the definition of M_K it is clear that in order to obtain a uniform bound on $\|M_K\|_{2\to 2}$ one can start with a suitable position $T(K)$ (where $T\in GL_n$) of K. Bourgain used the isotropic position; the property that played an important role in his argument was that when K is isotropic then $L_K\mathrm{Vol}_{n-1}(K\cap\theta^{\perp})\simeq 1$ for all $\theta\in S^{n-1}$. Bourgain mentioned the fact that $L_K\geqslant c$ and asked whether a reverse inequality holds true.

The result for $\|M_K\|_{2\to 2}$ was generalized to all $p>3/2$ by Bourgain [**99**] and, independently, by Carbery [**135**]. Afterwards, Müller [**480**] obtained dimension free maximal bounds for all $p>1$, which however depend on L_K and on the maximal volume of hyperplane projections of K. In the case of the cube, Bourgain [**102**] showed that for every $p>1$ there exists a constant $C_p>0$ such that $C_p(B_\infty^n)\leqslant C_p$ for all n.

Convex bodies associated with log-concave functions

The bodies $K_p(\mu)$ were introduced by K. Ball (see [**37**] and [**38**]) who established their convexity. One should also mention Busemann's paper [**128**], where the case of a density which is the indicator of a convex body (and, say, $p=1$) is proved. Ball [**38**] showed that if μ is an even isotropic log-concave measure then the body $T=K_{n+2}(\mu)$ is an isotropic symmetric convex body with $L_\mu\simeq L_T$. The observation that one can reduce the study of the behaviour of the isotropic constants of all log-concave measures to the class of centrally symmetric convex bodies is due to Klartag [**335**].

Bourgain's upper bound for the isotropic constant

Bourgain's bound $L_K=O(\sqrt[4]{n}\log n)$ appeared in [**100**]. We present a modification of his argument, which is due to Dar [**162**]. Bourgain showed in [**101**] that if K is a symmetric convex body and K is a ψ_2-body with constant b then one can improve this estimate to $L_K\leqslant cb\log(1+b)$. It was later proved by Klartag and E. Milman [**344**] that in this case one has $L_K\leqslant cb$ and Theorem 10.3.4 was significantly improved.

Partial answers

There are several results confirming the hyperplane conjecture for important classes of convex bodies. To be more precise, let us say that a class $\mathcal{C}$ of symmetric convex bodies satisfies the hyperplane conjecture uniformly if there exists a positive constant C such that $L_K\leqslant C$ for all $K\in\mathcal{C}$.

The fact that the isotropic constants of unconditional convex bodies are bounded by an absolute constant is due to Bourgain, see [**459**]; a different proof using the Loomis-Whitney inequality is given by Schmuckenschläger in [**549**]. One more proof, leading to the bound $L_K\leqslant 1/\sqrt{2}$, can be found in the article [**88**] of Bobkov and Nazarov. A more general result, with a different proof, can be found in Milman and Pajor [**459**]. Uniform bounds are known for the isotropic constants of some other classes of convex bodies: convex bodies whose polar bodies contain large affine cubes (see again [**459**]), the unit balls of 2-convex spaces with a given constant α (see Klartag and E. Milman [**343**]), bodies with small diameter (in particular, the class of zonoids) etc.

Uniform boundedness of the isotropic constants of the unit balls of the Schatten classes was established by König, Meyer and Pajor in [**360**]. One of the main ingredients of the proof is a formula of Saint-Raymond from [**532**]. Before the work of König, Meyer and Pajor, Dar had obtained the estimate $L_{B(S_p^n)} \leqslant C\sqrt{\log n}$ in [**163**] (see [**162**] for the case $p = 1$).

Upper bounds for the isotropic constant of polytopes, which depend on the number of their vertices or facets, follow from results of Ball [**40**], Junge [**314**] and [**315**] and E. Milman [**436**]. A more geometric approach, that covers the case of not necessarily symmetric polytopes too, was given of Alonso-Gutiérrez in [**15**].

Paouris' inequality

L_q-centroid bodies were introduced by Lutwak and Zhang [**407**] who used a different normalization. If K is a convex body in $\mathbb{R}^n$ then, for every $1 \leqslant q < \infty$, the body $\Gamma_q(K)$ was defined in [**407**] through its support function

$$h_{\Gamma_q(K)}(y) = \left(\frac{1}{c_{n,q}\mathrm{Vol}_n(K)}\int_K |\langle x, y\rangle|^q dx\right)^{1/q},$$

where

$$c_{n,q} = \frac{\kappa_{n+q}}{\kappa_2\kappa_n\kappa_{q-1}}.$$

In other words, $Z_q(K) = c_{n,q}^{1/q}\Gamma_q(K)$ if $\mathrm{Vol}_n(K) = 1$. The normalization of $\Gamma_q(K)$ is chosen so that $\Gamma_q(B_2^n) = B_2^n$ for every q. Lutwak, Yang and Zhang [**408**] have established the following L_q affine isoperimetric inequality (see Campi and Gronchi [**133**] for an alternative proof): For every $q \geqslant 1$,

$$\mathrm{Vol}_n(\Gamma_q(K)) \geqslant 1,$$

with equality if and only if K is a centered ellipsoid of volume 1.

Alesker's theorem is from [**5**]; it is the starting point of the work of Paouris. His study of the L_q-centroid bodies from an asymptotic point of view started with [**490**] and [**491**], where the parameter $q_*(\mu)$ is introduced. The main results of this section (Theorem 10.4.6) were proved by Paouris in [**493**] (see also [**492**] for a sketch of the proof) and [**494**].

In the particular case of unconditional isotropic convex bodies, the inequality of Paouris had been previously proved by Bobkov and Nazarov (see [**88**] and [**89**]). The origin of the work of Bobkov and Nazarov is in the work of Schechtman, Zinn and Schmuckenschläger on the volume of the intersection of two L_p^n-balls (see [**542**], [**544**], [**545**] and [**550**]). Before Paouris' theorem, Guédon and Paouris had studied in [**289**] the case of the unit balls of the Schatten classes.

Klartag's convex perturbations

Klartag's solution to the isomorphic slicing problem and his $O(\sqrt[4]{n})$ bound for the isotropic constant are from [**336**]. A second proof of the same estimate was given by Klartag and E. Milman in [**344**].

Klartag and E. Milman in [**344**] defined the "hereditary" variant

$$q_*^H(\mu) := n \inf_k \inf_{E\in G_{n,k}} \frac{q_*(\pi_E\mu)}{k} \tag{10.7.1}$$

of $q_*(\mu)$ and then, for every $q \leqslant q_*^H(\mu)$, they showed that $\mathrm{Vol}_n(Z_q(\mu))^{1/n} \geqslant c_3\sqrt{q/n}$ where $c_3 > 0$ is an absolute constant. An immediate consequence of this inequality and of the fact that $q_*^H(\mu) \geqslant c\sqrt{n}$ is an alternative proof of the bound $L_\mu = O(\sqrt[4]{n})$ for the isotropic constant. Vritsiou [**602**] introduced a new parameter $r_\sharp(\mu, A)$ that dominates

$q_*^H(\mu)$ and modified the argument of Klartag and E. Milman to show that the lower bound $\mathrm{Vol}_n(Z_p(\mu))^{1/n} \geqslant cA^{-1}\sqrt{p/n}$ continues to hold for all $p \leqslant r_\sharp(\mu, A)$.

The observation that one can use Klartag's approach in order to give a purely convex geometric proof of the reverse Santaló inequality is due to Giannopoulos, Paouris and Vritsiou [**242**]. Theorem 10.5.6 was proved by Milman and Pajor [**460**] in the centrally symmetric case. The argument presented in the text is due to Hartzoulaki [**301**].

Reductions of the hyperplane conjecture

Bourgain, Klartag and V. Milman proved in [**103**] a reduction of the hyperplane conjecture to the case of convex bodies whose volume ratio is bounded by some absolute constant. The same fact follows from Klartag's approach in [**335**].

Dafnis and Paouris introduced in [**160**] the parameter $q_{-c}(\mu,\zeta) := \max\{p \geqslant 1 : I_2(\mu) \leqslant \zeta I_{-p}(\mu)\}$; among other things they proved that a positive answer to the hyperplane conjecture is equivalent to the existence of two absolute constants $C, \xi > 0$ such that $q_{-c}(K,\xi) \geqslant Cn$ for every isotropic convex body K in $\mathbb{R}^n$.

Giannopoulos, Paouris and Vritsiou proposed in [**241**] a reduction of the hyperplane conjecture to the study of the parameter $I_1(K, Z_q^\circ(K)) = \int_K \|\langle\cdot, x\rangle\|_q dx$ in the sense that it immediately recovers a bound that is slightly worse than Bourgain's and Klartag's bounds and opens the possibility for improvements: an upper bound of the form $I_1(K, Z_q^\circ(K)) \leqslant C_1 q^s \sqrt{n} L_K^2$ *for some* $q \geqslant 2$ and $\frac{1}{2} \leqslant s \leqslant 1$ and *for all* isotropic convex bodies K in $\mathbb{R}^n$ leads to the estimate

$$L_n \leqslant \frac{C_2 \sqrt[4]{n}\,\mathrm{log} n}{q^{\frac{1-s}{2}}}.$$

The central limit problem

Sudakov applied the spherical isoperimetric inequality in [**575**] to show that for every $\delta > 0$ there exists $n(\delta)$ such that if $n \geqslant n(\delta)$ and if μ is a Borel probability measure on $\mathbb{R}^n$ that satisfies $\mathbb{E}_\mu(\langle x,\theta\rangle^2) \leqslant 1$ for all $\theta \in S^{n-1}$ then we may find $A_\delta \subset S^{n-1}$ with $\sigma(A_\delta) \geqslant 1-\delta$ and such that

$$\kappa(F_\theta, F) := \int_{-\infty}^{\infty} |F_\theta(t) - F(t)|\, dt \leqslant \delta$$

for all $\theta \in A_\delta$, where

$$F_\theta(t) = \mu(\{x : \langle x,\theta\rangle \leqslant t\}), \qquad t \in \mathbb{R}$$

and

$$F(t) = \int_{S^{n-1}} F_\theta(t)\, d\sigma(\theta).$$

Variants of the problem were studied by Diaconis and Freedman in [**168**], and by von Weizsäker in [**606**].

The case where μ_K is the Lebesgue measure on an isotropic centrally symmetric convex body K in $\mathbb{R}^n$ was studied by Anttila, Ball and Perissinaki in [**20**] who made use of the log-concavity of μ_K and showed that a thin-shell estimate implies an affirmative answer to the central limit problem. This work made the central limit problem widely known among people working in convex geometry. A clear exposition of both the general and the log-concave case can be found in Bobkov's article [**83**].

Around the same time, the same problem was studied by Brehm and Voigt [**118**], Brehm, Vogt and Voigt [**119**], Brehm, Hinow, Vogt and Voigt [**120**], who made sharp computations for specific convex bodies.

Bobkov and Koldobsky [**87**] stated the "variance hypothesis", which is an equivalent form of the thin shell conjecture. Using the subindependence of coordinate slabs for the ℓ_p^n-balls, $p \geqslant 1$, due to Ball and Perissinaki [**47**], Anttila, Ball and Perissinaki confirmed the variance hypothesis in this case.

In his breakthrough work [**339**] Bo'az Klartag first obtained a dimension dependent thin-shell estimate in full generality. He showed that, if μ is an isotropic log-concave measure then, for all $0 < \varepsilon \leqslant 1$ one has

$$\mu\left(\left\{\left|\frac{|x|}{\sqrt{n}} - 1\right| \geqslant \varepsilon\right\}\right) \leqslant Cn^{-c\varepsilon^2}$$

where $c, C > 0$ are absolute constants. This establishes the ε-hypothesis with $\varepsilon_n \simeq \sqrt{\frac{\log\log n}{\log n}}$. Subsequently, Klartag obtained in [**340**] power-type estimates for the ε-hypothesis by showing that

$$\mathbb{E}\Big(\frac{|x|^2}{n} - 1\Big)^2 \leqslant \frac{C}{n^\alpha}$$

with some $\alpha \simeq 1/5$, and gave an affirmative answer to the central limit problem: if μ is an isotropic log-concave measure on $\mathbb{R}^n$ then the density f_θ of $x \mapsto \langle x, \theta\rangle$ satisfies

$$\int_{-\infty}^{\infty} |f_\theta(t) - \gamma(t)|\, dt \leqslant \frac{1}{n^\kappa}$$

and

$$\sup_{|t| \leqslant n^\kappa} \left|\frac{f_\theta(t)}{\gamma(t)} - 1\right| \leqslant \frac{1}{n^\kappa},$$

for all θ in a subset A of S^{n-1} with measure $\sigma(A) \geqslant 1 - c_1\exp(-c_2\sqrt{n})$, where γ is the density of a standard Gaussian random variable, and c_1, c_2, κ are absolute constants. Multidimensional versions of the first result appear already in [**339**]. A generalization of the second result to higher dimensions appears in Eldan and Klartag [**184**].

Soon after Klartag's first proof of a thin-shell estimate, Fleury, Guédon and Paouris obtained in [**212**] a slightly weaker result. Starting from Klartag's method and ideas, but also using ideas from [**212**], Fleury [**211**] proved that if μ is an isotropic log-concave measure on $\mathbb{R}^n$, then for every $2 \leqslant q \leqslant c_1\sqrt[4]{n}$ one has

$$(\mathbb{E}_\mu|x|^q)^{1/q} \leqslant \left(1 + \frac{c_2 q}{\sqrt[4]{n}}\right)(\mathbb{E}_\mu|x|^2)^{1/2},$$

where $c_1, c_2 > 0$ are absolute constants. The work of Guédon and E. Milman [**288**] provides the best known estimates for the problem:

$$\mu\left(\{\big||x| - \sqrt{n}\big| \geqslant t\sqrt{n}\}\right) \leqslant C\exp(-c\sqrt{n}\min\{t^3, t\})$$

for every $t > 0$, where $C, c > 0$ are absolute constants. Their estimates recover the large deviations inequality of Paouris, improve the thin-shell estimate of Fleury and interpolate continuously between all scales of t. One can also check that if μ has a better ψ_α behaviour (for example, if μ is ψ_2) then the proofs can be easily modified to yield better bounds. See also [**345**], where Klartag and E. Milman introduced a regularization technique leading to improved small ball estimates.

Other contributions related to the central limit problem for convex sets (before or soon after Klartag's first general proof of the thin-shell estimate) were made by Koldobsky and Lifshits [**354**], Naor and Romik [**481**], Paouris [**491**], S. Sodin [**565**], E. Meckes and M. W. Meckes [**431**] and [**432**], E. Milman [**438**], Wojtaszczyk [**608**], Bastero and Bernués [**55**].

The thin shell conjecture for the class of unconditional log-concave measures was established by Klartag in [**341**]. Klartag's approach was extended by Barthe and Cordero-Erausquin in [**54**]; they adapted Klartag's techniques to provide spectral gap estimates for log-concave measures with many symmetries.

The inequality $L_n \leqslant C\sigma_n$ (where $\sigma_n = \sup_\mu \sigma_\mu$ and $\sigma_\mu^2 := \int_{\mathbb{R}^n}\big(|x| - \sqrt{n}\big)^2 d\mu(x)$) was proved by Eldan and Klartag in [**185**]. It should be compared with a result of Ball and Nguyen [**46**] that states that the (stronger) Kannan-Lovász-Simonovits conjecture implies the hyperplane conjecture. Another work of Eldan and Klartag [**186**] shows that stability estimates for the Brunn-Minkowski inequality are closely related to the thin shell conjecture.

Kannan-Lovász-Simonovits conjecture

The conjecture of Kannan, Lovász and Simonovits was stated in [**325**] in connection with randomized volume algorithms. Early lower bounds for Is_μ, that provide some estimates depending on the dimension, can be found in works of Payne and Weinberger [**495**], Li and Yau [**382**]. The main result in [**325**] states that $\sqrt{n}\mathrm{Is}_\mu \geqslant c$, where $c > 0$ is an absolute constant. Alternative proofs are given by Bobkov [**81**] and E. Milman [**439**]. Actually, Bobkov proved in [**85**] that

$$\sqrt[4]{n}\sqrt{\sigma_\mu}\mathrm{Is}_\mu \geqslant c;$$

this provides a direct link between the KLS-conjecture and the thin-shell conjecture: combined with the thin shell estimate of Guédon and E. Milman his result leads to the bound $n^{5/12}\mathrm{Is}_\mu \geqslant c$. E. Milman offers a different proof in [**439**]. S. Sodin proved in [**566**] that Is_μ is uniformly bounded for the ℓ_p^n-balls, $1 \leqslant p \leqslant 2$. The case $p \geqslant 2$ was settled by Latała and Wojtaszczyk [**370**].

Klartag [**341**] established a logarithmic in the dimension lower bound for the Poincaré constant Poin_K of an unconditional isotropic convex body K in $\mathbb{R}^n$; one has $\mathrm{Is}_K \simeq \sqrt{\mathrm{Poin}_K} \geqslant \frac{c}{\log n}$, where $c > 0$ is an absolute positive constant. Eldan [**183**] showed that there exists an absolute constant $C > 0$ such that

$$\frac{1}{\mathrm{Is}_n^2} \leqslant C \log n \sum_{k=1}^{n} \frac{\sigma_k^2}{k}.$$

Taking into account the result of Guédon and E. Milman, one gets the currently best known bound for Is_n: $\mathrm{Is}_n^{-1} \leqslant Cn^{1/3}\log n$.

Let $(\Omega, \mathcal{A}, P)$ be a probability space. The *Shannon entropy* of a non-negative measurable function $f : \Omega \to \mathbb{R}$, with the property that $\int f \log(1+f) < +\infty$, is defined as $\mathrm{Ent}(f) = -\int_\Omega f \log f + \int_\Omega f \,\log \int_\Omega f$. In the case of an isotropic random vector X with density f, the entropy of X takes the form

$$\mathrm{Ent}(X) = -\int_{\mathbb{R}^n} f \log f.$$

Ball observed that if X is an isotropic log-concave random vector in $\mathbb{R}^n$ and if

$$\mathrm{Ent}\left(\frac{X+Y}{\sqrt{2}}\right) - \mathrm{Ent}(X) \geqslant \delta(\mathrm{Ent}(G) - \mathrm{Ent}(X))$$

for some $\delta > 0$ and an independent copy Y of X, then the isotropic constant L_X of X satisfies $L_X \leqslant e^{1+\frac{2}{\delta}}$. Ball and Nguyen proved in [**46**] that if X is an isotropic log-concave random vector in $\mathbb{R}^n$ and its density f satisfies the Poincaré inequality with constant $\kappa > 0$, then

$$\mathrm{Ent}\left(\frac{X+Y}{\sqrt{2}}\right) - \mathrm{Ent}(X) \geqslant \frac{\kappa}{4(1+\kappa)}\left(\mathrm{Ent}(G) - \mathrm{Ent}(X)\right),$$

where Y is an independent copy of X. Thus, one can conclude that, for each individual isotropic log-concave random vector X, a bound for the Poincaré constant implies a bound for the isotropic constant: if the density of X satisfies the Poincaré inequality with constant $\kappa > 0$, then

$$L_X \leqslant e^{1+\frac{8(1+\kappa)}{\kappa}}.$$

In other words, the KLS-conjecture "strongly" implies the hyperplane conjecture.

Covering numbers of the centroid bodies

A question, originally posed by V. Milman in the framework of convex bodies, asks if there exists an absolute constant $C > 0$ such that every centered convex body K of volume 1 has at least one ψ_2 direction with constant C. Klartag, using again properties of the logarithmic Laplace transform, proved in [**338**] that for every log-concave probability measure μ on $\mathbb{R}^n$ there exists $\theta \in S^{n-1}$ such that

$$\mu\left(\{x : |\langle x, \theta\rangle| \geqslant ct\|\langle\cdot,\theta\rangle\|_2\}\right) \leqslant e^{-\frac{t^2}{(\log(t+1))^{2\alpha}}},$$

for all $1 \leqslant t \leqslant \sqrt{n}\log^{\alpha} n$, where $\alpha = 3$ (see also [**238**] for a first improvement). The best known estimate, is due to Giannopoulos, Paouris and Valettas who proved in [**239**] and [**240**] that one can always have $\alpha = 1/2$.

The main idea in all these works is to define the symmetric convex set $\Psi_2(\mu)$ whose support function is $h_{\Psi_2(\mu)}(\theta) = \|\langle\cdot,\theta\rangle\|_{\psi_2}$ and to estimate its volume. A logarithmic in the dimension bound on the volume radius of $\Psi_2(\mu)$ was first obtained by Klartag in [**338**] and then by Giannopoulos, Pajor and Paouris in [**238**]. The best known estimate $\text{v.rad}(\Psi_2(\mu)) \leqslant c\sqrt{\log n}$ is proved in [**239**]. The main tool in the proof of this result is estimates for the covering numbers $N(Z_q(K), sB_2^n)$.

Mean width and mean norm of isotropic convex bodies

The question to obtain an upper bound for the mean width of an isotropic convex body

$$w(K) := \int_{S^{n-1}} h_K(x)\, d\sigma(x),$$

that is, the L_1-norm of the support function of K with respect to the Haar measure on the sphere, was open for a number of years. The upper bound $w(K) \leqslant cn^{3/4}L_K$ appeared in the Ph.D. Thesis of Hartzoulaki [**301**]. Other approaches leading to the same bound can be found in Pivovarov [**511**] and in Giannopoulos Paouris and Valettas [**240**]. E. Milman showed in [**440**] that if K is an isotropic convex body in $\mathbb{R}^n$ then, for all $q \geqslant 1$ one has

$$w(Z_q(K)) \leqslant C\log(1+q)\max\left\{\frac{q\log(1+q)}{\sqrt{n}}, \sqrt{q}\right\} L_K$$

where $C > 0$ is an absolute constant. In particular,

$$w(K) \leqslant C\sqrt{n}(\log n)^2 L_K.$$

The dependence on n is optimal up to the logarithmic term.

An interesting related question is to determine the distribution of the function $\theta \mapsto \|\langle\cdot,\theta\rangle\|_{\psi_2}$ on the unit sphere; that is, to understand whether most of the directions have ψ_2-norm that is, say, logarithmic in the dimension. For a discussion and partial results see [**240**]. As a consequence of E. Milman's theorem, Brazitikos and Hioni showed in [**117**] that the answer is affirmative. More precisely, they showed that for any $a > 1$ one has

$$\|\langle\cdot,\theta\rangle\|_{L_{\psi_2}(K)} \leqslant C(\log n)^{3/2}\max\left\{\sqrt{\log n}, \sqrt{a}\right\} L_K$$

for all θ in a subset Θ of S^{n-1} with $\sigma(\Theta) \geqslant 1 - n^{-a}$, where $C > 0$ is an absolute constant.

The dual problem to estimate as the respective L_1-norm of the Minkowski functional of K,

$$M(K) := \int_{S^{n-1}} \|x\|_K\, d\sigma(x),$$

when K is a centrally symmetric isotropic convex body, had not been studied until recently; partial non-trivial results can be found in Giannopoulos, Stavrakakis, Tsolomitis and Vritsiou [**245**]. The currently best known estimate estimate $M(K) \leqslant \frac{C\log^{2/5}(e+n)}{\sqrt[10]{n}L_K}$ is due to Giannopoulos and E. Milman (see [**227**]).

Random polytopes

The literature on the approximation of convex bodies by random polytopes is very rich. We refer the reader to the books of Gruber [**277**] and Schneider and Weil [**557**], and to the survey articles of Bárány [**48**], Hug [**306**], Reitzner [**520**] and Schneider [**555**] for updated information.

The next results concern the asymptotic shape of the random polytope $K_N := \mathrm{conv}\{\pm X_1, \dots, \pm X_N\}$ that is spanned by N independent copies $X_1, \dots, X_N$ of an isotropic log-concave random vector X.; the emphasis is on the assumption that N is fixed in the range $[n, e^n]$ and one is interested in estimates which do not depend on the affine class of a convex body K. Natural questions are: to determine the asymptotic behaviour of the volume radius $\mathrm{Vol}_n(K)^{1/n}$, to understand the typical "asymptotic shape" of K_N and to estimate the isotropic constant of K_N. Dafnis, Giannopoulos and Tsolomitis proved in [**158**] that for any $n \leqslant N \leqslant \exp(n)$, the random polytope K_N satisfies, with high probability, the next two conditions: (i) $K_N \supseteq c\, Z_{\log(N/n)}(\mu)$ and (ii) for every $\alpha > 1$ and $q \geqslant 1$,

$$\mathbb{E}\left[\sigma(\{\theta : h_{K_N}(\theta) \geqslant \alpha h_{Z_q(\mu)}(\theta)\})\right] \leqslant N\alpha^{-q}.$$

Using this description of the shape of K_N and the theory of centroid bodies, one can determine the volume radius and the quermassintegrals of a random K_N; see Dafnis, Giannopoulos and Tsolomitis [**159**] and Giannopoulos, Hioni and Tsolomitis [**226**].

The study of the isotropic constant of K_N was initiated by Klartag and Kozma who proved in [**342**] that if $N > n$ and if $G_1, \dots, G_N$ are independent standard Gaussian random vectors in $\mathbb{R}^n$, then the isotropic constant of the random polytopes $K_N := \mathrm{conv}\{\pm G_1, \dots, \pm G_N\}$ and $C_N := \mathrm{conv}\{G_1, \dots, G_N\}$ is bounded by an absolute constant $C > 0$ with probability greater than $1 - Ce^{-cn}$. So far, there is essentially no other successful approach to this question; one can see that the same idea works in the case that the vertices x_j of K_N are distributed according to the uniform measure on an isotropic ψ_2-convex body, leading to a bound depending on its ψ_2-constant b. With a similar method, Alonso-Gutiérrez [**14**] has obtained a uniform bound in the situation where K_N or C_N is spanned by N random points uniformly distributed on the Euclidean sphere S_2^{n-1}. The unconditional case was settled by Dafnis, Giannopoulos and Guédon in [**157**].

Inequalities for log-concave functions

The fact that every integrable log-concave function $f : \mathbb{R}^n \to [0, \infty)$ decays exponentially as $|x| \to \infty$ appears in Bourgain's paper [**98**]. Fradelizi's inequality is from [**215**]. We present a different proof of the inequality $\|f\|_\infty \leqslant e^n f(0)$.

Theorem 10.6.3 is due to Latała [**367**]. For the Kahane-Khintchine inequality for negative exponents (Theorem 10.6.4) see Guédon [**287**].

Lemma 10.6.5 is a generalization of Klartag's [**336**, Lemma 2.6]; see Marshall, Olkin and Proschan [**416**] and Milman and Pajor [**459**] for a similar result in the case where f is, additionally, assumed decreasing. Lemma 10.6.6 is from [**459**, Lemma 2.1].

APPENDIX A

Elementary convexity

This first appendix includes classical and well known facts and theorems from convexity theory which we use throughout the book.

A.1. Basic convexity

A set $A \subset \mathbb{R}^n$ is called *convex* if $(1-\lambda)x + \lambda y \in A$ for any $x, y \in A$ and any $\lambda \in [0,1]$. Examples of convex sets include intervals, subspaces (also affine), half-spaces, Euclidean balls. A convex set will be called a convex body if in addition it is compact and has a non-empty interior. A set $A \subset \mathbb{R}^n$ is called centrally symmetric if $x \in A$ implies $-x \in A$ (sometimes one neglects the word "centrally" and calls these sets simply "symmetric"). Centrally symmetric convex bodies are in fact unit balls of norms on $\mathbb{R}^n$: the connection in one direction is given by Minkowski's functional

$$\|x\|_K := \inf\{r > 0 : x \in rK\},$$

which is a norm if K is a centrally symmetric convex body, while in the reverse direction, given a norm on $\mathbb{R}^n$, the corresponding unit ball

$$K_X = \{x : \|x\| \leqslant 1\}$$

is a centrally symmetric convex body. We mention that when 0 is in the interior of K, even if K is not assumed to be symmetric, we may consider the gauge function $\|\cdot\|_K$ defined above; we then call it a "generalized norm". In fact, we shall use this term also in cases when the body is not bounded, or when the origin belongs to the boundary of K, in which case the norm is allowed to assume the value $+\infty$ and may be 0 on non-zero vectors.

After fixing a scalar product on the space $\mathbb{R}^n$, denoted by $\langle \cdot, \cdot \rangle$, to every closed convex set K which includes the origin there corresponds a *polar* set

$$K^\circ = \{y \in \mathbb{R}^n : \sup_{x \in K} \langle x, y \rangle \leqslant 1\}.$$

When K is a centrally symmetric convex body which is the unit ball of the norm $\|\cdot\|_K$, the convex body K° is the unit ball of the dual norm, $\|\cdot\|^*$ defined by

$$\|y\|^* = \sup_{x \in K} \langle x, y \rangle.$$

The *affine hull* of a set of points is the smallest affine subspace they all belong to. A set of points is called *affinely independent* if none of them belongs to the affine hull of the rest. A *convex combination* of points $\{x_i\}_{i=1}^k \subset \mathbb{R}^n$ is a point of the form $\sum_{i=1}^k \lambda_i x_i$ where $\sum \lambda_i = 1$ and $\lambda_i \geqslant 0$. It is easy to check by induction that a set $A \subset \mathbb{R}^n$ is convex if and only if all convex combinations of points in A lie in A.

We shall recall in this section many simple and useful definitions and facts about convex sets which can be proved easily from the above definition of convexity.

An intersection of convex sets is convex. This enables to define the convex hull of a general set A by

$$\mathrm{conv}(A) = \bigcap\{B : B \supset A, B \text{ convex}\}.$$

Given $\lambda \in \mathbb{R}$ and $A, B \subset \mathbb{R}^n$ we define the λ-homothety of A and the Minkowski sum of A and B, as follows

$$\lambda A = \{\lambda x : x \in A\}, \quad A + B = \{a + b : a \in A, b \in B\}.$$

For convex sets, one has $\alpha A + \beta A = (\alpha + \beta)A$ for any $\alpha, \beta > 0$, and this characterizes convexity. Clearly convexity is preserved by both operations. It is also preserved by any linear or affine mapping. In particular, projections of convex sets are convex.

The intersection of finitely many closed half-spaces is called a *polyhedral set*, the convex hull of finitely many points is called a *polytope*, and the convex hull of $r + 1$ affinely independent points is called an *r-simplex*. When $r = n$ is the dimension of the whole space we simply say "a simplex". Bounded polyhedral sets are polytopes and vice versa. This will become even clearer after we discuss supporting half-spaces, see Corollary A.1.16 below.

The *vertices* of a polytope P are those points x for which $P \setminus \{x\}$ is convex; we denote the set of vertices of P by $v(P)$. Clearly polytopes are preserved by affine maps, the Minkowski sum of two polytopes is still a polytope (as $\mathrm{conv}(A + B) = \mathrm{conv}(A) + \mathrm{conv}(B)$), and the convex hull of two polytopes is again a polytope. The fact that the intersection of two polytopes is a polytope follows from the characterization of polytopes as bounded polyhedral sets.

The following three theorems are at the foundations of combinatorial convexity.

THEOREM A.1.1 (Radon). *Let $x_1, \ldots, x_m \in \mathbb{R}^n$ be affinely dependent. There is a partition into two disjoint sets $I \cup J = \{1, \ldots, m\}$, $I \cap J = \emptyset$ such that*

$$\mathrm{conv}\{x_i\}_{i \in I} \cap \mathrm{conv}\{x_j\}_{j \in J} \neq \emptyset.$$

Proof. Since $x_1, \ldots, x_m$ are affinely dependent, one can find $\alpha_i \in \mathbb{R}$, not all of them zero, with $\sum_{i=1}^m \alpha_i = 0$ and $\sum_{i=1}^m \alpha_i x_i = 0$. Partition with respect to whether the respective coefficient is positive or negative, and normalize. □

THEOREM A.1.2 (Helly). *Let $\{A_i\}_{i=1}^m$ be convex sets in $\mathbb{R}^n$ with $m \geqslant n + 1$. If each $n + 1$ of them have non-empty intersection, then the intersection of all the sets is non-empty.*

Proof. Proceed by induction over m. For $m = n+1$ it is trivial. Assume inductively that we may find an element x_i in the intersection of any $(m - 1)$ of the sets. Apply Radon's theorem to the resulting $m \geqslant n+2$ points to find without loss of generality

$$x \in \mathrm{conv}\{x_i\}_{i=1}^k \cap \mathrm{conv}\{x_i\}_{i=k+1}^m.$$

The first convex hull lies in the intersection of all A_i with $i = k+1, \ldots, m$, and the second in the intersection of the remaining A_i. □

We remark that if the sets are compact the above can be generalized to an infinite family by the finite intersection property.

THEOREM A.1.3 (Carathéodory). *Let $A \subset \mathbb{R}^n$ and $x \in \mathrm{conv}(A)$. Then there exist $\{x_i\}_{i=0}^n \subset A$ such that $x \in \mathrm{conv}\{x_i\}$.*

Proof. It is not hard to check that there exists *some* $k \in \mathbb{N}$ for which $x = \sum_{i=1}^k \alpha_i x_i$ with $x_i \in A$, $\sum \alpha_i = 1$, $\alpha_i > 0$, since the set of all such combinations includes A and is convex. Choose such a representation of x where k is least possible. It suffices to show that the x_i in such a representation are affinely independent, because then there could be at most $n+1$ of them. But if they were not affinely independent, there would be some $\sum \beta_i x_i = 0$ with $\sum \beta_i = 0$, in which case we could split the β_i into negative and positive ones, pick from the positive ones the one with the smallest ratio α_i/β_i, say i_0, and then

$$x = \sum_{i=1}^k (\alpha_i - \beta_i \alpha_{i_0}/\beta_{i_0}) x_i$$

would be a way of writing x as a convex combination of elements of A with just $k-1$ non-zero coefficients (all the coefficients in the combination above are non-negative), a contradiction to our choice of the least possible k. □

REMARK A.1.4. We mention without proof the following generalization: If the set $A \subset \mathbb{R}^n$ is known to be connected (or, more generally, the union of n connected sets) then each point of $\mathrm{conv}(A)$ can be written as a convex combination of n or fewer points from A.

Let us mention a simple but perhaps surprising consequence of Theorem A.1.2.

LEMMA A.1.5. *Let $K \subset \mathbb{R}^n$ be a non-empty compact convex set, and let $T \subset \mathbb{R}^n$ be some set of points. Then T is contained in a translate of K if and only if for every $(n+1)$ points $\{t_i\}_{i=1}^{n+1} \subset T$ there is a translate $x+K$ of K such that $\{t_i\}_{i=1}^{n+1} \subset x+K$.*

Proof. For every $t \in T$ the set of $x \in \mathbb{R}^n$ such that $t \in x+K$ is a convex set (it is simply the set $t-K$). Denote it by $K_{(t)}$. The fact that for every $(n+1)$ points $\{t_i\}_{i=1}^{n+1} \subset T$ there is a translate $x+K$ of K such that $\{t_i\}_{i=1}^{n+1} \subset x+K$ means that for every $(n+1)$ points $\{t_i\}_{i=1}^{n+1} \subset T$ we have $\bigcap_{i=1}^{n+1} K_{(t_i)} \neq \emptyset$. Thus by Helly's theorem, the intersection of any number of sets of the form $K_{(t)}$ is non-empty, and as the sets $K_{(t)}$ are compact, so is the (possibly infinite) intersection $\bigcap_{t\in T} K_{(t)} \neq \emptyset$. Let x belong to this intersection, then $T \subset x+K$ as needed. □

From the topological point of view, convex sets are simple. Indeed, letting $\mathrm{relint}(A)$ stand for the relative interior of a set A, that is, its interior with respect to its affine hull, we have

THEOREM A.1.6. *If $\emptyset \neq A \subset \mathbb{R}^n$ is convex then* $\mathrm{relint}(A) \neq \emptyset$.

Proof. One easily checks that the relative interior of a k-simplex is non-empty. For a general A, that is assumed of dimension k, take $k+1$ affinely independent points in A. Their convex hull contains relative interior points which are relative interior points of A as well. □

Once we have an interior point, convexity easily implies that for $x \in \bar{A}$ (here $\bar{A}$ denotes the closure of A) and $y \in \mathrm{relint}(A)$ one has $[y,x) \subset \mathrm{relint}(A)$. From this fact one may easily deduce the next lemma.

LEMMA A.1.7. *For a convex $A \subset \mathbb{R}^n$ both* $\mathrm{relint}(A)$ *and $\bar{A}$ are convex. In addition we have* $\mathrm{relint}(A) = \mathrm{relint}(\overline{A})$ *and* $\overline{\mathrm{relint}(A)} = \bar{A}$.

The convex hull of a closed set need not be closed, as the example of $\{(x, 1/x) : x > 0\} \cup \{(0,0)\}$ clearly shows. However, one may easily check that the convex hull of an open set is open and the convex hull of a compact set is compact.

The following claim is standard.

CLAIM A.1.8. *For a closed convex set $K \subset \mathbb{R}^n$ and a point x which is outside K, there is a unique closest point to x in K in the Euclidean metric, called the metric projection of x onto K and denoted by $P_K(x)$.*

Here existence comes from the fact that K is closed, and uniqueness from it being convex (together with the fact that we are using the Euclidean metric, which is strictly convex, that is, equality in the triangle inequality implies affine dependence between the points).

DEFINITION A.1.9 (supporting hyperplane). Let $K \subset \mathbb{R}^n$ be closed and convex. Let H be some affine hyperplane, that is, a set of the form $\{x : f(x) = c\}$ for some linear functional $f : \mathbb{R}^n \to \mathbb{R}$. A hyperplane determines two closed half-spaces $H^+ = \{f \geqslant c\}$ and $H^- = \{f \leqslant c\}$. We say that H is a *supporting hyperplane* for K if either H^- or H^+ contains K and $H \cap K \neq \emptyset$. The set $K \cap H$ is then called a support set for K (it may of course consist of just one point) and any $x \in K \cap H$ is called a support point of K. The corresponding half-space which contains K is called a supporting half-space for K.

The connection with the metric projection is as follows.

THEOREM A.1.10. *Let $K \subset \mathbb{R}^n$ be closed and convex. For any $x \notin K$ the hyperplane through $P_K(x)$ which is orthogonal to $x - P_K(x)$ is a supporting hyperplane for K.*

Proof. Denote this hyperplane by $H = P_K(x) + u^\perp$ with $u = x - P_K(x)$. Clearly by orthogonality, $x \notin H$. Let H^+ denote the half-space not containing x, so that H^+ can be written as $\{z : \langle z, v\rangle \geqslant \langle P_K(x), v\rangle\}$ where $v = -u$. We claim that $K \subset H^+$. Assume not, then there is some $y \in (H^- \setminus H) \cap K$. This means $\langle y - P_K(x), u\rangle > 0$. Consider the segment joining y and $P_K(x)$, which is included in K and forms an acute angle with the segment $[P_K(x), x]$. The points on this segment which are close to $P_K(x)$ are closer to x than $P_K(x)$: to see this, use the Pythagorean theorem, say, or simply write

$$\begin{aligned}
&\|x - (P_K(x) + \delta(y - P_K(x)))\|_2^2 \\
&\quad = \|x - P_K(x)\|_2^2 - 2\delta\langle x - P_K(x), y - P_K(x)\rangle + \delta^2\|y - P_K(x)\|_2^2 \\
&\quad < \|x - P_K(x)\|_2^2
\end{aligned}$$

for small δ. We thus get a contradiction, and conclude that $K \subset H^+$ as needed. □

The above implies the well known fact, that every closed convex set (other than $\emptyset$ or $\mathbb{R}^n$) is the intersection of its supporting half-spaces. Actually, it implies more - that every boundary point has a supporting hyperplane passing through it. Indeed, given a point $x \in \partial K$ one may find a point y outside K such that $x = P_K y$. We proceed with yet another "separation" result. We mention in passing that the following result does not hold in infinite dimensional space.

CLAIM A.1.11. *Let $A, B \subset \mathbb{R}^n$ be non-empty and convex, and assume* $\mathrm{relint}(A) \cap \mathrm{relint}(B) = \emptyset$. *Then there exists a hyperplane E separating them.*

Proof. Clearly by the assumption, $0 \notin \text{relint}(A) - \text{relint}(B) = \text{relint}(A - B)$. It can be in the closure (thus in the boundary of $A - B$) or outside. In both cases we have seen above that we may separate 0 from $A - B$ using a linear functional f such that $f(A - B) \geqslant f(0) = 0$. This exactly means that $f(a) \geqslant f(b)$ for all $a \in A$ and $b \in B$. □

In the class of polytopes, the support sets are called *faces*, and they can have dimension between 0 and $n - 1$. The $n - 1$ dimensional faces are called *facets*, and the 1-dimensional faces are called *edges*. Note that support sets are convex. It is easy to check that the 0-faces of P are precisely the vertices $v(P)$, and that any face of a polytope is the convex hull of the vertices which lie inside it:

CLAIM A.1.12. *Let $P \subset \mathbb{R}^n$ be a polytope, and let F be a face of P. Then $F = \text{conv}(v(P) \cap F)$. In particular, a polytope $P \subset \mathbb{R}^n$ has finitely many faces.*

Proof. Denote the supporting hyperplane associated with F by H, so that $P \subset H^-$ and $H \cap P = F$. Encode it by $\{f = c\}$, so that $f(P) \leqslant c$. For every vertex in F, $f(v) = c$ and for every vertex of P not in F, $f(v) < c$. Let δ be chosen so that for $v \notin F$, $f(v) \leqslant c - \delta$. Every point in F can be written as a convex combination of vertices from $v(P)$, but by applying f we see that the weights of all $v \notin F$ must be 0, so that the point is in $\text{conv}(v(P) \cap F)$. □

As a corollary of this claim, we get that a polytope is a polyhedral set, since we may take, for each of the (finitely many) faces, a supporting half-space and intersect these half-spaces. It is not hard to check that the resulting set must be the original polytope P.

Generalizing slightly the notion of vertices of polytopes, we define the extremal points of a convex set.

DEFINITION A.1.13. Let $K \subset \mathbb{R}^n$ be closed and convex.
(i) A point $x \in K$ is called extremal (for K) if whenever $x = \lambda y + (1 - \lambda)z$ with $z, y \in K$ and $\lambda \in (0, 1)$ one has $y = z = x$. The set of extremal points of K is denoted by $\text{ext}(K)$.
(ii) A point $x \in K$ is called an exposed point of K if there exists a supporting hyperplane for K such that $K \cap H = \{x\}$ (that is, $\{x\}$ is a support set). The set of exposed points of K is denoted by $\exp(K)$.

REMARK A.1.14. For a closed half-space, $\text{ext}(H) = \emptyset$, and for a closed convex set $\text{ext}(A) \neq \emptyset$ if and only if A does not contain any straight line. Clearly $\exp(K) \subset \text{ext}(K)$. The converse is not true by the example of a football stadium. As in polytopes, $v \in \text{ext}(K)$ if and only if $K \setminus \{x\}$ is convex. For polytopes $v(P) = \text{ext}(P) = \exp(P)$.

The infinite dimensional analogue of the following theorem is the Krein-Milman theorem. The finite dimensional one has a much simpler proof (and one does not need to consider the closure of the convex hull of the extreme points).

THEOREM A.1.15 (Minkowski). *Let $A \subset K \subset \mathbb{R}^n$ for a compact convex K. Then $K = \text{conv}(A)$ if and only if $\text{ext}(K) \subset A$. In particular, $K = \text{conv}(\text{ext}(K))$.*

Proof. If $K = \text{conv}(A)$ and x is an extreme point which is not in A, then $K \setminus \{x\}$ is convex and $\text{conv}(A) \subseteq K \setminus \{x\} \subsetneq K$, a contradiction. We thus only need to show that $K = \text{conv}(\text{ext}(K))$. The proof is by induction on dimension, as it clearly holds for dimension $n = 1$. Given $x \in K$ we may write it as a convex combination of two points in ∂K: if $x \notin \text{ext}(K)$ then simply take some line emanating from x and intersecting ∂K at some points $x_1 = x + tu$ and $x_2 = x - su$ for $s, t > 0$. To see that, say, x_1 is a convex combination of extremal points of K use some supporting hyperplane for K at x_1, say H, and consider the extremal points of the support set $K \cap H$. They are easily seen to be extremal also for K, and by the induction hypothesis, we are done. □

COROLLARY A.1.16. *A bounded polyhedral set is a polytope.*

Proof. Let P be a polyhedral set, $P = \bigcap_{j=1}^m H_j^+$. We need to show that $\text{ext}(P)$ is finite. We claim that any $x \in \text{ext}(P)$ has to be of a very specific form, of which there cannot be more than 2^m points.

Indeed, given x which is extremal, let us consider the intersection of those of the m hyperplanes H_j for which $x \in H_j$, and intersect the resulting set with all sets of the form $H_j^+ \setminus H_j$ for j such that $x \notin H_j$ (hence $x \in H_j^+ \setminus H_j$). This is a subset of P, it is relatively open, and it includes x which is an extreme point of P, thus it must be equal to $\{x\}$. □

THEOREM A.1.17. *For a compact convex* $K \subset \mathbb{R}^n$, $K = \overline{\text{conv}(\exp(K))}$.

Proof. We first show that $\exp(K)$ is non-empty. To this end consider any Euclidean ball which includes K and such that their boundaries intersect (for example, take the smallest R such that $K \subset x + RB_2^n$). Let $y \in \partial K \cap \partial B$ denote a tangency point. The point y is clearly exposed for K. Denote $\overline{\text{conv}(\exp)(K)} = L$. Assume there is a point $x \in K \setminus L$ and separate x from L by a Euclidean ball $x' + R'B_2^n$. This is clearly possible if R' is large enough. Now take the minimal $R > R'$ such that $K \subset x' + RB$. As before, we get that any point $y \in \partial K \cap \partial(x' + RB)$ is an exposed point of K. However, it is also separated from L, which is a contradiction. □

COROLLARY A.1.18 (Straszewicz). *For a compact convex* $K \subset \mathbb{R}^n$, $\text{ext}(K) \subset \overline{\exp(K)}$.

Proof. This follows by Theorem A.1.17 together with Theorem A.1.15 for $A = \exp K$ and the fact that $\overline{\text{conv}(B)} = \text{conv}(\overline{B})$ for every bounded $B \subset \mathbb{R}^n$. □

A.2. Convex functions

Convex functions are a natural extension of convex sets. They come up naturally in the theory of asymptotic geometric analysis, and are used throughout the book, so we would like to introduce them along with some of their simple and basic properties. The theory of convex functions is very rich. It includes many delicate facts (for example, about their regularity properties) most of which we do not need in this book, and these we state without proof.

DEFINITION A.2.1. (i) A function $f : \mathbb{R}^n \to (-\infty, +\infty]$ is called convex if its epigraph, defined by

$$\text{epi}(f) = \{(x, y) \in \mathbb{R}^n \times \mathbb{R} : f(x) \leqslant y\},$$

is a convex subset of $\mathbb{R}^{n+1}$.

(ii) A function $f : A \to (-\infty, \infty]$ is called convex if its extension to $\mathbb{R}^n$ as $+\infty$ outside A is convex, so in particular A must be convex.

(iii) The domain dom(f) of a convex function f is the (convex) set where f is finite.
(iv) The constant $+\infty$ function is called "improper".
(v) A function f is called concave if $-f$ is convex.

One easily checks that a function f is convex if and only if

$$f(\lambda x + (1-\lambda)y) \leqslant \lambda f(x) + (1-\lambda)f(y)$$

for all $x, y \in \mathbb{R}^n$ and $\lambda \in (0,1)$.

REMARK A.2.2. Affine functions are convex and concave. The sub-level sets of a convex function, that is, $\{x : f(x) \leqslant c\}$, are convex sets. Convex functions are closed under addition and multiplication by positive scalars. Convex functions are also closed under pointwise supremum (since the epigraphs are intersected), though properness (that is, the function not assuming the value $+\infty$ at every point) may be lost.

REMARK A.2.3 (Jensen inequality). If f is convex, $\lambda_i \geqslant 0$ and $\sum_{i=1}^m \lambda_i = 1$, then

$$f\Big(\sum_{i=1}^m \lambda_i x_i\Big) \leqslant \sum_{i=1}^m \lambda_i f(x_i).$$

More generally, given some probability measure μ on $\mathbb{R}^n$ we have

$$f\left(\int x d\mu(x)\right) \leqslant \int f(x) d\mu(x).$$

This can also be written as $f\left(\mathbb{E}_\mu(X)\right) \leqslant \mathbb{E}_\mu(f(X))$. The proof is simple and standard.

REMARK A.2.4. For positively homogeneous functions, convex means subadditive. Indeed, $f(\lambda x + (1-\lambda)y) \leqslant f(\lambda x) + f((1-\lambda)y) = \lambda f(x) + (1-\lambda)f(y)$ and vice versa. In particular, norms are convex.

Convex functions are important in optimization, mainly since a local minimum is global. Indeed, if x_0 is a local minimum, say in a ball around it of radius ρ, then take any y farther from x_0 and set

$$x = \frac{\rho}{|y-x_0|} y + \Big(1 - \frac{\rho}{|y-x_0|}\Big) x_0.$$

This point x satisfies $|x - x_0| = \rho$, thus $f(x) \geqslant f(x_0)$, while at the same time

$$f(x) \leqslant \frac{\rho}{|y-x_0|} f(y) + \Big(1 - \frac{\rho}{|y-x_0|}\Big) f(x_0),$$

hence we must have $f(x_0) \leqslant f(y)$.

DEFINITION A.2.5. A convex function f is called "closed" if the convex set epi$(f) \subset \mathbb{R}^n \times \mathbb{R}$ is closed. The closure of a convex function is the largest closed function below it, and its epigraph is $\overline{\mathrm{epi}(f)}$.

Note that, given a convex function, the closure of the epigraph will indeed be the epigraph of another convex function. One easily checks that for a convex function, being "closed" is the same as being lower semi-continuous, and that the points for which this property may fail are only the points at the boundary of the domain of f. In particular, changing f with its closure one only modifies (if needed) the values of f on the boundary of its domain.

DEFINITION A.2.6. The convex hull (or envelope, or convexification) of a function $f : \mathbb{R}^n \to (-\infty, +\infty]$ is defined as

$$\mathrm{conv}(f) = \sup\{g : g \leqslant f, g \text{ convex}\},$$

provided there exists some affine $g \leqslant f$. The convex hull, or convex infimum, of two, or many, functions is defined as the convex hull of the pointwise infimum, or equivalently by using the operation of convex hull on the epigraphs. This can also be written as

$$\mathrm{conv}(\inf_{\alpha \in I}(f_\alpha)) = \hat{\inf}_{\alpha \in I}(f_\alpha) = \sup\{g : g \leqslant f_\alpha \text{ for all } \alpha \in I, g \text{ convex}\}.$$

It exists if and only if there is an affine function below all f_α.

By the standard separation argument, adopted to the function setting, we have

THEOREM A.2.7. *Let $f : \mathbb{R}^n \to (-\infty, +\infty]$ be closed and convex. Then* $f = \sup\{h : h \leqslant f, h \text{ affine }\}$.

Another way to obtain the convex envelope of a function is by using the well known Legendre-Fenchel transform (we also call it simply "the Legendre transform").

DEFINITION A.2.8. Let $f : \mathbb{R}^n \to (-\infty, +\infty]$ be convex and proper. The Legendre transform $\mathcal{L}f$ of f is defined by

$$\mathcal{L}f(y) = f^*(y) := \sup_{x \in \mathbb{R}^n} (\langle x, y \rangle - f(x))$$

(again we identify $\mathbb{R}^n$ with its dual $(\mathbb{R}^n)^*$ via the scalar product).

THEOREM A.2.9. *For $f : \mathbb{R}^n \to (-\infty, +\infty]$ convex and proper, f^* is always closed, convex and proper. We have that $f^{**} = \overline{f}$.*

REMARK A.2.10. In fact, for any f we can take f^* (namely it makes sense to define f^* with the supremum in the definition taken over the domain of f), and we will have $f^{**} = \overline{\mathrm{conv}(f)}$.

Proof. Convexity follows by the way f^* is defined as the supremum of affine functions. Closedness also follows by the definition (the epigraph is an intersection of closed sets). Note that there is some affine function below f, say $\langle \cdot, y \rangle + c$ and thus $f^*(y) \leqslant \sup_x[\langle x, y \rangle - (\langle x, y \rangle + c)] = -c$ is proper.

Next we check that the Legendre transform is "an involution" on closed functions, and that applying it twice to a convex function f produces its closure $\overline{f}$. One direction is quite easy: $f^{**} \leqslant \overline{f}$. Indeed, we can show that $f^{**} \leqslant f$ by choosing $y = x$ in the "inner supremum":

$$f^{**}(x) = \sup_z (\langle x, z \rangle - f^*(z)) = \sup_z \inf_y (\langle x, z \rangle - \langle z, y \rangle + f(y)) \leqslant f(x).$$

As f^{**} is closed, we also have that $f^{**} \leqslant \overline{f}$.

For the other direction, we use Theorem A.2.7 which says that $\overline{f}$ is the supremum of affine functions below $\overline{f}$. Thus

$$\overline{f} = \sup\{(\langle \cdot, y \rangle - c) : \langle \cdot, y \rangle - c \leqslant \overline{f}\}.$$

The latter condition can be written as $c \geqslant \overline{f}^*(y)$. So,

$$\overline{f} = \sup\{(\langle \cdot, y \rangle - c) : c \geqslant \overline{f}^*(y)\} \leqslant \sup_y(\langle \cdot, y \rangle - \overline{f}^*(y)) = \overline{f}^{**}.$$

We see thus that for a closed function f, we have $f^{**} = f$. When f is not closed, given that $\overline{f} \leqslant f$, we have $\overline{f}^* \geqslant f^*$, and by applying the Legendre transform again we see that $\overline{f}^{**} \leqslant f^{**}$. Putting these together we get $\overline{f} \leqslant \overline{f}^{**} \leqslant f^{**} \leqslant \overline{f}$. The proof is thus complete. □

Convex functions satisfy good regularity properties. Regarding continuity:

THEOREM A.2.11. *Convex functions are continuous in the interior of their domain and Lipschitz continuous on compact subsets of the interior of their domain.*

Proof. Given x in the interior of the domain of a convex function f, we can restrict to a simplex P in the interior of the domain such that $x \in \text{int}(P)$. By convexity the values of $|f|$ on the whole simplex are bounded by the maximum on its vertices, which we denote by C say. Next we restrict to some neighbourhood $B_2^n(x,\rho) \subset P$ of x, and assume $|y| = \rho$.

For any $\varepsilon \in (0,1)$ write

$$f(x+\varepsilon y) = f((1-\varepsilon)x + \varepsilon(x+y)) \leqslant (1-\varepsilon)f(x) + \varepsilon C.$$

Thus $f(x+\varepsilon y) - f(x) \leqslant \varepsilon(C - f(x)) \leqslant 2\varepsilon C$. On the other hand, writing

$$x = \frac{1}{1+\varepsilon}(x+\varepsilon y) + \frac{\varepsilon}{1+\varepsilon}(x-y),$$

we have

$$f(x) \leqslant \frac{1}{1+\varepsilon}f(x+\varepsilon y) + \frac{\varepsilon}{1+\varepsilon}C.$$

Thus

$$f(x+\varepsilon y) - f(x) \geqslant \varepsilon(f(x) - C) \geqslant -2\varepsilon C.$$

Together,

$$|f(x+\varepsilon y) - f(x)| \leqslant 2\varepsilon C = |\varepsilon y| 2C/\rho$$

and this gives continuity as $\varepsilon \to 0$. Actually, this gives, for $z = x + \varepsilon y \in B_2^n(x,\rho)$, that

$$|f(z) - f(x)| \leqslant K|z - x|,$$

where $K = 2C/\rho$.

The fact that f is Lipschitz on compact subsets of the interior, is left to the reader. □

Differentiability is a slightly more complicated issue, which we shall not delve into too deeply. We shall start with $n = 1$. We then consider directional derivatives of higher dimensional functions, the subdifferential, and some explicit formulae when dealing with support functions (to be defined below).

Note that by Lebesgue's theorem, an absolutely continuous function on an interval is almost everywhere differentiable (and we have seen that convex functions are locally Lipshitz in the interior of their domain). In the convex case, however, almost everywhere differentiability becomes even stronger: there are at most countably many points at which the function is not differentiable. The following is a standard theorem from calculus:

THEOREM A.2.12. *Let $f : \mathbb{R} \to (-\infty, +\infty]$ be convex, $x \in \text{int}(\text{dom}f)$.*

(i) *The right and left derivatives exist and satisfy $f'_- \leqslant f'_+$.*
(ii) *Each is monotonously increasing, and they are equal except at countably many points.*

(iii) *The function* f'_+ *is continuous from the right, while* f'_- *is from the left, and* f *is the indefinite integral of each.*

Note that (iii) in particular implies that, if a convex function is differentiable, it is continuously differentiable.

REMARK A.2.13. For a convex $f : \mathbb{R} \to (-\infty, \infty]$ the connection between the directional derivatives and the supporting half-planes is straightforward:

$$f(y) \geqslant f(x) + u(y - x)$$

whenever $u \in [f'_-(x), f'_+(x)]$.

COROLLARY A.2.14. *Let* $f : \mathbb{R}^n \to (-\infty, \infty]$ *be convex,* $x \in \operatorname{int}(\operatorname{dom}(f))$. *Then for each* $u \in S^{n-1}$, *or, more generally, each* $u \in \mathbb{R}^n$, *the directional derivative, defined by*

$$f'(x; u) = \frac{\partial f}{\partial u^+}(x) := \lim_{t \to 0^+} \frac{f(x + tu) - f(x)}{t}$$

exists.

REMARK A.2.15. Note that even though the one sided directional derivatives exist, this need not be a linear operation with respect to u, in other words, f is not necessarily differentiable. However, as a function of u (defined for all $u \in \mathbb{R}^n$) the expression $\frac{\partial f}{\partial u^+}(x)$ is convex (and positively homogeneous):

$$\begin{aligned}
f'(x; \lambda u + (1-\lambda)v) &= \lim_{t \to 0^+} \frac{f(x + t(\lambda u + (1-\lambda)v)) - f(x)}{t} \\
&= \lim_{t \to 0^+} \frac{f(\lambda(x + tu) + (1-\lambda)(x + tv)) - f(x)}{t} \\
&\leqslant \lim_{t \to 0^+} \frac{\lambda(f(x + tu) - f(x)) + (1-\lambda)(f(x + tv) - f(x))}{t} \\
&= \lambda \lim_{t \to 0^+} \frac{f(x + tu) - f(x)}{t} + (1-\lambda) \lim_{t \to 0^+} \frac{f(x + tv) - f(x)}{t}.
\end{aligned}$$

REMARK A.2.16. Let us list some facts which are not hard to show and can be found in standard references for convex functions. Given a convex function f, differentiability at a point x is *equivalent* to existence of $\frac{\partial f}{\partial e_j}(x)$ for all $j = 1, \ldots, n$. Another equivalent fact is that the epigraph of f has a unique supporting affine hyperplane at the point $(x, f(x))$; in this case the supporting hyperplane is given by $a(y) = f(x) + \langle \nabla f(x), y - x \rangle$. It *is* true that any convex f is differentiable almost everywhere (with respect to Lebesgue measure). Actually, the second derivative also exists almost everywhere in the appropriate sense, and this is a theorem of Alexandrov. Finally, a convex differentiable function (on an open domain) is C^1, that is, all partial derivatives are continuous.

In the case of a differentiable function, or twice differentiable, there is a simple way to characterize convexity. We shall use $\nabla^2 f(x)$ to denote the Hessian matrix of f, namely the matrix whose elements are $\frac{\partial^2 f}{\partial x_i \partial x_j}(x)$ (sometimes also denoted $\operatorname{Hess}(f)$).

THEOREM A.2.17. *Let* $f : A \to \mathbb{R}$, *where* $A \subset \mathbb{R}^n$ *is convex and open.*

(i) *When* $n = 1$, *if* f *is differentiable,* f *is convex if and only if* f' *is increasing; if* f *is twice differentiable,* f *is convex if and only if* $f'' \geqslant 0$.

(ii) *For general n, if f is differentiable, f is convex if and only if ∇f is "monotone" in the following sense: for all $x, y \in A$*

$$\langle \nabla f(x) - \nabla f(y), x - y \rangle \geqslant 0.$$

Another equivalent condition is that $f(y) - f(x) \geqslant \langle \nabla f(x), y - x \rangle$ for all x, y in the domain of f.

(iii) *If f is twice differentiable, f is convex if and only if $(\nabla^2 f)(x)$ is a positive semidefinite matrix for all $x \in A$.*

In the non-differentiable case, we can substitute the gradient by another notion, defined as follows:

DEFINITION A.2.18. Let $f : \mathbb{R}^n \to (-\infty, +\infty]$ be convex and $x \in \operatorname{dom}(f)$. The subdifferential of f at x is defined by

$$\partial f(x) = \{u \in \mathbb{R}^n : f(y) \geqslant f(x) + \langle u, y - x \rangle \text{ for all } y\}.$$

Any $u \in \partial f(x)$ is called a subgradient of f at x. Note that, again, u is actually an element of the dual space, which we indentify with the original space via the fixed in advance Euclidean structure.

The subdifferential parametrizes the supporting hyperplanes of f. For $x \in \operatorname{int}(\operatorname{dom}(f))$, the set $\partial f(x)$ is non-empty, since there is at least one supporting hyperplane at $(x, f(x))$ which is the graph of an affine function which lies below f and touches it at $(x, f(x))$. Furthermore, the set $\partial f(x)$ is convex. For $x \in \operatorname{int}(\operatorname{dom}(f))$, it is also compact (since it is closed, convex, and it cannot include a ray because x is in the interior of the domain).

THEOREM A.2.19. *Let $f : \mathbb{R}^n \to (-\infty, +\infty]$ be convex and $x \in \operatorname{dom}(f)$. Then $u \in \partial f(x)$ if and only if $(u, -1)$ is an exterior normal to $\operatorname{epi}(f)$ at $(x, f(x))$, that is, $\operatorname{epi}(f) \subset (x, f(x)) + \{(u, -1)^{\perp}\}^{-}$.*

Proof. The condition $\operatorname{epi}(f) \subset (x, f(x)) + \{(u, -1)^{\perp}\}^{-}$ means that, whenever $(y, t) \in \operatorname{epi}(f)$, we have $\langle y - x, u \rangle - (t - f(x)) \leqslant 0$. This is equivalent to $\langle y - x, u \rangle \leqslant f(y) - f(x)$ for all y, which is the definition of the subgradient. □

REMARK A.2.20. Let $f : \mathbb{R}^n \to (-\infty, +\infty]$ be convex, let $x \in \operatorname{int}(\operatorname{dom}(f))$ and assume f is differentiable at x. Then $\partial f(x) = \{\nabla f(x)\}$. Indeed, let $u \in \partial f(x)$. Then for all $t > 0$ we have

$$\frac{f(x + tv) - f(x)}{t} \geqslant \langle u, v \rangle,$$

so that $\frac{\partial f}{\partial v^+}(x) \geqslant \langle u, v \rangle$. But under differentiability $-\frac{\partial f}{\partial v^+} = \frac{\partial f}{\partial (-v)^+} \geqslant -\langle u, v \rangle$, so we get that $\frac{\partial f}{\partial v^+} = \langle u, v \rangle$. Since this holds for all v, we have that u is unique, and equal to $\nabla f(x)$. The converse is also true, namely if $\partial f(x)$ is a single point then f is differentiable at x, but we omit the proof.

The subgradient is closely related to the Legendre transform. Let us mention only one of many such relations.

THEOREM A.2.21. *Let $f : \mathbb{R}^n \to (-\infty, +\infty]$ be convex, closed and proper. Then*

$$f^*(y) + f(x) \geqslant \langle x, y \rangle,$$

with equality if and only if $y \in \partial f(x)$ and if and only if $x \in \partial f^*(y)$.

Proof. The inequality holds by definition, $f^*(y) = \sup_z (\langle z, y\rangle - f(z)) \geqslant \langle x, y\rangle - f(x)$. In the differentiable case, the equality case can be elucidated by differentiation (by finding the maximum). In the general case:

Assume $y \in \partial f(x)$, then $f(z) \geqslant f(x) + \langle y, z - x\rangle$ for all z, and so

$$\begin{aligned} f^*(y) &= \sup_z (\langle z, y\rangle - f(z)) \leqslant \sup_z (\langle z, y\rangle - f(x) - \langle y, z - x\rangle) \\ &= \langle x, y\rangle - f(x) \leqslant f^*(y). \end{aligned}$$

Assume equality holds. Then $\sup_z (\langle z, y\rangle - f(z)) = \langle x, y\rangle - f(x)$, which means that for all z, $f(z) - f(x) \geqslant \langle z - x, y\rangle$.

Since, under the given conditions, $f^{**} = f$, we also get the second if and only if. □

Note that in particular we have just shown that for a convex f, $y \in \partial f(x)$ if and only if $x \in \partial f^*(y)$. In the differentiable setting this is sometimes stated as $(\nabla f^*)(\nabla f(x)) = x$, which by differentiating once more gives us the useful relation

$$(\nabla^2 f^*)(\nabla f(x)) \cdot (\nabla^2 f)(x) = \mathrm{Id}.$$

The notion of the polar of a convex body, which we defined at the beginning of the section, is best captured by the definition of the support function of a body.

DEFINITION A.2.22. For $K \subset \mathbb{R}^n$ non-empty and convex, we define the support function corresponding to K by

$$h_K(u) = \sup_{x \in K} \langle x, u\rangle$$

where $u \in \mathbb{R}^n$.

The properties listed in the claim below are easy to verify directly.

CLAIM A.2.23. *Let $K, T \subset \mathbb{R}^n$ be non-empty and convex. Then*

(i) *h_K is positively homogeneous, closed, and convex.*
(ii) *$h_K = h_{\overline{K}}$ and $\overline{K} = \{x : \langle x, u\rangle \leqslant h_K(u)$ for all $u\}$.*
(iii) *h_K is order preserving: $K \subset T$ implies $h_K \leqslant h_T$ and $h_K \leqslant h_T$ implies $\overline{K} \subset \overline{T}$.*
(iv) *h_K is finite if and only if K is bounded.*
(v) *$h_{\lambda K + \mu T} = \lambda h_K + \mu h_T$ for $\lambda, \mu \geqslant 0$ and $h_{-K}(u) = h_K(-u)$.*
(vi) *$h_{\mathrm{conv} K_i} = \sup h_{K_i}$ and $h_{\cap K_i} = \widehat{\inf h_{K_i}}$.*
(vii) *$(1_K^\infty)^* = h_K$ where 1_K^∞ stands for the convex indicator function of K, which equals 0 on K and $+\infty$ elsewhere.*

CLAIM A.2.24. *Every positively homogeneous closed convex $h : \mathbb{R}^n \to (-\infty, \infty]$ is the support function of some (unique) closed convex set K.*

Proof. Given h, we consider the function h^*. Note that for any $a > 0$ we have

$$h^*(u) = \sup_{x \in \mathbb{R}^n} (\langle ax, u\rangle - h(ax)) = ah^*(u),$$

which implies that h^* can only assume the values 0 or ∞. Letting K be the domain of h^*, which is a closed and convex set, we get that $h^* = 1_K^\infty$, and thus $h = h^{**} = h_K$. □

REMARK A.2.25. (i) $K = \{x\}$ if and only if $h_K = \langle x, \cdot \rangle$.
(ii) $h_{K+\{x\}} = h_K + \langle x, \cdot \rangle$.
(iii) $K = -K$ if $h_K(-u) = h_K(u)$ for all u.
(iv) $0 \in K$ if and only if $h_K \geqslant 0$.

In the case that $0 \in K$, the support function is positive and positively one-homogeneous, so we can consider it as a generalized norm on $\mathbb{R}^n$. It has a unit ball, which is exactly K°.

Let us revisit support sets, with the aid of the support function.

DEFINITION A.2.26. *Let $K \subset \mathbb{R}^n$ be non-empty closed and convex. We set*

$$H(u) = \{x : \langle x, u \rangle = h_K(u)\},$$
$$K(u) = K \cap H(u) = \{x \in K : \langle x, u \rangle = h_K(u)\}.$$

For bounded K, the hyperplane $H(u)$ is a supporting hyperplane for K with corresponding support set $K(u)$. For unbounded K, the set $K(u)$ might be empty, in which case $H(u)$ is not formally a supporting hyperplane. If $K(u) \neq \emptyset$, however, then indeed $K(u)$ is a support set and $H(u)$ a supporting hyperplane.

In this book we usually work in the setting of convex bodies, or in the more general setting of non-empty, compact, convex sets (without requiring non-empty interior).

The support function of polytopes is very simple: if $P = \mathrm{conv}\{x_i\}_{i=1}^m$, by the above remarks

$$h_P(u) = \max_{i=1,\ldots,m} \langle x_i, u \rangle.$$

We thus see that the support function is piecewise linear (that is, there is a partition of $\mathbb{R}^n$ in finitely many convex sets and on each one of them the function is linear). In fact we have that for a non-empty compact convex K, h_K is piecewise linear if and only if K is a polytope. Indeed, if h_K is piecewise linear, since it is positively homogeneous, its linearity sets must be cones, say some finitely many A_i. On each A_i, the function is given by $\langle \cdot, x_i \rangle$ (from homogeneity). Convexity of h_K implies that $h_K \geqslant \max \langle \cdot, x_i \rangle$, but since on each A_i the maximum is attained for the ith element, we have equality.

We conclude this section with a claim that will play a part in the next chapter.

PROPOSITION A.2.27. *Let $f : \mathbb{R}^n \to (-\infty, +\infty]$ be convex and let $x \in \mathrm{int}(\mathrm{dom}(f))$. The directional derivative in direction z is given by*

$$\frac{\partial f}{\partial z^+}(x) = h_{\partial f(x)}(z).$$

In particular, for convex $f, g : \mathbb{R}^n \to (-\infty, +\infty]$ and $x \in \mathrm{int}(\mathrm{dom}(f)) \cap \mathrm{int}(\mathrm{dom}(g))$ we have

$$\partial(f+g)(x) = \partial f(x) + \partial g(x)$$

Proof. We shall use the fact that, as a function of z, $\frac{\partial f}{\partial z^+}(x)$ is positively homogeneous and convex (hence sub-additive), as we saw in Remark A.2.15. Further, we may assume it is finite, since x is in the interior of the domain. Therefore, it is also closed (in fact, continuous).

We can thus use Theorem A.2.24 and find some convex set K_x such that $\frac{\partial f}{\partial z^+}(x) = h_{K_x}(z)$. We shall show that $K_x = \partial f(x)$.

Let $y \in K_x$. Then, by convexity, for $z \in S^{n-1}$,

$$f(x+tz) - f(x) \geqslant t\frac{\partial f}{\partial z^+}(x) = th_{K_x}(z) \geqslant t\langle y, z\rangle.$$

Thus, for any w, writing $w = x + tz$ with z being some unit vector, we get

$$f(w) - f(x) \geqslant \langle y, w - x\rangle.$$

This means that $y \in \partial f(x)$.

Conversely, if $y \in \partial f(x)$ then, for any z, $f(x+tz) \geqslant f(x) + t\langle y, z\rangle$. Taking the limit as $t \to 0^+$, we see that $h_{K_x}(z) = \frac{\partial f}{\partial z^+}(x) \geqslant \langle y, z\rangle$. This implies, by Theorem A.2.23 (ii), that $y \in K_x$.

The last claim follows from Theorem A.2.23 (v) and (iii). □

THEOREM A.2.28. *Let $K \subset \mathbb{R}^n$ be a non-empty compact convex set, and let $u \in \mathbb{R}^n \setminus \{0\}$. Then*

$$\partial h_K(u) = K(u)$$

and in particular, for $K, T \subset \mathbb{R}^n$ non-empty compact and convex sets, and any $z \in \mathbb{R}^n$,

$$(K+T)(u) = K(u) + T(u).$$

Proof. The second assertion follows from the first by the previous theorem. Note that

$$\begin{aligned}\partial h_K(u) &= \{v \in \mathbb{R}^n : h_K(y) \geqslant h_K(u) + \langle v, y-u\rangle \text{ for all } y\}\\ &= \{v \in \mathbb{R}^n : \sup_{x\in K}\langle x, y\rangle \geqslant \sup_{x\in K}\langle x, u\rangle + \langle v, y-u\rangle \text{ for all } y\}.\end{aligned}$$

If $v \in K(u)$ then $h_K(u) = \langle u, v\rangle$ and then, since $v \in K$, we have, for all y, $h_K(y) \geqslant \langle v, y\rangle = h_K(u) + \langle v, y-u\rangle$. So we see that $K(u) \subseteq \partial h_K(u)$.

If $v \in \partial h_K(u)$ then use the defining condition of the subdifferential with $y = 0$ to get $\langle v, u\rangle \geqslant h_K(u)$ and with $y = 2u$ to get $\langle v, u\rangle \leqslant h_K(u)$. This shows that there is equality. To see that $v \in K$ note that, for all y, $\langle v, y\rangle - h_K(y) \leqslant \langle v, u\rangle - h_K(u) = 0$, hence for all y, $\langle v, y\rangle \leqslant h_K(y)$. This by Theorem A.2.23 (ii) implies $v \in K$. □

COROLLARY A.2.29. *Let $K \in \mathbb{R}^n$ be a non-empty compact convex set and let $u \in \mathbb{R}^n \setminus \{0\}$. The set $K(u)$ contains only one point if and only if h_K is differentiable at u (in which case $h_{K(u)}$ is the support function of a point, thus linear, and so $\frac{\partial h_K}{\partial z^+}(u)$ is linear in z, and it is the gradient). In such a case, one has the useful formula*

$$\langle \nabla h_K(u), u\rangle = h_K(u).$$

A.3. Hausdorff distance

The space $\mathcal{K}^n$ of convex bodies in $\mathbb{R}^n$ is endowed with a natural topology. It can be introduced as the topology induced by the Hausdorff metric.

DEFINITION A.3.1. For $K, T \in \mathcal{K}^n$ we define

$$\delta^H(K,T) = \max\left\{\max_{x\in K}\min_{y\in T}|x-y|, \max_{x\in T}\min_{y\in K}|x-y|\right\}.$$

LEMMA A.3.2. *For $K, T \subset \mathbb{R}^n$ non-empty compact and convex sets, we have*

(i) $\delta^H(K,T) = \inf\{\delta \geqslant 0 : K \subset T + \delta B_2^n, T \subset K + \delta B_2^n\}$,

(ii) $\delta^H(K,T) = \max\{|h_K(u) - h_T(u)| : u \in S^{n-1}\}$.

Proof. The first assertion is clear from the definitions. The second claim is also simple: $K \subset T + \delta B_2^n$ if and only of $h_K \leqslant h_T + \delta h_B$ which, as an inequality for functions on S^{n-1}, means $h_K - h_T \leqslant \delta$. We see therefore that $\delta^H(K,T) \leqslant \delta$ if and only if $||h_K - h_T||_{C(S^{n-1})} \leqslant \delta$. □

Note that (ii) in particular means that the embedding $K \mapsto h_K$ from $\mathcal{K}^n$ to $C(S^{n-1})$ is an isometry between $(\mathcal{K}^n, \delta^H)$ and the subset $\mathcal{H}^n \subset C(S^{n-1})$ of the class of convex functions on the sphere endowed with the supremum norm (a function on S^{n-1} is called convex if its homogeneous extension to $\mathbb{R}^n$ is).

COROLLARY A.3.3. *Let $K, L, M \subset \mathbb{R}^n$ be non-empty compact and convex sets. We have*

$$\delta^H(K,L) = \delta^H(K+M, L+M).$$

In particular, we have the cancellation law

$$K + M = L + M \Longrightarrow K = L.$$

REMARK A.3.4. The isometries of the space of convex bodies with respect to the Hausdorff metric are exactly the mappings of the form

$$K \mapsto UK + L$$

for a fixed body L and a rigid motion U.

One method of proving results for convex bodies is to first prove them for polytopes and then use polytopes to approximate general convex bodies. To this end we can use

PROPOSITION A.3.5. *Let $K \subset \mathbb{R}^n$ be a non-empty compact and convex set, and let $\varepsilon > 0$.*

(i) *There exists a polytope $P \subset K$ with $\delta^H(P,K) \leqslant \varepsilon$.*

(ii) *There exists a polytope $P \supset K$ with $\delta^H(P,K) \leqslant \varepsilon$.*

(iii) *If $0 \in \operatorname{relint}(K)$ there exists a polytope P such that $P \subset \operatorname{relint}(K)$ and $K \subset \operatorname{relint}((1+\varepsilon)P)$.*

Proof. (i) We cover the boundary ∂K by a finite number of open balls of radius ε centered at the boundary, $\partial K \subset \bigcup_{i=1}^m (x_i + \varepsilon B_2^n)$. Let $P = \operatorname{conv}(x_i)_{i=1}^m$. Then $P \subset K$ and $\partial K \subset P + \varepsilon B_2^n$, so by convexity $K \subset P + \varepsilon B_2^n$ as needed.

(ii) We cover $\partial(K + \varepsilon B_2^n)$ by open halfspaces whose boundaries are supporting hyperplanes of K (this can be done since points on this boundary are necessarily not in K). By compactness, we find a finite sub-covering, $\partial(K + \varepsilon B_2^n) \subset \bigcup_{i=1}^m \operatorname{int}(H_i^+)$. We let $P = \cap_{i=1}^m H_i^-$. Then $K \subset P$ and $\partial(K + \varepsilon B_2^n) \subset \mathbb{R}^n \setminus P$, so $P \subset K + \varepsilon B_2^n$ as needed.

(iii) Is left to the reader. □

When approximating several sets simultaneously, one may impose some additional requirements regarding the structure of the approximating polytopes. To this end we define the notion of strongly isomorphic polytopes and of simple polytopes.

DEFINITION A.3.6. A polytope P in $\mathbb{R}^n$ is called *simple* if each of its vertices is contained in exactly n facets.

Two polytopes P_1, P_2 are called *strongly isomorphic* if

$$\dim P_1(u) = \dim P_2(u)$$

for every $u \in S^{n-1}$. Obviously this is an equivalence relation.

REMARK A.3.7. It is worth noting that, when we deal with a polytope P, every support set of P is again a polytope and its support sets are also support sets of the original polytope P. This verification is left to the reader, who might want to use that $(P(u))(v)$ for $v \in u^{\perp}$ is the same as $P(w)$ for $w = v + tu$ when $t > 0$ is sufficiently large.

PROPOSITION A.3.8. *If P_1, P_2 are strongly isomorphic polytopes then, for each $u \in S^{n-1}$, the support sets $P_1(u)$ and $P_2(u)$ are strongly isomorphic.*

Proof. Let us fix $u \in S^{n-1}$. For any $v \in S^{n-1}$ consider $(P_i(u))(v) =: Q_i'$, $i = 1, 2$. Without loss of generality we may assume that $0 \in Q_1' \cap Q_2'$. As we saw in the above remark, when η is sufficiently large, then the hyperplane $H = \{\langle z, \eta u + v\rangle = 0\}$ supports P_i and $H \cap P_i = Q_i'$, $i = 1, 2$. Since P_1 and P_2 are strongly isomorphic, we get $\dim Q_1' = \dim Q_2'$ and thus the assertion. □

PROPOSITION A.3.9. *For every integer $m \geqslant 2$, every $K_1, \ldots, K_m \in \mathcal{K}^n$ and every $\varepsilon > 0$, we can find simple strongly isomorphic polytopes $P_1, \ldots, P_m \in \mathcal{K}^n$ satisfying $\delta^H(K_i, P_i) < \varepsilon$ for all $i = 1, \ldots, m$.*

The proof of this proposition is very similar to the proof of Proposition 1.1.8. The simplicity can be assured by taking a small perturbation of the relevant directions, since not being simple means that there is some linear relation between the supporting hyperplanes. The strong isomorphism can be ensured if we begin with any approximating polytopes for the different sets, take all the possible normals, and change the approximating polytopes ever so slightly that they will include facets in all of these directions. We omit precise details here.

A.4. Compactness: Blaschke's selection theorem

THEOREM A.4.1. *Any sequence of convex bodies $K_j \subset \mathbb{R}^n$ of which all elements are contained in some fixed ball, has a convergent subsequence.*

Proof. One way is to use Ascoli-Arzela for the distance functions. We shall instead give a direct proof with subsequences: First we construct a "rectangle" of bodies $K_{i,j}$ where each row is a subsequence of the row above it, namely $\{K_{i,j}\}_{j=1}^{\infty}$ is a subsequence of $\{K_{i-1,j}\}_{i=1}^{\infty}$, and moreover, the distances between all elements in $\{K_{i,j}\}_{j=1}^{\infty}$ are at most $1/2^i$. This is done by covering the big ball by small ones of radius 2^{-i} (a finite number, N_i, using compactness) and saying that an infinite subsequence of the sequence $\{K_{i-1,j}\}_{j=1}^{\infty}$ has the same elements of the covering intersecting them. (That is, the same subset of $\mathcal{P}(N_i)$).

Next, take a diagonal, to get a subsequence $K_{j,j}$ with distance between each two at most $1/2^{\min(i,j)}$. We claim that taking $K_{n_j} = K_{j,j}$ we have $K_{n_j} + 1/2^{j-1}B_2^n$ is monotone decreasing, since $K_{n_{j+1}} \subset K_{n_j} + 1/2^j B_2^n$ and then we may add $1/2^j B_2^n$.

We thus may intersect all of its elements to get a set which we shall call

$$K_0 = \bigcap_j (K_{n_j} + 1/2^{j-1}B_2^n).$$

We claim this K is the limit of K_{n_j}.

Indeed, on the one hand for any $\varepsilon > 0$, for j large enough so that $2^{-(j-1)} < \varepsilon$, clearly K_0 is included in $K_{n_j} + \varepsilon B_2^n$.

On the other hand, take $K+\varepsilon B_2^n$ and let G denote its interior. We may subtract G from each $K_{n_j}+1/2^{j-1}B_2^n$ and get a decreasing sequence of compact sets

$$\cdots \supset K_{n_j}+1/2^{j-1}B_2^n \setminus G \supset K_{n_{j+1}}+1/2^j B_2^n \setminus G \supset \cdots$$

which are all included both in K (from its definition) and in $\mathbb{R}^n \setminus G$ and is thus empty (since K and $\mathbb{R}^n \setminus G$ do not intersect). From the finite intersection property, it must be empty from some point onwards, and in particular for large enough j, $K_{n_j} \subset G \subset K+\varepsilon B_2^n$. We thus get that $K_{n_j} \to K_0$, as needed. □

A.5. Steiner symmetrization properties in detail

To end the "elementary convexity" part of the appendix, we return to the notion of Steiner symmetrization which was discussed in the main text, and provide a full proof for the list of properties of this symmetrization which we have used.

PROPOSITION A.5.1. *Let $K, T \subset \mathbb{R}^n$ be non-empty compact and convex and let $u \in S^{n-1}$. Then*

(i) *$S_u(K)$ is a compact convex set.*
(ii) *$\lambda S_u K = S_u(\lambda K)$.*
(iii) *$S_u K + S_u T \subset S_u(K+T)$.*
(iv) *$K \subset T$ implies $S_u K \subset S_u T$.*
(v) *S_u is continuous with respect to Hausdorff metric on bodies with non-empty interior.*
(vi) $\mathrm{Vol}(S_u(K)) = \mathrm{Vol}(K)$.
(vii) *For all $1 \leqslant i \leqslant n$, $W_i(S_u(K)) \leqslant W_i(K)$, where W_i are the quermassintegrals, introduced in Chapter 1 and to be discussed in the next chapter.*
(viii) *$diam(S_u(K)) \leqslant diam(K)$.*
(ix) *$inradius(S_u(K)) \geqslant inradius(K)$, $circumrad(S_u K) \leqslant circumrad(K)$.*

Proof. (i) follows from trapezoids (Brunn's argument). Indeed, convexity is a two-dimensional notion, so it is enough to consider x and y in $u^\perp$ and check that for any $0<\lambda<1$ the interval centered at $z=(1-\lambda)x+\lambda y$ with length $l_z = |K \cap (z+\mathbb{R}u)|$ has length that is at least $(1-\lambda)l_x+\lambda l_y$, which in turn follows from the fact that $(1-\lambda)(K\cap(x+\mathbb{R}u))+\lambda(K\cap(y+\mathbb{R}u)) \subset K\cap(z+\mathbb{R}u)$.

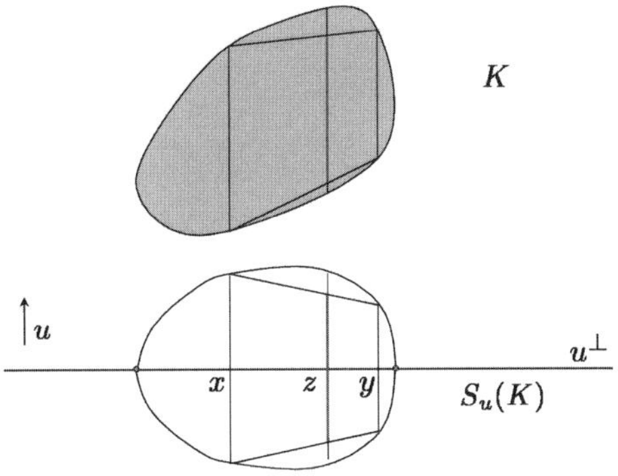

(ii) is trivial, as all notions are homogeneous.

(iii) follows by computation: as in (i), let $l_z^{(K)} = |K\cap(z+\mathbb{R}u)|$. Take an element in the set on the left hand side of (iii): this must be of the form $x+y$ where $x = x'+tu$

with $x' \in u^\perp$ and $|t| \leqslant \frac{1}{2}l_x^{(K)}$, and similarly $y = y' + su$ with $y' \in u^\perp$ and $|s| \leqslant \frac{1}{2}l_y^{(T)}$. Their sum is $x' + y' + (t+s)u$. To see that the latter belongs to the set on the right hand side, observe that the length of $(K+T) \cap (x+y+\mathbb{R}u)$ is at least $l_x^{(K)} + l_y^{(T)}$. This is true since

$$(K+T) \cap (x+y+\mathbb{R}u) \supset (K \cap (x+\mathbb{R}u)) + (T \cap (y+\mathbb{R}u))$$

and the right-hand-side set is an interval in $x+y+\mathbb{R}u$ of length $l_x^{(K)} + l_y^{(T)}$.

(iv) is immediate, since it is just an inequality of lengths.

(v) One should first note that a non-empty interior is important. One can easily construct a non-proper example with no continuity, as in the following picture:

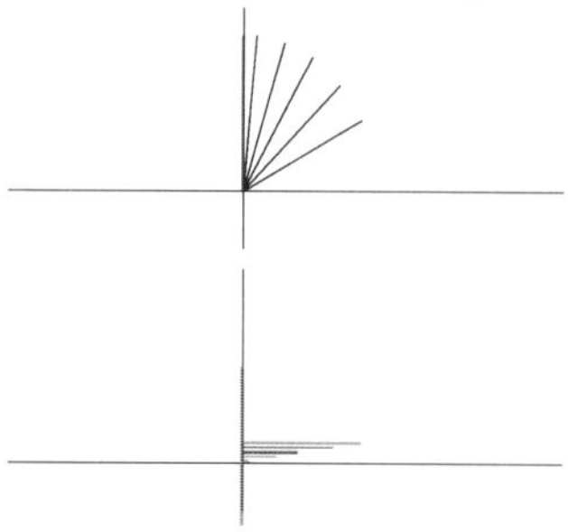

In the subclass of proper convex bodies: take a sequence of proper convex bodies that converges to a proper body, $K_j \to K$. We may assume without loss of generality that $0 \in \text{int}(K)$, since clearly Steiner symmetrization of a translation is a translation of the original symmetrization (by a vector which is the projection onto $u^\perp$ of the original translation). Then by definition of distance, and comparability of B_2 and K (since K is proper and $0 \in \text{int}(K)$), given any $\varepsilon > 0$ there is some index M such that, if $j > M$, we have $(1-\varepsilon)K \subset K_j \subset (1+\varepsilon)K$. (since 0 is in the interior we can do this "subtraction": $K \subset K_j + \varepsilon B_2^n$ means $h_K \leqslant h_{K_j} + \varepsilon h_{B_2^n} \leqslant h_{K_j} + \varepsilon' h_K$). Now use the homothety property (ii) to get that $(1-\varepsilon)S_uK \subset S_uK_j \subset (1+\varepsilon)S_uK$.

(vi) is just Fubini.

(vii) We shall prove it only for W_1, that is, for surface area. The proof for other quermassintergrals follows from the Rogers-Shephard scheme of moving shadows, see the notes and remarks. We use the equivalent definition of surface area as a limit of volume differences, and the "Brunn-Minkowski"-type property (iii) as follows:

$$\begin{aligned} S_n(S_uK) &= \lim_{\varepsilon \to 0^+} \frac{\text{Vol}_n(S_uK + \varepsilon B_2^n) - \text{Vol}_n(S_uK)}{\varepsilon} \\ &\leq \lim_{\varepsilon \to 0^+} \frac{\text{Vol}_n(S_u(K + \varepsilon B_2^n)) - \text{Vol}_n(S_uK)}{\varepsilon} \\ &= \lim_{\varepsilon \to 0^+} \frac{\text{Vol}_n(K + \varepsilon B_2^n) - \text{Vol}_n(K)}{\varepsilon} = S_n(K). \end{aligned}$$

(viii) is again a two-dimensional claim. The diameter is attained as $x - y$ for some $x, y \in K$. Consider the trapezoid, all of which is inside K, that is defined as the convex hull of $K \cap (x + \mathbb{R}u)$ and $K \cap (y + \mathbb{R}u)$. It has two diagonals, the larger of which is $x - y$.

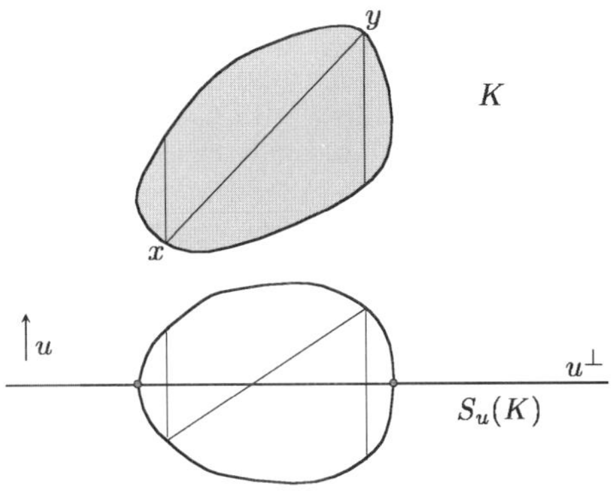

The large diagonal is only decreased after symmetrizing (one can deduce that from Pythagoras' theorem for example).

(ix) The symmetral of a ball is a ball with the same radius and inclusions are preserved by (iv); this applies both for inradius and circumradius. □

A.6. Notes and remarks

Recently, many excellent books on convexity were published, see for example the books of Schneider [**556**], of Gruber [**277**], of Schneider and Weil [**557**], of Gardner [**219**], of Groemer [**266**], of Thompson [**591**], of Burago and Zalgaller [**127**] and the recently translated Bonnesen and Fenchel [**93**]. We included in the first chapter and in both appendices some material we frequently use or that we think is of special interest to the readers of this book and may be useful in asymptotic geometric analysis.

For some of the history of the three classical combinatorial results (Radon, Helly and Caratheodory), see for example the survey by Danzer Grünbaum and Klee [**161**]. We mention Lutwak's containment theorem from [**406**] which is the opposite statement of Lemma A.1.5:

THEOREM A.6.1 (Lutwak's containment theorem). *Let K and L be convex bodies such that every simplex which contains K must contain a translate of L. Then the body K contains a translate of L.*

For the proof of the result mentioned in Remark A.1.4 and some further generalizations see Fenchel [**194**], and Hanner and Radstrom [**298**].

An example in infinite dimensions for which Claim A.1.11 does not hold is due to Klee; we reproduce such an example without details: one may take the space of all polynomials in a variable x, and the following two convex subsets: A is the subset of all polynomials with positive leading coefficiet, and B is the subset of all polynomials with negative leading coefficient. One easily sees that they are open, convex and do not intersect. However, it is easy to check that there is no linear functional separating them since if such a hyperplane existed then it would correspond to a linear functional which is, say, positive on A and negative on B, but one may construct elements $x \in A$ and $y \in B$ such that $x + ty \in A$ for all $t \in \mathbb{R}$ and this is clearly a contradiction. The correct formulation of Claim A.1.11 in the infinite dimensional setting is the famous Hahn-Banach Theorem and its relatives, see for example the book by Eidelman, Milman and Tsolomitis [**180**].

For more on exposed points see e.g. Gruber's [**277**]. To read on convex functions, we recommend Rockafellar [**521**], see also [**277**]. In particular, see there for the proofs of the facts listed in Remark A.2.16. Remark A.2.20 works in the opposite direction as well, but the proof needs a bit more effort, first deducing differentiability on every line, and then using that together with convexity this suffices. See e.g. [**556**].

It turns out that any map which is an order preserving isomorphism between the space of closed convex sets which contain the origin and positively one-homogeneous convex functions must be the support function (with respect to some Euclidean structure). In other words, properties (i) and (iii) essentially characterize the support map, and in particular force the map to satisfy all of the other properties listed in Claim A.2.23. Other properties of the support map can be used for this characterization as well. For more details on this somewhat unexpected result see the paper of Böröczky and Schneider [**96**] and the papers of Artstein-Avidan and Milman [**26**], [**27**] and the references therein.

For a proof of Remark A.3.4 see the paper of Gruber and Lettl [**278**], and also the related paper of Schneider [**553**]. For details on approximating bodies by strongly isomorphic and simple polytopes, see Schneider's book [**556**].

In relation with maps preserving convexity in $\mathbb{R}^n$, we mention an interesting class of non-linear maps which preserve convexity, called "fractional linear maps". These are maps defined on the complement of an affine hyperplane in $\mathbb{R}^n$, which are induced by a linear map on projective space, hence are also called "projective linear". They are of the form $x \mapsto A(x)/f(x)$ where $A : \mathbb{R}^n \to \mathbb{R}^n$ and $f(x) : \mathbb{R}^n \to \mathbb{R}$ are affine maps. It is easy to check that these maps preserve convexity on their domain, since they map intervals to intervals. See Shiffman [**562**] for generalizations of the fundamental theorem of affine geometry, and Artstein-Avidan, Florentin and Milman [**23**] for the use of these maps in mappings of convex functions.

We mention another construction from basic convexity which is useful in our field, that is the Steiner point of a convex body. This is a continuous (in the Hausdoff sense) mapping p from the class of convex bodies in $\mathbb{R}^n$ to $\mathbb{R}^n$ which satisfies $p(K) \in K$ (in fact, in the relative interior of K). It was defined by Steiner in $\mathbb{R}^2$ (for bodies sufficiently smooth, and for polygons) and later generalized to $\mathbb{R}^n$ as the following vector valued integral, which can also be interpreted as a curvature centroid of the body

$$s(K) = c_n \int_{S^{n-1}} h_K(u) u d\sigma(u).$$

Clearly it is additive with respect to Minkowski addition of bodies. The Steiner point is equivariant under rigid motions: $s(gK) = gs(K)$. It was shown by Schneider [**551**] that the Steiner point is unique in the following sense

THEOREM A.6.2 (Schneider). *Let $n \geqslant 2$ and let p be a mapping from all convex bodies to $\mathbb{R}^n$ such that p is Minkowski linear, equivariant under proper rigid motions and continuous. Then p is the Steiner point map.*

In fact, it is enough to assume continuity at B_2^n. There is an example showing that continuity is needed. See Schneider [**556**] for details.

APPENDIX B

Advanced convexity

B.1. Mixed volumes

In this section we introduce mixed volumes and we describe the proof of Minkowski's theorem on the volume of the sum of compact convex sets.

THEOREM B.1.1. *Let $K_1, \ldots, K_m$ be non-empty compact convex subsets of $\mathbb{R}^n$. There exist coefficients $V(K_{i_1}, \ldots, K_{i_n})$, $1 \leqslant i_1, \ldots, i_n \leqslant m$, that are symmetric with respect to the indices $i_1, \ldots, i_n$, such that*

$$\mathrm{Vol}_n\left(t_1K_1 + \cdots + t_mK_m\right) = \sum_{i_1,\ldots,i_n=1}^{m} V(K_{i_1}, \ldots, K_{i_n})t_{i_1} \cdots t_{i_n}$$

for all $t_1, \ldots, t_m \geqslant 0$. The coefficient $V(K_{i_1}, \ldots, K_{i_n})$ is called the ***mixed volume*** *of $K_{i_1}, \ldots, K_{i_n}$ (and depends only on the bodies $K_{i_1}, \ldots, K_{i_n}$).*

As a special case, we obtain a formula for $\mathrm{Vol}_n(K + sT)$, which is a polynomial of degree n in $s > 0$.

Below we discuss the proof of Theorem B.1.1 starting with the case of polytopes, and then using those to approximate arbitrary convex bodies. This is one of the canonical ways in which proofs are obtained in convexity theory. It provides some useful formulae for mixed volumes of polytopes. At the same time let us note that there are very short and ideologically simple ways to prove the same results (an analytic-geometric way) which we already hinted at in Chapter 1 Section 1.3, to which we shall return in Section B.4 below.

B.1.1. Volume and surface area

Both volume and surface area are notions which we are familiar with from the study of calculus, namely volume is Lebesgue measure on $\mathbb{R}^n$, and surface area is $(n-1)$ dimensional Lebesgue measure of ∂K. This notion coincides with a much more elementary one which can be recursively defined for polytopes (recursive with respect to the dimension), and then for arbitrary convex sets using approximation by polytopes. To this end we shall need to look at the support sets of a convex body as sets in a lower dimensional space - this is done, for example, by translating a support set $K(u)$ to $u^{\perp}$, which is then identified with $\mathbb{R}^{n-1}$ (any choice of an orthonormal basis there would work in the same way, since the only thing that matters is the Euclidean metric, and this is inherited from $\mathbb{R}^n$).

CLAIM B.1.2. Let $P \subset \mathbb{R}^n$ be a polytope. For $n = 1$, $\mathrm{Vol}_1([a, b]) = b - a$ and $S_1(P) = 2$ (surface area). For $n \geqslant 2$,

$$\mathrm{Vol}_n(P) = \frac{1}{n} \sum_{u} h_P(u)\mathrm{Vol}_{n-1}(P(u))$$

and

$$S_n(P) = \sum_u \mathrm{Vol}_{n-1}(P(u)).$$

Here u runs over all S^{n-1}, but clearly only finitely many terms in the sum are non-zero.

Note that if the polytope is of dimension $n-1$, its volume is 0 (since the only two support sets of dimension $n-1$ have opposite coefficients in the sum, $h_P(u) = -h_P(-u)$), as is its usual Lebesgue measure. Its surface area is twice its $(n-1)$-dimensional volume.

To prove the claim one should make sure that the formula coincides with usual Lebesgue measure for polytopes, and this can be done using induction on the dimension, and assuming that $0 \in P$ dissecting it into pyramids and summing up their volumes. A delicate point is why one may without loss of generality assume that $0 \in P$, before showing that this is indeed usual volume (thus invariant under rigid motions). To overcome this small obstacle one must make sure that the formula in the claim is invariant under translation, which amounts to

$$\sum \langle x_0, u_i\rangle \mathrm{Vol}_{n-1}(P(u_i)) = 0 \tag{B.1.1}$$

(or equivalently that $\sum u_i \mathrm{Vol}_{n-1}(P(u_i)) = 0$). But this can be verified by starting with some small ε, any $y_0 \in \mathrm{int}(P)$ and any x_0, in which case we do know $\mathrm{Vol}_n(P - y_0 + \varepsilon x_0) = \mathrm{Vol}_n(P - y_0)$ since we may dissect into pyramids and compare the formulae. This proves the desired equality and completes the (only non-trivial step in the) proof of the claim.

Volume is clearly monotone with respect to inclusion. Under convexity, so is surface area.

PROPOSITION B.1.3. *If $P \subset Q$ are polytopes, then $S_n(P) \leqslant S_n(Q)$.*

Proof. We need the following "triangle inequality": if $\{P(u_i)\}_{i=1}^m$ are all the facets of a given polytope P, then

$$\mathrm{Vol}_{n-1}(P(u_1)) \leqslant \sum_{i=2}^m \mathrm{Vol}_{n-1}(P(u_i)).$$

To prove this inequality, project all facets to the $u_1^\perp$ hyperplane, and note that projections do not increase $(n-1)$-dimensional volume. The union of the projections of $P(u_i)$ for $i \neq 1$ covers the facet $P(u_1)$, whence we are done. This formula implies that $S_n(Q \cap H^+) \leqslant S_n(Q)$ (since one is replacing all the facets of $Q \cap H^-$ but one, by just one facet). Assume then that $P \subset Q$ and, keeping in mind the above observation, inductively reduce Q to P by intersecting each time with a half-space that contains P. $\square$

REMARK B.1.4. We have just seen that, if P is a polytope with facets $P(u_i)$ of volume $V_i = \mathrm{Vol}_{n-1}(P(u_i))$, then $\sum u_i V_i = 0$. The opposite is also true and is called "Minkowski's theorem" which was discussed in Chapter 1.

Let $K \subset \mathbb{R}^n$ be a non-empty compact convex set. Then

$$\inf_{P \supset K} \mathrm{Vol}_n(P) = \sup_{P \subset K} \mathrm{Vol}_n(P) = \mathrm{Vol}_n(K).$$

Similarly

$$\inf_{P \supset K} S_n(P) = \sup_{P \subset K} S_n(P) =: S_n(K)$$

is the $(n-1)$-dimensional Lebesgue measure of $\partial K \subset \mathbb{R}^n$.

It is useful to note that $K \mapsto \mathrm{Vol}_n(K)$ is continuous (for the proof simply approximate by polytopes). We will see shortly that S_n is also continuous.

B.1.2. Mixed volumes: polytopes

It is useful to first give some intuition about the way one can prove polynomiality of volume with respect to Minkowski addition. We know that the volume of a polytope is given by

$$\mathrm{Vol}_n(P) = \frac{1}{n}\sum_u h_P(u)\mathrm{Vol}_{n-1}(P(u)),$$

where the sum is over all unit outer normals of facets of P.

Let us consider what happens when we sum two polytopes:

$$\mathrm{Vol}_n(sP_1 + tP_2) = \frac{1}{n}\sum_u (sh_{P_1}(u) + th_{P_2}(u))\mathrm{Vol}_{n-1}((sP_1 + tP_2)(u)),$$

since $h_{sP_1+tP_2} = sh_{P_1} + th_{P_2}$ for every $s, t \geqslant 0$. We now have to sum over unit normals of the sum, which is a set independent of s and t as we shall see below in Lemma B.1.5.

We could continue inductively, and use that $(sP_1 + tP_2)(u) = sP_1(u) + tP_2(u)$ (again, a fact that we shall see in Lemma B.1.5 below), and that by induction $\mathrm{Vol}_{n-1}(sP_1(u) + tP_2(u))$ is also a homogeneous polynomial in s, t, this time of degree $(n-1)$. In doing so, we would get a proof, but it is somewhat hard to keep track of the coefficients since we would be dealing with, for instance, $h_{(sP_1+tP_2)(u)}(v)$ in the inductive step. The second lemma below, Lemma B.1.7, helps us deal with these support functions and with the sets themselves, realizing the support sets as intersections of facets and recovering the vectors with respect to which these are support sets (certain combinations of the normals of the facets involved).

We shall start with an easy lemma regarding the support sets' behaviour with respect to Minkowski addition and multiplication by positive scalars.

LEMMA B.1.5. *Let $\lambda_i > 0$ and $P_i \subset \mathbb{R}^n$ polytopes, for $i = 1, \dots, m$. Let $u, v \in S^{n-1}$. Then*
(a) $(\sum_{i=1}^m \lambda_i P_i)(u) = \sum_{i=1}^m \lambda_i P_i(u)$.
(b) $\dim\left((\sum_{i=1}^m \lambda_i P_i)(u)\right) = \dim\left((\sum_{i=1}^m P_i)(u)\right)$.
(c) *If* $(\sum_{i=1}^m P_i)(u) \cap (\sum_{i=1}^m P_i)(v) \neq \emptyset$ *then*

$$\Big(\sum_{i=1}^m P_i\Big)(u) \cap \Big(\sum_{i=1}^m P_i\Big)(v) = \sum_{i=1}^m (P_i(u) \cap P_i(v)).$$

Proof. The first assertion follows from Theorem A.2.28. For (b) we first move all polytopes to have 0 in their relative interior, in which case their sum too has 0 in its relative interior, and then, letting $\lambda_{\min} = \min \lambda_i$ and $\lambda_{\max} = \max \lambda_i$, we have, using (a), that

$$\lambda_{\min}\sum P_i(u) \subset \sum_{i=1}^m \lambda_i P_i(u) \subset \lambda_{\max}\sum P_i(u). \tag{B.1.2}$$

Thus they have the same dimension (and in fact, the same spanned subspace after this translation).

Finally, assume the intersection in (c) is non-empty and take some x inside it. Thus $x = \sum x_i$ with $x_i \in P_i$. Clearly $h_{P_i}(u) = \langle x_i, u\rangle$ and $h_{P_i}(v) = \langle x_i, v\rangle$, otherwise x would not satisfy $h_{\sum P_i}(u) = \langle x, u\rangle$ and $h_{\sum P_i}(v) = \langle x, v\rangle$. Thus $x_i \in P_i(u) \cap P_i(v)$ for all i if (and only if) $x \in (\sum P_i)(u) \cap (\sum P_i)(v)$ (the "only if" part can be shown in much the same way, using (a) again). □

REMARK B.1.6. One consequence of the last lemma is that the normals of facets of P are also normals of facets of $P+Q$. Indeed, $(P+Q)(u) = P(u)+Q(u)$, so if $\dim P(u) = n-1$ then $\dim(P+Q)(u) = n-1$ as well.

We need a slightly more intricate fact about support sets that intersect.

LEMMA B.1.7. *Let $K \subset \mathbb{R}^n$ be a non-empty convex and compact set and let $u, v \in S^{n-1}$ be linearly independent. Let $w = \lambda u + \mu v$ where $\mu > 0$ and $\lambda \in \mathbb{R}$. Assume $K(u) \cap K(v) \neq \emptyset$. Then*

$$K(u) \cap K(v) = (K(u))(w).$$

In particular, for any $x \in K(u) \cap K(v)$ we have

$$h_{K(u)}(w) = \langle x, w\rangle = \lambda\langle x, u\rangle + \mu\langle x, v\rangle = \lambda h_K(u) + \mu h_K(v).$$

REMARK B.1.8. Note that we are not dealing with *simple* polytopes, so the dimension of the intersection is not necessarily $n-2$. Assume, however, that we are in the case of simple polytopes, and that u and v are unit normals of two intersecting facets, so that the intersection is of dimension $n-2$. In such a case, we could speak about the normal of the intersection with respect to one of the facets, say $P(u)$. We then get two unit normals, u and v, of the polytope P, and another normal w of $P(u) \cap P(v)$ with respect to the polytope $P(u)$, say. Geometrically we can see they all lie on a plane, and if we let θ stand for the angle between u and v, namely $\cos(\theta) = \langle u, v\rangle$, one sees that

$$v = u\cos(\theta) + w\sin(\theta)$$

(in other words, $w\sin(\theta) = \mathrm{Proj}_{u^\perp} v$).

Since the facet $P(u)$ lies in $u^\perp$, the facet of $P(u)$ in direction $w + \lambda u$ does not depend on $\lambda \in \mathbb{R}$, which is as claimed. Lemma B.1.7 above does not involve just polytopes and is quite general. In particular, the dimension of the intersection can be smaller than $n-2$.

Proof. Assume $z \in K(u) \cap K(v)$. Clearly $\langle z, u\rangle = h_K(u) = h_{K(u)}(u)$. Also clearly $h_{K(u)}(-u) = -\langle z, u\rangle$ since the inner product with u, and with $-u$, is fixed on $K(u)$. In particular, for all $\lambda \in \mathbb{R}$, $h_{K(u)}(\lambda u) = \lambda h_K(u) = \langle z, \lambda u\rangle$. Thus for $w = \lambda u + \mu v$, and given that $\langle z, v\rangle = h_K(v)$, we have that

$$\begin{aligned}\langle z, w\rangle &= h_{K(u)}(\lambda u) + \langle z, \mu v\rangle = h_{K(u)}(\lambda u) + h_K(\mu v)\\ &\geqslant h_{K(u)}(\lambda u) + h_{K(u)}(\mu v) \geqslant h_{K(u)}(w) \geqslant \langle z, w\rangle,\end{aligned}$$

hence $z \in (K(u))(w)$.

Similarly, if $z \in (K(u))(w)$, and if we take some $x \in K(u) \cap K(v)$, then x also belongs to $(K(u))(w)$ by above, so

$$\lambda\langle x, u\rangle + \mu\langle x, v\rangle = \langle x, w\rangle = \langle z, w\rangle = \lambda\langle z, u\rangle + \mu\langle z, v\rangle.$$

Since we know that z and x have the same inner product with u, we conclude by the above equality that $z \in K(v)$ too, as needed. □

We are finally in a position to define the mixed volume of n polytopes in $\mathbb{R}^n$. Usually one first says that the quantity $\mathrm{Vol}_n(\sum_{i=1}^m \lambda_i K_i)$ is a polynomial in the λ_i, then calls the coefficients "mixed volumes" and then gives a precise formula for them. We work in the opposite way, giving first the formula for the "mixed volumes" and then showing that $\mathrm{Vol}_n(\sum_{i=1}^m \lambda_i K_i)$ is a polynomial with these numbers as coefficients.

DEFINITION B.1.9. Let $P_1, \dots, P_n$ be polytopes in $\mathbb{R}^n$. For $n = 1$, and with $P_1 = [a, b]$, we define

$$V^{(1)}(P_1) = \mathrm{Vol}_1(P_1) = h_{P_1}(1) + h_{P_1}(-1) = b - a.$$

For $n \geqslant 2$ we define the mixed volume of $P_1, \dots, P_n$ inductively as

$$V^{(n)}(P_1, \dots, P_n) = \frac{1}{n} \sum_{u \in N(\sum_{i=2}^n P_i)} h_{P_1}(u) V^{(n-1)}(P_2(u), \dots, P_n(u)),$$

where $N(\sum_{i=2}^n P_i)$ is the set of unit normals of the $(n-1)$-dimensional faces of $\sum_{i=2}^n P_i$. When one of the sets is empty we set the mixed volume equal to 0.

REMARK B.1.10. Note that a-priori the expression for the mixed volume does not seem to be symmetric in the arguments, and this is one of the objectives of the next theorem. Also note that one easily sees by induction that $V^{(n)}(P, P, \dots, P) = \mathrm{Vol}_n(P)$.

THEOREM B.1.11. *Given polytopes $P_1, \dots, P_m \in \mathcal{K}^n$, we have that*

$$\mathrm{Vol}_n\Big(\sum_{i=1}^m \lambda_i P_i\Big) = \sum_{i_1, \dots, i_n = 1}^m \lambda_{i_1} \cdots \lambda_{i_n} V^{(n)}(P_{i_1}, \dots, P_{i_n}).$$

The function $V^{(n)}(P_1, \dots, P_n)$ is non-negative, translation invariant, symmetric in its arguments, and vanishes when the dimension of $\sum_{i=1}^n P_i$ is less than n.

REMARK B.1.12. To understand the result it may be useful to look at the special case $m = 2$. In such a case, we get that

$$\mathrm{Vol}_n(sP_1 + tP_2) = \sum_{j=0}^n s^j t^{n-j} \binom{n}{j} V^{(n)}(P_1[j], P_2[n-j]),$$

where $V^{(n)}(P_1[j], P_2[n-j])$ stands for P_1 appearing j times and P_2 appearing $(n-j)$ times.

We remark that in the proof we shall use inductively mixed volumes in lower dimensions for bodies that lie in a subspace (and are support sets). We thus also define for k polytopes in $\mathbb{R}^n$, say $Q_1, \dots, Q_k$, that satisfy $\dim(\sum_{i=1}^k Q_i) \leqslant k$, the quantity

$$V^{(k)}(Q_1, \dots, Q_k) = V^{(k)}(P_E(Q_1), \dots, P_E(Q_k))$$

where E is a k-dimensional subspace containing a translate of each of the polytopes. Because of translation invariance the choice of E will not be important.

Proof. We use induction on the dimension. In dimension 1, summing intervals is linear, and volume is linear, so everything holds:

$$\mathrm{Vol}_1\Big(\sum_{i=1}^m \lambda_i [a_i, b_i]\Big) = \mathrm{Vol}_1\Big(\Big[\sum \lambda_i a_i, \sum \lambda_i b_i\Big]\Big) = \sum \lambda_i (b_i - a_i).$$

Next assume the assertion is true in dimension $\leqslant n-1$. Let us first show that $V^{(n)}$ vanishes when the dimension of the Minkowski sum of the bodies is less than n. In such a case, the sum is contained in some translate of a hyperplane $u^\perp$. In particular, its $(n-1)$-dimensional faces with non-zero volume are either non-existent, in which case we sum over an empty set, or are in directions u and $-u$ for some vector, in which case $h_{P_1}(-u) = -h_{P_1}(u)$ and $P_j(u) = P_j(-u)$ for all j, hence $V^{(n)}$ vanishes then as well.

To prove polynomiality (with these given coefficients) it is sufficient to consider only positive λ_i, since if some of the coefficients are zero, we can simply reduce the value of m. We use the definition of volume and the inductive hypothesis to get

$$\begin{aligned}
\mathrm{Vol}_n\Big(\sum_{i=1}^m \lambda_i P_i\Big) &= \frac{1}{n} \sum_{u\in N(\sum_{i=1}^m \lambda_i P_i)} h_{\sum_{i=1}^m \lambda_i P_i}(u)\mathrm{Vol}_{n-1}\left(\Big(\sum_{i=1}^m \lambda_i P_i\Big)(u)\right) \\
&= \frac{1}{n} \sum_{u\in N(\sum_{i=1}^m \lambda_i P_i)} \sum_{i_1=1}^m \lambda_{i_1} h_{P_{i_1}}(u)\mathrm{Vol}_{n-1}\left(\sum_{i=1}^m \lambda_i (P_i(u))\right) \\
&= \sum_{i_1=1}^m \lambda_{i_1} \frac{1}{n} \sum_{u\in N(\sum_{i=1}^m \lambda_i P_i)} h_{P_{i_1}}(u)\mathrm{Vol}_{n-1}\left(\sum_{i=1}^m \lambda_i (P_i(u))\right).
\end{aligned}$$

Using the induction hypothesis

$$\mathrm{Vol}_{n-1}\left(\sum_{i=1}^m \lambda_i (P_i(u))\right) = \sum_{i_2,\ldots,i_n=1}^m \prod_{j=2}^n \lambda_{i_j} V^{(n-1)}(P_{i_2}(u),\ldots,P_{i_n}(u)).$$

Plugging this in, we get that

$$\begin{aligned}
&\mathrm{Vol}_n\Big(\sum_{i=1}^m \lambda_i P_i\Big) \\
&= \sum_{i_1=1}^m \sum_{i_2,\ldots,i_n=1}^m \lambda_{i_1} \prod_{j=2}^n \lambda_{i_j} \frac{1}{n} \sum_{u\in N(\sum \lambda_i P_i)} h_{P_{i_1}}(u) V^{(n-1)}(P_{i_2}(u),\ldots,P_{i_n}(u)) \\
&= \sum_{i_1,i_2,\ldots,i_n=1}^m \prod_{j=1}^n \lambda_{i_j} \frac{1}{n} \sum_{u\in N(\sum P_i)} h_{P_{i_1}}(u) V^{(n-1)}(P_{i_2}(u),\ldots,P_{i_n}(u)) \\
&= \sum_{i_1,i_2,\ldots,i_n=1}^m \prod_{j=1}^n \lambda_{i_j} V^{(n)}(P_{i_1},P_{i_2},\ldots,P_{i_n}).
\end{aligned}$$

We have exchanged the summation over $N(\sum_{i=1}^m \lambda_i P_i)$ first with $N(\sum_{i=1}^m P_i)$ by Lemma B.1.5 and then with $N(\sum_{j=2}^n P_{i_j})$ which is a smaller set (by Remark B.1.6). The latter is allowed because any u that is not in the set $N(\sum_{j=2}^n P_{i_j})$ will produce support sets $\{P_{i_j}(u)\}_{j=2}^n$ such that the dimension of the sum satisfies $\dim(\sum_{j=2}^n P_{i_j}(u)) \leqslant (n-2)$, and by the previous argument the $(n-1)$-dimensional mixed volumes of the corresponding support sets will equal 0 and not donate to the sum.

To show symmetry (it is important because we gave an explicit formula, otherwise we can just symmetrize any polynomial), we shall assume it inductively for dimension less than n, and show that

$$V^{(n)}(P_1,P_2,P_3,\ldots,P_n) = V^{(n)}(P_2,P_1,P_3,\ldots,P_n).$$

We may clearly assume the dimension of the sum $P = \sum_{i=1}^n P_i$ is n, otherwise both sides vanish. Let us write the expression for the left hand side above (in a way that is symmetric in P_1 and P_2): going back to the definition, we see that

$$\begin{aligned} &V^{(n)}(P_1, P_2, P_3, \dots, P_n) \\ &= \frac{1}{n} \sum_{u \in N(\sum_{i=2}^n P_i)} h_{P_1}(u) V^{(n-1)}(P_2(u), \dots, P_n(u)) \\ &= \frac{1}{n} \sum_{u \in N(\sum_{i=2}^n P_i)} \Big[h_{P_1}(u) \frac{1}{n-1} \\ &\qquad \times \sum_{w \in N(\sum_{i=3}^n P_i(u))} h_{P_2(u)}(w) V^{(n-2)}(P_3(u)(w), \dots, P_n(u)(w)) \Big]. \end{aligned}$$

In the inner sum, the vectors w summed upon are the normals of the facet $(\sum_{i=3}^n P_i)(u)$. By the same argument as before, we change all sums to sums on facets of $\sum_{i=1}^n P_i = P$ and $P(u)$ respectively. We see that

$$\begin{aligned} &V^{(n)}(P_1, P_2, P_3, \dots, P_n) = \\ &\frac{1}{n} \sum_{u \in N(P)} \Big[h_{P_1}(u) \frac{1}{n-1} \times \sum_{w \in N(P(u))} h_{P_2(u)}(w) V^{(n-2)}(P_3(u)(w), \dots, P_n(u)(w)) \Big]. \end{aligned}$$

Each facet of the form $(\sum_{i=1}^n P_i(u))(w)$ is an $(n-2)$-dimensional face of $P = \sum_{i=1}^n P_i$ contained in $P(u)$, thus it is given by some non-empty $P(u) \cap P(v)$. In particular, all these intersections are in $u^\perp$, and are the projections of vectors $v \in N(P)$ onto $u^\perp$. Thus, we are in fact summing over facets $P(v)$ that intersect $P(u)$, and we can refer to a previous lemma about intersections, Lemma B.1.5 (c), to continue. We first recall that, when $P(u)$ intersects $P(v)$ and an $(n-2)$-dimensional facet of $P(u)$ is produced, the normal of this facet with respect to $P(u)$, which is $w = P_{u^\perp} v$, is a linear combination of the form $w = \lambda u + \mu v$ with $\mu > 0$ (see Remark B.1.8 above), hence $P(u) \cap P(v) = P(u)(w)$. Even further, we know by Lemma B.1.5 that, if $\sum_{i=1}^n P_i(u) \cap \sum_{i=1}^n P_i(v)$ is such a face, then it is given as the sum $\sum_{i=1}^n (P_i(u) \cap P_i(v))$, so in particular these intersections are non-empty as well.

We then also have that $P_i(u)(w) = P_i(u) \cap P_i(v)$:

$$\begin{aligned} &V^{(n)}(P_1, P_2, P_3, \dots, P_n) \\ &= \frac{1}{n} \frac{1}{n-1} \sum_{u \in N(P)} \Big[h_{P_1}(u) \sum_{v \in N(P) \text{ s.t. } P(u) \cap P(v) \neq \emptyset, v \neq u} h_{P_2(u)}(P_{u^\perp} v) \\ &\qquad \times V^{(n-2)}(P_3(u) \cap P_3(v), \dots, P_n(u) \cap P_n(v)) \Big]. \end{aligned}$$

More precisely, since w is the (normalized) projection onto $u^\perp$ of v (as in Remark B.1.8), we have $w = \frac{v - u\langle u, v\rangle}{|v - u\langle u, v\rangle|}$. Similarly, $v = u\cos(\theta) + w \sin(\theta)$, with $\theta = \theta_{u,v}$ given by $\cos(\theta) = \langle u, v\rangle$. Thus $\sin(\theta) = |v - u\langle u, v\rangle|$ and

$$w = \frac{1}{\sin(\theta)} v - \frac{1}{\tan(\theta)} u.$$

Therefore, as in Lemma B.1.7, we have

$$h_{P_2(u)}(w) = \frac{h_{P_2}(v)}{\sin(\theta)} - \frac{h_{P_2}(u)}{\tan(\theta)}.$$

Going back to the main equation, we see that

$$\begin{aligned}
&V^{(n)}(P_1, P_2, P_3, \dots, P_n) \\
&= \frac{1}{n(n-1)} \sum_{u \in N(P)} \sum_{v \in N(P), P(v) \cap P(u) \neq \emptyset, v \neq \pm u} h_{P_1}(u) \Big[\frac{h_{P_2}(v)}{\sin(\theta_{u,v})} - \frac{h_{P_2}(u)}{\tan(\theta_{u,v})} \Big] \\
&\quad \times V^{(n-2)}(P_3(u) \cap P_3(v), \dots, P_n(u) \cap P_n(v)) \\
&= \frac{1}{n(n-1)} \sum_{u,v \in N(P), P(u) \cap P(v) \neq \emptyset} \Big[\frac{h_{P_1}(u) h_{P_2}(v)}{\sin(\theta_{u,v})} - \frac{h_{P_1}(u) h_{P_2}(u)}{\tan(\theta_{u,v})} \Big] \\
&\quad \times V^{(n-2)}(P_3(u) \cap P_3(v), \dots, P_n(u) \cap P_n(v)),
\end{aligned}$$

which clearly shows that the expression is symmetric in P_1, P_2.

What remains to be proven is that these numbers are independent of independent translations of the bodies: this follows from the respective property for Vol_n and uniqueness of polynomial decomposition, for example, when plugging in $n = m$. Indeed, write down the symmetric polynomial for $\mathrm{Vol}(\sum_{i=1}^n \lambda_i (P_i + x_i)) = \mathrm{Vol}(\sum_{i=1}^n \lambda_i P_i)$, and note that the coefficient of $\prod_{j=1}^n \lambda_{i_j}$ is both $V^{(n)}(P_{i_1}, \dots, P_{i_n})$ and $V^{(n)}(P_{i_1} + x_1, \dots, P_{i_n} + x_n)$; here we are also using the symmetry of the coefficients to have the relevant uniqueness of coefficients of a polynomial. □

As is usual in similar situations, there exists a reverse correspondence, namely one can get an expression for mixed volumes in terms of volumes of sums.

THEOREM B.1.13 (Polar decomposition). *Given polytopes $P_1, \dots, P_n$ in $\mathbb{R}^n$ we have*

$$V^{(n)}(P_1, \dots, P_n) = \frac{1}{n!} \sum_{k=1}^{n} (-1)^{n+k} \sum_{1 \leqslant j_1 < \cdots < j_k \leqslant n} \mathrm{Vol}_n \Big(\sum_{m=1}^{k} P_{j_m} \Big).$$

The proof of such a "polarization" result is standard; we include it here for the sake of completeness.

Proof. The right hand side is clearly some function of the polytopes, denote it by $F(P_1, \dots, P_n)$. Consider also the function $g(\lambda_1, \dots, \lambda_n) = F(\lambda_1 P_1, \dots, \lambda_n P_n)$ as a function of $(\lambda_1, \dots, \lambda_n)$. We already know that this is a homogeneous polynomial of degree n (on $(\mathbb{R}^+)^n$). Write it down with symmetric coefficients.

Plug in $P_1 = \{0\}$ to get a telescoping sum (for each term with P_1 there is a corresponding term without it, with an opposite sign, that cancels the former term). We thus have $g(0, \lambda_2, \dots, \lambda_n) = 0$. This means no terms without λ_1 have non-zero coefficients in the polynomial, and the same applies to λ_j for all j. As a consequence, the only term in the polynomial is of the form $C \prod_{j=1}^n \lambda_j$. Going back to the polynomial representation, we see that the only element that produces such a term (that is, the terms that correspond to it do not cancel with others) is $\mathrm{Vol}_n(\sum_{m=1}^n P_{j_m})$ with $j_1 = 1, \dots, j_n = n$. By the previous theorem we know the form of the coefficient of the product, which is simply $n! V^{(n)}(P_1, \dots, P_n)$. □

B.1.3. Mixed volumes: convex bodies

Naturally, we would like to extend the definition, and the theorem, from polytopes to general convex bodies.

THEOREM B.1.14. *For* $\{K_i\}_{i=1}^n \subset \mathbb{R}^n$ *non-empty compact convex sets, define their mixed volume by*

$$V(K_1,\ldots,K_n) = \lim V^{(n)}(P_1^{(m)},\ldots,P_n^{(m)})$$

where $P_i^{(m)} \to K_i$ *are any approximating sequences of polytopes. Then this function is well defined, and we have*

$$V(K_1,\ldots,K_n) = \frac{1}{n!}\sum_{k=1}^{n}(-1)^{n+k}\sum_{1\leqslant j_1<\cdots<j_k\leqslant n}\mathrm{Vol}_n\Big(\sum_{m=1}^{k} K_{j_m}\Big)$$

and, given $\lambda_i > 0$,

$$\mathrm{Vol}\Big(\sum_{i=1}^{m}\lambda_i K_i\Big) = \sum_{i_1,\ldots i_n=1}^{m}\prod_{j=1}^{n}\lambda_{i_j}V(K_{i_1},\ldots,K_{i_n}).$$

Furthermore, if $K, L, K_j \in \mathcal{K}^n$, *we have*

(i) $V(K,\ldots,K) = \mathrm{Vol}_n(K)$ *and* $nV(K,\ldots,K,B_2^n) = S_n(K)$.
(ii) V *is symmetric.*
(iii) V *is multilinear, that is, linear in each argument: for* $\lambda,\mu > 0$,

$$V(\lambda K + \mu L, K_2,\ldots,K_n) = \lambda V(K,K_2,\ldots,K_n) + \mu V(L,K_2,\ldots,K_n).$$

(iv) $V(K_1,K_2,\ldots,K_n) = V(K_1+\{x_1\},K_2+\{x_2\},\ldots,K_n+\{x_n\})$.
(v) $V(UK_1,UK_2,\ldots,UK_n) = V(K_1,K_2,\ldots,K_n)$ *for* $U \in SL(n)$.
(vi) V *is continuous with respect to each of its arguments.*
(vii) $V \geqslant 0$ *and is monotone increasing in each argument.*

Proof. We can use the formula for $V^{(n)}$ from Theorem B.1.13 which proves that the limit exists, is continuous and independent of the choice of approximating sequences, while it satisfies the corresponding equations. In particular we immediately get (iv), (v) and (vi). To show the first part of (i), it suffices to consider polytopes: for these (i) immediately follows by the main formula with $m = 1$. As for surface area, we first assume K to be a polytope. The definition requires writing B_2^n as $\lim P_m$. We know already that

$$nV(P,\ldots,P,P_m) = \sum_{u\in N(P)} h_{P_m}(u)\mathrm{Vol}_{n-1}(P(u)).$$

In the limit we have $h_{P_m}(u) \to 1$, so we see that

$$nV(P,\ldots,P,B_2^n) = \sum_{u\in N(P)} \mathrm{Vol}_{n-1}(P(u)) = S_n(P),$$

as needed. When K is a general convex body, we can approximate it from inside and outside by polytopes, and as the mixed volume is continuous, we are done. (We only use here the monotonicity of S_n, and we *get* its continuity.)

The symmetry with respect to permutations follows from symmetry for polytopes, as does non-negativity (which in the case of polytopes is clear to see if we move each summand so that 0 lies in the relative interior of it).

The linearity in each argument is also not hard to show by expanding in two different ways the polynomial $\mathrm{Vol}(\lambda_1(\lambda K + \mu L) + \lambda_2 K_2 + \cdots + \lambda_n K_n)$. Another way is to use additivity of the support function and the definition for polytopes, and then approximate a general body.

Monotonicity is also simple if one uses the definition with the support function, which is monotone. □

REMARK B.1.15. An alternative route towards mixed volumes is to restrict to classes of strongly isomorphic polytopes. Since we shall use strongly isomorphic polytopes in the next section, we illustrate the key steps in using these to define mixed volumes as well.

(1) Polytopes being strongly isomorphic implies the same for their corresponding faces. The property is preserved under Minkowski addition and under multiplication by positive scalars.
(2) Strongly isomorphic polytopes are dense in the following sense: given $K_1, \ldots, K_m$ and $\varepsilon > 0$ one may find strongly isomorphic $P_1, \ldots, P_m$ with $\delta^H(K_i, P_i) \leqslant \varepsilon$. In fact, one may also make P_i simple, that is, having each of its vertices belonging to exactly n facets (where n is the dimension of the space).
(3) Fixing a class of simple and strongly isomorphic polytopes, volume is a homogeneous polynomial of degree n in the heights $h_P(u_j)$.
(4) Fixing a class of simple and strongly isomorphic polytopes, the volume of $\sum_{i=1}^m \lambda_i P_i$ is a homogeneous polynomial of degree n in λ_i.
(5) From the above follows the polar decomposition of volume.
(6) The above steps allow to prove polynomiality of mixed volumes with *all* the needed properties.

B.2. The Alexandrov-Fenchel inequality

As explained in Chapter 1, one of the most beautiful, non-trivial and profound theorems in convexity, which is linked also with algebraic geometry and number theory, is the Alexandrov-Fenchel inequality for mixed volumes.

THEOREM B.2.1 (Alexandrov-Fenchel). *Let $K_1, K_2, K_3, \ldots, K_n$ be non-empty compact convex sets in $\mathbb{R}^n$. Then*

(B.2.1)
$$V(K_1, K_2, K_3, \ldots, K_n)^2 \geqslant V(K_1, K_1, K_3, \ldots, K_n) \cdot V(K_2, K_2, K_3, \ldots, K_n).$$

In (B.2.1) we have equality if K_1 and K_2 are homothetic, however there exist examples showing that there are other possibilities too for (B.2.1) to be an equality. In fact, the complete classification of the equality cases is still an open problem.

In the proof we shall make use of the special case where there are only two bodies involved. The proof of this case is a consequence of Brunn-Minkowski's theorem. This case is called "Minkowski's second inequality", and we state it here together with "Minkowski's first inequality" which will also be later used. Recall that $V(K[j], L[n-j])$ denotes the mixed volume of j copies of K and $(n-j)$ copies of L.

LEMMA B.2.2 (Minkowski's first and second inequality). *Let K, L be non-empty compact convex sets in $\mathbb{R}^n$. Then*

$$V(K, L[n-1])^2 \geqslant \mathrm{Vol}(K)^{\frac{1}{n}} \cdot \mathrm{Vol}(L)^{\frac{n-1}{n}}. \tag{B.2.2}$$

and

$$V(K, L[n-1])^2 \geqslant V(K, K, L[n-2]) \cdot \mathrm{Vol}(L). \tag{B.2.3}$$

Proof. Recall that

$$\mathrm{Vol}_n(\lambda K + (1-\lambda)L) = \sum_{i=0}^{n} \binom{n}{i} \lambda^i (1-\lambda)^{n-i} V(K[i], L[n-i]),$$

and that, by the Brunn-Minkowski inequality, the function

$$f(\lambda) := \mathrm{Vol}_n(\lambda K + (1-\lambda)L)^{1/n} - \lambda \mathrm{Vol}_n(K)^{1/n} - (1-\lambda)\mathrm{Vol}_n(L)^{1/n},$$

is concave on $[0,1]$. Note that $f(0) = f(1) = 0$, so for all $\lambda \in (0,1)$, $f(\lambda) \geqslant 0$ and if for some $\lambda_0 \in (0,1)$ we have $f(\lambda_0) > 0$, then $f(\lambda) > 0$ for every $\lambda \in (0,1)$. Differentiate once at the point $\lambda = 0$ to get

$$f'(0) = \mathrm{Vol}_n(L)^{-\frac{n-1}{n}} \left(V(K[1], L[n-1]) - \mathrm{Vol}_n(K)^{\frac{1}{n}} \cdot \mathrm{Vol}_n(L)^{\frac{n-1}{n}} \right) \geqslant 0,$$

with $f'(0) = 0$ if and only if $f \equiv 0$. The latter shows that the equality cases in (B.2.2) are exactly the equality cases in the Brunn-Minkowski inequality for convex bodies. Differentiate twice at the point $\lambda = 0$ to get

$$(n-1)\mathrm{Vol}_n(L)^{-\frac{2n-1}{n}} \left(V(K[2], L[n-2]) \cdot \mathrm{Vol}_n(L) - V(K[1], L[n-1])^2 \right) = f''(0) \leqslant 0,$$

which proves the desired inequality. □

We shall prove the Alexandrov-Fenchel inequality for classes of simple and strongly isomorphic polytopes, by induction on the dimension n. Then by continuity of volume and Proposition A.3.9 it will be proven for all n-tuples of bodies in $\mathbb{R}^n$.

We fix a class $\mathcal{A}$, of strongly isomorphic simple n-dimensional polytopes. Let $u_1, \dots, u_N$ denote the unit normal vectors corresponding to the facets of polytopes in $\mathcal{A}$. For a fixed i, the facets $P(u_i)$ of polytopes $P \in \mathcal{A}$ belong to a fixed class, call it $\mathcal{A}_i$, of strongly isomorphic polytopes in $u^\perp \simeq \mathbb{R}^{n-1}$. The directions of the facets of polytopes in $\mathcal{A}_i$ can be labelled by some $N_i < N$ labels.

We shall identify from here onwards a polytope $P \in \mathcal{A}$ with a vector $\overline{P} \in \mathbb{R}^N$ which is defined to be $\overline{P} = (h_1, \dots, h_N)$ with $h_i := h_P(u_i)$, the corresponding heights of the N facets of P. Note that the support vector $\overline{P}$ determines uniquely the polytope (and of course not any vector in $\mathbb{R}^N$ is a support vector of a polytope). Since we shall work by induction, it is already useful to mention that the support vector corresponding to the ith facet of P (as a member of $\mathcal{A}_i$), which is a vector in $\mathbb{R}^{N_i}$, is a linear image of the vector $\overline{P}$. This linear map depends only on the angles between the various normals, that is, on $\mathcal{A}$ and not on the specific polytope, see Remark B.1.8. We thus define $\Lambda_i : \mathbb{R}^N \to \mathbb{R}^{N_i}$ such that, if X is the support vector of $P \in \mathcal{A}$, then $\lambda_i(X)$ is the support vector of $P(u_i) \in \mathcal{A}_i$.

It is well known and easy to check that fixing a class of strongly isomorphic polytopes in $\mathbb{R}^n$, the volume of a positive linear combination of m of them is an n-homogeneous polynomial in the corresponding heights, and the coefficients are the mixed volumes of the corresponding polytopes; similarly, the mixed volume of an n-tuple of polytopes in $\mathcal{A}$ is a multi-linear form in their corresponding heights. Indeed, this was an alternative route to the proof of the existence, positivity and properties of mixed volumes. Let us show this briefly: Use the inductive assumption say that the mixed volume function for $(n-1)$-tuples of polytopes in $\mathcal{A}_i$ is a multi-linear function F_i in the corresponding heights $F_i : \mathrm{dom}(F_i) \subset (\mathbb{R}^{N_i})^{n-1} \to \mathbb{R}^+$ (for now the domain of F_i consists of these vectors which are support vectors of polytopes in $\mathcal{A}_i$, although assuming F_i is multi-linear means that in fact it can be defined on the whole linear space). Using the formula for mixed volumes we get

$$\begin{aligned} V(P_1, P_2, \dots, P_n) &= \frac{1}{n}\sum_{i=1}^{N} h_i^{P_1} \cdot V(P_2(u_i), \dots, P_n(u_i)) \\ &= \frac{1}{n}\sum_{i=1}^{N} (\overline{P_1})_i \cdot F_i(\Lambda_i(\overline{P_2}), \dots, \Lambda_i(\overline{P_n})) \\ &= \sum_{j_1,\dots,j_n=1}^{N} a_{j_1,\dots,j_n} (\overline{P_1})_{j_1} \cdots (\overline{P_n})_{j_n}. \end{aligned}$$

We see that the final expression is a multi-linear polynomial in the coordinates of the support vectors of $P_1, \dots, P_n$, as claimed.

Convinced that mixed volume is a multi-linear form in the heights, we may clearly extend it so that it is defined as a function $V : (\mathbb{R}^N)^n \to \mathbb{R}$, in such a way that on vectors of the form $(\overline{P_1}, \dots, \overline{P_n})$, which correspond to polytopes $P_1, \dots, P_n$, we will have that

$$V(\overline{P_1}, \dots, \overline{P_n}) = V(P_1, \dots, P_n).$$

In the rest of the proof we may use both polytopes and vectors in $\mathbb{R}^N$ as arguments of mixed volumes, and although we shall try to distinguish between the two usages, we do use the same notation.

Another useful fact is that translating a polytope $P \in \mathcal{A}$ by a vector $x \in \mathbb{R}^n$ amounts to adding the vector $(\langle x, u_i \rangle)_{i=1}^N$ to the corresponding support vector $\overline{P}$ of P. These translation vectors, which are support vectors of singletons in $\mathbb{R}^n$, form an n-dimensional subspace of $\mathbb{R}^N$ which will be of importance in the proofs below. More generally, of course, we have for $P, Q \in \mathcal{A}$ that $\overline{(P+Q)} = \overline{P} + \overline{Q}$.

Having fixed a class $\mathcal{A}$ of simple and strongly isomorphic polytopes in $\mathbb{R}^n$, and having introduced the identification between a polytope $P \in \mathcal{A}$ and its support vector $\overline{P}$ in $\mathbb{R}^N$, we shall prove the following.

THEOREM B.2.3. *Let $P, P_3, \dots, P_n \in \mathcal{A}$ and let $Z \in \mathbb{R}^N$. Then,*

$$V(Z, \overline{P}, \overline{P_3}, \dots, \overline{P_n})^2 \geqslant V(Z, Z, \overline{P_3}, \dots, \overline{P_n}) \cdot V(\overline{P}, \overline{P}, \overline{P_3}, \dots, \overline{P_n}).$$

Equality holds if and only if $Z = \lambda\overline{P} + p$, where $\lambda \in \mathbb{R}$, $\overline{P}$ is the support vector of P and p is the support vector of a point.

The proof of Theorem B.2.3 is based on the next proposition. Define a symmetric bilinear form Φ on $\mathbb{R}^N$ by

$$\Phi(X,Y) := V(X,Y,\overline{P_3},\ldots,\overline{P_n}), \quad \text{where } X,Y\in\mathbb{R}^N,$$

if $n \geqslant 2$ (with $\overline{P_3},\ldots,\overline{P_n}$ omitted if $n=2$).

PROPOSITION B.2.4. *Let $P, P_3,\ldots,P_n \in \mathcal{A}$, and Φ defined as above. If $\Phi(Z,\overline{P})=0$ then $\Phi(Z,Z)\leqslant 0$, and equality holds if and only if Z is the support vector of a point.*

Proof of Theorem B.2.3. Given $P, P_3,\ldots,P_n \in \mathcal{A}$ and $Z\in\mathbb{R}^N$, define Φ as above and let

$$\lambda := \frac{\Phi(Z,\overline{P})}{\Phi(P,P)} \quad \text{and} \quad Z' := Z-\lambda\overline{P}$$

(observe that $\Phi(\overline{P},\overline{P}) = V(P,P,P_3,\ldots,P_n)>0$). Then we have $\Phi(Z',\overline{P})=0$ and hence by Proposition B.2.4 it follows that $\Phi(Z',Z')\leqslant 0$, with equality if and only if Z' is the support vector of a point. Since

$$\Phi(Z',Z') = \Phi(Z,Z) - \frac{\Phi(Z,\overline{P})^2}{\Phi(\overline{P},\overline{P})},$$

the assertion of Theorem B.2.3 now follows. □

To prove Proposition B.2.4 we first restrict our attention to the special case $P_3=\cdots=P_n=P$ of Theorem B.2.3.

PROPOSITION B.2.5. *Given P and Z as in Theorem* B.2.3, *the inequality*

$$V(Z,\overline{P}[n-1])^2 \geqslant V(Z,Z,\overline{P}[n-2])\cdot \mathrm{Vol}_n(P) \tag{B.2.4}$$

holds. Moreover, equality holds if and only if $Z=\lambda\overline{P}+A$, where $\lambda\in\mathbb{R}$ and A is the support vector of a point.

REMARK B.2.6. In the proof that follows we only prove the inequality in Proposition B.2.5 in general, using induction, but establish the assertion of the equality case in this proposition only for $n=2$. We will then use this as base of the induction we need in order to prove Proposition B.2.4, and hence Theorem B.2.3; thus after completing the proofs we will have fully proven Proposition B.2.5 too, since the equality cases of Theorem B.2.3 cover the equality cases of the proposition.

Proof of Proposition B.2.5. Fixing $P\in\mathcal{A}$ and $Z\in\mathbb{R}^N$, if $|\varepsilon|>0$ is sufficiently small then $\overline{Q}=\overline{P}+\varepsilon Z$ is the support vector of a polytope $Q\in\mathcal{A}$. From Minkowski's inequality (B.2.3) in Lemma B.2.2 we infer that

$$\begin{aligned} 0 &\leqslant V(Q,P[n-1])^2 - V(Q,Q,P[n-2])\cdot\mathrm{Vol}_n(P) \\ &= \varepsilon^2[V^{(n)}(Z,\overline{P}[n-1])^2 - V^{(n)}(Z,Z,\overline{P}[n-2])\cdot\mathrm{Vol}_n(P)]. \end{aligned} \tag{B.2.5}$$

This proves (B.2.4).

If we now set $n=2$, then equality in (B.2.4) means that $V(Z,\overline{P})^2 = V(Z,Z)\cdot \mathrm{Vol}_2(P)$, which by (B.2.5) is equivalent to $V(Q,P)^2=\mathrm{Vol}_2(Q)\cdot\mathrm{Vol}_2(P)$. Given the equality cases in the Minkowski inequality (B.2.2), the latter holds if and only if Q and P are homothetic, which can happen if and only if $Z=\lambda\overline{P}+A$ with $\lambda\in\mathbb{R}$ and A the support vector of a point. □

We shall make use of the following:

PROPOSITION B.2.7. *Let $n \geqslant 3$. Fix a class $\mathcal{A}$, some $P_3, \ldots, P_n \in \mathcal{A}$ and define the bilinear form*

$$\Phi(X,Y) := V(X, Y, \overline{P_3}, \ldots, \overline{P_n})$$

as above. A vector $Z \in \mathbb{R}^N$ is an eigenvector of the bilinear form Φ with eigenvalue 0 if and only if Z is the support vector of a point.

Proof. The proof will be by induction on the dimension.

For each $i \in \{1, \ldots, N\}$ we shall consider $\mathcal{A}_i$ and define a symmetric bilinear form ϕ_i on $\mathbb{R}^{N_i}$ by

$$\phi_i(x,y) := V(x, y, \overline{P_4(u_i)}, \ldots, \overline{P_n(u_i)}), \quad \text{where } x, y \in \mathbb{R}^{N_i}$$

(with $P_4(u_i), \ldots, P_n(u_i)$ omitted if $n = 3$). Note that by the definitions we have that

$$\Phi(X,Y) = \frac{1}{n} \sum_{i=1}^{N} X_i \phi_i(\Lambda_i(Y), P_3). \tag{B.2.6}$$

Since $\phi(\cdot, \Lambda_i(\overline{P_3}))$ is linear, we see that (after composition with the linear Λ_i) we may write it in the form

$$\phi_i(\Lambda_i(Y), \Lambda_i(\overline{P_3})) = \sum_{j=1}^{N} b_{ij} Y_j$$

for some $b_{ij} \in \mathbb{R}$. Combining this with (B.2.6), we see that

$$\Phi(X,Y) = \frac{1}{n} \sum_{j=1}^{N} b_{ij} X_i \cdot Y_j, \tag{B.2.7}$$

where $b_{ij} = b_{ji}$ since Φ is symmetric. Saying that a vector $Z = (Z_1, \ldots, Z_N) \neq 0$ is an eigenvector of Φ corresponding to the eigenvalue 0 amounts to

$$\sum_{j=1}^{N} b_{ij} Z_j = 0 \quad \text{for every } i = 1, \ldots, N,$$

or equivalently to

$$\phi_i(\Lambda_i(Z), \Lambda_i(\overline{P_3})) = 0 \quad \text{for every } i = 1, \ldots, N. \tag{B.2.8}$$

If Z is the support vector of a point ζ, then for all i

$$\phi_i(\Lambda_i(Z), \Lambda_i(\overline{P_3})) = V(\{\zeta\}, P_3(u_i), P_4(u_i), \ldots, P_n(u_i)) = 0$$

so Z is indeed an eigenvector of Φ corresponding to the eigenvalue 0.

Suppose, conversely, that (B.2.8) holds. By the induction hypothesis, applied with respect to the class $\mathcal{A}_i$ of strongly isomorphic polytopes in $u^\perp$, we conclude that $\Lambda_i(Z)$ is the support vector of a point $\zeta_i \in u^\perp$ (this support vector calculated with respect to $\mathcal{A}_i$). To make sure that Z itself is a support vector of a point we select $\varepsilon \neq 0$ so that $\overline{P_3} + \varepsilon Z$ is a support vector of a polytope $Q \in \mathcal{A}$. Since $\Lambda_i(\overline{P_3} + \varepsilon Z) = \Lambda_i(\overline{P_3}) + \varepsilon \Lambda_i(Z)$ we have that for every i $Q(u_i) = P_3(u_i) + t_i$ (where the vector t_i is equal to $\varepsilon \zeta_i + a_i u_i$ for some $a_i \in \mathbb{R}$). Therefore, for all j such that $G_{ij} := Q(u_i) \cap Q(u_j)$ is an $(n-2)$-face, G_{ij} can be written as $(P_3(u_i))(u_j) + t_i$ as well as $(P_3(u_j))(u_i) + t_j$, hence $t_i = t_j$. Since any two facets of P_3 (or Q) can be joined by a chain of facets with the property that any two consecutive facets in the chain have an $(n-2)$-dimensional intersection, we conclude that $t_i = t_j$ for all

$i, j \in \{1, \ldots, N\}$. Therefore, Q is a translate of P_3, whence it follows that Z is the support vector of a point. This completes the proof of the proposition. □

For the proof of Proposition B.2.4 it will be useful to introduce, besides Φ, a second symmetric bilinear form Ψ on $\mathbb{R}^N$, which is in fact diagonal. To this end we shall assume without loss of generality that $h_P(u_i) > 0$ for all $i = 1, \ldots, N$, namely that $0 \in \text{int}(P)$. The second form will be given by

$$\Psi(X, Y) := \frac{1}{n} \sum_{i=1}^{N} \frac{\phi_i(\Lambda_i(\overline{P}), \Lambda_i(\overline{P_3}))}{h_P(u_i)} X_i Y_i, \quad \text{where } X, Y \in \mathbb{R}^N.$$

Since $\phi(P, P_3) > 0$, the form Ψ is positive definite. In analogy with (B.2.7) we can write

$$\Psi(X, Y) = \frac{1}{n} \sum_{j=1}^{N} c_{ij} X_i \cdot Y_j,$$

where

$$c_{ij} := \begin{cases} \frac{\phi_i(\Lambda_i(\overline{P}), \Lambda_i(\overline{P_3}))}{h_P(u_i)} & \text{if } i = j, \\ 0 & \text{if } i \neq j. \end{cases}$$

We shall consider the eigenvalues of Φ relative to Ψ. A vector $Z = (Z_1, \ldots, Z_N) \in \mathbb{R}^N$ is an eigenvector of Φ relative to Ψ with eigenvalue λ if and only if

$$\sum_{j=1}^{N} (b_{ij} - \lambda c_{ij}) Z_j = 0 \quad \text{for every } i = 1, \ldots, N,$$

or equivalently if and only if

$$\phi_i(Z, P_3) = \lambda \frac{\phi_i(\Lambda_i(\overline{P}), \Lambda_i(\overline{P_3}))}{h_P(u_i)} Z_i \quad \text{for every } i = 1, \ldots, N.$$

In particular, $\lambda = 1$ is an eigenvalue with corresponding eigenvector $Z = \overline{P}$. We next claim that it is the only positive eigenvalue, and that it is simple.

PROPOSITION B.2.8. *The only positive eigenvalue of Φ relative to Ψ is 1, and it is simple.*

Proof. First we assume that $P = P_3 = \cdots = P_n$. Suppose Proposition B.2.8 were false in this case: this would mean that either there is a positive eigenvalue $\mu \neq 1$ or that 1 is a multiple eigenvalue. If the former were true, we could find a non-zero vector $Z \in \mathbb{R}^N$ with $\Psi(Z, \overline{P}) = 0$ and $\Phi(Z, Z) = \mu \Psi(Z, Z) > 0$. If 1 were a multiple eigenvalue, then the corresponding eigenspace would be at least two-dimensional and hence it would contain a non-zero vector Z with $\Psi(Z, \overline{P}) = 0$ and $\Phi(Z, Z) = \Psi(Z, Z) > 0$. Thus in either case we would have a vector Z for which

$$V(Z, \overline{P}[n-1]) = \frac{1}{n} \sum_{i=1}^{N} \frac{\phi_i(\Lambda_i(\overline{P}), \Lambda_i(\overline{P}))}{h_P(u_i)} Z_i h_P(u_i) = \Psi(Z, \overline{P}) = 0,$$

while $V(Z, Z, \overline{P}[n-2]) = \Phi(Z, Z) > 0$. But this cannot hold because it contradicts Proposition B.2.5.

Now let $P_3, \ldots, P_n \in \mathcal{A}$ be arbitrary. Given $\theta \in [0, 1]$ let $P_r(\theta) := (1 - \theta)P + \theta P_r$, $r = 3, \ldots, n$. The coefficients of the corresponding forms Φ, Ψ depend in a continuous way on θ, hence the same must be true for the relative eigenvalues. By Proposition B.2.7, the number 0 is always an eigenvalue with multiplicity n. It follows that the sum of the multiplicities of the positive eigenvalues is independent of θ. Since this sum is equal to 1 when $\theta = 0$, it must also be equal to 1 when $\theta = 1$. This completes the proof of Proposition B.2.8. □

We finally get to the proof of our main Proposition.

Proof of Proposition B.2.4. We proceed by induction with respect to the dimension. When $n = 2$, the assertion follows by Proposition B.2.5. Let us assume now that $n \geqslant 3$ and that the assertion of Proposition B.2.4 is valid for dimensions lower than n. Proposition B.2.8 implies that the eigenspace corresponding to the eigenvalue 1 coincides with $\{\mathbb{R}\overline{P}\}$ and that the second eigenvalue is not positive, hence $\Phi(Z, Z) \leqslant 0$ for all Z such that $\Psi(Z, \overline{P}) = 0$. Note that $\Psi(Z, \overline{P}) = 0$ means $\sum_{i=1}^{N} \psi_i(P, P_3) Z_i = 0$ which is precisely $\Phi(Z, \overline{P}) = 0$. So in other words we have that $\Phi(Z, \overline{P}) = 0$ implies $\Phi(Z, Z) \leqslant 0$. Suppose that we have equality for some $Z \neq 0$. Since at this Z the maximal $\Phi(Z, Z)$ on the subspace $\Psi(Z, \overline{P}) = 0$ is attained, by the min-max principle, Z is an eigenvector with eigenvalue 0. By Proposition B.2.7, Z must be the support vector of a point. This completes the proof of Proposition B.2.4, and consequently of Theorems B.2.3 and B.2.1 too. □

B.3. More geometric inequalities of "hyperbolic" type

The Alexandrov-Fenchel inequalities are the most advanced representatives of a series of very important inequalities. They should perhaps be called "hyperbolic" inequalities in contrast to the more often used in analysis "elliptic" inequalities: Cauchy-Schwarz, Hölder, and their consequences (various triangle inequalities). The reason for using the name "hyperbolic" is connected with the fact that these inequalities are usually implied by the fact that the roots of some polynomials are real, as one may see in the topic which we next discuss. "Elliptic" inequalities are usually implied by complex roots of some polynomials. This is the same distinction as between hyperbolic and elliptic PDE.

A consequence of "hyperbolic" inequalities is *concavity* of some important quantities.

Let us start this short review by recalling some old and classical, but not well remembered, inequalities due to Newton. Let $x_1, \ldots, x_n$ be real numbers. We define the elementary symmetric functions $e_0(x_1, \ldots, x_n) = 1$, and

$$e_i(x_1, \ldots, x_n) = \sum_{1 \leqslant j_1 < \cdots < j_i \leqslant n} x_{j_1} x_{j_2} \ldots x_{j_i} \ , \quad 1 \leqslant i \leqslant n. \tag{B.3.1}$$

In particular, $e_1(x_1, \ldots, x_n) = \sum_{i=1}^{n} x_i$, $e_n(x_1, \ldots, x_n) = \prod_{i=1}^{n} x_i$. We then consider the normalized functions

$$E_i(x_1, \ldots, x_n) = \frac{1}{\binom{n}{i}} e_i(x_1, \ldots, x_n). \tag{B.3.2}$$

Newton proved that, for $k = 1, \ldots, n-1$,

$$E_k^2(x_1, \ldots, x_n) \geqslant E_{k-1}(x_1, \ldots, x_n) E_{k+1}(x_1, \ldots, x_n), \tag{B.3.3}$$

with equality if and only if all the x_i's are equal. An immediate corollary of (B.3.3), observed by Newton's student Maclaurin, is the string of inequalities

$$E_1(x_1,\dots,x_n) \geqslant E_2^{1/2}(x_1,\dots,x_n) \geqslant \cdots \geqslant E_n^{1/n}(x_1,\dots,x_n), \tag{B.3.4}$$

which holds true for any n-tuple $(x_1,\dots,x_n)$ of positive real numbers. Note the similarity between (B.3.3), (B.3.4) and the Alexandrov-Fenchel and Alexandrov inequalities respectively.

To prove (B.3.3) we consider the polynomial

$$P(x) = \prod_{i=1}^{n}(x - x_i) = \sum_{j=0}^{n}(-1)^j \binom{n}{j} E_j(x_1,\dots,x_n)x^{n-j}, \tag{B.3.5}$$

or in homogeneous form,

$$Q(t,\tau) = \tau^n P(\frac{t}{\tau}) = \sum_{j=0}^{n}(-1)^j \binom{n}{j} E_j(x_1,\dots,x_n)t^{n-j}\tau^j. \tag{B.3.6}$$

Since P has only real roots, by Rolle's theorem the same is true for the derivatives of P (with respect to t or τ) of any order. If we differentiate (B.3.6) $(n-k-1)$-times with respect to t and then $(k-1)$-times with respect to τ, we obtain the polynomial

$$\frac{n!}{2}E_{k-1}(x_1,\dots,x_n)t^2 - n!E_k(x_1,\dots,x_n)t\tau + \frac{n!}{2}E_{k+1}(x_1,\dots,x_n)\tau^2, \tag{B.3.7}$$

which has two real roots for fixed $\tau = 1$. This is exactly Newton's inequality (B.3.3).

A multidimensional, but still numerical, analogue of Newton's inequalities, with positive definite matrices, will be explained in the next section.

B.4. Positive definite matrices and mixed discriminants

As noted in the first chapters of the main text, positive definite matrices are a good model in which to search for similarities with convex bodies, where the determinant replaces volume, and usual addition corresponds to Minkowski summation for bodies. We have seen in Chapter 2, Section 2.1.2 a proof of the following inequality due to Minkowski.

LEMMA B.4.1. *Let $A_1, A_2 \in GL_n$ be real and positive definite. Then*

$$\det(A_1 + A_2)^{1/n} \geqslant \det(A_1)^{1/n} + \det(A_2)^{1/n}.$$

Equality holds if and only if $A_1 = \mu A_2$ for some $\mu > 0$. Equivalently,

$$\det((1-\lambda)A_1 + \lambda A_2) \geqslant \det(A_1)^{1-\lambda}\det(A_2)^{\lambda}$$

for all $\lambda \in [0,1]$. Here the equality condition is that $A_1 = A_2$.

It is useful to note that

$$\det(A)^{1/n} = \min\{\mathrm{tr}(AB)/n : \det(B) = 1\}$$

from which we can see another simple proof:

$$\begin{aligned}\min\{\mathrm{tr}(A_1 + A_2)B/n : \det(B) = 1\} &\geqslant \min\{\mathrm{tr}(A_1B)/n : \det(B) = 1\}\\ &+ \min\{\mathrm{tr}(A_2B)/n : \det(B) = 1\}.\end{aligned}$$

The original inequality can be proved as a consequence of the arithmetic-geometric means inequality, one only has to notice the fact from linear algebra that two positive definite linear transformations may be brought to diagonal form simultaneously by an SL_n transform. We gave in Chapter 2, Section 2.1.2 a different proof using Hölder's inequality.

Here is another similar inequality:

THEOREM B.4.2. *Let $A \in GL_n$ be a real and positive definite matrix. Then*

$$\det(A_{1,n}) \leqslant \det(A_{1,k}) \det(A_{k+1,n})$$

where $A_{r,s}$ stands for an $(s-r+1) \times (s-r+1)$ matrix indexed by $r, \ldots, s$ with $(i,j)^{th}$ entry the same as the $(i,j)^{th}$ entry of A. In particular,

$$\det(A) \leqslant \prod_{i=1}^{n} A_{i,i}.$$

The proof is very similar to that given for Lemma B.4.1 in the main text; we represent the determinant using the integral of $\exp(-\langle Ax, x\rangle)$, then note that the mixed terms are of the form $N+N^{-1}$ for some number N and use that $N+N^{-1} \geqslant 2$ for all N.

Next, we present an example of an inequality for "marginals", called Bergstrom's inequality.

THEOREM B.4.3. *Let $A, B \in GL_n$ be real and positive definite. Let A_i, B_i denote the submatrices obtained by deleting the ith row and column. Then*

$$\frac{\det(A+B)}{\det(A_i+B_i)} \geqslant \frac{\det(A)}{\det(A_i)} + \frac{\det(B)}{\det(B_i)}.$$

Proof. Given $A \in GL_n$ real and positive definite, we note that

$$\min\{\langle Ax, x\rangle : x_i = 1\} = \frac{\det(A)}{\det(A_i)}.$$

Indeed, this can be easily checked using Lagrange multipliers. Having this is hand, we clearly have

$$\min\{\langle (A+B)x, x\rangle : x_i = 1\} \geqslant \min\{\langle Ax, x\rangle : x_i = 1\} + \min\{\langle Bx, x\rangle : x_i = 1\}$$

□

Generalizing both this and the first lemma, Ky Fan proved the next theorem.

THEOREM B.4.4. *Given $A \in GL_n$ positive definite denote by $A_{(k)}$ the matrix obtained by deleting the first k rows and columns. Then*

$$\left(\frac{\det(A+B)}{\det(A_{(k)}+B_{(k)})}\right)^{1/(n-k)} \geqslant \left(\frac{\det(A)}{\det(A_{(k)})}\right)^{1/(n-k)} + \left(\frac{\det(B)}{\det(B_{(k)})}\right)^{1/(n-k)}.$$

We finally address the topic in the title of this section - mixed discriminants. Consider first the space of real symmetric $n \times n$ matrices. We polarize the function $A \to \det A$ to obtain the symmetric multilinear form

$$D(A_1, \ldots, A_n) = \frac{1}{n!} \sum_{\varepsilon \in \{0,1\}^n} (-1)^{n+\sum \varepsilon_i} \det\left(\sum \varepsilon_i A_i\right), \tag{B.4.1}$$

for symmetric $A_i \in GL_n$. Then, if $t_1, \ldots, t_m > 0$ and $A_1, \ldots, A_m \in S_n$, the determinant of $t_1 A_1 + \cdots + t_m A_m$ is a homogeneous polynomial of degree n in t_i of the form

$$\det(t_1 A_1 + \cdots + t_m A_m) = \sum_{1 \leqslant i_1 \leqslant \cdots \leqslant i_n \leq m} n! D(A_{i_1}, \ldots, A_{i_n}) t_{i_1} \cdots t_{i_n}. \tag{B.4.2}$$

The coefficient $D(A_1, \dots, A_n)$ is called the mixed discriminant of $A_1, \dots, A_n$.

The fact that the polynomial $P(t) = \det(A + tI)$ has only real roots for any $A \in S_n$ plays the central role in the proof of a number of very interesting inequalities connecting mixed discriminants, which are quite similar to Newton's inequalities. They were first discovered by Alexandrov in one of his approaches to the Alexandrov-Fenchel inequalities.

Alexandrov showed that for positive definite matrices the mixed discriminants are positive. In fact, we have:

THEOREM B.4.5. *Let $A_i, i = 1, \dots, n$ be positive definite matrices in GL_n. Then,*

$$D(A_1, A_2, \dots, A_n) \geqslant \prod_{i=1}^{n} [\det A]^{\frac{1}{n}}. \tag{B.4.3}$$

He also showed the following discriminant version for the Alexandrov-Fenchel inequality:

THEOREM B.4.6. *Let $A_i, i = 1, \dots, n$ be positive definite matrices in GL_n. Then,*

$$D^2(A_1, \dots, A_n) \geqslant D(A_1, A_1, A_3, \dots, A_n) D(A_2, A_2, A_3, \dots, A_n).$$

In fact, he showed that this remains true even if A_1 is not assumed positive definite.

We mention the useful fact that one may express the mixed discriminants in terms of the entries of the matrices A_k as follows

$$D(A_1, \dots, A_n) = \frac{1}{n!} \sum_{\sigma} \det \left(A_1^{(\sigma(1))} \cdots A_n^{(\sigma(n))} \right)$$

where the sum is over all permutations σ of $\{1, \dots, n\}$, and $A_i^{(k)}$ denotes the ith column of the matrix A_k.

It is worthwhile to mention how the above results relate, and enable us to prove, certain special cases and consequences of the Alexandrov-Fenchel inequalities for convex bodies. This was shortly mentioned in Chapter 1, Section 1.3; let us slightly elaborate here.

We start with the fact that the volume of $t_1 K_1 + \cdots + t_m K_m$ is a homogeneous polynomial in $t_i \geqslant 0$. Consider n fixed convex open bounded bodies K_i with normalized volume $\mathrm{Vol}_n(K_i) = 1$. As in Chapter 1, Section 1.3 consider the Brenier maps

$$\psi_i : (\mathbb{R}^n, \gamma_n) \to K_i,$$

where γ_n is the standard Gaussian probability density on $\mathbb{R}^n$. We have $\psi_i = \nabla f_i$, where f_i are convex functions on $\mathbb{R}^n$. By Caffarelli's regularity result (see [**130**], [**131**] and [**132**]) all the ψ_i's are smooth maps. Then, the image of $(\mathbb{R}^n, \gamma_n)$ by $\sum t_i \psi_i$ is the interior of $\sum t_i K_i$. Since each ψ_i is a measure preserving map, we have

$$\det \left(\frac{\partial^2 f_i}{\partial x_k \partial x_l} \right)(x) = \gamma_n(x) \quad , \quad i = 1, \dots, n.$$

It follows that

$$\begin{aligned}\mathrm{Vol}_n\left(\sum_{i=1}^n t_iK_i\right) &= \int_{\mathbb{R}^n}\det\left(\sum_{i=1}^n t_i(\frac{\partial^2 f_i}{\partial x_k\partial x_l})\right)dx\\ &= \sum_{i_1,\dots,i_n=1}^n t_{i_1}\dots t_{i_n}\int_{\mathbb{R}^n} D\left(\frac{\partial^2 f_{i_1}(x)}{\partial x_k\partial x_l},\dots,\frac{\partial^2 f_{i_n}(x)}{\partial x_k\partial x_l}\right)dx.\end{aligned}$$

In particular, we recover Minkowski's theorem on polynomiality of $\mathrm{Vol}_n(\sum t_iK_i)$, and see the connection between the mixed discriminants $D(\mathrm{Hess}f_{i_1},\dots,\mathrm{Hess}f_{i_n})$ and the mixed volumes

$$V(K_{i_1},\dots,K_{i_n}) = \int_{\mathbb{R}^n} D(\mathrm{Hess}f_{i_1}(x),\dots,\mathrm{Hess}f_{i_n}(x))dx.$$

The Alexandrov-Fenchel inequalities do not follow from the corresponding mixed discriminant inequalities, but the deep connection between the two theories is obvious. Also, some particular cases are indeed simple consequences. In some cases, this method allows to describe equality cases. For example, one can prove that

$$V(K_1,\dots,K_n) \geqslant \prod_{i=1}^n \mathrm{Vol}_n(K_i)^{1/n},$$

and characterize the equality cases.

B.5. Steiner's formula and Kubota's formulae

A special case of mixed volumes is when $m = 2$ and $K_2 = B_2^n$. In this case, which is a particular case of the polynomiality claim, the formula for $\mathrm{Vol}_n(K + \lambda B_2^n)$ is called "Steiner formula". Originally it was produced by directly studying the so-called "parallel body", $\{x : d(x,K) \leqslant \lambda\}$ which is the λ-neighborhood of the original body.

Here we introduce the following notation which is standard.

DEFINITION B.5.1. Given $K \in \mathcal{K}^n$ and $0 \leqslant j \leqslant n$, we let the quermassintegrals $W_j(K)$ be defined via the Steiner formula

$$\mathrm{Vol}_n(K + \lambda B_2^n) = \sum_{j=0}^n \binom{n}{j} W_j(K)\lambda^j.$$

Note that $W_j(K) = V^{(n)}(K\,[n-j], B_2^n\,[j])$ (for $\mu \geqslant 0$ we have $W_j(\mu K) = \mu^{n-j}W_j(K)$).

Note that the surface area $S_n(K)$ of K is given by

$$S_n(K) = nW_1(K) = \lim_{\varepsilon\to 0^+}\frac{1}{\varepsilon}\left(\mathrm{Vol}_n(K + \varepsilon B_2^n) - \mathrm{Vol}_n(K)\right).$$

This case of mixed volumes has been studied extensively. In this subsection we prove the Kubota formula for these numbers. Recall the Grassmannian manifold $G_{n,k}$ of k-dimensional subspaces of $\mathbb{R}^n$, on which we have a natural Haar measure, denoted by $\nu_{n,k}$ (which is normalized and rotation-invariant). The measure $\nu_{n,k}$ can be realized as follows: pick k independent vectors randomly on the sphere and take their span. With probability 1 this is an element in $G_{n,k}$, and this also gives the measure. Another option: if one agrees that there is a Haar measure on $O(n)$ (which can be obtained by taking a random element u_1 in S^{n-1}, then a random element in $S^{n-2} \subset u_1^\perp$ etc.), then one can pick a random element $U \in O(n)$, and consider the subspace $U(\text{span}\{e_1, \dots e_k\})$ that is an element in $G_{n,k}$. Different U can give rise to the same element in $G_{n,k}$, so that $G_{n,k}$ is a quotient space of $O(n)$ and inherits the measure from it. From uniqueness of Haar measure, all these constructions are identical.

We recall that $\kappa_n = \text{Vol}_n(B_2^n)$.

THEOREM B.5.2 (Kubota's formulae). *Let $K \in \mathcal{K}^n$ and $1 \leqslant j \leqslant n-1$. Then*

$$W_j(K) = \frac{\kappa_n}{\kappa_{n-j}} \int_{G_{n,n-j}} \text{Vol}_{n-j}(P_E(K)) d\nu_{n,n-j}(E).$$

Before proving the theorem in full, let us focus on just one of the cases, that is, when $j = 1$ (this will also serve as a base of induction later).

LEMMA B.5.3 (Cauchy's formula). *Let $K \in \mathcal{K}^n$. Then*

$$W_1(K) = \frac{\kappa_n}{\kappa_{n-1}} \int_{S^{n-1}} \text{Vol}_{n-1}(P_{u^\perp}(K)) d\sigma(u).$$

Recall that $W_1(K) = V^{(n)}(K, K, \dots, K, B_2^n) = \frac{1}{n} S_n(K)$, so this is in fact a formula for the surface area. Also, sometimes one uses integration on the sphere in place of integration with respect to the usual Lebesgue measure, the relation being $d\sigma(u) = du/\text{Vol}_{n-1}(S^{n-1}) = du/(n\kappa_n)$. Therefore, here we arrive at the formula

$$S_n(K) = \frac{1}{\kappa_{n-1}} \int_{S^{n-1}} \text{Vol}_{n-1}(P_{u^\perp}(K)) du,$$

which may be a more familiar form of the "Cauchy formula" to some readers.

Proof. We first work with a polytope P. For a generic $u \in S^{n-1}$ and for each facet F_i of P there is some angle $\theta_i \neq \pi/2$, between the outer normal v_i to the facet and u. The area of the projection of F_i onto $u^\perp$ is $|\cos(\theta_i)| \text{Vol}_{n-1}(F_i)$. The projection $P_{u^\perp}(P)$ is covered twice by the projections of the facets of P, and so we get that

$$\text{Vol}_{n-1}(P_{u^\perp}(P)) = \frac{1}{2} \sum_{i=1}^m |\cos(\theta_i)| \text{Vol}_{n-1}(F_i).$$

Integrating over the sphere, we see that

$$\int_{S^{n-1}} \text{Vol}_{n-1}(P_{u^\perp}(P)) d\sigma(u) = \frac{1}{2} \sum_{i=1}^m \text{Vol}_{n-1}(F_i) \int_{S^{n-1}} |\cos(\theta(v_i, u))| d\sigma(u).$$

By rotation invariance, the latter integral does not depend on v_i and is a constant depending only on the dimension, which we denote c_n. We have thus shown that for any polytope $P \subset \mathbb{R}^n$ we have

$$\int_{S^{n-1}} \text{Vol}_{n-1}(P_{u^\perp}(P)) d\sigma(u) = c_n S_n(P).$$

Since both sides are monotone, and since we have equality for all polytopes, we have equality for convex bodies as well: for all non-empty compact convex $K \subset \mathbb{R}^n$

$$\int_{S^{n-1}} \text{Vol}_{n-1}(P_{u^\perp}(K)) d\sigma(u) = c_n S_n(K).$$

To evaluate c_n set $K = B_2^n$ and get

$$\text{Vol}_{n-1}(B_2^{n-1}) = c_n S_n(B_2^n) = c_n n \text{Vol}_n(B_2^n).$$

Hence $c_n = \frac{1}{n} \frac{\kappa_{n-1}}{\kappa_n}$ as claimed. $\square$

We shall prove the Kubota formulae by induction. The induction step will be given by a formula which relates the quermassintegrals of a convex body in $\mathbb{R}^n$ with the quermassintergrals of its projections. Note that the function W_j depends on the dimension, namely if some K is a subset of E for some k-dimensional subspace of $\mathbb{R}^n$, there is a difference between its quermassintegrals when considered in $\mathbb{R}^n$ and when considered in $E \simeq \mathbb{R}^k$. For the purpose of the following lemma, denote by w_j the jth quermassintegral in dimension $n-1$. Note that if $K \subset \mathbb{R}^{n-1} \subset \mathbb{R}^n$ we have

$$\begin{aligned}\sum_{j=0}^{n}\binom{n}{j}W_j(K)\lambda^j &= \int_{-\lambda}^{\lambda}\mathrm{Vol}_{n-1}\left((K+\lambda B_2^n)\cap(\mathbb{R}^{n-1}+te_n)\right)dt\\ &= \int_{-\lambda}^{\lambda}\mathrm{Vol}_{n-1}\left(K+(\lambda^2-t^2)^{1/2}B_2^{n-1}\right)dt\\ &= \sum_{j=0}^{n-1}\binom{n-1}{j}w_j(K)\int_{-\lambda}^{\lambda}(\lambda^2-t^2)^{j/2}dt\\ &= \sum_{j=0}^{n-1}\binom{n-1}{j}w_j(K)\lambda^{j+1}\frac{\kappa_{j+1}}{\kappa_j},\end{aligned}$$

where in the last equality we have computed the volume of a $(j+1)$-ball using the Fubini theorem. We get by coefficient comparison that

$$W_j(K) = \frac{j}{n}\frac{\kappa_j}{\kappa_{j-1}}w_{j-1}(K).$$

LEMMA B.5.4. *Let $K \subset \mathbb{R}^n$ and $1 \leqslant j \leqslant n-1$. Then*

$$W_j(K) = \frac{\kappa_n}{\kappa_{n-1}}\int_{S^{n-1}} w_{j-1}(P_{u^\perp}(K))d\sigma(u).$$

(*When $j=1$ we have $w_0 = \mathrm{Vol}_{n-1}$ and this is simply the Cauchy formula.*)

Proof. We shall derive two polynomial formulae for $S_n(K+\lambda B_2^n)$, and compare coefficients. On the one hand we expand $P_{u^\perp}(K+\lambda B_2^n) = P_{u^\perp}(K) + \lambda(B_2^n \cap u^\perp)$ to get

$$\mathrm{Vol}_{n-1}(P_{u^\perp}(K+\lambda B_2^n)) = \sum_{j=0}^{n-1}\binom{n-1}{j}w_j(P_{u^\perp}(K))\lambda^j,$$

and then we integrate over the sphere to deduce by Cauchy formula that

$$S(K+\lambda B_2^n) = \frac{n\kappa_n}{\kappa_{n-1}}\sum_{j=0}^{n-1}\binom{n-1}{j}\lambda^j\int_{S^{n-1}} w_j(P_{u^\perp}(K))d\sigma(u).$$

On the other hand, using multi-linearity we can get the following polynomial expression in λ:

$$S(K+\lambda B_2^n) = nW_1(K+\lambda B_2^n) = \sum_{j=1}^{n}\binom{n}{j}W_j(K)j\lambda^{j-1}.$$

Comparing the coefficients of λ^{j-1} we find that

$$\frac{j}{n}\binom{n}{j}W_j(K) = \frac{\kappa_n}{\kappa_{n-1}}\binom{n-1}{j-1}\int_{S^{n-1}} w_{j-1}(P_{u^\perp}(K))d\sigma(u),$$

as claimed in the statement of the theorem. □

Proof of Theorem B.5.2. The proof follows by induction on j, where the case $j = 1$ is Cauchy's formula. Assume the claim holds for some $j < n - 1$, namely that it holds in all dimensions m such that $j \leqslant m - 1$. By Lemma B.5.4

$$W_{j+1}(K) = \frac{\kappa_n}{\kappa_{n-1}}\int_{S^{n-1}} w_j(P_{u^\perp}(K))d\sigma(u),$$

and by the inductive assumption the integrand can be written as

$$w_j(P_{u^\perp}(K)) = \frac{\kappa_{n-1}}{\kappa_{n-1-j}}\int_{G_{u^\perp,n-1-j}} \mathrm{Vol}_{n-1-j}(P_F(P_{u^\perp}K))d\nu_{n-1,n-1-j}(F)$$

(where each time we consider the Grassmanian manifold $G_{u^\perp,n-1-j}$ of $(n-1-j)$-dimensional subspaces F of $u^\perp$ with the corresponding Haar measure). By uniqueness of the Haar measure, putting the two formulae together we end up with

$$W_{j+1}(K) = \frac{\kappa_n}{\kappa_{n-1}}\frac{\kappa_{n-1}}{\kappa_{n-1-j}}\int_{G_{n,n-1-j}} \mathrm{Vol}_{n-1-j}(P_E K)d\nu_{n,n-1-j}(E)$$

as desired. □

B.6. Notes and remarks

Mixed volumes

For a thorough discussion of the theory of mixed volumes we refer to R. Schneider's book [**556**]. We follow closely, with his permission, the presentation of the proof of the Alexandrov-Fenchel inequality given there.

It may look as if Minkowski's theorem about the polynomiality in t_i of the volume of the Minkowski sum $\sum t_i K_i$ is simply a generalization of Steiner's polynomial formula for the volume of the t-extension $K + tB_2^n$ of a convex body. However, there is a very serious ideological difference between them. Steiner considers a t-extension, which may easily generalize to the setting of manifolds, or "tubes" around sets, see Weyl's paper [**607**]. However, it gives no clue to the fact that volume may be *polarized*, in the sense that there is a multi-linear form (with respect to Minkowski addition) defined on n-tuples of convex bodies (namely, mixed volumes), whose diagonal is volume.

We chose a "combinatorial" way to present mixed volumes and prove Minkowski's polarization result. This approach also leads us to the proof of the Alexandrov-Fenchel inequality, and follows, essentially, Alexandrov. A variant of this method using strongly isomorphic polytopes which we briefly explained can be read about in [**556**]. A very different and purely analytic approach to prove polynomiality was indicated in the text; it can be found for example in Gromov [**268**] and Alesker, Dar and Milman [**9**], and uses transportation of measure.

Some characterization theorems connected with mixed volumes were mentioned in the notes and remarks to Chapter 1. Here we mention a different kind of characterization: how to determine a body (up to translations) via its mixed volumes with other bodies. For example, W. Weil proved in [**605**] that if for some convex bodies A, B one has that

$$V(A, K_2, \dots, K_n) \leqslant V(B, K_2, \dots, K_n)$$

for all K_i, then there exists some $x_0 \in \mathbb{R}^n$ such that $A \subset x_0 + B$. If the bodies are centrally symmetric, it is enough to use $K_1 = \cdots = K_n$ and further, it is enough to consider only ellipsoids as this single body. Without a symmetry assumption on A and B, it is still possible to assume the condition only for $K_2 = \cdots = K_n$, and to consider only simplices as this single body. This follows from Theorem A.6.1 of E. Lutwak.

Notes and remarks regarding the history and significance of the Alexandrov-Fenchel inequality were given in Chapter 1.

An interesting developement in the study of mixed volumes came about recently. It was initiated by algebraic geometry, beginning with the works of Gusev [**292**] in the plane and continued by Esterov in [**190**]. Let us describe a particular case of the new formulea which they have discovered: Let K_1, K_2 and $C_3, \dots C_n$ be convex bodies in $\mathbb{R}^n$ and let $K = \mathrm{conv}(K_1 \cup K_2)$. Then

$$V(K_1, K_2, C_3, \dots, C_n) + V(K, K, C_3, \dots, C_n) =$$
$$V(K, K_1, C_3, \dots, C_n) + V(K, K_2, C_3, \dots, C_n).$$

We recomment reading Schneider's [**558**] where this developement is presented in a form standard for classical convex geometry. For the next stage of this developement see Kazarnovskii [**329**].

Valuations

Let us introduce and give a very short essay on valuation theory. The first example of valuations (and the most studied object) is based on the family of convex sets in $\mathbb{R}^n$. Actually, the notion of valuation was introduced by Dehn to solve one of the Hilbert Problems and was defined on the even smaller class of convex polytopes, however we shall consider the family of all convex sets $\mathcal{K}^n$. A function $V : \mathcal{K}^n \to \mathbb{R}$ is called a valuation if for $K_1, K_2 \in \mathcal{K}^n$ such that $K_1 \cup K_2 \in \mathcal{K}^n$ one has

$$V(K_1 \cup K_2) + V(K_1 \cap K_2) = V(K_1) + V(K_2).$$

Such a function may always be extended to finite unions of convex sets as a finitely additive function; sometimes one also allows complex valued functions. Of course, after such an extension a valuation is a finitely additive measure. However, the class of such measures is too large to control. In order to have a more restricted class of measures which still covers interesting examples from geometry and analysis, one could try to look for an extra condition of analytic nature which would determine how the measures of sets behave with respect to limits. In the classical measure theory of Lebesgue this condition is countable additivity. As is well known, this condition turned out to be extremely useful in numerous situations. However, in classical convexity theory in $\mathbb{R}^n$ a different and very useful condition is continuity of the valuation V on the class of convex compact sets with respect to the Hausdorff metric, namely if $(K_j)_{j \in \mathbb{N}}$ is a sequence of convex compact sets converging in the Hausdorff metric to another convex compact set K then $V(K_j)$ tends to $V(K)$. To the best of our knowledge, this condition of continuity was first introduced and systematically studied by Hadwiger, see his book [**295**].

We would like to emphasize that valuations may not be countably additive. Nevertheless countably additive measures are the simplest examples of valuations. Usually valuations cannot be defined on too broad a class of sets, say on Borel sets. The most interesting examples of valuations are derived from mixed volumes. Indeed, it is not difficult to check (although not completely trivial) that the functional $V(K) = V(K[n-j], A_1, ..., A_j)$ where A_i are convex sets, is a valuation (of homogeneity order $n-j$). Note that, say, the surface area of a convex bodies is a valuation, but it is not derived from a measure on $\mathbb{R}^n$. So, very different functionals, which are not measures are appearing when the closure of finite additive measures is taken in Hausdorff sense and not in the sense of sigma additivity. The first extremely deep result of this theory is the next theorem of Hadwiger [**295**].

THEOREM B.6.1 (Hadwiger). *Let V be a continuous valuation on $\mathcal{K}^n$ which is invariant under rigid motions. Then there exist $a_0, \dots, a_n \in \mathbb{R}$ such that*

$$V(K) = \sum_{j=0}^{n} a_i V_i(K)$$

for all $K \in \mathcal{K}^n$.

Many top experts, including Blaschke, put a lot of effort to try and characterize classes of valuations in the spirit of the Hadwiger theorem. A lot of results were obtained but, perhaps, the most exciting next step of this theory were two results by S. Alesker, one describing all continuous rotation invariant valuations, and the next one providing the complete characterization of all translation invariant valuations (see Alesker's [**7**]) which in particular solves a famous conjecture by McMullen [**428**] as follows:

THEOREM B.6.2 (Alesker). *Linear combinations of valuations given by mixed volumes in the form $V(K[i], A_{i+1}, \dots, A_n)$ where $0 \leqslant i \leqslant n$ and $A_i \in \mathcal{K}^n$ are dense (under a natural topology) in the space of all continuous translation invariant valuations.*

Actually, Alesker's result is much stronger than the consequence we described here, see [**556**, Chapter 6] for details. Of course, a lot of work was done in between, which helped Alesker to finish the task. Let us note the results by McMullen [**428**], and then with immediate and direct influence on Alesker, Klain [**332**] and Schneider [**552**]. We refer the reader to the surveys of McMullen and of McMullen and Schneider [**429**], [**430**], and the book [**556**]. More recent and modern developments include many beautiful results of M. Ludwig, for example see [**402**] and the references therein. Let us state one of them, belonging to Ludwig and Reitzner [**403**]:

THEOREM B.6.3 (Ludwig and Reitzner). *Let $\phi : \mathcal{K}^n \to \mathbb{R}$ be an upper semi-continuous valuation which is invariant under all volume preserving affine transformations. Then*

$$\phi = a \cdot \chi + b \cdot \mathrm{Vol}_n + c \cdot \Omega,$$

where χ is the Euler characteristic, Vol_n is the volume, Ω is the affine surface area, $a, b, c \in \mathbb{R}$, and $c \geqslant 0$. The converse is also true.

For the definition of affine surface area see Schneider [**556**]; we note that it is upper semi-continuous indeed, namely for any sequence $K_i \to K$ in $\mathcal{K}^n$ one has

$$\Omega(K) \geqslant \limsup_{i\to\infty} \Omega(K_i).$$

Let us end with a short remark on quermassintegrals. As we noted they are not defined by measures on $\mathbb{R}^n$ (besides the trivial one, namely volume). However, one may change the space on which the measure "lives", and they will be defined by such new spaces with measures on them. For example, consider the space of all affine one dimensional lines with the uniform measure. Then for every convex body its measure in this space is, after proper normalization, the surface area of this body (this is simply "Crofton's formula", see for example [**556**, page 245]). Similarly, the mean-width can be defined by another "affine Grassmannian": the uniform measure on all affine $(n-1)$-dimensional subspaces of $\mathbb{R}^n$. Here is an even more general fact proved by Alesker: Let $1 \leq i \leq n-1$. The following is an equivalent reformulation of the main theorem from [**8**, Proposition 6.1.5] which follows by elementary integration:

THEOREM B.6.4 (Alesker). *Let ϕ be a translation invariant even smooth i-homogeneous valuation. Then there exists a smooth measure ν on the affine Grassmannian $\bar{G}_{n-i}(\mathbb{R}^n)$ of $(n-i)$-subspaces such that ν is invariant with respect to $\mathbb{R}^n$-translations and*

$$\phi(K) = \int_{F \in \bar{G}_{n-i}} \chi(K \cap F)\, d\nu(F) \text{ for any } K \in \mathcal{K}^n.$$

Instead of giving a precise definition of a smooth valuation, let us give it only in the basic example of mixed volumes. The valuation

$$\phi(K) := V(K[i], A_1, \ldots, A_{n-i})$$

is smooth provided all A_j have infinitely smooth boundary with everywhere strictly positive Gauss curvature. Moreover ϕ is even provided all A_j are centrally symmetric. We remark that in the theorem above the measures μ, ν are unique for $i = 1, n-1$, and not unique for $1 < i < n$.

Mixed Discriminants

For the inequalities of Newton and some generalizations, see for example the book of Hardy Littlewood and Polya [**299**]. For more information on mixed discriminants and related inequalities, see for example Beckenbach and Bellman[**59**].

We mention that while many of the inequalities for mixed discriminants have their counterparts in convexity, it is not clear what should be the counterpart of Theorem B.4.2. The proof of theorem B.4.4 is due to Ky-Fan [**192**].

Using the fact that certain polynomials have only real roots, a proof of the inequality in Theorem B.4.6 can be found in the book of Hörmander [**305**, p. 63] and a very clear presentation of this is written up in Klartag [**337**]. In that paper one may also find some very interesting hyperbolic and elliptic type inequalities for mixed "volumes" of functions. The methods are not far from the ones discussed in Section B.3. Related inequalities, and some connections between mixed volumes and mixed discriminants, can be found in the papers of Frazelizi, Giannopoulos and Meyer [**216**], Giannopoulos Hartzoulaki and Paouris [**225**] and Artstein-Avidan, Florentin and Ostrover [**24**].

Bibliography

[1] A. D. Alexandrov, *On the theory of mixed volumes of convex bodies II: New inequalities between mixed volumes and their applications* (in Russian), Mat. Sb. N.S. **2** (1937), 1205-1238.

[2] A. D. Alexandrov, *On the theory of mixed volumes of convex bodies IV: Mixed discriminants and mixed volumes* (in Russian), Mat. Sb. N.S. **3** (1938), 227-251.

[3] A. D. Alexandrov, *Existence and uniqueness of a convex surface with a given integral curvature*, C. R. (Doklady) Acad. Sci. URSS (N.S.) **35** (1942), 131-134.

[4] A. D. Alexandrov, *Convex polyhedra*, Gosudarstv. Izdat. Techn.-Teor. Lit., Moscow-Leningrad 1950, Academie-Verlag, Berlin 1958, Springer-Verlag, Berlin 2005.

[5] S. Alesker, *ψ_2-estimate for the Euclidean norm on a convex body in isotropic position*, Geom. Aspects of Funct. Analysis (Lindenstrauss-Milman eds.), Oper. Theory Adv. Appl. **77** (1995), 1-4.

[6] S. Alesker, *Localization technique on the sphere and the Gromov-Milman theorem on the concentration phenomenon on uniformly convex sphere*, Convex geometric analysis (Berkeley, CA, 1996), MSRI Publications **34** (1999). 17-27.

[7] S. Alesker, *Description of translation invariant valuations on convex sets with solution of P. McMullen's conjecture*, Geom. Funct. Anal. **11** (2001), 244-272.

[8] S. Alesker, *A Fourier-type transform on translation-invariant valuations on convex sets*, Israel J. Math. 181 (2011), 189–294.

[9] S. Alesker, S. Dar and V. D. Milman, *A remarkable measure preserving diffeomorphism between two convex bodies in $\mathbb{R}^n$*, Geom. Dedicata **74** (1999), 201-212.

[10] N. Alon and V. D. Milman, *Embedding of ℓ_∞^k in finite-dimensional Banach spaces*, Israel J. Math. **45** (1983), 265-280.

[11] N. Alon and V. D. Milman, *Concentration of measure phenomena in the discrete case and the Laplace operator of a graph*, Seminar on Functional Analysis 1983/84, Publ. Math. Univ. Paris VII, 20, Univ. Paris VII, Paris, 55-68 (1984).

[12] N. Alon and V. D. Milman, Λ_1, *isoperimetric inequalities for graphs and superconcentrators*, J. Combinatorial Theory, Ser. B **38** (1985), 73-88.

[13] N. Alon and J. H. Spencer, *The Probabilistic Method*, 2nd. Edition, Wiley, (2000).

[14] D. Alonso-Gutiérrez, *On the isotropy constant of random convex sets*, Proc. Amer. Math. Soc. **136** (2008), 3293-3300.

[15] D. Alonso-Gutiérrez, *A remark on the isotropy constant of polytopes*, Proc. Amer. Math. Soc. **139** (2011), 2565-2569.

[16] D. Amir and V. D. Milman, *Unconditional and symmetric sets in n-dimensional normed spaces*, Israel J. Math. **37** (1980), 3-20.

[17] D. Amir and V. D. Milman, *A quantitative finite-dimensional Krivine theorem*, Israel J. Math. **50** (1985), 1-12.

[18] G. W. Anderson, *Integral Kashin splittings*, Israel J. Math. **138** (2003), 139-156.

[19] G. E. Andrews, R. Askey and R. Roy, *Special functions*, Encyclopedia of Mathematics and its Applications **71** (1999), Cambridge University Press.

[20] M. Anttila, K. M. Ball and I. Perissinaki, *The central limit problem for convex bodies*, Trans. Amer. Math. Soc. **355** (2003), 4723-4735.

[21] J. Arias de Reyna, K. M. Ball and R. Villa, *Concentration of the distance in finite dimensional normed spaces*, Mathematika **45** (1998), 245-252.

[22] S. Artstein, *Proportional concentration phenomena on the sphere*, Israel J. Math. **132** (2002), 337-358.

[23] S. Artstein-Avidan, D. Florentin and V. D. Milman, *Order isomorphisms on convex functions in windows*, Geometric aspects of functional analysis, Lecture Notes in Mathematics **2050**, Springer, Heidelberg, (2012), 61-122.

[24] S. Artstein-Avidan, D. Florentin and Y. Ostrover, *Remarks about Mixed Discriminants and Volumes*, to appear in Communications in Contemporary Mathematics.

[25] S. Artstein-Avidan and V. D. Milman, *Logarithmic reduction of the level of randomness in some probabilistic geometric constructions*, J. Funct. Anal. **235** (2006), 297-329.

[26] S. Artstein-Avidan and V. D. Milman, *A characterization of the support map*, Adv. Math. **223** (2010), no. 1, 379-391.

[27] S. Artstein-Avidan and V. D. Milman, *Hidden structures in the class of convex functions and a new duality transform*, J. Eur. Math. Soc. (JEMS) **13** (2011), no. 4, 975-1004.

[28] S. Artstein, V. D. Milman and S. J. Szarek, *Duality of metric entropy in Euclidean space*, C. R. Math. Acad. Sci. Paris **337** (2003), no. 11, 711-714.

[29] S. Artstein, V. D. Milman and S. J. Szarek, *More on the duality conjecture for entropy numbers*, C. R. Math. Acad. Sci. Paris **336** (2003), no. 6, 479-482.

[30] S. Artstein, V. D. Milman and S. J. Szarek, *Duality of metric entropy*, Annals of Math., **159** (2004), no. 3, 1313-1328.

[31] S. Artstein, V. D. Milman, S. J. Szarek and N. Tomczak-Jaegermann, *On convexified packing and entropy duality*, Geom. Funct. Anal. **14** (2004), no. 5, 1134-1141.

[32] S. Artstein-Avidan, D. Florentin, K. Gutman and Y. Ostrover, On Godbersen's Conjecture, to apear in *Geometria Dedicata* (2015).

[33] S. Artstein-Avidan and O. Raz, *Weighted covering numbers of convex sets*, Adv. Math. **227** (2011), no. 1, 730-744.

[34] S. Artstein-Avidan and B. Slomka *On weighted covering numbers and the Levi-Hadwiger conjecture*, preprint.

[35] K. Azuma, *Weighted sums of certain dependent random variables*, Tohoku Math. J. **19** (1967), 357-367.

[36] A. Badrikian and S. Chevet, *Mesures cylindriques*, in Espaces de Wiener et Fonctions Aléatoires Gaussiennes, Lecture Notes in Mathematics **379** (1974), Springer.

[37] K. M. Ball, *Isometric problems in ℓ_p and sections of convex sets*, Ph.D. Dissertation, Trinity College, Cambridge (1986).

[38] K. M. Ball, *Logarithmically concave functions and sections of convex sets in $\mathbb{R}^n$*, Studia Math. **88** (1988), 69-84.

[39] K. M. Ball, *Volumes of sections of cubes and related problems*, Lecture Notes in Mathematics **1376**, Springer, Berlin (1989), 251-260.

[40] K. M. Ball, *Normed spaces with a weak Gordon-Lewis property*, Lecture Notes in Mathematics **1470**, Springer, Berlin (1991), 36-47.

[41] K. M. Ball, *Volume ratios and a reverse isoperimetric inequality*, J. London Math. Soc. (2) **44** (1991), 351-359.

[42] K. M. Ball, *Ellipsoids of maximal volume in convex bodies*, Geom. Dedicata **41** (1992), 241-250.

[43] K. M. Ball, *An elementary introduction to modern convex geometry*, Flavors of Geometry, Math. Sci. Res. Inst. Publ. **31**, Cambridge Univ. Press (1997).

[44] K. M. Ball, *Convex geometry and functional analysis*, Handbook of the geometry of Banach spaces, Vol. I, North-Holland, Amsterdam, (2001), 161-194.

[45] K. M. Ball, *An elementary introduction to monotone transportation*, Geometric aspects of functional analysis, Lecture Notes in Math. **1850** Springer, Berlin (2004), 41-52.

[46] K. M. Ball and V. H. Nguyen, *Entropy jumps for isotropic log-concave random vectors and spectral gap*, Studia Math. **213** (2012), 81-96.

[47] K. M. Ball and I. Perissinaki, *The subindependence of coordinate slabs for the ℓ_p^n-balls*, Israel J. Math. **107** (1998), 289-299.

[48] I. Bárány, *Random points and lattice points in convex bodies*, Bulletin of the AMS, **45** (2008), 339-365.

[49] F. Barthe, *Inégalités fonctionelles et géomètriques obtenues par transport des mesures*, Thèse de Doctorat de Mathématiques, Université de Marne-la-Vallée (1997).

[50] F. Barthe, *Inégalités de Brascamp-Lieb et convexité*, C. R. Acad. Sci. Paris Ser. I Math. **324** (1997), no. 8, 885-888.

[51] F. Barthe, *On a reverse form of the Brascamp-Lieb inequality*, Invent. Math. **134** (1998), 335-361.

[52] F. Barthe, *An extremal property of the mean width of the simplex*, Math. Ann. **310** (1998), 685-693.

[53] F. Barthe, *Autour de l'inégalité de Brunn-Minkowski*, Ann. Fac. Sci. Toulouse **12** (2003), 127-178.

[54] F. Barthe and D. Cordero-Erausquin, *Invariances in variance estimates*, Proc. London Math. Soc. **106** (2013) 33-64.

[55] J. Bastero and J. Bernués, *Asymptotic behavior of averages of k-dimensional marginals of measures on $\mathbb{R}^n$*, Studia Math. **190** (2009), 1-31.

[56] J. Bastero and M. Romance, *John's decomposition of the identity in the non-convex case*, Positivity **6** (2002), 1-16.

[57] J. Bastero and M. Romance, *Positions of convex bodies associated to extremal problems and isotropic measures*, Adv. Math. **184** (2004), 64-88.

[58] J. Batson, D. Spielman and N. Srivastava, *Twice-Ramanujan Sparsifiers*, STOC 2009, SICOMP special issue (2012).

[59] E. F. Beckenbach and R. Bellman, *Inequalities*, Springer-Verlag (1971).

[60] F. Behrend, *Über einige Affininvarianten konvexer Bereiche*, Math. Ann. **113** (1937), 713-747.

[61] G. Bennett, *Probability inequalities for the sum of independent random variables*, J. Amer. Statistical Association **57** (1962), 33-45.

[62] G. Bennett, *Upper bounds on the moments and probability inequalities for the sum of independent, bounded random variables*, Biometrika **52** (1965), 559-569.

[63] G. Bennett, L. E. Dor, V. Goodman, W. B. Johnson and C. M. Newman, *On uncomplemented subspaces of $L_p, 1 < p < 2$*, Israel J. Math. **26** (1977), 178-187.

[64] Y. Benyamini and Y. Gordon, *Random factorization of operators between Banach spaces*, J. d'Analyse Math. **39** (1981), 45-74.

[65] C. Berg, *Corps convexes et potentiels sphériques*, Mat.-Fys. Medd. Danske Vid. Selsk. **37** (1969) no. 6.

[66] J. Bergh and J. Löfstrom, *Interpolation Spaces – An Introduction*, Springer-Verlag (1976).

[67] D. M. Bernstein, *The number of roots of a system of equations*, Funct. Analysis and its Applications **9** (1975), 183-185.

[68] S. Bernstein, *On a modification of Chebyshev's inequality and of the error formula of Laplace*, Ann. Sci. Inst. Sav. Ukraine, Sect. Math. **1** (1924), 38-49.

[69] S. Bernstein, *Theory of Probability*, Moscow, 1927.

[70] S. Bernstein, *On certain modifications of Chebyshev's inequality*, Dokl. AN SSSR **17** (1937), 275-277.

[71] K. Bezdek, *Classical topics in discrete geometry*, CMS Books in Mathematics/Ouvrages de Mathématiques de la SMC, Springer, New York, New York, 2010. ISBN: 978-1-4419-0599-4.

[72] K. Bezdek, A. E. Litvak, On the vertex index of convex bodies,, *Adv. Math.*, 215 (2007), 626–641.

[73] L. Bieberbach, *Über eine Extremaleigenschaft des Kreises*, J.-Ber. Deutsch. Math. Verein. **24** (1915), 247-250.

[74] B. J. Birch, *Homogeneous forms of odd degree in a large number of variables*, Mathematika **4** (1957), 102-105.

[75] W. Blaschke, *Kreis und Kugel*, Leipzig (1916).

[76] W. Blaschke, *Über affine Geometrie VII: Neue Extremaigenschaften von Ellipse und Ellipsoid*, Ber. Vergh. Sächs. Akad. Wiss. Leipzig, Math.-Phys. Kl. **69** (1917), 306-318, Ges. Werke **3**, 246-258.

[77] H. F. Blichfeldt, *The minimum value of quadratic forms and the closest packing of spheres*, Math. Ann. **101** (1929), 605-608.

[78] S. G. Bobkov, *Extremal properties of half-spaces for log-concave distributions*, Ann. Probab. **24** (1996), 35-48.

[79] S. G. Bobkov, *A functional form of the isoperimetric inequality for the Gaussian measure*, J. Funct. Anal. **135** (1996), 39-49.

[80] S. G. Bobkov, *An isoperimetric inequality on the discrete cube and an elementary proof of the isoperimetric inequality in Gauss space*, Ann. Probab. **25** (1997), 206–214.

[81] S. G. Bobkov, *Isoperimetric and analytic inequalities for log-concave probability measures*, Ann. Prob. **27** (1999), 1903-1921.

[82] S. G. Bobkov, *Remarks on the growth of L_p-norms of polynomials*, Geom. Aspects of Funct. Analysis (Milman-Schechtman eds.), Lecture Notes in Math. **1745** (2000), 27-35.

[83] S. G. Bobkov, *On concentration of distributions of random weighted sums*, Ann. Probab. **31** (2003), 195-215.

[84] S. G. Bobkov, *Spectral gap and concentration for some spherically symmetric probability measures*, Geom. Aspects of Funct. Analysis, Lecture Notes in Math. **1807**, Springer, Berlin (2003), 37-43.

[85] S. G. Bobkov, *On isoperimetric constants for log-concave probability distributions*, Geometric aspects of functional analysis, Lecture Notes in Math., **1910**, Springer, Berlin (2007), 81-88.

[86] S. G. Bobkov, F. Götze and C. Houdré, *On Gaussian and Bernoulli covariance representations*, Bernoulli **3** (2001), 439–451.

[87] S. G. Bobkov and A. Koldobsky, *On the central limit property of convex bodies*, Geom. Aspects of Funct. Analysis (Milman-Schechtman eds.), Lecture Notes in Math. **1807** (2003), 44-52.

[88] S. G. Bobkov and F. L. Nazarov, *On convex bodies and log-concave probability measures with unconditional basis*, Geom. Aspects of Funct. Analysis (Milman-Schechtman eds.), Lecture Notes in Math. **1807** (2003), 53-69.

[89] S. G. Bobkov and F. L. Nazarov, *Large deviations of typical linear functionals on a convex body with unconditional basis*, Stochastic Inequalities and Applications, Progr. Probab. **56**, Birkhäuser, Basel (2003), 3-13.

[90] V. Bogachev, *Gaussian measures*, Mathematical Surveys and Monographs **62**, American Mathematical Society, Providence, RI, 1998.

[91] E. D. Bolker, *A class of convex bodies*, Trans. Amer. Math. Soc. **145** (1969), 323-345.

[92] T. Bonnesen, *Les problèmes des isopérimètres et des isépiphanes*, Gauthiers? Villars, Paris, 1929.

[93] T. Bonnesen and W. Fenchel, *Theorie der konvexen Körper*, Springer, Berlin, 1934. Reprint: Chelsea Publ. Co., New York, 1948. English translation: BCS Associates, Moscow, Idaho, 1987.

[94] C. Borell, *Convex set functions in d-space*, Period. Math. Hungar. **6** (1975), 111-136.

[95] C. Borell, *The Brunn-Minkowski inequality in Gauss space*, Inventiones Math. **30** (1975), 207-216.

[96] K. J. Böröczky and R. Schneider, *A characterization of the duality mapping for convex bodies*, Geom. Funct. Anal. **18** (2008), no. 3, 657–667.

[97] J. Bourgain, *On martingales transforms in finite-dimensional lattices with an appendix on the K-convexity constant*, Math. Nachr. **119** (1984), 41-53.

[98] J. Bourgain, *On high dimensional maximal functions associated to convex bodies*, Amer. J. Math. **108** (1986), 1467-1476.

[99] J. Bourgain, *On the L^p-bounds for maximal functions associated to convex bodies*, Israel J. Math. **54** (1986), 257-265.

[100] J. Bourgain, *On the distribution of polynomials on high dimensional convex sets*, Lecture Notes in Mathematics **1469**, Springer, Berlin (1991), 127-137.

[101] J. Bourgain, *On the isotropy constant problem for ψ_2-bodies*, Geom. Aspects of Funct. Analysis (Milman-Schechtman eds.), Lecture Notes in Math. **1807** (2003), 114-121.

[102] J. Bourgain, *On the Hardy-Littlewood maximal function for the cube*, Israel J. Math. **203** (2014), 275-294.

[103] J. Bourgain, B. Klartag and V. D. Milman, *Symmetrization and isotropic constants of convex bodies*, Geom. Aspects of Funct. Analysis, Lecture Notes in Math. **1850** (2004), 101-115.

[104] J. Bourgain and J. Lindenstrauss, *Projection bodies*, Geometric aspects of functional analysis (1986/87), Lecture Notes in Mathematics **1317**, Springer, Berlin (1988), 250-270.

[105] J. Bourgain and J. Lindenstrauss, *Almost Euclidean sections in spaces with a symmetric basis*, Geometric aspects of functional analysis (1987-88), Lecture Notes in Mathematics **1376**, Springer, Berlin, (1989), 278-288.

[106] J. Bourgain, J. Lindenstrauss and V. D. Milman, *Minkowski sums and symmetrizations*, Geom. Aspects of Funct. Analysis (Lindenstrauss-Milman eds.), Lecture Notes in Math. **1317** (1988), 44-74.

[107] J. Bourgain, J. Lindenstrauss and V. D. Milman, *Estimates related to Steiner symmetrizations*, Geometric aspects of functional analysis (1987-88), Lecture Notes in Mathematics **1376**, Springer, Berlin, (1989), 264-273.

[108] J. Bourgain, J. Lindenstrauss and V. D. Milman, *Approximation of zonoids by zonotopes*, Acta Math. **162** (1989), no. 1-2, 73-141.

[109] J. Bourgain and V. D. Milman, *Distances between normed spaces, their subspaces and quotient spaces*, Integral Equations Operator Theory 9 (1986), no. 1, 31-46.

[110] J. Bourgain and V. D. Milman, *New volume ratio properties for convex symmetric bodies in* $\mathbb{R}^n$, Invent. Math. **88** (1987), 319-340.

[111] J. Bourgain, V.D. Milman and H. Wolfson, *On the type of metric spaces*, Trans. Amer. Math. Soc. **294** (1986), 295-317.

[112] J. Bourgain, A. Pajor, S. J. Szarek and N. Tomczak-Jaegermann, *On the duality problem for entropy numbers of operators*, Geometric aspects of functional analysis (1987-88), Lecture Notes in Mathematics **1376**, Springer, Berlin, (1989), 50-63.

[113] J. Bourgain and L. Tzafriri, *Invertibility of "large" submatrices with applications to the geometry of Banach spaces and harmonic analysis*, Israel J. of Math. **57** (1987), 137-224.

[114] J. Bourgain and L. Tzafriri, Embedding ℓ_p^k in subspaces of L_p for $p > 2$, Israel J. Math. **72** (1990), no. 3, 321-340.

[115] P. Brass, W. Moser and J. Pach, *Levi-Hadwiger covering problem and illumination*, in Research Problems in Discrete Geometry, Springer-Verlag (2005), 136-142.

[116] S. Brazitikos, A. Giannopoulos, P. Valettas and B-H. Vritsiou, *Geometry of isotropic convex bodies*, Mathematical Surveys and Monographs **196**, American Mathematical Society, Providence, RI, 2014.

[117] S. Brazitikos and L. Hioni, *Sub-Gaussian directions of isotropic convex bodies*, Preprint.

[118] U. Brehm and J. Voigt, *Asymptotics of cross sections for convex bodies*, Beiträge Algebra Geom. **41** (2000), 437-454.

[119] U. Brehm, H. Vogt and J. Voigt, *Permanence of moment estimates for p-products of convex bodies*, Studia Math. **150** (2002), 243-260.

[120] U. Brehm, P. Hinow, H. Vogt and J. Voigt, *Moment inequalities and central limit properties of isotropic convex bodies*, Math. Zeitsch. **240** (2002), 37-51.

[121] Y. Brenier, *Décomposition polaire et réarrangement monotone des champs de vecteurs*, C. R. Acad. Sci. Paris Sér. I Math., **305** (1987), 805-808.

[122] Y. Brenier, *Polar factorization and monotone rearrangement of vector-valued functions*, Comm. Pure Appl. Math. **44** (1991), 375-417.

[123] E. M. Bronstein, *ε-entropy of affine-equivalent convex bodies and Minkowski?s compactum*, Optimizatsiya **39** (1978), 5-11.

[124] H. Brunn, *Über Ovale und Eiflächen*, Inaugural Dissertation, München, 1887.

[125] H. Brunn, *Über Curven ohne Wendepunkte*, Habilitationsschrift, München, 1889.

[126] H. Brunn, *Referat über eine Arbeit: Exacte Grundlagen für eine Theorie der Ovale*, S.-B. Bayer. Akad. Wiss., (1894), 93-111.

[127] Y. D. Burago and V. A. Zalgaller, *Geometric Inequalities*, Springer Series in Soviet Mathematics, Springer-Verlag, Berlin-New York (1988).

[128] H. Busemann, *A theorem on convex bodies of the Brunn-Minkowski type*, Proc. Nat. Acad. Sci. U. S. A. **35** (1949), 27-31.

[129] P. Buser, *A note on the isoperimetric constant*, Ann. Sci. École Norm. Sup. **15** (1982), 213-230.

[130] L. A. Caffarelli, *A-priori estimates and the geometry of the Monge-Ampère equation*, Park City/IAS Mathematics Series **2** (1992).

[131] L. A. Caffarelli, *Boundary regularity of maps with a convex potential*, Comm. Pure Appl. Math. **45** (1992), 1141-1151.

[132] L. A. Caffarelli, *Regularity of mappings with a convex potential*, J. Amer. Math. Soc. **5** (1992), 99-104.

[133] S. Campi and P. Gronchi, *The L_p-Busemann-Petty centroid inequality*, Adv. in Math. **167** (2002), 128-141.

[134] S. Campi and P. Gronchi, *On volume product inequalities for convex sets*, Proc. Amer. Math. Soc. **134** (2006), 2393-2402.

[135] A. Carbery, *Radial Fourier multipliers and associated maximal functions*, Recent Progress in Fourier Analysis, North-Holland Math. Studies **111** (1985), 49-56.

[136] A. Carbery and J. Wright, *Distributional and L_q-norm inequalities for polynomials over convex bodies in* $\mathbb{R}^n$, Math. Res. Lett. **8** (2001), no. 3, 233-248.

[137] B. Carl, *Entropy numbers, s-numbers and eigenvalue problems*, J. Funct. Anal. **41** (1981), 290-306.

[138] B. Carl, *Entropy numbers of diagonal operators with an application to eigenvalue problems*, Journal Approx. Theory **32** (1981), 135-150.

[139] B. Carl, *Inequalities of Bernstein-Jackson type and the degree of compactness of operators in Banach spaces*, Ann. Inst. Fourier **35** (1985), 79-118.

[140] B. Carl and I. Stephani, *Entropy, compactness and the approximation of operators*, Cambridge Tracts in Mathematics **98**, Cambridge University Press, Cambridge, 1990.

[141] B. Carl and H. Triebel, *Inequalities between eigenvalues, entropy numbers and related quantities of compact operators in Banach spaces*, Math. Ann. **251** (1980), 129-133.

[142] G. Carlier, A. Galichon and F. Santambrogio, *From Knothe's transport to Brenier's map and a continuation method for optimal transport*, SIAM J. Math. Anal. **41** (2009), 2554-2576.

[143] G. D. Chakerian, *Inequalities for the difference body of a convex body*, Proc. Amer. Math. Soc. **18** (1967), 879-884.

[144] I. Chavel, *Riemannian geometry - A modern introduction*, Second edition, Cambridge Studies in Advanced Mathematics **98**, Cambridge University Press, Cambridge (2006).

[145] J. Cheeger, *A lower bound for the smallest eigenvalue of the Laplacian*, Problems in Analysis (Papers dedicated to Salomon Bochner, 1969), Princeton Univ. Press, Princeton (1970), 195-199.

[146] J. Cheeger and D. Ebin, *Comparison theorems in Riemannian geometry*, North Holland (1975).

[147] W. Chen, *Counterexamples to Knaster's conjecture*, Topology **37** (1998), 401-405.

[148] S. Y. Cheng and S. T. Yau, *On the regularity of the solution of the n-dimensional Minkowski problem*, Commun. Pure Appl. Math. **29** (1976), 495-516.

[149] G. Choquet, *Unicité des represéntations intégrales au moyen de points extrémaux dans les cônes convexes réticulés*, C. R. Acad. Sci. Paris **243** (1956), 555-557.

[150] E. B. Christoffel, *Über die Bestimmung der Gestalt einer krummen Oberfläche durch lokale Messungen auf derselben*, J. für die Reine und Angewandte Math. **64** (1865), 193-209.

[151] A. Colesanti, *Brunn-Minkowski inequalities for variational functionals and related problems*, Adv. Math. **194** (2005), 105-140.

[152] A. Colesanti, *Functional inequalities related to the Rogers-Shephard inequality*, Mathematika 53 (2006), no 1, 81–101 (2007).

[153] A. Colesanti and M. Fimiani, *The Minkowski problem for the torsional rigidity*, Indiana Univ. Math. J. **59** (2010), 1013-1039.

[154] D. Cordero-Erausquin, M. Fradelizi and B. Maurey, *The (B)-conjecture for the Gaussian measure of dilates of symmetric convex sets and related problems*, J. Funct. Anal. **214** (2004), 410-427.

[155] C. C. Craig, *On the Tchebycheff inequality of Bernstein*, Annals of Math. Statistics **4** (1933), 94-102.

[156] H. Cramér *Mathematical Methods of Statistics*, Princeton Mathematical Series, vol. 9, Princeton University Press, Princeton, N. J.,1946. xvi+575 pp.

[157] N. Dafnis, A. Giannopoulos and O. Guédon, *On the isotropic constant of random polytopes*, Advances in Geometry **10** (2010), 311-321.

[158] N. Dafnis, A. Giannopoulos and A. Tsolomitis, *Asymptotic shape of a random polytope in a convex body*, J. Funct. Anal. **257** (2009), 2820-2839.

[159] N. Dafnis, A. Giannopoulos and A. Tsolomitis, *Quermassintegrals and asymptotic shape of a random polytope in an isotropic convex body*, Michigan Math. Journal **62** (2013), 59-79.

[160] N. Dafnis and G. Paouris, *Small ball probability estimates, ψ_2-behavior and the hyperplane conjecture*, J. Funct. Anal. **258** (2010), 1933-1964.

[161] L. Danzer, B. Grünbaum and V. Klee, *Helly's theorem and its relatives*, 1963 Proc. Sympos. Pure Math., Vol. VII pp. 101-180 Amer. Math. Soc., Providence, R.I.

[162] S. Dar, *Remarks on Bourgain's problem on slicing of convex bodies*, in Geometric Aspects of Functional Analysis, Operator Theory: Advances and Applications **77** (1995), 61-66.

[163] S. Dar, *Isotropic constants of Schatten class spaces*, Convex Geometric Analysis, MSRI Publications **34** (1998), 77-80.

[164] S. Dar, *A Brunn-Minkowski-type inequality*, Geom. Dedicata **77** (1999), 1-9.

[165] S. Das Gupta, *Brunn-Minkowski inequality and its aftermath*, J. Multivariate Anal. **10** (1980), 296-318.

[166] J. S. Davidovič, B. I. Korenbljum and B. I. Hacet, *A certain property of logarithmically concave functions*, Dokl., Akad. Nauk Azerb. **185** (1969), 1215-1218.

[167] W. J. Davis, V. D. Milman and N. Tomczak-Jaegermann, *The distance between certain n-dimensional spaces*, Israel J. Math. **39** (1981), 1-15.

[168] P. Diaconis and D. Freedman, *Asymptotics of graphical projection pursuit*, Ann. of Stat. **12** (1984), 793-815.

[169] P. Diaconis and D. Freedman, *A dozen de Finetti-style results in search of a theory*, Ann. Inst. H. Poincaré Probab. Statist. **23** (1987), 397-423.

[170] S. J. Dilworth, *On the dimension of almost Hilbertian subspaces of quotient spaces*, J. London Math. Soc. **30** (1984), 481-485.

[171] S. J. Dilworth, *The dimension of Euclidean subspaces of quasi-normed spaces*, Math. Proc. Camb. Phil. Soc. **97** (1985), 311-320.

[172] S. J. Dilworth and S. J. Szarek, *The cotype constant and an almost Euclidean decomposition for finite-dimensional normed spaces*, Israel J. Math. **52** (1985), no. 1-2, 82-96.

[173] V. I. Diskant, *Bounds for convex surfaces with bounded curvature functions*, Siberian Math. J. **12** (1971), 78-89.

[174] V. I. Diskant, *Bounds for the discrepancy between convex bodies in terms of the isoperimetric difference*, Siberian Math. J. **13** (1972), 529-532.

[175] V. L. Dol'nikov and R. N. Karasev, *Dvoretzky type theorems for multivariate polynomials and sections of convex bodies*, Geom. Funct. Anal. **21** (2011), 301-318.

[176] R. M. Dudley, *The sizes of compact subsets of Hilbert space and continuity of Gaussian processes*, J. Funct. Anal. **1** (1967), 290-330.

[177] A. Dvoretzky, *A theorem on convex bodies and applications to Banach spaces*, Proc. Nat. Acad. Sci. U.S.A **45** (1959), 223-226.

[178] A. Dvoretzky, *Some results on convex bodies and Banach spaces*, in Proc. Sympos. Linear Spaces, Jerusalem (1961), 123-161.

[179] A. Dvoretzky and C. A. Rogers, *Absolute and unconditional convergence in normed linear spaces*, Proc. Nat. Acad. Sci., U.S.A **36** (1950), 192-197.

[180] Y. Eidelman, V. D. Milman and A. Tsolomitis, *Functional analysis, An introduction*, Graduate Studies in Mathematics, 66. American Mathematical Society, Providence, RI, 2004.

[181] H. G. Eggleston, B. Grünbaum and V. Klee, *Some semicontinuity theorems for convex polytopes and cell complexes*, Comment. Math. Helvet. **39** (1964), 165-188.

[182] A. Erhard, *Symétrisation dans l'espace de Gauss*, Math. Scand. **53** (1983), 281-301.

[183] R. Eldan, *Thin shell implies spectral gap up to polylog via a stochastic localization scheme*, Geom. Funct. Anal. **23** (2013), 532-569.

[184] R. Eldan and B. Klartag, *Pointwise estimates for marginals of convex bodies*, J. Funct. Anal. **254** (2008), 2275-2293.

[185] R. Eldan and B. Klartag, *Approximately Gaussian marginals and the hyperplane conjecture* Concentration, functional inequalities and isoperimetry, Contemp. Math. **545**, Amer. Math. Soc., Providence, RI (2011), 55-68.

[186] R. Eldan and B. Klartag, *Dimensionality and the stability of the Brunn-Minkowski inequality*, Ann. Sc. Norm. Super. Pisa (to appear).

[187] J. Elton, *Sign-embeddings of ℓ_1^n*, Trans. Amer. Math. Soc. **279** (1983), 113-124.

[188] P. Enflo, *On the nonexistence of uniform homeomorphisms between L_p-spaces*, Ark. Mat. **8** (1969), 103-105.

[189] P. Enflo, *Uniform homeomorphisms between Banach spaces*, Séminaire Maurey-Schwartz 75-76, Exposé no. 18, Ecole Polytechnique, Paris.

[190] A. Esterov, *Tropical varieties with polynomial weights and corner loci of piecewise polynomials*, Mosc. Math. J. 12, 55-76 (2012). arXiv:1012.5800v3

[191] K. J. Falconer, A result on the Steiner symmetrization of a compact set, J. London Math. Soc. **14** (1976), 385-386.

[192] K. Fan, *Some inequalities concerning positive definite hermitian matrices*, Proc. Cambridge Phil. Soc. **51** (1955), 414-421.

[193] G. Fejes Tóth and W. Kuperberg, *Packing and covering with convex sets*, Handbook of Convex Geometry (Gruber-Wills eds.), North-Holland, Amsterdam (1993), 799-860.

[194] W. Fenchel, *Über Krümmung und Windung geschlossener Raumkurven*, Math. Ann. **101** (1929), no. 1, 238-252.

[195] W. Fenchel, *Inégalités quadratiques entre les volumes mixtes des corps convexes*, C. R. Acad. Sci. Paris **203** (1936), 647-650.

[196] W. Fenchel, *Über die neuere Entwicklung der Brunn-Minkowskischen Theorie der konvexen Körper*, In Proc. 9th Congr. Math. Scand., Helsingfors 1938, pp. 249-272.

[197] W. Fenchel, *Convexity through the ages*, Convexity and its applications (Gruber-Wills eds.) Birkhauser, Basel (1983), 120-130.

[198] W. Fenchel and B. Jessen, *Mengenfunktionen und konvexe Körper*, Danske Vid. Selskab. Mat.-fys. Medd. **16** (1938), 31 pp.

[199] X. Fernique, *Régularité des trajectoires des fonctions aléatoires Gaussiennes*, Ecole d'Eté de St. Flour IV, 1974, Lecture Notes in Mathematics **480** (1975), 1-96.

[200] A. Figalli and D. Jerison, *How to recognize convexity of a set from its marginals*, J. Funct. Anal. **266** (2014), 1685-1701.

[201] A. Figalli, F. Maggi and A. Pratelli, *A refined Brunn-Minkowski inequality for convex sets*, Ann. Inst. H. Poincaré Anal. Non Linéaire **26** (2009), 2511–2519.

[202] A. Figalli, F. Maggi and A. Pratelli, *A mass transportation approach to quantitative isoperimetric inequalities*, Invent. Math. **182** (2010), 167–211.

[203] T. Figiel, *A short proof of Dvoretzky's theorem*, Comp. Math. **33** (1976), 297-301.

[204] T. Figiel and W. B. Johnson, *Large subspaces of ℓ_∞^n and estimates of the Gordon-Lewis constant*, Israel J. Math. **37** (1980), 92-112.

[205] T. Figiel, J. Lindenstrauss and V. D. Milman, *The dimension of almost spherical sections of convex bodies*, Acta Math. **139** (1977), 53-94.

[206] T. Figiel and N. Tomczak-Jaegermann, *Projections onto Hilbertian subspaces of Banach spaces*, Israel J. Math. **33** (1979), 155-171.

[207] W. J. Firey, *p-means of convex bodies* Mathematica Scandinavica, 10: 17-24, 1962.

[208] W. J. Firey, *The mixed area of a convex body and its polar reciprocal*, Israel J. Math. **1** (1963), 201-202.

[209] W. J. Firey, *The determination of convex bodies from their mean radius of curvature functions*, Mathematika **14** (1967), 1-13.

[210] W. J. Firey, *Christoffel's problem for general convex bodies*, Mathematika **15** (1968), 7-21.

[211] B. Fleury, *Concentration in a thin Euclidean shell for log-concave measures*, J. Funct. Anal. **259** (2010), 832-841.

[212] B. Fleury, O. Guédon and G. Paouris, *A stability result for mean width of L_p-centroid bodies*, Adv. Math. **214**, 2 (2007), 865-877.

[213] D. Florentin, V. D. Milman and R. Schneider, *A characterization of the mixed discriminant*, to appear.

[214] E. E. Floyd, *Real-valued mappings of spheres*, Proc. Amer. Math. Soc. **6** (1955), 957-959.

[215] M. Fradelizi, *Sections of convex bodies through their centroid*, Arch. Math. **69** (1997), 515-522.

[216] M. Fradelizi, A. Giannopoulos and M. Meyer, *Some inequalities about mixed volumes*, Israel J. Math. **135** (2003), 157-180.

[217] S. Gallot, D. Hulin and J. Lafontaine, *Riemannian Geometry*, Second Edition, Springer (1990).

[218] R. J. Gardner, *The Brunn-Minkowski inequality*, Bull. Amer. Math. Soc. (N.S.) **39** (2002), 355-405.

[219] R. J. Gardner, *Geometric Tomography*, Second Edition Encyclopedia of Mathematics and its Applications **58**, Cambridge University Press, Cambridge (2006).

[220] R. J. Gardner, D. Hug and W. Weil, *Operations between sets in geometry*, J. Eur. Math. Soc. (JEMS) **15** (2013), 2297-2352.

[221] A. Garnaev and E. D. Gluskin, *On diameters of the Euclidean sphere*, Dokl. A.N. U.S.S.R. **277** (1984), 1048-1052.

[222] A. Giannopoulos, *A note on the Banach-Mazur distance to the cube*, in Geometric Aspects of Functional Analysis, Operator Theory: Advances and Applications **77** (1995), 67-73.

[223] A. Giannopoulos, *A proportional Dvoretzky-Rogers factorization result*, Proc. Amer. Math. Soc. **124** (1996), 233-241.

[224] A. Giannopoulos and M. Hartzoulaki, *On the volume ratio of two convex bodies*, Bull. London Math. Soc. **34** (2002), 703-707.

[225] A. Giannopoulos, M. Hartzoulaki and G. Paouris, *On a local version of the Aleksandrov-Fenchel inequalities for the quermassintegrals of a convex body*, Proc. Amer. Math. Soc. **130** (2002), 2403-2412.

[226] A. Giannopoulos, L. Hioni and A. Tsolomitis, *Asymptotic shape of the convex hull of isotropic log-concave random vectors*, Preprint.

[227] A. Giannopoulos and E. Milman, *M-estimates for isotropic convex bodies and their L_q centroid bodies*, in Geometric Aspects of Functional Analysis, Lecture Notes in Mathematics **2116** (2014), 159-182.

[228] A. Giannopoulos and V. D. Milman, *Low M^*-estimates on coordinate subspaces*, Journal of Funct. Analysis **147** (1997), 457-484.

[229] A. Giannopoulos and V. D. Milman, *On the diameter of proportional sections of a symmetric convex body*, International Mathematics Research Notices (1997) **1**, 5-19.

[230] A. A. Giannopoulos and V. D. Milman, *How small can the intersection of a few rotations of a symmetric convex body be?*, C.R. Acad. Sci. Paris **325** (1997), 389-394.

[231] A. Giannopoulos and V. D. Milman, *Mean width and diameter of proportional sections of a symmetric convex body*, J. Reine angew. Math. **497** (1998), 113-139.

[232] A. Giannopoulos and V. D. Milman, *Extremal problems and isotropic positions of convex bodies*, Israel J. Math. **117** (2000), 29-60.

[233] A. Giannopoulos and V. D. Milman, *Concentration property on probability spaces*, Adv. in Math. **156** (2000), 77-106.

[234] A. Giannopoulos and V.D. Milman, *Euclidean structure in finite dimensional normed spaces*, Handbook of the Geometry of Banach spaces (Lindenstrauss-Johnson eds), Elsevier (2001), 707-779.

[235] A. Giannopoulos and V. D. Milman, *Asymptotic convex geometry: short overview*, Different faces of geometry, 87-162, Int. Math. Ser. **3**, Kluwer/Plenum, New York, 2004.

[236] A. Giannopoulos, V. D. Milman and M. Rudelson, *Convex bodies with minimal mean width*, Geometric Aspects of Functional Analysis (Milman-Schechtman eds.), Lecture Notes in Mathematics **1745** (2000), 81-93.

[237] A. Giannopoulos, V. D. Milman and A. Tsolomitis, *Asymptotic formulas for the diameter of sections of symmetric convex bodies*, Journal of Functional Analysis **223** (2005), 86-108.

[238] A. Giannopoulos, A. Pajor and G. Paouris, *A note on subgaussian estimates for linear functionals on convex bodies*, Proc. Amer. Math. Soc. **135** (2007), 2599-2606.

[239] A. Giannopoulos, G. Paouris and P. Valettas, *On the existence of subgaussian directions for log-concave measures*, Contemporary Mathematics **545** (2011), 103-122.

[240] A. Giannopoulos, G. Paouris and P. Valettas, *On the distribution of the ψ_2-norm of linear functionals on isotropic convex bodies*, in Geom. Aspects of Funct. Analysis, Lecture Notes in Mathematics **2050** (2012), 227-254.

[241] A. Giannopoulos, G. Paouris and B-H. Vritsiou, *A remark on the slicing problem*, Journal of Functional Analysis **262** (2012), 1062-1086.

[242] A. Giannopoulos, G. Paouris and B-H. Vritsiou, *The isotropic position and the reverse Santaló inequality*, Israel J. Math. **203** (2014), 1-22.

[243] A. Giannopoulos and M. Papadimitrakis, *Isotropic surface area measures*, Mathematika **46** (1999), 1-13.

[244] A. Giannopoulos, I. Perissinaki and A. Tsolomitis, *John's theorem for an arbitrary pair of convex bodies*, Geom. Dedicata **84** (2001), 63-79.

[245] A. Giannopoulos, P. Stavrakakis, A. Tsolomitis and B-H. Vritsiou, *Geometry of the L_q-centroid bodies of an isotropic log-concave measure*, Trans. Amer. Math. Soc. (to appear).

[246] H. Gluck, *Manifolds with preassigned curvature - a survey*, Bull. Amer. Math.Soc. **81** (1975), 313-329.

[247] E. D. Gluskin, *The diameter of the Minkowski compactum is approximately equal to n*, Funct. Anal. Appl. **15** (1981), 72-73.

[248] E. D. Gluskin, *Finite dimensional analogues of spaces without basis*, Dokl. Akad. Nauk USSR **216** (1981), 1046-1050.

[249] E. D. Gluskin, *On the sum of intervals*, Geometric aspects of functional analysis, Lecture Notes in Math. **1807** (2003), 122-130.

[250] E. D. Gluskin, A. E. Litvak, A remark on vertex index of the convex bodies, GAFA, *Lecture Notes in Math.*, 2050, 255–265, Springer, Berlin, 2012.

[251] H. J. Godwin, *On generalizations of Tchebychef's inequality*, J. Amer. Statist. Assoc. **50** (1955), 923-945.

[252] I. Gohberg and A. S. Markus, *A certain problem about the covering of convex sets with homothetic ones*, Izvestiya Mold. Fil. Akad. Nauk SSSR **10** (1960), 87-90.

[253] P. Goodey and W. Weil, *Centrally symmetric convex bodies and the spherical Radon transforms*, J. Diff. Geom. **35** (1992), 675-688.

[254] P. Goodey, V. Yaskin and M. Yaskina, *A Fourier transform approach to Christoffel's problem*, Trans. Amer. Math. Soc. **363** (2011), 6351-6384.

[255] Y. Gordon, *Some inequalities for Gaussian processes and applications*, Israel J. Math. **50** (1985), 265-289.

[256] Y. Gordon, *Elliptically contoured distributions*, Probab. Th. Rel. Fields **76** (1987), 429-438.

[257] Y. Gordon, *Gaussian processes and almost spherical sections of convex bodies*, Ann. Probab. **16** (1988), 180-188.

[258] Y. Gordon, *On Milman's inequality and random subspaces which escape through a mesh in $\mathbb{R}^n$*, Lecture Notes in Mathematics **1317** (1988), 84-106.

[259] Y. Gordon, O. Guédon and M. Meyer, *An isomorphic Dvoretzky's theorem for convex bodies*, Studia Math. **127** (1998), 191-200.

[260] Y. Gordon, H. König and C. Schütt, *On the duality problem for entropy numbers*, Texas functional analysis seminar 1984-1985 (Austin, Tex.), 141-150, Longhorn Notes, Univ. Texas Press, Austin, TX, 1985.

[261] Y. Gordon, H. König and C. Schütt, *Geometric and probabilistic estimates for entropy and approximation numbers of operators*, J. Approx. Theory **49** (1987), 219-239.

[262] Y. Gordon, A. E. Litvak, M. Meyer and A. Pajor, *John's decomposition in the general case and applications*, J. Differential Geom. **68** (2004), no. 1, 99-119.

[263] Y. Gordon, M. Meyer and S. Reisner, *Zonoids with minimal volume product - a new proof*, Proc. Amer. Math. Soc. **104** (1988), 273-276.

[264] J. W. Green, *Length and area of a convex curve under affine transformation*, Pacific J. Math. **3** (1953), 393-402.

[265] H. Groemer, *On the Brunn-Minkowski theorem*, Geom. Dedicata **27** (1988), 357-371.

[266] H. Groemer, *Geometric applications of Fourier series and spherical harmonics*, Encyclopedia of Mathematics and its Applications, **61**, Cambridge University Press, Cambridge (1996).

[267] M. Gromov, *Paul Lévy isoperimetric inequality*, I.H.E.S. Preprint (1980).

[268] M. Gromov, *Convex sets and Kähler manifolds*, in "Advances in Differential Geometry and Topology", World Scientific Publishing, Teaneck NJ (1990), 1-38.

[269] M. Gromov, *Metric Structures for Riemannian and Non-Riemannian Spaces*, based on "Structures métriques des variétés Riemanniennes" (L. LaFontaine, P. Pansu. eds.), English translation by Sean M. Bates, Birkhäuser, Boston-Basel-Berlin, 1999 (with Appendices by M. Katz, P. Pansu and S. Semmes).

[270] M. Gromov, *Isoperimetry of waists and concentration of maps*, Geom. Funct. Anal. **13** (2003), 178-215.

[271] M. Gromov and V. D. Milman, *A topological application of the isoperimetric inequality*, Amer. J. Math. **105** (1983), 843-854.

[272] M. Gromov and V. D. Milman, *Brunn theorem and a concentration of volume phenomenon for symmetric convex bodies*, GAFA Seminar Notes, Tel Aviv University (1984).

[273] M. Gromov and V. D. Milman, *Generalization of the spherical isoperimetric inequality to uniformly convex Banach spaces*, Compos. Math. **62** (1987), 263-282.

[274] W. Gross, *Die Minimaleigenschaft der Kugel*, Monatsh. Math. Phys. **28** (1917), 77-97.

[275] A. Grothendieck, *Sur certaines classes de suites dans les espaces de Banach et le theoreme de Dvoretzky-Rogers*, Bol. Soc. Mat. Sao Paulo **8** (1953), 83-110.
[276] P. M. Gruber, *Minimal ellipsoids and their duals*, Rend. Circ. Mat. Palermo (2) **37** (1988), 35-64.
[277] P. M. Gruber, *Convex and Discrete Geometry*, Grundlehren Math. Wiss. **336**, Springer, Heidelberg (2007).
[278] P. M. Gruber and G. Lettl, *Isometries of the space of convex bodies in Euclidean space*, Bull. London Math. Soc. **12** (1980), 455-462.
[279] B. Grünbaum, *Partitions of mass-distributions and of convex bodies by hyperplanes*, Pacific J. Math. **10** (1960), 1257-1261.
[280] B. Grünbaum, *Measures of symmetry for convex sets*, Proceedings of Symposia in Pure Mathematics, Vol. VII, AMS, 1963.
[281] B. Guan and P. Guan, *Convex hypersurfaces of prescribed curvatures*, Ann. of Math. **156** (2002), 655-673.
[282] P. Guan, C. Lin and X. Ma, *The Christoffel-Minkowski problem II. Weingarten curvature equations*, Chinese Ann. Math. Ser. B **27** (2006), 595-614.
[283] P. Guan and X. Ma, *The Christoffel-Minkowski problem I. Convexity of solutions if a Hessian equation*, Invent. Math. **151** (2003), 553-577.
[284] P. Guan and X. Ma, *Convex solutions of fully nonlinear elliptic equations in classical differential geometry. Geometric evolution equations*, Contemp. Math. **367** (2005), 115-127.
[285] P. Guan, X. Ma and F. Zhou, *The Christoffel-Minkowski problem III. Existence and convexity of admissible solutions*, Comm. Pure Appl. Math. **59** (2006), 1352-1376.
[286] O. Guédon, *Gaussian version of a theorem of Milman and Schechtman*, Positivity **1** (1997), 1-5.
[287] O. Guédon, *Kahane-Khinchine type inequalities for negative exponent*, Mathematika **46** (1999), 165-173.
[288] O. Guédon and E. Milman, *Interpolating thin-shell and sharp large-deviation estimates for isotropic log-concave measures*, Geom. Funct. Anal. **21** (2011), 1043-1068.
[289] O. Guédon and G. Paouris, *Concentration of mass on the Schatten classes*, Ann. Inst. H. Poincare Probab. Statist. **43** (2007), 87-99.
[290] V. Guruswami, J. Lee and A. Razborov, *Almost Euclidean subspaces of ℓ_1^n via expander codes*, Combinatorica **30** (2010), 47-68.
[291] V. Guruswami, J. Lee and A. Wigderson, *Euclidean sections with sublinear randomness and error-correction over the reals*, Approximation, randomization and combinatorial optimization, Lecture Notes in Comput. Sci., 5171, Springer, Berlin (2008), 444-454.
[292] G. Gusev, *Euler characteristic of the bifurcation set for a polynomial of degree 2 or 3*, arXiv:1011.1390v2. Translated from: Monodromy zeta-functions and Newton diagrams (in Russian); Ph.D. thesis, MSU, Moscow (2008)
[293] U. Haagerup, *The best constants in the Khintchine inequality*, Studia Math. **70** (1982), 231-283.
[294] H. Hadwiger, *Einfache Herleitung der isoperimetrischen Ungleichung für abgeschlossene Punktmengen*, Math. Ann. **124** (1952), 158-160.
[295] H. Hadwiger, *Vorlesungen über Inhalt, Oberfläche und Isoperimetrie*, Springer, Berlin (1957).
[296] H. Hadwiger, *Ungeloste Probleme Nr.20*, Elemente der Mathematik **12** (1957), 121.
[297] H. Hadwiger and D. Ohmann, *Brunn-Minkowskischer Satz und Isoperimetrie*, Math. Z. **66** (1956), 1–8.
[298] O. Hanner and H. Radström, *A generalization of a theorem of Fenchel*, Proc. Amer. Math. Soc. **2** (1951), 589-593.
[299] G. H. Hardy, J. E. Littlewood and G. Pólya, *Inequalities*, 2nd ed. London: Cambridge University Press (1964).
[300] L. H. Harper, *Optimal numberings and isoperimetric problems on graphs*, J. Combin. Theory **1** (1966), 385-393.
[301] M. Hartzoulaki, *Probabilistic methods in the theory of convex bodies*, Ph.D. Thesis (March 2003), University of Crete.
[302] R. Henstock and A. M. Macbeath, *On the measure of sum-sets, I: The theorems of Brunn, Minkowski and Lusternik*, Proc. London Math. Soc. **3** (1953), 182-194.

[303] D. Hilbert, *Beweis für die Darstellbarkeit der ganzen Zahlen durch eine feste Anzahl n^{ten} Potenzen (Waringsches Problem)*, Math. Ann. **67** (1909), no. 3, 281-300.

[304] W. Hoeffding, *Probability inequalities for sums of bounded random variables*, J. Amer. Statist. Assoc. **58** (1963), 13-30.

[305] L. Hörmander, *Notions of Convexity*, Progress in Math. **127**, Birkhäuser, Boston-Basel-Berlin (1994).

[306] D. Hug, *Random polytopes*, Stochastic Geometry, Spatial Statistics and Random Fields - Asymptotic Methods (ed. Evgeny Spodarev), Lecture Notes in Mathematics **2068** (2013), 205-238.

[307] A. Hurwitz, *Uber die Darstellung der ganzen Zahlen als Summen von n^{ten} Potenzen ganzer Zahlen.* (German) Math. Ann. **65** (1908), no. 3, 424-427.

[308] I. A. Ibragimov, V. N. Sudakov and B. S. Tsirelson, *Norms of Gaussian sample functions*, Proceedings of the Third Japan USSR Symposium on Probability Theory Lecture Notes in Mathematics Volume 550, 1976, 20–41.

[309] D. Jerison, *A Minkowski problem for electrostatic capacity*, Acta Math. **176** (1996), 1-47.

[310] D. Jerison, *The direct method in the calculus of variations for convex bodies*, Adv. Math. **122** (1996), 262-279.

[311] C. H. Jiménez and M. Naszódi, *On the extremal distance between two convex bodies*, Israel J. Math. **183** (2011), 103-115.

[312] F. John, *Polar correspondence with respect to a convex region*, Duke Math. J. **3** (1937), 355-369.

[313] F. John, *Extremum problems with inequalities as subsidiary conditions*, Courant Anniversary Volume, Interscience, New York (1948), 187-204.

[314] M. Junge, *On the hyperplane conjecture for quotient spaces of L_p*, Forum Math. **6** (1994), 617-635.

[315] M. Junge, *Proportional subspaces of spaces with unconditional basis have good volume properties*, in Geometric Aspects of Functional Analysis, Operator Theory: Advances and Applications **77** (1995), 121-129.

[316] W. B. Johnson and J. Lindenstrauss, *Extensions of Lipschitz mappings into a Hilbert space*, in Conference in modern analysis and probability (New Haven, Conn.) (1982), 189-206.

[317] P. Indyk, *Dimensionality reduction techniques for proximity problems*, Proceedings of the Eleventh Annual ACM-SIAM Symposium on Discrete Algorithms (San Francisco, CA, 2000), ACM, New York (2000), 371-378.

[318] P. Indyk, *Uncertainty principles, extractors and explicit embedding of ℓ_2 into ℓ_1*, STOC'07 Proceedings of the 39th Annual ACM Symposium on Theory of Computing, ACM, New York (2007), 615-620.

[319] P. Indyk and S. J. Szarek, *Almost-Euclidean subspaces of ℓ_1^N via tensor products: a simple approach to randomness reduction*, Approximation, randomization, and combinatorial optimization, Lecture Notes in Comput. Sci., 6302, Springer, Berlin (2010), 632-641.

[320] G. A. Kabatianski and V. I. Levenshtein, *Bounds for packings on a sphere and in space*, Problems Inform. Transmission **14** (1978), 1-17.

[321] M. I. Kadets and M. G. Snobar, *Certain functionals on the Minkowski compactum*, Mat. Zametki **10** (1971), 453-457.

[322] J.-P. Kahane, *Some Random Series of Functions*, Cambridge Studies in Advanced Mathematics **5**, Cambridge Univ. Press, Cambridge (1985).

[323] J.-P. Kahane, *Une inégalité du type de Slepian et Gordon sur les processus Gaussiens*, Israel J. Math. **55** (1986), 109-110.

[324] S. Kakutani, *A proof that there exists a circumscribing cube around any bounded closed convex set in $\mathbb{R}^3$*, Annals of Math. **43** (1942), 739-741.

[325] R. Kannan, L. Lovász and M. Simonovits, *Isoperimetric problems for convex bodies and a localization lemma*, Discrete Comput. Geom. **13** (1995), 541-559.

[326] B. S. Kashin, *Sections of some finite-dimensional sets and classes of smooth functions*, Izv. Akad. Nauk. SSSR Ser. Mat. **41** (1977), 334-351.

[327] B. S. Kashin and S. J. Szarek, *The Knaster problem and the geometry of high-dimensional cubes*, C. R. Acad. Sci. Paris Ser. I Math. **336** (2003), 931-936.

[328] B. S. Kashin and V. N. Temlyakov, *A remark on compressed sensing*, Math. Notes **82** (2007), 748-755.

[329] B. Ya. Kazarnovskii,*On the action of the complex Monge-Ampere operator on piecewise linear functions.* (Russian) Funktsional. Anal. i Prilozhen. 48 (2014), no. 1, 19–29; translation in Funct. Anal. Appl. 48 (2014), no. 1, 15-23 .

[330] A. Khintchine, *Über dyadische Brüche*, Math. Z. **18** (1923), 109-116.

[331] A. G. Khovanskii, *Newton polyhedra and the genus of complete intersections*, Funct. Analysis and its Applications **12** (1977), 55-67.

[332] D. Klain, *A short proof of Hadwiger's characterization theorem*, Mathematika **42** (1995), 329-339.

[333] B. Klartag, *A geometric inequality and a low M-estimate*, Proc. Amer. Math. Soc. **132** (2004), 2619-2628.

[334] B. Klartag, *Rate of convergence of geometric symmetrizations*, Geom. Funct. Anal. **14** (2004), no. 6, 1322-1338.

[335] B. Klartag, *An isomorphic version of the slicing problem*, J. Funct. Anal. **218** (2005), 372-394.

[336] B. Klartag, *On convex perturbations with a bounded isotropic constant*, Geom. Funct. Analysis **16** (2006), 1274-1290.

[337] B. Klartag, *Marginals of geometric inequalities*, Geom. Aspects of Funct. Analysis, Lecture Notes in Math. **1910**, Springer, Berlin (2007), 133-166.

[338] B. Klartag, *Uniform almost sub-gaussian estimates for linear functionals on convex sets*, Algebra i Analiz (St. Petersburg Math. Journal) **19** (2007), 109-148.

[339] B. Klartag, *A central limit theorem for convex sets*, Invent. Math. **168** (2007), 91-131.

[340] B. Klartag, *Power-law estimates for the central limit theorem for convex sets*, J. Funct. Anal. **245** (2007), 284-310.

[341] B. Klartag, *A Berry-Esseen type inequality for convex bodies with an unconditional basis*, Probab. Theory Related Fields **145** (2009), 1-33.

[342] B. Klartag and G. Kozma, *On the hyperplane conjecture for random convex sets*, Israel J. Math. **170** (2009), 253-268.

[343] B. Klartag and E. Milman, *On volume distribution in 2-convex bodies*, Israel J. Math. **164** (2008), 221-249.

[344] B. Klartag and E. Milman, *Centroid Bodies and the Logarithmic Laplace Transform - A Unified Approach*, J. Funct. Anal. **262** (2012), 10-34.

[345] B. Klartag and E. Milman, *Inner regularization of log-concave measures and small-ball estimates*, in Geom. Aspects of Funct. Analysis, Lecture Notes in Math. **2050** (2012), 267-278.

[346] B. Klartag and V. D. Milman, *Isomorphic Steiner symmetrization*, Invent. Math. **153** (2003), no. 3, 463-485.

[347] B. Klartag and O. Regev, *Quantum one-way communication can be exponentially stronger than classical communication*, Proceedings of the 43rd ACM Symposium on Theory of Computing (STOC'11), Assoc. Comput. Mach., (2011), 31-40.

[348] B. Klartag and R. Vershynin, *Small ball probability and Dvoretzky theorem*, Israel J. Math. **157** (2007), 193-207.

[349] B. Knaster, *Problem 4*, Colloq. Math. **30** (1947), 30-31.

[350] H. Kneser and W. Süss, *Die Volumina in linearen Scharen konvexer Körper*, Mat. Tidsskr. B (1932), 19-25.

[351] H. Knothe, *Contributions to the theory of convex bodies*, Michigan Math. J. **4** (1957), 39-52.

[352] A. Koldobsky, *Fourier analysis in convex geometry*, Mathematical Surveys and Monographs **116**, American Mathematical Society, Providence, RI, 2005.

[353] A. Koldobsky and H. König, *Aspects of the isometric theory of Banach spaces*, Handbook of the Geometry of Banach spaces, Vol. I, North-Holland, Amsterdam (2001), 899-939.

[354] A. Koldobsky and M. Lifshits, *Average volume of sections of star bodies*, Geom. Aspects of Funct. Analysis (Milman-Schechtman eds.), Lecture Notes in Math. **1745** (2000), 119-146.

[355] A. N. Kolmogorov, *Über die beste Annäherung von Functionen einer gegebenen Functionenklasse*, Ann. Math. **37** (1936), 107-110.

[356] A. N. Kolmogorov, *On certain asymptotic characteristics of completely bounded metric spaces*, Dokl. Akad. Nauk SSSR **108** (1956), 385-388.

[357] A. N. Kolmogorov and V. M. Tikhomirov, *ε-entropy and ε-capacity of sets in function spaces*, Uspehi Mat. Nauk **14** (1959), 3-86.

[358] H. König, *Eigenvalue distribution of compact operators*, Operator Theory: Advances and Applications, **16** (1986), Birkhauser Verlag, Basel.

[359] H. König, *Isometric imbeddings of Euclidean spaces into finite-dimensional lp-spaces* Panoramas of mathematics (Warsaw, 1992/1994), 79-87, Banach Center Publ., 34, Polish Acad. Sci., Warsaw, 1995.

[360] H. König, M. Meyer and A. Pajor, *The isotropy constants of the Schatten classes are bounded*, Math. Ann. **312** (1998), 773-783.

[361] H. König and V. D. Milman, *On the covering numbers of convex bodies*, Geometric Aspects of Functional Analysis (Lindenstrauss-Milman eds.), Lecture Notes in Math. **1267**, Springer (1987), 82-95.

[362] G. Kuperberg, *A low-technology estimate in convex geometry*, Internat. Math. Res. Notices, **9** (1992), 181-183.

[363] G. Kuperberg, *From the Mahler conjecture to Gauss linking integrals*, Geom. Funct. Anal. **18** (2008), 870-892.

[364] A. G. Kushnirenko, *Newton polyhedra and the number of solutions of a system of k equations with k unknowns*, Uspehi Mat. Nauk. **30** (1975), 302-303.

[365] S. Kwapien, *A theorem on the Rademacher series with vector coefficients*, Proc. Int. Conf. on Probability in Banach Spaces, Lecture Notes in Math. **526**, Springer, 1976.

[366] D. G. Larman and P. Mani, *Almost ellipsoidal sections and projections of convex bodies*, Math. Proc. Camb. Phil. Soc. **77** (1975), 529-546.

[367] R. Latała, *On the equivalence between geometric and arithmetic means for log-concave measures*, Convex geometric analysis (Berkeley, CA, 1996), Math. Sci. Res. Inst. Publ., **34**, Cambridge Univ. Press, Cambridge (1999), 123-127.

[368] R. Latała, *On some inequalities for Gaussian measures*, Proceedings of the International Congress of Mathematicians, Beijing, Vol. II, Higher Ed. Press, Beijing (2002), 813-822.

[369] R. Latała and K. Oleszkiewicz, *Small ball probability estimates in terms of widths*, Studia Math. **169** (2005), 305-314.

[370] R. Latała and J. O. Wojtaszczyk, *On the infimum convolution inequality*, Studia Math. **189** (2008), 147-187.

[371] M. Ledoux, *A simple analytic proof of an inequality by P. Buser*, Proc. Am. Math. Soc. **121** (1994), 951-959.

[372] M. Ledoux, *Isoperimetry and Gaussian Analysis*, Ecole d'Eté de Probabilités de St.-Flour 1994, Lecture Notes in Math. **1709** (1996), 165-294.

[373] M. Ledoux, *The concentration of measure phenomenon*, Mathematical Surveys and Monographs **89**, American Mathematical Society, Providence, RI (2001).

[374] M. Ledoux and M. Talagrand, *Probability in Banach spaces*, Ergeb. Math. Grenzgeb., 3. Folge, Vol. **23** Springer, Berlin (1991).

[375] K. Leichtweiss, *Über die affine Exzentrizität konvexer Körper*, Archiv der Mathematik **10** (1959), 187-199.

[376] K. Leichtweiss, *Konvexe Mengen*, Hochschulbucher fur Mathematik **81** VEB Deutscher Verlag der Wissenschaften, Berlin, 1980.

[377] L. Leindler, *On a certain converse of Hölderis inequality II*, Acta. Sci. Math. Szeged **33** (1972), 217-223.

[378] V. I. Levenshtein, *The maximal density of filling an n-dimensional Euclidean space with equal balls*, Mat. Zametki **18** (1975), 301-311; Math. Notes **18** (1976), 765-771.

[379] F. W. Levi, *Überdeckung eines Eibereiches durch Parallelverschiebung seines offenen Kerns* Arch. Math. (Basel), 369–370, vol. 6, 1955.

[380] P. Lévy, *Problèmes Concrets d'Analyse Fonctionelle*, Gauthier-Villars, Paris (1951).

[381] D. R. Lewis, *Ellipsoids defined by Banach ideal norms*, Mathematika **26** (1979), 18-29.

[382] P. Li and S. T. Yau, *On the parabolic kernel of the Schrödinger operator*, Acta Math. **156** (1986), 153-201.

[383] W. V. Li and W. Linde, *Small deviations of stable processes via metric entropy*, J. Theoret. Probab. **17** (2004), 261-284.

[384] J. Lindenstrauss, *Almost spherical sections, their existence and their applications*, Jber. Deutsch. Math.-Vereinig., Jubiläumstagung (Teubner, Stuttgart), (1990), 39-61.

[385] J. Lindenstrauss and V. D. Milman, *The Local Theory of Normed Spaces and its Applications to Convexity*, Handbook of Convex Geometry (edited by P.M. Gruber and J.M. Wills), Elsevier 1993, 1149-1220.

[386] J. E. Littlewood, *On bounded bilinear forms in an infinite number of variables*, Q. J. Math. (Oxford) **1** (1930), 164-174.

[387] A. Litvak, V. D. Milman and A. Pajor, *The covering numbers and "low M^*-estimate" for quasi-convex bodies*, Proc. Amer. Math. Soc. **127** (1999), no. 5, 1499-1507.

[388] A. E. Litvak, V. D. Milman, A. Pajor and N. Tomczak-Jeagermann, *Entropy extension*, (Russian) Funktsional. Anal. i Prilozhen. **40** (2006), no. 4, 65-71, 112; translation in Funct. Anal. Appl. **40** (2006), no. 4, 298-303.

[389] A. E. Litvak, V. D. Milman, A. Pajor and N. Tomczak-Jeagermann, *On the Euclidean metric entropy of convex bodies*, Geometric aspects of functional analysis, Lecture Notes in Math., **1910**, Springer, Berlin, (2007), 221-235.

[390] A. Litvak, V. D. Milman and G. Schechtman, *Averages of norms and quasi-norms*, Math. Ann. **312** (1998), 95-124.

[391] A. E. Litvak, V. D. Milman and N. Tomczak-Jeagermann, *Essentially-Euclidean convex bodies*, Studia Math., **196** (2010), 207-221.

[392] A. E. Litvak, A. Pajor and N. Tomczak-Jaegermann, *Diameters of sections and coverings of convex bodies*, J. Funct. Anal. **231** (2006), no. 2, 438-457.

[393] A. E. Litvak, A. Pajor, M. Rudelson, N. Tomczak-Jaegermann and R. Vershynin, *Random Euclidean embeddings in spaces of bounded volume ratio*, C. R. Acad. Sci. Paris **339** (2004), no. 1, 33-38.

[394] A. E. Litvak, A. Pajor, M. Rudelson, N. Tomczak-Jaegermann and R. Vershynin, *Euclidean embeddings in spaces of finite volume ratio via random matrices*, J. Reine Angew. Math. **589** (2005), 1-19.

[395] G. G. Lorentz, *Approximation of functions*, New York, Toronto, London: Academic Press (1966).

[396] L. Lovász, *On the ratio of optimal integral and fractional covers*, Discrete Mathematics **4**, (1975), 383–390.

[397] L. Lovász and M. Simonovits, *Random walks in a convex body and an improved volume algorithm*, Random Structures and Algorithms **4** (1993), 359-412.

[398] L. Lovász and S. Vempala, *The geometry of logconcave functions and sampling algorithms*, Random Structures Algorithms **30** (2007), 307-358.

[399] S. Lovett and S. Sodin, *Almost Euclidean sections of the n-dimensional cross-polytope using $O(n)$ random bits*, Commun. Contemp. Math. **10** (2008), 477-489.

[400] Y. Lyubarskii and R. Vershynin, *Uncertainty principles and vector quantization*, IEEE Trans. Inform. Theory **56** (2010), 3491-3501.

[401] E. Lucas, Nouv. Corresp. Math. **2** (1876), 101.

[402] M. Ludwig *Minkowski valuations*, Trans. Amer. Math. Soc. 357 (2005), 4191-4213.

[403] M. Ludwig and M. Reitzner, *A characterization of affine surface area*, Adv. Math. **147** (1999), 138-172.

[404] L. Lusternik, *Die Brunn-Minkowskische Ungleichung für beliebige messbare Mengen*, Doklady Akad. SSSR **3** (1935), 55-58.

[405] E. Lutwak, *The Brunn-Minkowski-Firey theory. I. Mixed volumes and the Minkowski problem*, J. Differential Geom. **38** (1993), 131-150.

[406] E. Lutwak, *Containment and circumscribing simplices*, Discrete Comput. Geom. **19** (1998), 229-235.

[407] E. Lutwak and G. Zhang, *Blaschke-Santaló inequalities*, J. Differential Geom. **47** (1997), 1-16.

[408] E. Lutwak, D. Yang and G. Zhang, *L_p affine isoperimetric inequalities*, J. Differential Geom. **56** (2000), 111-132.

[409] K. Mahler, *Ein Minimalproblem für konvexe Polygone*, Mathematika B (Zutphen) **7** (1938), 118-127.

[410] K. Mahler, *Übertragungsprinzip für konvexe Körper*, Casopis Pest. Mat. Fys. **68** (1939), 93-102.

[411] V. V. Makeev, *Some properties of continuous mappings of spheres and problems in combinatorial geometry*, Geometric questions in the theory of functions and sets (Russian), 75-85, Kalinin. Gos. Univ., Kalinin, 1986.

[412] Y. Makovoz, *A simple proof of an inequality in the theory of n-widths*, Constructive Theory of Functions '87, Sofia (1988).

[413] P. Mani, *Random Steiner symmetrizations*, Studia Sci. Math. Hungar. **21** (1986), 373-378.

[414] M. B. Marcus and G. Pisier, *Random Fourier Series with Applications to Harmonic Analysis*, Center for Statistics and Probability, Northwestern University, No. 44, 1980.

[415] E. Markessinis, G. Paouris and Ch. Saroglou, *Comparing the M-position with some classical positions of convex bodies*, Math. Proc. Cambridge Philos. Soc. **152** (2012), 131-152.

[416] A. W. Marshall, I. Olkin and F. Proschan, *Monotonicity of ratios of means and other applications of majorization*, Inequalities edited by O. Shisha, Acad. Press, New York, London (1967), 177-190.

[417] B. Maurey, *Constructions de suites symétriques*, C.R. Acad. Sci. Paris **288** (1979), 679-681.

[418] B. Maurey, *Some deviation inequalities*, Geom. Funct. Anal. **1** (1991), 188-197.

[419] B. Maurey, *Type, cotype and K-convexity*, Handbook of the Geometry of Banach Spaces (Johnson-Lindenstrauss eds.), Vol. 2, Elsevier (2001), 1299-1332.

[420] B. Maurey, *Inégalité de Brunn-Minkowski-Lusternik, et autres inégalités géométriques et fonctionnelles*, Séminaire Bourbaki, Astérisque **299** (2005), 95-113.

[421] B. Maurey and G. Pisier, *Series de variables aleatoires vectorielles independentes et proprietes geometriques des espaces de Banach*, Studia Math. **58** (1976), 45-90.

[422] V. G. Maz'ja, *The negative spectrum of the higher-dimensional Schrödinger operator*, Dokl. Akad. Nauk SSSR **144** (1962), 721-722.

[423] V. G. Maz'ja, *On the solvability of the Neumann problem*, Dokl. Akad. Nauk SSSR **147** (1962), 294-296.

[424] R. J. McCann, *A convexity theory for interacting gases and equilibrium crystals*, PhD Thesis, Princeton University (1994).

[425] R. J. McCann, *Existence and uniqueness of monotone measure preserving maps*, Duke Math. J. **80** (1995), 309-323.

[426] C. McDiarmid, *Concentration*, in Probabilistic Methods for Algorithmic Discrete Mathematics (Habib, McDiarmid, Ramirez-Alfonsin, Reed eds.), Algorithms and Combinatorics **16** (1991), Springer.

[427] P. McMullen, *Valuations and Euler-type relations on certain classes of convex polytopes*, Proc. London Math. Soc. **35** (1977), 113-135.

[428] P. McMullen, *Continuous translation invariant valuations on the space of compact convex sets*, Arch. Math. **34** (1980), 377-384.

[429] P. McMullen, *Valuations and dissections*, Handbook of convex geometry, Vol. A, B, 933-988, North-Holland, Amsterdam, 1993.

[430] P. McMullen and R. Schneider, *Valuations on convex bodies*, Convexity and its applications, 170-247, Birkhauser, Basel, 1983.

[431] E. Meckes and M. W. Meckes, *The central limit problem for random vectors with symmetries*, J. Theoret. Probab. **20** (2007), 697-720.

[432] M. W. Meckes, *Gaussian marginals of convex bodies with symmetries*, Beiträge Algebra Geom. **50** (2009), 101-118.

[433] M. Meyer, *Une characterisation volumique de certains éspaces normés*, Israel J. Math. **55** (1986), 317-326.

[434] M. Meyer and A. Pajor, *On Santaló's inequality*, in Geom. Aspects of Funct. Analysis (eds. J. Lindenstrauss, V. D. Milman), Lecture Notes in Math. **1376**, Springer, Berlin, 1989, 261-263.

[435] M. Meyer and A. Pajor, *On the Blaschke–Santaló inequality*, Arch. Math. **55** (1990), 82-93.

[436] E. Milman, *Dual mixed volumes and the slicing problem*, Adv. Math. **207** (2006), 566-598.

[437] E. Milman, *A remark on two duality relations*, Integral Equations Operator Theory **57** (2007), 217-228.

[438] E. Milman, *On Gaussian marginals of uniformly convex bodies*, J. Theoret. Probab. **22** (2009), no. 1, 256-278.

[439] E. Milman, *On the role of convexity in isoperimetry, spectral gap and concentration*, Invent. Math. **177** (2009), no. 1, 1-43.

[440] E. Milman, *On the mean width of isotropic convex bodies and their associated L_p-centroid bodies*, Int. Math. Research Notices (to appear).

[441] V. D. Milman, *New proof of the theorem of Dvoretzky on sections of convex bodies*, Funct. Anal. Appl. **5** (1971), 28-37.

[442] V. D. Milman, *Asymptotic properties of functions of several variables defined on homogeneous spaces*, Soviet Math. Dokl. **12** (1971), 1277-1281.

[443] V. D. Milman, *On a property of functions defined on infinite-dimensional manifolds*, Soviet Math. Dokl. **12** (1971), 1487-1491.

[444] V. D. Milman, *Geometrical inequalities and mixed volumes in the Local Theory of Banach spaces*, Astérisque **131** (1985), 373-400.

[445] V. D. Milman, *Random subspaces of proportional dimension of finite dimensional normed spaces: approach through the isoperimetric inequality*, Lecture Notes in Mathematics **1166** (1985), 106-115.

[446] V. D. Milman, *Almost Euclidean quotient spaces of subspaces of finite dimensional normed spaces*, Proc. Amer. Math. Soc. **94** (1985), 445-449.

[447] V. D. Milman, *Volume approach and iteration procedures in local theory of normed spaces*, Banach spaces Proceedings, Missouri 1984 (edited by N. Kalton and E. Saab), Lecture Notes in Math. **1166** (1985), 99-105.

[448] V. D. Milman, *Inegalité de Brunn-Minkowski inverse et applications à la théorie locale des espaces normés*, C.R. Acad. Sci. Paris **302** (1986), 25-28.

[449] V. D. Milman, *The concentration phenomenon and linear structure of finite-dimensional normed spaces*, Proceedings of the ICM, Berkeley (1986), 961-975.

[450] V. D. Milman, *Isomorphic symmetrization and geometric inequalities*, Lecture Notes in Mathematics **1317** (1988), 107-131.

[451] V. D. Milman, *A few observations on the connection between local theory and some other fields*, Lecture Notes in Mathematics **1317** (1988), 283-289.

[452] V. D. Milman, *The heritage of P. Lévy in geometrical functional analysis*, Astérisque **157-158** (1988), 273-302.

[453] V. D. Milman, *A note on a low M^*-estimate*, in "Geometry of Banach spaces, Proceedings of a conference held in Strobl, Austria, 1989" (P.F. Muller and W. Schachermayer, Eds.), LMS Lecture Note Series, Vol. 158, Cambridge University Press (1990), 219-229.

[454] V. D. Milman, *Spectrum of a position of a convex body and linear duality relations*, in Israel Math. Conf. Proceedings 3, Festschrift in Honor of Professor I. Piatetski-Shapiro, Weizmann Science Press of Israel (1990), 151-162.

[455] V. D. Milman, *Some applications of duality relations*, Lecture Notes in Mathematics **1469** (1991), 13-40.

[456] V. D. Milman, *Dvoretzky's theorem - Thirty years later*, Geom. Functional Anal. **2** (1992), 455-479.

[457] V. D. Milman, *Topics in asymptotic geometric analysis*, Proceedings of "Visions in Mathematics - Towards 2000", GAFA 2000, Special Volume (2000), 792-815.

[458] V. D. Milman, *Phenomena arizing from high dimensionality*, Special Volumes for 100 years of Kolmogorov, UMN, **59**, No. 1 (355) (2004), 157-168.

[459] V. D. Milman and A. Pajor, *Isotropic position and inertia ellipsoids and zonoids of the unit ball of a normed n-dimensional space*, Lecture Notes in Mathematics **1376**, Springer, Berlin (1989), 64-104.

[460] V. D. Milman and A. Pajor, *Cas limites dans les inégalités du type de Khinchine et applications géométriques*, C. R. Acad. Sci. Paris **308** (1989), 91-96.

[461] V. D. Milman and A. Pajor, *Entropy and asymptotic geometry of non-symmetric convex bodies*, Adv. Math. **152** (2000), no. 2, 314-335.

[462] V. D. Milman and A. Pajor, *Essential uniqueness of an M-ellipsoid of a given convex body*, Geom. Aspects of Funct. Analysis, Lecture Notes in Math., **1850**, Springer, Berlin (2004), 237-241.

[463] V .D. Milman and L. Rotem, *Characterizing addition of convex sets by polynomiality of volume and by the homothety operation*, CCM , to appear (2014)

[464] V. D. Milman and G. Schechtman, *Asymptotic Theory of Finite Dimensional Normed Spaces*, Lecture Notes in Mathematics **1200**, Springer, Berlin (1986).

[465] V. D. Milman and G. Schechtman, *An "isomorphic" version of Dvoretzky's theorem*, C.R. Acad. Sci. Paris **321** (1995), 541-544.

[466] V. D. Milman and G. Schechtman, *Global versus Local asymptotic theories of finite-dimensional normed spaces*, Duke Math. Journal **90** (1997), 73-93.

[467] V. D. Milman and R. Schneider, *Characterizing the mixed volume*, Adv. Geom. **11** (2011), 669-689.

[468] V. D. Milman and S. J. Szarek, *A geometric lemma and duality of entropy numbers*, in Geom. Aspects of Funct. Analysis, Lecture Notes in Mathematics **1745**, Springer, Berlin, (2000), 191-222.

[469] V. D. Milman and S. J. Szarek, *A geometric approach to duality of metric entropy*, C. R. Acad. Sci. Paris Ser. I Math. **332** (2001), no. 2, 157-162.

[470] V. D. Milman and R. Wagner, *Some remarks on a lemma of Ran Raz*, Geom. Aspects of Funct. Analysis (Milman-Schechtman eds.), Lecture Notes in Math. **1807** (2003), 158-168.

[471] V. D. Milman and H. Wolfson, *Minkowski spaces with extremal distance from Euclidean spaces*, Israel J. Math. **29** (1978), 113-130.

[472] F. Minding, *Über die Bestimmung des Grades einer durch Elimination hervorgehenden Gleichung*, J. Reine. Angew. Math. **22** (1841), 178-183.

[473] H. Minkowski, *Volumen und Oberfläche*, Math. Ann. **57** (1903), 447-495.

[474] H. Minkowski, *Geometrie der Zahlen*, Teubner, Leipzig, 1910.

[475] H. Minkowski, *Allgemeine Lehrsätze über die konvexen Polyeder*, Nachr. Ges. Wiss. Göttingen (1911), 198-219.

[476] B. Mityagin, *Approximative dimension and bases in nuclear spaces*, Uspehi Mat. Nauk **16** (1961), 63-132.

[477] B. Mityagin and A. Pelczynski, *Nuclear operators and approximative dimension*, Proc. of ICM, Moscow (1966), 366-372.

[478] G. Monge, *Mémoire sur la théorie des déblais et des remblais*, Histoire de l'Académie Royale des Sciences de Paris, avec les Mémoires de Mathématique et de Physique pour la mème année (1781), 666-704.

[479] I. Molchanov *Continued fractions built from convex sets and convex functions*, CCM, to appear.

[480] D. Müller, A geometric bound for maximal functions associated to convex bodies, Pacific J. Math. **142** (1990), 297-312.

[481] A. Naor and D. Romik, *Projecting the surface measure of the sphere of ℓ_p^n*, Ann. Inst. H. Poincaré Probab. Statist. **39** (2003), 241-261.

[482] F. L. Nazarov, *The Hörmander proof of the Bourgain-Milman theorem*, Geom. Aspects of Funct. Analysis, Lecture Notes in Math. **2050**, Springer, Heidelberg (2012), 335-343.

[483] F. L. Nazarov, M. Sodin and A. Volberg, *The geometric Kannan-Lovsz-Simonovits lemma, dimension-free estimates for the distribution of the values of polynomials, and the distribution of the zeros of random analytic functions*, Algebra i Analiz, **14** (2002), 214-234.

[484] A. Nemirovski, *Advances in convex optimization: conic programming*, International Congress of Mathematicians. Vol. I, 413-444, Eur. Math. Soc., Zürich, 2007.

[485] A. Pajor, *Metric entropy of the Grassmann manifold*, Convex geometric analysis (Berkeley, CA, 1996), Math. Sci. Res. Inst. Publ. **34**, Cambridge Univ. Press, Cambridge, (1999), 181-188.

[486] A. Pajor and N. Tomczak-Jaegermann, *Remarques sur les nombres d'entropie d'un opérateur et de son transposé*, C.R. Acad. Sci. Paris **301** (1985), 743-746.

[487] A. Pajor and N. Tomczak-Jaegermann, *Subspaces of small codimension of finite dimensional Banach spaces*, Proc. Amer. Math. Soc. **97** (1986), 637-642.

[488] A. Pajor and N. Tomczak-Jaegermann, *Volume ratio and other s-numbers of operators related to local properties of Banach spaces*, J. Funct. Anal. **87** (1989), no. 2, 273-293.

[489] O. Palmon, *The only convex body with extremal distance from the ball is the simplex*, Israel J. Math. **80** (1992), 337-349.

[490] G. Paouris, *On the ψ_2-behavior of linear functionals on isotropic convex bodies*, Studia Math. **168** (2005), 285-299.

[491] G. Paouris, *Concentration of mass and central limit properties of isotropic convex bodies*, Proc. Amer. Math. Soc. **133** (2005), 565-575.

[492] G. Paouris, *Concentration of mass on isotropic convex bodies*, C. R. Math. Acad. Sci. Paris **342** (2006), 179-182.

[493] G. Paouris, *Concentration of mass in convex bodies*, Geom. Funct. Analysis **16** (2006), 1021-1049.

[494] G. Paouris, *Small ball probability estimates for log-concave measures*, Trans. Amer. Math. Soc. **364** (2012), 287-308.

[495] L. E. Payne and H. F. Weinberger, *An optimal Poincaré inequality for convex domains*, Arch. Ration. Mech. Anal. **5** (1960), 286-292.

[496] V. G. Pestov, *Amenable representations and dynamics of the unit sphere in an infinite-dimensional Hilbert space*, Geom. Funct. Anal. **10** (2000), 1171-1201.

[497] V. G. Pestov, *Ramsey-Milman phenomenon, Urysohn metric spaces, and extremely amenable groups*, Israel J. Math. **127** (2002), 317-357.

[498] C. M. Petty, *Surface area of a convex body under affine transformations*, Proc. Amer. Math. Soc. **12** (1961), 824-828.

[499] A. Pietsch, *Theorie der Operatorenideale* (Zusammenfassung), Friedrich-Schiller-Universität Jena (1972).

[500] I. Pinelis *Extremal probabilistic problems and Hotelling's T^2 test under a symmetry condition*, Ann. Statist. **22** (1994), no. 1, 357-368.

[501] A. Pinkus, *n-widths in approximation theory*, Berlin, Heidelberg, New York: Springer Verlag (1985).

[502] G. Pisier, *Un nouveau théorème de factorisation*, C. R. Acad. Sc. Paris, S/erie A, **185** (1977), 715-718.

[503] G. Pisier, *Sur les espaces de Banach K-convexes*, Séminaire d'Analyse Fonctionnelle 1979-80, exp. 11, Ecole Polytechnique, Palaiseau.

[504] G. Pisier, *Un théorème sur les opérateurs linéaires entre espaces de Banach qui se factorisent par un espace de Hilbert*, Ann. E. N. S. **13** (1980), 23-43.

[505] G. Pisier, *Sur les espaces de Banach K-convexes*, Séminaire d'Analyse Fonctionnelle 79/80, Ecole Polytechniques, Palaiseau, Exp. 11 (1980).

[506] G. Pisier, *Holomorphic semi-groups and the geometry of Banach spaces*, Ann. of Math. **115** (1982), 375-392.

[507] G. Pisier, *Probabilistic methods in the geometry of Banach spaces*, Lecture Notes in Mathematics **1206** (1986), 167-241.

[508] G. Pisier, *A new approach to several results of V. Milman*, J. Reine Angew. Math. **393** (1989), 115-131.

[509] G. Pisier, *The Volume of Convex Bodies and Banach Space Geometry*, Cambridge Tracts in Mathematics **94** (1989).

[510] G. Pisier, *On the metric entropy of the Banach-Mazur compactum*, Preprint.

[511] P. Pivovarov, *On the volume of caps and bounding the mean-width of an isotropic convex body*, Math. Proc. Cambridge Philos. Soc. **149** (2010), 317-331.

[512] A. V. Pogorelov, *The Minkowski Multidimensional Problem*, Winston and Sons, Washington DC, 1978.

[513] L. Pontrjagin and L. Schnirelman, *Sur une propriete metrique de la dimension*, Annals of Mathematics, Vol 33, No 1, 156–162.

[514] A. Prékopa, *Logarithmic concave measures with applications to stochastic programming*, Acta. Sci. Math. Szeged **32** (1971), 301-316.

[515] A. Prékopa, *On logarithmic concave measures and functions*, Acta Sci. Math. Szeged **34** (1973), 335-343.

[516] R. Raz, *Exponential separation of quantum and classical communication complexity*, Proceedings of the 31st Annual ACM Symposium on Theory of Computing (STOC'99) ACM Press, Atlanta, Georgia (1999), 358-367.

[517] S. Reisner, *Random polytopes and the volume-product of symmetric convex bodies*, Math. Scand. **57** (1985), 386-392.

[518] S. Reisner, *Zonoids with minimal volume product*, Math. Z. **192** (1986), 339-346.

[519] S. Reisner, *Minimal volume-product in Banach spaces with a 1-unconditional basis*, J. London Math. Soc. **36** (1987), 126-136.

[520] M. Reitzner, *Random polytopes*, New perspectives in stochastic geometry, Oxford University Press, Oxford (2010), 45-76.

[521] R. T. Rockafellar, *Convex analysis*, Princeton Mathematical Series **28**, Princeton University Press, Princeton, NJ (1970).

[522] C. A. Rogers and G. C. Shephard, *The difference body of a convex body*, Arch. Math. (Basel) **8** (1957), 220-233.

[523] C. A. Rogers and G. C. Shephard, *Convex bodies associated with a given convex body*, J. London Soc. **33** (1958), 270-281.

[524] C. A. Rogers and G. C. Shephard, *Some extremal problems for convex bodies*, Mathematika **5** (1958), 93-102.

[525] C. A. Rogers and C. Zong, *Covering convex bodies by translates of convex bodies*, Mathematika 44 (1997), no. 1, 215–218.

[526] M. Rudelson, *Contact points of convex bodies*, Israel J. Math. **101** (1997), 93-124.

[527] M. Rudelson, *Random vectors in the isotropic position*, J. Funct. Anal. **164** (1999), no. 1, 60-72.

[528] M. Rudelson, *Distances between non-symmetric convex bodies and the MM^* estimate*, Positivity **4** (2000), no. 2, 161–178.

[529] M. Rudelson and R. Vershynin, *Combinatorics of random processes and sections of convex bodies*, Ann. of Math. **164** (2006), no. 2, 603-648.

[530] W. Rudin, *Trigonometric series with gaps*, J. Math. Mech. **9** (1960), 203-227.

[531] J. Saint-Raymond, *Sur le volume des corps convexes symétriques*, Initiation Seminar on Analysis: G. Choquet-M. Rogalski-J. Saint-Raymond, 20th Year: 1980/1981, Publ. Math. Univ. Pierre et Marie Curie, vol. **46**, Univ. Paris VI, Paris 1981, Exp. No. 11, 25 (French).

[532] J. Saint-Raymond, *Le volume des ideaux d'operateurs classiques*, Studia Math. **80** (1984), 63-75.

[533] L. A. Santaló, Un invariante afin para los cuerpos convexos del espacio de n dimensiones, Portugaliae Math. **8** (1949), 155-161.

[534] Ch. Saroglou, *Minimal surface area position of a convex body is not always an M-position*, Israel. J. Math. **30**, (2012), 1-15.

[535] G. Schechtman, *Lévy type inequality for a class of metric spaces*, Martingale theory in harmonic analysis and Banach spaces, Springer-Verlag, Berlin-New York (1981), 211-215.

[536] G. Schechtman, *Random embeddings of Euclidean spaces in sequence spaces*, Israel J. Math. **40** (1981), 187-192.

[537] G. Schechtman, *A remark concerning the dependence on ε in Dvoretzky's theorem*, Lecture Notes in Mathematics **1376** (1989), 274-277.

[538] G. Schechtman, *Concentration results and applications*, Handbook of the Geometry of Banach Spaces (Johnson-Lindenstrauss eds.), Vol. 2, Elsevier (2003), 1603-1634.

[539] G. Schechtman, *Special orthogonal splittings of L_1^{2k}*, Israel J. Math. **139** (2004), 337-347.

[540] G. Schechtman, *Two observations regarding embedding subsets of Euclidean spaces in normed spaces*, Adv. Math. **200** (2006), 125-135.

[541] G. Schechtman, *Euclidean sections of convex bodies*, Proc. of the Asymptotic Geom. Anal. Programme, Fields Institute Comm. **68** (2013).

[542] G. Schechtman and M. Schmuckenschläger, *Another remark on the volume of the intersection of two L_p^n balls*, Geom. Aspects of Funct. Analysis (1989-90), Lecture Notes in Math., **1469**, Springer, Berlin (1991), 174-178.

[543] G. Schechtman and M. Schmuckenschläger, *A concentration inequality for harmonic measures on the sphere*, in Geometric Aspects of Functional Analysis, Operator Theory: Advances and Applications **77** (1995), 255-273.

[544] G. Schechtman and J. Zinn, *On the volume of the intersection of two L_p^n balls*, Proc. Am. Math. Soc. **110** (1990), 217-224.

[545] G. Schechtman and J. Zinn, *Concentration on the ℓ_p^n ball*, Geom. Aspects of Funct. Analysis, Lecture Notes in Math., **1745**, Springer, Berlin (2000), 245-256.

[546] E. Schmidt, *Zum Hilbertschen Beweise des Waringschen Theorems*, Math. Ann. **74** (1913), no. 2, 271-274.

[547] E. Schmidt, *Die Brunn-Minkowski Ungleichung*, Math. Nachr. **1** (1948), 81-157.

[548] M. Schmuckenschläger, *On the dependence on ε in a theorem of J. Bourgain, J. Lindenstrauss and V. D. Milman*, Geometric aspects of functional analysis, Lecture Notes in Math. **1469** (1991), 166-173.

[549] M. Schmuckenschläger, *Die Isotropiekonstante der Einheitskugel eines n-dimensionalen Banachraumes mit einer 1-unbedingten Basis*, Arch. Math. (Basel) **58** (1992), 376-383.

[550] M. Schmuckenschläger, *Volume of intersections and sections of the unit ball of ℓ_p^n*, Proc. Amer. Math. Soc. **126** (1998), 1527-1530.

[551] R. Schneider, *On Steiner points of convex bodies*, Israel J. Math. **9** (1971), 241-249.

[552] R. Schneider, *Krümmungsschwerpunkte konvexer Körper*, II. Abh. Math. Sem. Univ. Hamburg **37** (1972), 204-217.

[553] R. Schneider, *Isometrien des Raumes der konvexen Körper*, Colloquium Math. **33** (1975), 219-224.

[554] R. Schneider, *Das Christoffel-Problem für Polytope*, Geom. Dedicata **6** (1977), 81-85.

[555] R. Schneider, *Recent results on random polytopes*, Boll. Un. Mat. Ital., Ser. (9) 1 (2008), 17-39.

[556] R. Schneider, *Convex Bodies: The Brunn-Minkowski Theory*, Second expanded edition. Encyclopedia of Mathematics and Its Applications 151, Cambridge University Press, Cambridge, 2014.

[557] R. Schneider and W. Weil, *Stochastic and Integral Geometry*, Springer, Berlin-Heidelberg (2008).

[558] R.Schneider, *A Formula for Mixed Volumes*, Geometric aspects of functional analysis, 423-426 , Lecture Notes in Math., 2116, Springer, Heidelberg, 2014; edited by B.Klartag and E.Milman.

[559] C. Schütt, *Entropy numbers of diagonal operators between symmetric Banach spaces*, J. Approx. Theory **40** (1984), 121-128.

[560] H. A. Schwarz, *Beweis des Satzes, daßdie Kugel kleinere Oberfläche besistzt als jeder andere Körper gleichen Volumens*, Nachr. Ges. Wiss. Göttingen (1884), 1-13; Ges. math. Abh. **2** (Berlin, 1890), 327-340.

[561] A. Segal, *Remark on stability of Brunn-Minkowski and isoperimetric inequalities for convex bodies*, Preprint.

[562] B. Shiffman, *Synthetic projective geometry and Poincaré's theorem on automorphisms of the ball*, Enseign. Math. (2) **41** (1995), no. 3-4, 201-215.

[563] V. M. Sidelnikov, *On the densest packing of balls on the surface of an n-dimensional Euclidean sphere and the number of binary code vectors with a given code distance*, Dokl. Akad. Nauk. SSSR **213** (1973), 1029-1032; Soviet Math. Dokl. **14** (1973), 1851-1855.

[564] D. Slepian, *The one-sided barrier problem for Gaussian noise*, Bell System Technical Journal (1962), 463-501.

[565] S. Sodin, *Tail-sensitive Gaussian asymptotics for marginals of concentrated measures in high dimension*, Geom. Aspects of Funct. Analysis (Milman-Schechtman eds.), Lecture Notes in Mathematics **1910**, Springer, Berlin (2007), 271-295.

[566] S. Sodin, *An isoperimetric inequality on the ℓ_p balls*, Ann. Inst. Henri Poincaré Probab. Stat. **44** (2008), 362-373.

[567] D. Spielman and N. Srivastava, *An elementary proof of the restricted invertibility theorem*, Israel J. Math **190** (2012).

[568] N. Srivastava, *On contact points of convex bodies*, in Geom. Aspects of Funct. Analysis, Lecture Notes in Mathematics **2050** (2012), 393-412.

[569] J. M. Steele, *Probability theory and combinatorial optimization*, CBMS-NSF Regional Conference Series in Applied Mathematics, SIAM **69** (1996).

[570] E. Stein, *The development of square functions in the work of A. Zygmund*, Bull. Amer. Math. Soc. **7** (1982), 359-376.

[571] J. Steiner, *Einfache Beweis der isoperimetrische Hauptsätze*, J. Reine Angew. Math. **18** (1838), 289-296.

[572] J. Steiner, *Über Maximum und Minimum bei den Figuren in der Ebene, auf der Kugelfläche und im Raume überhaupt*, J. Math. pures applic. **6** (1842) 105-170; J. reine angew. Math. **24** (1842), 93-152.

[573] D. W. Stroock, *Probability Theory, an Analytic View*, Cambridge University Press, Cambridge (1993).

[574] V. N. Sudakov, *Gaussian random processes and measures of solid angles in Hilbert spaces*, Soviet Math. Dokl. **12** (1971), 412-415.

[575] V. N. Sudakov, *Typical distributions of linear functionals in finite-dimensional spaces of high dimension*, Soviet Math. Dokl. **19** (1978), 1578-1582.

[576] V. N. Sudakov and B. S. Tsirelson, *Extremal properties of half-spaces for spherically invariant measures*, J. Soviet. Math. **9** (1978), 9-18; translated from Zap. Nauch. Sem. L.O.M.I. **41** (1974), 14-24.

[577] A. Szankowski, *On Dvoretzky's theorem on almost spherical sections of convex bodies*, Israel J. Math. **17** (1974), 325-338.

[578] S. J. Szarek, *On the best constants in the Khinchin inequality*, Studia Math. **58** (1976), no. 2, 197-208.

[579] S. J. Szarek, *On Kashin's almost Euclidean orthogonal decomposition of ℓ_1^n*, Bull. Acad. Polon. Sci. **26** (1978), no. 8, 691-694.

[580] S. J. Szarek, *Nets of Grassmann manifold and orthogonal group*, Proceedings of research workshop on Banach space theory (Iowa City, Iowa, 1981), Univ. Iowa, Iowa City, IA, (1982), 169-185.

[581] S. J. Szarek, *The finite dimensional basis problem, with an appendix on nets of Grassman manifold*, Acta Math. **159** (1983), 153-179.

[582] S. J. Szarek, *Condition numbers of random matrices*, J. Complexity **7** (1991), no. 2, 131-149.

[583] S. J. Szarek, *Metric entropy of homogeneous spaces*, Quantum probability (Gdańsk, 1997), Banach Center Publ., **43**, Polish Acad. Sci., Warsaw, (1998), 395-410.

[584] S. J. Szarek, *Convexity, complexity, and high dimensions*, International Congress of Mathematicians, Vol. II, Eur. Math. Soc., Zürich, (2006), 1599-1621.

[585] S. J. Szarek and M. Talagrand, *An isomorphic version of the Sauer-Shelah lemma and the Banach-Mazur distance to the cube*, Lecture Notes in Mathematics **1376** (1989), 105-112.

[586] S. J. Szarek and N. Tomczak-Jaegermann, *On nearly Euclidean decompositions of some classes of Banach spaces*, Compositio Math. **40** (1980), 367-385.

[587] M. Talagrand, *An isoperimetric theorem on the cube and the Kintchine-Kahane inequalities*, Proc. Amer. Math. Soc. **104** (1988), 905-909.

[588] M. Talagrand, *A new isoperimetric inequality and the concentration of measure phenomenon*, Geometric aspects of functional analysis (1989–90), Lecture Notes in Math. **1469**, Springer, Berlin (1991), 94-124.

[589] M. Talagrand, *A new look at independence*, Ann. Probab. **24** (1996), 1-34.

[590] B. Teissier, *Du théorème de Hodge aux inéfalités isopérimètriques*, C. R. Acad. Sci. Paris **288** (1979), 287-289.

[591] A. C. Thompson, *Minkowski Geometry*, Encyclopedia of Mathematics and its Applications **63**, Cambridge University Press, Cambridge (1996).

[592] N. Tomczak-Jaegermann, *Dualité des nombres d'entropie pour des opérateurs à valeurs dans un espace de Hilbert*, C.R. Acad. Sci. Paris **305** (1987), 299-301.

[593] N. Tomczak-Jaegermann, *Banach-Mazur Distances and Finite Dimensional Operator Ideals*, Pitman Monographs **38** (1989), Pitman, London.

[594] H. Triebel, *Interpolationseigenschaften von Entropie - und Durchmesseridealen kompakter Operatoren*, Studia Math. **34** (1970), 89-107.

[595] B. Uhrin, *Curvilinear extensions of the Brunn-Minkowski-Lusternik inequality*, Advances in Mathematics **109** (1994), 288-312.

[596] P. S. Urysohn, *Mittlere Breite und Volumen der konvexen Körper im n-dimensionalen Raume*, Mat. Sb. SSSR **31** (1924), 313-319.

[597] R. Vershynin, *Isoperimetry of waists and local versus global asymptotic convex geometries* (with an appendix by M. Rudelson and R. Vershynin), Duke Math. Journal **131** (2006), no. 1, 1-16.

[598] C. Villani, *Topics in Optimal Transportation*, Graduate Texts in Mathematics **58**, Amer. Math. Soc. (2003).

[599] C. Villani, *Optimal transport – Old and new*, Grundlehren der Mathematischen Wissenschaften **338** Springer-Verlag, Berlin (2009).

[600] A. Volčič, *Random Steiner symmetrizations of sets and functions*, Calc. Var. Partial Differential Equations **46** (2013), 555-569.

[601] J. von Neumann, *Approximative properties of matrices of high finite order*, Portugaliae Math. **3** (1942), 1-62.

[602] B-H. Vritsiou, *Further unifying two approaches to the hyperplane conjecture*, International Math. Research Notices (2014), no. 6, 1493-1514.

[603] R. Wagner, *On the relation between the Gaussian orthogonal ensemble and reflections, or a self-adjoint version of the Marcus-Pisier inequality*, Canadian Math Bulletin **49** (2006), 313-320.

[604] D. L. Wang and P. Wang, *Extremal configurations on a discrete torus and a generalization of the generalized Macaulay theorem*, SIAM J. Appl. Math. **33** (1977), 55-59.

[605] W. Weil *Decomposition of convex bodies*, Mathematika 21 (1974), 19-25.

[606] H. von Weizsäcker, *Sudakov's typical marginals, random linear functionals and a conditional central limit theorem*, Probab. Theory Relat. Fields **107** (1997), 313-324.

[607] H. Weyl, *On the volume of tubes*, Amer. J. Math. **61** (1939), 461-472.

[608] J. O. Wojtaszczyk, *The square negative correlation property for generalized Orlicz balls*, Geom. Aspects of Funct. Analysis, Lecture Notes in Math., **1910**, Springer, Berlin (2007) , 305-313.

[609] H. Yamabe and Z. Yujobo, *On continuous functions defined on a sphere*, Osaka Math. J. **2** (1950), 19-22.

[610] V. V. Yurinskii, *Exponential bounds for large deviations*, Theor. Probability Appl. **19** (1974), 154-155.

[611] V. V. Yurinskii, *Exponential inequalities for sums of random vectors*, J. Multivariate Anal. **6** (1976), 473-499.

[612] C. Zong, *Strange phenomena in convex and discrete geometry*, Universitext, Springer (2003).

Subject Index

Author Index